COMPUTATIONAL STRATEGIES FOR SPECTROSCOPY

COMPUTATIONAL STRATEGIES FOR SPECTROSCOPY

From Small Molecules to Nano Systems

Edited by

VINCENZO BARONE

WILEY

A JOHN WILEY & SONS, INC., PUBLICATION

Published by John Wiley & Sons, Inc., Hoboken, New Jersey
Published simultaneously in Canada

For general information on our other products and services or for technical support, please contact our Customer Care Department within the United States at 877-762-2974, outside the United States at 317-572-3993 or fax 317-572-4002.

Wiley also publishes its books in a variety of electronic formats. Some content that appears in print may not be available in electronic formats. For more information about Wiley products, visit our web site at www.wiley.com.

Library of Congress Cataloging-in-Publication Data:

Computational Strategies for Spectroscopy: From Small Molecules to Nano Systems / Edited by: Vincenzo Barone.
p. cm.
Includes index.
ISBN 978-0-470-47017-6 (cloth)
1. Spectrum analysis–Data processing. I. Barone, Vincenzo, Dr.
QD95.5.D37C66 2011
543′.50285–dc22

2010041044

ePDF ISBN: 978-1-118-00870-6
oBook ISBN: 978-1-118-00872-0
ePub ISBN: 978-1-118-00871-3

10 9 8 7 6 5 4 3 2 1

CONTENTS

CONTRIBUTORS

Hans Ågren, Theoretical Chemistry, School of Biotechnology, Royal Institute of Technology, SE10044 Stockholm, Sweden

Vincenzo Barone, Scuola Normale Superiore, Piazza dei Cavalieri 7, I56126 Pisa, Italy

Maurizzio Becucci, LENS and Dipartimento di Chimica "Ugo Schiff," Polo Scientifico e Tecnologico, Università degli Studi di Firenze, Via N. Carrara 1, I50019 Sesto Fiorentino, Florence, Italy

Malgorzata Biczysko, Scuola Normale Superiore, Piazza dei Cavalieri 7, I56126 Pisa, Italy and Dipartimento di Chimica "Paolo Corradini," Università di Napoli Federico II, Complesso Univ. Monte S. Angelo, Via Cintia, I80126 Naples, Italy

Julien Bloino, Scuola Normale Superior, Piazza dei Cavalieri 7, I56126 Pisa, Italy and Dipartimento di Chimica "Paolo Corradini," Università di Napoli Federico II, Complesso Univ. Monte S. Angelo, Via Cintia, I80126 Naples, Italy

Giuseppe Brancato, Italian Institute of Technology, IIT@NEST Center for Nanotechnology Innovation, Piazza San Silvestro 12, I56125 Pisa, Italy

Marina Brustolon, Dipartimento di Scienze Chimiche, Università degli Studi di Padova, Via Marzolo 1, 35131 Padova, Italy

Chiara Cappelli, Dipartimento di Chimica e Chimica Industriale, Università di Pisa, Via Risorgimento 35, I56126 Pisa, Italy

Vincenzo Carravetta, CNR—Consiglio Nazionale delle Ricerche, Istituto per i Processi Chimico-Fisici (IPCF), Via G. Moruzzi, I56124 Pisa, Italy

Sonia Coriani, Dipartimento di Scienze Chimiche, Università degli Studi di Trieste Via L. Giorgieri 1, I34127 Trieste, Italy and Centre for Theoretical and Computational Chemistry (CTCC), University of Oslo, Blindern, N0315 Oslo, Norway

Orlando Crescenzi, Dipartimento di Chimica "Paolo Corradini," Università di Napoli Federico II, Complesso Univ. Monte S. Angelo, Via Cintia, I80126 Naples, Italy

Wolfgang Domcke, Department of Chemistry, Technische Universität München, D-85747 Garching, Germany

Sabine Van Doorslaer, Department of Physics, University of Antwerp, Universiteitsplein 1 (N 2.16), B-2610 Antwerp, Belgium

Dassia Egorova, Institute of Physical Chemistry, Universität Kiel, D-24098 Kiel, Germany

Jack H. Freed, Department of Chemistry and Chemical Biology, Cornell University, Ithaca, New York 14853

Maxim F. Gelin, Department of Chemistry, Technische Universität München, D-85747 Garching, Germany

Roberto Improta, CNR—Consiglio Nazionale delle Ricerche, Istituto Biostrutture e Bioimmagini, Via Mezzocannone 16I, I80134 Naples, Italy

Alessandro Lami, CNR—Consiglio Nazionale delle Ricerche, Istituto di Chimica dei Composti OrganoMetallici, UOS di Pisa, Via G. Moruzzi, I56124 Pisa, Italy

Ermelinda S. M. Maçôas, Centro de Química-Física Molecular (CQFM) and Institute of Nanoscience and Nanotechnology (IN), Instituto Superior Técnico, 1049-001 Lisbon, Portugal

Alfonso Pedone, Scuola Normale Superiore, Piazza dei Cavalieri 7, I56126 Pisa, Italy

Giangaetano Pietraperzia, LENS and Dipartimento di Chimica "Ugo Schiff," Polo Scientifico e Tecnologico, Università degli Studi di Firenze, Via N. Carrara 1, I50019 Sesto Fiorentino, Florence, Italy

Antonino Polimeno, Dipartimento di Scienze Chimiche, Università degli Studi di Padova, Via Marzolo 1, I35131 Padova, Italy

Cristina Puzzarini, Dipartimento di Chimica "G. Ciamician", Università degli Studi di Bologna, Via Selmi 2, I40126 Bologna, Italy

Nadia Rega, Dipartimento di Chimica "Paolo Corradini", Università di Napoli Federico II, Complesso Univ. Monte S. Angelo, Via Cintia, I80126 Naples, Italy

Antonio Rizzo, CNR—Consiglio Nazionale delle Ricerche, Istituto per i Processi Chimico-Fisici (IPCF),Via G. Moruzzi, I56124 Pisa, Italy

Kenneth Ruud, Centre for Theoretical and Computational Chemistry (CTCC), Department of Chemistry, University of Tromsø, N9037 Tromsø, Norway

Fabrizio Santoro, CNR—Consiglio Nazionale delle Ricerche, Istituto di Chimica dei Composti OrganoMetallici, UOS di Pisa, Via G. Moruzzi, I56124 Pisa, Italy

Fabio Trani, Scuola Normale Superiore, Piazza dei Cavalieri 7, I56126 Pisa, Italy

PREFACE

Within the plethora of modern experimental techniques, vibrational, electronic, and resonance spectroscopies are uniquely suitable to probe the static and dynamic properties of molecular systems under realistic environmental conditions and in a noninvasive fashion. Indeed, the impact of spectroscopic techniques in practical applications is huge, ranging from astrophysics to drug design and biomedical studies, from the field of cultural heritage to characterizations of materials and processes of technological interest, and so on. However, the development of more and more sophisticated experimental techniques poses correspondingly stringent requirements on the quality of the models employed to interpret spectroscopic results and on the accuracy of the underlying chemical–physical descriptions. As a matter of fact, spectra do not provide direct access to molecular structure and dynamics, and interpretation of the indirect information that can be inferred from analysis of the experimental data is seldom straightforward. Typically these complications arise from the fact that spectroscopic properties depend on the subtle interplay of several different effects, whose specific roles are not easy to separate and evaluate.

In such a complex scenario, theoretical studies can be extremely helpful, essentially at three different levels: (i) supporting and complementing the experimental results to determine structural, electronic, and dynamical features of target molecule(s) starting from spectral properties; (ii) dissecting and quantifying the role of different effects in determining the spectroscopic properties of a given molecular/supramolecular system; and (iii) predicting electronic, molecular, and spectroscopic properties for novel/modified systems. For this reason, computational spectroscopy is rapidly evolving from a highly specialized research field into a versatile and fundamental tool for the assignment of experimental spectra and their interpretation in terms of basic physical–chemical processes.

The predictive and interpretative ability of computational chemistry experiment can be clearly demonstrated by state-of-the-art quantum mechanical approaches to spectroscopy, which at present yield results comparable to the most accurate experimental measurements. In this book several examples of such highly accurate studies with particular reference to rotational spectroscopy or electronic transitions for small molecular systems showing complex nonadiabatic interactions will be presented. However, the highly accurate approaches available for small molecular systems are not the main scope of the present work, as they are not transferable directly to the study of large, complex molecular systems. Clearly, the definition of efficient computational approaches aimed at spectroscopic studies of macrosystems is in general a nontrivial task, and the basic requirement is that such effective models need to reflect a correct physical picture. Then, as will be presented, appropriate schemes can be introduced even for challenging cases, retaining the reliability of more demanding computational approaches for molecular systems of, for example, drug design, materials science, and nanotechnology.

The main aim of the book is the presentation and analysis of several examples illustrating the current status of computational spectroscopy approaches applicable to medium-to-large molecular system in the gas phase and in more complex environments. Particular attention is devoted to theoretical models able to provide data as close as possible to the results directly available from experiment in order to avoid ambiguities in the interpretation of the latter. Additionally, the main focus is on approaches easily accessible to nonspecialists, possibly through integrated computational strategies available in standard computational packages.

In fact, one of the objectives of the book is to introduce nonexpert readers to modern computational spectroscopy approaches. In this respect, the essential basic background of the described theoretical models is provided, but for the extended description of concepts related to theory of molecular spectra readers are referred to the widely available specialized volumes. Similarly, although computational spectroscopy studies rely on quantum mechanical computations, only necessary aspects of quantum theory related directly to spectroscopy will be presented. Additionally, we have chosen to analyze only those physical–chemical effects which are important for molecular systems containing atoms from the first three rows of the periodic table, while we will not discuss in detail effects and computational models specifically related to transition metals or heavier elements. Particular attention has been devoted to the description of computational tools which can be effectively applied to the analysis and understanding of complex spectroscopy data. In this respect, several illustrative examples are provided along with discussions about the most appropriate computational models for specific problems.

The book has been set as a joint effort of members of an Italian network devoted to applications of computational approaches to molecular and supramolecular problems (http://m3village.sns.it) along with some international collaborators and is organized as follows. After the preface by the editor, short chapters (authored by Brustolon, Van Doorslaer, Maçôas, Becucci, and Pietraperzia) summarize the point of view of experimental spectroscopists about the status and many interesting perspectives in the field. Then, different topics of computational spectroscopy are examined starting with a section devoted to transitions between electronic and spin states within a static framework. This part starts with a chapter (by Improta) describing electronic spectroscopy in the ultraviolet (UV)–visible region with particular attention to environmental effects; the following chapters deal with response function theory applied to linear and nonlinear optical spectroscopy (by Rizzo, Coriani, and Ruud), X-ray spectroscopy (by Carravetta and Agren), magnetic resonance spectroscopies (by Pedone and Crescenzi), and photoluminescence in nanocrystals (by Trani). Then, time-independent approaches to nuclear motions, with special reference to rotational, vibrational, and electronic spectroscopies, are analyzed by Puzzarini Cappelli, Biczysko, Bloino, and Santoro. The last section is devoted to time-dependent approaches and includes a contribution by Domcke and co-workers concerning the computation of ultrafast time- and frequency-resolved spectroscopic signals; followed by two chapters devoted to quantum, semiclassical, and classical dynamical approaches authored by Lami, Santoro, Rega, and Brancato; and closed by a chapter by Freed and Polimeno devoted to the application of stochastic techniques to "slow" spectroscopies like nuclear magnetic resonance (NMR) and electron paramagnetic resonance (EPR).

Computational spectroscopy is a rapidly evolving field that is producing versatile and widespread tools for the assignment of experimental spectra and their interpretation in terms of basic chemical–physical effects. We hope that the topics covered in the book and related to the computation of infrared (IR), UV–visible, NMR, and EPR spectral parameters, with reference to the underlying vibronic and environmental effects, will allow computational chemists, analytical chemists and spectroscopists, physicists, materials scientists, and graduate students to benefit from this thorough resource.

VINCENZO BARONE

Note: Color versions of selected figures are available at ftp://ftp.wiley.com/public/sci_tech_med/computational_strategies.

INTRODUCTION TO ELECTRON PARAMAGNETIC RESONANCE

MARINA BRUSTOLON
Dipartimento di Scienze Chimiche, Universita degli Studi di Padova, Padova, Italy
SABINE VAN DOORSLAER
Department of Physics, University of Antwerp, Antwerp, Belgium

Ever since its first observation in 1944, electron paramagnetic resonance (EPR) has offered a unique tool to investigate paramagnetic systems. In the first four decades of EPR, continuous-wave (cw) EPR at X-band frequencies (9.5 GHz) was the main technique. Although a lot of information can sometimes already be determined from these cw-EPR experiments, in many cases these spectra consist of very uninformative single lines. The introduction of the cw electron nuclear double resonance (cw-ENDOR) technique in 1956 [1] offered a first way of obtaining more detailed information about the interactions of the unpaired electrons with the surrounding magnetic nuclei (and hence about the electronic state of the system). However, a revolutionary new area of EPR started in the 1980s with the development of pulsed EPR spectrometers and more recently with the introduction of high-frequency (HF) EPR. The EPR toolbox has now moved from the X-band cw-EPR analysis to a large number of different EPR methods, each of them resulting in particular information. Combined multi frequency cw and pulsed EPR techniques allow characterizing in detail the dynamics and the electronic and geometric structure of long-living and transient paramagnetic species.

In the following the potential of some of these methods will be outlined further. At the same time we will point out the increasing importance of a synergic

Computational Strategies for Spectroscopy: From Small Molecules to Nano Systems, First Edition.
Edited by Vincenzo Barone.

development of theoretical models and computational tools for a full interpretation of the new EPR experiments.

EPR: DYNAMICS AND SPIN RELAXATION

The analysis of cw-EPR spectral profiles for radicals in solution to obtain information on their intermolecular and intramolecular dynamics has been a customary procedure since chemists began to use this spectroscopy in the 1960s. For motions sufficiently fast to average the magnetic anisotropies, the spectral profile can be simulated as a collection of Lorentzian lines at the resonance frequencies, with linewidths proportional to the spin–spin relaxation rates. This type of simulation is based on the Redfield theory, in which the random magnetic interactions are considered as a time-dependent perturbation of the static spin Hamiltonian. The linewidths are obtained as functions of the magnetic anisotropies (g and hyperfine tensors for radicals) and of the correlation times of the motions [2]. The border between "fast" and "slow" motions is defined by the comparison between the correlation times and the inverse of magnetic anisotropies. Therefore, in a multifrequency EPR approach (MF-EPR), each experiment requires an adjustment of the time scale on the basis of the frequency of the spectrometer, since the Zeeman anisotropy depends on the external magnetic field. The use of MF-EPR therefore facilitates the study of complex dynamics. HF-EFR is a fast time-scale technique "freezing out" the slower motions and giving lineshapes affected by the faster motions only, whereas EPR at lower frequencies is sensitive to slower motions. The combined approach thus permits the separation of different types of motion [e.g., 3].

Today the spectral profiles can be simulated for any motional regime by a numerical integration of the stochastic Liouville equation, as discussed in Chapter 12 and in the references therein. The noticeable improvement in the techniques of calculation of the magnetic parameters and their dependence on the solvent, and of the minimum energy conformation of the molecules, have opened the possibility of an integrated computational approach. Since it gives calculated spectral profiles completely determined by the molecular and physical properties of the radical and of the solvent at a given temperature, this method is a step forward in the direction of a sound interpretation of complex spectra.

EPR spectral profiles of paramagnetic probes in solid phases are determined by the anisotropic distribution of spin packets due to the nonaveraged magnetic anisotropies. The inhomogeneously broadened bands are scarcely informative, and the so-called hyperfine methods (see below) are used in these cases to extract information on the hyperfine parameters.

A viable method to extract further information on the residual motions of paramagnetic probes in solids is echo-detected EPR (ED-EPR), where the spectral profile versus the magnetic field is given by the integrated electron spin echo (ESE). The echo experiment probes directly the homogeneous linewidth at each spectral position affected by the motional process, bypassing the inhomogeneous broadening [4]. This method has been applied conveniently when residual motions of paramagnetic probes are present in glasses [e.g., 5], in biological systems [6], and in complex solids. Multifrequency ED-EPR has been applied to study nitroxide spin probes in a supramolecular compound [7]. In another example, ED-EPR has been used for the detection of a photoexcited dye triplet in glassy solution and in an inclusion channel compound; the different types of tumbling motions in the two environments give ED–time resolved EPR spectra with strongly different spectral profiles [8].

Finally, pulsed EPR experiments in solids allow determination of spin–lattice relaxation and phase memory times (T_1, T_M), whose dependence on temperature and nature of the environment can be analyzed to give information on the collective relaxation phenomena due to nuclear spin diffusion, electron–electron dipolar interaction, and instantaneous diffusion [9]. Note that the electron–electron dipolar interaction is also exploited in the DEER (PELDOR) technique for measuring the distance between paramagnetic centers. Instantaneous diffusion can give information on the microconcentration of radicals produced by high-energy irradiation in solids [e.g., 10].

The studies on spin dynamics and spin relaxation properties of paramagnetic species in solids are particularly interesting today in the perspective of a development of spintronics. In this respect the relaxation properties of the relatively simple, well-known and studied organic radicals in solids can help in understanding more complex behaviors [e.g., 11].

PROBING ELECTRONIC STRUCTURE THROUGH HYPERFINE SPECTROSCOPY

In paramagnetic systems, the unpaired electron(s) can interact with the surrounding magnetic nuclei (hyperfine interaction) [2, 12]. This interaction reflects the spin density distribution (and hence the electronic energy), and the electron spin–nuclear spin distances. Furthermore, for a nuclear spin larger than $\frac{1}{2}$, the nuclear quadrupole interaction reflects the electric field gradient experienced by the nucleus [12, 13]. The measurement of these hyperfine and nuclear quadrupole interactions thus completes the electronic information obtained from determining the electron Zeeman interaction and (for $S > \frac{1}{2}$) the zero-field interaction and, hence, allows in principle for a full description of the electronic structure of the paramagnetic species at hand. The majority of the hyperfine and nuclear quadrapole interactions unfortunately remain unresolved in the field-swept EPR spectra, explaining the need of other EPR approaches, the so-called hyperfine spectroscopy techniques, to obtain this information.

A large variety of hyperfine spectroscopy methods exist that allow the detection of hyperfine and nuclear quadrupole interactions: electron spin-echo envelope modulation (ESEEM), ENDOR, and ELDOR-detected NMR (electron–electron double-resonance detected nuclear magnetic resonance) [13]. Although there are cases in which ESEEM and ENDOR perform equally well, ESEEM-like methods tend to be favorable for the detection of small nuclear frequencies (<10 MHz), whereas ENDOR and ELDOR-detected NMR methods are more appropriate for the larger nuclear transition frequencies. At high microwave frequencies, the use of ELDOR-detected over ENDOR techniques may be advantageous [14].

All ESEEM pulse sequences have in common that they consist only of microwave pulses with one microwave frequency and that they generate an electron spin echo, whose intensity is detected as a function of one or two interpulse distances. The detected modulation of this echo intensity reflects the nuclear interactions. Details on the physical origin of these nuclear modulations can be found in the literature [13]. Although a large variety of ESEEM methods are available, the HYSCORE (hyperfine sublevel correlation) spectroscopy method [15] has become the most favored technique. This four-pulse ESEEM method leads to a two-dimensional (2D) time-domain signal. In the frequency-domain spectrum obtained after 2D Fourier transformation of this signal the cross peaks directly link the nuclear frequencies of different electron spin manifolds. In this way, spectral interpretation is largely facilitated. While HYSCORE experiments are easy to perform at lower microwave frequencies (2–35 GHz), high-frequency HYSCORE is hard to realize, needing high-power microwave pulses, which are at present only attainable in prototype spectrometers [16, 17].

This limitation is not there for pulsed ENDOR methods, which can be used at all microwave frequencies. In ENDOR, the sample is irradiated with a combination of microwaves and radio waves. Continuous-wave ENDOR was already introduced in 1956 by Feher [1] and for a long time remained an important tool to determine the hyperfine and nuclear quadrupole interactions. However, nowadays this technique is largely replaced by the pulsed counterparts, which are more versatile. Two of the most commonly used ENDOR pulse sequences are Davies ENDOR [18] and Mims ENDOR [19]. In these techniques a combination of microwave pulses and a π radio frequency (RF) pulse with variable RF is used. A first set of microwave pulses creates electron polarization. When the RF matches one of the nuclear transitions, the populations of the different energy levels will be affected. This will change the electron polarization that is read out by a last sequence of microwave pulses, usually via electron spin echo detection as a function of the radio frequency. In this way, the nuclear frequencies can be directly detected.

In some cases, it may be beneficial to use ELDOR-detected NMR [20] instead of ENDOR to obtain the larger nuclear transitions. This technique bears similarities with the above-mentioned ENDOR techniques, but instead of using an RF pulse, a microwave pulse with variable microwave frequency is used to affect the electron populations. The nuclear transition frequencies follow from monitoring the polarization changes as a function of the difference between the two microwave frequencies.

In practice, the EPR spectroscopist will thus be using a variety of multifrequency cw-EPR and hyperfine spectroscopy techniques to determine the different spin Hamiltonian parameters. This has led to new insights in biological systems as well as in material sciences and it would be impossible to give an exhaustive account of all these applications. Here, we merely give a few examples to give the reader an idea about the wide applicability of the hyperfine techniques. ENDOR and HYSCORE spectroscopy allowed for instance for the direct detection of a hydrogen ligand in the [NiFe] center of the regulatory H_2-sensing hydrogenase of *Ralstonia eutropha* in its reduced state [21], a ligand about which a lot had been speculated, but no hard proof could be provided by other (spectroscopic) means. Similarly, hyperfine spectroscopies were found to be key in the identification of a stable Rh-coordinated aminyl radical metal complex obtained after one-electron oxidation of an amide complex [22] and in the detection of a Co (III)-bound phenoxyl radical formed during the activation of a chiral cobalt salen catalyst [23]. Pulsed EPR and ENDOR of N@C_{60} in polycrystalline C_{60} allowed the detection of the freezing of cage rotation and the symmetry lowering induced by a phase transition [24]. High-field EPR and ENDOR even allowed, in a unique way, the probing of the wave function of shallow Li and Na donors in ZnO nanoparticles [25].

One of the big challenges is the translation of the different spin Hamiltonian parameters in a structural model. Indeed, the g values, zero-field splittings, and hyperfine and nuclear quadrupole tensors depend in a complicated way on the geometric structure and electronic ground state. Although some simplified approaches, such as the point dipolar approximation to extract distance information from the dipolar part of the hyperfine interaction [26], are readily used, in most cases these approaches are not sufficient to determine all information from the experimental parameters. Therefore, EPR analyses are increasingly combined with quantum chemical computation, whereby density functional theory (DFT) plays an extremely important role. While DFT reproduces the EPR parameters of organic radicals very well [27, 28], the computations of these parameters for paramagnetic transition metal complexes is often still problematic [29–31]. Especially challenging is the computation of zero-field parameters of high-spin systems and the g and metal hyperfine values of many transition metal systems. Nevertheless, it is by now clear that quantum chemical methods are pivotal for the correct interpretation of the EPR findings and that an improvement in the quality of these computations will have a huge impact on the application potential of hyperfine spectroscopy.

INTERELECTRON SPIN DISTANCES: PROBING NANOMETER DISTANCES WITH EPR

As mentioned earlier, dipolar electron spin–spin interactions can be probed with EPR techniques. If the interspin distance is smaller than 2 nm, considerable broadening of the cw-EPR spectrum occurs that can be interpreted in terms of the spin–spin distance [32]. Recent advances in pulsed EPR, most importantly the introduction of the four-pulse DEER (or four-pulse PELDOR) method [33] and the double quantum coherence techniques [34], have extended the electron interspin distance range accessible by EPR from 1 to 8 nm. However, only few natural systems contain two paramagnetic sites, limiting the applications. Wayne Hubbel and co-workers [35, 36] overcame this limitation in their pioneering work on site-directed spin labeling (SDSL) of proteins. This method allows attaching paramagnetic spin labels in proteins at specific points. In this way, any diamagnetic protein can be turned into an EPR-active molecule. By careful selection of the spin label insertion points and subsequent distance determination, a lot of structural information can be obtained about the protein. The spin labels are usually nitroxides, with the (1-oxyl-2,2,5,5-tetramethylpyrroline-3-methyl) methanethiosulfonate label (MTSSL) [37] being the most popular. Similarly, routes to spin label DNA or RNA have been introduced. Spin label EPR is now a booming area, and the number of systems that can be addressed by this technique is growing continuously. We refer the reader to a number of excellent reviews on this matter [38–43]. It is important to mention that the spin label orientation typically will show a certain amount of flexibility around the tether attaching the label to the protein. This spread in orientations will influence the distance determination. Hence, spin label EPR results need to be combined with molecular dynamics computations to link the experimentally observed distances to the molecule's structure [44].

TIME-RESOLVED EPR: DETECTION OF SHORT-LIVING PARAMAGNETIC STATES

Time-resolved EPR (TR-EPR) allows the investigation of shortly living excited states of molecules with lifetimes down to hundreds of nanoseconds. It is used to study paramagnetic states produced by photophysical as well as photochemical events. The spin populations shortly after a laser pulse are far from equilibrium, and the system is spin polarized. The acquisition of the EPR spectrum after a short delay via cw-EPR or pulsed EPR then allows the recording of spin-polarized lines, in emission or enhanced absorption, with a much better signal-to-noise (S/N) ratio than for equilibrium spin populations. The first observation of spin polarization was made for radical pairs produced in solution by photolysis, the CIDEP (chemically induced dynamic electron polarization) effect [45]. Radicals produced in pairs are spin correlated (SCRP); for example, they "remember" whether they are deriving by dissociation from a molecule in a singlet or in an excited triplet state. Studies on the effect of spins in chemical reactions have led to a new branch of chemistry, spin chemistry, involving specific studies on reaction yield detected magnetic resonance (RYDMR), magnetic field effects (MFE), in chemistry and so on [46]. Short-lived radical pairs (RP) are formed also during the primary energy conversion steps of natural photosynthesis [e.g., 47, 48] and in donor–acceptor pairs, or donor–bridge–acceptor triads via electron or charge transfer in materials mimicking natural photosynthesis [49]. The search is for materials allowing intramolecular electron transfer events ultimately resulting in a charge transfer (CT) excited state with a relatively long lifetime [50].

Photoexcited triplet states of many organic molecules have been characterized also by TR-EPR, obtaining information on the electron distribution in the triplet (via the zero-field splitting parameters), on the photophysical path leading to the triplet (as the spin populations depend on the previous history), and on the evolution of the excited state in time [51]. The spin evolution of systems of higher multiplicity, such as an organic excited triplet on a dye and spin exchanging with the unpaired electron on a stable radical, has also been studied [52]. This has been performed also on a series of fullerenes linked to a stable nitroxide radical [e.g., 53].

Time-resolved EPR observations have played a pivotal role in obtaining information about the properties and evolution of spin states in natural biological events driven by light and in complex materials. This information is of relevance not only in relation to solar energy conversion but also in relation to perspectives such as molecular electronics, molecular photonics, and molecular spintronics, where the spin manipulation by light is a very challenging topic. As dipolar electron–electron interactions and spin exchange interactions are fundamental parameters for the interpretation of these experiments, any further improvement in the ability in calculating zero field splitting (ZFS) parameters and spin exchange energy is very important. Reliable calculations of ZFS parameters are also of relevance for PELDOR experiments mentioned in the previous part, since the distance measured via this method between pairs of paramagnetic species is based on the value of their electron–electron dipolar interaction [54, 55].

EPR IMAGING

EPR is not only a spectroscopic tool but can also be used for imaging. This branch of EPR focuses on the imaging and location of paramagnetic species in different samples. One of the most interesting recent developments in this field is the electron spin resonance microscopy aimed at the observation of paramagnetic species in small (solid-state) samples with (sub-)micrometer-scale resolution [56]. In case of in vivo imaging, the strong absorption of the microwaves by the water results in large-dimension constraints. As a result, L-band (1–1.5 GHz) EPR spectrometers are used for studies of animals no larger than mice, whereas EPR spectrometers working at RFs (200–300 MHz) allow examination of rats [57]. Despite this limitation, EPR imaging is finding growing medical applications, such as in the monitoring of drug delivery processes [57], skin and hair research [58, 59], and cardiac research [60]. Further information can be found elsewhere [57, 58, 60].

CONCLUSION

In the last decades, EPR has become a very versatile research field, with different subdisciplines. Recent developments, including the introduction of high-field EPR, different pulsed EPR methodologies, spin-labeling techniques, and miniaturization, have enormously increased the number of problems that can be addressed with EPR. At the same time, there is a strong need of new and dedicated theoretical models and calculation tools in order to extract the maximum of information from the obtained data.

REFERENCES

1. G. Feher, *Phys. Rev.* **1956**, *103*, 834.
2. N. M. Atherton, *Principles of Electron Spin Resonance*, Ellis Horwood, New York, 1993.
3. B. Dzikovski, D. Tipikin, V. Livshits, K. Earle, J. H. Freed, *Phys. Chem. Chem. Phys.* **2009**, *11*, 6676.
4. G. L. Millhauser, J. H. Freed, *J. Chem. Phys.* **1984**, *81*, 37.
5. S. A. Dzuba, E. P. Kirilina, E. S. Salnikov, *J. Chem. Phys.* **2006**, *125*.
6. I. V. Borovykh, P. Gast, S. A. Dzuba, *Appl. Magn. Reson.* **2007**, *31*, 159.
7. E. G. Bagryanskaya, D. N. Polovyanenko, M. V. Fedin, L. Kulik, A. Sehnegg, A. Savitsky, K. Möbius, A. W. Coleman, G. S. Ananchenkof J. A. Ripmeesterf, *Phys. Chem. Chem. Phys.* **2009**, *11*, 6700.
8. A. Barbon, M. Bortolus, M. Brustolon, A. Comotti, A. L. Maniero, U. Segre, P. Sozzani, *J. Phys. Chem. B* **2003**, *107*, 3325.
9. S. S. Eaton, G. R. Eaton, *Biol. Magn. Reson.* **2000**, *19*, 29.
10. M. Marrale, M. Brai, A. Barbon, M. Brustolon, *Rad. Res.* **2009**, *171*, 349.
11. M. Mannini, L. Sorace, L. Gorini, F. M. Piras, A. Caneschi, A. Magnani, S. Menichetti, D. Gatteschi, *Langmuir* **2007**, *23*, 2389.
12. J. A. Weil, J. R. Bolton, J. E. Wertz, *Electron Paramagnetic Resonance*, Wiley, New York, 1994.
13. A. Schweiger, G. Jeschke, *Principles of Pulse Electron Paramagnetic Resonance*, Oxford University Press, Oxford, 2001.
14. M. Fittipaldi, I. García-Rubio, F. Trandafir, I. Gromov, A. Schweiger, A. Bouwen, S. Van Doorslaer, *J. Phys. Chem. B* **2008**, *112*, 3859.
15. P. Höfer, A. Grupp, H. Nebenführ, M. Mehring, *Chem. Phys. Lett.* **1986**, *132*, 279.
16. P. A. S. Cruickshank, D. R. Bolton, D. A. Robertson, R. I. Hunter, R. J. Wylde, G. M. Smith, *Rev. Sci. Instrum.* **2009**, *80*, 103102.
17. D. Goldfarb, Y. Lipkin, A. Potapov, Y. Gorodetsky, B. Epel, A. M. Raitsimring, M. Radoul, I. Kaminker, *J. Magn, Reson.* **2008**, *194*, 8.
18. E. R. Davies, *Phys. Lett. A* **1974**, *47*, 1.
19. W. B. Mims, *Proc. R. Soc.* **1965**, *283*, 452.
20. P. Schosseler, Th. Wacker, A. Schweiger, *Chem. Phys. Lett.* **1994**, *224*, 319.
21. M. Brecht, M. Van Gastel, T. Buhrke, B. Friederich, W. Lubitz, *J. Am. Chem. Soc.* **2003**, *125*, 13075.
22. T. Buttner, J. Geier, G. Frison, J. Harmer, C. Calle, A. Schweiger, H. Schonberg, H. Grutzmacher, *Science* **2005**, *307*, 235.
23. E. Vinck, D. M. Murphy, I. A. Fallis, R. R. Strevens, S. Van Doorslaer, *Inorg. Chem.* **2010**, *49*, 2083.
24. N. Weiden, H. Kass, K. P. Dinse, *J. Phys. Chem. B* **1999**, *103*, 9826.
25. S. B. Orlinskii, J. Schmidt, P. G. Baranov, D. M. Hofmann, C. D. Donega, A. Meijerink, *Phys. Rev. Lett.* **2004**, *92*, 47603.
26. G. C. Hurst, T. A. Henderson, R. W. Kreilick, *J. Am. Chem. Soc.* **1985**, *107*, 7294.
27. V. Barone, P. Cimino, O. Crescenzi, M. Pavone, *J. Mol. Struct: Theochem* **2007**, *811*, 323.

28. F. De Proft, E. Pauwels, P. Lahorte, V. Van Speybroeck, M. Waroquier, P. Geerlings, *Magn. Reson. Chem.* **2004**, *42*, S3.
29. F. Neese, *Coord. Chem. Rev.* **2009**, *253*, 526.
30. S. Riedel, M. Kaupp, *Coord. Chem. Rev.* **2009**, *253*, 606.
31. A. Bencini, *Inorg. Chim. Acta* **2008**, *361*, 3820.
32. C. Altenbach, K. J. Oh, R. J. Trabanino, K. Hideg, W. L. Hubbell, *Biochemistry* **2001**, *40*, 15471.
33. M. Pannier, S. Veit, A. Godt, G. Jeschke, H. W. Spiess, *J. Magn. Reson.* **2000**, *142*, 331.
34. P. P. Borbat, J. H. Freed, *Chem. Phys. Lett.* **1999**, *313*, 145.
35. C. Altenbach, S. L. Flitsch, H. G. Khorana, W. L. Hubbell, *Biochemistry* **1989**, *28*, 7806.
36. C. Altenbach, T. Marti, H. G. Khorana, W. L. Hubbel, *Science* **1990**, *248*, 1088.
37. L. J. Berliner, J. Grunwald, H. O. Hankovsky, K. Hideg, *Anal. Biochem.* **1982**, *119*, 450.
38. J. P. Klare, H.-J. Steinhoff, *Photosynthesis Res.* **2009**, *102*, 377.
39. Y. D. Tsvetkov, Y. A. Grishin, *Instr. Exp. Techniques* **2009**, *52*, 615.
40. O. Schiemann, *Methods Enz. Ser.* **2009**, *469* (Part B), 329.
41. O. Schiemann, T. F. Prisner, *Q. Rev. Biophys.* **2007**, *40*, 1.
42. P. P. Borbat, A. J. Costa-Filho, K. A. Earle, J. K. Moscicki, J. H. Freed, *Science* **2001**, *291*, 266.
43. G. Jeschke, Y. Polyhach, *Phys. Chem. Chem. Phys.* **2007**, *9*, 1895.
44. Y. Polyhach, A. Godt, C. Bauer, G. Jeschke, *J. Magn, Reson.* **2007**, *185*, 118.
45. P. W. Atkins, K. A. McLauchlan, A. F. Simpson, *Nature* **1968**, *219*, 927.
46. K. A. McLauchlan, *J. Chem. Soc. Perkin Trans.* **1997**, *2*, 2465.
47. G. Kothe, M. C. Thurnauer, *Photosynth. Res.* **2009**, *102*, 349.
48. (a) A. Savitsky, K. Möbius, *Helv. Chim. Acta.* **2006**, *89*, 2544; (b) K. Möbius, A. Savitsky, *High-Field EPR Spectroscopy on Proteins and Their Model Systems: Characterization of Transient Paramagnetic States*, Royal Society of Chemistry, Cambridge, 2009.
49. Z. E. X. Dance, Q. Mi, D. W. McCamant, M. J. Ahrens, M. A. Ratner, M. R. Wasielewski, *J. Phys. Chem. B* **2006**, *110*, 25163.
50. J. W. Verhoeven, *J. Photochem. Photobiol. C Photochem. Rev.* **2006**, *7*, 40.
51. N. Hirota, S. Yamauchi. *J. Photochem. Photobiol. C Photochem. Rev.* **2003**, *4*, 109.
52. G. I. Likhtenstein, K. Ishii, S. Nakatsuji, *Photochem. Photobiol.* **2007**, *83*, 871.
53. L. Franco, M. Mazzoni, C. Corvaja V. P. Gubskaya, L. S. Berezhnaya, I. A. Nuretdinov, *Mol. Phys.* **2006**, *104*, 1543.
54. S. Sinnecker, F. Neese, *J. Phys. Chem. A* **2006**, *110*, 12267.
55. C. Riplinger, J. P. Y. Kao, G. M. Rosen, V. Kathirvelu, G. R. Eaton, S. S. Eaton, A. Kutateladze, F. Neese, *J. Am. Chem. Soc.* **2009**, *131*, 10092.
56. A. Blank, E. Suhovoy, R. Halevy, L. Shtirberg, W. Harneit, *Phys. Chem. Chem. Phys.* **2009**, *11*, 6689.
57. D. L. Lurie, K. Mäder, *Adv. Drug Deliv. Rev.* **2005**, *57*, 1171.
58. P. M. Plonka, *Exp. Dermatol.* **2009**, *18*, 472.
59. E. Vanea, N. Charlier, J. De Wever, M. Dinguizli, O. Feron, J. F. Baurain, B. Gallez, *NMR Biomed.* **2008**, *21*, 296.
60. P. Kuppusamy, J. L. Zweier, *NMR Biomed.* **2004**, *17*, 226.

CHALLENGE OF OPTICAL SPECTROSCOPIES

ERMELINDA M. S. MAÇÔAS

Centro de Química-Física Molecular (CQFM) and Institute of Nanoscience and Nanotechnology (IN), Instituto Superior Técnico, Lisbon, Portugal

INTRODUCTION

Optical spectroscopy reveals information about the perturbation of the oscillating charged particles of atoms and molecules (nuclei and electrons) induced by the oscillating electric field of light. The extent of the perturbation depends on the structure of the material giving insight into key structure–reactivity relationships. The interaction of the electric field with matter can lead to a series of events that are better understood considering either the wavelike (refraction and reflection) or particlelike (absorption, emission, scattering) nature of radiation and matter.

A light wave can be refracted (bent) and reflected when crossing the boundary between media of different density (different index of refraction). Refraction is a very important phenomenon when dealing with high-power, short light pulses. The magnitude of the wave vector (k) defining the direction of propagation of a light wave depends on the index of refraction (n) of the medium ($k = 2\pi/\lambda = \omega n/c$), which is frequency and intensity dependent. This dependence results in the time delay, duration broadening, and frequency variation in time (chirp) experienced by short pulses propagating through a transparent medium [1].

Resonant light–matter interaction occurs when the photon energy (or frequency of the oscillating field) matches the energy difference (transition frequency) between rotational, vibrational, or electronic states of the system leading to absorption or

Computational Strategies for Spectroscopy: From Small Molecules to Nano Systems, First Edition. Edited by Vincenzo Barone.

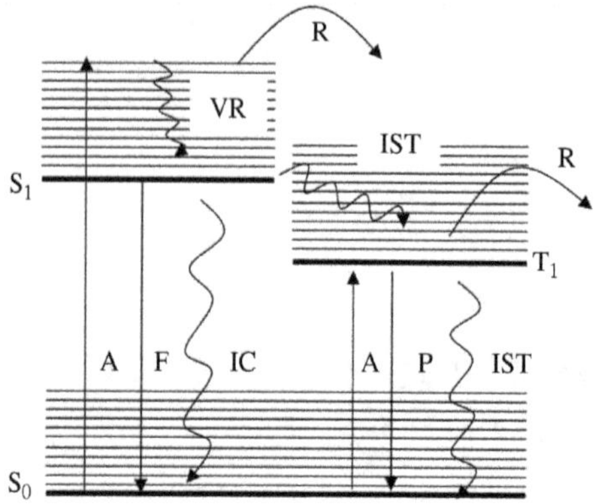

Figure 1 State energy diagram representing possible photochemical and photophysical processes triggered by absorption of a photon: A, absorption; VR, vibrational relaxation; F, fluorescence; IC, internal conversion; P, phosphorescence; IST, intersystem crossing; R, reaction.

emission of a photon. The excess energy introduced in a molecular system by interaction with the electric field can be relaxed by different photochemical and photophysical processes (Figure 1). The energy can be redistributed among the vibrational states of the system and further relaxed by intermolecular energy transfer. It can lead to transitions between states of the same spin (internal conversion, IC) or states of different spin (intersystem crossing, IST) as well as reactions in the excited state (R). It can also be deactivated radiatively by singlet–singlet emission (fluorescence, *F*) and triplet–singlet emission (phosphorescence, *P*) [2]. Light scattering is a nonresonant phenomenon that is best understood by considering the periodic oscillating charge induced within the system due to perturbation of the electron cloud by the oscillating electric field. The induced oscillating dipole moment is by definition a source of radiation, thereby resulting in emission of light. The light is emitted at either the same or different frequencies from the incident light in the so-called elastic and inelastic scattering interaction, respectively. Raman spectroscopy probes the inelastic scattering due to molecular rotations and vibrations.

The molecular response to a perturbation induced by an incident radiation field can be generalized by the polynomial expansion of the macroscopic electric polarizability vector (**P**) as follows:

$$\mathbf{P} = \varepsilon_0 \left[\chi^{(1)}\mathbf{E} + \chi^{(2)}\mathbf{E}.\mathbf{E} + \chi^{(3)}\mathbf{E}.\mathbf{E}.\mathbf{E} + \chi^{(4)}\mathbf{E}.\mathbf{E}.\mathbf{E}.\mathbf{E} + \chi^{(5)}\mathbf{E}.\mathbf{E}.\mathbf{E}.\mathbf{E}.\mathbf{E} + \ldots \right] \quad (1)$$

where ε_0 is the electric permittivity, **E** is the complex electric field function of the incident light wave, and $\chi^{(n)}$ are the $(n + 1)$-rank tensors of the macroscopic susceptibility. For weak radiation fields interacting with the material, the induced perturbation scales linearly with the magnitude of the electric field due to the vanishingly small contributions from higher order terms than the first order. When two or more independent radiation fields interact with the medium or intense radiation fields are used, the higher order terms must be taken into account. In this situation, the response of the material becomes nonlinear. The imaginary parts of the odd terms are used to express the linear, two-photon and three-photon absorption cross sections. The even terms are at the origin of phenomena like second-harmonic generation, sum-and-difference frequency mixing, optical rectification, and linear electro-optical effects. They are relevant for surfaces and interfaces but vanish in isotropic centro-symmetric media. Depending on the particular orientation and delay between independent radiation fields acting on the material, the third-order term can be used to express many of the nonlinear optical spectroscopies commonly used in photo-physic and dynamic studies, like coherent Raman, pump–probe, transient grating, and photon echo spectroscopies.

The optical susceptibility $\chi^{(n)}$ is a property of the medium that relates to its microscopic structure. It can be measured experimentally, but a fundamental understanding of the underlying complex structure–optical properties relationship necessarily relies on the development of theoretical models able to predict the nth-order nonlinear optical effects. Even though optical susceptibility can be best evaluated by a full quantum mechanical treatment, for practical purposes, interpretation of the experimental results is based on some sort of simplified models. The following sections in this chapter are a brief illustration of the application of linear and third-order spectroscopies to study the photophysics and dynamics of molecular systems pointing out the interplay between experiments and computational modeling. The focus will be mainly on third-order spectroscopies that have largely superseded the more conventional linear spectroscopy approach due to the great detail of information that can be extracted from spectroscopic methods based on nonlinear interaction of the electric field with matter.

STEADY-STATE LINEAR SPECTROSCOPY

In steady-state linear optical spectroscopies, resonance frequencies, amplitudes, and lineshapes are measured that provide valuable information about structure, reactivity, and photophysics. Interpretation of the observed spectral features is greatly aided by computational simulations. Vibrational spectra of small- to medium- size molecules and aggregates are routinely interpreted based on computed harmonic vibrational transitions and oscillator strength for optimized ground-state geometries, which are readily available from different computational programs at various levels of theoretical approximation. Spectroscopic investigation of the conformational landscape and reaction pathways in prototypical molecules and supramolecular assemblies relevant in biochemistry (e.g., biomolecular building blocks, physiologically active compounds, models for ligand–receptor interactions) are extensively supported by ab initio and density functional theory (DFT) calculations of ground-state potential energy surface, allowing us to understand at a fundamental level intra- and intermolecular interactions like hydrogen-bonding effects, isomerization, and specific solute–solvent interactions [3–5]. Due to the high computational cost associated with anharmonic calculations, vibrational spectra of medium- to large-size molecules are typically done under the harmonic approximation. In conformationally flexible systems characterized by highly anharmonic large-amplitude modes, this treatment has significant limitations. This is usually the case in carboxylic acids and amino acid groups so ubiquitous in all sorts of chemical and biological processes (e.g., molecular recognition, self-assembly, protein folding, formation of micellar carriers for drug delivery, and smart polymers) [6–11]. In order to achieve spectroscopic accuracy in these types of systems, different strategies are being explored to introduce realistic anharmonic potential. Efforts have been made in finding suitable computationally tractable solutions of the vibrational Schrödinger equation, including anharmonic potentials defined at different levels of approximation, in introducing anharmonic corrections via a perturbation approach based on the harmonic oscillator states, and in the parallelization of the calculations of both the anharmonic potential and the vibrational eigenvalues and eigenvectors [12–14].

Computation of excited-state energies, structures, and electronic properties has also become an indispensable tool in the interpretation of electronic spectra and excited-state dynamics. Electron density based time-dependent DFT (TD-DFT) methods have become very popular in the simulation of excited-state properties due to their ability to give impressive quantitative results at a moderate computational cost for relevant medium- to large-size molecule, as opposed to the more time-consuming correlated wavefunction-based methods, such as multireference configuration interaction (MRCI), multiconfigurational self-consistent field (MCSCF), complete active-space SCF (CASCF), and complete active-space perturbation theory (CASPT) [15]. Every week several reports are published demonstrating the usefulness of the TD-DFT methods in increasingly complex systems. TD-DFT has been systematically applied in photophysical studies addressing metal complexes and dyes as potential sensitizers in solar cells [16, 17], common fluorescent probes used in spectroscopy and microscopy [18, 19], nucleobases present in RNA and in DNA [20], conjugated polymers with applications in photonic devides [21], and many other complex systems. The recent simulations of the electronic spectra of carbonyl-protected gold cluster are but one example of how far this method can go [22]. Implementation of TD-DFT in electronic structure calculation programs was a great driving force for routine application of these methods in photochemistry and photophysical studies. In addition, the set of programs nowadays available to visualize molecular orbitals can give a very intuitive picture of the electronic transitions. TD-DFT also has its own limitations that one should be aware of in order to extract reliable information. The choice of a functional and basis set should be critically accessed against known experimental data about the particular system under study. The failure of common functionals to describe long-range dispersion forces of significant importance to supramolecular systems and the limited accuracy in describing delocalized excited states (charge transfer state, valence states of molecules with extended π systems, Rydberg states, and doubly excited states) are known shortfalls in the performance of TD-DFT [15, 23, 24]. Even within the same molecule it is not unreasonable to question the validity of a particular choice of functional to address different electronic states. Nevertheless, development of corrected exchange-correlation functional is still a very active area of research.

The apparent structureless bandshape of broad electronic transitions observed in ultraviolet (UV)–visible absorption or emission spectra hides a complicated set of vibronic transitions that can only be unraveled by excited-state electronic structure calculation methods [18, 25]. Even though the formulation exists to include vibrational couplings and anharmonic corrections in computing electronic transitions that would allow for a quantitative comparison with experimental data, their implementation has so far not been trivial [18, 25]. However, this picture is most likely about to change as an integrated set of computational tools easy to handle by nonspecialists and able to run on standard personal computers has been recently reported [26].

For molecules in the condensed phase, including environment effects in the simulations of electronic and vibrational spectra is of utmost importance. The environment can stabilize preferentially specific molecular conformations and excited states either due to polarity effects or via specific interactions. Indeed, information about the environmental polarity can be retrieved from solvatochromic effects observed in electronic absorption spectra due to differences in the electric dipole moment of the states involved in the electronic transition [27]. Several approaches are available to account for environmental effects, from the simplest model representing the solvent by a continuum characterized by a dielectric constant to the supramoleular approach where the solute plus solvent molecules introduced as near neighbors are treated quantum mechanically within an external potential created by the solvent [28, 29].

STEADY-STATE THIRD-ORDER SPECTROSCOPY

In the last two decades many creative technological applications of nonlinear absorption have been demonstrated stimulated by the development of suitable powerful pulsed lasers operating at near-IR wavelengths. Frequency up-conversion lasing; optical power limiting, stabilization, and reshaping; optical 3D microfabrication; drug delivery; high-density optical data storage; and laser scanning multiphoton excited fluorescence microscopy are among the areas where exploration of different events triggered by multiphoton absorption has led to significant technological developments [30]. Multiphoton absorption is here understood as the simultaneous absorption of more than one photon by the same molecule (Figure 2). It provides a mean to access a given excited state by absorption of lower energy photons when compared to the corresponding one-photon-induced transition. The need for high-power pulsed lasers stems from the fact that multiphoton absorption cross sections, σ_n, are considerably lower than linear absorption cross sections (e.g., $\sigma_1 \sim 10^{-19}$–$10^{-16}\,\mathrm{cm}^2$, $\sigma_2 \sim 10^{-51}$–$10^{-46}\,\mathrm{cm}^4\,\mathrm{s}$, and $\sigma_3 \sim 10^{-78}$–$10^{-80}\,\mathrm{cm}^8\,\mathrm{s}^2$). The high-order dependence of the multiphoton absorption probability on the light intensity is responsible for an intrinsic confinement of the absorption to a smaller volume when compared to the linear absorption. Elimination of undesirable out-of-focus absorption when conjugated with a greater depth of penetration of the longer wavelengths used in multiphoton excitation constitutes the fundamental advantages of nonlinear absorption over the corresponding linear process. Under diffraction-limited focusing conditions,

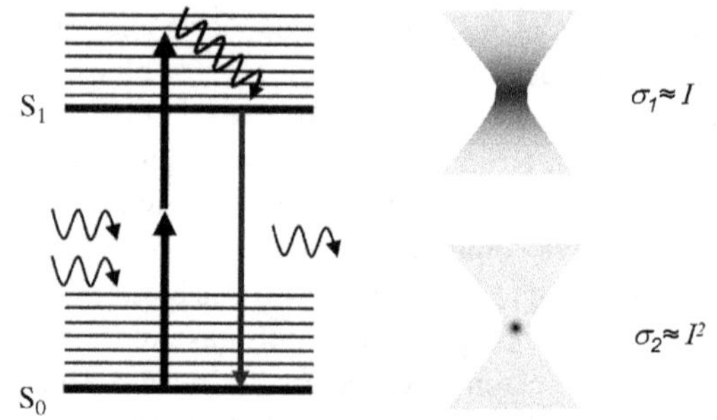

Figure 2 Schematic representation of simultaneous two-photon absorption followed by vibrational energy relaxation and emission of higher energy photon. The quadratic dependence of the two-photon absorption cross section on the excitation intensity leads to a confinement of the absorption process to a smaller volume when compared with linear absorption.

the absorption can be confined to a nanometric volume by high-numerical-aperture objectives enabling the activation of chemical or physical processes with high 3D spatial resolution.

Due to the ease of implementation, connected with the widespread availability of high-peak-power pulsed lasers tunable in the range of 700–1000 nm (Ti–sapphire oscillators), two-photon absorption of molecular and supramolecular systems with linear absorption in the range of 350–500 nm has been by far the most extensively explored process. The complexity of optimized two-photon absorption chromophores range from simple dipolar molecules to hyperbranched dendritic structures, polymers, conjugated polymer dots, semiconductor, and up-conversion nanoparticles. In cases where the optimized system to perform a given task (e.g., delivery of a given drug, singlet oxygen release) is not a particularly good two-photon absorption chromophore, ingenious architectures have been devised where good two-photon absorption chromophores are used as antenna that later will channel the excitation energy into the desired reaction center [30, 31].

The two-photon absorption cross section is expressed in terms of the third-order macroscopic susceptibility ($\chi^{(3)}$ in expression 1), which is an orientational average of the microscopic polarizabilities induced by the effective local field [32]. The microscopic polarizabilities can be expanded in terms of the transition dipole moments associated with many-body excited states, so that the two-photon absorption cross section is proportional to the sum over the transition dipole moments between the ground- and excited-state wavefuncions [33]. The general requirements for successful two-photon absorption systems are quite well understood [30]. These systems will preferably have either a permanent ground-state dipole or resonant charge transfer excited states. Thus, strong electron-donating (D) and electron-withdrawing (A) groups bridged by polarizable π-conjugated systems are needed. Many different geometries have been explored in order to establish the relevant structure–activity relationships, from simple dipolar molecules (D–π–A), quadrupolar (D–π–A–π–D, A–π–D–π–A), and octupolar [D(–π–A)$_3$, A(–π–D)$_3$] systems to dendrimers, cyclic oligomers, and polymers. Systematic studies have brought insight into how factors like the nature of the D and A groups, type and extent of the bridge, branching, symmetry, and solvation affect the two-photon absorption cross section. Nowadays, there are records of supramolecular systems with extremely high two-photon absorption cross sections (200,000 GM, 1 GM $= 10^{-50}$ cm^4 s, more than three orders of magnitude higher than the two-photon active molecules known 10 years ago) [34]. Much of this achievement is due to an extremely huge effort of molecular engineering in the quest to maximize two-photon absorption cross sections. Nevertheless, as the complexity of the structures increases, simple structure–activity relationships became less clear as the factors affecting the two-photon activity became strongly entangled.

The guidance that computational estimations can bring to the molecular engineering strategies designed to tackle specific applications is precious. Just as in the case of linear absorption, wave function-based correlated ab initio methods are currently of little use in the treatment of nonlinear optical properties of molecules of practical interest. Two-photon absorption properties have been most successfully modeled by TD-DFT extensions for the calculation of molecular nonlinear optical properties based on the residues of the quadratic response function [35], and on the quasi-particle formalism of the TD-DFT equations for arbitrary frequency-dependent nonlinear optical polarizability [36, 37]. Fortunately, for organic molecules only a limited number of states are optically active in the UV–visible spectral region. These are usually states with strong transition dipole moments associated with a charge transfer nature of the excited state and the delocalized π-electron systems. Indeed, the complete sum-of-all-states expression providing expansions of the molecular polarizability into dipolar contributions from different many-body excited states can be simplified under certain assumptions to an actual sum over few states. These simplified expressions have been applied with great success to understand the trends in two-photon absorption activity of dipolar and quadrupolar molecules [33]. Branched molecules with several chromophores are usually simulated by a Frenkel exciton model [33]. Nevertheless, it has become clear that electrostatic interactions alone cannot provide an accurate description of the two-photon absorption process in these systems and electronic correlation as to be explicitly introduced [33, 38].

Despite the remarkable progresses made in the last decade, it should be noted that while many of the identified potential applications require high-energy excited states that can efficiently transfer the excitation energy into desired reaction centers, the high two-photon absorption cross sections reported so far are usually connected with low-energy excited states. Materials that can be excited by simultaneous absorption of two-photons at higher energies are needed. In addition, for two-photon active material to be of practical use, optimization of secondary properties key to specific applications must also be carefully considered (e.g, solubility, cytotoxicity). Future developments can be envisaged in the exploration of different types of systems (charged and open-shell molecules) and different types of processes connected with two-photon absorption (e.g., circular dichroism). Simplified theoretical models that include electronic correlation are needed to give clear indication as the way to proceed avoiding the up-to-now trial-and-error search for improved chromophores.

TRANSIENT THIRD-ORDER SPECTROSCOPY

The linewidth characteristic of steady-state spectroscopic features contains contributions from coherent dephasing, inhomogeneous dephasing, and relaxation. Nonlinear transient spectroscopy is able to isolate these contributions giving precious insight into the evolution of coherences in the excited state, lifetimes, and solvation dynamics. All of those are crucial to understand the dynamics of ultrafast events such as bond breaking/bond formation, photoisomerization, electron and proton transfer, exciton relaxation, and the dynamics of energy relaxation in general. The widespread availability of powerful femtosecond laser pulses together with the development of tools to effectively shape and control these pulses (tuning of central frequencies, bandwidth, pulsewidth, phase and polarization direction) has led to major developments in ultrafast spectroscopic techniques in the last two decades.

Time-correlated single-photon counting (TCSPC) and fluorescence up-conversion are the most popular techniques for real-time measurements of fluorescence emission [39]. In TCSPC, the arrival time of single photons emitted by a sample excited by a train of ultrafast pulses (femtoseconds to picoseconds) is recorded. After accumulation of enough single-photon events, a histogram of detected times allows for reconstruction of the waveform of the emitted signal [40]. Even when using the fastest avalanche photodiodes as detector, time resolutions are limited to 20–50 ps. In the less sensitive fluorescence up-conversion technique, the fluorescence lifetime can be recorded with a resolution that is only limited by the duration of the laser pulse (typically less than 100 fs). An ultrafast laser beam is split so that part of it is used for excitation of the sample and part of it is used for optical gating of the emission signal. The gate pulse is sum–frequency mixed with the photons emitted by the sample in a nonlinear crystal with an appropriate phase-matching geometry. The up-converted signal, which is separated in space from both the gate pulse and the fluorescence pulse, can be measured by a "slow" detector. The fluorescence up-conversion signal is measured as a function of a time delay between the excitation event and the optical gate [39]. Just like most of the transient spectroscopy methods, the fluorescence up-conversion technique is a variation of the general pump–probe method [41, 42]. In pump–probe methods a strong pump beam interacts with the sample and a weaker beam arriving at a variable delay time is used to probe the time-dependent changes induced by the pump pulse (Figure 3). The frequencies of the two pulses and the time delays between them are the controllable variables of the setup. There are many variations of the pump–probe scheme depending on which processes are being monitored (absorption and emission) and on which type of wavelength is used for the pump and probe beams (UV–vis, UV–IR, vis–IR, IR–IR, etc). Other popular third-order transient spectroscopics used to interrogate coherent dynamics are transient grating (TG), coherent anti-Stokes–Raman (CARS), and photon echo (PE) spectroscopies [43]. All of those are based on noncollinear four-wave mixing arrangements differing in the pulse sequence and the number of controllable wavelengths (Figure 3). Several time-resolved wave-mixing experiments can be performed with minimum modifications on the setup. The type of response observed

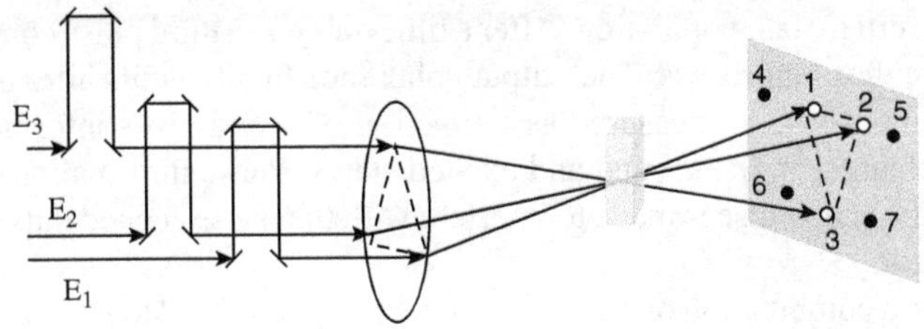

Phase-matching conditions:

$$\mathbf{k}_4 = 2\mathbf{k}_1 - \mathbf{k}_2 \qquad \mathbf{k}_6 = \mathbf{k}_1 - \mathbf{k}_2 + \mathbf{k}_3$$

$$\mathbf{k}_5 = -\mathbf{k}_1 + \mathbf{k}_2 \qquad \mathbf{k}_7 = -\mathbf{k}_1 + \mathbf{k}_2 + \mathbf{k}_3$$

Figure 3 Folded BOXCARS geometry applied in several transient nonlinear optical spectroscopies. In pump–probe spectroscopy, one of the three beams is blocked and the intensity of one of the incoming beams is monitored as a function of the time delay between the remaining two beams (e.g., beam 3 is blocked and beam 2 is monitored as a function of its delay with respect to beam 1, phase-matching condition would be $\mathbf{k}_2 = \mathbf{k}_1 - \mathbf{k}_1 + \mathbf{k}_2$). Beams 4 and 5 are photon echo signals generated from beams 1 and 2. Beams 6 and 7 can be stimulated photon echo or transient grating signals generated from beams 1, 2, and 3. In transient grating two of the beams are time coincident. In coherent anti-Stokes–Raman spectroscopy, beams 1 and 3 are time coincident and carry the same frequency; the difference between this frequency and that of beam 2 (so-called Stokes beam) matches a vibrational frequency of the system; and beam 6 will correspond to the anti-Stokes emission.

is determined by the phase-matching geometry based on momentum conservation of the involved photons. In TG spectroscopy two time-coincident pulses cross at the sample with some angle [44]. The two beams create a spatially varying interference pattern along the transverse direction that interacts with the sample. The optical grating can be written by changes in absorption (excited-state grating), subsequent relaxation can lead to a periodic temperature profile (thermal grating), and nonresonant scattering can create a spatial modulation of the refractive index. A third pulse approaching the grating at an angle that satisfies the Bragg condition is diffracted off the grating probing the time evolution of the system. In CARS spectroscopy the two-time coincident pulses have different colors and the energy difference matches a vibrational frequency of the sample creating a vibrational coherence [45]. The probe measures the degree of vibrational coherence remaining after a certain time delay. PE spectroscopy is the optical analogue of spin echo spectroscopy in NMR [46]. It is the method of excellence to disentangle homogeneous and inhomogeneous line broadening. A laser pulse creates a coherence between ground (g) and excited state (e), which is allowed to evolve freely with a frequency $\omega_{eg} = (E_e - E_g)/\hbar$ for a given time interval (τ). A second pulse switches the coherence to evolve at a frequency $-\omega_{eg}$ restoring the initial phase after a time 2τ. The emitted coherent signal recorded as a function of the evolution time bares only the homogeneous relaxation dynamics. In the three-pulse version of PE, also known as stimulated photon echo, the second pulse converts the ω_{eg} coherence into a ground- or excited-state population. After a time delay T, a third pulse comes that will initiate the rephasing process. The output coherence finally emits after a time $t < \tau$. The polarization echo is measured as a function of T and gives information about population changes in the ground and excited states. Phase information can also be collected if the echo pulse is made to interfere with the phase-locked pulse, called the local oscillator.

Generally, a combination of the described techniques is used to interrogate in great detail the dynamics of the systems. The femtosecond time resolution provided by this set of spectroscopic techniques is the key for observation of atomic-scale dynamics [47]. Molecular rotations and vibrations can be observed in real time rather than in energy spectra. Following directly the motion of the system along the potential energy surface while it moves from the reactant well trough the transition state into the product well became possible with femtosecond spectroscopies. Dissociation of the NaI molecule is a paradigmatic example of chemistry probed in real time [47]. With dephasing times slower than the nuclear motion, the wavepacket dynamics and quantum beats could be investigated by femtosecond spectroscopy. As expected for reactive systems with slow dephasing, the wavepacket repeatedly sampled the reactive curve crossing, each time crossing with some probability into the product side. The kinetics of product formation showed a stepwise rising curve with a frequency corresponding to the period of the wavepacket motion [48].

The theoretical treatment of transient nonlinear spectroscopic features is based on the density matrix and the perturbative solution of the quantum mechanical Liouville equations describing the temporal evolution of the system under the influence of the incident electromagnetic fields as well as the intrinsic relaxation processes [32, 49]. Generally, the signals observed are understood in terms of a diagrammatic representation on double-sided Feynman diagrams of the time evolution of the density matrix operator and its transformations by interaction with a particular time order and geometric arrangement (phase-matching conditions in Figure 3) of the electric fields involved [50]. The corresponding mathematical expressions can then be written in a very intuitive way following some general rules that translate the relevant diagrammatic representations [32]. Frequently, a reduction of the high dimensionality of the problem is needed in order to obtain computationally tractable expressions. Simulations at diverse levels of theoretical approximation, from classical mechanics to mixed quantum mechanical/molecular mechanical (QM/MM) models to full quantum mechanics, are applied depending on the nature of the problem at hands [32, 47, 51]. The computation of electronic energies as a function of molecular geometry can be used to obtain adiabatic transition energies and identify state crossings important in ultrafast nonradiative decay probed by transient spectroscopies. Such calculations at a high level of accuracy have located conical intersections pointed out to be responsible for the ultrafast energy relaxation observed in DNA base pairs providing the mechanism for protection of UV-induced phodamage in DNA [52, 53]. Much of our present understanding of the rich photophysics and photochemistry observed in transition metal complexes, with relevance to their application in solar cells, organic light-emitting diodes (LEDs), and photoswitches, results from a combined effort of transient spectroscopic studies and TD-DFT simulations [54]. Modeling of transient spectroscopy signals has to account also for transition frequency fluctuations due to solute–solvent interactions. The environment plays a key role in the excited-state dynamics. Besides providing a bath for energy relaxation [55], environment fluctuations can aid intramolecular energy redistribution by providing frequency components that equal the energy mismatch between two eigenstates [56], and it can assist in barrier crossing by modifying the intramolecular potential energy surface [57, 58].

Ultrafast spectroscopies have opened the door not only to understand early time dynamics of the intrinsically fast processes but also to control the outcome of the reaction in real time [59, 60]. Quantum control necessarily relies on very detailed a priori knowledge of the potential energy surface and dynamics of the system for designing pulse shapes and pulse sequences that will ultimately promote the molecules into the desired reaction path with minimal energy losses.

All the above-described techniques have potential to reveal much more and in a more intuitive way if the coherent higher order polarization induced by the sequence of optical trains can be projected in multidimensions. Indeed, coherent multidimensional optical spectroscopy represents the state of the art in optical spectroscopies. It is a hectic fast-evolving field of research that was experimentally realized for the first time nearly 10 year ago [50]. In two-dimensional (2D) pump–probe vibrational and electronic spectroscopies the pump pulse is tunable and the probe pulse is a spectrally broad pulse dispersed by an array detector [61]. The molecular response is projected over the two frequency axis resulting in a 2D spectrum analogue to the well-known 2D nuclear magnetic resonance (NMR) spectra. The availability of double-array detectors in the visible has made single-shot 2D electronic spectroscopy possible [62, 63]. The pump pulse is spectrally dispersed along a given direction (say the x axis) perpendicular to the propagation direction (z axis) before interacting with the sample, and the generated pump–probe signal is spectrally dispersed in a different direction (y axis) originating a 2D signal directly imaged at the array detector. Imaging of frequency up-converted infrared signals by the same double-array visible detectors was also proposed for single-shot 2D IR [62]. The 2D-PE method is based on the stimulated photon echo pulse sequence described above. This time the echo is recorded interferometrically by a monochromator as a function of the time between the first two pulses (τ). The first Fourier transform is perform experimentally by the interferometric detection of the echo ($t \rightarrow \omega_t$) giving one of the axes in the 2D plot, while a second numerical Fourier transform with respect to τ gives the second axis ($\tau \rightarrow \omega_\tau$) [64, 65].

Regardless of being at a very insipient stage of development, these coherent multidimensional nonlinear spectroscopies have already proven their potential for becoming a complementary technique to multidimensional NMR for resolving molecular structures, with the added value of increased time resolution and wide dynamic range. In a 2D spectrum, the off-diagonal peaks give detailed information about molecular couplings and coherences, which can be traced back to structural properties. Important correlations between secondary-structure proteins and 2D IR spectroscopic features in the amide I band have been established and used in the interpretation of 2D IR spectra of different globular proteins [51]. The amide I band in model peptides has been studied by femtosecond IR pump–probe spectroscopy revealing the delocalized nature of this mode [66]. The lineshapes in the 2D spectra were correctly simulated by a simple excitonic coupling model. Nonlinear transient 1D IR and 2D IR photon echo experiments together with an extensive computational modeling have been systematically applied to study hydrogen-bonding dynamics [67]. Studies in water using the reporter HOD molecule have shown ultrafast spectral diffusion at the sub-100-fs time scale and hydrogen-bonding lifetimes of about 1 ps [67]. In carboxylic acid dimers, OH stretch dephasing times of 200 fs and multiple quantum beats associated with coherent wavepacket motions along low-frequency hydrogen bond modes have been observed [67]. Recently, a surface-specific 2D IR method has been demonstrated able to study the dynamics in self-assembled monolayers [68]. Very interestingly, 2D electronic photon echo has shown evidences of the wave like nature of energy transfer from the antenna chromophore to the reaction center in photosynthetic systems [69, 70].

Understanding the origin of the off-diagonal signals in 2D spectra, their amplitude and lineshape evolution, and the observed beats relies heavily on theoretical modeling and computations [51]. Depending on the complexity of the particular system under study, different simplifications of the many-body electronic problem have been proposed. Multichromophoric complex systems have been quite successfully approximated by the Frenkel exciton model [51]. For electronic transitions the chromophores are described by a two-level system (ground and excited state), while for vibrational transitions the chromophores are considered to be three-level anharmonically corrected systems (the ground state and the two excited states involved in the fundamental and overtone transitions) [51].

Multidimensional nonlinear spectroscopy is a huge technological revolution that is still in its infancy. Up to now only a handful of laboratories are sufficiently well equipped to carry on these technically demanding experiments. The obtained results are encouraging, and one can only hope that in the future turnkey 2D spectrometers will be made available. At this point there is room for much improvement at both the technological and theoretical levels. For instance, circularly polarized light could be applied that would discriminate between chiral molecules so ubiquitous in biologically relevant systems. Exploring higher dimensionalities can be anticipated for the near future [71]. More realistic, computationally tractable theoretical models accounting for the effects of electron correlation and anharmonic couplings for strongly interacting chromophores are needed as well as an integrated set of computational tools that can be applied to any type of molecular system irrespective of its size and complexity.

REFERENCES

1. C. Rulliére, Ed., *Femtosecond Laser Pulses*, Springer Science+Business Media, New York, 2005.
2. N. J. Turro, *Modern Molecular Photochemistry*, University Science Books, Sausalito, 1991.
3. J. Kubelka, T. A. Keiderling, *J. Am. Chem. Soc.* **2001**, *123*, 12048.
4. J. R. Carney, T. S. Zwier, *J. Phys. Chem. A* **2000**, *104*, 8677.
5. P. Butz, R. T. Kroemer, N. A. Macleod, J. P. Simons, *J. Phys. Chem. A* **2001**, *105*, 544.

6. V. Munoz, P. A. Thompson, J. Hofrichter, W. A. Eaton, *Nature* **1997**, *390*, 196.
7. W. I. Weis, K. Drickamer, *Annu. Rev. Biochem.* **1996**, *65*, 441.
8. R. P. Sijbesma, F. H. Beijer, L. Brunsveld, B. J. B. Folmer, J. Hirschberg, R. F. M. Lange, J. K. L. Lowe, E. W. Meijer, *Science* **1997**, *278*, 1601.
9. M. Matsumura, D. H. Fremont, P. A. Peterson, I. A. Wilson, *Science* **1992**, *257*, 927.
10. U. Manna, S. Patil, *J. Phys. Chem. Lett.* **2010**, *1*, 907.
11. S. J. Chen, J. L. Hu, C. W. Yuen, L. K. Chan, *Polym. Int.* **2010**, *59*, 529.
12. D. Begue, N. Gohaud, R. Brown, C. Pouchan, *J. Math. Chem.* **2006**, *40*, 197.
13. V. Barone, *J. Chem. Phys.* **2005**, *122*, 10.
14. O. Christiansen, *Phys. Chem. Chem. Phys.* **2007**, *9*, 2942.
15. A. Dreuw, M. Head-Gordon, *Chem. Rev.* **2005**, *105*, 4009.
16. J. E. Monat, J. H. Rodriguez, J. K. McCusker, *J. Phys. Chem. A* **2002**, *106*, 7399.
17. M. P. Balanay, D. H. Kim, *Phys. Chem. Chem. Phys.* **2008**, *10*, 5121.
18. R. Improta, V. Barone, F. Santoro, *Angew. Chem.-Int. Edit.* **2007**, *46*, 405.
19. D. Jacquemin, E. A. Perpete, I. Ciofini, C. Adamo, *Acc. Chem. Res.* **2009**, *42*, 326.
20. T. Gustavsson, A. Banyasz, E. Lazzarotto, D. Markovitsi, G. Scalmani, M. J. Frisch, V. Barone, R. Improta, *J. Am. Chem. Soc.* **2006**, *128*, 607.
21. J. F. Briere, M. Cote, *J. Phys. Chem. B* **2004**, *108*, 3123.
22. O. Lopez-Acevedo, J. Rintala, S. Virtanen, C. Femoni, C. Tiozzo, H. Grnbeck, M. Pettersson, H. Hkkinen, *J. Am. Chem. Soc.* **2009**, *131*, 12573.
23. S. Grimme, J. Antony, T. Schwabe, C. Mück-Lichtenfeld. *Org. Biomol. Chem.* **2007**, *5*, 741.
24. Y. Zhao, D. G. Truhlar, *Phys. Chem. Chem. Phys.* **2005**, *7*, 2701.
25. M. Dierksen, S. Grimme, *J. Phys. Chem. A* **2004**, *108*, 10225.
26. V. Barone, J. Bloino, M. Biczysko, F. Santoro, *J. Chem. Theory Comput.* **2009**, *5*, 540.
27. C. Reichardt, *Chem. Rev.* **1994**, *94*, 2319.
28. J. Tomasi, B. Mennucci, R. Cammi, *Chem. Rev.* **2005**, *105*, 2999.
29. M. M. Karelson, M. C. Zerner, *J. Phys. Chem.* **1992**, *96*, 6949.
30. G. S. He, L. S. Tan, Q. Zheng, P. N. Prasad, *Chem. Rev.* **2008**, *108*, 1245.
31. M. A. Oar, W. R. Dichtel, J. M. Serin, J. M. J. Frechet, J. E. Rogers, J. E. Slagle, P. A. Fleitz, L. S. Tan, T. Y. Ohulchanskyy, P. N. Prasad, *Chem. Mat.* **2006**, *18*, 3682.
32. S. Mukamef, *Principles of Nonlinear Optical Spectroscopy*, Oxford, Oxford University Press, 1995.
33. F. Terenziani, C. Katan, E. Badaeva, S. Tretiak, M. Blanchard-Desce, *Adv. Mater.* **2008**, *20*, 4641.
34. M. Pawlicki, H. A. Collins, R. G. Denning, H. L. Anderson, *Angew Chem.-Int. Edit.* **2009**, *48*, 3244.
35. P. Salek, O. Vahtras, J. D. Guo, Y. Luo, T. Helgaker, H. Agren, *Chem. Phys. Lett.* **2003**, *374*, 446.
36. O. Berman, S. Mukamel, *Phys. Rev. A* **2003**, *67*, 13.
37. S. Tretiak, V. Chernyak, *J. Chem. Phys.* **2003**, *119*, 8809.
38. C. Katan, S. Tretiak, M. H. V. Werts, A. J. Bain, R. J. Marsh, N. Leonczek, N. Nicolaou, E. Badaeva, O. Mongin, M. Blanchard-Desce, *J. Phys. Chem. B* **2007**, *111*, 9468.

39. J. R. Lakowicz, *Principles of Fluorescence Spectroscopy*, 3rd ed., Springer, New York, 2006.
40. W. Becker, *Advanced Time-Correlated Single Photon Counting Techniques*, Springer, 2005.
41. P. Hamm, M. Lim, R. M. Hochstrasser, *J. Chem. Phys.* **1997**, *107*, 10523.
42. F. C. De Schryver, S. De Feyter, G. Schweitzer, Eds., *Femtochemistry*, Wiley-VCH, Weinheim, 2001.
43. T. H. Joo, Y. W. Jia, J. Y. Yu, D. M. Jonas, G. R. Fleming, *1 Phys. Chem.* **1996**, *100*, 2399.
44. E. J. Brown, Q. G. Zhang, M. Dantus, *J. Chem. Phys.* **1999**, *110*, 5772.
45. R. F. Begley, A. B. Harvey, R. L. Byer, *Appl. Phys. Lett.* **1974**, *25*, 387.
46. W. P. de Boeij, M. S. Pshenichnikov, D. A. Wiersma, *Annu. Rev. Phys. Chem.* **1998**, *49*, 99.
47. A. H. Zewail, *J. Phys. Chem. A* **2000**, *104*, 5660.
48. T. S. Rose, M. J. Rosker, A. H. Zewail, *J. Chem. Phys.* **1988**, *88*, 6672.
49. Y. R. Shen, *The Principles of Nonlinear Optics*, Wiley, New York, 1984.
50. S. Mukamel, Y. Tanimura, P. Hamm, *Acc. Chem. Res.* **2009**, *42*, 1207.
51. M. H. Cho, *Chem. Rev.* **2008**, *108*, 1331.
52. C. E. Crespo-Hernández, B. Cohen, Patrick M. Hare, B. Kohler, *Chem Rev.* **2004**, *104*, 1977.
53. S. B. Nielsen, T. I. Sølling, *Chem. Phys. Chem.* **2005**, *6*, 1276.
54. J. E. Monat, J. H. Rodriguez, J. K. McCusker, *J. Phys. Chem. A* **2002**, *106*, 7399.
55. V. E. Bondybey, *Annu. Rev. Phys. Chem.* **1984**, *35*, 591.
56. P. Hamm, *Chem. Phys.* **2008**, *347*, 503.
57. M. Pettersson, E. M. S. Macoas, L. Khriachtchev, J. Lundell, R. Fausto, M. Rasanen, *J. Chem. Phys.* **2002**, *117*, 9095.
58. J. Hicks, M. Vandersall, Z. Babarogic, K. B. Eisenthal, *Chem. Phys. Lett.* **1985**, *116*, 18.
59. T. Brixner, G. Gerber, *Chem. Phys. Chem.* **2003**, *4*, 418.
60. M. Dantus, *Annu. Rev. Phys. Chem.* **2001**, *52*, 639.
61. P. Hamm, R. M. Hochstrasser, in M. D. Fayer, Ed., Marcel Dekker, 2001, p. 273.
62. M. F. DeCamp, L. P. DeFlores, K. C. Jones, A. Tokmakoff, *Opt. Express* **2007**, *15*, 233.
63. M. F. DeCamp A. Tokmakoff, *Opt. Lett.* **2006**, *31*, 113.
64. J. Zheng, K. Kwak, M. D. Fayer, *Acc. Chem. Res.* **2067**, *40*, 75.
65. M. C. Asplund, M. T. Zanni, R. M. Hochstrasser, *Proc. Natl. Acad. Sci. U.S.A.* **2000**, *97*, 8219.
66. P. Hamm, M. H. Lim, R. M. Hochstrasser, *J. Phys. Chem. B* **1998**, *102*, 6123.
67. T. Elsaesser, *Acc. Chem. Res.* **2009**, *42*, 1220.
68. J. Bredenbeck, A. Ghosh, H. K. Nienhuys, M. Bonn, *Acc. Chem. Res.* **2009**, *42*, 1332.
69. G. S. Engel, T. R. Calhoun, E. L. Read, T. K. Ahn, T. Mancal, Y. C. Cheng, R. E. Blankenship, G. R. Fleming, *Nature* **2007**, *446*, 782.
70. G. D. Scholes, *J. Phys. Chem. Lett.* **2010**, *1*, 2.
71. S. Garrett-Roe, P. Hamm, *Acc. Chem. Res.* **2009**, *42*, 1412.

QUEST FOR ACCURATE MODELS: SOME CHALLENGES FROM GAS-PHASE EXPERIMENTS ON MEDIUM-SIZE MOLECULES AND CLUSTERS

Maurizio Becucci and Giangaetano Pietraperzia
LENS and Dipartimento di Chimica "Ugo Schiff", Polo Scientifico e Tecnologico, Università degli Studi di Firenze, Florence, Italy

INTRODUCTION

Spectroscopic experiments have always required a strong theoretical support for the interpretation of data in terms of equilibrium molecular geometry, vibrational force fields, energy and nature of the different electronic excited states, and so on. Nowadays the trend toward the study of molecular systems of ever-increasing complexity in terms of dimensionality and/or flexibility makes the theoretical support to experiments even more important. This results in the development of integrated computational approaches that combine accurate determination of structural and electronic properties and/or explicit description of their time evolution, to be compared to state-of-the-art experimental results, either in the frequency or time high resolution.

Contemporary spectroscopic methods that are now available to modern researchers open new possibilities in terms of systems and problems that can be studied. As a consequence such experiments have disclosed a complex world with peculiar

Computational Strategies for Spectroscopy: From Small Molecules to Nano Systems, First Edition. Edited by Vincenzo Barone.

behaviors which, not surprisingly, need to be interpreted by new ideas. In turn, theoretical interpretation via computational modeling of spectroscopic observables in solutions, complex materials, and biosystems is becoming an invaluable tool to gather information on the structure, dynamics, and reactivity, on par with the technical progress of the spectroscopic apparatus. On these grounds, the final goal from a theoretical point of view is to provide coherent and integrated computational tools, thoroughly validated, robust, and effective enough to be of practical use in the spectroscopic investigation of realistic molecular systems or in the design of new ones. Additionally it is crucial to make them available to a large scientific community, including researchers who do not specialize in theory and computation. In order to help these developments, the game for experimentalists is to provide new data from systems that can represent a challenge for the theories and methods lying beyond quantum calculations.

Frequency-resolved experiments in the gas phase provide detailed information on the molecular geometries averaged over different quantum states, thus representing direct tests for the construction of the true molecular Hamiltonian and effective wavefunctions. Already the early time-resolved experiments were probing the evolution of population between the different excited states and the nature of some transient species. Recent progress of novel ultrafast nonlinear spectroscopic techniques yielded a dramatic increase in structural and time resolution. At the same time, high-spectral-resolution methodologies allow for an enormous precision in the evaluation of spectroscopic parameters. Furthermore, details on reaction dynamics provided by experimental studies on gas-phase photodissociation give a complete picture of the reaction kinematics, based on the identification of the fragments, together with their quantum states and the full vectorial determination of their speed. Interpretation of such accurate modern spectroscopy experiments stands as a true challenge for theoretical treatments.

As mentioned above, the development of theoretical tools appropriate for computational spectroscopy studies capable of describing complex molecular systems and still endowed with the accuracy needed to gain full insight into important real-world processes requires a novel vision: old paradigms employed for most of the last century and based on clear-cut space-and-time separations are no longer valid. For example, discussing emission/absorption spectroscopy of photoactive biomolecules or the electron spin resonance signal of labeled systems considered as isolated objects is not a realistic option anymore, and the multiscale nature of these observations must be taken into account at a rather sophisticated level in order to simulate/reproduce/interpret available experimental data.

In order to consolidate the bases of theoretical treatments and understand clearly the effective limits of approximate models, molecular beam gas-phase experiments are an ideal choice to study molecular systems under well-controlled conditions [1]. They allow making a clear picture of isolated system properties and then introducing the perturbation due to the environment in a controlled way. Given the characteristics of the supersonic expansion, it is possible to obtain cold isolated molecules distributed in very few quantum states as well as complexes that are held together by weak interactions including clusters that are not stable under the usual static gas-cell conditions. Moreover, information on the system under study obtained in molecular beam conditions is not mediated by the environment statistical fluctuations. The results from spectroscopic experiment are then the best training set to test high-level quantum calculations. In return, the combination of high-resolution experiments and high-level quantum calculations gives us the best chance to describe properties of complex systems in fine details. In particular, the design and characterization of complex molecular systems triggers novel research trends in the fields of soft matter and material sciences; spectroscopy is always a powerful tool for their investigation.

In this short chapter we will present some recent experimental results in which quantum computation support was desired to extract the correct information from experimentally acquired spectra.

The chosen systems (see Figure 1) include very floppy molecules (as an example we report some recent results about 1,3-benzodioxole [2, 3]) and/or weakly interacting molecular and supramolecular complex structures whose properties and functionality are ruled by the interplay of many subtle effects (as an example the anisole clusters [4–9]) for which accurate determination of structural and energetic properties represent today's challenge for the computational methods. Additionally, an example of the need for theoretical support in understanding reaction dynamics results [10, 11] is also reported.

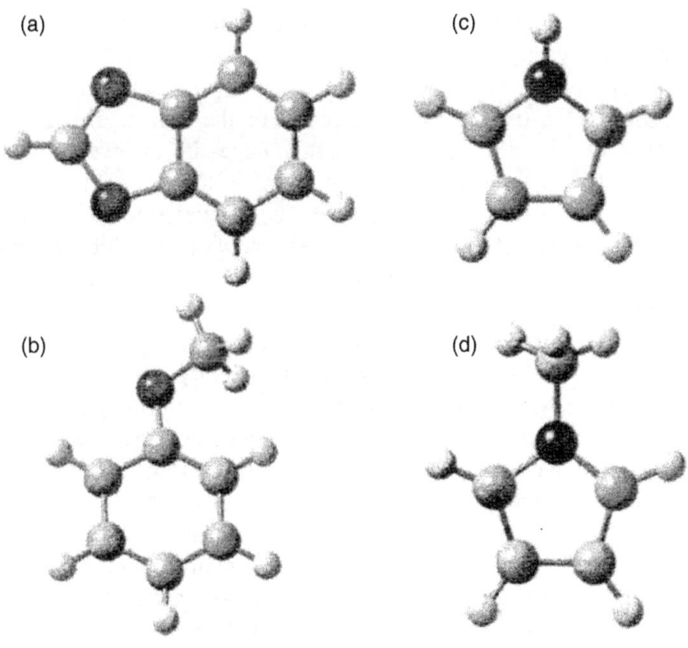

Figure 1 Pictorial view of the equilibrium geometry structure of (a) 1,3-benzodioxole (BDO), (b) anisole, (c) pyrrole, and (d) *N*-methylpyrrole (NMP).

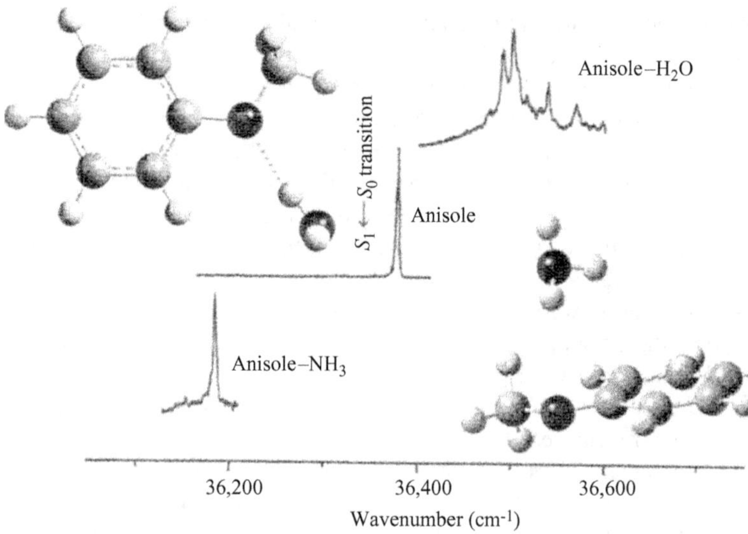

Figure 2 Relative shift of origin band for S_0–S_1 electronic transition for clusters with respect to isolated anisole molecule. Data are derived from refs. [5–7].

MOLECULAR STRUCTURE AND PROPERTIES

Let us start by discussing problems related to accurate determination of molecular structures in floppy systems. In fact, the elucidation of the molecular conformation is of fundamental importance in chemistry and biochemistry, as the molecular properties depend mostly on the geometric structure of the molecule. Conformational studies are of particular relevance for medium- to large-size molecules characterized by their possibility to have different local minima in the potential energy surface, which correspond to various conformations and give rise to different chemical properties of the molecule itself. This is a well-known fact in biochemistry as the efficiency of a protein in performing its function strongly depends on the extent of the folding of the macromolecule. However, prior to trying to understand the above-stated phenomena in very complicated systems, it is generally useful to consider simpler molecular models. For this purpose four- and five-membered rings are particularly important [12]. The ring frames can have several stable conformations as a result of a delicate balance of different factors, mainly the ring angle strain and the torsional forces [13]. So, for a five-membered ring, there will be the ring angle strain tending to give the molecule a planar structure; on the other hand torsional forces may lead to nonplanarity of the ring. Conformation-related problems can be effectively discussed with the example of BDO, a floppy molecule in which the presence of very low frequency and large-amplitude vibrational modes causes issues in the determination of the equilibrium structure and vibrational assignments. The BDO can serve as a model system for conformational studies of larger molecules of biological interest, and for this reason it has been intensively studied for a number of years. Following the general arguments on the structure of five-membered ring molecules, BDO should be planar (ring angle strain, but no torsional forces due to the lack of eclipsed hydrogen atoms); however, a more complex picture has been revealed by spectroscopy studies in the ground [14–17] and excited [18] electronic states.

As to the electronic ground state, Caminati et al. [14] used pure rotational spectroscopy to study the isolated molecule. The ground-state rotational spectrum was recorded and some thermally populated vibrationally excited states were observed as well. Combining a bidimensional flexible model, which describes molecular deformations, with the highly detailed rotational information, the authors concluded that the molecule is nonplanar due to the coupling of two vibrational modes, that is, the ring puckering of the five-membered ring and the butterfly motion of the two rings. An assignment scheme for the first vibrational levels was also proposed. Sakurai et al. [15] reinvestigated the S_0 electronic state of the BDO molecule using the data obtained by far-infrared, Raman, and dispersed fluorescence spectroscopy. Their assignment of the vibrational spectrum was fairly similar to that proposed by Caminati et al. [14], the major difference being the assignment of the butterfly mode. However, they confirmed the hypothesis that the ring-puckering and the butterfly modes strongly interact. In fact, an analysis of the obtained potential energy surface showed an intimate coupling between these two vibrational modes and it was also found that the molecule has a puckered conformation with a small (164-cm^{-1}) barrier to planarity. In their study Sakurai et al. introduced for the first time the concept of the anomeric effect to explain the bent equilibrium structure of the molecule. This has been defined as a stereoelectronic effect that involves the donation of electron density from a lone pair on one oxygen atom to the adjacent carbon–oxygen bond. Such an interaction, believed to be the result of $n - \sigma^*$ overlap [16, 17], reaches its maximum value when the –C–O–C–O torsional angle is at 90°, whereas it vanishes for a planar ring. Since the ring puckering increases the magnitude of the anomeric effect, this can result in the stabilization of a nonplanar structure.

The excited (S_1) electronic state of BDO has been studied by Laane and co-workers [18] through the analysis of the vibronic spectrum measured in a supersonic jet laser spectrometer. They made a complete assignment of the vibronic structure of the spectrum, and, with the help of data from dispersed fluorescence and UV absorption in a cell, concluded that the molecule is nonplanar in both the S_0 and the S_1 electronic states. Also in this case the anomeric effect was invoked in order to justify the nonplanarity due to the puckering of the five-membered ring and the butterfly motion of the two rings.

Next, some of the S_1–S_0 vibronic bands have been studied by us with rotationally resolved laser-induced fluorescence experiments, with the goal of providing a more precise determination of the BDO S_1-state equilibrium structure and its deformation in the first vibrational excited levels. In this study, based on analysis of the rotational structure of the different vibronic bands, some problems with the vibrational assignment of the vibronic spectrum became immediately clear [18]. In fact, on the basis of previous studies some of the bands under investigation should have a level in common, either in the S_0 or in the S_1 state, but the experimentally determined rotational constants for those states were not equal within the experimental error.

More recently, some quantum calculations on the BDO molecule by Orlandi and his group were published [19, 20], but they did not provide a completely satisfactory solution of the problem. At present some issues remain unclear as none of the applied models is able to describe at the same time molecular deformation and vibrational energy levels for the different harmonics and combination low-frequency bands, large-amplitude motions of BDO, in both the S_0 and in the S_1 states. As a consequence, further theoretical and experimental investigations are foreseen to disclose the full picture.

Flexibility and a large number of possible coordination modes are even more important factors in the case of molecular complexes where many contact points and different reciprocal orientations of the fragments could exist. Experiments on molecular complexes stand also as a viable route to study the fundamental forces tuning the physical–chemical properties and processes of bulk or solvated systems as, for example, molecular recognition and interaction between large molecules. In this respect, anisole molecule and its clusters with different solvent molecules are very convenient model systems allowing the study of several types of weak intermolecular interactions. Because of the coexistence in anisole of prototypical chemical functional groups acting as H-bond donors or acceptors, presence of both a π- and n-electron systems and a permanent electric dipole moment, anisole can participate in different interaction schemes. Additionally its absorption spectrum is in an easily accessible spectral region (near UV), facilitating experimental studies. For this reason, anisole complexes formed with different small partner molecules like CO_2 [17], NH_3 [18, 19], and H_2O [20] have been extensively studied. Very recently, detailed spectroscopy studies on anisole dimer have also been reported [8, 9]. Studies on the complexes formed by anisole with the above-mentioned small molecules demonstrated the large variety of interaction mechanisms possible for anisole while involved in supramolecular systems: two different planar structures for the complex, in anisole–CO_2 and anisole–H_2O, and a nonplanar structure of the anisole–NH_3 complex have been observed.

The anisole–CO_2 complex is stabilized by quadrupolar interactions with the carbon atom of CO_2 close to the n orbitals of oxygen from anisole and the two oxygen atoms of CO_2 involved in improper hydrogen bonds with the H atoms from the methyl and from the aromatic ring of anisole. At variance, for the anisole–H_2O complex the principal interaction is the hydrogen bond between one of the water hydrogen atoms and the oxygen atom of the anisole frame. The water center of mass lies practically in the symmetry plane of anisole while the second H atom of water stays above that plane. However, as shown by the results from the rotationally resolved spectra, two H atoms of water are statistically equivalent. This issue has been resolved by quantum mechanical calculations which pointed out the possibility of tunneling between two equivalent conformations with the free H atom above or below the anisole symmetry plane. Additionally, changes in the average position of water in the cluster when its ^{1}H atoms are exchanged with ^{2}H atoms have been observed experimentally. Quantum mechanical calculations have shown that the bending motion of water, a vibration that modulates the distance of water between the methyl group and the nearest aromatic H atom, is related to a very anharmonic and asymmetric potential surface. Thus, as isotopic substitution leads to changes of the zero-point vibrational energy, it also allows the water molecule to explore different regions of the potential energy surface, and such an effect is responsible for the different effective geometries of anisole complexes with ^1H_2O and ^2H_2O. For anisole–ammonia the experimentally determined nonplanar arrangement can be attributed to existence of a number of different contact points between anisole and ammonia. These are the nitrogen electron lone pairs involved in the improper hydrogen bond with the H atoms of the anisole methyl group, one H atom from ammonia pointing toward the oxygen *n* electron density and a second H atom toward the aromatic electron density.

Additional insights on the structure of the complex and the nature underlying the intermolecular interaction can be obtained by analysis of the red or blue shift of the origin band for the S_0–S_1 electronic transition for the clusters with respect to the isolated anisole molecule reported in Figure 2. This effect is related to the arrangement of the two units together with the changes in electron density on the different nonhydrogen atoms in the anisole upon electronic excitation. A detailed description of the properties of anisole complexes was provided by quantum mechanical calculations with methods based on density functional theory (DFT) and functionals able to account for dispersion interactions. It has been shown that the S_0–S_1 electronic transition in anisole is related to the transfer of electron density from oxygen to the aromatic ring leading to the weakening of interaction energy in the excited state for the anisole–water complex and strengthening in the anisole–ammonia complex, in perfect agreement with experimentally observed blue and red shifts, respectively.

Very recently we also reported on combined experimental and theoretical studies of the anisole dimer [8, 9]. In considering the experimentally observed red shift of the origin band for the $S_1 \leftarrow S_0$ electronic transition of this cluster with respect to the isolated anisole molecule, suppose that the interaction between the two units is ruled by the π-electron system. However, direct analysis of the rotationally resolved data of high-spectral-resolution experiments was quite difficult to determine the rotational constants, mainly due to the small values of the rotational constants that caused a relevant superimposition between the rotational components of the experimental spectrum. Support from theoretical investigations (i.e., from molecular mechanics calculations, further refined with DFT geometric optimizations) was then very useful to generate reliable trial sets for the rotational constants. In this way sound data analysis led to the determination of a center-symmetric structure of the cluster with the two units on parallel planes separated by 3.4 Å and the π-electron systems facing one another. This spatial arrangement allows for a superimposition of the π orbitals from the two units leading to a possible excitonic coupling. The existence of excitonic coupling was revealed in experiments on different isotopically substituted species of anisole. It turned out that, as the excitonic coupling is small, the difference in the zero-point vibrational energy for anisole units with different H-atom isotopic composition on the aromatic ring is sufficient to displace the energy levels of the two aromatic rings in such a way that the excitonic coupling does not become effective. Thus, excitonic coupling occurs only for the cluster with the two units of exactly the same H-atom isotopic distribution on the aromatic ring. This is an example in which a purely electronic coupling effect is modulated by the vibrational properties of the system: Such a case requires special care for an appropriate quantum mechanical description, Additionally, it has been found that changes in the isotopic H-atom distribution on the methyl group do not affect the excitonic coupling. This demonstrates that isotopic substitution in the methyl group has a negligible effect with respect to the electronic properties of the aromatic ring leading to the conclusion that effectively separated different fragments can be present even in a small molecule like anisole.

PHOTOCHEMICAL PROCESSES

Photochemical reactions on electronically excited states represent another group of spectroscopic studies where strong connection between experimental and computational methods is clearly needed. Usually several different photochemical reactions can take place upon electronic excitation, and their efficiency (total and relative) is controlled by the balance between the rates of the different relaxation processes from the excited state. Some of these processes involve direct or multistep mechanisms for energy release to the environment, such as direct or relaxed radiative decays or nonradiative degradation of the excited state up to complete thermalization. The number and the efficiency of the relaxation channels depend on both the internal properties of the molecule and the effect of the surrounding environment. Some relevant examples of photochemical processes are represented by studies of heteroaromatic molecules, such as those present in DNA, in which electronic excitation often leads to true chemical reactions that are extremely important for all biological processes. In particular, such photoprocesses may lead to reactions that are extremely dangerous due to potentially mutagenic effects. However, such undesired effects can be reduced through effective relaxation pathways that transform the dangerous electronic excitation into heat, a primary mechanism for the protection for DNA. In biological systems the electronic states populated by UV radiation originate from the coupling of the nucleobase π states. Effective decay mechanisms depend on the DNA molecular structure and its slow fluctuations, which rule stacking and pairing interactions, along with solvent effects and charges carried by the phosphate moieties. Such mechanisms are ultrafast and multiexponential, encompassing femto- to nanosecond time scales: They may operate on a single nucleobase or be collective, involving variable-length excitons and charge transfers, with the latter triggering slow and large-amplitude structure rearrangements. Photodamage competes with protection mechanisms, causing, as an example, thymine dimerization, which is ultrafast but is ruled by stacking-geometry fluctuations. A secondary mechanism for the protection (and therefore the effectiveness) of DNA bases is related to simple reactions, such as N–H or O–H bond cleavage, that make use of the excess energy upon UV photon absorption without the occurrence of potentially more dangerous reactions [10]. Deeper understanding of these processes in the condensed phase and then in complex biological systems requires a clear picture of the isolated molecule properties. As an example, we discuss here the case of pyrrole and *N*-methylpyrrole (NMP), molecules exhibiting N–X (X = H, CH_3) photocleavage, the photoreaction which has been extensively studied by the use of ion imaging–velocity mapping experiments and by theoretical calculations [10, 11, 21, 22]. These studies pointed out the role of the $\pi\sigma_{N\text{-}X}{}^*$ dissociative state (along the N–X bond) as a relevant decay channel for the photoexcited pyrrole and substituted pyrroles. Let us start from the photodissociation dynamics of pyrrole [10, 21] for which both slow and fast H atoms have been observed upon excitation above the S_1 state. Ion–imaging experiments performed at different wavelengths between 240 and 193 nm show that fast H atoms are produced only at the longest wavelengths upon photodissociation of pyrrole, while slow H atoms are observed in all the different conditions. Fast H atoms are associated with fast and direct reaction mechanisms involving the population of the N–H repulsive $\pi\sigma^*$ state while slow H atoms are related to statistical dissociation processes that proceed, via internal conversion, in the ground electronic state. The different nature of the two reaction mechanisms has also been probed by the angular distribution of the H atoms released where slow H atoms are produced with an isotropic distribution with respect to the laser beam polarization axis while fast H atoms exhibit a strongly anisotropic distribution. This behavior has been described in the case of pyrrole based on the presence of the S_1 state with $\pi\sigma_{NH}{}^*$ character and an S_2 state with $\pi\pi^*$ character. The S_0 and S_1 states possess different symmetry (A_1 and A_2, respectively, in the C_{2v} symmetry group) so the $S_1 \leftarrow S_0$ transition is only weakly allowed. Thus a molecule prepared in the S_1 state can evolve to the S_0 surface by the extension of the N–H bond along the repulsive $\pi\sigma_{NH}{}^*$ surface or with a passage, mediated by internal conversion (IC), followed by a statistical NH or CH bond fission. However, once the energy of the exciting photon is high enough to go into the S_2 state, the situation changes drastically as the $S_2 \leftarrow S_0$ transition is optically allowed and S_2 is a bound state. Thus, the direct reaction pathway is closed and only IC-mediated statistical processes possibly exist.

The situation is somewhat different in the case of NMP electronic excitation [11, 22] where the N–CH_3 cleavage represents the only possible source for the CH_3 fragment. For NMP, experiments carried out at different excitation energy show that slow CH_3 fragments were always observed, while fast CH_3 fragments were observed only for excitation very close to the $S_1 \leftarrow S_0$ band origin or for excitation around 193 nm. This oscillatory behavior has been understood only through the detailed picture of the nature and energy order of the different electronic excited states. In the case of pyrrole the energetic ordering of the five π molecular orbitals is the following: starting from lower energies, there is first a single MO and then two couples of degenerate MOs. Such a picture can be already derived from the ordinary Hückel theory and has been confirmed by standard excited-state quantum mechanical calculations. Instead, in the case of NMP the interaction between the methyl σ_{CH} MO and the π orbitais (usually referred to as hyperconjugation) leads to a complete removal of degeneracy for the π MOs. There are two different low-energy $\pi\sigma_{N\text{-}X}{}^*$ MOs, originating from electron excitation from the two upper π MOs. It must be noticed that the energy separation for these two MOs does not vanish only when the methyl group is close to the pyrrolydil system. Once the CH_3 fragment is removed, the two $\pi\sigma_{N\text{-}X}{}^*$ states become degenerate, as in pyrrole. Therefore, as the final state for photodissociation starting from the two different excited states is the same, the difference in the kinetic energy released for the dissociation process in the two regimes must equal the difference in excitation photon energy. That is exactly the experimental finding from the ion imaging–velocity mapping experiments at 242 and 193 nm [11]. Additionally quantum mechanical calculations [11] have shown that the S_1 and S_2 states of NMP are almost degenerate and have $\pi\pi^*$ and $\pi\sigma_{N-X}{}^*$ character. Then, the second $\pi\pi^*$ state and the second $\pi\sigma_{N-X}{}^*$ state are present while increasing the excitation energy. Thus, in analogy with the case of pyrrole, a direct dissociation channel exists around the origin of the electronic absorption spectrum, which is quenched by the presence of the S_3 $\pi\pi^*$ excited state and finally a new direct reaction channel exists when the dissociative S_4 $\pi\sigma_{N-X}{}^*$ state is reached. This picture, fully consistent with other experimental evidences, is further corroborated by the very low quantum yield of the second direct reaction channel. This low yield is caused by the competition with the S_3 $\pi\pi^*$ state, which is lying just below and is characterized by the very strong quantum yield. In fact, other experimental evidence as the angular distribution of the CH_3 fragments fits nicely in this scheme.

Further insight into the nature of decay processes added by highly energy resolved experiments on the photodissociation of pyrrole (by H-atom Rydberg tagging methods) reveals a mode-specific energy distribution in the pyrrolidyl fragment; this suggests a relevant role for specific excitation of the parent molecule in the promotion of the dissociation process [21]. Such results lead to the almost unexplored idea of exploiting ultrafast pulse-shaping techniques for selectively controlling the reaction quantum yields. Clearly new adventures in this area would require a detailed description of the involved excited sates and reaction channels as those presently obtained for isolated molecules.

CONCLUSIONS

Laser spectroscopy methods applied to molecules prepared in the gas phase allows for the determination of rotational, vibrational, and electronic excited-state energy levels. Moreover, due to time-resolved experiments direct measurement of the population transfer dynamics between different levels is possible, including the excited-state evolution with a chemical change in the system, identification of the reaction products, and, if fragmentation occurs, the measure of their recoil energy. In this chapter we have shown how the outcome of such spectroscopic studies might be interpreted with the help of computations providing a link between electronic structure and spectroscopic properties. However, as real-life processes span a wide range of space and time scales, satisfactory understanding of the underlying phenomena requires integrated approaches that combine the description of local/short-timed and global/slow-motion regimes. As was discussed intrinsically fast processes such as electronic excitations are often followed by slow processes related to structural relaxation. Beyond that, most relaxation mechanisms undergo intermediate steps where actual characteristic times of the evolution of the system and the environment cannot be separated and a proper and balanced consideration of all the involved effects is required when one aims to explore an overall mechanism.

REFERENCES

1. G. Scoles, Ed., *Atomic and Molecular Beam Methods,* Oxford University Press, 1992.
2. G. Pietraperzia, A. Zoppi, M. Becucci, E. Droghetti, E. Castellucci, *Chem. Phys. Lett.* **2004**, *385*, 304.
3. Z. Kisiel, L. Pszczolkowski, G. Pietraperzia, M. Becucci, W. Caminati, R. Meyer, *Phys. Chem. Chem. Phys.* **2004**, *6*, 5469.
4. C. G. Eisenhardt, M. Pasquini, G. Pietraperzia, M. Becucci, *Phys. Chem. Chem. Phys.* **2002**, *4*, 5590.
5. G. Piani, M. Pasquini, G. Pietraperzia, M. Becucci, A. Armentano, E. Castellucci, *Chem. Phys. Lett.* **2007**, *434*, 25.
6. M. Biczysko, G. Piani, M. Pasquini, N. Schiccheri, G. Pietraperzia, M. Becucci, M. Pavone, V. Barone, *J. Chem. Phys.* **2007**, *127*, 144303.
7. M. Becucci, G. Pietraperzia, M. Pasquini, G. Piani, A. Zoppi, R. Chelli, E. Castellucci, W. Demtroeder, *J. Chem. Phys.* **2004**, *120*, 5601.
8. G. Pietraperzia, M. Pasquini, N. Schiccheri, G. Piani, M. Becucci, E. Castellucci, M. Biczysko, J. Bloino, V. Barone, *J. Phys. Chem. A* **2009**, *113*, 14343.
9. N. Schiccheri, M. Pasquini, G. Piani, G. Pietraperzia, M. Becucci, M. Biczysko, J. Bloino, V. Barone, *PCCP* **2010**, *12*, 13547.
10. A. L. Sobolewski, W. Domcke, C. Dedonder-Lardeux, C. Jouvet, *Phys. Chem. Chem. Phys.* **2002**, *4*, 1093.
11. G. Piani, L. Rubio-Lago, M. A. Collier, T. N. Kitsopoulos, M. Becucci, *J. Phys. Chem. A* **2009**, *113*, 14554.
12. A. C. Legon, *Chem. Rev.* **1980**, *80*, 231.
13. J. Laane, *J. Chem. Phys.* **1969**, *50*, 776.
14. W. Caminati, S. Melandri, G. Corbelli, L. B. Favero, R. Meyer, *Mol. Phys.* **1993**, *80*, 1297.
15. S. Sakurai, N. Meinander, K. Morris, J. Laane, *J. Am. Chem. Soc.* **1999**, *121*, 5056.
16. A. J. Kirby, *Stereoelectronic Effects*, Oxford University Press, Oxford, 1996.
17. E. Cortez, R. Verastegui, J. Villareal, J. Laane, *J. Am. Chem. Soc.* **1993**, *115*, 12132.
18. J. Laane, E. Bondoc, S. Sakurai, K. Morris, N. Meinander, J. Choo, *J. Am. Chem. Soc.* **2000**, *122*, 2628.
19. E. Emanuele, F. Negri, G. Orlandi, *Chem. Phys.* **2006**, *321*, 75.
20. E. Emanuele, G. Orlandi, *J. Phys. Chem. A* **2005**, *109*, 6471.
21. M. N. R. Ashfold, B. Cronin, A. L. Devine, R. N. Dixon, M. G. D. Nix, *Science* **2006**, *312*, 1637.
22. A. G. Sage, M. G. D. Nix, M. N. R. Ashfold, *Chem. Phys.* **2008**, *347*, 300.

PART I

ELECTRONIC AND SPIN STATES

1

UV–VISIBLE ABSORPTION AND EMISSION ENERGIES IN CONDENSED PHASE BY PCM/TD-DFT METHODS

Roberto Improta
CNR—Consiglio Nazionale della Ricerche, Istituto Biostrutture e Bioimmagini, Naples, Italy

Computational Strategies for Spectroscopy: From Small Molecules to Nano Systems, First Edition.
Edited by Vincenzo Barone.

1.1 INTRODUCTION

The absorption and emission of light in the ultraviolet–visible (UV–vis) energy range (200–750 nm, 6.2–1.6 eV) are probably the 'most popular' spectroscopic processes which are continuously experienced by living beings in every-day life. For example, the existence of colors and the possibility of seeing depend on these processes.

From a scientific perspective, absorption and fluorescence spectra are a source of fundamental information in many chemical–physical molecular features [1–3]. First, they are two of the most straightforward methods of identifying a compound and thus represent key analytical techniques. Furthermore, advances in pump–probe time-resolved experiments have made absorption and fluorescence spectroscopy ideal tools to monitor the time evolution and outcome of many reactive processes [4]. These techniques can also provide indications of the electronic structure of molecules, their ground- and excited-state geometry, and the most relevant vibrational features, just to name a few basic properties.

From a complementary point of view, absorption spectra are often recorded to investigate the interactions between a molecule and its environment, and part of these interactions are indeed revealed by modifications of diagnostic molecular vibrations [5, 6]. Measuring absorption and fluorescence spectra is thus one of the first and key steps in many different areas of molecular and biological physics and nanotechnology science.

UV–vis absorption and emission are also extremely important from a technological and industrial point of view, as shown, for example, by the huge efforts made in the design of new dyes [7] or new photovoltaic materials [8], in the fields of light-emitting diodes [9] and phototropic materials, or, concerning pharmaceutical research, in the fields of photodynamic therapy [10] and molecular imaging [11].

The absorption and emission of light by molecules obviously depend on a basic quantum mechanical property, that is, the existence of discrete energy levels. It is not surprising that these processes have also been investigated from a theoretical/ computational point of view. Computation of absorption and emission spectra is thus a very well developed research field, and an increasing number of theoretical and computational advances have made computational approaches a fundamental complement of experiments [12–32]. Schematically, we can say that computations have been shown to be able to:

1. Significantly help in the assignment of experimental spectra. Consider, for example, a complex organic reaction with multiple possible products (or a mixture of products), exhibiting different, though similar, absorption (or fluorescence) spectra. Comparison between the spectra computed for the most likely products and the experimental ones often provides fundamental information for identifying the products [33].
2. Increase the amount of information derived from experiments. Due to the large number and complexity of the effects that can potentially influence absorption and emission spectra, it is not always easy to disclose all the information thereby contained, limiting the potentialities of these techniques. This is true

especially in the condensed phase, where many fundamental features are often somewhat hidden in broad and structureless spectra. In such a scenario, being able to accurately compute spectra (possibly vibrationally resolved) is extremely useful.

3. Shed light on the relationship between the structure of a compound and its spectral properties. Computations can more easily dissect the different chemical–physical effects (intrinsic and environmental) modulating the spectral properties of molecule. They can thus help the design of new compounds with specifically tailored features (e.g., new dyes), with significant benefits for industrial research.

On the other hand, notwithstanding the huge theoretical and computational efforts, a reliable calculation of absorption and emission spectra, especially when dealing with sizable molecules, is not a trivial task. In fact, while it is relatively easy to approach the so-called chemical accuracy for molecules in the ground electronic states, an accurate determination of the energy of the excited electronic states is computationally and theoretically much more demanding. Indeed, excited electronic states are often quite close in energy: The energy ordering often depends on fine details of the adopted computational method (e.g., the size of the basis set). The breaking of the Born–Oppenheimer approximation is rather common, and nonadiabatic couplings cannot be safely neglected [13]. Furthermore, not only for computing emission spectra but also for obtaining the Franck–Condon factors between the ground and the electronic excited states, the excited-state stationary points are needed, and, thus, performing excited-state geometry optimizations is often necessary. In this respect, analytical excited-state gradients are available only for a limited number of methods (CIS, CASSCF, CASPT2, SAC-CI, CC2) [29–40].[1] Finally, the most accurate ab initio methods are too computationally expensive for studying the excited states of most systems of technological and biological processes, which are usually medium/large-size molecules.

The usefulness of computational methods would of course be quite limited if environmental effects could not be taken into proper account, since almost all of the above-mentioned processes occur in solution. As a consequence, even a qualitative agreement with experiments requires the use of a suitable solvation model. The inclusion of environmental effects involves additional difficulties: Not only should the solvation model be able to provide an accuracy comparable to that attained in vacuo, but in solution any problem involving excited states becomes intrinsically dynamic [41]. The solvent reaction field couples the ground-state density with the density correction and the orbital relaxation arising from the electronic transition. Furthermore, the coupling is modulated by the solvent relaxation times [41].

However, thanks to very recent methodological and computational advances [especially in the methods rooted in the time-dependent density functional theory

[1] Configuration interaction singles (CISs); complete active-space self-consistent field (CASSCF) and its second-order perturbation counterpart (CASPT2); symmetry-adapted cluster configuration interaction (SAC-CI); double coupled cluster (CC2).

(TD-DFT)][17, 18] it is nowadays possible to compute remarkably accurate absorption and emission spectra of sizable molecules in solution. Furthermore, the inclusion of these methods in user-friendly computational methods has made these techniques available to nonspecialists.

In this chapter we shall thus discuss some suitable strategies for computing the absorption and emission spectra of medium–large systems in "complex" environment, focusing on the computation of the vertical excitation energies and the vertical emission energies. Treatment of the lineshape can be found in the Chapters 8 and 10.

1.2 QUANTUM MECHANICAL METHODS FOR STUDY OF ELECTRONIC EXCITED STATES

A real "panoplia" of quantum mechanical methods for computing the properties (energy, oscillator strengths, eventually the minima, etc.) of the excited electronic states of medium-size molecules is currently available. A detailed discussion of the main features of each computational approach is obviously outside the scope of this chapter, and many interesting books and review articles are already available within the literature [12–32, 42, 43]. Very schematically, we can distinguish between wavefunction-based methods and electron-density-based methods. In some methods of the former family, such as single-reference configuration interaction (CI) or multiconfigurational-based ones—multiconfigurational SCF (MCSCF), for example, the CASSCF, CASPT2 [14–16], or multireference CI [19]—any electronic state is described as the combination of several Slater determinants corresponding to different electronic configurations (i.e., different occupation schemes of the molecular orbitals, MOs). The expansion coefficients of the different Slater determinants and, in multiconfigurational SCF approaches, the expansion coefficients of the MOs in the Slater determinants are then variationally computed.

A different approach is followed by other wavefunction-based methods as those belonging to the coupled-cluster (CC) family [20, 21, 28]. CC models are based on the single-reference wavefunction and allow us to compute excitation energies within the equation-of-motion [24–26] and linear response [27] CC formalisms (EOM-CC and LR-CC, respectively). Accuracy of CC results depends on the level of truncation in the CC expansion. In this respect, extensions of CC theory for excited states to include triplet excitations within both iterative and noninterative schemes [26, 27] allow for very accurate computations of excited-state properties. However, due to their high computational cost, these methods cannot be applied to large molecules. Recently, promising results have been obtained by an approximated single and double CC method, which, exploiting the resolution of the identity (RI) approximation for two electron integrals, can be applied to fairly large systems with a good degree of accuracy [22, 23]. Additionally, the SAC-CI approach [34], being formally equivalent to EOM-CC and LR-CC models, introduces some approximations by neglecting the unimportant unlinked terms and the perturbation selection of the linked operators. By doing so, the computations become more efficient and allow studies of larger and more complex systems, which are further facilitated by the availability of SAC-CI analytical gradients [35].

The simpler wavefunction method is CIS, where the excited states are computed by considering only single excited Slater determinants using the HF MO to describe the ground state. Despite exhibiting some advantages (limited computational cost, size consistency, availability of analytical first and second derivatives), CIS excitation energies are not very accurate, being usually overestimated by 0.5–2 eV (see Dreuw and Head-Gordon [29] for a more detailed discussion).

A second class of methods is instead based on the knowledge of the electron density in a theoretical framework similar to that of the DFT. In DFT the effects of the exchange and of the electronic correlation are included by the so-called exchange-correlation (xc) functional, which is obtained by an empirical fit of experimental data or by imposing some physical constrains based on the behavior of some "limit" model systems [44]. For the excited states, the TD-DFT [17, 18] recently emerged as a very effective tool, since, when coupled to suitable density functionals, it often reaches an accuracy comparable to that of the most sophisticated (but expensive) post-Hartree–Fock methods [45–59], with a much more limited computational cost. As a consequence in the last years an increasing number of TD-DFT applications have appeared in the literature, also because this method can be used as a "blackbox" and is thus also easily accessible to nonspecialists.

TD-DFT has often been criticized for not being a first-principle nonempirical method (in analogy with its "parent" DFT), not showing uniform accuracy in treating electronic transitions with different characters, and delivering a qualitatively wrong description of the crossing region between different electronic excited states (see below). On the other hand, although wavefunction-based methods have known impressive methodological advances, significantly increasing their range of applicability, they suffer from very high computational costs. As a consequence, some compromises concerning the basis set or, for MC methods, the active space employed are often necessary, not only decreasing the expected accuracy but also introducing a significant degree of arbitrariness in formally rigorous methods. Furthermore, semiempirical parameters are often also present in sophisticated calculations, such as CASPT2, which often include the so-called IPEA shift. This empirical parameter has been introduced to compensate for systematic errors in CASPT2 ionization potentials (IPs) and electron affinities (EAs). Its application typically increases the computed CASPT2 excitation energies by about 0.1–0.3 eV and corrects for the known tendency of CASPT2 to slightly underestimate excitation energies.

In our opinion, each method has its own advantages and limitations, and the choice of quantum mechanical (QM) method depends on the phenomena/system investigated. As discussed in the introduction, a method coupling accuracy and computational feasibility is necessary when treating systems of biological and technological interest. On the grounds of our experience, we think that TD-DFT probably represents the best compromise between accuracy and computational cost for describing the excited-state behavior in medium/large-size molecules. As a consequence it will be used as reference computational method in this chapter.

1.3 TIME-DEPENDENT DFT

1.3.1 Foundations of Time-Dependent DFT

TD-DFT has been reviewed in several excellent papers [17, 18]. We thus limit ourselves to its basic foundations.

TD-DFT is rooted in the Runge–Gross theorem [60] (which is not valid for the degenerate ground state), allowing the extension of the Hohemberg–Kohn–Sham formulation of the TD-DFT theory to the treatment of time-dependent phenomena:

$$\Psi(t) = e^{-i\phi(t)}\Psi[\rho, \Psi_0](t) \tag{1.1}$$

The Runge–Gross theorem implies that:

(a) A functional of the time dependent charge density, ρ, determines the wavefunction up to a time-dependent phase factor.
(b) All the observables can be calculated with knowledge of the time-dependent one-body density.

When dealing with time-independent processes, we can determine the ground state of a given system by minimizing its total energy. In time-dependent systems no variational principle based on the energy can be exploited, since the total energy is not a conserved quantity. It is possible, however, to resort to another quantity, that is, the quantum mechanical action:

$$[\Psi] = \int_{t_0}^{t_1} dt \langle \Psi(t)| \frac{\partial}{\partial t} - \hat{\mathcal{H}}(t)|\Psi(t)\rangle \tag{1.2}$$

The true time-dependent density is that making the action stationary:

$$\frac{\delta A}{\delta \rho(\mathbf{r}, t)} = 0 \tag{1.3}$$

On the grounds of the Runge–Grosse theorem, it is then quite straightforward to derive the time-dependent Kohn–Sham equation, in analogy with the procedure used in TD-DFT. For adiabatically introduced perturbations, whose frequency does not exceed the highest occupied–lowest occupied molecular orbital (HOMO–LUMO) gap, it is possible thus to assume the existence of a potential v_{eff} ($\mathbf{r}$, t) for an independent particle system whose orbitals ψ_i ($\mathbf{r}$, t) yield the same charge density ρ ($\mathbf{r}$, t) as for the interacting system,

$$\rho(\mathbf{r}, t) = \sum_i f_i |\psi_i(\mathbf{r}, t|^2 \tag{1.4}$$

Minimizing the action, we then obtain

$$\left[-\frac{1}{2}\nabla^2 + v_{\text{eff}}(\mathbf{r}, t)\right]\psi_i(\mathbf{r}, t) = i\frac{\partial \psi_i(\mathbf{r}, t)}{\partial t} \tag{1.5}$$

The first term on the left-hand side of Eq. 1.5 represents the kinetic energy of the electrons, whereas the second one, v_{eff}, is given as

$$v_{\text{eff}} = v_{\text{ext}}(\mathbf{r}, t) + \int \frac{\rho(\mathbf{r}, t)}{|r - r'|} dr' + v_{\text{xc}}(\mathbf{r}, t) \tag{1.6}$$

where v_{ext} is the external potential (electron–nuclei interaction), the second term gives account of the electron–electron Coulomb interaction, and the third one, v_{xc}, represents the time-dependent counterpart of the stationary exchange–correlation functionals

$$v_{\text{xc}}(\mathbf{r}, t) = \frac{\delta A_{\text{xc}}[\rho]}{\delta \rho(\mathbf{r}, t)} \tag{1.7}$$

$$A_{\text{xc}} = \int_{t_0}^{t_1} E_{\text{xc}}[\rho_t] dt \tag{1.8}$$

Up to this point, no approximation has been made in TD-DFT: As in DFT, the only problem is the knowledge of the xc functional. However, in the most commonly used approaches, the determination of the time-dependent xc functional involves additional approximations. The simplest approximation is the adiabatic approximation, which can be applied if the external potential varies slowly in time [61]. It is a local approximation in time, assuming that v_{xc} is determined only by the density $\rho(t)$ at the same time,

$$v_{\text{xc}}[\rho](\mathbf{r}, t) = \frac{\partial A_{\text{xc}}[\rho]}{\partial \rho(\mathbf{r}, t)} \simeq= \frac{\partial E_{\text{xc}}[\rho]}{\partial \rho_t(\mathbf{r})} = v_{\text{xc}}[\rho_t](\mathbf{r}) \tag{1.9}$$

From the physical point view, the adiabatic approximation assumes that the reaction of the SCF to temporal changes in ρ is instantaneous, neglecting all the retardation effects.

Most of the TD-DFT implementations compute the excitation energies relying on a second approximation, based on the linear response theory.

For a small external time-dependent potential it is not necessary to solve the full time-dependent KS equation, but most time-dependent properties can be calculated from the first-order variation of the density:

$$\delta P_{ij}(\omega) = \frac{n_j - n_i}{\omega - (\epsilon_i - \epsilon_j)} [\delta \mu_{\text{ext}}(\omega) + \delta v_{\text{scf}}(\omega)] \tag{1.10}$$

where δP_{ij} is the linear response of the density matrix in the frequency domain and δv_{scf} is the linear response of the SCF due to the change in the charge density.

On the other hand, δv_{scf} depends on the response of the density matrix,

$$\delta v_{\text{scf}}^{ij} = \sum_{k,l} K_{ij,kl}(\omega)\, \delta P_{kl} \tag{1.11}$$

The above equations are usually expressed in matrix from:

$$\begin{bmatrix} \mathbf{A} & \mathbf{B} \\ \mathbf{B} & \mathbf{A} \end{bmatrix} \begin{bmatrix} \mathbf{X} \\ \mathbf{Y} \end{bmatrix} = \omega \begin{bmatrix} 1 & 0 \\ 0 & -1 \end{bmatrix} \begin{bmatrix} \mathbf{X} \\ \mathbf{Y} \end{bmatrix} \tag{1.12}$$

where vectors **X** and **Y** represent the linear response of the density matrix to the time-dependent perturbation,

$$A_{ai,bj} = \delta_{ab}\delta_{ij}(\epsilon_a - \epsilon_i) + K_{ai,bj} \tag{1.13}$$

and

$$B_{ai,bj} = K_{ai,jb} \tag{1.14}$$

where ϵ is the energy of the time-independent KS orbitals (ψ), i, j referring to the occupied and a, b to the virtual orbitals.

Excitation energies are thus computed as poles of the dynamic polarizability, that is, as the values of ω leading to zero eigenvalues on the left-hand side of the matrix of Eq. 1.12. In the framework of the above equations, an efficient "fast" iterative solution for the lowest eigenvalue/excitation energies can be attained [62]. Oscillator strengths can also be obtained by the eigenvectors of Eq. 1.12, as explained by Casida [17].

The recent implementation of TD-DFT analytical gradients [38–40] allows for the determination of the excited-state stationary points and their properties (e.g., the multipole moments). Harmonic frequencies can be obtained by performing numerical differentiation of the analytic gradients, enabling us to perform the same kind of vibrational analysis performed in the ground electronic state [45–50].

TD-DFT calculations thus allow determining the energy and the properties of the excited states with a limited computational cost. As a consequence, it is usually not necessary to impose any symmetry constraint, very large basis sets can be used, and no ad hoc choice (see, e.g., the active space in CASSCF/CASPT2 calculations) is usually necessary, also when dealing with large-size systems. These are important features: On the one hand, it is possible to treat different systems (e.g., a supramolecular system and its component) and different kinds of transitions (e.g., $n\pi^*$ and $\pi\pi^*$) with a similar degree of accuracy, putting the analysis of the computational results on firmer ground. On the other hand, they make TD-DFT a very user-friendly method easily accessible to nonspecialists. As a consequence in the last years the number of TD-DFT studies has continuously increased, allowing important advances in our knowledge of the potentialities and the limitations of this method. It is nowadays well assessed that TD-DFT, when employing a suitable density functional, can provide fairly accurate results (with 0.2–0.3 eV the experimental results) in several classes of systems, despite the limited computational cost [51–58].

It is important, however, to remember that TD-DFT cannot be considered a blackbox method, since the accuracy of its results depends on several factors (functional, basis set, etc.) which have to be properly addressed. Furthermore, as discussed in the next section, there are several systems/processes for which TD-DFT has often shown significant failures.

1.3.2 Limitations of Time-Dependent DFT

TD-DFT is a mono-determinantal method, and thus it cannot be applied to electronic states with an intrinsic multireference character [2]. Analogously, TD-DFT can exhibit deficiencies in treating electronic transitions with substantial contributions from double excitations [63–66], although interesting attempts to overcome the above limitations have been proposed [65–71]. In several cases, however, an electronic transition exhibits a multireference character (or a significant contribution from double excitations) just because of a poor description of the ground-state MO by HF orbitals, and such features are not present when using MOs computed at the DFT level.

Another traditional failure of TD-DFT concerns the treatment of long-range charge transfer (CT) transitions between zero-overlap donor–acceptor pair. Standard functionals significantly underestimate the transition energy and fail to reproduce the correct $1/R$ trend when the donor/acceptor distance (R) increases [29].

However, in the last five years new functionals have been developed, which are able to deliver a correct estimate also of the long-range CT transitions [72–76]. Furthermore, it is worth noting that for electronic transitions which involve only partial CT character, the underestimate of the excitation energies by TD-DFT may be controlled by the use of hybrid functionals, whereas the performances of pure functionals are much poorer (see next sections). Therefore, some of the deficiencies ascribed to TD-DFT are not intrinsic features of the method but depend on the choice of the functional.

Also "long-range" corrected functionals are not able, however, to accurately treat other classes of compounds such as cyanines, especially when the length of the ethylenic bridge connecting the two NH_2^+/NH_2 moieties increases. This failure is likely due to the intrinsic multideterminantal character of the electronic transitions in such molecules [58, 77].

1.4 SOLVATION MODELS

As anticipated in the introduction, since most of the UV–vis spectra are recorded in the condensed phase, suitable theoretical models, able to include the effect of the solvent on the absorption and the emission spectra, are necessary. This topic has been discussed in detail in several reviews, and thus, also in this case, we limit our discussion to some basic aspects [41, 78]. The most direct procedure to compute the spectra of a given molecule (the solute) in solution consists in including in the calculations a certain number of explicit solvent molecules [79, 80]. However, this approach has to face two severe difficulties: (i) the number of solvent molecules necessary to reproduce the bulk properties of a liquid (say, its macroscopic dielectric constant) is very large; (ii) a dynamical treatment averaging all the possible configurations of the solvent molecules is in principle necessary. As a consequence, this approach has a large computational cost, especially when used for studying excited states, which, as discussed above, need rather expensive QM computational methods. As a consequence, this "purely supramolecular" approach usually relies on an approximate description of the solvent molecules by using classical force fields within a mixed quantum mechanics/molecular mechanics (QM/MM) mixed approach. However, correct placement of the first solvation shell molecule is required, which can often be a difficult task for the standard molecular mechanics force field [81]. Furthermore, a large number of QM/MM calculations are necessary to reach converged excitation energies.

Complementing the results obtained for the study of ground electronic states in solution, many computational studies indicate that approaches exploiting continuum solvation models are very effective tools for evaluating the solvent effect on the excited-state properties. Among continuum models, the polarizable continuum model (PCM) is probably the one most commonly used. In the following, we thus focus mainly on this method [78, 82].

In the PCM the solvent is described as a homogeneous dielectric which is polarized by the solute. The latter is placed within a cavity in the solvent medium (built as the envelope of spheres centered on the solute atoms) and the proper electrostatic problem at the cavity surface is solved using a boundary element approach [78]. In the PCM framework, the solvent loses its molecularity and, especially in hydrogen-bonding solvents, the explicit inclusion of solute–solvent interactions is very important for getting accurate results. In these cases, as discussed in detail in the next section, mixed discrete/continuum models, where a limited number of solvent molecules are included in the computational model, usually provide accurate results.

Inclusion of the solvent effect within a continuum model in time-dependent processes as absorption or emission poses several problems. On the one hand, the solution of the electrostatic problem is highly nonlinear. Indeed, the solvent reaction field should be variationally determined together with all the other parameters in the electronic method used (MO coefficients, CI coefficients, excitation amplitudes, etc), but all those parameters do depend on the solvent reaction field. Different approaches have been envisaged to tackle the above problem. The most used can be classified in two classes, that is, state-specific (SS) and linear response (LR) approaches [83–86]. In SS methods (e.g., CASSCF/PCM) a different effective Shrödinger equation is solved for each state of interest, achieving a fully variational formulation of solvent effect on the excited-state properties. In the methods exploiting the LR response formalism (as TD-DFT) the excitation energies are "directly" determined without computing the exact excited electron density. As discussed in more detail in the next sections, SS and LR methods can provide very different estimates of the solvent effect on the excited-state properties and dynamical solvation effects. This point should thus be treated very carefully when using PCM in excited-state calculations.

Dynamical Solvent Effect Another critical topic to be considered when studying excited states in solution is the dynamical solvent effect. Electron excitation is an intrinsically dynamic process: The full equilibration of solvent degrees of freedom to the excited-state density requires a finite time. It is thus fundamental that the characteristic times of solvent degrees of freedom are properly taken into account.

Several theoretical approaches to the description of dynamical solvent effects have been proposed within the framework of PCM of other continuum models [41]. The simplest, and most commonly used, treatment involves the definition of two limit time regimes: equilibrium (EQ) and nonequilibrium (NEQ). In the former all the solvent degrees of freedom are in equilibrium with the electron density of the excited-state density, and the solvent reaction field depends on the static dielectric constant of the embedding medium. In the latter, only solvent electronic polarization (fast degrees of freedom) is in equilibrium with the excited-state electron density of the solute, while the slow solvent degrees of freedom remain equilibrated with the ground-state electron density. In the NEQ time regime the "fast" solvent reaction field is ruled by the dielectric constant at optical frequency (ϵ_{opt}, usually related to the square of the solvent refractive index).

The NEQ limit is the most suitable to the treatment of the absorption process. The study of the fluorescence process is instead more complex, since in this case dynamical solvent effects cannot be rigorously decoupled from the intramolecular effects due to the motion of the wave-packet (WP) on the excited-state surface. However, it is possible to define some limit reference models, and intuitive consideration of the properties of the solvent and/or the excited potential energy surface is often sufficient to define what is the most suitable to the treat the case under study (see next sections). PCM can be used in conjunction with all the most important excited-state electronic methods. Since we selected TD-DFT as our reference electronic method, we shall treat PCM/TD-DFT in more detail in the next sections.

1.4.1 SS-PCM/TD-DFT

The solvent reaction field contribution to the solute free energy (G) can be expressed in PCM as

$$G = \tfrac{1}{2}\mathbf{V}^{\dagger}\mathbf{q} \tag{1.15}$$

where vector $\mathbf{V}$ collects the values of the solute's electrostatic potential and $\mathbf{q}$ is the apparent surface charge placed at the center of the surface *tesserae* (i.e., the small tiles which the cavity surface is finely subdivided in), where also $\mathbf{V}$ is computed.

In the latest version of the method [78] the polarization charges depend on the solute's electrostatic potential and, thus, on its density through a general relationship of the form

$$\mathbf{q} = -\mathbf{DV} \tag{1.16}$$

where the square matrix $\mathbf{D}$ is related to cavity geometric parameters and the solvent dielectric constant ϵ.

Consider a generic excited electronic state (2) together with the corresponding ground state (1). As discussed by Improta et al. [86], in SS methods the excited-state equilibrium [$\mathcal{G}_{\mathrm{eq}}(2)$] free energy in solution thus explicitly depends on the excited state (2) density:

$$\mathcal{G}_{\text{eq}}^{(2)} = \frac{1}{2}\sum_i q_i^{(2)} V_{i,\rho}^{(2)} = \frac{1}{2}\sum_i q_{i,f}^{(2)} V_{i,\rho}^{(2)} + \frac{1}{2}\sum_i q_{i,s}^{(2)} V_{i,\rho}^{(2)} \tag{1.17}$$

The nonequilibrium ($\mathcal{G}_{\text{neq}}^{(2)}$) free energy in solution involves an explicit dependence on the density of the ground state (1):

$$\mathcal{G}_{\text{neq}}^{(2)} = \frac{1}{2}\sum_i q_{i,f}^{(2)} V_{i,\rho}^{(2)} + \left(\sum_i q_{i,s}^{(1)} V_{i,\rho}^{(2)} - \frac{1}{2}\sum_i q_{i,s}^{(1)} V_{i,\rho}^{(1)}\right) + \left(\frac{1}{2}\sum_i q_{i,s}^{(1)} V_{i,f}^{(2)} - \frac{1}{2}\sum_i q_{i,s}^{(1)} V_{i,f}^{(1)}\right) \tag{1.18}$$

In the above equations q_f/q_s and V_f/V_s are the polarization charges and the corresponding potentials relative to the "fast" and "slow" solvent degrees of freedom. The potential generated by the density of state (n) is given as $V_\rho^{(n)}$.

The absorption process is ruled by nonequilibrium solvation, and thus the solvent contribution to the vertical excitation energy can be computed by using the following relationship:

$$\Delta G_{\text{abs}} = \mathcal{G}_{\text{neq}}^{(2)} - \mathcal{G}_{\text{eq}}^{(1)} \tag{1.19}$$

Soon after the electronic transition has occurred, the system starts evolving on the excited-state potential energy surface (PES) toward its energy minimum. At the same time, slow solvent degrees of freedom start equilibrating on the excited-state electron density. These two processes cannot be rigorously decoupled, especially when they exhibit similar time scales, and we cannot thus expect that a single strategy is suitable to all the possible emission processes. In many cases, however, this complex scenario can be simplified by means of qualitative considerations on the properties of the solvent and/or the excited potential energy surface. The equilibration of intramolecular degrees of freedom is faster than solvent equilibration except for very flat PESs or when many low-frequency motions are involved (large-amplitude torsional motions, inversions, etc.). This is especially true in polar solvents and for electronic transitions involving significant variations of the excited-state electron density. In such cases, indeed, time-resolved experiments suggest that the equilibration of the slow solvent degrees of freedom occurs on the picosecond time scale. This time should be long enough to assume that the excited electronic state has reached its minimum. A simple limiting case is that of ultrafast excited-state decay when only fast solvent degrees of freedom are expected to be in equilibrium with the excited-state density. In this limit, ΔG_{em} can be computed exactly in the same way as ΔG_{abs} [86]:

$$\Delta G_{\text{em}} = \mathcal{G}_{\text{neq}}^{(2)} - \mathcal{G}_{\text{eq}}^{(1)} \tag{1.20}$$

Of course, in this case excited-state geometry optimizations should also be performed in the nonequilibrium limit.

Another simple limit is that of "very long" excited-state lifetimes, which characterize, for instance, strongly fluorescent species. In this case, we can assume that all the solvent degrees of freedom are in equilibrium with the excited-state density. The ground-state ($\mathcal{G}^{(1)}_{\text{neq}}$) nonequilibrium free energy in solution describing the emission process can thus be obtained from Eq. 1.18, interchanging labels 1 and 2. The fast solvent degrees of freedom are equilibrated with the ground-state electron density, whereas the slow ones are kept frozen at the value obtained in the equilibrium calculation of the excited state.

In this limit, excited-state geometry optimizations should be performed with the solvent equilibrium limit, and the solvent contribution to the fluorescence energy (ΔG_{em}) is given as

$$\Delta G_{\text{em}} = \mathcal{G}^{(2)}_{\text{eq}} - \mathcal{G}^{(1)}_{\text{neq}} \tag{1.21}$$

The above relationship is the most suitable for treating phosphorescence, where states (2) and (1) correspond to T_1 and S_0, respectively.

The computation of the quantities involved in Eqs. 1.20 and 1.21 is straightforward using a generalization of the SS-PCM/TD-DFT method presented elsewhere [86], where the nonlinear problem of determining the polarization charges corresponding to the excited-state density is solved by using a self-consistent iterative procedure. Starting from a TD-DFT calculation, a first approximation to the state-specific reaction field is computed using the electron density of the state of interest by solving Eq. 1.16. In the next step, a TD-DFT calculation is performed in the presence of this first set of polarization charges, providing an updated excited-state density and, consequently, a new set of polarization charges. This iterative procedure is continued until convergence on the reaction field is achieved. In the cases examined until now, 4/5 iterations are usually sufficient to reach a convergence $\leq$0.0001 a.u. on the final energy.

The final equilibrium and nonequilibrium energies of the state of interest are easily determined by adding the corrections obtained by Eq. 1.17 or 1.18 to the excited-state energy provided by the TD calculation.

1.4.2 LR-PCM/TD-DFT

In LR-PCM/TD-DFT the excitation energies are "directly" determined without computing the excited-state density by plugging in a PCM contribution in the TD-DFT equations reported in Section 3.1 [87]. The coupling matrix **K** of Eq. 1.11 can thus be decomposed in two terms, the former being related to the gas-phase calculation, the second related to PCM:

$$K_{st,uv} = K^{0}_{ai,bj} + K^{\text{PCM}}_{ai,bj} \tag{1.22}$$

The PCM contributions to the time-dependent Kohn–Sham equations depend [87] on the term $\delta\phi(\mathbf{s}', \omega)$,

$$\delta\phi(\mathbf{s}', \omega) = \int_{\mathbf{R}^3} \frac{\delta\rho^{\mathrm{el}}(\mathbf{r}', \omega)}{|\mathbf{s}' - \mathbf{r}'|} d\mathbf{r}' \tag{1.23}$$

which formally corresponds to an electrostatic potential computed using the electron density variation ($\delta\rho^{\mathrm{el}}$) associated with the electronic transition in place of a specific electron density (ρ^{el}). The contribution from the PCM operator is then defined as

$$v^{\mathrm{PCM}}\left[\delta\rho^{\mathrm{el}}\right](\mathbf{r}) = \int_\Gamma \int_\Gamma \delta\phi(\mathbf{s}', \omega) \mathcal{Q}(\epsilon_{\mathrm{opt}}; \mathbf{s}', \mathbf{s}) \frac{1}{|\mathbf{s} - \mathbf{r}|} ds\, ds' \tag{1.24}$$

if only the fast solvation degrees of freedom are equilibrated with the excited-state density of the solute, and

$$v^{\mathrm{PCM}}\left[\delta\rho^{\mathrm{el}}\right](\mathbf{r}) = \int_\Gamma \int_\Gamma \delta\phi(\mathbf{s}', \omega) \mathcal{Q}(\epsilon; \mathbf{s}', \mathbf{s}) \frac{1}{|\mathbf{s} - \mathbf{r}|} ds\, ds' \tag{1.25}$$

when treating equilibrium solvation. In the nonequilibrium case, the PCM response matrix $\mathcal{Q}$ depends on the dielectric constant at optical frequency (ϵ_{opt}), whereas in the equilibrium case it depends on ϵ.

A significant part of solvent effect on the excited-state energies is recovered by LR approaches using MO orbitals computed in solution (and, thus, including the polarization due to the solvent reaction field) [83]. However, since the exact excited-state electron density is never computed, all the solvent contributions depending on the variation of the multipole moment upon excitation are missing in LR computations. An additional PCM correction is instead introduced which depends on the ground–excited state transition density [83]. As a consequence, the treatment of dynamical solvent effects is completely different with respect to that made by SS-PCM (in Eq. 1.18 explicit reference is made to the excited-state density). Furthermore, in the LR-PCM method the ground state is thus always fully equilibrated with the solvent degrees of freedom. As a consequence, in the nonequilibrium case the solvation contribution to the emission energy is computed as

$$\Delta G_{\mathrm{em}}(\mathrm{LR}) = \mathcal{G}^{(2)}_{\mathrm{neq}}(\mathrm{LR}) - \mathcal{G}^{(1)}_{\mathrm{eq}} \tag{1.26}$$

whereas for long living excited state it is computed as

$$\Delta G_{\mathrm{em}}(\mathrm{LR}) = \mathcal{G}^{(2)}_{\mathrm{eq}}(\mathrm{LR}) - \mathcal{G}^{(1)}_{\mathrm{eq}} \tag{1.27}$$

that is, in standard LR-PCM the equilibrium solvation energy for the ground state is used in both cases, making this method not suitable to an accurate treatment of the solvent effect on the emission process.

1.5 COMPUTING SPECTRA: THEORY

1.5.1 Choice of Functional

As anticipated above, results of the many computational studies of excited states exploiting TD-DFT have clearly highlighted that the accuracy of the computed vertical excitation energy (VEE) remarkably depends on the adopted functional. Several studies show that the local density approximation (LDA) (the exchange/correlation energy is a function of only the local value of the electron density) underestimates the VEE of valence transitions in organic molecules [89]. In particular, there is a marked deterioration of the results for high-lying bound states [51–58].

Better results are obtained by functionals exploiting the generalized gradient approximation (GGA, the functional depends on both the density and its gradient), such as BLYP [90], PBE [91], or HCTH [92]. These functionals have limited computational requirements, they can be easily implemented in ab initio molecular dynamics methods like Car–Parrinello, and in some systems, such as metals, they can provide rather reliable excitation energies. However, a huge number of computational studies indicate that they are usually outperformed by their "hybrid" analogues, that is, functionals, including a fraction of "exact" exchange, computed at the Hartree–Fock level of theory. The most popular hybrid functionals, such as B3LYP [93] or PBE0 [94–96], contain, in each point of the space, the same percentage of HF exchange, and they can thus be labeled as "global hybrids". These functionals, especially B3LYP, are surely the most commonly used functionals not only for ground-state DFT but also for TD-DFT excited-state calculations [53].

Local hybrid (LH) functionals are characterized by a mixing of HF exchange that depends on the spatial electronic coordinate (as LH-BLYP) [97]. Finally, range-separated hybrids (RSHs) use a growing fraction of exact exchange as the interelectronic distance increases, giving a long-range correction (LC) to the original DFT scheme [72, 73]. Functionals such as CAM-B3LYP [72] or LC-ωPBE [73] are representative of this class of functionals. Functionals that depend explicitly on the semilocal information in the Laplacian of the spin density or of the local kinetic energy density also have been developed [98]. Such functionals (see, e.g., the functionals of the M05 or M06 family) [51] are generally referred to as meta-GGA functionals.

Obviously, a thorough comparison among the performances of the different functionals is outside the scope of the present chapter (see refs. 51–58 for recent reviews). On the other hand, it is possible to draw some general guidelines for the selection of the density functional to use in TD-DFT calculations, summarizing the requirements for an "optimal" density functional. A first critical feature of a functional concerns the philosophy underlying its development. According to a first broad school of thought, once an appropriate functional form has been selected, the functional is heavily parameterized by reference to experimental data or data from explicitly correlated ab initio calculations [92]. Alternatively, the exact properties of the functional (determined to fulfill a series of physical conditions) can be used to determine both its structure and, eventually, the parameters in its functional form.

Based on our experience in the field, we think that a density functional with a small number of (ideally zero) adjustable parameters, and thus tailored according to this latter approach, would be preferable. Although it can deliver worse performances than heavily parameterized functionals in several cases, it is safer for studying systems that could not belong to the class of compounds used in the parameterization procedure.

A suitable functional should obviously also provide reliable results, namely (i) accurate ground-state geometries/vibrational frequencies, (ii) balanced description of valence and Rydberg states, (iii) balanced description of bright and dark excited states, (iv) good excited-state geometries/vibrational frequencies, and (v) a correct description of charge transfer states.

It is clear that a "perfect" functional fulfilling all the above requirements probably does not exist. Several different analyses have appeared in the literature comparing the performance of different functionals, sometimes with contradictory results [51–58]. Actually, it is necessary to pay a lot of attention when analyzing the studies assessing the performance of the different functionals (or, obviously, of different QM methods), especially when they are based on comparison with experimental results. In fact, (i) the VEE is not an observable and it does not necessarily coincide with the band maximum; (ii) as a consequence, absolute error is not necessarily the most significant property; (iii) reproducing trends (i.e., substituent effect) can be more important; and (iv) most of the experimental results are obtained in solution but most of the computational studies are performed in vacuo.

However, it is nowadays well assessed that hybrid functionals usually outperform nonhybrid ones and that the most commonly used hybrid functional, B3LYP, is not the best one (vide infra) [51–58].

Our previous experience suggests that the PBE0 hybrid functional is a "well-balanced" functional, usually providing very accurate results. In PBE0 the amount of exact exchange has been determined in order to fulfill a number of physical conditions without resorting to any fitting procedure [94]. PBE0, obeying both the Levy condition [99] and the Lieb–Oxford bound [100], provides a fairly accurate description of the regions characterized by a low electron density but, at the same time, by a high value of the electron density gradient. A reliable description of this region is important not only for a system dominated by dispersions interactions (e.g., van der Waals complexes) but also for higher lying excited states, as, for example, Rydberg states. For these reasons, TD-PBE0 excitation energies are, on the average, more accurate than those provided by other commonly used hybrid functionals, such as B3LYP [96]. The Becke exchange functional, not obeying the Levy condition [99] or the Lieb–Oxford bound [100], exhibits an incorrect asymptotic limit. Despite the absence of adjustable parameters, besides providing a reliable description of the ground-state properties of several classes of compounds, TD-PBE0 results have shown an overall degree of accuracy comparable with that of the best last-generation functionals in the description of both bright and dark excitation and both valence and Rydberg states [53–58]. Furthermore, the vibrational analysis performed on the ground of TD/PBE0 results (excited-state minima geometry and vibrational frequencies) turned out in remarkable agreement with the experimental indications for several classes of compounds [45–50].

On the other hand, PBE0 (like all the standard density functionals) exhibits important failures in describing zero-overlap CT transitions. In these cases, a check using purposely tailored functionals (CAM-B3LYP, M06-2X, LCω-PBE) is highly recommended.

1.5.2 Choice of Basis Set

As already anticipated, one of advantages of using TD-DFT is that this method does not exhibit any dramatic dependence on the size of the basis set. For valence transitions, many studies indicate that a medium-size basis set (valence double-zeta or triple-zeta adding polarization and diffuse functions) provides VEE close to convergence [53–58].

Based on our experience, the small 6-31G(d) basis sets provide sufficiently accurate equilibrium geometries and vibrational frequencies, and 6-31G(d) energy ordering is usually qualitatively correct, but for Rydberg states. Although the computed trend depends on the kind of transition considered, the VEE obtained at the 6-31+G(d,p) level are close to convergence (representing a good compromise between accuracy and computational cost). Finally, 6-311+G(2d,2p) results can be considered converged in the great majority of the systems.

The above considerations do not hold when studying excited states with Rydberg character: These transitions require more extended basis sets [6-31+G(d,p) is the minimal basis set possible], the inclusion of diffuse functions is mandatory, and their energies converge more slowly with the size of the basis set.

Analogously, the study of charged species, especially anions, in their electronic excited states requires, on average, more extended basis sets than that of their neutral counterpart. In this case, for example, the energy ordering obtained at the 6-31G(d) level is often misleading.

Particular attention has to be paid to the basis set adopted in PCM/TD-DFT calculations. The PCM cavity radii most commonly adopted have been parameterized in order to reproduce solvation energies (i.e., a ground-state property) at a specific level of theory [HF/6-31G(d), HF/6-31+G(d,p) for the anions, PBE0/6-31G(d), etc.] [101]. When using different basis sets, different results can be obtained, especially when thinking that VEE involves virtual molecular orbitals, which are usually more diffuse than the occupied ones and whose behavior "optimal" radii are likely different with respect to those used for the ground state. To compute the solvent effect on the excited states, the safer procedure is using a basis set not too large, similar to that used for the radii optimizations 6-31G(d), 6-31+G(d,p) for charged species.

1.5.3 Choice of Solvation Model

The two basic choices concern (i) the solvation model and (ii) the method to use. As discussed above, a continuum model such as PCM offers several advantages over purely supramolecular methods. On the other hand, PCM is expected to provide a good estimate of the electrostatic contribution to the solute–solvent interaction, whereas less accurate results can be obtained when dealing with nonpolar or hydrogen-bonding solvents. In the former case, the nonelectrostatic contribution (dispersion interactions, electronic repulsion) to the solute–solvent interaction is larger than the electrostatic one. Solvent shift in a nonpolar solvent, for example, can depend on the variation of the polarizability associated with the electronic transition more than on the dipole moment shift. Interesting attempts to treat this kind of problem within the framework of PCM/TD-DFT calculations have been made [102], but it cannot be taken for granted that standard calculations are able reproduce solvent effect in nonpolar solution (dielectric constant <4) or when dealing with nonpolar solutes. For hydrogen-bonding solvents, as anticipated above, the solute–solvent interactions can be highly directional and have a strongly molecular nature, that is, that cannot be properly described by simple electrostatic interactions. Consider, for example, the hydrogen bond formed by a water molecule and the lone pair of a carbonyl group. As we shall verify when reporting the results of our study on uracil, an electronic transition involving the carbonyl lone pair would significantly perturb the hydrogen bonds in which it participates. An accurate estimate of the effects of the electronic transition to the hydrogen bond interaction can be obtained only at the QM level, that is, explicitly including the solvent molecule in the computational model. Although PCM alone can provide useful hints on the solvent shift in hydrogen-bonding solvents, an accurate determination of the absorption and emission spectra requires that the most important solute–solvent interactions, say, the first solvation shell, are considered in the calculations by means of a mixed discrete-continuum approach [81]. Although there is no general rule concerning the number and geometry of the solvent molecules to be considered, as we shall see in the next sections, combining chemical intuition with experimental results, it is relatively easy to design a computational model able to provide accurate results.

As discussed above, PCM/TD-DFT being a reference method, it is still necessary to choose between LR-PCM/TD-DFT and SS-PCM/TD-DFT calculations. It is however important to remind that several of the following considerations are valid for many of the LR-PCM and SS-PCM methods.

LR-PCM/TD-DFT is surely the simplest and fastest method for computing absorption and emission energies in solution, its computational cost being only marginally larger than the corresponding gas-phase TD-DFT calculation. The availability of LR-PCM/TD-DFT analytical gradients gives relatively easy access to excited-state minima, other excited-state properties as the multipole moment, as well as all the different population analyses available for the ground state. Vibrational frequencies in solution can be obtained by numerical second derivatives of the energy, increasing their computational cost. The $6 \times N$ PCM/TD-DFT single-point calculations are indeed necessary to obtain the vibrational frequencies of a molecule containing N atoms. Furthermore, it is necessary that the procedure is performed with a lot of caution when in the presence of two close-lying excited states.

The results of a large number of studies indicate that LR-PCM is able to provide a fairly accurate estimate of the VEE in solution, especially for what concerns the bright excited states [53–58]. On the other hand, LR-PCM has been shown to overestimate solvent effects on the intensities, especially at the EQ level [85, 86]. The two most significant limitations of LR-PCM concerns (i) the treatment of the emission process (and in general the dynamical solvation effect) and (ii) the study of the electronic transitions involving a substantial electron density shift [85, 86].

For the emission process, we recall that in LR-PCM calculations solvent degrees of freedom are always equilibrated with the ground-state density and in general, this method provides a less rigorous treatment of dynamical solvent effect than SS-PCM. For example, the solvent reorganization energy (λ), that is, the difference between the energy of a state in the NEQ and in EQ limits, is not correctly computed by LR-PCM. In this method λ is proportional to the square of the transition dipole moment: It is thus larger for bright transitions, being zero for dark states. Furthermore, in LR-PCM the excited-state energy does not exhibit any dependence on the excited-state dipole moment, and thus the energy of the excited states with a large dipolar character (e.g., the CT transitions) is significantly underestimated, especially at the EQ level [85, 86].

On the other hand, most of the above deficiencies are not present in SS approaches, as SS-PCM/TD-DFT, which instead gives a balanced description of strong and weak electronic transitions (see also the results reported below). Several studies (see below) indicate that SS-PCM/TD-DFT provides accurate estimates of dynamical solvent effects on the absorption and emission processes, of solvent reorganization energy, and thus of inhomogeneous broadening. For example, in SS-PCM/TD-DFT λ is indeed proportional to the square of the dipole moment shift associated with the transition, which is indeed expected to be the leading term in a polar solvent [85, 86].

Unfortunately, SS-PCM/TD-DFT excited-state analytical gradients are not available. Furthermore, SS-PCM has an iterative implementation which not only increases the possibility of convergence failures but would also make geometry optimizations rather cumbersome.

An effective strategy is thus complementing the results of LR-PCM/TD-DFT geometry optimizations by single-point SS-PCM calculations. In any case, notwithstanding the above caveat, in many systems a good agreement between LR-PCM/TD-DFT results and experimental absorption and emission spectra has been found [53–58, 78]. It has indeed been shown [83] that at the zero order, related to the interaction with the slow solvation degrees of freedom, LR and SS approaches are identical [83].

It is necessary, finally, to comment on another factor potentially affecting the computed spectral parameters, that is, the cavity radii [88]. As we have anticipated above, the solute cavity within PCM has built as envelope of spheres centered on the atoms or atomic groups. Different parameters are involved in the building of the cavity; one of the most important obviously is the radii associated with each sphere, which rules the volume of the cavity and, thus, the distance between the atoms and the cavity surface. It is well known that the computed properties depend on the adopted atomic radii, and this is obviously true also for the excited-state energies. Both SS-PCM and LR-PCM results depend on the cavity model adopted, especially on the radii used. Additional studies are necessary to shed light on this latter point. Several studies indicate that using the cavity models developed for ground-state computations usually provides fairly accurate results [53–58]. However, this feature has to be considered with particular attention when comparing different kinds of transitions. Furthermore, since the choice of the cavity affects the relative energy of the transition, some critical features (e.g., the presence of a crossing) can depend on the cavity.

1.6 COMPUTING SPECTRA: APPLICATIONS

Before examining in detail the results of some TD-DFT applications to the computation of the spectra, it is useful to provide some hints about the expected accuracy of this method and of the different functionals. Although the huge number of studies makes it almost impossible to give a comprehensive picture, very interesting indications are provided by the very careful studies of Jacquemin, Adamo, and colleagues, performed a systematic comparison among the experimental absorption spectra of several classes of organic dyes of industrial interest (azobenzene, anthraquinone, indigos, arylethenes) and the VEE computed at the TD-DFT level using different density functionals [53–58].

As expected, their analysis shows that the performances of TD-DFT, in general, and of the different functionals depend on the class of compounds examined. For 189 anthraquinone derivatives solvated in various media (CH_2Cl_2, methanol, and ethanol), the mean absolute errors (MAEs) are 0.10 eV for both PBE0 and B3LYP [53]. For 304 indigoid derivatives, PCM-TD-PBE0/6-311+G(2d,p)//PCM-PBE0/6-311G(d,p) calculations provide excitation energies with a MAE limited to 0.04 eV with respect to the experimental band maxima [53].

Including other class of compounds in the analysis, such as azobenzene, coumarins, diaryl-ethenes, and diphenylamine derivatives, does not change significantly the picture: The best performances are delivered by PBE0, with an average absolute deviation limited to 0.14 eV/22 nm, which has been proposed to be the expected PCM/TD-PBE0 accuracy for low-lying excited states of conjugated organic compounds. The second best approach, CAM-B3LYP, suffers larger deviations (0.26 eV/38 nm) but appears particularly well suited for studying dyes with a very delocalized excited state. Furthermore, CAM-B3LYP provides a very good estimate of substituent effect on the absorption spectra [54].

When class of compounds that are known to be ill-treated by TD-DFT, for example, cyanine, are included in the set of the experimental data ($\sim$500 compounds and more than 700 excited states), the average error of TD-PBE0 VEE increases up to 0.24 eV, and a similar value is obtained when the comparison is made with the best theoretical estimates computed for a smaller set of compounds in the gas phase (104 singlet state). In any case, such a value is an average between that expected for electron transitions with a monodeterminantal nature, for which PBE0 (and other hybrid functionals including 20/30% of HF exchange) is remarkably accurate (expected error $\sim$0.15 eV), and those with a strong multideterminantal nature (e.g., cyanine, triphenylmethane, and acridine derivatives) for which TD-DFT is inadequate [54].

Interestingly, the analysis of Jacquemin et al. [54] shows that the accuracy of PBE0 (and of other hybrid functionals containing a similar percentage of HF exchange) for $n \rightarrow \pi^*$ transitions is even larger than that found for $\pi \rightarrow \pi^*$ transitions, the MAE with respect to the experiments being close to zero. This results, as highlighted by Jacquemin et al. is probably due to the more local character of $n\pi^*$ transitions in the examined set.

Rydberg excitations are another class of transitions that have been traditionally considered "not reliably" treated at the TD-DFT level. Actually, several studies have shown that, besides the use of at least one diffuse function of the heavy atoms, the accuracy of TD-DFT calculations dramatically depends on the adopted functional. Indeed, in order to correctly describe low-density/high-density gradient regions as those involved in Rydberg excitations, it is necessary to use an exchange-correlation functional exhibiting a correct asymptotic behavior:

$$\lim_{r \to \infty} v_{xc}(r) = -\frac{1}{r} \tag{1.28}$$

When functionals, such as PBE0, fulfilling the above condition are employed, fairly accurate VEEs are obtained for Rydberg excitations, the MAE being <0.4 eV for several organic compounds [54].

For metallo-organic systems, to the best of our knowledge, no such systematic comparisons between TD-DFT and experimental results exist in the literature. Accurate results have been obtained on several porphyrin-like metal complexes [103]. TD-PBE0 delivers reliable results in the study of several Ru–polypyridine complexes, although its accuracy is lower than that found for organic molecules (errors in the range 0.15–0.4 eV) [53]. Standard functionals such as B3LYP have often been successfully applied to interpret the spectra of Ru–dye complexes of potential interest for the development of solar cells [104]. On the other hand, studies of $[Ni(H_2O)_6]^{2+}$ complexes show that TD-B3LYP are not fully reliable, even qualitatively, due to the lack of transitions of double-excitation character and the wrong treatment of CT transitions (the energy of the LMCT transitions is significantly underestimated) [105]. In summary, additional studies are probably necessary to assess the general reliability of TD-DFT for the study of the visible spectra of metallo-organic systems, since the quality of the results can depend on several additional factors (proper inclusion of relativistic effects, pseudopotential employed, spin–orbit coupling) with respect to those examined for organic molecules.

Both for organic and metallo-organic systems, studies in the literature show that TD-DFT, also employing a standard hybrid functional such as PBE0, can provide a remarkably accurate description of transitions with partial CT character (see also Section 1.6.1.2). In fact, we remind the reader that the failures of TD-DFT calculations are usually associated with long-range CT between two partners whose molecular orbitals have a vanishing overlap. On the other hand, CT transitions in which donor and acceptor have a small but nonzero overlap (e.g., the metal to ligand charge transfer (MLCT) transitions in several inorganic complexes) can be treated in a fairly accurate way without resorting to long-range corrected functionals [53].

1.6.1 Selected Examples

To better illustrate the general concepts summarized above, it can be useful to make explicit reference to the calculation of the vertical absorption and emission energies of some representative compounds in different environments. Based on previous experience, we discuss in some detail the results obtained in the study of nucleic acid components and one of the most used solvent probes, coumarin C153.

1.6.1.1 Bases of Nucleic Acids

Uracil in Gas Phase Absorption of UV–vis radiation by DNA and its constituents is a phenomenon of fundamental biological interest. Indeed, since many DNA photolesions are triggered by the population of singlet excited electronic states, the availability of a fast radiationless decay channel is very important to life [106]. As a consequence, the excited states of the nucleic acid bases (nucleobases) have been thoroughly investigated by many experimental and computational studies, both in the gas phase and in solution, providing several indications of the performances of the different computational approaches. Uracil (see Figure 1.1) is probably the nucleobase examined more in detail [107–123]. Experiments reveal the presence of a strong absorption band at ∼4.8–5.0 eV (depending on the solvent), and computations ascribe this feature to a π/π^* transitions (hereafter $S\pi$), which mainly corresponds to a HOMO → LUMO excitation [107]. Different computational methods agree in predicting that $S\pi$ is not the lowest energy excited state in the FC region: Indeed, an underlying dark state is predicted, with $n\pi^*$ character (hereafter Sn), which mainly arises from the excitation of an electron from the lone pair (HOMO-1) of the C4–O8 carbonyl group (see Figure 1.1) to the π^* LUMO [108]. The energy difference between Sn and $S\pi$ is thus another quantity of relevant interest, especially when

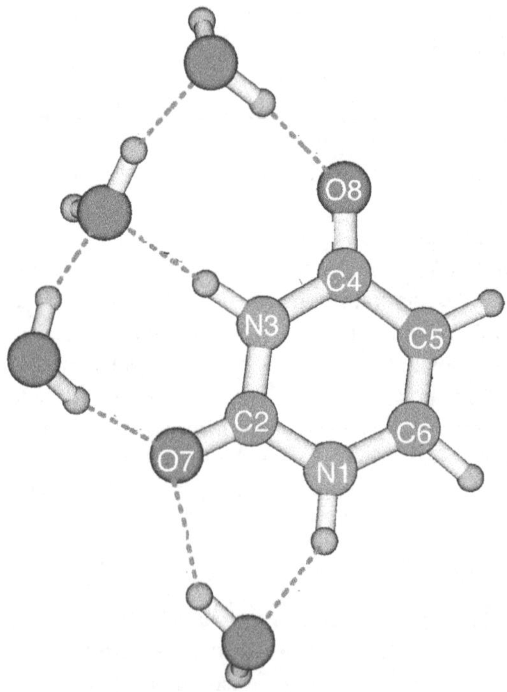

Figure 1.1 Schematic description (and atom labeling) of uracil·4 H_2O model.

thinking that, according to both experiments and computations, a significant $S\pi \rightarrow Sn$ population transfer is possible [109, 110].

A thorough study of the uracil in the gas phase has been performed by Matsika et al., which compared the results obtained by the equation-of-motion coupled-cluster and multireference configuration interaction methods with those provided by other computational methods [108]. The completely renormalized EOM-CCSD with the noniterative triples CR-EOM-CCSD(T) method, based on the methods of moments of coupled-cluster equations, used with the aug-cc-pVTZ basis set, predicts that $S\pi$ and Sn VEEs are 5.25 and 5.00 eV, respectively. MRCI(12,9)/aug-ANO-DZ calculations, including the Davidson correction (the expansion consists of about 330 million configuration state functions), predicts that $S\pi$ VEE is 5.32 eV, whereas, at this level of theory, the authors did not succeed in obtaining the Sn energy due to technical difficulties. CASPT2(14,10)/6-31G(d,p) calculations provide similar indications, $S\pi$ VEE = 5.18 eV and Sn VEE = 4.93 eV. Inspection of Table 1.1 clearly shows that the results provided by the TD-PBE0 method are extremely close to those obtained by using much more computationally demanding methods and much more accurate than other ab initio methods as CC2. The only significant discrepancy with respect to CCSD(T) predictions concerns the energy difference between $S\pi$ and Sn, which is 0.2 eV larger in PBE0, and, in this respect, this prediction looks more consistent with the experimental indications in water (vide infra). Furthermore, TD-PBE0 results are much less dependent on the basis set size than those provided by EOM/CCSD(T) or CASPT2. These two latter methods also exhibit a marked dependence on the size of the active space included in the calculations. In the wave function method these features make it much more difficult to treat large-size systems, for which the use of

Table 1.1 Comparison between VEE Computed for Two Lowest Energy Excited States of Uracil Using Different Computational Approaches

Method	Basis Set	$n\pi^*$	$\pi\pi^*$
CR-EOM-CCSD(T)	aug-cc-pVTZ	5.00	5.25
CR-EOM-CCSD(T)	6-31G(d)	5.19	5.65
EOM-CCSD	aug-cc-pVTZ	5.23(0.00)	5.59(0.20)
CASPT2(14,10)	6-31G(d,p)	4.93(0.00)	5.18(0.20)[a]
MS(3)-CASPT2(12,10)[b]	6-31G(d)	5.16(0.00)	5.52
CASPT2//CASSCF(14,10)[c]	ANO		5.02
MRCISD+Q(12,9)	aug-ANO-DZ	not conv	5.32
CC2	TZVP	4.91	5.52
PBE0	6-311+G(2d,2p)	4.81(0.00)	5.26(0.14)
PBE	6-311+G(2d,2p)	4.00(0.00)	4.76(0.06)
LC-ωPBE	6-311+G(2d,2p)	5.06(0.00)	5.40(0.18)
B3LYP	6-311+G(2d,2p)	4.68(0.00)	5.15(0.13)
CAM-B3LYP	6-311+G(2d,2p)	5.08(0.00)	5.40(0.18)
M052X	6-311+G(2d,2p)	5.02(0.00)	5.51(0.20)

[a] From ref. 111.
[b] From ref. 112.
[c] From ref. 113.

large basis sets and active spaces is rather cumbersome. Inspection of Table 1.1 shows that, on average, PBE0 exhibits the best performances among the functionals examined. Long-range corrected functionals, CAM-B3LYP or LC-ωPBE, overestimate the VEE of $S\pi$ by at least 0.2 eV. A pure density functional such as PBE significantly underestimates both $S\pi$ and Sn VEE. B3LYP provides similar results as PBE0, but Sn VEE exhibits a larger discrepancy with respect to the most accurate WF-based methods than that provided by PBE0.

Interestingly, the maximum of the absorption band of uracil in the gas phase is 5.08 eV, that is, 0.1–0.2 eV red shifted with respect to the "best" computational estimates. In any case, it is important to remember that the shape and maximum of the experimental absorption band can be significantly modulated by vibrational and vibronic effects.

Uracil in Water Solvent noticeably affects the energy of the excited states and their relative stability. For example, the absorption band maximum of uracil in water is ~0.25 eV red shifted with respect to the vapor phase [114]. Sn is known to be significantly destabilized in water, and solvent has been shown to modulate the excited-state dynamics in uracil derivatives [107, 115].

Our studies in the field have already unambiguously shown that proper description of solvent shifts of uracil excited states in aqueous solution requires that both bulk solvent effect and solute–solvent hydrogen bond interactions have to be considered [107, 109, 112, 115, 116]. As shown in Table 1.2, in fact, inclusion of bulk solvent effects by the PCM model decreases by 0.1 eV $S\pi$ VEE and increases by 0.35 eV that of Sn. When four water molecules of the first solvation shell are included in the model (see Figure 1.1), and gas-phase calculations provide similar trends. When both effects are considered, Sn is strongly destabilized (by ~0.5 eV), whereas the computed solvent red shift for $S\pi$ (0.2 eV) approaches that predicted by the experiments (~0.25–0.3 eV). Inclusion of water molecules is particularly important to correctly reproduce the solvent effect on Sn: This electronic transition involves the transfer of an electron from the oxygen LP, which can potentially act as hydrogen bond acceptors, toward the more diffuse π^* molecular orbital, leading to a decrease of the solute–solvent hydrogen bond strength.

The VEEs of uracil in water solution have also been studied by other approaches [117–120]. MRCI/cc-pVDZ/MM calculations (considering 257 water molecules as fixed charges, whose position is averaged by means of MD simulation) predicts that in water $S\pi$ VEE is red shifted by ~0.05–0.1 eV and Sn VEE blue shifted

Table 1.2 Comparison between the VEE Computed for Two Lowest Energy Excited State of Uracil Using Different Models

Excited State	Gas Phase	Gas Phase + 4 H_2O	Water PCM	Water PCM + 4 H_2O
Sn	4.81(0.00)	5.11(0.00)	5.15(0.00)	5.32(0.00)
$S\pi$	5.26(0.14)	5.19(0.15)	5.16(0.20)	5.10(0.20)

Note: PBE0/6-311+ G(2d,2p)//PBE0/6-31G(d) calculations.

by ~0.4 eV with respect to the gas phase [117]. The use of the fragment molecular orbital multiconfigurational self-consistent field (FMO-MCSCF), which partially includes quantum effects in the description of the solute–solvent interactions, provides solvent shifts closer to the experimental estimates and to our computational results: for $S\pi \sim -0.15/-0.18$ eV, for $Sn \sim +0.42/+0.47$ eV. EOM-CCSD(T)/MM calculations provide a +0.44 blue shift for *Sn* and a +0.05 blue shift for $S\pi$, and this latter result is in qualitatively disagreement with experiments. Using Monte Carlo simulations on 200 water molecules to generate solvent configurations around uracil, which have then been used for intermediate neglect of differential overlap (INDO) excited-state calculations, indicate +0.5-eV and −0.19-eV solvent shifts for *Sn* and $S\pi$, respectively [119]. On the balance, it seems that a mixed discrete–continuum approach, notwithstanding by far the computationally less demanding, provides the most accurate predictions. On average, we can thus estimate that, when going from the gas phase to water, *Sn* is destabilized by ~0.7 eV with respect to $S\pi$. As a consequence, in the FC region *Sn* is less stable than $S\pi$ by ~0.2 eV only. Such a difference is fully compatible with the partial $S\pi \rightarrow Sn$ population transfer evidenced by time-resolved experiments for uracil in water [109, 110]. On the other hand, if we assume that in the gas phase *Sn* is less stable than $S\pi$ by ~0.25 eV only (as predicted by EOM-CCSD(T) or CASPT2 calculations), the energy difference in water would be ~0.5 eV, a value which would make the population transfer much less likely. Those considerations clearly show that careful study of the solvent effect is mandatory in order to use computational results obtained in the gas phase to study processes occurring in solution or just to assess the reliability of a given computational method.

Actually, our mixed discrete–continuum model is not limited to the study of UV–vis spectra, but it has been already successfully employed to model solvent effects on several different spectral properties, such as electron paramagnetic resonance (EPR) hyperfine coupling constants, nuclear magnetic resonance (NMR) chemical shifts, and so on [45, 121].

The only point not unambiguously defined in our model concerns the number and position of the solvent molecules to be explicitly included in the calculations. Although it is not possible to define strict and general rules, a suitable combination between experimental and computational indications and chemical intuition in most of cases allows for obtaining reliable results. For uracil in water, for example, NMR experiments indicate that no water molecule is strongly bonded to C5 and C6 carbon atoms and that O7 and O8 are coordinated by two and one water molecules, respectively [122]. Car–Parrinello dynamics suggest that the first coordination shell of uracil (up to 2.5 Å) is formed by six water molecules, four in the molecular plane (as in Figure 1.1) and two more or less perpendicular to it [123]. Although a full description of the first solvation shell in solution requires, of course, a proper dynamic treatment, a number of studies have confirmed that the PCM is able to accurately account for the effect of water molecules that are more distant and/or not directionally bound to the carbonyl oxygen lone pairs. The model depicted in Figure 1.1 thus appears a reasonable guess, to be further refined by means of PCM geometry optimizations. Furthermore, our previous experience suggests that inclusion of bulk solvent effects by means of PCM significantly decreases the dependence of the computed VEE on the "exact" coordination geometry of the first solvation shell, and, in many cases, also on the number of solvent molecules considered. Geometry arrangements that in the gas phase are predicted to have significantly different energy and/or spectral properties appear almost isoenergetic at the PCM level, which is able to correctly reproduce the energetic effect of the outer solvation shells.

Guanine in Water at Strongly Acidic pH According to our model, in both ground, and excited-state geometry, optimizations, the first solvation shell is fully optimized, that is, it is treated like solute degrees of freedom. Since outer solvation shells are not included in our calculations, this choice is expected to overestimate the conformational flexibility of the water molecules of the cybotactic region, suggesting that particular attention has to be paid when investigating dynamical solvation effects. In some cases the number of solvent molecules included in the model can also affect the computed excitation energies. An example of this feature is provided by our recent study of the excited-state dynamics of guanine monophosphate (G) [124]. In this study it is shown that protonation of nitrogen N7 (see Figure 1.2) in strongly acidic conditions noticeably affects the behavior of the lowest energy bright state (L_a). We have compared the results obtained by including five (G-Wa5) or 6 (G-Wa6) water molecules in our computational model. While for the neutral compound no minimum is found in the path leading from the FC to the CoI with S_0, for the protonated

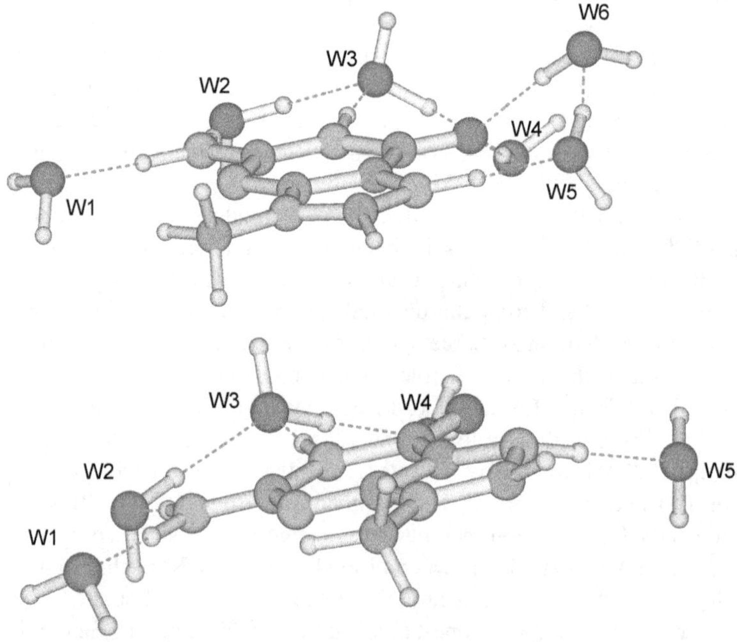

Figure 1.2 Schematic description of the minima of the lowest excited state of 9-methylguanine in water in strongly acidic conditions, computed in water at the PCM/TD-PBE0/6-31G(d) level by using a model including five (up) or six (down) water molecules.

compound a minimum is found, where the N7 and C8 hydrogens are out of the molecular plane (see Figure 1.2), with concomitant motion of water molecules in the first solvation shell. Interestingly, the out-of-plane distortion is larger for the model with G-Wa5, since in G-Wa6, due the presence of an additional hydrogen bond between the solvent molecules of the first solvation shell, the motion of the water molecule hydrogen bonded to the N7-H group is more restrained. The different geometry of the excited-state minimum is obviously mirrored in the computed emission energy, which is larger by $\sim$2200 cm^{-1} for G-Wa5, since distortion from planarity significantly destabilizes S_0. It is thus clear that, especially for excited-state geometry optimizations, there are situations where different choices of the solvation shell can noticeably affect the computed spectra. On the other hand, it is important to remind that both G-Wa5 and G-Wa6 models are able to reproduce the effect of the protonation on the emission spectra, disclosing the underlying chemical–physical effects. In fact, experiments show that the Stokes shift is significantly larger (by $\sim$4000 cm^{-1}) for the protonated compound, in line with computational predictions.

1.6.1.2 Coumarins The results obtained in the study of coumarin, especially that of C153, allows getting additional insights on the treatment of non-hydrogen-bonding solvents. In coumarins (see Figure 1.3), the $S_0 \rightarrow S_1$ transition has a $\pi \rightarrow \pi^*$ character, and it essentially corresponds to a HOMO $\rightarrow$ LUMO transition [46, 47]. However, while the HOMO is delocalized on the whole molecule, with significant contribution by the π orbitals of the "central" benzene ring and of the nitrogen atom, the LUMO is mainly localized on the "quinone like" terminal ring with significant contribution of the π^* orbital of the carbonyl group. As a consequence, the $S_0 \rightarrow S_1$ transition has a partial intramolecular charge transfer character (from the nitrogen atom to the carbonyl group), and S_1 has a partial zwitterionic character (with the nitrogen

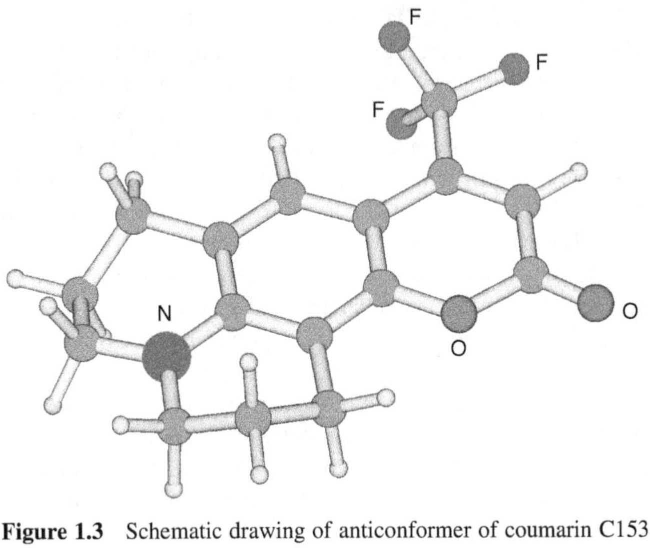

Figure 1.3 Schematic drawing of anticonformer of coumarin C153.

atom and the oxygen of the carbonyl group bearing a formal positive and negative charge, respectively). This results in a significant solvent shift of this transition, which, coupled to the molecular rigidity, makes coumarin and, in particular, C153 ideal tools for investigating the different solvation rates by time-resolved spectroscopic techniques, being the most used "molecular probes" of crucial dynamical aspects of solvation [125, 126].

Starting our analysis from the gas phase, the picture of the C153 excited states obtained at the TD-PBE0 level is in remarkable agreement with the experimental one. The 0–0 transition energies computed in the gas phase by the PBE0 functional for the nearly isoenergetic syn and anti isomers of C153 (25,680 and 25,660 cm^{-1}) are indeed in almost quantitative agreement with the experimental values (25,898 and 25,710 cm^{-1}, respectively) [127]. The calculated values for the oscillator strength (0.36 and 0.37, respectively) match their experimental counterpart (0.37) [128]. Finally, PCM/TD-PBE0 calculations predict that in both dimethyl sulfoxide (DMSO) and cyclohexane the dipole moment shift ($\Delta\mu$) associated with the electronic transition is ≈6.0 D (see also ref. 85). This value is in good agreement with the estimates based on electroabsorption experiments [129] and slightly underestimated with respect to the experimental determination in the gas phase [130].

The very good agreement between experimental and computational results could seem at a first sight surprising due to the partial CT character of the S_0–S_1 transition. On the other hand, the C153 lowest energy transition involves only partial intramolecular CT and can be correctly treated by hybrid functionals, as shown by the results obtained on dimethylaminobenzonitrile [131]. In this respect, it is noteworthy that for C153 conventional GGA functionals provide much less accurate results than their hybrid counterpart: For instance, the PBE functional [91] provides gas-phase 0–0 transition energies of 19,845 and 19,931 cm^{-1} for the syn and anti isomer, with a significant underestimation (≈6000 cm^{-1}) with respect to experimental results.

Comparison between vertical excitation and emission energies computed at the SS-PCM/TD-PBE0 and LR-PCM/TD-PBE0 levels confirms that the former method is significantly more accurate than the latter, especially for the emission energy in DMSO and, consequently, for estimating both the solvatochromic shift and the Stokes shift in DMSO. In fact, the predicted Stokes shift in DMSO, which is underestimated by 1200 cm^{-1} at the SS level, is underestimated by 3000 cm^{-1} at the LR level and, in the same way, the solvatochromic effect DMSO–cyclohexane for fluorescence (overestimated by 400 cm^{-1} at the SS level) is underestimated by 2200 cm^{-1} at the LR level [47]. These results confirm that it is important to resort to SS-PCM single-point calculations for an accurate evaluation of the emission process and of the dynamical solvent effects. On the other hand, the very good agreement between the computed and experimental lineshapes supports the general accuracy of the equilibrium structures and vibrational frequencies provided by LR-PCM/TD-DFT calculations [45–50]. From the general point of view, our studies on coumarins (not only on C153 but also on other coumarin derivatives) [86] shows that, as already stated, solvent models based on a polarizable continuum, at least when dealing with non-hydrogen-bonding solvents, are able to capture the "physics" of time-dependent phenomena in solution.

1.6.2 Dealing with Supramolecular Interactions: Optical Properties of DNA

In the examples considered until now, we have tackled the study of a medium-size molecule (a single chromophore) in solution. The theoretical/computational framework we have outlined can be rather easily applied to other several systems, such as, for example, a chromophore embedded in a protein. In this latter case, the chromophore can be treated by using the methods described above, and the only difference concerns the inclusion of environmental effects. Mixed QM/MM methods are indeed usually more suitable than continuum models for treating highly nonisotropic environments as proteins [132]. Many processes of biological or technological interest, however, depend on multiple chromophores arranged in supramolecolar structures. The optical properties of those systems are traditionally treated by semiempirical methods, where the interaction among the chromophores is included at a simplified level (e.g., by considering the dipolar/dipolar coupling between the excited-state transitions in the isolated chromophore). However, it is nowadays possible to attain a fully quantum mechanical description of the optical properties of complex supramolecular structures, as shown by the results obtained on DNA single and double strands [133–136].

The computation of nucleic acid UV–vis spectra is an ideal test case for theoretical and computational models. Not only are a large number of experimental results (also time resolved) available [106], but, since the DNA monomers have been thoroughly studied, it is possible to assess the accuracy of the different theoretical approaches in describing the effect that the subtle balance of supramolecular interactions (such as π stacking and hydrogen bonds) has on the excited-state behavior.

The study of π-stacked systems in their excited states is particularly challenging. Indeed, simpler models describing multichromophore spectra on the basis of excitations localized on the chromophore, weakly coupled by exciton coupling, are not fully adequate to describe strongly interacting systems, that is, face-to-face π-stacked systems close to their van der Waals minimum (intermonomer distance ~3–4 Å). In these latter systems, the frontier orbitals can have a substantial overlap and the resulting electronic transitions delocalized over multiple bases. This is indeed the case of the lowest energy bright electronic transitions of stacked adenine nucleobases in polyAde single and double strands [133–136]. In this latter case, our studies in the field provide encouraging indications about the reliability of TD-DFT calculations when based on a suitable functional for the treatment of the excited states in π-stacked nucleobases. Both in stacked adenine dimer and in $Ade_2 \cdot Thy_2$ stacked tetramer [133–136], PCM/TD-PBE0 calculations (at both the SS-PCM and LR-PCM levels) fully reproduce all the features that experiments show polyAde·polyThy oligomers always exhibit when compared to the spectrum of equimolar mixtures of the corresponding monomers [137, 138]:(i) a weak blue shift of the band maximum, (ii) a strong decrease of the absorption intensity, and (iii) a noticeable shoulder in its redwing. Interestingly different functionals are able to reproduce the above features, but PCM/TD-PBE0 VEE are closer to the experimental band maximum. Those results have been obtained also on a system explictly including the phosphoribose backbone (see Figure 1.4), providing useful indications on the effect of the backbone on the absorption spectrum of DNA.

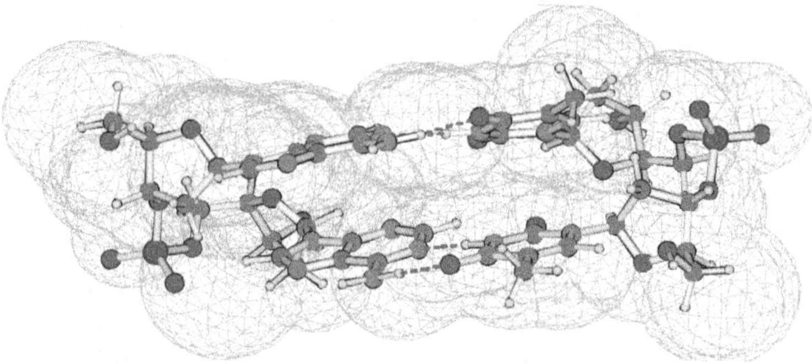

Figure 1.4 Schematic drawing of the $(dT)_2{\cdot}(dA)_2$ tetramer, adopting the B-DNA conformation. The PCM cavity used for calculations in solution is also shown.

The transition energy of several dark-excited states falls within the absorption band of the tetramer and the Ade dimer. Some of them have a clear-cut Ade → Ade or Ade → Thy CT character, and their energy is significantly underestimated by PBE0 with respect to CAM-B3LYP or M052X. In order to investigate this point in more detail, we have optimized the (9Methyl-Ade)·(1Methyl-Thy) Watson–Crick hydrogen-bonded pair in the gas phase, computing the lowest energy transitions by using different density functionals [133]. M052X and CAM-B3LYP functionals predict that the A → T CT transition is less stable by ~0.8 eV than the bright excited state localized on T. This result agrees within 0.15 eV with predictions of ab initio CC2 methods [139, 140].

PBE0 overestimates the stability of intrastrand CT transitions, although the amount of the overestimation is smaller than what was found for Ade → Thy interstrand CT transitions. We recall indeed that the most dramatic failures of "standard" density functionals in treating CT transitions occur in the case of zero overlap between the MOs of the donor and the acceptor molecule [17]. We have computed the absorption spectrum of $(9\text{Methyl-Ade})_2$ in the gas phase by using the B-DNA-like ground-state minimum, optimized in aqueous solution, as reference geometry. CAM-B3LYP and M05-2X predict that the lowest energy 9Methyl-Ade → 9Methyl-Ade CT transition is ~0.45 eV less stable than the excited states corresponding to the most intense electronic transition. This estimate is in a good agreement with the 0.64 eV value provided by CC2/TZVP calculations on a similar system (two A molecules adopting the experimental B-DNA structure) [139]. As anticipated above, PBE0 instead predicts that the CT transition is ~0.25 eV more stable than the bright excited state, with a smaller discrepancy with respect to the CC2 results than that found for the interstrand CT transition.

Although the above results confirm that PBE0 suffers from the same deficiencies as standard functionals in properly treating CT transitions, it is expected to deliver a better description of the stacked systems (in both the ground and excited states) than other commmonly used functionals, such as B3LYP, which do not exhibit the correct asymptotic limit. In fact, when applied to the study of cytosine-stacked dimer, TD-PBE0 provides a very accurate description of the dependence of the S_1 energy on the intermonomer distance, in very good agreement with that obtained at the CASPT2 level [141].

As a consequence, a suitable approach for the study of large-size systems could be that of using TD-PBE0 for predicting and interpreting the behavior of the bright excited states or of the "local" dark transitions (e.g., $n \rightarrow \pi^*$ excitations) but correcting the estimate of the CT transitions energy using results from such long-range corrected functionals as CAM-B3LYP and M062X. An alternative could be that of resorting only to these latter functionals, although the description of the bright excited states could be less accurate than that obtained at the TD-PBE0 level.

In any case, it is clear that proper treatment of the solvent effect, both static and dynamical, is fundamental for reliable evaluation of the CT transition's stability in the condensed phase. When using continuum solvation models, a state-specific approach combined with an accurate description of the excited-state electron density (averaging procedures of the excited-state density such as those usually employed in CASPT2 should be avoided) is mandatory, since LR-PCM/TD-DFT strongly underestimates the stability of transition with even partial CT character.

1.7 CONCLUDING REMARKS

In this chapter we have tried to provide a methodological and computational framework for the calculation of the vertical excitation and emission energies of medium/large molecules in solution. Despite its limitations, TD-DFT appears to be a very promising method. It has a very limited computational cost, it is "user friendly," and, when using a suitable functional (PBE0, CAM-B3LYP, M062X, depending on the system/process under investigation), it provides absorption and emission energies within the 0.2–0.3-eV range of the corresponding experimental peaks. Contemporarily, PCM has shown to be a very effective method for including the solvent effect in the calculation of the optical properties. On this ground, we can sketch the basic steps to follow when computing the absorption and emission spectra of a given compound:

1. Ground-state geometry optimization and frequency calculations. DFT/6-31G (d) (in the gas phase) and PCM/DFT/6-31G(d) (in solution) calculations, employing a global hybrid density functional, usually, provide reliable results.

For Absorption Process

2. TD-PBE0/6-31+G(d,p) (in the gas phase) and LR-PCM/TD-PBE0/6-31+G(d,p) (in solution) should provide a good estimate of the VEE, but for Rydberg states where a TZV basis is needed.
3. When looking for a more accurate estimate of the energy difference between the different excited states in solution: (i) SS-PCM/TD-DFT/6-31+G(d,p) calculations and, for hydrogen bonding solvent, (ii) including in the calculation the molecules of the first solvation shell.

For Emission Process

4. Computing the excited-state geometry. TD-PBE0/6-31G(d) (in the gas phase) and LR-PCM/TD-PBE0/6-31G(d) calculations should provide a reliable geometry.

5. TD-PBE0/6-31+G(d,p) (in the gas phase) and LR-PCM/TD-PBE0/6-31+G(d,p) (in solution) should provide a good estimate of the emission energy, although, in solution, a SS-PCM/TD-PBE0/6-31+G(d,p) check is highly recommended.
6. For supramolecular systems TD-PBE0 can understimate the VEE of CT transition. A check using other functionals (as CAM-B3LYP or M06-2X) or computational approaches is highly recommended.
7. When treating hydrogen-bonding solvents, accurate absorption and emission energies can be obtained only when the cybotactic region is explicitly included in the calculations. Mixed continuum–discrete approaches are very effective. Sometimes the choice of the model is not unambiguous but, in any case, for reasonable choices the results are much more accurate than that obtained by using only PCM or purely supramolecular approaches in the gas phase.

In most cases, the above recipes can be directly applied to supramolecular systems containing up to ~150 atoms, providing fairly accurate results. On the other hand, it is important to be aware that any computational approach, when used as a blackbox, can lead to disappointing results, especially when dealing with complex systems/processes. The approaches we have described in this chapter do not yet enable, instead, the calculation of the optical and electronic properties of very large supramolecular systems at the nanoscale. However, the possibility of accurately computing the properties of "meaningful" building blocks (see, e.g., the $Ade_2 \cdot Thy_2$ tetramer for AT DNA) is an important step toward the definition of effective and reliable multiscale models.

REFERENCES

1. N. Tkachenko, *Optical Spectroscopy Methods and Instrumentations*, Elsevier, Amsterdam, 2006.
2. (a) J. Laane, H. Takahashi, A. D. Bandrauk, Eds., *Structure and Dynamics of Electronic Excited States*, Springer, Berlin, 1998. (b) M. Klessinger, J. Michl, Eds., *Excited States and Photo-Chemistry of Organic Molecules*, VCH, Mannheim, Berlin, 1995.
3. (a) C. H. Wang, *Spectroscopy of Condensed Media*, Academic, New York, 1985. (b) G. R. Fleming, *Chemical Applications of Ultrafast Spectroscopy*, Oxford University, New York, 1986.
4. A. H. Zewail, *Femtochemistry: Ultrafast Dynamics of the Chemical Bond*, World Scientific, Singapore, 1994.
5. M. L. Horbg, J. A. Gardecki, A. Papazyan, M. Maroncelli, *J. Phys. Chem.* **1995**, *99*, 17311.
6. L. P. Lustres, S. A. Kovalenko, M. Mosquera, T. Senyushkina, W. Flasche, N. P. Ernsting, *Angew. Chem. Int. Ed.* **2005**, *44*, 5635.
7. R. M. Christie, *Colour Chemistry*, Royal Society of Chemistry, Cambridge, 1971.
8. M. Grätzel, *Acc. Chem. Res.* **2009**, *42*, 1788–1798.

9. K. Walzer, B. Maennig, M. Pfeiffer, K. Leo, *Chem. Rev.* **2007**, *107*, 1233.
10. S. Verma, G. M. Watt, Z. Mai, T. Hasan, *Photochem. Photobiol.* **2007**, *83*, 996.
11. E. Terreno, D. Delli Castelli, A. Viale, S. Aime, *Chem. Rev.* **2010**, *110*, 3019.
12. M. Olivucci, A. Sinicropi, Computational photochemistry, in *Computational Photochemistry*, Vol. 16, M. Olivucci, Ed., Elsevier, Amsterdam, 2005.
13. W. Domcke, D. R. Yarkony, H. Koppel, Eds., *Conical Intersections: Electronic Structure, Dynamics & Spectroscopy.* (Advanced Series in Physical Chemistry), World Scientific Publishing, Singapore, 2004.
14. B. O. Roos, *Acc. Chem. Res.* **1999**, *32*, 137.
15. B. O. Roos, K. Andersson, M. P. Fülscher, P. A. Malmqvist, L. Serrano-Andres, K. Pierloot, M. Merchán, Multiconfigurational perturbation theory: Applications in electronic spectroscopy, *Adv. Chem. Phys.* **1996**, *93*, 219.
16. L. Gagliardi, B. Roos, *Chem. Soc. Rev.* **2007**, *36*, 893.
17. M. K. Casida, in *Recent Advances in Density Functional Methods*, Part I, D. P. Chong, Ed., World Scientific, Singapore, 1995.
18. K. Burke, J. Werschnik, E. K. U. Gross, *J. Chem. Phys* **2005**, *123*, 62206.
19. R. J. Buenker, S. D. Peyerimhoff, P. J. Bruna, in *Computational Theoretical Organic Chemistry*, I. G. Csizmadia, R. Daudel, Eds., Reidel, Dordrecht, The Netherlands, 1981, p. 91.
20. J. Gauss, in *Encyclopedia of Computational Chemistry*, P. v. R. Schleyer, N. L. Clark, J. Gasteiger, H. F. Schaefer III, P. R. Schreiner, Eds., Wiley, Chichester, 1998, p. 615.
21. R. J. Bartlett, Ed., *Modern Ideas in Coupled-Cluster Methods*, World Scientific, Singapore, 1997.
22. O. Christiansen, H. Koch, P. Jorgensen, *Chem. Phys. Lett.* **1995**, *243*, 409.
23. C. Hättig, F. Weigend, *J. Chem. Phys.* **2000**, *113*, 5154.
24. A. I. Krylov, *Annu. Rev. Phys. Chem.* **2008**, *59*, 433.
25. A. I. Krylov, *Acc. Chem. Res.* **2006**, *39*, 8391.
26. K. Kowalski, P. Piecuch, *J. Chem. Phys.* **2004**, *120*, 1715.
27. S. P. A. Sauer, M. Schreiber, M. R. Silva-Junior, W. Thiel, *J. Chem. Theory Comput.* **2009**, *5*, 555.
28. O. Christiansen, *Theor. Chim. Acta* **2006**, *116*, 106.
29. A. Dreuw, M. Head-Gordon, *Chem. Rev.* **2005**, *105*, 4009.
30. (a) C. J. Cramer, *Essentials of Computational Chemistry*, Wiley, Hoboken, NJ, 2002. (b) F. Jensen, *Introduction to Computational Chemistry*, Wiley, Hoboken, NJ, 2007.
31. A. Dreuw, *Chem. Phys. Chem.* **2006**, *7*, 2259.
32. S. Grimme, *Rev. Comput. Chem.* **2004**, *20*, 153.
33. A. Pezzella, L. Panzella, O. Crescenzi, A. Napolitano, S. Navaratman, R. Edge, E. J. Land, V. Barone, M. d'Ischia, *J. Am. Chem. Soc.* **2006**, *128*, 15490.
34. H. Nakatsuji, K. Hirao, *J. Chem. Phys.* **1978**, *68*, 2053.
35. K. Toyota, I. Mayumi, M. Ehara, M. J. Frisch, H. Nakatsuji, *Chem. Phys. Lett.* **2003**, *367*, 730.
36. A. Köhn, C. Hättig, *J. Chem. Phys.* **2003**, *119*, 5021.
37. P. Celani, H.-J. Werner, *J. Chem. Phys.* **2003**, *119*, 5044.

38. F. Furche, R. Ahlrichs, *J. Chem. Phys.* **2002**, *117*, 7433.
39. (a) C. V. Caillie, R. D. Amos *Chem. Phys. Lett.* **1999**, *308*, *249;* (b) **2000**, *317*, 159.
40. G. Scalmani, M. J. Frisch, B. Mennucci, J. Tomasi, R. Cammi, V. Barone, *J. Chem. Phys.* **2006**, *124*, 094107.
41. B. Mennucci, R. Cammi, Eds., *Continuum Solvation Models in Chemical Physics: From Theory to Applications*, Wiley, New York, 2007.
42. M. Schreiber, M. R. Silva-Junior, S. P. A. Sauer, W. Thiel, *J. Chem. Phys.* **2008**, *128*, 134110.
43. C. M. Marian, N. Gilka, *J. Chem. Theoret. Comp.* **2008**, *4*, 1501.
44. E. K. U. Gross, R. M. Dreizler, *Density Functional Theory*, Springer, Berlin, 1995.
45. V. Barone, R. Improta, N. Rega, *Acc. Chem. Res.* **2008**, *41*, 605.
46. R. Improta, V. Barone, F. Santoro, *Angew. Chem. Int. Ed.* **2007**, *46*, 405.
47. R. Improta, V. Barone, F. Santoro, *J. Phys. Chem. B* **2007**, *111*, 14080.
48. F. Santoro, R. Improta, A. Lami, J. Bloino, V. Barone, *J. Chem. Phys.* **2007**, *126*, 084509.
49. F. Santoro, A. Lami, R. Improta, V. Barone, *J. Chem. Phys.* **2007**, *126*, 184102.
50. F. Santoro, A. Lami, R. Improta, J. Bloino, V. Barone, *J. Chem. Phys.* **2008**, *128*, 224311.
51. Y. Zhao, D. G. Truhlar, *Acc. Chem. Res.* **2008**, *41*, 157.
52. Y. Zhao, D. G. Truhlar, *J. Chem. Theory Comput.* **2005**, *1*, 415.
53. D. Jacquemin, E. A. Perpete, I. Ciofini, C. Adamo, *Acc. Chem. Res.* **2009**, *42*, 326.
54. D. Jacquemin, V. Wathelet, E. A. Perpete, C. Adamo, *J. Chem. Theory Comput.* **2009**, *5*, 2420.
55. D. Jacquemin, E. A. Perpete, I. Ciofini, C. Adamo, *J. Chem. Theory Comput.* **2010**, *6*, 1532.
56. D. Jacquemin, V. Wathelet, E. A. Perpete *J. Phys. Chem. A* **2006**, *110*, 9145.
57. D. Jacquemin, E. A. Perpete, O. A. Vydrov, G. E. Scuseria, C. Adamo, *J. Chem. Phys.* **2007**, *127*, 094102.
58. D. Jacquemin, E. A. Perpete, G. Scalmani, M. J. Frisch, R. Kobayashi, C. Adamo, *J. Chem. Phys.* **2007**, *127*, 094102.
59. M. Dierksen, S. Grimme, *J. Chem. Phys.* **2005**, *122*, 244101.
60. E. Runge, E. K. U. Gross, *Phys. Rev. Lett.* **1984**, *52*, 997.
61. (a) C. Jamorski, M. E. Casida, D. R. Salahub, *J. Chem. Phys.* **1996**, *104*, 5134. (b) M. Petersilka, U. J. Gossmann, E. K. U. Gross, *Phys. Rev. Lett.* **1996**, *76*, 1212. (c) R. Bauernschmitt, R. Ahlrichs, *Chem. Phys. Lett.* **1996**, *256*, 454.
62. R. E. Stratmann, G. E. Scuseria, M. J. Frisch, *J. Chem. Phys.* **1998**, *109*, 8218.
63. M. Wanko, M. Garavelli, F. Bernardi, T. A. Niehaus, T. Frauenheim, M. Elstner, *J. Chem. Phys.* **2004**, *120*, 1674.
64. D. J. Tozer, R. D. Amos, N. C. Handy, B. O. Roos, L. Serrano-Andres, *Mol. Phys.* **1999**, *97*, 859.
65. A. Dreuw, J. L. Weisman, M. Head-Gordon, *J. Chem. Phys.* **2003**, *119*, 2943.
66. A. Dreuw, M. Head-Gordon, *J. Am. Chem. Soc.* **2004**, *126*, 4007.
67. K. Burke, J. Werschnik, E. K. U. Gross, *J. Chem. Phys.* **2005**, *123*, 62206.
68. O. Gritsenko, E. J. Baerends, *J. Chem. Phys.* **2004**, *121*, 655.
69. Y. Tawada, T. Tsuneda, S. Yanagisawa, T. Yanai, K. Hirao, *J. Chem. Phys.* **2004**, *120*, 8425.

70. T. N. Maitra, F. Zhang, R. J. Cave, K. Burke, *J. Chem. Phys.* **2004**, *120*, 5932.
71. N. T. Maitra, *J. Chem. Phys.* **2005**, *122*, 234104.
72. T. Yanai, D. Tew, N. Handy, *Chem. Phys. Lett.* **2004**, *393*, 51.
73. O. A. Vydrov, G. E. Scuseria, *J. Chem. Phys.* **2006**, *125*, 234109.
74. M. A. Rohrdanz, K. M. Martins, J. M. Herbert, *J. Chem. Phys.* **2009**, *130*, 054112.
75. Y. Zhao, D. G. Truhlar, *Theor. Chem. Acc.* **2008**, *120*, 215.
76. Y. Zhao, N. E. Schultz, D. G. Truhlar, *J. Chem. Theoret. Comp.* **2006**, *2*, 364.
77. M. Schreiber, V. Bub, M. P. Fulscher, *Phys. Chem. Chem. Phys.* **2001**, *3*, 3906.
78. J. Tomasi, B. Mennucci, R. Cammi, *Chem. Rev.* **2005**, *105*, 2999.
79. M. Orozco, F. J. Luque, *Chem. Rev.* **2000**, *100*, 4187.
80. (a) J. Gao, *Acc. Chem. Res.* **1996**, *29*, 298. (b) G. Monard, K. M. Merz, Jr., *Acc. Chem. Res.* **1999**, *32*, 904.
81. M. Pavone, O. Crescenzi, G. Morelli, N. Rega, V. Barone, *Theor. Chem. Acc.* **2006**, *116*, 456.
82. S. Miertus, E. Scrocco, J. Tomasi, *J. Chem. Phys.* **1981**, *55*, 117.
83. (a) R. Cammi, S. Comi, B. Mennucci, J. Tomasi, *J. Chem. Phys.* **2005**, *122*, 104513. (b) S. Comi, R. Cammi, B. Mennucci, J. Tomasi, *J. Chem. Phys.* **2005**, *123*, 134512.
84. M. Caricato, B. Mennucci, J. Tomasi, F. Ingrosso, R. Cammi, S. Comi, G. Scalmani, *J. Chem. Phys.* **2006**, *124*, 24520.
85. R. Improta, V. Barone, G. Scalmani, M. Frisch, *J. Chem. Phys.* **2006**, *125*, 54103.
86. R. Improta, G. Scalmani, M. J. Frisch, V. Barone, *J. Chem. Phys.* **2007**, *127*, 74504.
87. M. Cossi, V. Barone, *J. Chem. Phys.* **2001**, *115*, 4708.
88. R. Improta, V. Barone, *J. Mol. Struct. Theochem.* **2009**, *914*, 87.
89. P. Hohenberg, W. Kohn, *Phys. Rev. B* **1964**, *136*, 864.
90. (a) A. D. Becke, *Phys. Rev. A* **1988**, *38*, 3098–100. (b) C. Lee, W. Yang, R. G. Parr, *Phys. Rev. B* **1988**, *37*, 785.
91. J. P. Perdew, K. Burke, M. Ernzerhof, *Phys. Rev. Lett.* **1996**, *77*, 3865.
92. A. D. Boese, N. C. Handy, *J. Chem. Phys.* **2001**, *114*, 5497.
93. A. D. Becke, *J. Chem. Phys.* **1993**, *98*, 5648.
94. C. Adamo, V. Barone, *J. Chem. Phys.* **1999**, *110*, 6158.
95. M. Enzerhof, G. E. Scuseria, *J. Chem. Phys.* **1999**, *110*, 5029.
96. C. Adamo, G. E. Scuseria, V. Barone, *J. Chem. Phys.* **2000**, *111*, 2889.
97. J. Jaramillo, G. E. Scuseria, M. Ernzerhof, *J. Chem. Phys.* **2003**, *118*, 1068.
98. (a) S. K. Ghosh, R. G. Parr, *Phys. Rev. A* **1986**, *34*, 785. (b) J. Tau, J. P. Perdew, V. N. Staroverov, E. Scuseria, *Phys. Rev. Lett.* **2003**, *91*, 146401.
99. M. Levy, J. P. Perdew, *Phys. Rev. B* **1993**, *48*, 11638.
100. E. H. Lieb, S. Oxford, *Int. J. Quantum Chem.* **1981**, *19*, 427.
101. V. Barone, M. Cossi, J. Tomasi, *J. Chem. Phys.* **1997**, *107*, 3210.
102. M. Cossi, V. Barone, *J. Chem. Phys.* **2000**, *112*, 2427.
103. E. J. Baerends, G. Ricciardi, A. Rosa, S. J. A. van Gisbergen, *Coord. Chem. Rev.* **2002**, *230*, 5.
104. M. Pastore, E. Mosconi, F. De Angelis, M. Gratzel, *J. Phys. Chem. C* **2010**, *114*, 7205.

105. (a) F. Neese, *Coord. Chem. Rev.* **2009**, *253*, 526. (b) F. Neese, T. Petrenko, D. Ganyushin, G. Olbrich, *Coord. Chem. Rev.* **2007**, *205*, 288.
106. (a) C. E. Crespo-Hernandez, B. Cohen, P. M. Hare, B. Kohler, *Chem. Rev.* **2004**, *104*, 1977. (b) C. T. Middleton, K. de La Harpe, C. Su, Y. K. Law, C. Crespo-Hernadez, B. Kohler, *Annu. Rev. Phys. Chem.* **2009**, *60*, 13.
107. T. Gustavsson, A. Banyasz, E. Lazzarotto, D. Markovitsi, G. Scalmani, M. J. Frisch, V. Barone, R. Improta, *J. Am. Chem. Soc.* **2006**, *128*, 607.
108. E. Epifanovsky, K. Kowalski, P.-D. Fan, M. Valiev, S. Matsika, A. I. Krylov, *J. Phys. Chem. A* **2008**, *112*, 9983.
109. R. Improta, V. Barone, A. Lami, F. Santoro *J. Phys. Chem. B* **2009**, *113*, 14491.
110. P. M. Hare, C. Crespo-Hernandez, B. Kohler, *Proc. Natl. Acad. Sci. U.S.A.* **2007**, *104*, 435.
111. T. Climent, R. Gonzalez-Luque, M. Merchan, L. Serrano-Andres, *Chem. Phys. Lett.* **2007**, *441*, 327.
112. Y. Mercier, F. Santoro, M. Reguero, R. Improta, *J. Phys. Chem. B* **2008**, *112*, 10769.
113. M. Merchan, R. Gonzalez-Luque, T. Climent, L. Serrano-Andres, E. Rodriguez, M. Reguero, D. J. Pelez, *J. Phys. Chem. B* **2006**, *110*, 26471.
114. L. B. Clark, I. Tinoco, Jr., *J. Am. Chem. Soc.* **1965**, *87*, 11.
115. R. Improta, V. Barone, *J. Am. Chem. Soc.* **2004**, *126*, 14320.
116. F. Santoro, V. Barone, T. Gustavsson, T. Improta, *J. Am. Chem. Soc.* **2006**, *128*, 16312.
117. K. A. Kistler, S. Matsika, *J. Phys. Chem. A* **2009**, *113*, 12396.
118. C. Zazza, A. Amadei, N. Sanna, A. Grandi, G. Chillemi, A. Di Nola, M. D'Abramo, M. Aschi, *Phys. Chem. Chem. Phys.* **2006**, *8*, 1385.
119. V. Ludwig, K. Coutinho, S. Canuto, *Phys. Chem. Chem. Phys.* **2007**, *9*, 4907.
120. J. M. Olsen, K. Aidas, K. V. Mikkelsen, J. Kongsted, *J. Chem. Theoret. Comp.* **2010**, *6*, 249.
121. R. Improta, V. Barone, *Chem. Rev.* **2004**, *104*, 1231.
122. M. Chahinian, H. B. Seba, B. Ancian, *Chem. Phys. Lett.* **1998**, *285*, 337.
123. M. P. Gaigeot, M. Sprik, *J. Phys. Chem. B* **2004**, *108*, 7458.
124. V. Karunakaran, K. Kleinermanns, R. Improta, S. A. Kovalenko, *J. Am. Chem. Soc.* **2009**, *131*, 5839.
125. M. Maroncelli, G. R. Fleming, *J. Chem. Phys.* **1987**, *86*, 6221.
126. (a) Y. Kimura, *J. Chem. Phys.* **1999**, *111*, 5474. (b) T. Gustavsson, L. Cassara, V. Gulbinas, G. Gurzadyan, J. C. Mialocq, S. Pommeret, M. Sorgius, P. van der Meulen, *J. Phys. Chem. A* **1998**, *102*, 4229. (c) F. Ingrosso, B. M. Ladanyi, B. Mennucci, G. Scalmani, *J. Phys. Chem B* **2006**, *110*, 4953.
127. B. A. Pryor, P. M. Palmer, Y. Chen, M. R. Topp, *Chem. Phys. Lett.* **1999**, *299*, 536.
128. A. Muhlpfordt, R. Schanz, N. P. Ernsting, V. Farzdtinov, S. Grimme, *Phys. Chem. Chem. Phys.* **1999**, *1*, 3209.
129. A. Chowdhury, S. A. Locknar, L. L. Premvardham, L. A. Peteanu, *J. Phys. Chem. A* **1999**, *103*, 9614.
130. R. Kanya, Y. Ohshima, *Chem. Phys. Lett* **2003**, *370*, 211.
131. S. Carlotto, A. Polimeno, C. Ferrante, C. Benzi, V. Barone, *J. Phys. Chem. B* **2008**, *112*, 8106.

132. L. M. Frutos, T. Andruniow, F. Santoro, N. Ferré, M. Olivucci, *Proc. Nalt. Acad. Sci. U.S.A.* **2007**, *104*, 7764.

133. F. Santoro, V. Barone, R. Improta, *J. Am. Chem. Soc.* **2009**, *131*, 15232.

134. R. Improta, *Phys. Chem. Chem. Phys.* **2008**, *10*, 2656.

135. F. Santoro, V. Barone, R. Improta, *Chem. Phys. Chem.* **2008**, *9*, 2531.

136. F. Santoro, V. Barone, R. Improta, *Proc. Natl. Acad. Sci. U.S.A.* **2007**, *104*, 9931.

137. D. Markovitsi, T. Gustavsson, F. Talbot, *Photochem. Photobiol. Sci.* **2007**, *6*, 717.

138. D. Onidas, T. Gustavsson, E. Lazzarotto, D. Markovitsi, *J. Phys. Chem. B* **2007**, *111*, 9644.

139. A. W. Lange, J. M. Herbert, *J. Am. Chem. Soc.* **2009**, *131*, 3913.

140. S. Perun, A. L. Sobolewski, W. Domcke, *J. Phys. Chem. A.* **2006**, *110*, 9031.

141. F. Santoro, V. Barone, R. Improta, *J. Comp. Chem.* **2008**, *29*, 957.

2

RESPONSE FUNCTION THEORY COMPUTATIONAL APPROACHES TO LINEAR AND NONLINEAR OPTICAL SPECTROSCOPY

ANTONIO RIZZO

CNR—Consiglio Nazionale delle Ricerche, Istituto per i Processi Chimico-Fisici (IPCF), Area della Ricerca, Pisa, Italy

SONIA CORIANI

Dipartimento di Scienze Chimiche e Farmaceutiche, Università degli Studi di Trieste, Trieste, Italy, and Centre for Theoretical and Computational Chemistry (CTCC), University of Oslo, Blindern, Norway

KENNETH RUUD

Centre for Theoretical and Computational Chemistry (CTCC), Department of Chemistry, University of Tromsø, Tromsø, Norway

Computational Strategies for Spectroscopy: From Small Molecules to Nano Systems, First Edition. Edited by Vincenzo Barone.

Spectroscopy studies the transitions between quantum-mechanical stationary states of atoms or molecules, and it encompasses energies ranging from those involved in the process of reorienting nuclear spins in external magnetic fields (nuclear magnetic resonance) to those arising in subnuclear processes (γ-rays). In this chapter we will be concerned with the optical spectroscopies exploiting ultraviolet (UV), visible, and infrared (Raman) radiation, typically involving transitions of outer-shell electrons and/or vibrational states of molecules. This is the area where the potential of molecular spectroscopies was first realized—the observations by Fraunhofer of transitions in atomic flames in the first decade of the nineteenth century—and it encompasses a range of wavelengths from roughly 200 nm (near UV) to 5000 nm (near infrared)—that is, from 6 to 0.25 eV. Within this rather wide area of study, we will concentrate on a few spectroscopic techniques involving linear and nonlinear interactions between electromagnetic radiation and matter—in some cases in the presence of external static fields—where direct absorption of one or more photons or the differential absorption of quanta of linearly or circularly polarized light (linear or circular dichroism) occur. We will discuss in particular the connections and implications, as well as the practical aspects, of the computational approaches related to the use of modern response theory in ab initio studies of UV, visible, and infrared spectroscopies involving transitions between bound stationary states. Transitions involving photoemission of electrons will be treated elsewhere in this book. Also, other processes involving the interaction of excited states with the dynamic fields associated with electromagnetic radiation, with or without external static fields, such as double refringence and the related phenomena of optical rotation and the building of ellipticities in dispersive (nonabsorptive) regions of the spectrum will not be considered in this chapter.

2.1 OVERVIEW OF RESPONSE FUNCTION THEORY APPROACHES

One of the main objectives of response function theory (RFT) [1–5] has from the beginning been the calculation of spectroscopic properties. RFT can be regarded as a convenient way to reformulate time-dependent perturbation theory [6] to a form where it is particularly well suited for parametrization and implementation within approximate wavefunction/density-based approaches. RFT provides a "recipe" for computing the response of a given (molecular) system to an external, time-dependent perturbation such as an electromagnetic field. Among the most important features of RFT are the possibility to straightforwardly compute linear as well as nonlinear response properties (thus describing a plethora of optical effects) and the ability to address excited-state properties (excitation energies from the ground state and between excited states, transition moments, excited-state dipole moments, excited-state structures, and so on) without a direct knowledge (i.e., nonlinear optimization) of the excited-state wavefunction(s). For both ground- as well as excited-state properties, no sum-over-state procedure is required, since the explicit summations over intermediate excited states appearing in the time-dependent perturbation theory expressions are replaced by the solution of linear sets of equations [4, 5]. RFT is today the only way to address excited-state properties in the context of density functional theory (DFT), where it is commonly referred to as time-dependent density functional theory (TD-DFT) [7, 8].

2.1.1 Ehrenfest versus Quasienergy Formulations

In terms of the formal derivations, we can pragmatically classify RFT approaches into two main groups: (1) Ehrenfest approaches and (2) quasi-energy approaches. The starting point is, in both cases, the time evolution of the expectation value of a time-independent operator when the system is subject to a (relatively weak) time-dependent perturbing field V^t, consistent with the fact that in the presence of a dynamic field there are no well-defined energy eigenvalues, whereas there are well-defined expectation values. The (Hermitian) perturbing field is usually expanded in the frequency domain as

$$V^t = \int_{-\infty}^{+\infty} d\omega\, V^{\omega} \exp\{(-i\omega + \delta)t\} \tag{2.1}$$

or as sum of periodic perturbations

$$\begin{aligned} V^t &= \sum_{j=-N}^{N} \exp\{(-i\omega_j)t\} V^{\omega_j} = \sum_{j=-N}^{N} \exp\{(-i\omega_j)t\} \sum_B \epsilon_B(\omega_j) B \\ &\equiv \sum_b \exp\{(-i\omega_b)t\} \epsilon_b(\omega_b) B \end{aligned} \tag{2.2}$$

In these equations, ω (Eq. 2.1) and ω_j (Eq. 2.2) indicate the circular frequencies associated with the perturbing field, δ is a small positive number that ensures that the perturbation vanishes for $t \rightarrow -\infty$, $\epsilon_{\mathrm{B}}(\omega_j)$ is the field strength parameter and B the associated perturbation operator. The expectation value of the time-independent operator A is written as [1]

$$\begin{aligned} \langle 0(t)|A|0(t)\rangle &= \langle 0|A|0\rangle + \int_{-\infty}^{+\infty} d\omega_1 e^{(-i\omega_1 + \delta)t} \langle\langle A; V^{\omega_1}\rangle\rangle_{\omega_1} \\ &+ \frac{1}{2!} \int_{-\infty}^{+\infty} d\omega_1\, d\omega_2 e^{[-i(\omega_1 + \omega_2) + 2\delta]t} \langle\langle A; V^{\omega_1}, V^{\omega_2}\rangle\rangle_{\omega_1,\omega_2} + \cdots \end{aligned} \tag{2.3}$$

or, alternatively, for the periodic perturbation [9]

$$\langle 0(t)|A|0(t)\rangle = \langle 0|A|0\rangle + \sum_{b;B} e^{-i\omega_b t}\langle\langle A;B\rangle\rangle_{\omega_b}\epsilon_B(\omega_b) + \frac{1}{2!}\sum_{(b;B),(c;C)} e^{-i(\omega_b+\omega_c)t}\langle\langle A;B,C\rangle\rangle_{\omega_b,\omega_c}\epsilon_B(\omega_b)\epsilon_C(\omega_c) + \cdots \tag{2.4}$$

where the expansion coefficients to each order in the perturbation are called "response functions": the *linear response function* $\langle\langle A;V^{\omega_1}\rangle\rangle_{\omega_1}$ [or $\langle\langle A;B\rangle\rangle_{\omega_b}$] collects all terms of first order in the perturbation, the *quadratic response function* $\langle\langle A;V^{\omega_1},V^{\omega_2}\rangle\rangle_{\omega_1,\omega_2}$ [or $\langle\langle A;B;C\rangle\rangle_{\omega_b,\omega_c}$] those of second order in the perturbation, and so on; $|0(t)\rangle$ is the time-dependent wavefunction and $|0\rangle$ the eigenfunction of the unperturbed Hamilton operator H_0. The time-dependent wavefunction $|0(t)\rangle$ evolves in time according to the time-dependent Schrödinger equation, and it is normalized

$$H(t)|0(t)\rangle = i\frac{\partial}{\partial t}|0(t)\rangle \quad \langle 0(t)|0(t)\rangle = 1 \quad \frac{d}{dt}\langle 0(t)|0(t)\rangle = 0 \tag{2.5}$$

where $H = H_0 + V^t$, and it is conveniently rewritten in the phase-isolated form $|0(t)\rangle = e^{-iF}|\tilde{0}(t)\rangle$, where the so-called regular wavefunction $|\tilde{0}(t)\rangle$ reduces to the unperturbed wavefunction $|0\rangle$ in the static limit and satisfies the phase-isolated time-dependent Schrödinger equation

$$\left(H - i\frac{\partial}{\partial t} + i\dot{F}\right)|\tilde{0}(t)\rangle = 0 \tag{2.6}$$

The phase factor F is found by projection of Eq. 2.6 against $\langle\tilde{0}(t)|$ and time integration

$$\dot{F} \equiv Q(t) = \langle\tilde{0}(t)|H - i\frac{\partial}{\partial t}|\tilde{0}(t)\rangle \quad F = \int_{-\infty}^{t} Q(t)\,dt \tag{2.7}$$

and it can immediately be proved that

$$\langle 0(t)|A|0(t)\rangle = \langle\tilde{0}(t)|A|\tilde{0}(t)\rangle \tag{2.8}$$

According to Eq. 2.3 (or Eq. 2.4), one can obtain explicit expressions for the response functions by assuming a perturbation expansion for the wavefunction, for example,

$$\begin{aligned}|0(t)\rangle &= |0^{(0)}\rangle + |0^{(1)}(t)\rangle + \cdots \\ &= |0\rangle + \int_{-\infty}^{+\infty} d\omega_1\, e^{(-i\omega_1+\delta)t}|0^{(\omega_1)}\rangle \\ &\quad + \int_{-\infty}^{+\infty} d\omega_1\, d\omega_2\, e^{[-i(\omega_1+\omega_2)+2\delta]t}|0^{(\omega_1,\omega_2)}\rangle + \cdots\end{aligned} \tag{2.9}$$

introducing it into the expression for the time-dependent expectation value and collecting all terms to a given order in the applied perturbation. The *response equations* from which the wavefunction response to each order is obtained are derived by rewriting the time-dependent Schrödinger equation as a time-dependent variation principle—even though it does not correspond to a variational condition on the expectation value of the perturbed Hamiltonian H—after having assumed a convenient parametrization for the time evolution of the (regular) wavefunction $|\tilde{0}(t)\rangle$. One popular choice of time evolution is the exponential parametrization [4]

$$|\tilde{0}(t)\rangle = e^{X(t)}|0\rangle \quad X(t) = \sum_p x_p(t)R_p \tag{2.10}$$

where the $x_p(t)$ are the time-dependent parameters to be determined and R_p the associated one-electron operators. For the exact case, R_p corresponds to the state transfer operator

$$R_p = \begin{cases} |p\rangle\langle 0| & p > 0 \\ |0\rangle\langle p| & p < 0 \end{cases} \tag{2.11}$$

Note that different variational conditions have been proposed [6, 10, 11] and that they are not necessarily equivalent for approximate wavefunction methods.

Within the Ehrenfest framework [12], valid for exact as well as variational approximate methods, either the time-variational principle of Frenkel [11],

$$\langle\delta\tilde{0}(t)|\left(H - i\frac{\partial}{\partial t} + \dot{F}\right)|\tilde{0}(t)\rangle \quad \langle\delta\tilde{0}(t)|\tilde{0}(t)\rangle + \langle\tilde{0}(t)|\delta\tilde{0}(t)\rangle = 0 \tag{2.12}$$

or the equivalent one of Langhoff, Epstein, and Karplus [6],

$$\Re\langle\delta\tilde{0}(t)|\left(H - i\frac{\partial}{\partial t} + \dot{F}\right)|\tilde{0}(t)\rangle \quad \langle\delta\tilde{0}(t)|\tilde{0}(t)\rangle + \langle\tilde{0}(t)|\delta\tilde{0}(t)\rangle = 0 \tag{2.13}$$

is chosen, which can be shown to yield a set of generalized Ehrenfest equations for the excitation/deexcitation operators R_p,

$$\frac{d}{dt}\langle\tilde{0}(t)|R_p|\tilde{0}(t)\rangle = \langle\tilde{0}(t)|\left(\frac{\partial}{\partial t}R_p\right)|\tilde{0}(t)\rangle - i\langle\tilde{0}(t)|\left[R_p, H\right]|\tilde{0}(t)\rangle \tag{2.14}$$

The latter are then solved to each order in the perturbation to obtain the perturbed amplitudes $x_p(t)$ to each order.

In the quasi-energy framework, on the other hand, it is recognized that the time derivative $Q(t)$ of the time-dependent phase factor $F(t)$ is reminiscent of the energy in the static case (thus the name *quasi-energy* or *pseudoenergy*) and it can be proven that its time average

$$\{Q(t)\}_T = \frac{1}{T}\int_{-T/2}^{T/2} Q(t)\, dt \tag{2.15}$$

over a common period T of the perturbation operators in Eq. 2.2 satisfies both a time-averaged variational principle

$$\delta\{Q(t)\}_T = \delta\left\{\langle\tilde{0}(t)|\left(H - i\frac{\partial}{\partial t}\right)|\tilde{0}(t)\rangle\right\}_T = 0 \tag{2.16}$$

and a generalized time-averaged Hellman–Feynman theorem [13, 14]

$$\frac{d\{Q(t)\}_T}{d\epsilon_A(\omega_a)} = \left\{\langle\tilde{0}(t)|\frac{\partial H}{\partial\epsilon_A(\omega_a)}|\tilde{0}(t)\rangle e^{-i\omega_a t}\right\}_T = \{\langle\tilde{0}(t)|A|\tilde{0}(t)\rangle e^{-i\omega_a t}\}_T \tag{2.17}$$

which, when introduced into the expansion in Eq. 2.4, allows us to rewrite the nth-order response functions as $(n+1)$th-order derivatives with respect to the perturbation strengths of the time-averaged quasi-energy

$$\langle\langle A; B, \ldots, N\rangle\rangle_{\omega_b,\ldots,\omega_n} = \frac{d^{n+1}\{Q(t)\}_T}{d\epsilon_A(\omega_a)\, d\epsilon_B(\omega_b)\ldots d\epsilon_N(\omega_n)} \tag{2.18}$$

with $\omega_a = -\sum_{i=b}^{n}\omega_i$. The time-averaged variational principle, Eq. 2.16, is used to determine the time evolution of the wavefunction after having chosen the preferred parametrization of the time-dependent wavefunction—for example, an exponential parametrization—and after having expanded its time-dependent parameters in orders of the perturbation. It turns out that also the response equations at each order can be rewritten as derivatives of Eq. 2.16 (at each order) with respect to the perturbation strengths and the time-dependent parameters. The quasi-energy approach is clearly similar to the energy derivative approach used for static properties [15–17], to which it naturally reduces in the limit of static perturbations.

The Ehrenfest approach has been followed in the seminal paper by Olsen and Jørgensen [4] to derive formal expressions for the linear and nonlinear response functions of exact, self-consistent field (SCF), and multiconfigurational SCF (MCSCF) wavefunctions and in the consequent implementations at the linear [18, 19], quadratic [20, 21], and cubic [22–24] response theory level for SCF and MCSCF wavefunctions; see also Olsen and Jørgensen [5]. An extension to DFT going beyond the local density approximation (LDA) is described elsewhere [25]. The quasi-energy approach has become increasingly popular since the end of the 1990s [9, 26] and it allows for, in combination with a Lagrangian method [9], a unique definition of response functions also for nonvariational parametrizations of the (time-dependent) wavefunction, such as approximate coupled-cluster methods. The quasi-energy approach therefore provides a unified framework for treating variational and nonvariational wavefunctions in a manner analogous to time-independent theory. Another advantage over the Ehrenfest method is that in the quasi-energy method the permutational symmetries with respect to the exchange of operators become inherent in the equation; see the discussion by Cristiansen, Hättig and Jørgensen [9].

Independently of the approach used, in the exact case one obtains the well-known spectral representation [also known as the sum-over-states (SOS) expression] of the linear response function (LRF)

$$\langle\langle A;B\rangle\rangle_{\omega} = \sum_{j\neq n}\left\{\frac{\langle n|A|j\rangle\langle j|B|n\rangle}{\omega-\omega_{jn}} - \frac{\langle n|B|j\rangle\langle j|A|n\rangle}{\omega+\omega_{jn}}\right\} \tag{2.19}$$

where $\hbar\omega_{jn} = E_j - E_n$. A pole and residue analysis of the LRF [4, 9] allows for the identification of the excitation energies $\hbar\omega_{fn}$ from the ground state $|n\rangle$ to the final states $|f\rangle$ and the one-photon transition strength

$$\lim_{\omega\to\omega_{fn}}(\omega-\omega_{fn})\langle\langle A;B\rangle\rangle_{\omega} = \langle n|A|f\rangle\langle f|B|n\rangle \tag{2.20}$$

For the quadratic response function (QRF)

$$\langle\langle A;B,C\rangle\rangle_{\omega_b,\omega_c} = -\mathcal{P}(a,b,c)\sum_{j,k\neq n}\left\{\frac{\langle n|A|j\rangle\langle j|\bar{B}|k\rangle\langle k|C|n\rangle}{(\omega_a+\omega_{jn})(\omega_c-\omega_{kn})}\right\} \tag{2.21}$$

where $\mathcal{P}(a,b,c)$ generates all terms obtained by permuting the operators A, B, C and their corresponding frequencies ω_a, ω_b, and ω_c, with $\omega_a = -(\omega_b+\omega_c)$. Barred operators are defined as $\bar{A} = A - \langle n|A|n\rangle$. The QRF has poles when the absolute values of the individual frequencies ω_b and ω_c (or their sum $-\omega_a$) match an excitation energy. The so-called single residues are obtained as

$$\lim_{\omega_c\to\omega_{fn}}(\omega_c-\omega_{fn})\langle\langle A;B,C\rangle\rangle_{\omega_b,\omega_c} = -\mathcal{P}(a,b)\left[\sum_{j}\frac{\langle n|A|j\rangle\langle j|\bar{B}|f\rangle}{\omega_a+\omega_{jn}}\right]\langle f|C|0\rangle \tag{2.22}$$

which is in the form of a two-photon transition moment times a one-photon transition moment. Double residues are obtained at those values of the remaining frequencies that make the above two-photon transition moment resonant,

$$\lim_{\omega_a\to-\omega_{in}}(\omega_a+\omega_{in})\left[\lim_{\omega_c\to\omega_{fn}}(\omega_c-\omega_{fn})\langle\langle A;B,C\rangle\rangle_{\omega_b,\omega_c}\right] = -\langle n|A|i\rangle\langle i|\bar{B}|f\rangle\langle f|C|n\rangle \tag{2.23}$$

which is suggestive of a route to obtain transition moments between excited states as well as excited-state first-order properties [4, 9].

The spectral representation of the cubic response function (CRF) is [5]

$$\langle\langle A; B, C, D\rangle\rangle_{\omega_b,\omega_c,\omega_d} = \mathcal{P}(a,b,c,d)$$
$$\times \left[-\sum_{k,l,m\neq n} \frac{\langle n|A|k\rangle\langle k|\bar{B}|l\rangle\langle l|\bar{C}|m\rangle\langle m|D|n\rangle}{(\omega_a+\omega_{kn})(\omega_c+\omega_d-\omega_{ln})(\omega_d-\omega_{mn})} \right.$$
$$\left. + \sum_{k\neq n} \frac{\langle n|A|k\rangle\langle k|B|n\rangle}{(\omega_a+\omega_{kn})(\omega_b-\omega_{kn})} \sum_{l} \frac{\langle n|C|m\rangle\langle m|D|n\rangle}{(\omega_d-\omega_{ln})} \right] \tag{2.24}$$

The residue analysis of the CRF yields different types of excited-state quantities such as three-photon transition matrix elements (three-photon absorption) [27], the two-photon matrix elements between excited states (the cross section for second-order transitions), and the excited-state polarizability (dynamic second-order property).

The spectral representations above are not computationally efficient, as they would require knowledge of all intermediate excited states. Computationally tractable formulas for the response functions within the various approximate methods are obtained instead through the following steps: (1) choose a time-independent reference wavefunction; (2) choose a parametrization of its time-development, for instance an exponential parametrization; (3) set up the appropriate equations for the time development of the chosen wavefunction parameters; (4) solve these equations in orders of the perturbation to obtain the wavefunction (parameters); (5) insert the solutions of these equations into the expectation value expression and obtain the RTFs; and (6) identify the excited-state properties from the poles and residues. The computationally tractable formulas for the response functions therefore differ depending on the electronic structure method at hand, and the true spectral representations given above are only valid in the limit of a full-configuration interaction (FCI) wavefunction. For approximate methods (i.e., where electron correlation is only partially included), matrix equations appear instead of the SOS expressions, for example,

$$\langle\langle A; B\rangle\rangle_{\omega} = -\mathbf{V}_A^{[1]T}\mathbf{N}^B(\omega) \quad \text{where} \quad \mathbf{N}^B(\omega) = (\mathbf{E}^{[2]}-\omega\mathbf{S}^{[2]})^{-1}\mathbf{V}_B^{[1]} \tag{2.25}$$

where $\mathbf{E}^{[2]}$ is a generalized Hessian matrix, $\mathbf{S}^{[2]}$ is called the metric matrix, and $\mathbf{V}^{[1]}$ is the property-gradient matrix. The explicit expressions for these matrices depend on the chosen approximate method. Excitation energies are obtained by solving a generalized eigenvalue equation such as

$$(\mathbf{E}^{[2]}-\omega\mathbf{S}^{[2]})\mathbf{X} = 0 \tag{2.26}$$

During the last decade, much effort has been directed toward the development of response theory methods that scale linearly with the size of the system. One essential requirement to achieve linear scaling is then to work in the local, atomic orbital (AO) basis, and a few AO-based formulations of response theory have therefore appeared [e.g., 28–31]. Larsen et al. [28] used an unconstrained exponential parametrization of the density matrix in the AO basis to determine response functions for standard basis sets following to a large extent the approach of Olsen and Jørgensen [4, 5] for the molecular orbital basis. Coriani et al. [29] used the Löwdin orthogonal AO basis to obtain an AO-based response solver which allows for linear-scaling calculations of excitation energies, transition strengths, and frequency-dependent linear response functions (polarizabilities) at the Hartree–Fock and Kohn–Sham (TD-DFT) levels of theory. The method was later extended to the general frequency-dependent quadratic response function and its residues by Kjærgaard et al. [30]. Another important reference is the work of Ochsenfeld and co-workers [31].

Thorvaldsen et al. [32] have presented a general method for the calculation of molecular properties to arbitrary order, in which the quasi-energy and Lagrangian formalisms are combined to derive response functions by differentiation of the quasi-energy derivative Lagrangian using the elements of the density matrix in the AO representation as variational parameters and the idempotency relation for the density matrix as a constraint. The authors demonstrated that, even if in exact theory the quasi-energy and the density operator (or density matrix) are fundamentally different quantities with respect to operations on the bra and ket vectors—and therefore the quasi-energy cannot be expressed in terms of the density operator or the density matrix in the same way as the energy in time-independent theory—the perturbation-strength derivative of the quasi-energy may be expressed in terms of the density matrix and its time derivative and one can use this to identify molecular response functions by differentiation of the quasi-energy perturbation-strength derivative $\{L^a(\mathbf{D}, t)\}_T$,

$$\langle\langle A; B\rangle\rangle_{\omega_b} = \left.\frac{d\{L^a(\mathbf{D}, t)\}_T}{d\epsilon_B(\omega_b)}\right|_{\epsilon=0} = L^{ab} \quad \omega_a = -\omega_b \tag{2.27}$$

The response equations obtained were shown to have the same structure as the ones obtained using the exponential parametrization [28, 29] and may therefore be solved using the same linear response solver [29]. The formalism was derived with emphasis on an easy extension to higher-order molecular properties and to provide a uniform framework for the calculation of static and frequency-dependent molecular properties. It focuses on defining generic building blocks, making it straightforward to implement molecular properties of arbitrary order. Standard basis sets and basis sets that depend explicitly on the applied frequency-dependent perturbations—perturbation-dependent basis sets such as frequency-dependent London atomic orbitals (LAOs) [33, 34]—could be treated on an equal footing. We refer the interested reader to the original paper [32] for a thorough discussion.

2.1.2 Infinite lifetime versus finite lifetime Response Approaches

Physically, the excited states possess a finite lifetime due to spontaneous emission, which is one of the causes of the broadening of the absorption bands associated with the various electronic transitions in an experimental absorption spectrum. In standard response theory, however, the excited states are assumed to have an infinite lifetime, and the response functions are therefore singular when one or more frequencies (or a sum of them) approach an excitation energy. These singularities result in divergencies in the dispersion curve for the given response function and infinitely narrow absorption peaks. Thus, to mimic an experimental absorption spectrum within standard response theory, one usually carries out a pole and a residue calculation to determine the excitation energies and the corresponding oscillator strengths and then attach a lineshape functions to each stick line; see Section 2.2. If one could somehow incorporate the finite lifetime of the excited state in the response functions from the very beginning, one would be able to reproduce their correct physical behavior at resonance frequencies and directly simulate the absorption spectrum. This can be done by introducing an empirical, effective lifetime parameter or "damping term" in order to remove the unphysical behavior of the molecular property in the resonant region.

In the "complex polarization propagator" (CPP) approach of Norman and co-workers [35–40], the damping term is introduced as a modification to the Ehrenfest theorem governing the time development of the molecular system based on an analogy with the Liouville equation and it effectively introduces finite lifetimes in the excited states. The approach represents a generalization of the approach by Olsen and Jørgensen [4] for standard response theory, and at the linear response level it is equivalent to the pragmatic addition of a damping term in the SOS expression of the undamped linear response function [41, 42]. A slightly different route is followed in the so-called "damped response theory" (DRT) method of Kristensen et al. [43], where the damping is obtained by introducing complex excitation energies into the standard response function expressions for exact theory and then generalized to variational Hartree–Fock, DFT, and configuration interaction (CI) (the matrix structure of the response equations is similar to exact theory) within a quasi-energy formulation. An ad hoc introduction of damping factors into the standard expression for the properties has been used by Krykunov and co-workers specifically for magnetic circular dichroism (MCD) [44] in a TD-DFT context. We refer to the original papers for details.

One clear advantage of the damped approaches compared to the standard formulations of response theory (residue analysis coupled with the lineshape convolution) is the fact that they allow for the exploration of high-energy regions of the spectrum. They are therefore particularly relevant for X-ray absorption phenomena and for systems with a high density of excited states, as expected for very large molecular systems [43]. Applications of the CPP approach in the literature include optical rotation dispersion and electronic circular dichroism [39, 45], magnetic-circular dichroism [40, 46], X-ray absorption and dichroism [47–51], along with dispersion effects and scattering phenomena [36, 37, 52] at the Hartree–Fock, MCSCF, and TD-DFT levels of theory.

2.1.3 Response Functions in the Vibrational Domain

The response theory approach discussed in Section 2.1.1 can also be extended to the vibrational domain. The formal derivation of a vibrational response theory will follow along the lines discussed in Section 2.1.1, but there will be differences in the working equations due to the differences in the representation of the vibrational states compared to the electronic states, as well as due to the fact that the vibrational state is characterized by a number of different quantum numbers corresponding to the occupation of the different normal modes, often referred to as modals in the case of vibrational self-consistent field-optimized wavefunctions. Vibrational (and vibronic) response theory has been developed by the group of Christiansen [53–55] (see also refs. 56 and 57 for recent reviews of this field).

In contrast to electronic structure theory, for the vibrational wavefunction there exists an exactly solvable model which in many cases leads to results that are of sufficiently high accuracy to allow for a good approximation to the more elaborate vibrational self-consistent field or vibrational configuration interaction models, namely the harmonic oscillator model. Within the Born–Oppenheimer approximation [58], we can consider the nuclei to move in an effective potential generated by the electron density of the molecule. By assuming the change in the potential to be well represented by a harmonic potential, we can recover a set of equations that in the normal coordinate basis can be solved analytically, corresponding to the well-known harmonic oscillator problem for each normal coordinate in the molecule. We will not discuss the derivation of the normal coordinate basis and the vibrational eigenfunctions any further here, as this is treated in other chapters in this book. Here we only note that the vibrational wavefunction is separable, the total wavefunction being a product of harmonic oscillator wavefunctions for each normal mode, and that energy can be written as the corresponding sum of the harmonic oscillator energies for the different normal modes.

Assuming that we can do a Taylor expansion of the dependence of a molecular property P on the nuclear displacements along the normal coordinates $\mathbf{Q}_k$, we may write

$$P(\mathbf{Q}) = \hat{P}|_{\mathbf{Q}_e} + \sum_k \left.\frac{\partial P}{\partial \mathbf{Q}_k}\right|_{\mathbf{Q}_e} \Delta\mathbf{Q}_k + \frac{1}{2}\sum_{k,l} \left.\frac{\partial^2 P}{\partial \mathbf{Q}_k \partial \mathbf{Q}_l}\right|_{\mathbf{Q}_e} \Delta\mathbf{Q}_k\, \Delta\mathbf{Q}_l \ldots \tag{2.28}$$

where we have used $\mathbf{Q}_e$ to indicate that the derivative of the property P is calculated at a reference molecular geometry for the given normal coordinate (in most cases the equilibrium geometry, although other reference geometries have been proposed [59, 60]). The property derivative is therefore independent of the nuclear coordinates. Within the harmonic oscillator approximation, the evaluation of the transition matrix elements that appear in the response functions and their residues is therefore reduced to the calculation of geometric derivatives of molecular properties and the evaluation of matrix elements between harmonic oscillator wavefunctions of the form $\langle \nu_i^k | Q_k^n | \nu_j^k \rangle$, where the superscript k indicates normal mode k, n the order of the geometric distortion, and i and j the quantum number of the vibrational state. Using the properties of the harmonic oscillator wavefunctions [61], these matrix elements can be evaluated analytically and, furthermore, only a limited number of these matrix elements are nonzero. In contrast to RFT for electronic states, where the summations run over all excited electronic states of the molecule, only a very limited number of vibrational states actually contribute to the vibrational response function within the harmonic oscillator approximation, and the SOS expressions can therefore in general be evaluated exactly.

At room temperature, only the vibrational ground state is in general populated. Truncating the expansion of the geometry dependence of the dipole moment in Eq. 2.28 at the linear term (often referred to as the double-harmonic approximation [62] since we are also using a harmonic approximation for the vibrational wavefunction), there will be only one surviving term in the SOS expression for the vibrational wavefunction, and we can write the vibrational polarizability as

$$\alpha^{\text{vib}} = \tfrac{1}{2}\mathcal{P}(\alpha,\beta)\sum_k \frac{d\mu_\alpha}{d\mathbf{Q}_k}\frac{d\mu_\beta}{d\mathbf{Q}_k}\left(\omega_k^2-\omega_a^2\right)^{-1} \tag{2.29}$$

As for the electronic response functions, a pole and residue analysis can be performed, and from this we would recover the conventional transition dipole moments and vibrational frequencies as observed in infrared spectroscopy and discussed in more detail in other chapters of this book.

Before leaving the topic of vibrational RFT, let us note that when optical wavelengths are used for the incoming light, the energy associated with the radiation is in general much larger than the energy difference between two different vibrational states, meaning that we in general will not observe divergences in the response functions for optical frequencies and that the denominators in Eq. 2.29 will be significantly larger than if only vibrational energy differences were used, thus significantly reducing the magnitude of the vibrational polarizability compared to applied static (frequency-independent) perturbations. (Several reviews on pure vibrational contributions to nonlinear optical properties exist, and we refer the interested reader to refs. 63–65 for more details.)

2.2 CONNECTION BETWEEN OBSERVABLES AND COMPUTABLE QUANTITIES

The International System of Quantities is adopted in this section for the choice of units in our equations [66]. Also, so-called conventional units are introduced for those quantities for which other units are more frequently used (in place of the standard recommended) due to old habits and practice.

2.2.1 One-Photon Electronic Absorption and Circular Dichroism

The connection between experimental and computed quantities in the case of optical spectroscopies is given by the Beer–Lambert equation, describing the attenuation of the intensity (or irradiance) I of a monochromatic laser beam from its original value I_0 after passing through a homogeneous isotropic medium where the absorber has a molar concentration c along an optical path z (throughout this section z indicates the propagation direction of the electromagnetic beam),

$$I = I_0 \times 10^{-\epsilon(\omega)cz} \tag{2.30}$$

It is customary to express the concentration in Eq. 2.30 in the form of molarity (moles per liter) and the optical path in centimeter. Here, $\epsilon(\omega)$ is the so-called decadic molar extinction coefficient, depending on the circular frequency of the radiation ω, and it is the typical measure of absorption, especially in the UV region of the spectrum. Its commonly used units are liter per mole per centimeter and is related to the absorption cross section

$$\sigma(\omega) = \frac{1000 \ln(10)\epsilon(\omega)}{N_A} \tag{2.31}$$

and to the absorption coefficient

$$\alpha(\omega) = \ln(10)\epsilon(\omega)c_0 \tag{2.32}$$

where c_0 is the speed of light in vacuo. The concentration c implies that the number of molecules per cubic centimeter in the sample is $N = cN_A/1000$, with N_A denoting Avogadro's number. In samples with moderate absorption, that is, when the exponent in Eq. 2.30 is small enough to allow the exponential to be expanded in a Taylor series, and keeping only the linear term, one can write

$$-\Delta I = I - I_0 \cong -I_0 \ln(10)\epsilon(\omega)cz = -\frac{1000 N I_0 \ln(10)\epsilon(\omega)z}{N_A} \tag{2.33}$$

The expressions connecting the observable (the decadic molar extinction coefficient) to the microscopic quantities that can be computed can be derived within a semiclassical [42, 67–74] or a quantum electrodynamic [75, 76] approach.

For a transition from state i to state j, assuming propagation of a monochromatic electromagnetic wave in an isotropic medium with refractive index n, and neglecting the contributions arising from the interaction of the magnetic dipole with the magnetic field and the electric quadrupole with the spatial gradient of the electric field of the wave, one obtains an expression for the one-photon absorption (OPA) decadic molar extinction coefficient,

$$\epsilon(\omega) = \frac{4\pi^2 N_A \omega}{3 \times 1000 \times \ln(10)(4\pi\epsilon_0)n\hbar c_0} \sum_{\text{transitions, } ij} g_{ij}(\omega)|\mu_{ij}|^2 \tag{2.34}$$

$$= \frac{2e^2\pi^2 N_A \omega}{1000 \times \ln(10)(4\pi\epsilon_0)n m_e c_0} \sum_{\text{transitions, } ij} g_{ij}(\omega)\frac{f_{ij}}{\omega_{ij}} \tag{2.35}$$

where the matrix element of the electric dipole operator $\boldsymbol{\mu}$ has been introduced,

$$\boldsymbol{\mu} = \sum_j q_j \mathbf{r}_j \tag{2.36}$$

involving the charges q_j and the spatial coordinates $\boldsymbol{r}_j$ of all particles, ϵ_0 is the electric constant, and $f_{ij} = 2m_e\omega_{ij}/3\hbar e^2$ is the oscillator strength.

Let us at this stage discuss the matter of the lineshape function $g_{ij}(\omega)$ [68, 77], already mentioned in Section 2.1.2. In theoretical derivations of Eq. 2.34, requiring the conservation of the energy in the transition from state i to state j leads to the introduction of the Dirac delta function $\delta(\omega_{ij} - \omega)$ involving the excitation energy $E_{ij} = \hbar\omega_{ij}$. This neglects the fact that in actual spectroscopic conditions the radiation is polychromatic and the energies of the states are broadened by several processes, among them collisions and the interaction with light. In molecules, moreover, assuming validity of the Born–Oppenheimer approximation, transitions between electronic states must take care of the complexity of the vibrational and rotational manifolds, belonging both to the ground and to the excited electronic states. Therefore, the infinitely sharp transition lines predicted by theory usually turn out to be rather broad bands in experiment, resulting from a large number of ro-vibrational excitations, and from the complex collisional dynamics occurring in gas or condensed phase. The effects of vibrations and rotations on electronic absorption and dichroism spectra in molecules will be discussed later in this chapter as well as elsewhere in this book. For the time being, we assume that we can take care of the broadening of the absorption spectra, as most often done, in a phenomenological way by substituting the Dirac delta function with an appropriate lineshape function $g_{ij}(\omega)$ centered at ω_{ij} and which, for each given transition $i \rightarrow j$, is normalized to unity over the given absorption band, that is,

$$\int_{\text{band}} g_{ij}(\omega)\, d\omega = 1 \tag{2.37}$$

Typical functions suitable to represent bandshapes are the Lorentzian function, usually employed to describe homogeneous broadening,

$$g_{ij}(\omega) = L_{ij}(\omega) = \frac{1}{\pi}\frac{\gamma/2}{(\omega - \omega_{ij})^2 + (\gamma/2)^2} \tag{2.38}$$

with value at the maximum $L(\omega_{ij}) = 2/(\pi\gamma)$ and with γ as full width at half maximum (FWHM), or the Gaussian function, mainly used to model inhomogeneous broadening,

$$g_{ij}(\omega) = G_{ij}(\omega) = \frac{2}{\gamma\sqrt{2\pi}} e^{-2(\omega-\omega_{ij})^2/\gamma^2} \tag{2.39}$$

with value at the maximum $G(\omega_{ij}) = 2/(\gamma\sqrt{2\pi})$ and FWHM $= \gamma\sqrt{2\ln 2}$ ($\gamma\sqrt{2}$ is the full width at the points where the Gaussian is at a fraction $1/e$ of the maximum). The conversion to wave lengths ($\lambda = 2\pi c_0/\omega$)—that is, a transition from $\omega g_{ij}(\omega)$ to $\lambda g_{ij}(\lambda)$—yields, by straightforward substitution,

$$\omega L_{ij}(\omega) = \frac{\omega}{\pi} \frac{(\gamma/2)}{(\omega-\omega_{ij})^2 + (\gamma/2)^2} \to \frac{\lambda_{ij}}{\pi} \frac{\gamma'/2}{(\lambda-\lambda_{ij})^2 + (\gamma'/2)^2} \approx \frac{\lambda}{\pi} \frac{\gamma''/2}{(\lambda-\lambda_{ij})^2 + (\gamma''/2)^2}$$
$$= \lambda L_{ij}(\lambda) \tag{2.40}$$

$$\omega G_{ij}(\omega) = \frac{2\omega}{\gamma\sqrt{2\pi}} e^{-2(\omega-\omega_{ij})^2/\gamma^2} \to \frac{2\lambda_{ij}}{\gamma'\sqrt{2\pi}} e^{-2(\lambda-\lambda_{ij})^2/(\gamma')^2} \approx \frac{2\lambda}{\gamma''\sqrt{2\pi}} e^{-2(\lambda-\lambda_{ij})^2/(\gamma'')^2}$$
$$= \lambda G_{ij}(\lambda) \tag{2.41}$$

where $\gamma' = \lambda\lambda_{ij}\gamma/(2\pi c_0)$, whereas $\gamma'' = \lambda_{ij}^2\gamma/(2\pi c_0)$ and $\lambda_{ij} = 2\pi c_0/\omega_{ij}$. The use of the intermediate expressions above yields inhomogeneous lineshapes, non-symmetric with respect to the maximum, due to the dependence of γ' on λ. These functions (divided by λ_{ij}) are by necessity not normalized to unity when integration with respect to λ is performed over the whole band. Substituting $\omega g_{ij}(\omega)$ with $\lambda g_{ij}(\lambda)$, with γ being replaced by γ'', yields instead lineshapes with proper normalization in the respective domains. It is straightforward to prove that replacing $\omega g_{ij}(\omega)$ with $\nu g_{ij}(\nu)$ and γ by $\gamma' = \gamma/(2\pi)$ yields properly normalized lineshapes in the frequency $\omega = 2\pi\nu$ variable. The FWHM is the measure of the broadening of the spectroscopic lines, and it is often treated as an empirical parameter in ab initio simulations, in most cases keeping the same value, independent therefore of the frequency, for all computed transitions. Furthermore, bands due to different electronic or vibrational transitions overlap, and this interaction between different lines in the spectrum is almost invariably treated by combining linearly the lineshapes of the various transitions.

Before leaving the topic of lineshape functions, let us just note that the use of the CPP method discussed in Section 2.1.2 will remove the need to add empirical linehape functions to the computed data. However, the lifetime remains an empirical parameter also in the CPP calculations and the computed spectra will always get a bandshape consistent with a Lorentzian function, thus always corresponding to homogeneous broadening of the absorption bands.

Equation 2.34 is valid for linearly polarized as well as unpolarized light. Linear dichroism, arising in oriented samples, is also dealt with straightforwardly by taking the difference $\epsilon_x(\omega)-\epsilon_y(\omega)$ for the two directions perpendicular to the direction of propagation. This will not be discussed further here.

Circular dichroism (ECD) is defined as the difference in extinction coefficients for left (*L*) and right (*R*) circularly polarized light, $\Delta\epsilon(\omega)$,

$$\Delta\epsilon(\omega) = \epsilon_L(\omega)-\epsilon_R(\omega) \tag{2.42}$$

Keeping only the electric field interaction in the perturbing operator, that is, neglecting magnetic and higher-order electric multipolar interactions, yields a vanishing circular dichroism, since in the difference taken between absorption of right and left circularly polarized light, the electric dipole moment contribution cancels out, being identical for left and right circularly polarized light. It is therefore mandatory to include the leading higher-order terms in the expansion of the interaction operator, those involving the magnetic dipole (**m**) and the electric quadrupole (**q**) operators. The latter, which in its traced form can be written as

$$\mathbf{q} = \sum_j q_j \mathbf{r}_j \mathbf{r}_j \tag{2.43}$$

can be proven to yield contributions to the ECD that vanish for isotropic samples. We will not consider this contribution further at this time. We note, however, that these contributions will be important also for isotropic samples for other chiroptical spectroscopies such as Raman optical activity (ROA) (see Section 2.3.4.2). As far as the magnetic dipole operator is concerned, it involves in principle both the orbital and the spin electronic angular momenta. The effect of the latter in ECD studies is neglected here since we will only be concerned with closed-shell molecules, and we therefore define it as

$$\mathbf{m} = \sum_j \frac{q_j}{2m_j} (\mathbf{r}_j \times \mathbf{p}_j) = \sum_j \frac{q_j}{2m_j} \mathbf{l}_j \tag{2.44}$$

where m_j indicates the mass of particle j and both the linear momentum operator of each particle $\mathbf{p}_i = -i\hbar\nabla_i$ and the orbital angular momentum $\mathbf{l}_j = (\mathbf{r}_j \times \mathbf{p}_j)$ are introduced. Adopting a form of the interaction operator involving only the electric dipole and the orbital magnetic dipole operators yields, for the difference in extinction coefficients,

$$\Delta\epsilon(\omega) = \epsilon_L(\omega) - \epsilon_R(\omega) = \frac{16\pi^2 N_A \omega}{3 \times 1000 \times \ln(10)(4\pi\epsilon_0)\hbar c_0^2} \sum_{\text{transitions, } ij} g_{ij}(\omega) \Im(\boldsymbol{\mu}_{ij} \cdot \mathbf{m}_{ij}) \tag{2.45}$$

This expression connects the observed anisotropy of the decadic molar extinction coefficient to the rotatory strength, R_{ij}, the scalar product of the electric dipole and magnetic dipole transition matrix elements,

$$R_{ij} = \Im(\boldsymbol{\mu}_{ij} \cdot \mathbf{m}_{ij}) \tag{2.46}$$

The ratio of ECD to OPA intensities is obtained as

$$\frac{\Delta\epsilon(\omega)}{\epsilon(\omega)} = \frac{4n}{c_0} \frac{\sum_{\text{transitions},ij} g_{ij}(\omega) \Im(\boldsymbol{\mu}_{ij} \cdot \mathbf{m}_{ij})}{\sum_{\text{transitions},ij} g_{ij}(\omega) (\boldsymbol{\mu}_{ij})^2} \tag{2.47}$$

Assuming that the lineshapes of ordinary absorption and circular dichroism are the same, a good approximation to this ratio, for an individual transition $i \to j$, is

$$\frac{\Delta\epsilon_{ij}}{\epsilon_{ij}} \approx \frac{4n}{c_0}\frac{\Im(\boldsymbol{\mu}_{ij}\cdot\mathbf{m}_{ij})}{(\boldsymbol{\mu}_{ij})^2} \approx 2.91894 \times 10^{-2}\frac{\Im(\boldsymbol{\mu}_{ij}[\text{a.u.}]\cdot\mathbf{m}_{ij}[\text{a.u.}])}{(\boldsymbol{\mu}_{ij}[\text{a.u.}])^2} \quad (2.48)$$

The rotatory strength vanishes when the transition is either electric dipole ($\boldsymbol{\mu}_{ij}=0$) or magnetic dipole ($\mathbf{m}_{ij}=0$) forbidden or when the two transition moments are perpendicular. One of these three instances always occurs in molecules belonging to point groups with at least an improper axis. Electronic circular dichroism (in isotropic samples without external fields) therefore manifests itself only for chiral molecules, that is (by definition), in molecules belonging to point groups missing improper axes, and it is therefore an example of a chiroptical property.

Turning to wavelengths and reversing Eqs. 2.34 and 2.45, one obtains an expression which is commonly used when comparing experiment and ab initio studies. Taking into account the fact that $\int_{\text{band}} g_{ij}(\lambda)d\lambda = 1$ (see above)

$$\sum_{\text{transitions},ij} |\boldsymbol{\mu}_{ij}|^2 \approx \frac{3\times 1000\times \ln(10)(4\pi\epsilon_0)n\hbar c_0}{4\pi^2 N_A}\int_{\text{spectrum}}\frac{\epsilon(\lambda)}{\lambda}d\lambda \quad (2.49)$$

$$\sum_{\text{transitions},ij} \Im(\boldsymbol{\mu}_{ij}\cdot\mathbf{m}_{ij}) \approx \frac{3\times 1000\times \ln(10)(4\pi\epsilon_0)\hbar c_0^2}{16\pi^2 N_A}\int_{\text{spectrum}}\frac{\Delta\epsilon(\lambda)}{\lambda}d\lambda = 0 \quad (2.50)$$

The last equality in Eq. 2.50 can be proven to apply to all molecules, independent of their chirality. These relationships are most useful when the bands are well separated, so that an individual transition can be identified. In this case

$$|\boldsymbol{\mu}_{ij}|^2 \approx \frac{3\times 1000\times \ln(10)(4\pi\epsilon_0)n\hbar c_0}{4\pi^2 N_A}\int_{ij\text{ band}}\frac{\epsilon(\lambda)}{\lambda}d\lambda \quad (2.51)$$

$$\Im(\boldsymbol{\mu}_{ij}\cdot\mathbf{m}_{ij}) \approx \frac{3\times 1000\times \ln(10)(4\pi\epsilon_0)\hbar c_0^2}{16\pi^2 N_A}\int_{ij\text{ band}}\frac{\Delta\epsilon(\lambda)}{\lambda}d\lambda \quad (2.52)$$

We can summarize the equations above as follows[1]:

$$\epsilon(\omega)[\text{L mol}^{-1}\text{ cm}^{-1}] = 9.78404\times 10^{60}\omega[\text{s}^{-1}]\sum_{\text{transitions},ij} g_{ij}(\omega[\text{s}])|\mu_{ij}[\text{C m}]|^2 \quad (2.53)$$

[1] Note that $\omega g_{ij}(\omega)$ can be replaced by $\lambda g_{ij}(\lambda)$ or $\nu g_{ij}(\nu)$, keeping the numerical constants unchanged. The frequency-dependent refractive index is approximated as $n = n(\lambda) \approx 1$ from now on.

$$\Delta\epsilon(\omega)[\text{L mol}^{-1}\text{ cm}^{-1}] =$$
$$1.30544 \times 10^{53}\omega[\text{s}^{-1}] \sum_{\text{transitions},ij} g_{ij}(\omega[\text{s}])\Im(\boldsymbol{\mu}_{ij}[\text{C m}] \cdot \mathbf{m}_{ij}[\text{J T}^{-1}]) \tag{2.54}$$

$$\epsilon(\omega)[\text{L mol}^{-1}\text{ cm}^{-1}] =$$
$$1.08862 \times 10^{38}\omega[\text{s}^{-1}] \sum_{\text{transitions},ij} g_{ij}(\omega[\text{s}])\left|\mu_{ij}[\text{Fr cm}]\right|^2 \tag{2.55}$$

$$\Delta\epsilon(\omega)[\text{L mol}^{-1}\text{ cm}^{-1}] =$$
$$4.35449 \times 10^{38}\omega[\text{s}^{-1}] \sum_{\text{transitions},ij} g_{ij}(\omega[\text{s}])\Im(\boldsymbol{\mu}_{ij}[\text{Fr cm}] \cdot \mathbf{m}_{ij}[\text{erg G}^{-1}]) \tag{2.56}$$

$$\epsilon(\omega)[\text{L mol}^{-1}\text{ cm}^{-1}] =$$
$$7.03301 \times 10^{2}\omega[\text{a.u.}] \sum_{\text{transitions},ij} g_{ij}(\omega[\text{a.u.}])|\mu_{ij}[\text{a.u.}]|^2 \tag{2.57}$$

$$\Delta\epsilon(\omega)[\text{L mol}^{-1}\text{ cm}^{-1}] =$$
$$2.05289 \times 10^{1}\omega[\text{a.u.}] \sum_{\text{transitions},ij} g_{ij}(\omega[\text{a.u.}]\Im(\boldsymbol{\mu}_{ij}[\text{a.u.}] \cdot \mathbf{m}_{ij}[\text{a.u.}]) \tag{2.58}$$

We also have that

$$|\mu_{ij}[\text{C m}]|^2 \approx 1.02207 \times 10^{-61} \times \int_{ij\text{ band}} \frac{\epsilon(\lambda)}{\lambda}\,d\lambda \tag{2.59}$$

$$|\mu_{ij}[\text{Fr cm}]|^2 \approx 9.18593 \times 10^{-39} \times \int_{ij\text{ band}} \frac{\epsilon(\lambda)}{\lambda}\,d\lambda \tag{2.60}$$

$$|\mu_{ij}[\text{a.u.}]|^2 \approx 1.42187 \times 10^{-3} \times \int_{ij\text{ band}} \frac{\epsilon(\lambda)}{\lambda}\,d\lambda \tag{2.61}$$

$$\Im(\boldsymbol{\mu}_{ij}[\text{C m}] \cdot \mathbf{m}_{ij}[\text{J T}^{-1}]) \approx 7.66024 \times 10^{-54} \times \int_{ij\text{ band}} \frac{\Delta\epsilon(\lambda)}{\lambda}\,d\lambda \tag{2.62}$$

$$\Im(\boldsymbol{\mu}_{ij}[\text{Fr cm}] \cdot \mathbf{m}_{ij}[\text{erg G}^{-1}]) \approx 2.29648 \times 10^{-39} \times \int_{ij\text{ band}} \frac{\Delta\epsilon(\lambda)}{\lambda} d\lambda \tag{2.63}$$

$$\Im(\boldsymbol{\mu}_{ij}[\text{a.u.}] \cdot \mathbf{m}_{ij}[\text{a.u.}]) \approx 4.87117 \times 10^{-2} \times \int_{ij\text{ band}} \frac{\Delta\epsilon(\lambda)}{\lambda} d\lambda \tag{2.64}$$

Before moving further, we mention here that the decadic molar extinction coefficient, also known as molar absorptivity, is not the only quantity employed by experimentalists to report their measurements of circular dichroism. Two other quantities often employed are the "specific ellipticity" $[\psi]$ and the "molar ellipticity" $[\Theta]$. The former is given in degrees $\text{cm}^3\,\text{dm}^{-1}\,\text{g}^{-1}$ and is related to the absolute "ellipticity" ψ, usually given in degrees, by the relationship

$$[\psi](\omega) = \frac{\psi}{dc_g} \tag{2.65}$$

where d is the optical path given in decimeters, and c_g is the concentration given in grams per liter. It can be shown that

$$[\psi](\omega) = \frac{180 \times 10^4 \ln(10)}{4\pi M} \Delta\epsilon(\omega) \tag{2.66}$$

with the decadic molar extinction coefficient being given in $\text{L mol}^{-1}\,\text{cm}^{-1}$ and where M is the molar mass in g mol^{-1}. The molar ellipticity $[\Theta]$, given in degrees $\text{dm}^3\,\text{mol}^{-1}\,\text{cm}^{-1}$, is obtained from the specific ellipticity as

$$[\Theta](\omega) = \frac{M}{100}[\psi] \tag{2.67}$$

and therefore

$$[\Theta](\omega) = \frac{18{,}000 \ln(10)}{4\pi} \Delta\epsilon(\omega) \approx 3298 \times \Delta\epsilon(\omega) \tag{2.68}$$

The reference to ellipticity hints at the change of polarization status of radiation coming out of a chiral sample. The differential absorption of the two circular components which, when in phase and of equal intensity represent a linearly propagating incident beam, yields an outgoing elliptically polarized beam, and the measure of the ellipticity induced by the sample has traditionally been a way of quantifying optical activity.

The expression given above for the electronic dichroism can be employed, in connection with Kramers–Kronig transforms [78, 79], to determine the optical rotation, which is the dispersive analogue of the ellipticity observed in regions of absorption. The procedure is made somewhat affordable if the dichroism is described in terms of the lineshape functions (Dirac delta, Lorentzian, or Gaussian functions) mentioned above (see refs. 67 and 80 for further details).

We finally note that Eqs. 2.34 and 2.45 can be rewritten in the so-called velocity gauge form by exploiting the relationship existing between the position **r** and linear momentum **p** for an exact (local) Hamiltonian $\mathcal{H}$ [5],

$$[\mathbf{r}_i, \mathcal{H}] = \imath \frac{\hbar}{m_i} \mathbf{p}_i \tag{2.69}$$

which also implies [81, 82]

$$[\mathbf{r}_i \mathbf{r}_i, \mathcal{H}] = \imath \frac{\hbar}{m_i} (\mathbf{r}_i \mathbf{p}_i + \mathbf{p}_i \mathbf{r}_i) \tag{2.70}$$

This yields "velocity" formulations for our properties, which have both advantages (the most interesting one being that an inherently origin-invariant formulation of electronic circular dichroism is obtained in this formulation) and disadvantages (e.g., that the use of the velocity operator in calculations carried out with limited basis set might suffer from slower convergence with respect to the size of the basis set expansion). "Velocity" formulations are generally slightly less popular than the corresponding "length" formulations, which are equivalent to the adoption of the so-called dipole approximation. The implications in practical calculations, resorting to approximate Hamiltonians and wavefunction models, will be discussed later in this chapter.

2.2.2 Nonlinear Absorption and Dichroism: Two-Photon Absorption and Circular Dichroism

The possibility that more than one photon may be absorbed simultaneously is a fact known since the early days of quantum theory, having been discussed already by Göppert-Mayer in 1931 [83]. Experimental observation required the advent of coherent intense sources of radiation (lasers) becoming available at the beginning of the 1960s. Since then the field has evolved rapidly, and the number of papers on the subject of multiphoton processes, as for the whole area of nonlinear optical processes, is growing at a very fast pace. References for this section are (in particular for the direct absorption part) the books by Bloembergen [84], Shen [85], Butcher and Cotter [86], Faisal [87], and Lambropoulos and Eberly [88]; see also the classical work of McClain [89]. For recent reviews of the area of two-photon absorption see also Terenziani et al. [90] and He et al. [91]. In the following we will treat only two-photon processes. The extension to more than two-photon phenomena is rather straightforward (although as the order of nonlinearity increases, the formulas become increasingly more complicated).

The quantity of interest in studies of two-photon absorption (TPA), in the case of two laser sources of circular frequency ω_1 and ω_2 with associated wavelengths λ_1 and

λ_2, is the two-photon transition strength $\delta(\omega_1, \omega_2)$, defined in a way somewhat similar to the decadic molar extinction coefficient $\epsilon(\omega)$, through the relationship connecting the attenuation (ΔP) of the optical power (P) of the laser beam(s) to the optical power itself. The latter is defined as the number of photons in the beam per second, and the following relationship applies [89]:

$$-\Delta P_1 = -\Delta P_2 \approx P_1 P_2 \delta(\omega_1, \omega_2) C z A^{-1} \tag{2.71}$$

Here C is the concentration (in number of molecules per cubic centimeter), z is the optical path (in centimeters, see above), and A is the area of the beam (in square centimeters). Therefore, $\delta(\omega_1, \omega_2)$ has units of $\text{cm}^4\,\text{s}\,\text{molecule}^{-1}\,\text{photon}^{-1}$. The unit $10^{-50}\,\text{cm}^4\,\text{s}\,\text{molecule}^{-1}\,\text{photon}^{-1}$, which is the usual order of magnitude of the two-photon transition strength, is often referred to as one "Göppert-Mayer," or one GM. With the appropriate manipulations it is possible to rewrite Eq. 2.71 as

$$-\frac{\Delta I_1}{z} \approx \frac{I_1 I_2}{\hbar\omega_1} 1000 N_A c \delta(\omega_1, \omega_2) \tag{2.72}$$

where the irradiance I_1 (I_2), the lost irradiance ΔI_1, and the concentration now are defined as in Eq. 2.33.

The derivation of an expression for $\delta(\omega_1,\omega_2)$ via semiclassical or quantum electrodynamic approaches, can be found in several places. Time-dependent perturbation theory must be carried out to at least second order to describe the process where two photons impinge on the molecule within a time shorter than the lifetime of a "virtual state"—a superposition of the whole manifold of excited stationary states—and are both being absorbed.

By adopting, as for OPA, an approximation for the interaction operator including only the electric dipole perturbation, the isotropically averaged two-photon transition strength $\delta(\omega_1,\omega_2)$ can be written as

$$\delta(\omega_1, \omega_2) = \frac{1}{15} \frac{(2\pi)^3 \omega^2}{c_0^2 (4\pi\epsilon_0)^2} \times \sum_{ij,\text{transitions}} g_{ij}(\omega_1, \omega_2) \delta_{ij}^{\text{TPA}}(\omega_1, \omega_2) \tag{2.73}$$

where, as for the one-photon case, the Dirac delta function has been replaced by the appropriate lineshape function $g_{ij}(\omega_1,\omega_2)$, again usually a Lorentzian or a Gaussian function:

$$\begin{aligned} \delta_{ij}^{\text{TPA}}(\omega_1, \omega_2) = {} & A \times S_{\alpha\alpha}^{ij}(\omega_1, \omega_2) S_{\beta\beta}^{ij,*}(\omega_1, \omega_2) + B \times S_{\alpha\beta}^{ij}(\omega_1, \omega_2) S_{\alpha\beta}^{ij,*}(\omega_1, \omega_2) \\ & + C \times S_{\alpha\beta}^{ij}(\omega_1, \omega_2) S_{\beta\alpha}^{ij,*}(\omega_1, \omega_2) \end{aligned} \tag{2.74}$$

where implicit Einstein summation over repeated Greek indices is implied and A, B, and C are numbers depending on the polarization state of the two photons and the geometric setup (relative orientations of the beams). If the two (generally complex) unitary polarization vectors are $\mathbf{i}_1$ and $\mathbf{i}_2$, then [75, 89]

$$A = 4|\mathbf{i}_1 \cdot \mathbf{i}_2|^2 - 1 - |\mathbf{i}_1 \cdot \mathbf{i}_2^*|^2 \tag{2.75}$$

$$B = -|\mathbf{i}_1 \cdot \mathbf{i}_2|^2 + 4 - |\mathbf{i}_1 \cdot \mathbf{i}_2^*|^2 \tag{2.76}$$

$$C = -|\mathbf{i}_1 \cdot \mathbf{i}_2|^2 - 1 + 4|\mathbf{i}_1 \cdot \mathbf{i}_2^*|^2 \tag{2.77}$$

In Eq. 2.74, in the dipole approximation

$$\mathcal{S}^{ij}_{\alpha\beta}(\omega_1,\omega_2) = \frac{1}{\hbar}\sum_n \left\{ \frac{(\mu_\alpha)_{in}(\mu_\beta)_{nj}}{\omega_{ni}-\omega_1} + \frac{(\mu_\beta)_{in}(\mu_\alpha)_{nj}}{\omega_{ni}-\omega_2} \right\} \tag{2.78}$$

is the second-rank, two-photon tensor which, contrary to what was happening to the transition moment for one-photon processes, is now a function of both frequencies. The energy conservation relationship applies, $\omega_1 + \omega_2 = \omega_{ij}$, and the use of expression 2.78 implies off-resonance conditions—that is, frequencies ω_1 and ω_2 sufficiently far off the values at which the denominators in Eq. 2.78 vanish. Resonant two-photon absorption or dichroism will not be mentioned further in this chapter. The Greek indices α and β refer to the Cartesian components of the operator $\boldsymbol{\mu}$ in the molecular frame, and the summation runs over the whole set of excited states n, including in principle also the states i and j. The tensor is nonsymmetric in the exchange of the two frequencies except for $\omega_1 = \omega_2$, and its symmetry properties are described in detail elsewhere [75]. Symmetry enforces that $\delta_{ij}(\omega_1, \omega_2)$ depends on three polarization/geometry parameters for the general case of two beams with different frequencies and on two parameters only in the degenerate case. In the latter case, which is the one we will focus on from now on [75] ($\omega = \omega_1 = \omega_2$ and $\mathbf{i}_1 = \mathbf{i}_2 = \mathbf{i}$),

$$\delta^{\mathrm{TPA}}_{ij}(\omega) = \{(2|\mathbf{i}\cdot\mathbf{i}|^2 - 1) \times [\mathcal{A}_1(\omega)]_{ij} - (|\mathbf{i}\cdot\mathbf{i}|^2 - 3) \times [\mathcal{A}_2(\omega)]_{ij}\} \tag{2.79}$$

where

$$[\mathcal{A}_1(\omega)]_{ij} = \sum_{\rho\sigma} \mathcal{S}^{ij}_{\rho\rho}(\omega)\mathcal{S}^{ij,*}_{\sigma\sigma}(\omega) \tag{2.80}$$

$$[\mathcal{A}_2(\omega)]_{ij} = \sum_{\rho\sigma} \mathcal{S}^{ij}_{\rho\sigma}(\omega)\mathcal{S}^{ij,*}_{\rho\sigma}(\omega) \tag{2.81}$$

The process analogous to one-photon ECD in the two-photon case was discussed first by Tinoco in the mid-1970s [92]. The author derived expressions for the difference between the two-photon absorption strengths measured for left $\delta_L(\omega)$ and right $\delta_R(\omega)$ circularly polarized light—two-photon circular dichroism (TPCD)—by an extension carried out with a semiclassical approximation to the TPA formulas, along the lines of the procedures described for ECD in the previous section. Work on the subject was carried out in the same period by Power [93], who got the same expression for the property by employing a quantum electrodynamic formalism. TPCD has also been discussed by Andrews [94] and by Meath and Power [95–98]. The latter presented extensions of the formulas to the general n-photon circular dichroism process [97] and discussed the effect of elliptical polarization [98]. Nonlinear optical activity has also been discussed by Wagniere [99, 100], as well as multiphoton optical rotation, mentioned by Power [101], and it has been the subject of experimental [102, 103] and computational [104, 105] work for the last 20 years and more.

Fluorescence-detected two-photon circular dichroism of lanthanides, discussed in 1986 [106], was observed nine year later [107]. Radiation-induced nonlinear circular dichroism was seen in solutions of ruthenium bipyridil salts in the last decade [108–112]. Upper limits for the two-photon dichroism that could be detected in experiments were determined by Markowicz and co-workers by employing a modified Z-scan technique [113]. Very recently, Hernández and his group developed a double L-scan technique which permits the experimental detection of well-resolved TPCD and two-photon circular-linear dichroism (TPCLD) (see below) spectra for chiral samples in solution [114]. The experimental determination of nonlinear circular dichroism has proven to be especially challenging, mostly due to the weakness of the effect. On the other hand, computational activity in the field has been rather intense, and reliable approaches to the calculation of differential two-photon absorption strengths were developed mostly in our group [82, 115–123].

In addition to the original papers by Tinoco [92], Power [93], and Andrews [94], derivations of the expression for the TPCD can be found in the book by Lin et al. [124]. Contrary to the one-photon case, as noted above, the two-photon absorption strength depends on the choice of the polarization of the photons, but—within the dipole approximation—it remains the same for left- and right-circularly polarized photons. Higher-order multipoles (in particular the magnetic dipole and the electric quadrupole) must be employed in the perturbing operator to obtain a nonvanishing dichroism $\delta_L(\omega) - \delta_R(\omega)$. Following Tinoco [92], who employed a velocity form for the perturbing operator within a second-order semiclassical perturbation theory approach, the difference in TPA cross section for left- and right-circularly polarized photons can be expressed as

$$\Delta\delta(\omega) = \delta_L(\omega) - \delta_R(\omega) = \frac{4}{15} \frac{(2\pi)^3 \omega^2}{c_0^3 (4\pi\epsilon_0)^2} \times \sum_{ij,\text{transitions}} g_{ij}(2\omega) R_{ij}^{\text{TPCD}}(\omega) \tag{2.82}$$

$$R_{ij}^{\text{TPCD}}(\omega) = -b_1[\mathcal{B}_1(\omega)]_{ij} - b_2[\mathcal{B}_2(\omega)]_{ij} - b_3[\mathcal{B}_3(\omega)]_{ij} \tag{2.83}$$

Due to the use of the velocity form of the operators, the TPCD given by this expression is gauge-origin independent. The parameters b_1, b_2 and b_3 are numbers, simple combinations of the analogous A, B, and C coefficients in Eqs. 2.75–2.77. Tinoco [92] gave a list of different triplets (b_1, b_2, b_3) for a variety of combinations of polarization and relative orientations for the two photons, of which at least one must be left/right circularly polarized. The "molecular" parameters $\mathcal{B}_1$, $\mathcal{B}_2$, and $\mathcal{B}_3$ take the form

$$[\mathcal{B}_1(\omega)]_{ij} = \frac{1}{\omega^3} \sum_{\rho\sigma} \mathcal{M}^{p,ij}_{\rho\sigma}(\omega) \mathcal{P}^{p*,ij}_{\rho\sigma}(\omega) \tag{2.84}$$

$$[\mathcal{B}_2(\omega)]_{ij} = \frac{1}{2\omega^3} \sum_{\rho\sigma} \mathcal{T}^{+,ij}_{\rho\sigma}(\omega) \mathcal{P}^{p*,ij}_{\rho\sigma}(\omega) \tag{2.85}$$

$$[\mathcal{B}_3(\omega)]_{ij} = \frac{1}{\omega^3} \sum_{\rho\sigma} \mathcal{M}^{p,ij}_{\rho\rho}(\omega) \mathcal{P}^{p*,ij}_{\sigma\sigma}(\omega) \tag{2.86}$$

and they are therefore appropriate contractions of the generalized two-photon, second-rank tensors such as the one given in Eq. 2.78. Indeed, these tensors are defined (for the general case of two photons of different frequency) as

$$\mathcal{P}^{p,ij}_{\alpha\beta}(\omega_1, \omega_2) = \frac{1}{\hbar} \sum_n \left\{ \frac{(\mu^p_\alpha)_{in} \left(\mu^p_\beta\right)_{nj}}{\omega_{ni} - \omega_1} + \frac{\left(\mu^p_\beta\right)_{in} (\mu^p_\alpha)_{nj}}{\omega_{ni} - \omega_2} \right\} \tag{2.87}$$

$$\mathcal{M}^{p,ij}_{\alpha\beta}(\omega_1, \omega_2) = \frac{1}{\hbar} \sum_n \left\{ \frac{(\mu^p_\alpha)_{in} (m_\beta)_{nj}}{\omega_{ni} - \omega_1} + \frac{(m_\beta)_{in} (\mu^p_\alpha)_{nj}}{\omega_{ni} - \omega_2} \right\} \tag{2.88}$$

$$\mathcal{T}^{+,ij}_{\alpha\beta}(\omega_1, \omega_2) = \frac{1}{\hbar} \epsilon_{\beta\rho\sigma} \sum_n \left\{ \frac{\left(T^+_{\alpha\rho}\right)_{in} (\mu^p_\sigma)_{nj}}{\omega_{ni} - \omega_1} + \frac{(\mu^p_\sigma)_{in} \left(T^+_{\alpha\rho}\right)_{nj}}{\omega_{ni} - \omega_2} \right\} \tag{2.89}$$

Here the velocity dipole operator

$$\boldsymbol{\mu}^p = \sum_j \frac{q_j}{m_j} \mathbf{p}_j \tag{2.90}$$

and the "velocity" form of the electric quadrupole operator

$$\mathbf{T}^+ = \sum_j \frac{q_j}{m_j} (\mathbf{r}_j \mathbf{p}_j + \mathbf{p}_j \mathbf{r}_j) \tag{2.91}$$

appear and $\epsilon_{\alpha\beta\gamma}$ is the Levi-Civita alternating tensor.

As mentioned above, the formulation presented here has the advantage of yielding results independent of the choice of origin for the multipolar expansion and therefore of the electric and magnetic fields, not only (as it should) formally, in the "exact" case, but also in approximate treatments. Adopting the "length" form of the interaction yields alternative formulations [82] which on the other hand are not origin invariant in approximate calculations. One of these formulations was labelled ω_0 [82], and it gives the same results as those yielded by Eq. 2.82 in the limit of a complete one-electron basis set. This formulation can be obtained via the following substitutions:

$$\mathcal{P}^{p,ij}_{\alpha\beta}(\omega) \rightarrow -\omega^2 \mathcal{S}^{ij}_{\alpha\beta}(\omega) \tag{2.92}$$

$$\mathcal{M}^{p,ij}_{\alpha\beta}(\omega) \rightarrow -\imath\omega \mathcal{M}^{ij}_{\sigma\beta}(\omega) \tag{2.93}$$

$$\mathcal{T}^{+,ij}_{\alpha\beta}(\omega) \rightarrow \omega^2 \mathcal{Q}^{ij}_{\alpha\beta}(\omega) \tag{2.94}$$

yielding

$$\mathcal{R}^{\text{TPCD}}_{ij} = -\imath b_1 \left[\mathcal{B}^{\omega_0}_1(\omega)\right]_{ij} + b_2 \left[\mathcal{B}^{\omega_0}_2(\omega)\right]_{ij} + \imath b_3 \left[\mathcal{B}^{\omega_0}_3(\omega)\right]_{ij} \tag{2.95}$$

$$\left[\mathcal{B}^{\omega_0}_1(\omega)\right]_{ij} = \sum_{\rho\sigma} \mathcal{M}^{ij}_{\rho\sigma}(\omega) \mathcal{S}^{ji*}_{\rho\sigma}(\omega) \tag{2.96}$$

$$\left[\mathcal{B}^{\omega_0}_2(\omega)\right]_{ij} = \frac{\omega}{2} \sum_{\rho\sigma} \mathcal{Q}^{ij}_{\rho\sigma}(\omega) \mathcal{S}^{ji*}_{\rho\sigma}(\omega) \tag{2.97}$$

$$[\mathcal{B}^{\omega_0}_3(\omega)]_{ij} = \left[\sum_{\rho} \mathcal{M}^{ij}_{\rho\rho}(\omega)\right] \left[\sum_{\sigma} \mathcal{S}^{ij}_{\sigma\sigma}(\omega)\right] \tag{2.98}$$

The tensors $\mathcal{M}^{ij}_{\alpha\beta}(\omega_1, \omega_2)$ and $\mathcal{Q}^{ij}_{\alpha\beta}(\omega_1, \omega_2)$ are defined as

$$\mathcal{M}^{ij}_{\alpha\beta}(\omega_1, \omega_2) = \frac{1}{\hbar} \sum_n \left\{ \frac{(\mu_\alpha)_{in}(m_\beta)_{nj}}{\omega_{ni} - \omega_1} + \frac{(m_\beta)_{in}(\mu_\alpha)_{nj}}{\omega_{ni} - \omega_2} \right\} \tag{2.99}$$

$$\mathcal{Q}^{ij}_{\alpha\beta}(\omega_1, \omega_2) = \frac{1}{\hbar} \epsilon_{\beta_{\rho\sigma}} \sum_n \left\{ \frac{(q_{\alpha\rho})_{in}(\mu_\sigma)_{nj}}{\omega_{ni} - \omega_1} + \frac{(\mu_\sigma)_{in}(q_{\alpha\rho})_{nj}}{\omega_{ni} - \omega_2} \right\} \tag{2.100}$$

TPCD arises therefore from the interaction of light with the electric and magnetic multipoles of matter, and, contrary to what happens in isotropic samples for ECD, also the molecular quadrupole plays a role, although its importance is in practice limited [115, 117, 119, 120, 123].

The ratio between TPCD and TPA for two photons of the same frequency is

$$\frac{\Delta\delta(\omega)}{\delta} = \frac{8}{c_0} \frac{\sum_{ij,\text{transitions}} g_{ij}(2\omega) R_{ij}^{\text{TPCD}}(\omega)}{\sum_{ij,\text{transitions}} g_{ij}(2\omega) \delta_{ij}^{\text{TPA}}(\omega)} \tag{2.101}$$

or, for a given transition and assuming the same lineshape for both properties,

$$\frac{\Delta\delta_{ij}(\omega)}{\delta_{ij}} \approx \frac{8}{c_0} \frac{R_{ij}^{\text{TPCD}}(\omega)}{\delta_{ij}^{\text{TPA}}(\omega)} \approx 5.83788 \times 10^{-2} \frac{R_{ij}^{\text{TPCD}}(\omega)[\text{a.u}]}{\delta_{ij}^{\text{TPA}}(\omega)[\text{a.u.}]} \tag{2.102}$$

As for the rotatory strength of ECD, it can be shown that the two-photon rotatory strength of TPCD vanishes for isotropic samples of achiral molecules, appearing therefore as another manifestation of optical activity in chiral systems.

The main equations of this section can be summarized as follows:

$$\delta(\omega)[\text{GM}] = 8.35150 \times 10^{-4} \times (\omega[\text{a.u}])^2 \times \sum_{ij,\text{transitions}} g_{ij}(2\omega)[\text{a.u.}] \times \delta_{ij}^{\text{TPA}}(\omega)[\text{a.u.}] \tag{2.103}$$

$$\Delta\delta(\omega)[\text{GM}] = 4.87555 \times 10^{-5} \times (\omega[\text{a.u.}])^2 \times \sum_{ij,\text{transitions}} g_{ij}(2\omega)[\text{a.u.}] \times R_{ij}^{\text{TPCD}}(\omega)[\text{a.u.}] \tag{2.104}$$

$$\delta(\omega)[\text{GM}] = 2.02015 \times 10^{-20} \times (\omega[\text{s}^{-1}])^2 \times \sum_{ij,\text{transitions}} g_{ij}(2\omega)[\text{s}] \times \delta_{ij}^{\text{TPA}}(\omega)[\text{a.u.}] \tag{2.105}$$

$$\Delta\delta(\omega)[\text{GM}] = 1.17934 \times 10^{-21} \times (\omega[\text{s}^{-1}])^2 \times \sum_{ij,\text{transitions}} g_{ij}(2\omega)[\text{s}] \times R_{ij}^{\text{TPCD}}(\omega)[\text{a.u.}] \tag{2.106}$$

These expressions permit the computation of the experimental observables, the two-photon absorption strength or its anisotropy, in their standard units (GM) starting from the variables (frequency, lineshapes, and molecular parameters) in the units given in square brackets.

Quite recently, attention has been drawn to TPCLD, defined as [125]

$$\Delta\delta^{\text{TPCLD}}(\omega) = \frac{\delta_{\text{CPL}}^{\text{TPA}}(\omega) - \delta_{\text{LPL}}^{\text{TPA}}(\omega)}{\delta_{\text{CPL}}^{\text{TPA}}(\omega) + \delta_{\text{LPL}}^{\text{TPA}}(\omega)} = \frac{R_{\text{CL}}^{\text{TPA}}(\omega) - 1}{R_{\text{CL}}^{\text{TPA}}(\omega) + 1} \tag{2.107}$$

where LPL and CPL are acronyms for linearly and circularly polarized light, respectively, and $R_{\text{CL}}^{\text{TPA}}(\omega) = \delta_{\text{CPL}}^{\text{TPA}}(\omega)/\delta_{\text{LPL}}^{\text{TPA}}(\omega)$. The quantity in Eq. 2.107 is well defined in both chiral and achiral systems, and its dispersion yields in both cases a signature of the type and general symmetry features of each excited state. For chiral molecules, the TPCLD is different for the two enantiomers, and the difference is a measure of TPCD. Indeed, using the formulas given above (Eqs. 2.79, 2.82, and 2.83), the latter specialized to the case of two circularly polarized photons, it is easy to recast Eq. 2.107 for chiral response as [126]

$$\Delta\delta_{L/R}^{\text{TPCLD}}(\omega) = \frac{-2[\mathcal{A}_1(\omega)]_{ij} + [\mathcal{A}_2(\omega)]_{ij} \pm (4/c_0)\{3[\mathcal{B}_1(\omega)]_{ij} + [\mathcal{B}_2(\omega)]_{ij} - [\mathcal{B}_3(\omega)]_{ij}\}}{5[\mathcal{A}_2(\omega)]_{ij} \pm (4/c_0)\{3[\mathcal{B}_1(\omega)]_{ij} + [\mathcal{B}_2(\omega)]_{ij} - [\mathcal{B}_3(\omega)]_{ij}\}} \tag{2.108}$$

where the molecular parameters were defined in Eqs. 2.80, 2.81 and 2.84 to 2.86 and the *L*/*R* subscript applies to left- and right-circularly polarized light, respectively. If the TPCD can be considered as a small perturbation (see Eq. 2.102), then [126]

$$\Delta\delta_{L/R}^{\text{TPCLD}}(\omega) \approx \frac{1}{5} - \frac{2}{5}\frac{[\mathcal{A}_1(\omega)]_{ij}}{[\mathcal{A}_2(\omega)]_{ij}} \pm \frac{4[\mathcal{A}_2(\omega)]_{ij} + 2[\mathcal{A}_1(\omega)]_{ij}}{25[\mathcal{A}_2(\omega)]_{ij}^2} \times \left\{3[\mathcal{B}_1(\omega)]_{ij} + [\mathcal{B}_2(\omega)]_{ij} - [\mathcal{B}_3(\omega)]_{ij}\right\} \tag{2.109}$$

TPCLD has been reported for BINOL, that is, 1,1-bi(2-naphtol); see Toro et al. [126], who compared theoretical predictions and experiment.

The atomic units of some of the quantities of relevance in this section are

$$S_{\alpha\beta}^{ij} \rightarrow \frac{e^2 m_e a_0^4}{\hbar^2} \tag{2.110}$$

$$\delta_{ij}^{\text{TPA}} \rightarrow \frac{e^4 m_e^2 a_0^8}{\hbar^4} \tag{2.111}$$

$$\mathcal{M}_{\alpha\beta}^{ij} \rightarrow \frac{e^2 a_0}{m_e} \tag{2.112}$$

$$\mathcal{P}_{\alpha\beta}^{ij} \rightarrow \frac{e^2}{m_e} \tag{2.113}$$

$$\mathcal{T}_{\alpha\beta}^{+,ij} \rightarrow \frac{e^2 a_0}{m_e} \tag{2.114}$$

$$R_{ij}^{\text{TPCD}} \rightarrow \frac{e^4 m_e a_0^7}{\hbar^3} \tag{2.115}$$

2.2.3 Magnetic Circular Dichroism

Circular dichroism can be observed in all fluids, independently of whether they are optically active (chiral) or not, by applying a static external magnetic field with a component parallel to the direction of propagation to a sample traversed by linearly polarized light in the setup of the classical Faraday experiment [127, 128]. The different refractive indices between the two circular components of the linearly polarized beam in regions of transparency yield the well-known magnetic field-induced optical rotation (MOR) and its frequency dispersion (MORD), the oldest manifestation of (circular) birefringence. The corresponding differences in the absorption coefficients lead to an ellipticity of the outgoing beam, commonly referred to as magnetic field–induced circular dichroism (MCD). An account of these matters is given in the literature [42, 68, 129]; see also Rizzo and Coriani [130]. Here we will focus on MCD, which is treated by Mason [131] and Piepho and Schatz [132] as well as in a few dedicated reviews [133–138].

The measurement of MCD involves, as for ECD, the detection of the molar, $[\Theta]$, Eq. 2.67, or specific, $[\psi]$, Eq. 2.65 ellipticity induced by the magnetic field. From this the decadic extinction coefficient $\epsilon(\omega)$, which in this case depends also on the strength of the applied magnetic field, being therefore usually given as molar absorptivity per unit magnetic field strength ($\mathrm{L\,mol^{-1}\,cm^{-1}\,T^{-1}}$) is determined. The dichroism can be explained by considering the effect of the external magnetic field, having a component lying along the direction of propagation, on the absorption spectra arising in its absence.

In the classical treatment of MCD, the possible degeneracies of both the i and j states are usually taken into account. These components, which all have the same energy in the absence of the external longitudinal magnetic field, are split due to the Zeeman effect when the field is switched on. In the expression for the molar decadic extinction coefficient, the degeneracy introduces a summation over the Boltzmann-weighted components of the ground state. The external field, modifying the electronic energy, modifies the weights as well as influences the position of the absorption peaks, the transition matrix elements—through the effect on both the excitation energies and the (complex) wavefunctions—and the form of the lineshape. The latter is usually treated within the "rigid-shift" approximation, which assumes that the external magnetic field does not modify the actual shape of the function g_{ij} but simply shifts it in the abscissa by an amount related to the Zeeman shift. The rigid-shift approximation is a rather crude assumption which is abandoned when a proper vibronic perturbation theory is adopted for MCD.

At the end of the theoretical development, the following classical expression is obtained for the anisotropy of the molar decadic extinction coefficient in an MCD experiment where the strength of the external magnetic field is B_{ext},

$$\Delta\epsilon(\omega) = -\frac{8\pi^2 N_A \omega}{3 \times 1000 \times \ln(10)(4\pi\epsilon_0)\hbar c_0} B_{\text{ext}} \times \sum_{ij,\text{transitions}} \left\{ \frac{1}{\hbar}\frac{\partial g_{ij}(\omega)}{\partial \omega}\mathcal{A}_{ij} + g_{ij}(\omega)\left[\mathcal{B}_{ij} + \frac{\mathcal{C}_{ij}}{kT}\right] \right\} \tag{2.116}$$

where the $\mathcal{A}$, $\mathcal{B}$, and $\mathcal{C}$ MCD terms are defined, for an isotropic sample, as

$$\mathcal{A}_{ij} = \frac{1}{2N_i} \sum_{i_\alpha j_\beta} \sum_{i'_\alpha j'_\beta} \left[\mathbf{m}_{j'_\beta j_\beta} \delta_{i_\alpha i'_\alpha} - \mathbf{m}_{i_\alpha i'_\alpha} \delta_{j_\beta j'_\beta} \right] \cdot \boldsymbol{\mu}_{i'_\alpha j'_\beta} \times \boldsymbol{\mu}_{j_\beta i_\alpha} \tag{2.117}$$

$$\mathcal{B}_{ij} = \frac{1}{\hbar N_i} \sum_{i_\alpha j_\beta} \Im \left\{ \sum_{r \neq i} \frac{\mathbf{m}_{r i_\alpha}}{\omega_{ri}} \cdot \boldsymbol{\mu}_{j_\beta r} \times \boldsymbol{\mu}_{i_\alpha j_\beta} + \sum_{r \neq j} \frac{\mathbf{m}_{j_\beta r}}{\omega_{rj}} \cdot \boldsymbol{\mu}_{r i_\alpha} \times \boldsymbol{\mu}_{i_\alpha j_\beta} \right\} \tag{2.118}$$

$$\mathcal{C}_{ij} = \frac{1}{2N_i} \sum_{i_\alpha j_\beta} \sum_{i'_\alpha} \mathbf{m}_{i_\alpha i'_\alpha} \cdot \boldsymbol{\mu}_{i'_\alpha j_\beta} \times \boldsymbol{\mu}_{j_\beta i_\alpha} \tag{2.119}$$

Here N_i is the dimension of the degenerate manifold for state i, with i_α and j_β indicating the components of the ith and jth states; T is the temperature and k is Boltzmann's constant. The $\mathcal{A}$ term vanishes if both i and j states are nondegenerate, whereas the $\mathcal{C}$ term vanishes for nondegenerate i states. It can be seen that the former is connected to the difference in Zeeman shifts of the two electronic states involved in the transition, whereas the latter arises from the changes in population of the degenerate initial state due to the effect of the external magnetic field. The $\mathcal{B}$ term is the only term surviving in the case of nondegenerate initial and final states, and it is the result of the changes induced in the transition moments due to the electric field perturbation by the external longitudinal magnetic field. The $\mathcal{B}$ and $\mathcal{C}$ terms are both associated with the usual absorption lineshape function $g_{ij}(\omega)$, with the profile of a Lorentzian or a Gaussian. The $\mathcal{A}$ term is instead associated with the ω-derivative of the absorption lineshape, with the typical dispersive profile

$$\frac{\partial L_{ij}(\omega)}{\partial \omega} = -\frac{\gamma}{\pi} \frac{\omega - \omega_{ij}}{[(\omega - \omega_{ij})^2 + (\gamma/2)^2]^2} \tag{2.120}$$

$$\frac{\partial G_{ij}(\omega)}{\partial \omega} = -\frac{8}{\gamma^3 \sqrt{2\pi}} (\omega - \omega_{ij}) e^{-2(\omega - \omega_{ij})^2/\gamma^2} \tag{2.121}$$

In rewriting Eq. 2.116 using wavenumbers ($\tilde{\nu} = \nu/c_0$), frequencies (ν), or wavelengths (λ) in place of angular frequencies (ω), it is important to note that while it is possible to interchange the three variables easily in the combination $\omega g_{ij}(\omega) \leftrightarrow \nu g_{ij}(\nu) \leftrightarrow \tilde{\nu} g_{ij}(\tilde{\nu}) \leftrightarrow \lambda g_{ij}(\lambda)$, see above, and that any of the four products are indeed unitless, some complication arises with the derivative of the shape function. The relationship is

$$\omega \frac{\partial g_{ij}(\omega)}{\partial \omega} \leftrightarrow \frac{\tilde{\nu}}{2\pi c_0} \frac{\partial g_{ij}(\tilde{\nu})}{\partial \tilde{\nu}} \leftrightarrow \frac{\nu}{2\pi} \frac{\partial g_{ij}(\nu)}{\partial \nu} \leftrightarrow \frac{\lambda_{ij}^2}{2\pi c_0} \lambda \frac{\partial g_{ij}(\lambda)}{\partial \lambda} \tag{2.122}$$

Note that the product $\sigma(\partial g_{ij}(\sigma)/\partial \sigma)(\sigma = \omega, \nu, \tilde{\nu}$ or $\lambda)$ has the units of inverse σ.

As noted above, in the case of nondegenerate initial and final electronic states, only the $\mathcal{B}$ term survives in Eq. 2.116, and therefore, as for the regular CD, we can write (in the λ domain)

$$\sum_{\text{transitions},ij} \mathcal{B}_{ij} = -\frac{3 \times 1000 \times \ln(10)(4\pi\epsilon_0)\hbar c_0}{8\pi^2 N_A B_{\text{ext}}} \int_{\text{spectrum}} \frac{\Delta\epsilon(\lambda)}{\lambda} d\lambda \tag{2.123}$$

The method commonly employed to determine separately the three terms in the case where all three are nonvanishing is known as the "method of moments", and it is based on the following relationships that apply to the lineshape functions and their first derivatives:

$$\int_{\text{spectrum}} \frac{\partial g_{ij}(\lambda)}{\partial \lambda} d\lambda = 0 \tag{2.124}$$

$$\int_{\text{spectrum}} g_{ij}(\lambda) d\lambda = 1 \tag{2.125}$$

$$\int_{\text{spectrum}} \lambda \frac{\partial g_{ij}(\lambda)}{\partial \lambda} d\lambda = -\int_{\text{spectrum}} g_{ij}(\lambda) d\lambda = -1 \tag{2.126}$$

$$\int_{\text{spectrum}} \lambda g_{ij}(\lambda) d\lambda = \lambda_{ij}^{\max} \tag{2.127}$$

where $\lambda_{ij}^{\max}$ is the wavelength at the center of the band. Based on these relationships, which are the consequence of the vanishing of the lineshape function at the extremes of the spectrum, we can use Eq. 2.116 [with $\lambda g_{ij}(\lambda)$ replacing $\omega g_{ij}(\omega)$, see also Eq. 2.122], divide both sides by λ, and integrate to obtain

$$\sum_{\text{transitions},ij} \left\{ \mathcal{B}_{ij} + \frac{\mathcal{C}_{ij}}{kT} \right\} = -\frac{3 \times 1000 \times \ln(10)(4\pi\epsilon_0)\hbar c_0}{8\pi^2 N_A B_{\text{ext}}} \int_{\text{spectrum}} \frac{\Delta\epsilon(\lambda)}{\lambda} d\lambda \tag{2.128}$$

In particular, if a given transition yields a dichroic band which is well separated from the rest of the spectrum,

$$\left\{ \mathcal{B}_{ij} + \frac{\mathcal{C}_{ij}}{kT} \right\} = -\frac{3 \times 1000 \times \ln(10)(4\pi\epsilon_0)\hbar c_0}{8\pi^2 N_A B_{\text{ext}}} \int_{\text{band}} \frac{\Delta\epsilon(\lambda)}{\lambda} d\lambda \tag{2.129}$$

The use of Eq. 2.129 allows for the determination of the combination of the $\mathcal{B}_{ij}$ and $\mathcal{C}_{ij}$ terms. By performing measurements at different temperatures and doing a linear regression, a determination of the individual terms is possible. To obtain $\mathcal{A}_{ij}$, one can proceed to a direct integration of Eq. 2.116 using Eq. 2.129 above to get rid of the other two terms,

$$\sum_{\text{transitions},ij} \mathcal{A}_{ij} = \frac{3 \times 1000 \times \ln(10)(4\pi\epsilon_0)\hbar^2 c_0^2}{4\pi N_A \lambda_{ij}^2 B_{\text{ext}}} \int_{\text{spectrum}} \frac{\Delta\epsilon(\lambda)}{\lambda}(\lambda - \lambda_{ij}^{\max})\, d\lambda \tag{2.130}$$

or, for a well-separated band,

$$\mathcal{A}_{ij} = \frac{3 \times 1000 \times \ln(10)(4\pi\epsilon_0)\hbar^2 c_0^2}{4\pi N_A \lambda_{ij}^2 B_{\text{ext}}} \int_{\text{band}} \frac{\Delta\epsilon(\lambda)}{\lambda}(\lambda - \lambda_{ij}^{\max})\, d\lambda \tag{2.131}$$

Equation 2.116 is translated into the following when the units of the magnetic field, wavelength, shape function, temperature, and MCD terms are those given in square brackets:

$$\begin{aligned}
\frac{\Delta\epsilon(\lambda)}{B_{\text{ext}}}\left[\text{L mol}^{-1}\,\text{cm}^{-1}\text{T}^{-1}\right] = &\\
& -1.31339 \times 10^{-4} \times \lambda[\text{nm}] \sum_{ij,\text{transitions}} (\lambda_{ij}[\text{nm}])^2 \frac{\partial g_{ij}(\lambda)}{\partial\lambda}\left[\text{nm}^{-2}\right]\mathcal{A}_{ij}[\text{a.u.}]\\
& -5.98442 \times 10^{-3} \times \lambda[\text{nm}] \sum_{ij,\text{transitions}} g_{ij}(\lambda)[\text{nm}^{-1}]\mathcal{B}_{ij}[\text{a.u.}]\\
& -1.88967 \times 10^{+3} \times \frac{\lambda[\text{nm}]}{T[\text{K}]} \sum_{ij,\text{transitions}} g_{ij}(\lambda)[\text{nm}^{-1}]\mathcal{C}_{ij}[\text{a.u.}]
\end{aligned} \tag{2.132}$$

If the magnetic field is given in Gauss, the prefactors above need to be multiplied by 10^{-4}. The atomic units of the three MCD terms are

$$\mathcal{A}_{ij} \rightarrow \frac{e^3\hbar a_0^2}{m_e} \tag{2.133}$$

$$\mathcal{B}_{ij} \rightarrow \frac{e^2 a_0^4}{\hbar} \tag{2.134}$$

$$\mathcal{C}_{ij} \rightarrow \frac{e^3\hbar a_0^2}{m_e} \tag{2.135}$$

2.3 RESPONSE THEORY AND SPECTROSCOPY

In the following we will discuss the application of RFT methods for the determination of well-known spectroscopic properties, connecting the expressions reported in

Section 2.1 with those in Section 2.2 for one- and two-photon absorption and dichroism and for magnetic circular dichroism. In the second part of the section, attention will be focused on some vibrational spectroscopies, namely vibrational circular dichroism (VCD), ROA, and coherent anti-Stokes Raman scattering (CARS), which have become increasingly popular in the last few years.

2.3.1 One-Photon Absorption and Circular Dichroism

As mentioned in Section 2.1, (vertical) excited-state energies and transition moments (and consequently the oscillator strengths) entering Eqs. 2.34 and/or 2.35 can be obtained from the poles and residues of the (infinite-lifetime) linear response function; see Eqs. 2.19 and 2.20. For singlet excitations, one usually computes poles and residues of the electric dipole polarizability, that is, the linear response function of the electric dipole operator. For triplet excitations (out of a singlet ground state), an explicitly spin-dependent operator must be used, for instance the spin–orbit operator [139]. If both the poles and the residues have been determined, the absorption spectrum can be simulated by lineshape convolution of each peak. Implementations of UV spectra simulations based on this approach at various levels of theory, from (time-dependent) Hartree–Fock (HF) to multiconfigurational self-consistent field to coupled-cluster [including the equation-of-motion (EOM) approach] to (time-dependent) density functional theory (in the so-called adiabatic approximation) can be found in many quantum chemistry packages such as DALTON [140], TURBOMOLE [141, 142], ADF [143–145], GAUSSIAN [146], MOLCAS [147], and CADPAC [148].

Alternatively, one can generate the absorption spectrum using complex polarization propagator/damped response approaches. Implementations of the CPP/DRT exist at the HF and TD-DFT levels of theory [35, 43].

The response theory approach in general only addresses purely electronic excitations and (strongly) dipole-allowed transitions. To account for the vibrational structure (vibrational progression) of the electronic spectrum obtained from response theory, one often invokes the Frank–Condon (FC) [149–151] and Herzberg–Teller (HT) [152] approximations (within the Born–Oppenheimer adiabatic approximation [58]). This subject is treated in detail elsewhere in this book. The FC alone is usually employed to interpret strongly dipole-allowed electronic bands, whereas the HT approximation is used for weakly dipole-allowed transitions, in which vibronically induced transition moments are also taken into account via the linear term in a Taylor series expansion of the electronic transition dipole moment around the ground-state equilibrium geometry [153]. Thus, the FC approximation "only" requires the overlap between the ground- and excited-state vibrational wavefunctions, whereas for HT it is also necessary to compute the derivatives with respect to the nuclear displacements of the transition strengths. The latter can still be determined by response theory, as described for instance by Coriani et al. [154], by combining the RFT approach with geometry derivative techniques.

Another spectroscopy which can be addressed by (linear) response theory is electronic circular dichroism (ECD). The ECD spectrum contains a wealth of information on the absolute configuration of optically active molecules as well as on their conformation, and it is therefore an important resource not only for the assignment of the absolute configuration of chiral species but also for conformational studies of, for example, biomolecules. For an account of recent progress in the calculation of ECD spectra, with emphasis on RTF approaches, see Pecul and Ruud [155].

The key quantity for the computational simulation of an ECD spectrum is the scalar rotational strength for a transition from the ground state $|0\rangle$ to an excited state $|n\rangle$, which can be evaluated from the residue of a linear response function. In the so-called *length-gauge* (LG) formulation, see also Eqs. 2.46, we have

$$R_{0n}^{\mathrm{LG}} = \sum_{\alpha} \lim_{\omega \to \omega_{0n}} (\omega - \omega_{0n}) \langle\langle \mu_\alpha ; m_\alpha \rangle\rangle_\omega \tag{2.136}$$

whereas in the *velocity-gauge* (VG) formalism [156] we have

$$R_{on}^{\mathrm{VG}} = \frac{1}{\omega_{0n}} \sum_{\alpha} \lim_{\omega \to \omega_{0n}} (\omega - \omega_{0n}) \langle\langle \mu_\alpha^P ; m_\alpha \rangle\rangle_\omega \tag{2.137}$$

The linear response function of relevance is here the mixed electric dipole–magnetic dipole polarizability tensor $G'_{\alpha\beta}$, the trace of which is proportional to the specific optical rotation [155]. For an exact wavefunction in the limit of a complete basis set, the length and velocity expressions are equivalent. For approximate methods in the length-gauge form, special care is required to ensure gauge-origin independent results. For variational methods, the most successful approach is based on the use of LAOs [33, 34], as proposed by Bak et al. in the 1990s [157]. For nonvariational methods such as the coupled-cluster approximation, the use of LAOs is not sufficient to obtain gauge invariance even in the limit of a complete basis set, and the velocity and length formulations are not equivalent. Pedersen, Koch, and co-workers proposed the use of a coupled-cluster Lagrangian including orbital rotations explicitly [158, 159]. When employed with relatively small basis sets, the VG formalism leads to results which can be quite unreliable, as it is well known that convergence with the extension and quality of the basis set is faster in the length than in the velocity formalism.

Most of the calculations of ECD spectra are nowadays carried out employing TD-DFT [7, 8], which allows for studies of rather large molecules, unaffordable by more sophisticated ab initio methods. TD-DFT implementations of ECD spectra are available in most of the commonly used molecular structure and property codes, such as DALTON [140], GAUSSIAN [146], and ADF [143–145]. In most cases, both the LG and VG formulations of ECD are implemented. Nevertheless, in the LG, only DALTON is able to carry out simulations of ECD spectra employing LAOs [160]. A wide range of exchange–correlation (XC) functionals is available. We just mention here the good performance observed for the Coulomb attenuated method B3LYP (CAM-B3LYP) spectra [161–163] in the calculation of ECD spectra [164]. The CPP approach has been applied quite recently to the calculation of ECD spectra [39] within a TD-DFT framework, allowing for a direct determination of the circular dichroism spectrum.

Solvent effects have been shown to be rather important, and a polarizable continuum model (PCM) [165–167] implementation has been reported [168]; see also Rizzo et al. [120]. Kongsted and co-workers have implemented a technique coupling coupled cluster and a dielectric continuum model for the calculation of rotatory strength tensors [169, 170]. Cavity field effects on ECD spectra have been analyzed very recently, again within PCM [171]. Conformational flexibility is also quite important, as discussed elsewhere [120, 122, 172, 173]. As far as the effect of molecular vibrations is concerned, there is an increasing amount of interest in the subject, and we leave full details to other chapters of this book. We however briefly mention that studies of vibronic effects on the ECD spectra at the FC level [174] or including HT effects [175, 176] have been carried out rather recently. Also, the linear coupling model approach (LCM) approach [177–179] and the adiabatic Franck–Condon or adiabatic full harmonic (AFH) model approach have been applied to the analysis of the vibrational structure of the ECD spectrum of gaseous *R*-(+)3-methyl-cyclopentanone [180, 181].

2.3.2 Response Theory Formulation of MCD

MCD spectroscopy [68, 131, 132, 182] is widely used in inorganic chemistry—often focusing on elucidating the electronic structure of porphyrins and phthalocyanines [183, 184]—and in biological chemistry [185–188] for probing the electronic structure of metalloenzyme active sites [185].

After a few decades of relatively scarce attention, the computational determination of MCD parameters and spectra is experiencing a renewed interest, both from the methodological and the applicational point of view, with a noticeable increase in the number of publications on this subject during the last five years [40, 44, 46, 189–202]. Early methods for calculating the MCD terms, and the $\mathcal{B}$ term in particular, have in general been based on SOS procedures [203–209], often at a semiempirical level. An SOS method has also been proposed in recent years by Seth et al. [196] for the $\mathcal{B}$ term within TD-DFT in a molecular orbital–based formulation. A SOS method is however intrinsically hampered by the errors due to truncations in the number of excited states that are considered.

Let us consider first the $\mathcal{B}$ term; see also Eq. 2.118. The possibility to obtain the $\mathcal{B}$ term of MCD from response theory was already mentioned by Olsen and Jørgensen [4], but it was only in 1998 [189] that the response function formalism was actually applied to compute the $\mathcal{B}$ term of ethylene and *para*-benzoquinone from the first residue of the frequency-dependent QRF at the Hartree–Fock and MCSCF levels of theory,

$$\mathcal{B}_{nj} \equiv \mathcal{B}(n \rightarrow j) = -i\epsilon_{\alpha\beta\gamma} \lim_{\omega_b \rightarrow 0} \left(\lim_{\omega_c \rightarrow \omega_{jn}} (\omega_c - \omega_{jn}) \langle\langle \mu_\beta; m_\gamma, \mu_\alpha \rangle\rangle_{\omega_b,\omega_c} \right) \tag{2.138}$$

Coriani et al. [189] showed that the Hartree–Fock method may not be able to reproduce the correct sign for the $\mathcal{B}$ term, and hence electron correlation effects can be essential in order to correctly determine the $\mathcal{B}$ term. The inability of Hartree–Fock to reproduce the sign of the $\mathcal{B}$ term was confirmed by Solheim et al. [193], where the response method was extended to time-dependent density functional theory and to the PCM to account for the effects of solvation.

In 2000, a slightly different strategy, still rooted in RFT, was presented for calculating the $\mathcal{B}$ term at the coupled-cluster singles and doubles (CCSD) level, as the magnetic field derivative of the one-photon dipole transition strength [190],

$$\mathcal{B}(n \to j) = \frac{1}{2}\epsilon_{\alpha\beta\gamma}\left(\frac{d\Im\{\langle n|\mu_\alpha|j\rangle\langle j|\mu_\beta|n\rangle\}}{dB_\gamma}\right)_{\mathbf{B}=0} \equiv \frac{1}{2}\epsilon_{\alpha\beta\gamma}\Im\left(\frac{dS_{nj}^{\alpha\beta}}{dB_\gamma}\right)_{\mathbf{B}=0} \tag{2.139}$$

where B_γ is a Cartesian component of the magnetic field (induction) **B**. This formulation allowed for an easy inclusion of the magnetic field dependence in the orbitals when using LAOs [33, 190] and thus to obtain a gauge-origin independent expression for the $\mathcal{B}$ term even for the nonvariational CCSD approximation. The use of LAOs to remove the gauge-origin dependence in the calculation of the $\mathcal{B}$ term had been originally proposed in 1972 by Seaman and Linderberg and implemented within a finite-perturbation (FP) method [210]. The CCSD approach of Coriani et al. [190] was recently applied to a number of molecules [192] and gave $\mathcal{B}$ terms of high quality, hampered however by the computational cost of the CCSD approximation and thus limited to molecules of the size of pyrimidine and phosphabenzene. The CCSD study of Kjærgaard et al. [192] also showed the difficulties that are encountered in benchmarking the computational results against experiment due to the cancellation of positive and negative contributions.

One way to efficiently include electron correlation effects for large molecules is via DFT, which is known to give reasonable accuracy at a low computational cost [211–213], and linear scaling implementations make Kohn–Sham DFT (and Hartree-Fock) applicable to molecules consisting of more than 1000 atoms [29, 214]. The renewed interest in MCD goes to some extent along with the development in recent years of DFT-based procedures and their applicability to large systems of interest in MCD [e.g., 44, 46, 193–199, 201, 202]. Seth et al. [198] apply a magnetically-perturbed TD-DFT method [215] to calculate the $\mathcal{B}$ term of MCD and describe both a SOS approach and a direct approach. The direct approach is similar to the RFT approach presented by Coriani et al. [189] for the Hartree–Fock and MCSCF approximations.

Kjærgaard and co-workers [202], presented a gauge origin-independent formulation of both the Verdet constant of MOR and the $\mathcal{B}$ term of MCD within the linear scaling framework of Hartree–Fock/Kohn–Sham–DFT response theory of Coriani et al. [29]. Their approach is based on a Lagrangian technique where each property is expressed as a total derivative—with respect to the strength of the external magnetic field—of an appropriate LRF whose magnetic field dependence is parameterized by means of LAOs. By choosing an AO-based expression for lower-order properties as a starting point for the derivation [28, 29], the authors obtained working equations that will lead to linear scaling for sufficiently sparse matrices. When LAOs are not used, they recover the standard expression for the QRF $\langle\langle\mu_\alpha; \mu_\beta, m_\gamma\rangle\rangle_{\omega,0}$ and its residues—that is, the standard quadratic response expressions for the Verdet constant and the $\mathcal{B}$ term of MCD [189]. Note that when evaluating the $\mathcal{B}$ term they also ensured that any unphysical divergencies related to singularities in the response equations are projected out. Benchmark calculations against CCSD results [190, 192] proved to be of particular importance in order to assess the reliability of these calculations in the assignment of the excited states for large molecules.

Turning our attention to the $\mathcal{A}$ term (Eq. 2.117), we note that although it is possible to rewrite this expression in terms of a double residue of the QRF

$$\begin{aligned}\mathcal{A}_{nj} &\equiv \mathcal{A}(n \rightarrow j) \propto \epsilon_{\alpha\beta\gamma} \lim_{\omega_a \rightarrow -\omega_{jn}} (\omega_a + \omega_{jn}) \left[\lim_{\omega_c \rightarrow \omega_{jn}} (\omega_c - \omega_{jn}) \langle\langle \mu_\alpha; m_\beta, \mu_\gamma \rangle\rangle \omega_b, \omega_c \right] \\ &= \epsilon_{\alpha\beta\gamma} \langle n|\mu_\alpha|j_\sigma\rangle \langle j_\sigma|\overline{m_\beta}|j_\sigma\rangle \langle j_\sigma|\mu_\gamma|n\rangle \equiv \epsilon_{\alpha\beta\gamma} \langle n|\mu_\alpha|j_\sigma\rangle \left\{ \langle j_\sigma|m_\beta|j_\sigma\rangle - \langle n|m_\beta|n\rangle \right\} \langle j_\sigma|\mu_\gamma|n\rangle\end{aligned} \tag{2.140}$$

most calculations published so far for the $\mathcal{A}$ term have used SOS-based approaches [216, 217]. A similar approach could possibly also be used for the $\mathcal{C}$ term (Eq. 2.119), assuming that the response approach at hand is capable of dealing with degenerate ground states. We shall not concern us with calculations of the $\mathcal{C}$ term in this chapter, limiting ourselves to refer to the work of Ganyushin and Neese [200], Bolvin [218], and Ziegler and co-workers [195, 219]. Seth and co-workers [194] showed that the $\mathcal{A}$ term alternatively can be derived as the magnetic field derivative of the excitation energy,

$$\mathcal{A}(n \rightarrow j) = -\frac{1}{2} \epsilon_{\alpha\beta\gamma} \sum_\sigma \left(\frac{\partial \omega_{jn}}{\partial B_\gamma} \right)_{\mathbf{B}=0} \Im \left\{ \langle n|\mu_\alpha|j_\sigma\rangle \langle j_\sigma|\mu_\beta|n\rangle \right\} \tag{2.141}$$

where ω_{jn} is the excitation frequency from the ground state $|n\rangle$ to the degenerate excited state $|j_\sigma\rangle$, with σ running over the number of degenerate states.

MCD spectra can also be computed using damped approaches such as the CPP approach of Solheim et al. [40], the damped RFT approach of Kristensen et al. [43], or the ad hoc approach based on the imaginary part of the Verdet constant using damped time-dependent density functional theory of Krykunov et al. [44]. Following Solheim et al. [40, 46], the MCD spectrum is computed directly as

$$\text{MCD}(\omega) = \chi \hbar \omega \frac{1}{\mu_B} \epsilon_{\alpha\beta\gamma} \Re \langle\langle \mu_\alpha; \mu_\beta, m_\gamma \rangle\rangle_{\omega,0} \tag{2.142}$$

without the need to use an explicit separation into $\mathcal{A}$ and $\mathcal{B}$ terms χ is here a factor that depends on the quantities actually measured in experiment (see the book by Mason [131]) and μ is the Bohr magneton. Actually, comparing their results for three metal porphyrins with those obtained by Peralta, Seth, and Ziegler [199] via separate calculations of the $\mathcal{A}$ and $\mathcal{B}$ terms and lineshape convolution, Solheim et al. [46] question whether it is convenient and safe to apply such a separation. Even if the molecules under investigation have similar overall spectra (the only difference is the metal ion), the contributions from the $\mathcal{A}$ term obtained for each of them by Peralta et al. [199] are qualitatively different. However, this difference in the contributions of the $\mathcal{A}$ and $\mathcal{B}$ terms appears primarily to be due to a mixing of metal orbitals into the orbitals involved in the excitation process for nickel porphyrin for the SAOP functional used by Peralta and Seth [199], a metal orbital mixing which is neither observed in the Zn- nor Mg-porphyrin nor the calculations presented by Solheim et al. [46], where a different functional was used.

2.3.3 Two-Photon Absorption and Circular Dichroism

TPA, and more generally multiphoton absorption, exhibits characteristics such as 3D spatial selectivity and higher resolution, reduced energy usage for increased wavelengths—which ensures deeper penetration into matter and tissues, reduced scattering loss, photodamage, and background fluorescence—thus providing major advantages with respect to conventional one-photon spectroscopy [90, 91]. These advantages are of particular importance in areas such as optoelectronics, microscopy, optical data storage and power limiting, upconversion lasing, imaging of tissues, photodynamic therapy, and more. The development of computational approaches and strategies for the calculation of two-photon absorption has in recent years given valuable support to the development of protocols for the design of molecules with large nonlinear responses. These approaches range from semiempirical methods able to yield qualitative information, through effective few-state or exciton models, to highly accurate, wavefunction-based ab initio correlated methods, applicable to molecules from small to medium size, and therefore not easily extendible to large systems of biological interest or to nanostructures. An account of the progress made in the field can be found in the excellent reviews by Terenziani and co-workers [90] and by Prasad and his group [91]. Among the wide selection of computational tools, an important role is taken by RFT techniques. Time-dependent density functional theory (TD-DFT) [7, 8] is particularly widely employed for calculations of TPA spectra of molecules of ever-increasing size.

As noted in Section 2.1, the second-rank two-photon tensor Eq. 2.78, involving a sum over the manifold of excited electronic states, can be obtained from the single residue of the appropriate QRF,

$$\langle\langle \hat{\mu}_\alpha; \hat{\mu}_\beta, V^{\omega_j} \rangle\rangle_{\omega_1,\omega_2} \Rightarrow \mathcal{S}^{0j}_{\alpha\beta}(\omega_1, \omega_2) \tag{2.143}$$

where V^{ω_j} is an arbitrary operator, corresponding to the excitation vector to state j. This approach yields the most popular route, based on analytic RFT, to the calculation of TPA spectra. An alternate technique rarely employed in calculations of TPA spectra calls for the direct calculation of the squared two-photon transition amplitude that enters the expression for the two-photon transition rate (Eq. 2.74) from the residue of the cubic response function [16]. In cases where the electronic spectrum is dominated by one or more low-lying transitions, the use of few-, often two-state models, amounting to restricting the sum-over-states in Eq. 2.22 to the few states that yield major contributions, has been quite widespread [220, 221]. This approach also resorts to the techniques of RFT, since the transition matrix elements (between ground and excited states and between excited states) appearing in the numerator of Eq. 2.22 and the excitation energies appearing in the denominator are obtained from residues and poles, respectively, of the appropriate response functions; see Section 2.1 and Eq. 2.23. Incidentally, in what is the first use of response techniques to the calculation of TPA in atoms and molecules, dating to the mid-1980s [222–225] and preceding the landmark papers by Jørgensen, Olsen, and co-workers which made the calculation of frequency-dependent quadratic response functions and their residues possible [4, 20, 25], equation-of-motion techniques [226, 227] were employed to approximate the excited state–excited state transition electric dipole matrix elements by employing the results of linear response calculations (excitation vectors and one-electron transition matrix elements).

The vast majority of the literature on RFT studies of TPA relies on TD-DFT [25, 228, 229] and in particular on the general implementation in the DALTON program [140]; see Cronstrand et al. [221] for a review of this field. The effect of electron correlation [230] and conjugation length in linear and multibranched polymeric chains [231, 232], proper choice of exchange–correlation functional [229], the effect of solvation [122, 233–235], molecular vibrations [177–179], and aggregation in organic nonlinear optical (NLO) chromophores [236, 237] have been studied and find increasing interest in the literature; see again Terenziani et al. [90] for a recent summary and for pertinent references. The computational requirements for a meaningful application of RFT to the calculation of TPA spectra are comparable to those of the response functions supplying the transition amplitudes. Therefore, roughly speaking, the basis set employed in calculations of two-photon amplitudes must have the same characteristics as the basis sets used in calculations of first electric dipole hyperpolarizabilities, that is, a suitable extension and an appropriate supply of diffuse and high-cardinal-number basis functions. For the choice of basis set, as for the choice of the appropriate exchange–correlation functional in TD-DFT calculations, the nature of the excited state, its localized (valence) or diffuse (Rydberg) character, must in addition be taken into account. Our experience, based on a number of studies of TPA spectra of organic chromophores of different size and complexity, is that a popular functional such as the B3LYP [238–240] functional performs quite well for valence-excited states in most cases, whereas a functional such as CAM-B3LYP [161–163], developed to account for long-range charge transfer excitations, works better for charge transfer and Rydberg-type excited states. Condensed-phase effects in studies of TPA employing a RFT approach resort mostly to the PCM [165–167], in particular in its integral-equation formulation (IEF) [241, 242] for the description of the interactions occurring between solute and solvent. PCM has proven in some cases to be particularly effective in reproducing some solvatochromic effects in TPA spectra, otherwise poorly described in the isolated molecule model [122, 233–235]. The effect of molecular vibrations and the coupling of electronic and vibrational models have been described, in connection with a response theory approach, in the linear coupling model approach [177–179]. Very recently, an adiabatic Franck–Condon, or adiabatic full harmonic model approach, which permits quite successful analysis of the vibrational structure of the TPA spectra, has been presented [121].

For relatively small molecules, multiphoton transition moments are in principle available with high accuracy resorting to a coupled-cluster response theory approach employing variational transition moment functionals [16, 243]. Applications are known for atoms [243] and small molecules such as formaldehyde, diacetylene, and water [163, 244]. A coupled-cluster/molecular mechanics response theory approach was exploited to study the solvent effect on the two-photon absorption of microsolvated formaldehyde and liquid water [245]. Relativistic effects in TPA of noble gases have been studied due to the implementation of the single residue of the quadratic response function at the four-component Hartree–Fock approximation [246].

The three-photon transition amplitude has been shown to be obtainable from the single residue of the appropriate CRF [27], although the technique has been employed only rarely. Lin et al. [247] analyzed for instance, solvent effects (accounted for by PCM) on the three-photon absorption spectrum of a symmetric charge transfer molecule using TD-DFT.

The computational route to the calculation of two-photon circular dichroism spectra was opened only recently [115], when it was recognized that the molecular parameters entering the TPCD rotatory strength equation 2.83 can be obtained, as for the two-photon transition amplitude in Eq. 2.78, as single residues of appropriate quadratic response functions [82]:

$$\langle\langle \hat{\mu}^{\mathrm{p}}_{\alpha}; \hat{\mu}^{\mathrm{p}}_{\beta}, V^{\omega_n} \rangle\rangle_{\omega_1,\omega_2} \Rightarrow \mathcal{P}^{\mathrm{p},0n}_{\alpha\beta}(\omega_1,\omega_2) \tag{2.144}$$

$$\langle\langle \hat{\mu}^{\mathrm{p}}_{\alpha}; \hat{m}_{\beta}, V^{\omega_n} \rangle\rangle_{\omega_1,\omega_2} \Rightarrow \mathcal{M}^{\mathrm{p},0n}_{\alpha\beta}(\omega_1,\omega_2) \tag{2.145}$$

$$\epsilon_{\beta\rho\sigma}\langle\langle \hat{T}^{+}_{\alpha\rho}; \hat{\mu}^{\mathrm{p}}_{\sigma}, V^{\omega_n} \rangle\rangle_{\omega_1,\omega_2} \Rightarrow \mathcal{T}^{+,0n}_{\alpha\beta}(\omega_1,\omega_2) \tag{2.146}$$

$$\langle\langle \hat{\mu}_{\alpha}; \hat{m}_{\beta}, V^{\omega_n} \rangle\rangle_{\omega_1,\omega_2} \Rightarrow \mathcal{M}^{0n}_{\alpha\beta}(\omega_1,\omega_2) \tag{2.147}$$

$$\epsilon_{\beta\rho\sigma}\langle\langle \hat{q}_{\alpha\rho}; \hat{\mu}_{\sigma}, V^{\omega_n} \rangle\rangle_{\omega_1,\omega_2} \Rightarrow \mathcal{Q}^{0n}_{\alpha\beta}(\omega_1,\omega_2) \tag{2.148}$$

As such, they can be obtained by a rather straightforward generalization of the standard two-photon tensor, and they are available in the distribution of the DALTON suite of programs. The calculation of the quantity determining the TPCD observable involves three (in both Tinoco's origin-independent formulation and the origin-dependent ω_0 dipole length formulation) or four (in the alternative mixed dipole/velocity origin independent formulations discussed elsewhere [82]) tensors, compared to the single two-photon tensor needed in studies of TPA. The computational cost of a TPCD calculation is therefore roughly three to four times higher than a regular TPA analysis. Note also that an origin-dependent calculation carried out along the lines of the ω_0 formulation has the advantage of yielding as a byproduct the two-photon transition amplitudes, allowing for the simultaneous determination of both TPCD and TPA spectra.

In the computational studies of two-photon circular dichroism that have been published to date [115, 117, 119–123] TD-DFT has been used, the size of the systems preventing more refined ab initio structure models to be used, in part due to the lack of symmetry in systems for which TPCD can be observed. The choice of the XC functional has been limited to B3LYP and CAM-B3LYP, and in the only instance to date where a comparison with experiment was made possible, the study on BINOL [123], which extended also to TPCLD [126], see Eq. 2.107, it was found that the former reproduces better the position of the peaks of the TPCD spectrum with respect to the latter. An analysis of the signatures of the excited states, in terms of molecular orbitals and single excitations, appears to confirm the localized character of the excitations in this case.

Basis sets of triple-zeta (when affordable) or double-zeta quality have been used for TPCD studies. It is well known that the convergence of the property with the extension of the basis set, in terms of both cardinal number and diffuse functions, is more difficult when employing the velocity form of the transition multipole operators, as is done in Tinoco's formulation of TPCD. This aspect has been treated by Rizzo et al. [82], where a detailed analysis of the dependence of the spectroscopic signal on the quality of the basis set was carried out on a chiral frozen structure of hydrogen peroxide using a hierarchy of augmented correlation-consistent aug-cc-pVXZ (X=D, T,Q,5) basis sets [248]. The conclusions of the study were that for this small system, basis, sets of at least triple-zeta quality are needed to reach a good degree of convergence for the TPA and TPCD spectra. Nevertheless, in other studies on extended systems, such as amino acids [117, 122], helicenes [119], or organic chromophores [120, 121, 123], it was found that reasonably converged results can also be obtained using smaller basis sets (e.g., of double-zeta quality) relying on the cooperative effects observed in many-center expansions, where basis functions on different centers help overcome the shortcomings of the reduced set of functions placed on each individual center.

As for TPA, condensed-phase effects on TPCD spectra are in some cases important, and they are currently accounted for by resorting to PCM [120, 122, 123]. In the case of BINOL, where theory could be compared to experiment, including the effect of solvation through PCM proves to be essential in order to recover the features observed in the range of wavelengths where the strongest peaks appear [123, 126]. Recently, also vibronic effects on the TPCD spectra have been analyzed [121], employing both the LCM and the AFH approximations. It was shown that the Hertzberg–Teller couplings might be strong enough to heavily influence the shape and strength of the TPCD signal, to the point of inducing in principle a change of sign within the vibronic envelope of a peak assigned to a specific excited electronic state.

2.3.4 Examples of Chiroptical and Nonlinear Vibrational Spectroscopies

Even if response theory is conventionally formulated for electronic properties, there are several spectroscopies whose molecular theory involves the nuclei that can be addressed by combining response theory approaches with derivative techniques as discussed in Section 2.1.1. We consider here, as examples, VCD, ROA, and CARS.

2.3.4.1 *Vibrational Circular Dichroism* Vibrational circular dichroism is the vibrational analogue of ECD, corresponding to the differential absorption of left- and right-circularly polarized light in the infrared wavelength region, thus corresponding to excitations in the vibrational manifold rather than between different electronic states. Phenomenologically, we can therefore write the rotational strength for a transition between two different vibrational states i and j for normal mode k as

$$R^{k;ij}_{\alpha\beta} = \Im\left(\langle \nu^k_i|\mu_\alpha|\nu^k_j\rangle\langle \nu^k_i|m_\beta|\nu^k_j\rangle\right) \tag{2.149}$$

Using harmonic oscillator wavefunctions as basis functions, truncating the Taylor expansion of the geometry dependence of the electric and magnetic dipole moments at linear order as done in Eq. 2.28 (in some contexts referred to as the Placzek approximation [249]) and following the discussion in Section 2.1.3 for the two vibrational transition moments, the rotational strength can be shown to be

$$R^k_{\alpha\beta} = \Im\left(\langle \nu_0\left|\left.\frac{\partial\mu_\alpha}{\partial\mathbf{Q}_k}\right|_{\mathbf{Q}_e}\mathbf{Q}_k\right|\nu_k\rangle\langle\nu_k\left|\left.\frac{\partial m_\beta}{\partial\mathbf{Q}_k}\right|_{\mathbf{Q}_e}\mathbf{Q}_k\right|\nu_0\rangle\right) = \frac{\hbar^2}{2}\left.\frac{\partial\mu_\alpha}{\partial\mathbf{Q}_k}\right|_{\mathbf{Q}_e}\left.\frac{\partial m_\beta}{\partial\mathbf{Q}_k}\right|_{\mathbf{Q}_e} \tag{2.150}$$

where we have assumed that only the vibrational ground state is sufficiently populated at normal experimental conditions. Within the double-harmonic approximation, only transitions to the first vibrationally excited state are possible for each vibrational normal mode.

Whereas the dipole moment gradient (the *atomic polar tensor*) is well defined and it is the same quantity that determines conventional infrared absorption spectra (see the chapter on vibrational spectroscopy), the gradient of the magnetic dipole moment is zero within the Born–Oppenheimer approximation. This is due to the fact that the magnetic dipole moment for a closed-shell molecule is quenched (since it corresponds to an expectation value of an imaginary operator), making the rotational strength in Eq. 2.150 zero.

Stephens showed that, by including also nonadiabatic effects, a vibrationally induced magnetic moment can arise that will lead to a nonvanishing rotational strength [250]. Vibrational circular dichroism is thus an example of a nonadiabatic effect, in a similar manner as properties such as the nuclear spin–rotation constant or the rotational g tensor [251, 252].

Before proceeding with some more details regarding the evaluation of the magnetic dipole moment gradient (the *atomic axial tensor*, M_A) in the context of a nonadiabatic wavefunction, let us first note that it is from a computational point of view advantageous to derive equations with respect to Cartesian displacements of a nucleus A rather than with respect to the normal modes. These two different representations of the nuclear motion are related by a linear transformation

$$\Delta\mathbf{R}_{A,\alpha} = \sum_k S_{A,\alpha;k}\mathbf{Q}_k \tag{2.151}$$

where $S_{A,\alpha k}$ is the transformation matrix from normal coordinate Q_k to Cartesian displacement along the α direction for nucleus A.

Following Stephens, we can write the vibronic wavefunction as a combination of a pure Born–Oppenheimer zero-order wavefunction, and the first-order perturbed wavefunction due to the nonadiabatic coupling matrix elements as

$$\psi_{Ii}^{\text{pert}} = \psi_{Ii}^{(0)} + \sum_{(J,j)\neq(0,0)} \frac{\langle \psi_{Ii}^{(0)} | \mathcal{H}^{\text{el}} + \mathcal{T}_N | \psi_{Jj}^{(0)} \rangle}{E_{Ii} - E_{Jj}} \left| \psi_{Jj}^{(0)} \right\rangle \tag{2.152}$$

where $\psi_{Ii}^{(0)}$ denotes a vibronic wavefunction corresponding to the ith vibrational level of the Ith electronic state, $\mathcal{H}^{\text{el}}$ the electronic Hamiltonian and $\mathcal{T}_N$ the operator for the kinetic energy of the nuclei. We do not go into the details of the derivation here, referring the interested readed instead to the original work of Stephens [250], but note that the atomic axial tensors as evaluated using this nonadiabatic wavefunction can, to leading order in the geometry dependence, be expressed as the overlap of the magnetic field–perturbed and geometry-perturbed wavefunctions

$$M_{A,\alpha\beta} = I_{A,\alpha\beta} + J_{A,\alpha\beta} \tag{2.153}$$

$$I_{A,\alpha\beta} = \left\langle \frac{\partial \psi_0^{\text{el}}}{\partial R_{A,\alpha}} \middle| \frac{\partial \psi_0^{\text{el}}}{\partial B_\beta} \right\rangle \Bigg|_{\mathbf{R}=\mathbf{R}_0,\mathbf{B}=0} \tag{2.154}$$

$$J_{A,\alpha\beta} = \frac{i}{4} \epsilon_{\alpha\beta\gamma} Z_A R_{A\gamma} \tag{2.155}$$

We note that the atomic axial tensor cannot be expressed directly as an energy derivative since it involves the overlap over two perturbed wavefunctions. However, the perturbed wavefunctions can be obtained as the dot product of the solution vectors $\mathbf{N}^{\mathbf{B}}$ defined by the linear response function in Eq. 2.26 for a magnetic field and nuclear displacement perturbation, respectively. It is important here to note that some care needs to be exercised in the way the orbital rotations, in the language of Section 2.1.1, are defined [253, 254].

In contrast, the *atomic polar tensor* (the dipole gradient) can be expressed as an energy derivative, being the second derivative of the electronic energy, once with respect to an external electric field and once with respect to a nuclear distortion. We will not adress any further here the calculation of the atomic axial and polar tensors from ab initio wavefunctions, referring instead to the original works describing these implementations at the Hartree–Fock [253, 255–257], DFT [258, 259], and multi-configurational SCF [253] levels of theory.

The development of theoretical methods for the calculation of VCD spectra has been reviewed on several occasions [260, 261]. We here only note that two major breakthroughs were necessary for making the calculations reliable and routine: (1) the introduction of London atomic orbitals for ensuring that the results were independent of the choice of the global gauge origin and for ensuring fast basis set convergence [253] and (2) the use of density functional theory for calculating the force field and rotational strengths [258]. Stephens and co-workers have in several studies demonstrated the powers of this computational approach [261] and how the combined use of experimental VCD spectra with ab initio calculations is a very efficient and reliable approach for determining the absolute configurations of chiral molecules [262, 263]. It is particularly noteworthy that VCD has also been shown to lead to a reassessment of earlier determinations of the absolute configuration of chiral molecules based on ECD [264] as well as X-ray crystallography [265]. Density functional theory has been the dominating correlated approach for calculating VCD spectra, and the only other fully correlated approach we are aware of for calculating the atomic axial and polar tensors is the multiconfigurational self-consistent field approach of Bak et al. [253].

The cost of a VCD calculation is of the same order of magnitude as that of determining the force field itself, and indeed there is very little computational overhead relative to a force field calculation (three response equations due to the external magnetic field induction need to be solved in addition to the $3N$ equations that need to be solved for the nuclear distortions). However, basis set requirements for the atomic axial and polar tensors may differ from that of the force field, potentially leading to a higher computational cost than the force field alone. Stephens and co-workers have demonstrated that basis sets of TZ2P or cc-pVTZ quality in combination with hybrid density functionals such as B3LYP and B3PW91 give results of an accuracy sufficiently high to unambigously assign experimental spectra, both for the force field and the rotational strengths [261].

Extensions of the basic VCD calculations to also decribe solvent effects have been presented by Cappelli et al. [266], including also nonequilibrium effects in the solvent due to the vibrational excitations in the solute. Although solvent effects have been shown to be nonnegligible, these methods have not been much utilized, largely due to the fact that only very rarely does the solvent as modeled by a dielectric continuum model lead to a reversal of the sign of the VCD rotational strengths. The use of the continuum model is therefore in general not important in terms of determining the absolute configuration of chiral molecules in solution. There are, however, a couple of noteworthy exceptions. For conformationally flexible molecules, the population fractions of different conformers may depend strongly on the solvent, and for such systems, proper account of solvation effects is mandatory [267, 268]. Another area where more elaborate models may be required is when specific interactions such as hydrogen bonding are important for the calculated spectra [269, 270].

2.3.4.2 Raman Optical Activity A vibrational chiroptical spectroscopy complementary to VCD, in particular in terms of the frequency range covered, is ROA, the chiroptical analogue of Raman spectroscopy in the same manner as VCD is the chiroptical version of infrared spectroscopy. In ROA, the key quantity is therefore the differential scattering of right- and left-circularly polarized light. For the most common setup of backscattering, the ROA circular intensity differences (CIDs) and the Raman intensity are given by [42, 271]

$$I^R_{\mathrm{SCP}}(\xi)-I^L_{\mathrm{SCP}}(\xi) = \frac{1}{45c_0}\left(\frac{\omega^2\mu_o}{4\pi}\right)^2 \frac{\hbar}{2\omega_i}\left[24\beta^2_{G_i} + 8\beta^2_{A_i}\right] \tag{2.156}$$

$$I^{R}_{\mathrm{SCP}}(\xi) + I^{L}_{\mathrm{SCP}}(\xi) = \frac{1}{90}\left(\frac{\omega^2 \mu_o}{4\pi}\right)^2 \frac{\hbar}{2\omega_i}\left[90a_i^2 + 14\beta_i^2\right] \tag{2.157}$$

where we have introduced a number of Raman and ROA invariants, defined respectively as

$$a_i^2 = \frac{1}{9}\left(\frac{\partial \alpha_{\alpha\alpha}}{\partial \mathbf{Q}_k}\right)\bigg|_{\mathbf{Q}_e}\left(\frac{\partial \alpha_{\beta\beta}}{\partial \mathbf{Q}_k}\right)\bigg|_{\mathbf{Q}_e} \tag{2.158}$$

$$\beta_i^2 = \frac{1}{2}\left[3\left(\frac{\partial \alpha_{\alpha\beta}}{\partial \mathbf{Q}_k}\right)\bigg|_{\mathbf{Q}_e}\left(\frac{\partial \alpha_{\alpha\beta}}{\partial \mathbf{Q}_k}\right)\bigg|_{\mathbf{Q}_e} - \left(\frac{\partial \alpha_{\alpha\alpha}}{\partial \mathbf{Q}_k}\right)\bigg|_{\mathbf{Q}_e}\left(\frac{\partial \alpha_{\beta\beta}}{\partial \mathbf{Q}_k}\right)\bigg|_{\mathbf{Q}_e}\right] \tag{2.159}$$

$$\beta_{G_i}^2 = \frac{1}{2}\left[3\left(\frac{\partial \alpha_{\alpha\beta}}{\partial \mathbf{Q}_k}\right)\bigg|_{\mathbf{Q}_e}\left(\frac{\partial G'_{\alpha\beta}}{\partial \mathbf{Q}_k}\right)\bigg|_{\mathbf{Q}_e} - \left(\frac{\partial \alpha_{\alpha\alpha}}{\partial \mathbf{Q}_k}\right)\bigg|_{\mathbf{Q}_e} - \left(\frac{\partial G'_{\beta\beta}}{\partial \mathbf{Q}_k}\right)\bigg|_{\mathbf{Q}_e}\right] \tag{2.160}$$

$$\beta_{A_i}^2 = \frac{\omega}{2}\left[\left(\frac{\partial \alpha_{\alpha\beta}}{\partial \mathbf{Q}_k}\right)\bigg|_{\mathbf{Q}_e}\left(\frac{\epsilon_{\alpha\gamma\delta}\partial A_{\gamma,\delta\beta}}{\partial \mathbf{Q}_k}\right)\bigg|_{\mathbf{Q}_e}\right] \tag{2.161}$$

where $\alpha_{\alpha\beta}$, $G'_{\alpha\beta}$ and $A_{\gamma,\delta\beta}$ are, respectively, the electric dipole–electric dipole, electric dipole–magnetic dipole, and electric dipole–electric quadrupole polarizabilities, as discussed in previous sections of this chapter. Note, however, that in our definition of the mixed electric dipole–electric quadrupole polarizability, we here use the traceless form of the operator, defined as

$$\Theta_{\alpha\beta} = \frac{1}{2}\sum_i q_i\left[3r_{i,\alpha}r_{i,\beta} - \delta_{\alpha\beta}r_{i,\gamma}r_{i,\gamma}\right] \tag{2.162}$$

Until recently, ROA CIDs were calculated by finite geometric differences of the (mixed) electric polarizability tensors, making the calculation ROA CIDs computationally expensive. During the last five years, several analytic implementations have been presented [271–273], making the calculation of ROA CIDs much more feasible. Indeed, with the most recent advances in which no response equations are solved for the geometric perturbations [271, 273], the calculation of the ROA CIDs actually becomes less expensive than the calculation of the force field itself, making the calculation of ROA spectra more computationally tractable than the calculation of VCD spectra, because for ROA one may choose different basis sets for the force field and the ROA CIDs, thus reducing the basis set requirements for both properties. As already noted, this is in many cases not possible when calculating the VCD spectra.

An exhaustive account of existing approaches to calculate ROA spectra was recently given by Thorvaldsen and Ruud [271], and we will not discuss any of these approaches in detail here. However, to illustrate the additional complications that arise when perturbation-dependent basis sets are introduced in a quasi-energy formalism, we briefly discuss the quasi-energy expressions for the gradients of the polarizability, mixed electric dipole–magnetic dipole polarizability and mixed electric dipole–electric quadrupole polarizabilities, in order to highlight the differences compared to the case where only one-electron perturbing operators are involved, as was discussed in Section 2.1.1.

Thorvaldsen and Ruud showed that the gradients of the (mixed) electric and magnetic polarizabilities that appear in the ROA CIDs can be written as [271]

$$\begin{aligned} Q^{gf^*f} = \{&\mathrm{Tr}\mathbf{V}^{gf^*}\mathbf{D}^{f} + \mathrm{Tr}\mathbf{V}^{gf}\mathbf{D}^{f^*} + \mathrm{Tr}(\mathbf{h}^{g} + \mathbf{G}^{g}(\mathbf{D}))\mathbf{D}^{f^*f} \\ &+ \mathrm{Tr}\mathbf{G}^{g}(\mathbf{D}^{f^*})\mathbf{D}^{f} - \mathrm{Tr}\mathbf{S}^{g}\mathbf{W}^{f^*f}\}_T \end{aligned} \tag{2.163}$$

$$\begin{aligned} Q^{gf^*q} = \{&\mathrm{Tr}\mathbf{V}^{gf^*}\mathbf{D}^{q} + \mathrm{Tr}\mathbf{V}^{gq}\mathrm{D}^{f^*} + \mathrm{Tr}(\mathbf{h}^{g} + \mathbf{G}^{g}(\mathbf{D}))\mathbf{D}^{f^*q} \\ &+ \mathrm{Tr}\mathbf{G}^{g}(\mathbf{D}^{q})\mathbf{D}^{f^*} - \mathrm{Tr}\mathbf{S}^{g}\mathbf{W}^{f^*q}\}_T \end{aligned} \tag{2.164}$$

$$\begin{aligned} Q^{gf^*b} = \{&\mathrm{Tr}\mathbf{V}^{gf^*}\mathbf{D}^{b} \\ &+ \mathrm{Tr}(\mathbf{h}^{gb} + \mathbf{V}^{gb} - \tfrac{i}{2}\mathbf{T}^{gb} + \mathbf{G}^{gb}(\mathbf{D}))\mathbf{D}^{f^*} \\ &+ \mathrm{Tr}(\mathbf{h}^{g} + \mathbf{V}^{g} - \tfrac{i}{2}\mathbf{T}^{g} + \mathbf{G}^{g}(\mathbf{D}))\mathbf{D}^{f^*b} \\ &+ \mathrm{Tr}\mathbf{G}^{g}(\mathbf{D}^{b})\mathbf{D}^{f^*} - \mathrm{Tr}\mathbf{S}^{gb}\mathbf{W}^{f^*} - \mathrm{Tr}\mathbf{S}^{g}\mathbf{W}^{f^*b}\}_T \end{aligned} \tag{2.165}$$

In these equations, the superscripts f, q, b, and g correspond to derivatives of the quasi-energy equation 2.27 with respect to the perturbation strength of the electric field, the electric field gradient, the magnetic field, and a geometric distortion of a nucleus, respectively. The perturbed densities $\mathbf{D}^f$ are determined by solving linear sets of equations such as those given in Eq. 2.26. In a similar manner, $\mathbf{h}^g$, $\mathbf{G}^g$, and $\mathbf{S}^g$ denote the (geometric) derivatives of the one-electron, the two-electron, and the overlap matrix, respectively. The derivatives of the (frequency-dependent) interaction operator

$$\begin{aligned} V(t) &= -F(t)\mu - Q(t)\Theta - B(t)m \\ &= -[f\exp(-i\omega t) + f^*\exp(i\omega t)]\mu \\ &\quad -[q\exp(-i\omega t) + q^*\exp(i\omega t)]\Theta \\ &\quad -[ib\exp(-i\omega t) + ib^*\exp(i\omega t)]m \end{aligned} \tag{2.166}$$

are defined in a similar manner, and we have also introduced the antisymmetric time derivative overlap matrix

$$T_{\mu\nu} = \langle \chi_\mu | \dot{\chi}_\nu \rangle - \langle \dot{\chi}_\mu | \chi_\nu \rangle \quad \dot{\chi}_\mu = \frac{d}{dt}\chi_\mu \tag{2.167}$$

with χ_μ denoting a spherical or Cartesian Gaussian atomic basis fucntion, and a dot over the matrix being used to indicate a time derivative. Finally, we have introduced the energy–frequency-weighted density matrix

$$\mathbf{W} = \mathbf{D}\tilde{\mathbf{F}}\mathbf{D} + \frac{i}{2}\dot{\mathbf{D}}\mathbf{S}\mathbf{D} - \frac{i}{2}\mathbf{D}\mathbf{S}\dot{\mathbf{D}} \quad \tilde{\mathbf{F}} = \frac{\partial \tilde{E}(\mathbf{D})}{\partial \mathbf{D}^T} = \mathbf{h} + \mathbf{V} - \frac{i}{2}\mathbf{T} + \mathbf{G}(\mathbf{D}) \tag{2.168}$$

The perturbed densities appearing in the equations above can be obtained by solving linear response equations as in Eq. 2.26 [32, 274].

By computing the force fields using basis sets and density functionals suited for these kinds of calculations (in many cases B3LYP/6-311G** may give very accurate force fields), the calculation of the ROA CIDs hinges on the basis set requirements needed for the geometric derivatives of the electronic tensors in Eqs. 2.163–2.165. Hug and co-workers [273, 275] have developed very compact rarified basis sets that have been shown to give very good results at a very low computation [276], making analytic ROA calculations computationally feasible.

An important technique for calculating ROA spectra of large molecules has been the tensor transfer technique of Bouř et al. [277]. In this approach, the exact origin dependence of the different polarizabilities is used to translate the polarizability gradients calculated for a molecular fragment in a given orientation into a larger molecular complex. In this way one can "synthesize" the full ROA spectrum from the polarizability gradients calculated for the different fragements, translated and rotated into the full molecular structure. The method has proven to give very good approximations to the full ROA spectra [278] and has enabled calculations on large molecules using ab initio methodology [279, 280]. It is worth noting that the harmonic force field fulfills similar translational and rotational relations and that the tensor transfer technique thus also can be used to generate accurate force fields for large molecules from high-quality force fields calculated for smaller fragments [279, 281].

Several studies have combined force fields calculated in the presence of a dielectric continuum, most often the PCM [166, 167, 282, 283] or the COSMO model [284] in order to get an approximate treatment of the solvent effects on the ROA spectra. To the best of our knowledge, the only studies in which the PCM has been applied to all quantities determining the ROA spectrum are those of Pecul and co-workers [285, 286].

The solvent most commonly used in ROA experiments is water, for which explicit intermolecular interactions can be expected to be important. These specific solute–solvent interactions cannot be expected to be well modeled by continuum approaches. Jalkanen and co-workers have devoted much attention to the description of specific solute–water interactions in the modeling of ROA spectra, optimizing the structure of the dominating conformation of the solute with the innermost coordination sphere [287–289].

2.3.4.3 Coherent Anti-Stokes–Raman Scattering Ab initio calculations of nonlinear vibrational spectroscopies is a field that is very much in its infancy, and we will therefore not dwell much on this topic here, limiting ourselves to a recent example from our own research group, namely the four-wave mixing processes that lead to CARS [290].

In the CARS process, which formally is a four-wave mixing process in which two of the incoming frequencies are the same, the observed CARS signal is determined by the square of the CARS susceptibility tensor [290]:

$$S_{\text{CARS}}(\omega_1;\omega_2) = |\chi^{(3)}(-\omega_\sigma;\omega_1,-\omega_2,\omega_1)|^2 \tag{2.169}$$

Here $\chi^{(3)}(-\omega_\sigma;\omega_1,\omega_2,\omega_3)$ is the conventional fourth-order susceptibility tensor with $\omega_\sigma = \omega_1 + \omega_2 + \omega_3$. At the microscopic level, this susceptibility tensor is governed by the second hyperpolarizability, which in a sum-over-states formalism can be written as [291]

$$\begin{aligned}\gamma_{\alpha\beta\gamma\delta}(-\omega_\sigma;\omega_1,\omega_2,\omega_3) &= \frac{1}{\hbar^3}\sum \mathcal{P}_{1,2,3}\\ &\times \sum_{nmp}\left[\frac{\langle 0|\mu_\alpha|n\rangle\langle n|\mu_\beta|m\rangle\langle m|\mu_\gamma|p\rangle\langle p|\mu_\delta|0\rangle}{(\omega_{n0}-\omega_\sigma-i3\delta)(\omega_{m0}-\omega_2-\omega_3-i2\delta)(\omega_{p0}-\omega_3-i\delta)}\right.\\ &+\frac{\langle 0|\mu_\beta|n\rangle\langle n|\mu_\alpha|m\rangle\langle m|\mu_\gamma|p\rangle\langle p|\mu_\delta|0\rangle}{(\omega_{n0}+\omega_1+i\delta)(\omega_{m0}-\omega_2-\omega_3-i2\delta)(\omega_{p0}-\omega_3+i\delta)}\\ &+\frac{\langle 0|\mu_\gamma|m\rangle\langle m|\mu_\beta|n\rangle\langle n|\mu_\alpha|p\rangle\langle p|\mu_\delta|0\rangle}{(\omega_{n0}+\omega_1+\omega_2+i2\delta)(\omega_{m0}+\omega_2+i\delta)(\omega_{p0}-\omega_3+i\delta)}\\ &\left.+\frac{\langle 0|\mu_\delta|p\rangle\langle p|\mu_\gamma|m\rangle\langle m|\mu_\beta|n\rangle\langle n|\mu_\alpha|0\rangle}{(\omega_{n0}+\omega_\sigma+i3\delta)(\omega_{m0}+\omega_2+\omega_3+i2\delta)(\omega_{p0}+\omega_3+i\delta)}\right]\end{aligned} \tag{2.170}$$

We have here chosen to retain the factor $i\delta$ in order to avoid divergencies when the denominators go to zero. As discussed in Section 2.1.2, $i\delta$ can also be considered as a phenomenological finite lifetime for the, in this case, vibrationally excited state.

In the CARS process, the dominant signal arises from resonances between the difference frequency $\omega_1 - \omega_2$ and a specific vibrational transition (and in a similar manner for $\omega_3 - \omega_2$ since $\omega_3 = \omega_1$). From Eq. 2.170, it can be realized that these vibrational resonances can only occur for the state $|m\rangle$ in the SOS expression. Making the approximation $\omega_\sigma = 2\omega_1 - \omega_2 \approx \omega_1$ and thus that $\omega_2 \approx \omega_1$, which is reasonable considering that the vibrational excitation energy is much smaller than the energy of the optical laser field, Eq. 2.170 can be approximated as

$$\gamma_{\alpha\beta\gamma\delta}^{\text{CARS, Res}}(-\omega_\sigma;\omega_1,-\omega_2,\omega_1) \approx \frac{2}{\hbar}\sum_k \left[\frac{\langle 0|\alpha_{\alpha\beta}(\omega_1)|k\rangle\langle k|\alpha_{\gamma\delta}(\omega_1)|0\rangle}{\omega_{k0}-(\omega_1-\omega_2)+i2\epsilon} + \frac{\langle 0|\alpha_{\alpha\delta}(\omega_1)|k\rangle\langle k|\alpha_{\gamma\beta}(\omega_1)|0\rangle}{\omega_{k0}+(\omega_1-\omega_2)+i2\epsilon}\right] \tag{2.171}$$

where k now corresponds to a vibrationally excited state of the electronic ground state and $\hbar\omega_{k0}$ is the corresponding vibrational excitation energy.

The strategy for evaluating these vibrational matrix elements is now clear, and invoking the double-harmonic approximation, we obtain the resonance contribution to the CARS intensity as

$$\gamma_{\alpha\beta\gamma\delta}^{\text{CARS, Res}}(-\omega_\sigma;\omega_1,-\omega_2,\omega_1) \approx \frac{2}{\hbar}\sum_k \left\{\left(\frac{\partial\alpha_{\alpha\beta}(\omega_1)}{\partial\mathbf{Q}_k}\right)\bigg|_{\mathbf{Q}_e}\left(\frac{\partial\alpha_{\gamma\delta}(\omega_1)}{\partial\mathbf{Q}_k}\right)\bigg|_{\mathbf{Q}_e}[\omega_{k0}-(\omega_1-\omega_2)+i2\epsilon]^{-1} + \left(\frac{\partial\alpha_{\alpha\delta}(\omega_1)}{\partial\mathbf{Q}_k}\right)\bigg|_{\mathbf{Q}_e}\left(\frac{\partial\alpha_{\gamma\beta}(\omega_1)}{\partial\mathbf{Q}_k}\right)\bigg|_{\mathbf{Q}_e}[\omega_{k0}+(\omega_1-\omega_2)+i2\epsilon]^{-1}\right\} \tag{2.172}$$

Although the contribution in Eq. 2.172 dominates in the vicinity of a resonance, the CARS signal also has contributions from the nonresonant part of the second hyperpolarizability tensor, both electronic and vibrational, and we will denote these contributions as γ_e. Denoting furthermore the real and imaginary parts of the tensor in Eq. 2.172 as γ_υ^R and γ_υ^I, respectively we can write our final expression for the CARS signal as

$$|\gamma|^2 = [(\gamma_\upsilon^R + \gamma_\upsilon^I) + \gamma_e]^2 = (\gamma_\upsilon^R)^2 + 2\gamma_\upsilon^R\gamma_e + (\gamma_e)^2 + (\gamma_\upsilon^I)^2 \tag{2.173}$$

where we have performed an isotropic averaging of the hyperpolarizability tensor

$$\gamma = \frac{1}{15}(\gamma_{\xi\eta\eta\xi} + \gamma_{\xi\eta\xi\eta} + \gamma_{\xi\xi\eta\eta}) \tag{2.174}$$

We note that the evaluation of CARS requires largely the same quantity as is needed for conventional Raman spectroscopy as well as ROA, namely the polarizability gradient. If the nonresonant contributions are also wanted, then the electronic second hyperpolarizability is also needed, as well as the off-resonant pure vibrational contributions [292]. Ab initio studies of CARS have to date been very limited [292, 293].

Experimentally, an important advantage of CARS over convential Raman spectroscopy is the fact that the resonance process increases the scattering cross section by

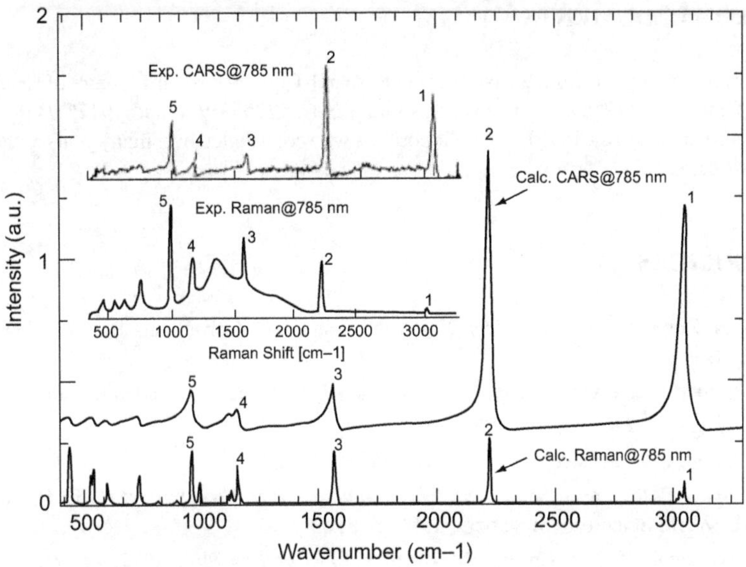

Figure 2.1 Comparison of theoretical and experimental Raman and CARS spectra for benzonitrile. (Data taken from ref. 293.)

several orders of magnitude, in addition to less problem with fluorescence due to the anti-Stokes scattering. However, the resonance conditions may also lead to different selection rules in terms of the intensity of the calculated CARS signal. This is illustrated in Figure 2.1 where both theoretical and experimental Raman and CARS spectra are compared for benzonitrile. We note that the calculations reproduce the experimentally observed CARS enhancement of the vibrational bands at approximately 2250 and 3050 cm^{-1}. We refer to Mohammed et al. [293] for computational details and more information on the experimental data shown in Figure 2.1.

In addition to the lifetime response methods cited in Section 2.1.2, we would like to mention here that Jensen, Autschbach and coworkers have also presented linear response functions using finite lifetimes at the time-dependent density-functional level of theory [294].

Also, two additional references worth mentioning on the calculation of ECD spectra by response theory methods are authored by Crawford and coworkers [295,296].

Finally, since the submission of this contribution, it has become recognized that the atomic axial tensor (AAT) appear in VCD can be expressed as a derivative, specifically the frequency derivative at zero frequency of a linear response function for operators referencing a nuclear displacement and a magnetic field [297]. Using such definition within the density matrix-based quasienergy derivative Lagrangian approach of Thorvaldsen et al. [32] allows to express the AAT in a form where the need to solve response equations for the nuclear displacements is removed, significantly reducing the computation cost compared to existing formulations.

ACKNOWLEDGMENTS

KR has received support from the Norwegian Research Council through a Centre of Excellence Grant (Grant No. 179568/V30) as well as Grant Nos. 177558/V30 and 191251/V00. Many thanks are also due to the long list of collaborators who contributed over many years to the work reviewed here.

REFERENCES

1. D. N. Zubarev, *Nonequilibrium Statistical Mechanic*, Consultants Bureau, New York, 1974.
2. J. Linderberg, Y. Öhrn, *Propagators in Quantum Chemistry*, Academic, London, 1973.
3. E. Dalgaard, *Phys. Rev. A* **1982**, *26*, 26.
4. J. Olsen, P. Jørgensen, *J. Chem. Phys.* **1985**, *82*, 3235.
5. J. Olsen, P. Jørgensen, in *Modern Electronic Structure Theory, Part II*, D. R. Yarkony, Ed., World Scientific, Singapore, 1995, p. 857.
6. P. W. Langhoff, S. T. Epstein, M. Karplus, *Rev. Modern Phys.* **1972**, *44*, 602.
7. E. Runge, E. K. U. Gross, *Phys. Rev. Lett.* **1984**, *52*, 997.
8. M. A. L. Marques, E. K. U. Gross, *Annu. Rev. Phys. Chem.* **2004**, *55*, 427.
9. O. Christiansen, C. Hättig, P. Jørgensen, *Int. J. Quant. Chem.* **1998**, *68*, 1.
10. R. Moccia, *Int. J. Quant. Chem.* **1973**, *7*, 779.
11. J. Frenkel, *Wave Mechanics, Advanced General Theory*, Clarendon, Oxford, 1934.
12. P. Ehrenfest, *Z. Phys.* **1927**, *45*, 455.
13. H. Hellman, *Einführung in die Quantenchemie*, Franz Deuticke, Leipzig, 1937.
14. R. P. Feynman, *Phys. Rev.* **1939**, *56*, 340.
15. T. Helgaker, P. Jørgensen, in *Methods in Computational Molecular Physics*, S. Wilson, G. H. F. Diercksen, Eds., Plenum, New York, 1992, 353.
16. C. Hättig, O. Christiansen, P. Jørgensen, *J. Chem. Phys.* **1998**, *108*, 8331.
17. C. Hättig, P. Jørgensen, *J. Chem. Phys.* **1998**, *109*, 9219.
18. H. J. Aa. Jensen, H. Koch, P. Jørgensen, J. Olsen, *Chem. Phys.* **1988**, *119*, 297.
19. J. Olsen, H. J. Aa. Jensen, P. Jørgensen, *J. Comp. Phys.* **1988**, *74*, 265.
20. H. Hettema, H. J. Aa. Jensen, P. Jørgensen, J. Olsen, *J. Chem. Phys.* **1992**, *97*, 1174.
21. O. Vahtras, H. Ågren, P. Jørgensen, H. J. Aa. Jensen, T. Helgaker, J. Olsen, *J. Chem. Phys.* **1992**, *97*, 9178.
22. P. Norman, D. Jonsson, O. Vahtras, H. Ågren, *Chem. Phys. Lett.* **1995**, *242*, 7.
23. D. Jonsson, P. Norman, H. Ågren, *J. Chem. Phys.* **1996**, *105*, 6401.
24. P. Norman, D. Jonsson, H. Ågren, P. Dahle, K. Ruud, T. Helgaker, H. Koch, *Chem. Phys. Lett.* **1996**, *253*, 1.
25. P. Sałek, O. Vahtras, T. Helgaker, H. Ågren, *J. Chem. Phys.* **2002**, *117*, 9630.
26. K. Sasagane, F. Aiga, R. Itoh, *J. Chem. Phys.* **1993**, *99*, 3738.
27. P. Cronstrand, Y. Luo, P. Norman, H. Ågren, *Chem. Phys. Lett.* **2003**, *375*, 233.
28. H. Larsen, P. Jørgensen, J. Olsen, T. Helgaker, *J. Chem. Phys.* **2000**, *113*, 8908.

29. S. Coriani, S. Høst, B. Jansík, L. Thøgersen, J. Olsen, P. Jørgensen, S. Reine, F. Pawlowski, T. Helgaker, P. Sałek, *J. Chem. Phys.* **2007**, *126*, 154108.
30. T. Kjærgaard, P. Jørgensen, J. Olsen, S. Coriani, T. Helgaker, *J. Chem. Phys.* **2008**, *129*, 054106.
31. J. Kussmann, C. Ochsenfeld, *J. Chem. Phys.* **2007**, *127*, 204103.
32. A. J. Thorvaldsen, K. Ruud, K. Kristensen, P. Jørgensen, S. Coriani, *J. Chem. Phys.* **2008**, *129*, 214108.
33. F. London, *J. Phys. Radium* **1937**. *8*, 397.
34. M. Krykunov, J. Autschbach, *J. Chem. Phys.* **2005**, *123*, 114103.
35. P. Norman, D. Bishop, H. Jensen, J. Oddershede, *J. Chem. Phys.* **2001**, *115*, 10323.
36. P. Norman, A. Jiemchooroj, B. E. Sernelius, *J. Chem. Phys.* **2003**, *118*, 9167.
37. A. Jiemchooroj, P. Norman, B. Sernelius, *J. Chem. Phys.* **2005**, *123*, 1.
38. P. Norman, D. Bishop, H. Jensen, J. Oddershede, *J. Chem. Phys.* **2005**, *123*, 1.
39. A. Jiemchooroj, P. Norman, *J. Chem. Phys.* **2007**, *126*, 134102.
40. H. Solheim, K. Ruud, S. Coriani, P. Norman, *J. Chem. Phys.* **2008**, *128*, 094103.
41. B. J. Orr, J. F. Ward, *Mol. Phys.* **1971**, *20*, 513.
42. L. D. Barron, *Molecular Light Scattering and Optical Activity*, Cambridge University Press, Cambridge, 2004.
43. K. Kristensen, J. Kauczor, T. Kjærgaard, P. Jørgensen, *J. Chem. Phys.* **2009**, *131*, 044112.
44. M. Krykunov, M. Seth, T. Ziegler, J. Autschbach, *J. Chem. Phys.* **2007**, *127*, 244102.
45. P. Norman, K. Ruud, T. Helgaker, *J. Chem. Phys.* **2004**, *120*, 5027.
46. H. Solheim, K. Ruud, S. Coriani, P. Norman, *J. Phys. Chem. A* **2008**, *112*, 9615.
47. U. Ekström, P. Norman, V. Carravetta, H. Ågren, *Phys. Rev. Lett.* **2006**, *97*, 143001.
48. U. Ekström, P. Norman, *Phys. Rev. A* **2006**, *74*, 042722.
49. G. Tu, Z. Rinkevicius, O. Vahtras, H. Ågren, U. Ekström, P. Norman, V. Carravetta, *Phys. Rev. A* **2007**, *76*, 022506.
50. A. Jiemchooroj, U. Ekström, P. Norman, *J. Chem. Phys.* **2007**, *127*, 165104.
51. A. Jiemchooroj, P. Norman, *J. Chem. Phys.* **2008**, *128*, 234304.
52. A. Jiemchooroj, P. Norman, B. Sernelius, *J. Chem. Phys.* **2006**, *125*, 124306.
53. O. Christiansen, *J. Chem. Phys.* **2005**, *122*, 194105.
54. O. Christiansen, J. Kongsted, M. J. Paterson, J. M. Luis, *J. Chem. Phys.* **2006**, *125*, 214309.
55. M. B. Hansen, O. Christiansen, C. Hättig, *J. Chem. Phys.* **2009**, *131*, 154101.
56. O. Christiansen, *Phys. Chem. Chem. Phys.* **2007**, *9*, 2942.
57. J. M. Bowman, T. Carrington, H.-D. Meyer, *Mol. Phys.* **2008**, *106*, 2145.
58. M. Born, R. Oppenheimer, *Ann. der Phys.* **1927**, *84*, 457.
59. P.-O. Åstrand, K. Ruud, D. Sundholm, *Theor. Chem. Acc.* **2000**, *103*, 365.
60. P.-O. Åstrand, K. Ruud, P. R. Taylor, *J. Chem. Phys.* **2000**, *112*, 2655.
61. E. B. Wilson, Jr., J. C. Decius, P. C. Cross, *Molecular Vibrations: The Theory of Infrared and Raman Vibrational Spectra*, Dover, Mineola, NY, 1980.
62. D. M. Bishop, B. Kirtman, *J. Chem. Phys.* **1991**, *95*, 2646.
63. D. M. Bishop, *Rev. Mod. Phys.* **1990**, *62*, 343.

64. D. M. Bishop, *Adv. Chem. Phys.* **1998**, *104*, 1.
65. D. M. Bishop, P. Norman, in *Handbook of Advanced Electronic and Photonic Materials*, H. S. Nalwa, Ed., Academic, San Diego, 2000.
66. E. R. Cohen, T. Cvitaš, J. G. Frey, B. Holmström, K. Kuchitsu, R. Marquardt, I. Mills, F. Pavese, M. Quack, J. Stohner, H. L. Strauss, M. Takami, A. J. Thor, *Quantities, Units and Symbols in Physical Chemistry*, 3rd ed., IUPAC, RCS Publishing, Mineola, NY, 2007.
67. J. A. Schellman, *Chem. Rev.* **1975**, *75*, 323.
68. J. Michal, E. W. Thulstrup, *Spectroscopy with Polarized Light*, VCH Publishers, New York, 1986.
69. W. Moffitt, A. Moscowitz, *J. Chem. Phys.* **1959**. *30*, 648.
70. N. Berova, K. Nakanishi, R. W. Woody, Eds., *Circular Dichroism: Principles and Applications*, 2nd ed., Wiley, New York, 2000.
71. A. Kosłowski, N. Sreerama, R. W. Woody, in *Circular Dichroism: Principles and Applications*, 2nd ed., N. Berova, K. Nakanishi, R.W. Woods, Eds., Wileys, New York, 2000, 55.
72. A. Moscowitz, in *Optical Rotatory Dispersion*, C. Djerassi, Ed., McGraw-Hill, New York, 1960, p. 151.
73. R. W. Woody, in *Circular Dichroism and the Conformational Analysis of Biomolecules*, G. D. Fasman, Ed., Plenum, New York, 1996, p. 26.
74. D. Keller, in *Circular Dichroism and the Conformational Analysis of Biomolecules*, G. D. Fasman, Ed., Plenum, New York, 1996, p. 413.
75. D. P. Craig, T. Thirunamachandran, *Molecular Quantum Electrodynamics: An Introduction to Radiation Molecule Interaction*, Dover, Mineola, NY, 1984.
76. W. Heitler, *The Quantum Theory of Radiation*, Dover, Mineola, NY, 1984.
77. I. E. Levine, *Molecular Spectroscopy*, Wiley, New York, 1975.
78. H. A. Kramers, *Atti Cong. Intern. Fisica (Transactions of Volta Centenary Congress) Como* **1927**, *2*, 545.
79. R. de L. Kronig, *J. Opt. Soc. Am.* **1926**, *12*, 547.
80. P. L. Polavarapu, *J. Phys. Chem. A* **2005**, *109*, 7013.
81. T. B. Pedersen, A. E. Hansen, *Chem. Phys. Lett.* **1995**, *246*, 1.
82. A. Rizzo, B. Jansík, T. B. Pedersen, H. Ågren, *J. Chem. Phys.* **2006**, *125*, 064113.
83. M. Göppert-Mayer, *Ann. Phys. (Leipzig)* **1931**, *9*, 273.
84. N. Bloembergen, *Nonlinear Optics*, 4th ed., World Scientific, Singapore, 1996.
85. Y. R. Shen, *The Principles of Nonlinear Optics*, Wiley Classic Library Ed., Wiley Interscience, Hoboken, NJ, 2003.
86. P. N. Butcher, D. Cotter, *The Elements of Nonlinear Optics*, Cambridge University Press, Cambridge, 1990.
87. F. H. M. Faisal, *Theory of Multiphoton Processes*, Plenum, New York and London, 1987.
88. P. Lambropoulos, J. H. Eberly, Eds., *Multiphoton Processes*, Wiley, New York, 1978.
89. W. M. McClain, *Acc. Chem. Res.* **1974**, *7*, 129.
90. F. Terenziani, C. Katan, E. Badaeva, S. Tretiak, M. Blanchard-Desce, *Adv. Mater.* **2008**, *20*, 4641.
91. G. S. He, L.-S. Tan, Q. Zheng, P. N. Prasad, *Chem. Rev.* **2008**, *108*, 1245.

92. I. Tinoco, Jr., *J. Chem. Phys.* **1975**, *62*, 1006.
93. E. A. Power, *J. Chem. Phys.* **1975**, *63*, 1348.
94. D. L. Andrews, *Chem. Phys.* **1976**, *16*, 419.
95. W. J. Meath, E. A. Power, *J. Phys. B At. Mol. Phys.* **1984**, *17*, 763.
96. W. J. Meath, E. A. Power, *Mol. Phys.* **1984**, *51*, 585.
97. W. J. Meath, E. A. Power, *J. Phys. B At. Mol. Phys.* **1987**, *20*, 1945.
98. W. J. Meath, E. A. Power, *J. Mod. Opt.* **1989**, *36*, 977.
99. G. Wagniére, *J. Chem. Phys.* **1982**, *77*, 2786.
100. G. H. Wagnière, *Linear and Nonlinear Optical Properties of Molecules*, Verlag Helvetica Chimica Acta, Basel, 1993.
101. E. A. Power, in *New Frontiers in Quantum Electrodynamics and Quantum Optics*, O. A. Barut, Ed., Plenum, New York, 1990, p. 253.
102. A. Gedanken, M. Tamir, *Rev. Sci. Instrum.* **1987**, *58*, 950.
103. R. Cameron, G. C. Tabisz, *Mol. Phys.* **1997**, *90*, 159.
104. W. Qu, G. C. Tabisz, *J. Chem. Phys.* **2006**, *124*, 184305.
105. R. Cameron, G. C. Tabisz, *J. Chem. Phys.* **2007**, *126*, 224507.
106. J. Szłucki, W. Stręk, *J. Chem. Phys.* **1986**, *85*, 5547.
107. K. E. Gunde, F. S. Richardson, *Chem. Phys.* **1995**, *194*, 195.
108. F. Hache, H. Mesnil, M. C. Schanne-Klein, *Phys. Rev. B* **1999**, *60*, 6405.
109. H. Mesnil, F. Hache, *Phys. Rev. Lett.* **2000**, *85*, 4527.
110. H. Mesnil, M. C. Schanne-Klein, F. Hache, M. Alexandre, G. Lemercier, C. Andraud, *Chem. Phys. Lett.* **2001**, *338*, 269.
111. H. Mesnil, M. C. Schanne-Klein, F. Hache, *Phys. Rev. A* **2002**, *66*, 013802.
112. M. Alexandre, G. Lemercier, C. Andraud, H. Mesnil, M. C. Schanne-Klein, F. Hache, *Synth. Metals* **2002**, *127*, 135.
113. P. P. Markowicz, M. Samoc, J. Cerne, P. N. Prasad, A. Pucci, G. Ruggeri, *Opt. Expr.* **2004**, *12*, 5209.
114. L. De Boni, C. Toro, F. E. Hernández, *Opt. Lett.* **2008**, *33*, 2958.
115. B. Jansík, A. Rizzo, H. Ågren, *Chem. Phys. Lett.* **2005**, *414*, 461.
116. A. Baranowska, A. Rizzo, B. Jansík, S. Coriani, *J. Chem. Phys.* **2006**, *125*, 054107.
117. B. Jansík, A. Rizzo, H. Ågren, *J. Phys. Chem. B* **2007**, 111, 446; Erratum, ibid, **2007**, 111, 2409.
118. A. Rizzo, in *AIP Conference Proceedings, Vol. 963, Computational Methods in Science and Engineering: Theory and Computation: Old Problems and New Challenges.* Lectures Presented at the International Conference on Computational Methods in Science and Engineering 2007 (ICCMSE 2007), G. Maroulis, T. E. Simos, Eds., American Institute of Physics, New York, 2007, p. 379.
119. B. Jansík, A. Rizzo, H. Ågren, B. Champagne, *J. Chem. Theor. Comp.* **2008**, *4*, 457.
120. A. Rizzo, N. Lin, K. Ruud, *J. Chem. Phys.* **2008**, *128*, 164312.
121. N. Lin, F. Santoro, A. Rizzo, Y. Luo, X. Zhao, V. Barone, *J. Phys Chem. A* **2009**, *113*, 4198.
122. M. Guillaume, K. Ruud, A. Rizzo, S. Monti, Z. Lin, X. Xu, *J. Phys. Chem. B* **2010**, *114*, 6500.

123. C. Toro, L. De Boni, N. Lin, F. Santoro, A. Rizzo, F. E. Hernández, *Chem. Eur. J.* **2010**, *16*, 3504.

124. S. H. Lin, Y. Fujimura, H. J. Neusser, E. W. Schlag, *Multiphoton Spectroscopy of Molecules*, Academic, New York, 1984.

125. D. Wanapun, R. D. Wampler, N. J. Begue, G. J. Simpson, *Chem. Phys. Lett.* **2008**, *455*, 6.

126. C. Toro, L. De Boni, N. Lin, F. Santoro, A. Rizzo, F. E. Hernández, *Chirality*, **2010**, 22, E202.

127. M. Faraday, *Philos. Mag.* **1846**, *28*, 294.

128. M. Faraday, *Philos. Trans. Roy. Soc.* **1846**, *136*, 1.

129. D. J. Caldwell, H. Eyring, *The Theory of Optical Activity*, Wiley Interscience, New York, 1971.

130. A. Rizzo, S. Coriani, *Adv. Quant. Chem.* **2005**, *50*, 143.

131. W. R. Mason, *A Practical Guide to Magnetic Circular Dichroism Spectroscopy*, Wiley, New York, 2007.

132. S. B. Piepho, P. N. Schatz, *Group Theory in Spectroscopy: With Applications to Magnetic Circular Dichroism*, Wiley, New York, 1983.

133. A. D. Buckingham, P. J. Stephens, *Ann. Rev. Phys. Chem.* **1966**, *17*, 399.

134. P. N. Schatz, A. J. McCaffery, *Q. Rev. Chem. Soc.* **1969**, *23*, 552.

135. P. J. Stephens, *Ann. Rev. Phys. Chem.* **1974**, *25*, 201.

136. P. J. Stephens, *Adv. Chem. Phys.* **1976**, *35*, 197.

137. P. J. Stephens, *Chem. Phys. Lett.* **1968**, *2*, 241.

138. P. J. Stephens, *J. Chem. Phys.* **1970**, *52*, 3489.

139. O. Vahtras, H. Ågren, P. Jørgensen, H. J. Aa. Jensen, T. Helgaker, J. Olsen, *J. Chem. Phys.* **1992**, *96*, 2118.

140. DALTON, a molecular electronic structure program, Release 2.0 (2005), available: http://www.kjemi.uio.no/software/dalton/dalton.html.

141. R. Ahlrichs, M. Bär, M. Häser, H. Horn, C. Kölmel, *Chem. Phys. Lett.* **1989**, *162*, 165.

142. TURBOMOLE V6.1 2009. A development of University of Karlsruhe and Forschungszentrum Karlsruhe GmbH, 1989–2007, TURBOMOLE GmbH, since 2007, available: http://www.turbomole.com.

143. E. J. Baerends, et al. ADF2009.01, SCM, Theoretical Chemistry, Vrije Universiteit, Amsterdam, The Netherlands, http://www.scm.com.

144. G. te Velde, F. M. Bickelhaupt, E. J. Baerends, C. Fonseca Guerra, S. J. A. van Gisbergen, J. G. Snijders, T. Ziegler, *J. Comp. Chem.* **2001**, *22*, 931.

145. C. Fonseca Guerra, J. G. Snijders, G. te Velde, E. J. Baerends, *Theor. Chem. Acc.* **1998**, *99*, 391.

146. M. J. Frisch et al., Gaussian 03, Revision C.02, Gaussian, Wallingford, CT, **2004**.

147. G. Karlström, R. Lindh, P.-Å. Malmqvist, B. O. Roos, U. Ryde, V. Veryazov, P.-O. Widmark, M. Cossi, B. Schimmelpfennig, P. Neogrady, L. Seijo, *Comput. Mater. Sci.* **2003**, *28*, 222.

148. CADPAC: The Cambridge Analytic Derivatives Package Issue 6, Cambridge, 1995, A suite of quantum chemistry programs developed by R. D. Amos with contributions from I. L. Alberts, J. S. Andrews, S. M. Colwell, N. C. Handy, D. Jayatilaka, P. J. Knowles,

R. Kobayashi, K. E. Laidig, G. Laming, A. M. Lee, P. E. Maslen, C. W. Murray, J. E. Rice, E. D. Simandiras, A. J. Stone, M.-D. Su, and D. J. Tozer.

149. J. Franck, *Trans. Faraday Soc.* **1926**, *21*, 536.
150. E. Condon, *Phys. Rev.* **1926**, *28*, 1182.
151. E. Condon, *Phys. Rev.* **1928**, *32*, 858.
152. G. Herzberg, E. Teller, *Z. Phys. Chem. B Chem. E* **1933**, *21*, 400.
153. T. Petrenko, F. Neese, *J. Chem. Phys.* **2007**, *127*, 164319.
154. S. Coriani, T. Kjærgaard, P. Jørgensen, K. Ruud, J. Huh, R. Berger, *J. Chem. Theory Comp.* **2010**, *6*, 1028.
155. M. Pecul, K. Ruud, *Adv. Quant. Chem.* **2005**, *50*, 185.
156. T. B. Pedersen, H. Koch, K. Ruud, *J. Chem. Phys.* **1999**, *110*, 2883.
157. K. L. Bak, A. E. Hansen, K. Ruud, T. Helgaker, J. Olsen, P. Jørgensen, *Theor. Chim. Acta* **1995**, *90*, 441.
158. T. B. Pedersen, H. Kock, C. Hättig, *J. Chem. Phys.* **1999**, *110*, 8318.
159. T. B. Pedersen, B. Fernandez, H. Kock, *J. Chem. Phys.* **2001**, *114*, 6983.
160. M. Pecul, K. Ruud, T. Helgaker, *Chem. Phys. Lett.* **2004**, *388*, 110.
161. Y. Yanai, D. P. Tew, N. C. Handy, *Chem. Phys. Lett.* **2004**, *393*, 51.
162. M. J. G. Peach, T. Helgaker, P. Sałek, T. W. Keal, O. B. Lutnæs, D. J. Tozer, N. C. Handy, *Phys. Chem. Chem. Phys.* **2006**, *8*, 558.
163. M. J. Paterson, O. Christiansen, F. Pawłowski, P. Jørgensen, C. Hättig, T. Helgaker, P. Sałek, *J. Chem. Phys.* **2006**, *124*, 054322.
164. D. Shcherbin, K. Ruud, *Chem. Phys.* **2008**, *349*, 234.
165. S. Miertuš, E. Scrocco, J. Tomasi, *Chem. Phys.* **1981**, *55*, 117.
166. R. Cammi, J. Tomasi, *J. Comp. Chem.* **1995**, *16*, 1449.
167. J. Tomasi, B. Mennucci, R. Cammi, *Chem. Rev.* **2005**, *105*, 2999.
168. M. Pecul, D. Marchesan, K. Ruud, S. Coriani, *J. Chem. Phys.* **2005**, *122*, 024106.
169. J. Kongsted, A. E. Hansen, T. B. Pedersen, A. Osted, K. V. Mikkelsen, O. Christiansen, *Chem. Phys. Lett.* **2004**, *391*, 259.
170. J. Kongsted, T. B. Pedersen, A. Osted, A. E. Hansen, K. V. Mikkelsen, O. Christiansen, *J. Phys. Chem. A* **2004**, *108*, 3632.
171. S. Pipolo, R. Cammi, A. Rizzo, C. Cappelli, B. Mennucci, J. Tomasi, *Int. J. Quant Chem.* **2011**, *111*, 826.
172. T. Mori, Y. Inoue, S. Grimme, *J. Org. Chem.* **2006**, *71*, 9797.
173. S. Tartaglia, D. Padula, P. Scafato, L. Chiummiento, C. Rosini, *J. Org. Chem.* **2008**, *73*, 4865.
174. J. Neugebauer, E. J. Baerends, M. Noojien, J. Autschbach, *J. Chem. Phys.* **2005**, *122*, 234305.
175. M. Noojien, *Int. J. Quant. Chem.* **2006**, *106*, 2489.
176. M. Dierksen, S. Grimme, *J. Chem. Phys.* **2006**, *124*, 174301.
177. P. Makac, Y. Luo, H. Ågren, *Chem. Phys. Lett.* **2000**, *330*, 447.
178. P. Makac, Y. Luo, P. Norman, H. Ågren, *J. Chem. Phys.* **2000**, *113*, 7055.
179. N. Lin, X. Zhao, A. Rizzo, Y. Luo, *J. Chem. Phys.* **2007**, *126*, 244509.
180. N. Lin, Y. Luo, F. Santoro, X. Zhao, A. Rizzo, *Chem. Phys. Lett.* **2008**, *464*, 144.

181. N. Lin, F. Santoro, X. Zhao, A. Rizzo, V. Barone, *J. Phys. Chem. A* **2008**, *112*, 12401.

182. E. W. Thulstrup, *Aspects of the Linear Magnetic Circular Dichroism of Planar Organic Molecules*, Springer-Verlag, Berlin, 1980.

183. N. Kobayashi, K. Nakai, *Chem. Commun.* **2007**, 4077.

184. A. Gorski, E. Vogel, J. L. Sessler, J. Waluk, *J. Phys. Chem. A* **2002**, *106*, 8139.

185. E. Solomon, E. Pavel, K. Loeb, C. Campochiaro, *Coord. Chem. Rev.* **1995**, *144*, 369.

186. M. Kirk, K. Peariso, *Curr. Opin. Chem. Biol.* **2003**, *7*, 220.

187. M. Cheeseman, C. Greenwood, T. J. Thomson, in *Advances in Inorganic Chemistry*, Vol. 36, A. G. Sykes, Ed., Academic, San Diego, 1991, p. 201.

188. L. L. Pearce, E. L. Bominaar, J. Peterson, *Biochem. Biophys. Res. Commun.* **2002**, *297*, 220.

189. S. Coriani, P. Jørgensen, A. Rizzo, K. Ruud, J. Olsen, *Chem. Phys. Lett.* **1999**, *300*, 61.

190. S. Coriani, C. Hättig, P. Jørgensen, T. Helgaker, *J. Chem. Phys.* **2000**, *113*, 3561.

191. Y. Honda, M. Hada, M. Ehara, H. Nakatsuji, J. Michl, *J. Chem. Phys.* **2005**, *123*, 164113.

192. T. Kjærgaard, B. Jansík, P. Jørgensen, S. Coriani, J. Michl, *J. Phys. Chem. A* **2007**, *111*, 11278.

193. H. Solheim, L. Frediani, K. Ruud, S. Coriani, *Theor. Chem. Acc.* **2008**, *119*, 231.

194. M. Seth, T. Ziegler, A. Banerjee, J. Autschbach, S. van Gisbergen, E. J. Baerends, *J. Chem. Phys.* **2004**, *120*, 10942.

195. M. Seth, T. Ziegler, J. Autschbach, *J. Chem. Phys.* **2005**, *122*, 094112.

196. G. A. Peralta, M. Seth, T. Ziegler, *J. Chem. Theory Comput.* **2007**, *3*, 434.

197. M. Seth, M. Krykunov, T. Ziegler, J. Autschbach, *J. Chem. Phys.* **2008**, *128*, 234102.

198. M. Seth, M. Krykunov, T. Ziegler, J. Autschbach, A. Banerjee, *J. Chem. Phys.* **2008**, *128*, 144105.

199. G. A. Peralta, M. Seth, T. Ziegler, *Inorg. Chem.* **2007**, *46*, 9111.

200. D. Ganyushin, F. Neese, *J. Chem. Phys.* **2008**, *128*, 114117.

201. M. Seth, T. Ziegler, *Inorg. Chem.* **2009**, *48*, 1793.

202. T. Kjærgaard, P. Jørgensen, A. J. Thorvaldsen, P. Sałek, S. Coriani, *J. Chem. Theory Comp.* **2009**, *5*, 1997.

203. D. W. Miles, H. Eyring, *Proc. Natl. Acad. Sci. U.S.A.* **1973**, *70*, 3754.

204. A. R. Meier, G. H. Wagnière, *Chem. Phys.* **1987**, *113*, 287.

205. D. J. Shieh, S. H. Lin, H. Eyring, *J. Phys. Chem.* **1972**, *76*, 1844.

206. E. Goldstein, S. Vijaya, G. A. Segal, *J. Am. Chem. Soc.* **1980**, *102*, 6198.

207. G. Marconi, *Chem. Phys. Lett.* **1988**, *146*, 259.

208. J. Michl, *Tetrahedron* **1974**, *30*, 4215.

209. J. Fleischhauer, J. Michl, *J. Phys. Chem. A* **2000**, *104*, 7776.

210. L. Seamans, J. Linderberg, *Mol. Phys.* **1972**, *24*, 1393.

211. E. K. U. Gross, J. F. Dobson, M. Petersilka, in *Topics in Current Chemistry*, Vol. 181, R. F. Nalewajski, Ed., Springer, New York, 1996, p. 81.

212. R. van Leeuwen, *Int. J. Mod. Phys. B* **2001**, *15*, 1969.

213. W. Koch, M. C. Holthausen, *A Chemist's Guide to Density Functional Theory*, 2nd ed., Wiley-VCH, Weinheim, Germany, 2001.

214. P. Sałek, S. Høst, L. Thøgersen, P. Jørgensen, P. Manninen, J. Olsen, B. Jansík, S. Reine, F. Pawlowski, E. Tellgren, T. Helgaker, S. Coriani, *J. Chem. Phys.* **2007**, *126*, 114110.

215. M. Seth, T. Ziegler, *J. Chem. Phys.* **2007**, *127*, 134108.

216. Y. Honda, M. Hada, M. Ehara, H. Nakatsuji, J. Downing, J. Michl, *Chem. Phys. Lett.* **2002**, *355*, 219.

217. J. O. Jensen, G. F. Adams, C. Chabalowski, *Chem. Phys. Lett.* **1990**, *172*, 379.

218. H. Bolvin, *Inorg. Chem.* **2007**, *46*, 417.

219. M. Seth, T. Ziegler, *J. Chem. Phys.* **2006**, *124*, 144105.

220. P. Cronstrand, Y. Luo, H. Ågren, *Chem. Phys. Lett.* **2002**, *352*, 262; Erratum ibid. **2002**, *363*, 198.

221. P. Cronstrand, Y. Luo, H. Ågren, *Adv. Quant. Chem.* **2005**, *50*, 1.

222. R. Moccia, A. Rizzo, *Mol. Phys.* **1984**, *51*, 1095.

223. R. Moccia, A. Rizzo, *J. Phys. B At. Mol. Phys.* **1985**, *18*, 3319.

224. I. Cacelli, V. Carravetta, R. Moccia, A. Rizzo, *Mol. Phys.* **1986**, *59*, 403.

225. I. Cacelli, V. Carravetta, R. Moccia, A. Rizzo, *Chem. Phys.* **1986**, *109*, 227.

226. D. L. Yeager, M. A. C. Nascimento, V. McKoy, *Phys. Rev. A* **1975**, *11*, 1168.

227. D. Lynch, M. F. Herman, D. L. Yeager, *Chem. Phys.* **1982**, *64*, 69.

228. B. Jansík, P. Sałek, D. Jonsson, O. Vahtras, H. Ågren, *J. Chem. Phys.* **2005**, *122*, 054107.

229. P. Sałek, O. Vahtras, J. Guo, Y. Luo, T. Helgaker, H. Ågren, *Chem. Phys. Lett.* **2003**, *374*, 446.

230. C. B. Nielsen, S. Rettrup, S. P. A. Sauer, *J. Chem. Phys.* **2006**, *124*, 114108.

231. Y.-H. Sun, K. Zhao, C.-K. Wang, Y. Luo, Y. Ren, X.-T. Tao, M.-H. Jiang, *J. Mol. Struct. THEOCHEM* **2004**, *682*, 185.

232. J.-X. Yang, C.-X. Wang, L. Li, N. Lin, X.-T. Tao, Y.-P. Tian, X. Zhao, M.-H. Jiang, *Chem. Phys.* **2009**, *358*, 39.

233. Y.-Z. Songa, D.-M. Lib, X.-N. Songa, X.-M. Huanga, C.-K. Wang, *J. Mol. Struct. THEOCHEM* **2006**, *772*, 75.

234. L. Frediani, Z. Rinkevicius, H. Ågren, *J. Chem. Phys.* **2005**, *122*, 244104.

235. K. Zhao, L. Ferrighi, L. Frediani, C.-K. Wang, Y. Luo, *J. Chem. Phys.* **2007**, *126*, 204509.

236. F. Terenziani, A. Painelli, *Phys. Rev. B* **2003**, *68*, 165405.

237. F. Terenziani, V. Parthasarathy, A. Pla-Quintana, T. Maishal, A.-M. Caminade, J.-P. Majorai, M. Blanchard-Desce, *Angew. Chem.* **2009**, *121*, 8847.

238. A. D. Becke, *J. Chem. Phys.* **1993**, *98*, 5648.

239. A. D. Becke, *Phys. Rev. A* **1988**, *38*, 3098.

240. C. Lee, W. Yang, R. G. Parr, *Phys. Rev. B* **1988**, *37*, 785.

241. E. Cancès, B. Mennucci, J. Tomasi, *J. Chem. Phys.* **1997**, *107*, 3032.

242. B. Mennucci, E. Cancès, J. Tomasi, *J. Phys. Chem. B* **1997**, *101*, 10506.

243. C. Hättig, O. Christiansen, P. Jørgensen, *J. Chem. Phys.* **1998**, *108*, 8355.

244. C. L. Thomsen, D. Madsen, S. R. Keiding, J. Thøgersen, O. Christiansen, *J. Chem. Phys.* **1999**, *110*, 3453.

245. M. J. Paterson, J. Kongsted, O. Christiansen, K. V. Mikkelsen, C. B. Nielsen, *J. Chem. Phys.* **2006**, *125*, 184501.

246. J. Henriksson, P. Norman, H. J. Aa. Jensen, *J. Chem. Phys.* **2005**, *122*, 114106.

247. N. Lin, L. Ferrighi, X. Zhao, K. Ruud, A. Rizzo, Y. Luo, *J. Phys. Chem. B* **2008**, *112*, 4703.
248. T. H. Dunning, Jr., *J. Chem. Phys.* **1989**, *90*, 1007.
249. G. Placzek, *Handbuch der Radiologie* **1934**, *6*, 205.
250. P. J. Stephens, *J. Phys. Chem.* **1985**, *89*, 748.
251. W. H. Flygare, *Chem. Rev.* **1974**, *74*, 653.
252. J. Gauss, K. Ruud, T. Helgaker, *J. Chem. Phys.* **1996**, *105*, 2804.
253. K. L. Bak, P. Jørgensen, T. Helgaker, K. Ruud, H. J. Aa. Jensen, *J. Chem. Phys.* **1993**, *98*, 8873.
254. K. L. Bak, P. Jørgensen, T. Helgaker, K. Ruud, *Faraday Discuss.* **1994**, *99*, 121.
255. R. D. Amos, N. C. Handy, K. J. Jalkanen, P. J. Stephens, *Chem. Phys. Lett.* **1987**, *133*, 21.
256. Aa. E. Hansen, P. J. Stephens, T. D. Bouman, *J. Phys. Chem.* **1991**, *95*, 4255.
257. K. J. Jalkanen, P. J. Stephens, P. Lazzeretti, R. Zanasi, *J. Chem. Phys.* **1989**, *90*, 3204.
258. J. R. Cheeseman, M. J. Frisch, F. J. Devlin, P. J. Stephens, *Chem. Phys. Lett.* **1996**, *252*, 211.
259. V. P. Nicu, J. Neugebauer, S. K. Wolff, E. J. Baerends, *Theor. Chem. Acc.* **2008**, *119*, 245.
260. L. A. Nafie, T. B. Freedman, *Enantiomer* **1998**, *3*, 283.
261. P. J. Stephens, F. J. Devlin, J. J. Pan, *Chirality* **2008**, *20*, 643.
262. F. J. Devlin, P. J. Stephens, J. R. Cheeseman, M. J. Frisch, *J. Am. Chem. Soc.* **1996**, *118*, 6327.
263. F. J. Devlin, P. J. Stephens, J. R. Cheeseman, M. J. Frisch, *J. Phys. Chem.* **1997**, *101*, 6322.
264. A. Aamouche, F. J. Devlin, P. J. Stephens, *J. Chem. Soc. Chem. Commun.* **1999**, 361.
265. F. J. Devlin, P. J. Stephens, P. Besse, *Tetrahedron Asymm.* **2005**, *16*, 1557.
266. C. Cappelli, S. Corni, B. Mennucci, R. Cammi, J. Tomasi, *J. Phys. Chem. A* **2002**, *106*, 12331.
267. D. Marchesan, S. Coriani, C. Forzato, P. Nitti, G. Pitacco, K. Ruud, *J. Phys. Chem. A* **2005**, *109*, 1449.
268. C. Cappelli, B. Mennucci, *J. Phys. Chem. B* **2008**, *112*, 3441.
269. W.-G. Han, K. J. Jalkanen, M. Elstner, S. Suhai, *J. Phys. Chem. B* **1998**, *102*, 2587.
270. E. Tajkhorshid, K. J. Jalkanen, S. Suhai, *J. Phys. Chem. B* **1998**, *102*, 5899.
271. K. Ruud, A. J. Thorvaldsen, *Chirality* **2009**, *21*, S54.
272. V. Liégeois, K. Ruud, B. Champagne, *J. Chem. Phys.* **2007**, *127*, 244103.
273. A. J. Thorvaldsen, B. Gao, K. Ruud, M. Fedorovsky, G. Zuber, W. Hug, unpublished.
274. R. Bast, A. J. Thorvaldsen, M. Ringholm, K. Ruud, *Chem. Phys.* **2009**, *356*, 177.
275. G. Zuber, W. Hug, *J. Phys. Chem. A* **2004**, *108*, 2108.
276. M. Reiher, V. Liégeois, K. Ruud, *J. Phys. Chem. A* **2005**, *109*, 7567.
277. P. Bouř, J. Sopková, L. Bednárová, P. Maloň, T. A. Keiderling, *J. Comp. Chem.* **1997**, *18*, 646.
278. M. Fedorovsky, Raman optical activity: From analyzing to synthesizing vibrational spectra, thesis, Universität Fribourg, **2008**.
279. J. Kapitán, F. Zhu, L. Hecht, J. Gardiner, D. Seebach, L. D. Barron, *Angew. Chem. Int. Ed.* **2008**, *47*, 6392.
280. J. Kapitán, V. Baumruk, P. Bouř, *J. Am. Chem. Soc.* **2006**, *128*, 2438.

281. V. Andrushchenko, H. Wieser, P. Bouř, *J. Phys. Chem. A* **2007**, *111*, 9714.

282. E. Cancés, B. Mennucci, *J. Math. Chem.* **1998**, *23*, 309.

283. R. Cammi, L. Frediani, B. Mennucci, J. Tomasi, K. Ruud, K. V. Mikkelsen, *J. Chem. Phys.* **2002**, *117*, 13.

284. A. Klamt, G. Schüürmann, *J. Chem. Soc. Perkin Trans. 2* **1993**, *5*, 799.

285. M. Pecul, E. Lamparska, C. Cappelli, L. Frediani, K. Ruud, *J. Phys. Chem. A* **2006**, *110*, 2807.

286. M. Pecul, C. Deillon, A. J. Thorvaldsen, K. Ruud, *J. Raman Spectrosc.*, **2010**, *41*, 1200.

287. M. Knapp-Mohammady, K. J. Jalkanen, F. Nardi, R. C. Wade, S. Suhai, *Chem. Phys.* **1999**, *240*, 63.

288. K. J. Jalkanen, V. W. Jürgensen, A. Claussen, A. Rahim, G. M. Jensen, R. C. Wade, F. Nardi, C. Jung, I. M. Degtyarenko, R. M. Nieminen, F. Herrmann, M. Knapp-Mohammady, T. A. Niehaus, K. Frimand, S. Suhai, *Int. J. Quant. Chem.* **2006**, *106*, 1160.

289. E. Deplazes, W. van Bronswijk, F. Zhu, L. D. Barron, S. Ma, L. A. Nafie, K. J. Jalkanen, *Theor. Chem. Acc.* **2009**, *119*, 155.

290. S. Mukamel, *Principles of Nonlinear Optical Spectroscopy*, Oxford University Press, New York, 1995.

291. P. Norman, K. Ruud, in *Non-Linear Optical Properties of Matter: From Molecules to Condensed Phases*, Challenges and Advances in Computational Chemistry and Physics, Vol. 1, M. G. Papadopoulos; A. J. Sadlej and J. Leszczynski, Eds., Kluwer, Dordrecht, 2006, p. 1.

292. A. J. Thorvaldsen, L. Ferrighi, K. Ruud, H. Ågren, S. Coriani, P. Jøgensen, *Phys. Chem. Chem. Phys.* **2009**, *11*, 2293.

293. A. Mohammed, H. Ågren, A. J. Thorvaldsen, K. Ruud, *Chem. Phys. Lett.* **2010**, *485*, 320.

294. L. Jensen, J. Autschbach, G. C. Schatz *J. Chem. Phys.* **2005**, *122*, 224115.

295. T. D. Crawford, *Theor. Chem. Acc.* **2006**, *115*, 227.

296. T. D. Crawford, M. C. Tam, M. L. Abrams, *J. Phys. Chem. A* **2007**, *111*, 12057.

297. S. Coriani, A. J. Thorvaldsen, K. Kristensen, P. Jørgensen, *Phys. Chem. Chem. Phys.*, **2011**, *13*, 4224.

3

COMPUTATIONAL X-RAY SPECTROSCOPY

VINCENZO CARRAVETTA

CNR—Consiglio Nazionale delle Ricerche, Instituto per i Processi Chimico-Fisici (IPCF), Pisa, Italy

HANS ÅGREN

Theoretical Chemistry, School of Biotechnology, Royal Institute of Technology, Stockholm, Sweden

Computational Strategies for Spectroscopy: From Small Molecules to Nano Systems, First Edition. Edited by Vincenzo Barone.

3.1 INTRODUCTION

X rays, electromagnetic waves situated between ultraviolet light and gamma rays on the wavelength scale from 10 to 0.01 nm, were discovered more than a century ago by W. C. Röntgen and have ever since been intensively applied in the medical and natural sciences. This owes primarily to the ability of X-ray photons to penetrate soft matter and to the fact that their wavelengths are comparable to atomic dimensions, making it possible to obtain diffractional fingerprints of ordered materials. The first and most common source of X rays is the vacuum tube where an electron beam emitted by a filament (cathode) hits a metallic anode after being accelerated by a voltage difference of several kilovolts. In addition to discrete emission spectral lines, generated by the electron impact ionization and characteritic for each element, the deceleration of electrons penetrating the anode gives rise to a continuous spectrum of X rays by a phenomenon known as *bremsstrahlung* [1]. Basically the same principle lies behind the production of X rays by the more modern sources, synchrotrons, where the electrons are accelerated, instead of decelerated, by magnetic fields to a speed close to that of the light in a large vacuum ring. Synchrotron radiation has the characteristics of high brightness, tunability, and polarization, which in the last few decades has vastly promoted the extensive and successful use of X rays in the field of spectroscopy, laying the ground for much of our contemporary understanding of the electronic structure of the atom. Intense and wavelength-tunable X rays are nowadays typically generated at synchrotron facilities equipped with high-energy resolution monochromators and electron and photon detectors.

In the X-ray region of the electromagnetic spectrum there is sufficient energy to excite different electronic states, including transitions from the lowest energy (core) orbitals. Due to the small spatial extent of core orbitals and the wide separation of their energies, it is possible to select by X-ray excitation a certain atomic species of interest and even distinguish among atoms of the same species in different local environments, that is, X-ray spectroscopy is chemically selective. Furthermore, effective local selection rules for X-ray transitions ensure a considerable potential for the determination of electronic density around the emitting atoms. The interaction of X rays with the core electrons of an atom or a molecule can be classified in a number of main processes pictorially represented by the independent electron diagrams in Figure 3.1.

The absorption of a photon of sufficiently high energy, from a few hundred to a few thousand electron-volts for most light atoms, may excite the core electron to either an empty orbital (photoabsorption) or the electronic continuum (photoemission or ionization), eventually exciting/ionizing a secondary electron in the valence shell (shake-up/off). The high energy of the different final states reached by the excitation/ionization processes may be dissipated by electronic decay processes that have radiative (X-ray emission or fluorescence) or nonradiative (Auger emission) character.

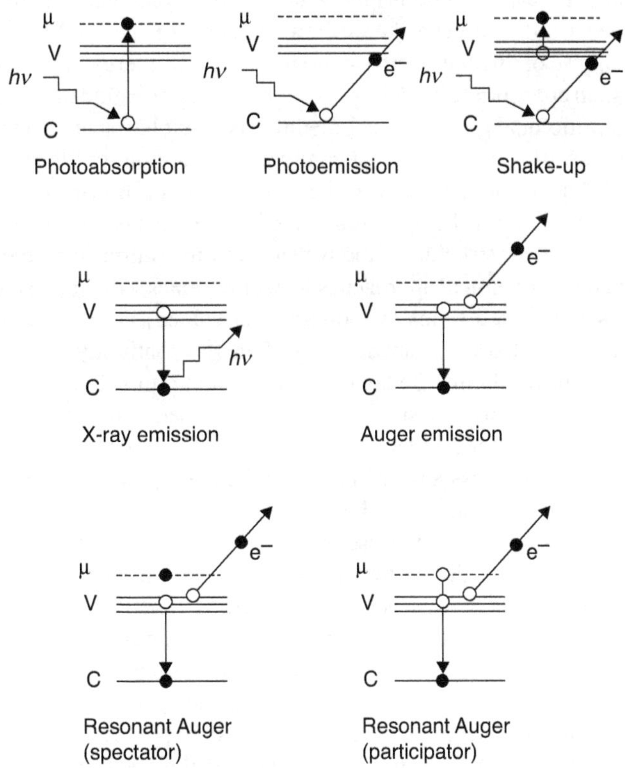

Figure 3.1 Diagrams of the main processes involving or originating from the interaction of core electrons and X-ray photons.

When core excitation and Auger decay occur simultaneously, resonant Auger decay, either spectator or participator, is realized. The processes diagrammatically represented in Figure 3.1 form the physical ground of a large variety of X-ray spectroscopies, some of which will be discussed in the following. Our illustrations of the computatonal methods and their related theories will focus on molecular systems but will have high relevance also for other phases: liquids and solids as well as polymers, surfaces, and surface adsorbates, for which X-ray spectroscopies have been extensively applied throughout the years.

3.2 DESCRIPTION BY QUANTUM CHEMICAL METHODS

Most of the electronic processes behind X-ray spectroscopies involve the electronic continuum where the ionized/excited core states appear as electronic resonances with a lifetime in the range of femtoseconds. Due to their quasi stationary character the core ionized/excited states can in principle be described by a number of quantum chemical methods developed for the low-energy excited electronic states. This is generally realized in variational or response methods by controlling the (single) occupancy of the core orbital in wavefunction approaches or the hole index in the excitation operator space in order to keep the desired character of the final state avoiding a variational collapse or, in other words, a mixing with low-energy valence excitations. While under such appropriate boundary conditions the application of a relatively large number of ab initio quantum chemical methods is feasible for the determination of accurate core transition energies, most commonly the simplest independent particle approximation has been applied for the calculation of transition intensities, a fundamental ingredient for the interpretation of the experimental spectroscopic data in terms of the electronic structure of the system. A further intrinsic problem in the use of standard quantum chemical approaches to excited states well above the ionization threshold (or superexcited states as they are also named) is the description of the electronic continuum that has characteristics of energy continuity but also of infinite degeneracy and an oscillating behavior of the continuum orbitals that are hardly represented by projections on standard finite basis sets commonly employed in quantum chemical methods. Unconventional basis functions (STOCOS [2], splines [3], oscillating Gaussians [4]) exposing the appropriate oscillating character of the electronic continuum have been proposed and sometimes employed to overcome such problems, but their use is generally limited to small-size molecular systems due to numerical difficulties. Most of the calculations have instead been performed by projection on basis sets of standard L^2 functions, mostly Gaussians, based on the observation that the local representation of the continuum wavefunction in the molecular region is sufficient, in many cases, for the evaluation of the transition moments involving the continuum. This is not true, however, for properties like the angular distribution of electrons in a photoemission process, which are strictly dependent by the infinitely degenerate character of the electronic continuum and by quantities, like the phase shift, that depend on the asymptotic behavior of the continuum orbital.

Scattering-based approaches, as, for instance, the reactance matrix or K matrix [5] projected on an L^2 basis sets, are found more appropriate for computing such kinds of properties.

3.3 STATIC EXCHANGE AND RANDOM PHASE APPROXIMATION

Despite the fact that the core excited/ionized state has a quasi-stationary character and, as such, can be computed by quantum chemistry methods using appropriate "boundary conditions," it remains the problem of an appropriate description of the emitted electron in the calculations of cross sections of the core excitation process. Standard quantum chemistry methods provide, in general, a representation of only the lowest virtual orbitals below the ionization threshold, although, as for instance in the case of Hartree–Fock (HF) orbitals, they are not the optimal ones from a physical point of view. Atomic one-center models have been frequently employed for the calculation of molecular high-energy transition intensities [6, 7]. They are based on an atomic approximation of the molecular continuum orbital, decomposed in partial waves, and on the approximation of the bound orbitals by using only the linear combination of atomic orbitals (LCAO) expansion coefficients pertaining to the atomic functions centered on the site of the core excitation/ionization. This approach neglects, of course, multicenter effects in the physical process and gives the intensity as an incoherent sum of atomic intensities.

Virtual molecular orbitals projected on multicenter basis sets and improved with respect to the HF canonical orbitals are provided by a rather simple but effective approximation called static exchange (STEX) [8]. By this method the virtual orbitals, both below and above the ionization threshold, are variationally optimized for the Coulomb plus exchange potential of the ionic target. The ionic wave function is assumed, instead, not to be affected by the presence of the excited electron but can otherwise be independently optimized; this means that in the case of core ionization the important electron relaxation around the core hole can be fully accounted for (in reality a bit overcounted due to the neglected interaction with the excited electron). By writing the excited state as a linear combination of single excited configurations corresponding to the antisymmetrized product of an ionic target wavefunction $|\Phi_j^{(N-1)}\rangle$ and of a virtual orbital $|\phi_{\epsilon c}\rangle$, we get, in the case of a closed-shell ground state, the singlet multichannel wavefunction

$$\left|\Psi_E^N\right\rangle = \sum_c \frac{1}{\sqrt{2}}\left(\hat{a}^\dagger_{\epsilon_c\alpha}\hat{a}_{j_c\alpha} + \hat{a}^\dagger_{\epsilon_c\beta}\hat{a}_{j_c\beta}\right)\left|\Psi_0^N\right\rangle = \sum_c \left|\Phi_{j_c}^{(N-1)}\phi_{\epsilon_c}\right\rangle \tag{3.1}$$

where the index c runs over all the channels included in the model, $|\Psi_0^N\rangle$ is the reference (ground) state, and α, β are spin indices.

The projected Schrödinger equation

$$\langle \Phi_{j_a}\delta\phi_{\epsilon_a}|\hat{H}-E|\sum_c \Phi_{j_c}\phi_{\epsilon_c}\rangle = 0 \tag{3.2}$$

for all possible variations $\delta\phi_{\epsilon\alpha}$ of the orbital $\phi_{\epsilon\alpha}$, while keeping frozen the ionic wavefunctions, allows us to derive the one-particle equation

$$\langle \hat{P}\delta\phi_{\epsilon_a}|\hat{F} - E_a - \hat{J}_{ja,ja} + 2\hat{K}_{ja,ja}|\hat{P}\phi_{\epsilon_a}\rangle + \sum_{c\neq a}\langle \hat{P}\delta\phi_{\epsilon_a}| - \hat{J}_{ja,jc} + 2\hat{K}_{ja,jc}|\hat{P}\phi_{\epsilon_c}\rangle = 0 \tag{3.3}$$

where the projector

$$\hat{P} = \sum_{i}^{\mathrm{occ}} \phi_i\rangle\langle\phi_i$$

imposes orthogonality between the virtual orbital and the occupied ones and $\hat{F}$, $\hat{J}$, and $\hat{K}$ are the Fock, Coulomb, and exchange operators, respectively, defining the STEX Hamiltonian for channel a:

$$\hat{h}_a^{N-1} = \hat{F} - \hat{J}_a + 2\hat{K}_a \tag{3.4}$$

For a different reference state and/or different ionic states, appropriate STEX Hamiltonians can be easily derived from an appropriate projected Schrödinger equation similar to Eq. 3.2. In the case, for instance, of a closed-shell ground state and a double-ionized target state with holes in orbitals x and j (ionic state in a shake-up/off process) the STEX Hamiltonian has the general expression [9]

$$\hat{h}_{x,j,s}^{N-2} = \hat{F} + c_{J_x}\hat{J}_x + c_{K_x}\hat{K}_x + c_{J_j}\hat{J}_j + c_{K_j}\hat{K}_j \tag{3.5}$$

where the coupling coefficients c_J, c_K depend on the adopted spin coupling (S) scheme (see Table 3.1) for the three open shells x, j, ϵ.

The STEX approximation is an independent-particle, independent-channel model and the corresponding Hamiltonian includes only the first term of Eq. 3.3, while channel interaction is introduced by the second term. The final state energy is obtained by adding the ionic target energy $E_a = \langle\Phi_{ja}|\hat{H}|\Phi_{ja}\rangle$ to a STEX eigenvalue. If the ground state is taken as the reference state, electron correlation effects may be

Table 3.1 Coupling Coefficients for Different Shake Channels

	x		j/ϵ	
	c_J	c_K	c_J	c_K
Singlet coupling of x and j orbitals first	−1.0	0.5	−1.0	0.5
Triplet coupling of x and j orbitals first	−1.0	1.5	−1.0	1.5
Singlet coupling of j and ϵ orbitals first	−1.0	0.5	−1.0	2.0
Triplet coupling of j and ϵ orbitals first	−1.0	1.5	−1.0	0.0

introduced at three levels: ionic target relaxation, interchannel coupling, and initial (ground) and final (ionic) state correlation. The first level is the one more important and easily implemented in the case of a core excited state. The main advantages of the STEX approximation are the following: (1) an optimized description of the excited/ionized electron motion in an ab initio nonlocal potential, derived from a physically meaningful self-consistent field (SCF) electron density, that can be improved by including electronic relaxation in the ionic wavefunction; (2) the one-electron Hamiltonian can be easily projected on L^2 basis functions and fully diagonalized, so including all the discrete-continuum intrachannel coupling; and (3) a direct calculation of the STEX Hamiltonian matrix on a large, almost complete, basis set is possible and the method [10, 11], implemented in the code STEX is easily applicable to systems with a large number of electrons. A similar approach is adopted, but not in a direct form and then limited to smaller size molecules, in the code GSCF3 [12, 13]. On the other hand, in the STEX approximation, the electronic relaxation around the core hole is taken into account by SCF optimization of the ionic wavefunction, and it is assumed to be independent from the excited electron, while other electron correlation effects and, in particular, the screening effect of the excited electron are neglected. As a main consequence, the energy scale of the bound virtual orbitals may become compressed against the ionization threshold; in the calculation of core photoabsorbtion spectra it is generally assumed that such an error is roughly constant for the different excitation channels.

The STEX approximation can be seen as the basic one in a series of methods of increasing complexity. The excited state in Eq. 3.1, including the coupling of several excitation channels, corresponds to the so-called Tamn–Dancoff approximation. A slightly more complex expression of the excitation operator, including also deexcitation terms,

$$\hat{\mathbf{T}}_I^\dagger = \sum_{j,\epsilon} \left(\mathbf{X}_{j\epsilon,I} \hat{a}_\epsilon^\dagger \hat{a}_j + \mathbf{Y}_{j\epsilon,I} \hat{a}_j^\dagger \hat{a}_\epsilon \right) \tag{3.6}$$

is the one corresponding to the random phase approximation (RPA) [14], also known as the time-dependent Hartree–Fock (here spin labels are not explicitly indicated). The coefficients **X** and **Y** are determined by the RPA equation

$$\begin{pmatrix} \mathbf{M} & \mathbf{Q} \\ \mathbf{Q}* & \mathbf{M}* \end{pmatrix} \begin{pmatrix} \mathbf{X}_{.,I} \\ \mathbf{Y}_{.,I} \end{pmatrix} = \omega_I \begin{pmatrix} \mathbf{I} & 0 \\ 0 & -\mathbf{I} \end{pmatrix} \begin{pmatrix} \mathbf{X}_{.,I} \\ \mathbf{Y}_{.,I} \end{pmatrix} \tag{3.7}$$

with

$$\mathbf{M}_{l\nu,\,j\mu} = \delta_{lj}\delta_{\nu\mu}(\epsilon_\mu - \epsilon_j) + (1 - \delta_{lj})(2[\nu l|\mu j] - [jl|\mu\nu]) \tag{3.8}$$

$$Q_{l\nu,j\mu} = (2[\nu l|\mu j] - [\mu l|\nu j]) \tag{3.9}$$

and ω_I is the excitation energy.

It can easily be shown that RPA corresponds to the linear response of the ground-state SCF wavefunction and that, by comparison with Eq. 3.4, the diagonal blocks of the matrix **M**, except for the occupied orbital energy ϵ_j, correspond to the STEX Hamiltonian matrix of the different excitation channels. The RPA excitation operator in Eq. 3.6 involves not only coupling of different channels but also, by the **Y** component, some electron correlation. It has, moreover, the convenient property of making RPA results gauge invariant in the limit of a complete basis set. For such positive characteristics and for the limited computational effort needed to solve its equation by matrix direct [15] and integral direct algorithms [16, 17], the RPA approach has been largely applied to the study of the lowest excited states of medium-size molecules. However, because the RPA equation is projected on the "frozen" orbitals of the ground state, RPA cannot properly handle the effect of the electron relaxation for the highly excited states. For this reason excitations of the core shell electrons cannot be described quantitatively. For such processes the RPA excitation energies are several electron-volts away from the experimental results, and the intensities can also be strongly affected by neglecting the electron relaxation so easily taken into account, instead, by the STEX approximation.

RPA belongs to the class of polarization propagator methods that form a universal approach to determining spectroscopic properties in the optical and ultraviolet regions. As discussed above, it has several formal and practical advantages in that it explicitly optimizes the ground-state wavefunction and ensures orthogonality among states as well as gauge operator invariance, sum rules, and general size consistency. It has been implemented for a variety of electronic structure methods, variational or perturbational, wavefunction or density based. In a recent work on the use of polarization propagators in X-ray absorption spectroscopy (XAS) [18, 19], it was demonstrated that it is indeed possible to construct an electronic polarization propagator which is applicable not only in the traditional optical region but also in the X-ray region. This was achieved through the formulation of a resonant-convergent first-order complex polarization propagator (CPP) [20, 21], where the imaginary part of the molecular polarizability is included. This makes it possible to directly calculate the photoabsorption cross section at a particular frequency without explicitly addressing the excited-state spectrum. With a correction of self-interaction energy of the core orbital the approach provides fully correlated density functional theory (DFT), absolute-scale X-ray spectra that are obtained based solely on the optimization of the electronic ground state. This propagator approach has the same qualities in X-ray spectroscopies as those long shown in optical spectroscopies.

3.4 CORE ELECTRON BINDING ENERGIES

Calculations of core hole states and core electron binding energies have mostly concerned state-by-state optimization using open-shell Hartree–Fock, multiconfiguration self-consistent field, and configuration interaction techniques, while propagator techniques have been of restricted use owing to the "relaxation problem" alluded to above. This can to some extent be seen as a consequence of the breakdown of Koopmans theorem due to the large orbital and energy relaxation and at the same time the need to apply an operator manifold that contains core–valence transfer operators. Furthermore, some core electron processes are delicately dependent on the precise nature of the electron correlation and would require higher order propagator methods if the unrelaxed ground state is used as reference state [22]. Most Green's function or propagator applications to X-ray spectra have so far rested on the fact that the relative X-ray or Auger emission energies are accessible as single- and double-ionization potentials, respectively, and they treat these processes as ionization events with the neutral ground state as reference [23].

The calculation of the X-ray energies and their chemical shifts apply to core hole states that are embedded in electronic continua and therefore metastable, subject to variational collapse. The large orbital relaxation following the creation of core hole states is entailed with an nonorthogonality and state interaction problem in the calculation of transition moments if a self-consistent field procedure is applied. The embedding of the core hole state with the continuum implies interactions with scattered electron waves at high energies; the interaction itself leads though only to a marginal shift (well below 0.1 EV for first-row molecules) of the core electron IP and can surely be neglected in practical problems. Another difference with respect to low-lying valence hole states is that for the calculations of the latter states the root index, that is, the number of negative Hessian eigenvalues, is known a priori, while this is usually not the case for core hole states or other superexcited states. When a finite basis is used to describe these states, the continuum has a discrete representation and the states will have a finite, but in general unpredictable, root index. One therefore needs criteria other than those based on explicit knowledge of the root index to converge to states embedded in the continuum.

The self-consistent field (SCF) solutions for core hole states have traditionally been optimized within the open shell restricted Hartree–Fock (OSRHF) approximations, nowadays also by DFT, by a series of diagonalizations of the (open-shell) Fock (Kohn–Sham) operator [24, 25], a procedure which is linear with respect to the residual error vector. A maximum overlap criterion has often been used to identify the singly occupied core orbital from iteration to iteration [25, 26]. Since the core orbital has negligible overlap with any other orbital (the core hole state is also almost completely orthogonal to any other lower-lying self-consistent state), this procedure does not collapse variationally. Various algorithms have been employed to improve convergence for such calculations. Quadratic SCF gives secure convergence for the states lowest of its symmetry. These procedures can be step restricted for global convergence and quadratically and sharply convergent locally. With the tools developed in direct second-order SCF or multiconfiguration SCF (MCSCF) [27, 28] it has been possible to perform such optimization also for core hole states without variational collapse, often in a two-step procedure where in the first step the singly occupied core orbital is kept frozen. A Newton–Raphson algorithm with full coupling between configuration and orbital variables (with zero-level shift) does straightforwardly converge to the relaxed core hole state as this algorithm does not require knowledge of the correct Hessian index in the local region [29].

The main computational engine has up to now thus been the ΔSCF method [24], which has found applicability for core electron spectroscopy partly owing to the fact that there are systematic cancellations of electron correlation effects for core ionization, that is, the correlating valence pairs remain unbroken upon core ionization. Although comparatively small, the correlation error of core binding energies is still of the order of the shift values measured for organic compounds. A fine tuning of the computations through explicitly correlated methods remain, however, elusive for somewhat larger compounds owing to the poor particle scaling of such methods. Modern DFT has shown a way out of this dilemma, especially in conjunction with the so-called transition potential method. Chong has applied transition potential DFT techniques in a Kohn–Sham (KS) orbital formulation in a series of papers [30–32]; see also other contributions in the literature [33–39]. A transition state [40] implementation of the KS theory means that the ionization potential is directly identified as the energy of a core KS orbital optimized for a transition state with a fractional $\left(\frac{1}{2}\right)$ occupation of the core orbital. In SCF theory the half-occupied transition state (or the equivalent transition potential) energy is a strict approximation of ΔSCF (the relaxation energy is obtained through second order). Although further refinements of the transition-state theory can be obtained through combinations of several transition-state energies, it still remains a strict approximation of ΔSCF. From these arguments one can argue that a ΔKS method [39, 41] should improve upon the transition-state KS procedure.

Despite the fact that DFT is essentially a ground-state theory for the total energy of the system, the similarity of the KS and SCF equations suggests to attach a physical meaning to the KS orbitals and orbital energies, rather than that they are just considered as mathematical quantities in the solution of the DFT problem. While "Koopman's theorem" is not strictly valid for KS orbital energies, the experience shows that the (negative) orbital energies of the valence shell may be, in practice, close to the experimental IPs [38, 39], while the deviation for the core orbitals is usually quite large. The pioneering work of Perdew and Zunger [42] demonstrated that the origin of this large deviation for core shells can be traced to an error due to the spurious self-interaction for a singly occupied orbital, which is much enhanced for strongly localized orbitals like core orbitals and therefore undermines the results of the DFT-KS model in different practical applications. It was also shown that, by correcting for the self-interaction error (SIE) in some appropriate way, the corrected KS orbital energies will account not only for the electron correlation effects, as it could be hoped from the character of DFT, but also for the electron relaxation effects. The KS orbital energies of the core shells of a molecule can be estimated by an appropriate self-interaction correction (SIC) that may allow an estimation of core IPs with good accuracy from DFT calculations for the ground state of a molecule. Such an approach [43] can be applied in a single calculation to reproduce the X-ray photoemission spectra (XPS) at different K edges and is particularly efficient for applications to large molecular systems. There are indeed several possibilities for SIC DFT that adds a further correction of the potential coupling to the Perdew–Zunger method. An empirical parameter α that weighs the strength of the coupling term with respect to the one-electron potential was introduced in Tu et al. [43], where it was shown that α has, in practice, only a slight variation from the best value for the core $1s$ orbital energies of C, N, O.

3.5 CALCULATIONS OF CORE ELECTRON CHEMICAL SHIFTS

An important historic aspect of computational X-ray spectroscopy refers to the notion of the core photoelectron chemical shift and the development of simplifying models for calculations of the chemical shift and for its structural or stoichiometric interpretation. After its discovery in 1958 by Kai Siegbahn and his co-workers [44], the core ionization chemical shift became the most intriguing aspect of photoelectron spectroscopy and formed the ground for the notion of electron spectroscopy for chemical analysis (ESCA) [45, 46]. Although, by that time, chemical shifts were known in X-ray emission spectroscopy for several decades, the revelation of a chemical shift also in photoelectron spectroscopy brought about completely new prospects for chemical analysis. This was owed to the fact that the photoelectron shift is an order of magnitude larger than the X-ray emission shift and it has a more direct relation to chemical structure. Since the discovery great efforts have been made in using chemical shifts to gain information on stoichiometric and structural properties of matter, and many models were proposed and evaluated for the prediction and interpretation of the shifts. Relations between the chemical shifts and other properties or quantities were explored, such as charge distributions, reactivities [47], heats of formation [48, 49], proton affinities [50, 51], nuclear magnetic resonance (NMR) shifts [52], IR and Mössbauer shifts [53, 54, 66], and X-ray [55] and Auger core electron shifts [56], to mention a few examples. It was realized that successful use of the shifts for quantitative predictions of chemical and physical properties relies on good models, and along with the intense instrumental development giving ever higher precision and resolution, concomitant efforts were made in deriving such models. It was realized that these must be simple, yet accurate, in order to be useful.

3.5.1 Potential Models

Potential models became instrumental in the classical analysis of chemical shifts [45]. The motivation behind the potential models has been reviewed rather extensively, [e.g., 45, 57]. The derivation of these models follows rather straightforwardly from electrostatic considerations where the charge of the core ionized atom plus the ligand charge distributions add up to a potential at the core electron in the ground state according to

$$\delta E_i = k_{AqA} + \sum_{A \neq B} \frac{q_B}{R_{AB}} + l_a \qquad (3.10)$$

where q_B is the effective charge of an atom at a distance R_{AB} from the site (A) of ionization. At this site the effective charge q_A is distributed over the valence orbitals at an average radial distance $\langle r \rangle = 1/k_A$ from the core electron to be ionized. A quantum chemical derivation of this expression starts out from the fact that the orbital energy gives the dominating contribution to the total ionization shift, considering, for example, the water molecule, the different contributions relate approximately as orbital energy 559.52 eV, relaxation energy −20.37 eV, correlation energy 0.49 eV, and relativistic energy 0.48 eV as derived in Carravetta and Ågren [58]. The shift of the core orbital energy is

$$\Delta\epsilon_i = \delta\left[\left\langle i|\hat{h}|i\right\rangle + \sum_j (2J_{ij} - K_{ij}) - \sum_{B\neq A}\left\langle i\left|\frac{Z_B}{r_B}\right|i\right\rangle\right] \tag{3.11}$$

where $\langle i|\hat{h}|i\rangle$ is the kinetic and one-center nuclear attraction energy for core orbital i situated on atom A and J_{ij} and K_{ij} are direct and exchange integrals between core orbital i and other orbitals. Successive approximations in this expression leads to the potential model. The constant k_A takes here the form of a one-center repulsion integral between the core and valence charge distributions. Linear relationships have been established between shifts and charges for many classes of compounds [45, 46]. Large shifts, up to 10 eV, result from substitution with atoms or groups with strong electropositive or electronegative character. For unpolar compounds the shifts are comparably small, typically below 2 eV, and with moderate charging or decharging the ground-state model becomes more limited. In order to improve the ground-state potential models, final state relaxation needs to be taken into account. One can here mention the transition operator formalism of Goscinski et al. [59] and the two-potential formalism of Hedin and Johansson [60]. In the transition operator method the binding energy is related to an orbital energy obtained as an eigenvalue of a Fock operator where half an electron has been removed from the orbital. It corresponds nearly to the self-consistent field value for the ionization energy. With arguments similar to those for the ground-state model, the transition orbital energies are decomposed to a charge model expression as

$$-\epsilon_A^T = k_A^T q_A^T + \sum_{B\neq A}\frac{q_B^T}{R_{AB}} + l_A^T = k_A^T q_A^T + V_A^T + l_A^T \tag{3.12}$$

By taking the difference orbital energies $-\epsilon_A^T$ and $-\epsilon_A^G$ for the transition- and ground-state models the relaxation energy is directly expressed in terms of atomic charges and extra-atomic potentials [61, 62]. These expressions expose different contributions to the shift, a contraction of valence charge around the core, a charge transfer term representing the increase of charge around the ionization site, and a rearrangement of the extra-atomic potential due to relaxation. For homologous organic compounds it is in general found that the relaxation potential covaries with the measured shift, while the screening roughly remains constant [62]. The simple potential model for ESCA shifts is useful for predicting not only the change of ionization energy through the charging but also the ionization energy gradient with respect to nuclear–nuclear coordinates through the charge displacements. In fact, from the ground-state potential model the vibronic coupling constants can be derived via the charging and decharging of the core ionized atom and, hence, the Franck–Condon factors [63]. For instance, the potential model predicts shortening of the CO bond upon C_{1s} ionization, but a lengthening of the bond upon O_{1s} ionization with a vibronic coupling that reasonably matches the high-resolution ESCA band profiles [63].

3.5.2 Dielectric and Polarization Models

Applications of photoelectron spectroscopy to molecular crystals, liquids, and solutions [64–67] inspired efforts to generalize the simple ESCA potential model to chemical shifts due to solvation [68–72]. One has thus looked for formulations of the chemical shift for extended systems, other than when substituents have strong electropositive or electronegative character, and studied the response of the full spectrum upon condensation [68, 72]. One has addressed intermolecular interactions behind such shifts of both short- and long-range types, like exchange, electrostatic, polarization, and dispersion interactions. The theoretical models covering the polarization response can be classified as the dielectric [68], microscopic polarization [73], and reaction field models [71]. The latter can be viewed as a supermolecular extension of the dielectric model.

The Born model for the solvation of a charged sphere in a uniform dielectric gives an estimate of the photoionization energy shift,

$$\Delta E = \frac{q^2}{2a}\frac{1}{1-\epsilon}$$

where a is the sphere radius, q the charge, and ϵ the dielectric constant of the solvent. As any molecular properties or interactions are neglected in this expression, it is quite coarse, especially for smaller solvent species which are charged or have a large permanent dipole moment. Polarization models form one generalization of the dielectric model. These models treat the polarization response upon photoionization, its dependence on local structure, and effective local polarizabilities. Such models can be formulated based on atom and bond decomposition schemes of the external polarizabilities [73].

In the reaction field model, the atom, molecule, or supermolecule is enclosed by a cavity and surrounded by a linear, homogeneous, dielectric medium described by its macroscopic dielectric constant. The reaction field model can with advantage be applied as a semicontinuum model when the first solvation shell is accounted for explicitly. In addition to long-range polarization effects, it then also describes short-range effects (charge transfer and exchange) by accounting for the electronic structure of the first solvation shell. It may thus be seen as a mix of the dielectric model, improved by means of a reaction field, with the supermolecular model. Charge transfer within a molecular complex has a self-consistent coupling with the solvent through the reaction field. For core ionization there is a considerable charge rearrangement to screen the core hole, and the reaction field contribution is therefore conceivably significant for the solvent shift. The generalization [74] of the self-consistent reaction field method to open-shell and multiconfiguration states allowed applications on spectroscopic problems in general and photoionization in particular, such as for ionic [71] or molecular solutions [75]. The model allows rather detailed conclusions concerning the origin of solvent effects on chemical shifts in solution, such as the role of localization, dielectric polarization and reaction field contributions, and the role of optical versus static dielectric response of the medium. For polar crystals the reaction field model can be corrected by a potential term describing the Madelung summation of the electrostatic contributions.

3.5.3 Statistical Models

In liquids and solutions a chemical shift model should ideally account for the dynamical disordering of the solvent structures. This calls for models that are based on a decomposition of the intermolecular contributions to the shift and a parameterization of these contributions in terms of solvent structure, for example, atom–atom distribution functions. Such models should ideally account for the dependence of shift on temperature and pressure. From the distribution functions the shifts can be derived as well as the full photoelectron spectral function, including shift, width, and asymmetry, upon condensation. A basic assumption is that photoionization is vertical, meaning that both initial and final states can be associated with the same nuclear conformation. This approximation is well grounded considering the time scales between the photoelectron process and the rearrangement of the solvent molecules, which means that the solvent is not in equilibrium with respect to the final state. A common assumption behind such models is also that the internal solute nuclear motion is decoupled from external forces. This means that the spectral function f^{tot} can be written as a convolution of internal and external parts, f^{free} and f^{solv}, respectively,

$$f^{\text{tot}}(\omega) = \int f^{\text{free}}(\omega')f^{\text{solv}}(\omega - \omega')\, d\omega' \tag{3.13}$$

where f^{solv} is infinitely sharp for each photionization transition. The spectral function results from a large number of photoionization events with distribution $F^{\text{solv}}(\omega)$ of energies ω. The moments of the energy distribution function characterize the photoelectron spectral function, as shown elsewhere [72].

Considering that the photoionization interaction energy ω depends only on the solvent structure which is denoted by a collective variable Ω, that is, $\omega = G(\Omega)$, the shift and change of width can be expressed as

$$\Delta E_{BE} = \int G(\Omega)P(\Omega)\, d\Omega \tag{3.14}$$

$$\sigma^2_{\Delta E_{BE}} = \int G^2(\Omega)P(\Omega)\, d\Omega - \left[\int G(\Omega)P(\Omega)\, d\Omega\right]^2 \tag{3.15}$$

The weighting function $P(\Omega)$ is the geometric distribution function of the system and is accessible from a statistical Monte Carlo or molecular dynamics simulation. The function $G(\Omega)$ can be decomposed by means of perturbation theory for intermolecular interactions, applied either to the transition state or separately for initial and final states. Having established the dependence of each term in such a perturbational expansion on Ω, one integrates over all structures using the appropriate statistical weights $P(\Omega)$. A few simplifications, motivated in Ågren et al. [69, 76], that is, neglect of dispersion contributions, point ionization, spherical averaging of solvent polarizabilities, and neglect of charge transfer and exchange effects, give the spectral functions in terms of radial [$g(r)$] and angular [$\Xi(r, \theta)$] distribution functions of the solvent. The shift, obtained as a sum of electrostatic and polarization interactions with the solvent, is then expressed as the sum of the following two terms:

$$\int \omega_{\text{pol}} F_1(\omega)\, d\omega_{\text{pol}} = \int g(r) E_{\text{pol}}(r)\, dr = -\frac{1}{2}(2Q_i + 1) \int \frac{\tilde{P}}{r^4} g(r)\, dr \tag{3.16}$$

$$\int \omega_{\text{Coul}} F_2(\omega)\, d\omega_{\text{Coul}} = \int \Xi(r, \theta) E_{\text{Coul}}(r, \theta)\, dr\, d\theta = -\int \frac{\mu \cos\theta}{r^2}\, dr\, d\theta \tag{3.17}$$

For solutions the polarization interaction is mainly responsible for the shift, while the width is mostly given by the electrostatic interaction. This interaction can often be localized to the first solvation shell, for instance, in the case of water and probably for other polar liquids as well. The angular distribution of the molecules in the first solvation shell thus represents the first moment of the energy distribution well, while the longer range electrostatic contribution largely cancels due to disorder. One can see the two terms above as an extension of the potential model with relaxation, where the terms are obtained from integration of distribution functions, which in turn are dependent on temperature and pressure.

3.5.4 Thermochemical Models

Thermochemical cycles have been used for studies of binding energy shifts and related solvation and reorganization energies for many different types of systems such as metallic elements [77], free molecules [47], and solutions [64, 65]. One particular application is to derive proton affinities from thermochemical cycles, including core ionization potentials [78] using relations that connect the core electron chemical shift with thermochemical energies such as heats of formation and proton affinities. The first proposal to use core electron binding energies for predictions of thermochemical energies was due to Jolly and Hendrickson [48] and was later followed by several applications [49, 78–81]. Through the equivalent core approximation a property like proton affinity can be linked to the core ionization energy of an isoelectronic compound. A correlation of this type was observed by Martin and Shirley [50] and further studied by Davis and Rabalais [82]. Later, Nordfors et al. [51] formalized a thermochemical model, containing core-ionized versus valence-ionized reference compounds, for predicting proton affinities from core binding energy shifts.

Core ionization energy shifts and the equivalent core approximation have frequently also been used in Born–Haber studies of condensation effects on photoelectron spectra; see the work by Mårtensson and Johansson [77]. In the Born–Haber cycle the substituent shift in the thermochemical cycle is replaced by the shift due to condensation and solvation enthalpies are replaced by heats of formation or proton affinities. In the Born–Haber cycle solvent photoionization is equivalent to the following steps: dissociating an atom from the ground-state compound, core ionizing the free atom, substituting the core ion for the valence-ionized equivalent core species, and reentering the ion into the compound. For metallic elements this scheme provides accurate predictions of ionization potentials for free atoms [77]. Processing of the Born–Haber cycle for solutions uncovers the reorganizational part of the solvation energy for the initial-state species. The reorganization term emerges since the solvation process is adiabatic, while photoionization is vertical. Siegbahn et al. [64, 65] used this fact to investigate reorganization contributions to the solvation of alkali, alkaline, and halogen ions in solutions and unravelled some regular trends for the reorganization versus charge state and effective ionic radius.

The thermochemical models thus mostly make use of the equivalent core approximation, which relates electronic structures and molecular geometries for the core and the resonance valence states. The equivalent core error is due to changes between valence and core charge penetration due to the different surroundings. This may contribute to half an electron-volt or more to the core ionization potential, which is much larger than localization effects and is the reason that core levels are more stabilized upon condensation than valence levels, as observed, for instance, in water [83] and benzene [84].

3.5.5 Empirical Models

Potential or statistical models make it possible to interpret the physical origin of the shift as they are based on rational principles. Common for them is, however, that although they establish a relation between shift and structure, this structure dependency is given only *implicitly*; the structure must be supplied to the model in order to produce the shift. The inverse problem of using the model for obtaining structure from shift is harder since one sole or a few shift values can be produced by a number of different charge distributions and structures. They can, often at best, distinguish between a few possible alternatives. Using the statistical models the large number of structural parameters can be reduced to a small number of "effective" parameters, such as those of the solvent distribution functions.

Empirical models can give more explicit relations for deriving structure. One class of empirical models is based on the assumption that the ground-state-induced effects together with the final state relaxation effects can be described in a common polarizability expression [85], which can be parameterized to display the structure dependency explicitly. Considering the situation when the position and polarizability of all atoms and bonds are known, one may write the shift as

$$\Delta E_B^{\text{pol}} = -\frac{1}{2}\sum_j \left[\sum_i \bar{\epsilon}_j^i \mathbf{P}_j \bar{\epsilon}_j^i + \sum_{i \neq k} \bar{\epsilon}_j^i \mathbf{P}_j \bar{\epsilon}_j^k\right]$$

where i denotes the charge centers and j the polarization centers. Using clamped charges for the ionized molecule and screened polarization, with the screening taking place through the chemical bonds and with the ionic field assumed homogeneous, one obtains at large distances

$$\Delta E_B^{\text{pol}} = -\frac{Q_f - Q_i}{2\chi}\sum_j \frac{P_j}{R_j^4} \tag{3.18}$$

Here χ is the dielectric constant. For more distant-lying centers the effect of molecular bonding is small and the screened interaction with the ionic field can be considered as homogeneous. Smearing out the polarization centers to a dielectric, $\chi = 1 + gP$, the summation in Eq. 3.18 gives

$$\Delta E_B^{\mathrm{pol}} = -\left[\frac{2Q_i + 1}{2a}\right]\left(1 - \frac{1}{\chi}\right) \tag{3.19}$$

where a is the lower limit of integration. In the other extreme one thus retains the usual Born model for the solvation energy of a uniformly charged sphere without any structure dependency. Parameterizing the polarizability in a suitable way one can make the structure dependency explicit, for instance, Gasteiger and Hutchings introduced an effective polarizability P_d which took the smallest number of bonds (n_j) between the charge and each atomic site into consideration [86]. The formula was determined empirically:

$$P_d = \frac{4}{N_c}\left(\sum_j (0.75)^{n_i - j} P_j\right)^2 \tag{3.20}$$

This entity was shown to correlate well with state relaxation energies derived from experimental results. The connectivity number N_c was introduced and is determined by a simple bond-counting procedure. The formula does not take atomic types into consideration, only the number of bonds b_n in the coordination spheres weighted by their orders n:

$$N_c = \sum_n b_n (0.5)^{n-1} \tag{3.21}$$

It was shown that core ionization shifts correlated very well with the connectivity number for homologous compounds [85, 86]. The model has been generalized by Nordfors to incorporate molecules containing many different types of atoms and different hybridizations [85]:

$$I^c = a_0 + \sum_i N_c^i a_i \tag{3.22}$$

$$N_c^i = \sum_n b_i (0.5)^{n-1} \tag{3.23}$$

The constants a_i are considered typical for each element i and are determined by a least-squares fit to already available experimental data. Connectivity numbers have found applications also in topological and graphical theories to describe chemical phenomena [87].

3.6 CALCULATIONS OF VIBRONIC COUPLING IN X-RAY SPECTRA

Core orbitals have been attributed a strictly nonbonding function as they are both localized and nonoverlapping with other orbitals. They are furthermore only marginally dependent on the change of nuclear configuration. In the mid 1970s this view was significantly disturbed by the demonstration that X-ray spectra involving strictly nonbonding valence orbitals show fine structure [88]. Furthermore, XPS measurements using monochromatized X rays showed that core photoelectron bands can be asymmetric [7, 89], even exhibiting structure at very high resolution. The early results were interpreted in terms of a geometric relaxation of the involved core hole states. Physical explanations were proposed by applying the concepts of the potential model for chemical shifts [63] and the equivalent core approximation model. Several investigations [63, 90–94] were devoted to the explanation of these features also in terms of electronic structure theory and ab initio calculations [92]. With the aid of Franck–Condon analysis the changes in the force fields and geometries upon the electronic transitions could be indicated [91, 95], giving rationalizations in terms of relaxation effects, localization, exchange interaction, and vibrational interference effects. Certain patterns were unraveled such that the bond length is shortened upon C_{ls} ionization, lengthened upon O_{ls} ionization, whereas ${F_{1s}}^{-1}$ states are dissociative.

Vibronic analysis of X-ray spectra is also intimately connected to the localization problem. Bagus and Schaeffer [96] showed that SCF calculations of the core ionization potentials of symmetric molecules were much in error when full molecular symmetry was imposed, whereas the error decreased to only normal error of core IPs in the case of symmetry breaking. The difference was related to a relaxation effect, or error, in the high-symmetry case. Further use of localized solutions was advocated by studies of other properties, such as energy gradients, vibronic coupling constants, electronic transition moments, photoelectron angular distributions, and satellites [97–101]. For example, only broken-symmetry OSRHF solutions could provide energy gradients that reproduce band profiles observed in XPS spectra of homonuclear species.

In practice it is simple to remedy the symmetry dilemma for core hole states by a limited configuration interaction calculation in the low-symmetry point group that retains a quasi-degenerate symmetry-adapted solution. Using MCSCF wavefunctions of high symmetry with a limited configurational expansion the missing relaxation energy at the Hartree–Fock level can largely be compensated [98]. The relaxation in the low symmetry can thus be described by correlation in the high symmetry. Of course, the symmetry breaking does not refer to the exact wavefunctions in the case of diatomics. For polyatomic molecules containing an element of symmetry the situation is different as symmetry breaking can follow vibronic coupling over antisymmetric modes. Thus core hole states quasi-degenerate in the high symmetry can couple through nontotally symmetric vibrational modes in a Jahn–Teller or pseudo-Jahn–Teller-like fashion. Domcke and Cederbaum showed that the ${O_{1s}}^{-1}$ Σ_u and Σ_g states of CO_2^+ couple over the antisymmetric stretching modes, thereby lowering the symmetry from $D_{\infty h}$ to $C_{\infty v}$ [102]. A diabatic electronic representation which diagonalizes the nuclear kinetic operator instead of the electronic Hamiltonian also displays such effects.

While for X-ray absorption-type spectra, the short lifetime manifests itself merely as a broadening of the bands, the effect of the short lifetime emerges differently for X-ray or Auger emission in that the lifetime broadening introduces interference among the various vibronic progressions. Several theoretical studies have been devoted to the effect of lifetime vibrational interference in vibronic decay of molecular core hole states [103–108]. Cesar et al. [109] approached the problem by using Åberg's formulation of atomic X-ray and Auger decay as multichannel resonance-scattering processes [110]. The expressions derived contain contributions due to both direct and interference terms. The latter are cross terms that correspond to second-order-like contributions for the combined process of formation and decay of the core hole state where virtual vibrational transitions are possible during the time of the existence of the intermediate state.

The analysis of vibronic contributions to X-ray spectra has mostly relied on Franck–Condon (FC) analysis. Two approaches emerge as important special cases, the vertical and the adiabatic approaches. These refer to the expansion of the excited-state potential energy surfaces at the equilibrium geometry of the ground state or at the equilibrium geometry of the excites states, respectively. Several methods have been advocated for multidimensional FC analysis of vibronic spectra within the adiabatic approach, for example, the generating function methods [111] and the use of efficient recurrence relations [112–114] and autocorrelation functions [115]. The adiabatic approach is advantageous for finer details of the spectra containing small vibrational progressions. For long vibrational progressions with many modes excited the treatment of the coupling force constants, and anharmonicity may become crucial, in particular for emission-type spectra. With an expansion of the potential curve around the excited state equilibrium, anharmonicity may be poorly described in the FC region. The vertical approach, although expanding the potential curve only to first order, does take into account the anharmonicity contribution to the intensities. The use of the vertical first-order coupling constants was early put forward by Domcke and Cederbaum [116–119]. In its simplest form the vertical approach gives FC factors for M modes as a product of Poisson distributions:

$$F(0, n_1, n_M) = \exp\left(-\sum_{s=1}^{M} a_s\right) \prod_{s=1}^{M} \frac{a_s^{n_s}}{n_s!} \tag{3.24}$$

where $a_s = (k_s/\omega_s)^2$ and k_s is the first-order vibronic coupling constant and ω_s the ground-state frequency for the mode. There are some advantages of using a vertical expansion as it better describes the local features of the excited-state potential surface where the gross of the intensity of a vibronic band resides. Whenever the geometry of the molecule in the excited state remains within the FC region, this approach produces an accurate overall bandshape, that is, the envelope of the vibronic band, but may poorly reproduce finer spectral details developed by the molecular dynamics in longer time periods.

One can also obtain bandshapes of X-ray spectra in a time-dependent fashion through the use of autocorrelation functions which are obtained from wavepackets propagated on the potential surfaces of the states participating in the X-ray transition. However, such methods are limited to a few degrees of freedom and therefore, practically, to diatomic or triatomic molecules. This follows from that, with $3N - 6$ degrees of freedom, the potential surface cannot be computed globally. With current analytic techniques the multidimensional gradients and Hessians can be obtained at the vertical and adiabatic points (stationary points for ground and excited states, respectively). These thus provide a local expansion of potentials and form basic quantities for the vibronic methods that in principle allow calculations of autocorrelation functions for the motion of polyatomic systems. Time autocorrelation functions then describe molecular motions on the multidimensional potential energy surface and can be obtained by solving the quantum equations of motions. A Fourier transform of the autocorrelation function gives a time-evolved interpretation of the formation of the vibronic spectrum. A general formalism based on the notion of autocorrelation functions which includes the vertical and adiabatic approaches as special cases has been derived by Cesar [115].

3.7 SHAKE-UP SPECTRA

3.7.1 Sudden Approximation

Shake-up states involve a valence electron excitation following a core hole creation. The process has mostly been considered as a dipole transition in an N-electron system where one assumes the creation of the shake-up states to be decoupled from their decay and neglects their interaction with continua of doubly or higher charged states. In a one-particle language the shake process takes place between a bound orbital i and an empty orbital v (bound for shake-up and continuum for shake-off) in the presence of an opened core orbital x. The final shake-up state thus has a configuration with three open shells in case the ground state is a closed shell. Such a system gives rise to one quartet and two doublet spin states. Using spin-restricted open-shell theory (OSRHF in the case of Hartree–Fock optimization), the quartet state is expressed in determinants of spin orbitals as

$$^{4}\Psi = \{\phi_x \phi_i \phi_v\} \tag{3.25}$$

for the $M_s = \frac{3}{2}$ component and the two doublets as

$$^{2}\Psi^{1} = \frac{1}{\sqrt{2}}(\{\phi_x \bar{\phi}_i \phi_v\} - \{\phi_x \phi_i \bar{\phi}_v\}) \tag{3.26}$$

$$^{2}\Psi^{3} = \frac{1}{\sqrt{6}}(2\{\bar{\phi}_x \phi_i \phi_v\} - \{\phi_x \bar{\phi}_i \phi_v\} - \{\phi_x \phi_i \bar{\phi}_v\}) \tag{3.27}$$

for the $M_s = \frac{1}{2}$ components. The barred spin orbitals have β spin, the unbarred ones α spin. These states are also called configuration state functions (CSFs) because they constitute the smallest linear combination of determinants that fulfills spin and spatial symmetry. The quartet state can be discarded due to the monopole selection rules that govern the overlap between the final shake state and the core-annihilated ground state that is a doublet. In the first CSF in Eq. 3.26 the i and v spin orbitals have been coupled to a singlet and then coupled with the core shell x to a doublet. The second of the two doublet states in Eq. 3.27 is formed by coupling the i and v shells to a triplet state, which in turn is coupled to the open core shell to form a final doublet state. The second CSF is called the spin-polarized CSF [120]. The short-lived character of the shake-up states emerges then only as an added Lorentzian broadening of the bands.

With the application of some approximations for the outgoing photoelectron one can further simplify the shake-up picture to the calculation of transition moments and energies between bound states. Most calculations of the shake-up process have been carried out with wavefunction approaches, in particular the configuration interaction method. The mainstream of applications have neglected postcollision interaction and adopted the strong orthogonality condition for the photoelectron in the Born–Oppenheimer approximation. With these assumptions cross sections for shake-up intensities have been derived by several researchers. Martin and Shirley [121] derived such cross sections for wavefunctions for initial and final states that are both correlated and relaxed with sets of mutually nonorthogonal orbitals. Equivalent expressions using second quantization were later derived by Arneberg et al. [100]. The three basic operational rules in this derivation are given by the strong orthogonality condition, the anticommutation relations for creation and annihilation operators(a_r^+ and a_r), and the nonorthogonality transformation between initial- ($\{\phi_q\}$) and final-state ($\{\phi_p\}$) orbitals imposing the operator relations $\hat{a}_p = \sum_q \langle \phi_p | \phi_q \rangle \hat{a}_q$ and $\hat{a}_p^+ = \sum_q \hat{a}_q^+ \langle \phi_q | \phi_p \rangle$. The strong orthogonality condition for the outgoing electron implies

$$\Psi_f(N) = \hat{a}_\epsilon^+ |\Psi_f(N-1)\rangle \tag{3.28}$$

and thus

$$\hat{a}_\epsilon |\Psi_f(N-1)\rangle = 0 \tag{3.29}$$

where Φ_ϵ represents the outgoing electron orbital and $\psi_f(N-1)$ the bound shake-up state of the ion. With this condition one neglects the correlation between continuum and bound electrons and ensures that the total final-state wavefunction $\psi_f\,(N)$ is normalized if $\psi_f(N-1)$ is normalized. The transition moments and intensities (cross sections) are obtained from the Fermi golden rule using the many-electron dipole operator

$$\hat{T} = \sum_{p,q} \langle p | \hat{t}_1 | q \rangle \hat{a}_p^+ \hat{a}_q \tag{3.30}$$

giving the transition moments

$$T_{f0} = \langle \Psi_f(N) | \hat{T} | \Psi_0(N) \rangle = \sum_{p,q} \langle p | \hat{t}_1 | q \rangle \langle \Psi_f(N) | \hat{a}_p^+ \hat{a}_q | \Psi_0(N) \rangle \tag{3.31}$$

Using the three operational rules stated above the transition moment is reduced to [100]

$$T_{f0} = \sum_q \langle\epsilon|\hat{t}_1|q\rangle\langle\Psi_f(N-1)|\hat{a}_q\Psi_0(N)\rangle + \sum_q \langle\epsilon|q\rangle\langle\Psi_f(N-1)|\hat{T}|\hat{a}_q\Psi_0(N)\rangle \tag{3.32}$$

The particular form of this expression with one direct term, the first one, and one conjugate term, the second one, is a direct consequence of the strong orthogonality condition for the outgoing photoelectron. The direct term describes a one-electron bound–continuum dipole transition weighted by a many-electron overlap element between bound states. The conjugate term is guided by a bound–continuum orbital overlap element weighted by a bound–bound state many-electron dipole transition element. The summation index q runs over the occupied orbitals of the initial state. For core electron shake states this summation can, to a very good approximation, be restricted to one index, the one corresponding to the vacated core orbital in the final state. All other terms are close to zero due to orbital orthogonality. Thus hole-mixing effects common in (inner) valence spectra [73] are not present for the core electron shake-up phenomenon and the intensity variation with excitation energy is guided by one continuum overlap element $\langle\epsilon\,|\,q\rangle$ and one dipole orbital matrix element $\langle\epsilon\,|\,t_1\,|\,q\rangle$ only. For low-energy excitation close to the ionization threshold these two orbital matrix elements are strongly varying with the excess energy ϵ and are furthermore of comparable magnitude. This means that the total cross section has a strong energy dependence and also that the conjugate term contributes much to this cross section.

For high-energy excitation the energy region is so narrow that the dipole matrix elements $\langle\epsilon|\hat{t}_1|q\rangle$ do not significantly vary for different shake-up states, while the overlap $\langle\epsilon|q\rangle$ is negligible due to the highly oscillating character of the continuum orbital. For X-ray excitation the shake-up intensity is thus well guided by the many-electron overlap element only X_q^f:

$$I_f \propto |T_{f0}|^2 \propto |X_q^f|^2 = |\langle\Psi_f(N-1)|\hat{a}_q\Psi_0(N)\rangle|^2 \tag{3.33}$$

The set of approximations leading to Eq. 3.33 is often summarized as the sudden approximation. The thumb rule described above concerns an independent particle description of the shake phenomenon. With this description one can thus associate each pair of occupied and unoccupied orbital indices to two specific shake-up states, a triplet and a singlet parent coupled state. Systems that can be described this way receive shake intensity through the effect of orbital relaxation — the more relaxation the more intensity. The effect of relaxation on intensities is clearly also seen from the overlap element in Eq. 3.33. However, in order to obtain quantitative prediction for shake-up intensities, more sophisticated models, including the effect of electron correlation, must be considered. With the definition of *correlation energy* as the difference between one-determinant, Hartree–Fock, energy and the exact expectation value of the electronic nonrelativistic Hamiltonian, final-state correlation is always needed for shake-up since a correct spin-coupled shake-up state needs at least two determinants. A better definition of correlation energy might be obtained by replacing a single determinant by a proper spin- and symmetry-adapted least linear combination of determinants. With such a definition both correlation and relaxation energy contributions to shake states are model dependent, since there is no unique way to define a many-open-shell Hartree–Fock energy. Including also correlation in the initial state there will be more contributions to the final overlap amplitude than that from the core-annihilated ground-state Hartree–Fock determinant $\hat{a}_x|\phi_0\rangle$. The general effect of initial-state correlation can be seen if we write the hole–particle expansion of the wavefunctions of the initial and final states in a common form:

$$|\Psi^A\rangle = c_0^A|\Phi_0^A\rangle + \sum_{i,j} c_i^{j,A}|\Phi_i^{j,A}\rangle + \text{higher order excitations} \tag{3.34}$$

Here A equals gs (ground state), 1s (main-hole state); or ss (shake-up state). The following obvious coefficient relations apply: $c_0^{gs} \gg c_i^{j,gs}$ (the Hartree–Fock determinant dominates for the ground-state wavefunction); $c_0^{1s} \gg c_i^{j,1s}$ (the 1s annihilated Hartree–Fock determinant dominates for the main core hole state wavefunction); and $c_0^{ss} \ll c_i^{j,ss}$ [the shake-up configuration (or CSF) is dominating for the shake-up state wavefunction]. Using the Slater–Condon rules in a frozen orbital approximation only contribution to the overlap amplitude is allowed from pairs of (initial- and final-state) determinants with no change in orbitals indices. The intensity

$$I = |g_{1s}|^2 = |\langle\Psi^{A=1s,ss}|a_{1s}|\Psi^{gs}\rangle|^2 \tag{3.35}$$

then simplifies to

$$I = |c_0^{gs} * c_0^A + \sum_{i,j} c_i^{j,gs} * c_i^{j,A}|^2 \tag{3.36}$$

Considering the relations of the coefficients given above, one finds that for $A = 1s$, main-core electron ionization, the intensity is well guided by $|c_0^{gs} * c_0^A|^2$. However, for the shake-up states ($A = ss$) *at least* one coefficient product in the second sum is of the same order as $c_0^{gs} * c_0^A$. Initial-state correlation is thus decisive for state-specific intensities, but also for total shake cross sections.

3.7.2 Calculations of Shake-Up Spectra

Optimization of large and accurate wavefunctions of shake-up states and the calculation of generalized overlap amplitudes from these wavefunctions are made possible by several program implementations. Many of these implementations contain features that are helpful for core ionization problems in general, including the shake phenomenon, such as algebraic constructions of higher order Green's functions, unitary parametrization of orbital rotations, algorithms for efficient core hole optimization, unconstrained optimization techniques, perturbation selection of wavefunctions, and evaluation of matrix elements between wavefunctions with mutually nonorthogonal orbitals, to mention a few of many examples with impact for the core electron shake-up problem [122].

As for most electron spectra, the calculation of shake-up spectra can be subdivided into two main categories; Green's function methods of algebraic diagrammatic construction (ADC3 and ADC4 [22, 123]) and wavefunction methods [here OSRHF, configuration interaction (CI), and MCSCF]. The latter comprise explicit Hamiltonian methods, active-space methods, configuration selection methods, and amplitude selection methods. The shake-off calculations can be divided into those using so-called L^2 (Stieltjes imaging) and scattering matrix (K-matrix) methods, which are described in the following in terms of primary photoionization.

The explicit Hamiltonian methods are straightforward in that the Hamiltonian elements over an N-electron basis of determinants (antisymmetrized product of spin orbitals) or CSFs are set up in core and diagonalized completely. The explicit Hamiltonian techniques have the advantage of being flexible for generation of the configurations; those with importance can be selected by inspection or by some perturbational argument. On the negative side one notes that the size of the CI Hamiltonian puts limits on the system that can be handled. For larger configurational expansions the CI Hamiltonian must be formed and diagonalized by direct techniques [124].

In the active-space methods the wavefunctions are determined by orbital divisions into subspaces of inactive, active, and secondary orbitals with double, fractional, and empty occupations, respectively. The determination of the orbital spaces are made by means of perturbation theory analysis or by natural orbital analysis from pre-calculations. There are two main types of active-space methods, the complete (CAS) and restricted (RAS) active-space methods [122, 125, 126]. With the former, configurations are generated by all possible excitations within the active space, fulfilling space and spin symmetry requirements; with the latter, a further orbital division is assumed with restrictions in the complete expansions. Since many shake-up states are governed by multiple excitations and the correlation of these implies further excitations with respect to the ground-state configuration, a complete expansion scheme has some conceptual merit. The practical problem is that the number of configuration state functions increases steeply with the number of active orbitals, and the active spaces must for computational reasons therefore be limited. There are, however, two other advantages with active-space techniques; the wavefunction can be self-consistently optimized also with respect to orbitals, that is, implemented as MCSCF, and the sudden approximation shake-up cross sections can be obtained very efficiently. Self-consistent optimizations though introduces nonorthogonality and interaction between states, something that both complicates the numerical procedure for the shake-up amplitudes and leads to poorer quality of the results. Postprocessing of the wavefunctions can alleviate this problem [127].

As indicated above the correlation of multiply excited wavefunctions describing shake-up states requires hole–particle expansions of high (maybe unknown) order [128, 129]. Computations of wavefunctions containing all CSFs generated by higher excitation levels soon become very costly. Similar problems are encountered in applications of configuration interaction for low-energy excitation states and spectra. Several methods have been proposed to deal with the problem, so-called contracted CI [e.g., 130], so-called multireference CI [e.g., 131], or selection CI techniques in general [132]. In large-scale calculations the configurational selection is automatized by perturbational, occupational, or other criteria.

Early multireference and selection CI calculations of shake-up spectra were carried out by Rodwell et al. [133, 134] and Barth et al. [135]. With multireference determinant CI (MRDCI) [131, 136], a space of determinants of small dimension is employed as reference for external (single and double) excitations. All configurations which appear in the CI expansion with expansion weight above a certain threshold are included. The inclusion of the individual external CSFs excited out of the reference space may be tested on the basis of their Hamiltonian matrix element with each of the reference CSFs. This procedure can then be repeated by successively lowering the thresholds or, iteratively, by successive solutions of the secular equation. A related approach, the so-called "extended root set approach", was applied by Rodwell et al. [133, 134]. A list of CSFs generated from a smaller set of important CSFs are obtained by perturbational criteria. Generally, a ΔSCF precalculation for the main-core-hole state generates a set of orbitals for all shake-up states. Improved orbitals were obtained by either MCSCF calculations in (part of) the reference space [134] or by natural orbital transformations of precalculated CI wavefunctions [134, 135].

The Green's function, or propagator, methods (see above) are often advocated due to the formal advantage that the physical information, namely, ionization energies and spectral intensities, is obtained directly without the need for separate calculations of initial and final states. Green's function methods have therefore been used extensively for characterizing photoelectron spectra in the valence region. They have served as improvements of Koopmans theorem predictions in the outer valence region or for identifying correlation state satellites and breakdown of the orbital picture in the inner valence region. Green's function calculations of core electron shake-up spectra have mostly used so-called algebraic diagrammatic construction, ADC3 and ADC4, methods [22, 123], which are respectively complete third- and fourth-order approximation schemes to the one-particle Green's function [123]. Within these schemes the ionization energies and spectroscopic factors are obtained as eigenpairs of a Hermitian secular matrix and computed with techniques similar to those used in diagonalizing of large CI matrices. The construction and solution of the secular matrix are actually simplified in the core electron problem since one is allowed to neglect the core–valence correlation (cf. single-occupancy restriction, neglect of semi-internal CI, which leads to a decoupling of the secular problem [22, 137]). The secular matrix is then defined by a configuration space comprising $1h$ (one-hole), $2h$–$1p$ (two-hole, one-particle) and $3h$–$2p$ (three-hole, two-particle) states with exactly one core level vacancy. The low-order particle states are treated at higher level ($1h$ states by fourth perturbation order) than higher order particle states (e.g., $3h$–$2p$ states by first perturbation order). Lower shake-up states are thus better described than higher lying ones [22]. The merits and limitations of ADC(3) and ADC(4) have been discussed in conjunction with applications to the shake-up spectra of the N_2 and CO molecules [22, 123].

3.8 AUGER SPECTRA

Auger spectra have extensively been used for sample analysis of elements and for surface structure analysis by means of scanning and surface-imaging techniques. For molecules the Auger experiment has mostly been used as a spectroscopic tool to obtain information on dicationic states. Auger spectra have thus been analyzed in terms of electronic and conformational structures of such states but have also been used as probes for the dynamics of electron–molecule scattering processes, mostly for the study of electronic structures in terms of molecular orbitals (MOs) and many-body character. A variety of methods have been proposed with chemical applications that cover symmetry, delocalization, hybridization, and bonding. They have also concerned more subtle effects such as vibronic couplings, fine structures, and the associated information on force fields and equilibrium structures of ionic states [138]. Effects due to electron correlation, such as breakdown of the molecular orbital picture [139, 140] and hole-mixing effects [141], frequently analyzed in the context of valence photoelectron spectra, all have counterparts in Auger emission but with different manifestations.

Molecular Auger spectra show considerable structure over a wide energy range. The high density of final states and the lack of strict selection rules make their analysis rather complicated. For small species the outermost high-kinetic-energy part shows resolved structures that can be described in terms of MO theory and even in terms of vibronic excitations, while the spectra grow progressively more complicated at lower kinetic energies. The final states of molecular Auger transitions are naturally divided into three classes, comprising outer–outer, inner–outer, and inner–inner valence states. These three groups of states represent fairly nonoverlapping energy regions and have rather different characteristics with respect to relaxation energy, electron correlation, transition amplitudes and vibrational (dissociative) broadenings, and so on. Most of the intensity is gathered into the outer–outer group of states, while for the other groups structure is often blurred by dissociative broadening and overlapping satellites. From a theoretical point of view these three spectral regions can roughly be described as independent particle states, hole-mixing states, and states with breakdown of MO theory, respectively. Final state correlation effects and the breakdown of the orbital picture is of great importance in many molecular spectra. It means that one finds more lines than two-hole combinations of orbitals in these spectra. These effects are due to energetic quasi-degeneracies between two-hole states with and without hole–particle (shake-up) excitations, which frequently occur in the inner and intermediate valence regions. Such interactions happen to be more pronounced for molecules than for atoms, the more so the more unsaturated the bonding.

Modern theoretical formulation of the Auger process follows many-channel resonant scattering theory [142] with roots in the classical work of Fano [143] on the theory for configuration interaction among continuum states. The scattering theory formulation of the Auger process in molecules [109] follows that for the atomic case. The main difference between the two cases is that the nuclear degrees of freedom increase the number of available open scattering channels and create additional vibronic coupling between these channels. The generalization of the conventional analysis of molecular Auger spectra from a two-step model, where the ionization/excitation process is considered decoupled from the following electronic decay, to a one-step model is necessary for several reasons. An explicit consideration of the excitation process is required for describing the vibrational and state interference effects and for the calculation of the Auger lines profiles. There will in general be a contribution from the nuclear degrees of freedom to the core hole state lifetimes and the discrete–continuum interaction energies. There will be inter- and intrachannel couplings in the electronic wavefunction depending on the nuclear motion. Furthermore, the total scattering cross section also includes the possibility of the direct scattering events for which vibronic coupling may play a role.

The generalization of the scattering theory formulation of Auger to include molecules was earlier given by Cesar et al. [109], emphasizing the consequences for the vibronic cross sections and the fine-structure analysis. A theory was presented where an explicit assumption of the validity of the Born–Oppenheimer approximation was made for the asymptotic vibronic states of the initial and final states but where the true electronic–vibrational states were assumed to represent the true continuum molecular wavefunctions. Attention was given to how the vibronic channel coupling is included in the transition amplitudes of the direct and resonant scattering events and its relative importance for the observed spectral functions.

In most practical applications, the starting point for interpretation of molecular Auger spectra has mostly been given by the Fermi golden rule expression

$$I = |\langle\Psi_i|(\hat{H}-E)\Psi_f\rangle|^2 = \left|\langle\Psi_0|a_q^+(\hat{H}-E)a_\epsilon^+ a_r a_s\Psi_0\rangle\right|^2 \tag{3.37}$$

assuming the nonradiative decay of core hole states occurs as a two-step process, with the deexcitation uncoupled from the excitation and independent of the interaction between primary photoelectrons and Auger electrons.

With a strong orthogonality condition imposed on the scattered electron orbital, fulfilling the killer condition $a_\epsilon^+|\Psi_0\rangle = 0$ the Auger amplitude resulting from the Fermi golden rule expression (Eq. 3.37) takes the form

$$M_{if} = \sum_x \tau_x\chi_x \tag{3.38}$$

that is a sum of terms in which each term constitutes a product of a molecular orbital (MO) factor τ_x and a many-body factor χ_x. For Auger this sum resolves as

$$\sum_x \tau_x\chi_x = \sum_{qrs}\langle\phi_\epsilon\phi_q|\frac{1-P}{r_{12}}|\phi_r\phi_s\rangle\langle\Psi_f(N-2)|\hat{a}_q^+\hat{a}_s\hat{a}_r|\Psi_i(N-1)\rangle \tag{3.39}$$

for which there is a threefold summation of indices: $x = q, r, s$. However, for all cases it is well motivated to limit the core index q to one item due to the almost perfect orthogonality ($\chi_x \simeq 0$) between states with holes in different core orbitals or with holes in core and valence orbitals. One can note that with, effectively, two indices for annihilation of the initial-state orbitals, there is a great number of possibilities for near degeneracies of the final states of an Auger transition.

With the single-channel and strong orthogonality condition imposed on the continuum electron orbital, a many-body factor is obtained that is only dependent on the characteristics of the residual bound states. A conventional MO analysis of the spectra is entailed only if there is just one large many-body term in the summation and if that term is close to 1. In that case one further reduces the MO factor in terms of, for example, local decomposition of symmetries and charges. One notes that for Auger a continuum orbital enters explicitly in the MO factor (as it does also in the photoionization case); the calculation of the MO factor will be discussed in a following section.

3.8.1 Many-Body Effects in Auger Spectra

Correlation, or many-body effects, can be classified according to the many-body factor χ_x. If χ_x is close to 1, the MO picture, the *aufbau* principle, a "Koopmans theorem" and the quasi-particle picture hold. The analysis of the Auger spectrum can then be conducted solely in terms of MO theory. When more than one χ_x enters in the wavefunction, we have hole-mixing effects and electronic interference in the transition cross sections, in analogy to the case of photoelectron spectra. When only one χ_x is large, but this χ_x is present in more than one state, one can then not associate a one-to-one correspondence between MOs (or MO factors in Eq. 3.39) and spectral bands (states). The states in question are thus associated with a breakdown of the MO picture. It could, finally, also be that no χ_x is large, in which case we talk about a correlation-state satellite.

Although there obviously is no rigorous distinction between these cases, this classification has turned out to be useful to analyze observed features in molecular Auger spectra. The correlation energy contribution to Auger states is model dependent in so far as there is no unambiguous way to define a many-open shell Hartree–Fock energy. However, often there is no one-to-one correspondence between states and spin-coupled MOs and the MO picture clearly breaks down. In contrast to closed-shell ground states where the correlation is classified by external single-, double-, excitation schemes, the correlation schemes of one-, two-, or many-open-shell states must include internal and semi-internal excitations. In fact, the configuration interaction due to such excitations is most often the dominant one.

The final states of the Auger transitions can often be characterized with a valid MO description in the outer–outer valence region, as hole-mixing states in the inner–outer region, and as states with breakdown of the MO picture in the inner–inner valence region. The energy limits for these effects are, of course, system dependent; in "atomlike" systems such as first-row hydrides there is a MO breakdown only in the inner–inner valence spectra and one witnesses breakdown effects only in the inner–inner part of the Auger spectrum, while the rest are comparatively well described at an independent-particle (MO) level. Already for a small species like CO there are hole-mixing states in the outer–outer valence region [106, 144], and for larger compounds the role of hole mixing or breakdown of the MO picture is pronounced in most parts of the spectra.

The correlation can often be seen as near degeneracies, either between Koopmans configurations (hole mixing) or between final eigenstates (MO breakdown). A *static* correlation is then entailed, which in the wavefunction picture is described by *internal* and *semi-internal* hole–particle excitations. For Auger states the internal CI is obtained by redistributing the two holes among the occupied valence levels, while the semi-internals are obtained as coupled internal–externals, that is, redistributing one hole in the occupied space coupled by an external hole–particle excitation. Since the number of two-hole states increases with the square of the number of electrons, it is quite clear that the MO breakdown effect due to semi-internal CI (SEMICI) increases steeply with the size of the molecule. The role of SEMICI is also pronounced, especially for electron-rich molecules with low symmetry. In contrast, to, for example, valence photoionization states, the hole-mixing states for Auger are often due to static correlation for which there are large many-body factors and hence large intensities. There are of course a great number of hole-mixing states where the dynamic correlation is leading and with small χ_x's, corresponding to those appearing in X-ray emission or valence photoionization spectra. However, such weak satellites are generally not observed in the Auger case.

For the correlation-state satellites the *dynamical* correlation effect is dominating. This is, in the wave function picture, described by *external* hole–particle excitations. The wavefunction is then characterized by only one small intermixed χ_x but is dominated by configurations generated by the hole–particle excitations. It can be noted that also for the outermost Koopmans states, the correlation (rather the correlation correction) is of dynamical, that is, external, origin. In this case, however, it acts only as a modulation to the intermixing of the main Koopmans configuration, with a large χ, and the quasi-particle picture still holds. Since static correlation, just like relaxation, always gives rise to a positive error, a Koopmans theorem, here the single ionization expression of Eq. 3.40 with $R_{ij}=0$ and $I_i=\epsilon_i$, fails badly for transitions to double-hole inner valence states. In contrast, the dynamical correlation error is negative, thereby counteracting the relaxation error for Koopmans states.

3.8.2 Calculation of Molecular Auger Spectra

As for several other types of eletronic spectra the computational methods that apply to Auger spectra fall into two major classes; wavefunction and Green's function methods. The former can be further distinguished into independent-particle, open-shell self-consistent field (OSSCF, OSRHF), MCSCF, and SEMICI approaches. Many calculations of molecular Auger spectra have been performed by ab initio wavefunction methods [e.g., 106, 139, 141]. Apart from ab initio calculations semiempirical approaches have also been frequently employed [145]. Green's function or propagator methods have been widely used in quantum chemical calculations of ionization potentials and excitation energies [146–152]. However, while particle–particle Green's functions are well-known tools in quantum physics [153–155], ab initio correlation calculations of double ionization potentials (and thus relative Auger energies) of finite electronic systems by Green's function methods seem to be relatively scarce [156–162]. Analysis of Auger spectra generally use the fact that the relative Auger kinetic energies are accessible by the double ionization potentials of the system. The kinetic energy is given by $E_{\text{kin}} = \text{IP}^c - \text{DIP}(n)$, where IP^c is the core ionization potential pertinent to the creation of the initial state and DIP(n) is the double ionization potential for the final state under consideration.

3.8.3 Independent-Particle Methods

The independent-particle methods have the decisive advantage that they appeal to our inclination to picture electronic processes as promoting one electron at a time and, using a MO interpretation, to assign individual molecular orbitals, or a simple combination of them, to each line in the electronic spectrum. The independent-particle MO analysis of Auger spectra starts out from the simple expression for double electron ionization energies:

$$E_{ij} = I_i + I_j - R_{ij} + V_{ij}^S \tag{3.40}$$

where I_i and I_j denote single ionization energies with holes in orbitals i and j; R_{ij} is a relaxation term and V_{ij}^S a hole–hole interaction term which is dependent on the spin–coupling of the two-hole state. In a first approximation, corresponding to the Koopmans level of approximation for photoelectron spectra, R_{ij} is assumed zero, which is correct in the limit of an infinite number of electrons. The double-hole ionization potentials are then simply given by sums of two (negative) orbital energies corrected by the hole–hole interaction parameter. For an ordinary molecule this parameter takes values in the order of electron-volts. Within the validity of this approximation the Auger spectrum is assigned directly from the photoelectron spectrum provided the hole–hole interaction parameters can be estimated.

Various versions of the independent-particle model expressed by Eq. 3.40 have been tested, for example, using ΔSCF energies or using experimental energies for I_i and I_j. In the latter case the correlation energy is implicitly included for the single ionization steps; however, the change in correlation (and relaxation) going from the first to the second ionization step is not. Also the hole–hole interaction parameters themselves can be obtained in different ways. The intrinsic errors, or relaxation energies, in expression 3.40 can be formulated as

$$S_{ij} = -\epsilon_i - \epsilon_j - E_{ij}^{\text{SCF}} + V_{ij}^{\epsilon} \tag{3.41}$$

where ϵ_i and ϵ_j are the ground-state orbital energies and V_{ij}^{ϵ} is obtained from ground-state orbitals, or as

$$D_{ij} = I_i^{\text{SCF}} + I_j^{\text{SCF}} - E_{ij}^{\text{SCF}} + V_{ij}^{\text{SCF}} \tag{3.42}$$

where I_i^{SCF} and I_j^{SCF} denote SCF single ionization potentials and V_{ii}^{SCF} now obtained self-consistently and state specifically [and for appropriate spin-coupling, singlet (S)

or triplet (T)]. It can be argued that orbitals optimized for one state in each of the three main inner–inner, inner–outer, and outer–outer groups of states can be used to construct D's and V's for a "one-particle" spectrum. It mimics a full ΔSCF solution quite well, although the correct ordering of states is not always obtained. Even though the model based on single ionization potentials cannot be used to recapitulate the correct ordering of the different Auger states, it can be used, together with the one-center intensity model recapitulated below, to grossly assign intensity to different parts of the spectra.

3.8.4 Wavefunction Methods

The self-consistent optimization of two-open-shell states with low-spin coupling (singlets) present in Auger spectra is afflicted by some complications that are not present in the Hartree–Fock description of single-hole states and which have been coped with by different orthogonality-constrained approaches. The first is the Roothaan coupled Fock operator open-shell SCF with orthogonality constraint obtained by the means of Lagrangian multipliers [163]. This method fulfills the so-called generalized Brillouin theorem [164]; however, the normal form of the Brillouin theorem is not fulfilled; that is, singly excited configurations are interacting with the Hartree–Fock reference state. An alternative Brillouin condition can be formulated imposing that second-order contributions to the energy from single replacements to the reference state cancel, which leads to improvements over the calculations using the generalized Brillouin theorem for two open-shell singlets of the same symmetry [165, 166]. These are also the two basic open-shell SCF methods that have been applied to molecular Auger spectra.

In general, open-shell Hartree–Fock has been found appropriate for a gross characterization of an Auger spectrum. However, this model becomes progressively poorer toward the low-kinetic-energy end of the spectrum, as explained above. With unconstrained MCSCF techniques two-hole states can routinely be optimized for any configurational state function. The orthogonal orbital parametrization allows for full optimization of large classes of wavefunctions, in particular those which describe two-hole states. The MCSCF method has therefore found several applications in the analysis of Auger spectra, in particular for the outer parts of the spectra where the root index is sufficiently low to guarantee convergence of the MCSCF cycles.

Computationally, semi-internal CI is an effective tool for analyzing double-hole states in Auger spectra [139, 140]. However, it can pose problems concerning the generation of the (semi-internal) CSFs and concerning the calculation of higher lying roots in the Hamiltonian CI matrix. The latter problem arises because those intensity-carrying CSFs with large intermixing of semi-internals are higher in energy than many multiply excited CSFs without any Auger intensity [small generalized overlap amplitudes (GOAs)]. These problems are best handled by explicit Hamiltonian approaches, in which the semi-internal configurations can be generated by flexible selection schemes. If the CI Hamiltonian is of moderate size, it can of course be diagonalized completely and all roots corresponding to main and satellite states in the Auger spectra can be obtained therefrom. Although flexible the small size of the CI

Hamiltonian puts limits on the system that can be handled. For larger configurational expansions the CI Hamiltonian has to be diagonalized with iterative techniques, that is, by direct techniques including linear transformations of the CI Hamiltonian on trial vectors. It is, however, often too cumbersome with this technique to diagonalize for all roots in the spectrum. Lanczos methods modified to only extract those roots with large intensities are suitable for constructing a full spectrum [100]. For very large sizes of the configurational expansion, active-space techniques have been used with complete (CAS) or restricted (RAS) active spaces. However, such wavefunctions will contain a very large number of configurations in case they have to include the appropriate semi-internal excitations, and with a large number of configurations these methods are again limited to the few lowest roots covering only a part of the Auger spectrum. Computationally efficient variants are given by so-called contracted CI techniques [130] and called multireference selection CI [167] and have been applied with some success to Auger. In the latter case the configurational selection is automatized by perturbational or other criteria, which often is an efficient route to calculations of many-hole states or higher excited states.

3.8.5 Green's Function Methods

The second major branch of calculations of molecular valence Auger spectra is given by the two-particle Green's function method. Two-particle Green's functions are well-known tools in quantum physics and have, after the pioneering work of Liegener [144, 168–170], been widely applied in calculations of double ionization potentials and thus relative Auger energies of finite electronic systems, especially in the so-called renormalized form of the Bethe–Salpeter equation and other approximation schemes. In the Green's function method the double ionization potentials (DIPs) are obtained as poles of a two-particle Green's function, namely the particle–particle Green's function (2p-GF), which is defined as

$$\mathcal{G}_{klmn}(\omega) = \int_{-\infty}^{\infty} dt\, e^{i\omega t}(-i)\langle \Psi_0^N | T[a_l(t)a_k(t)a_m^+ a_n^+] | \Psi_0^N \rangle \tag{3.43}$$

where Ψ_0 is the correlated ground state of the neutral molecule in the Heisenberg representation, T is Wick's time-ordering operator, and a_l^+ and a_l are the usual Heisenberg creation and annihilation operators for the canonical Hartree–Fock spinorbitals.

The particle–particle Green's function has the following spectral resolution, known as the Lehmann representation:

$$\mathcal{G}_{klmn}(\omega) = \sum_{\nu} \frac{\langle \Psi_0^N | a_l a_k | \Psi_\nu^{N+2} \rangle \langle \Psi_\nu^{N+2} | a_m^+ a_k^+ | \Psi_0^N \rangle}{\omega + \mathrm{DEA}_\nu + i\eta} - \sum_{\nu} \frac{\langle \Psi_0^N | a_m^+ a_n^+ | \Psi_\nu^{N-2} \rangle \langle \Psi_\nu^{N-2} | a_l a_k | \Psi_0^N \rangle}{\omega + \mathrm{DIP}_\nu - i\eta} \tag{3.44}$$

where η is the positive infinitesimal tending to zero in the distributional sense and the sums run over all $N+2$ or $N-2$ electron states as indicated by the superscripts. The Lehmann representation shows that it is possible to obtain $-\text{DIP}\nu$ as poles of the 2p-GF. The 2p-GF is accessible by time-independent perturbation theory. This means that the interacting many-electron ground state is generated by an adiabatic switching process and the terms arising from the series expansion of the time evolution operator are symbolized by diagrams. Partial summation of the diagrammatic series may lead to factorizable equations for the 2p-GF. Instead of using the diagrammatic approach, one can as well write the 2p-GF as a superoperator resolvent and use algebraic methods for the construction of appropriate approximations or one can derive matrix equations for the advanced and retarded part of the 2p-GF separately and derive suitable perturbation schemes for them. Diagrammatic expansions can effectively be summed via the Bethe–Salpeter equation, which enables us to establish an explicit connection between one particle, for example, photoelectron spectra and Auger spectra, and the breakdown phenomena manifested in both of them. The treatment of second-order irreducible particle–particle vertex parts via the Bethe–Salpeter equation for time-ordered diagrams [144, 168–170] represents a specific choice of a partial summation of diagrams for the particle–particle Green's function. Other choices of partial summations are possible by the algebraic diagrammatic construction (ADC) scheme [157] or the superoperator technique [158]. In the ADC scheme suggested by Schirmer and Barth [157] one starts by decomposing the advanced part G^+ of the 2p-GF (the second term in the Lehmann representation given in Eq. 3.44) in the following way:

$$G^+(\omega) = f^+(\omega - D - C)^{-1}f \tag{3.45}$$

where C is a Hermitian matrix called the effective interaction matrix, D a diagonal matrix containing the zero-order DIPs, and f the matrix of modified transition amplitudes. One assumes to have expansions of the expressions f and C in powers of the electron–electron interaction, inserts those expansions into the binomial expansion of the above expression for the 2p-GF, and orders the resulting products according to the order of electron–electron interaction. In this way one obtains, for a given order, an expression for the 2p-GF which can be compared to the corresponding term of the diagrammatic expansion. One can then determine the quantities f and C such that the two expressions become equal at a given order. Thus at that order one can use the above expression for G^+, which means that the poles can be determined by solving the eigenvalue problem

$$(D+C)\,Y = Y\omega \tag{3.46}$$

where Y denotes the eigenvector matrix and ω the diagonal matrix of eigenvalues. It should be mentioned that using a first-order irreducible vertex part and unrenormalized one-particle data and restricting the orbital indices to the occupied space correspond to the first-order ADC. This is also called the Tamm–Dancoff approximation. Beyond this simplest case ADC will lead to somewhat different levels of approximation than discussed in the previous section. A renormalization seems not yet to have been incorporated in the ADC framework, but the expressions have been formulated for the second [157] and third orders [161].

A purely algebraic approach to the 2p-GF suggested by Tarantelli and Cederbaum [160] can be obtained by rewriting the above equation for G^+ as

$$G^+(\omega) = f^+(\omega - E + H)^{-1} f \tag{3.47}$$

where E is a constant term (the ground-state energy of the reference system) and H is a Hermitian matrix called the effective Hamiltonian matrix. One can derive closed-form expressions for a unitary transformation which is constructed in such a way as to allow for an evaluation of the DIPs by means of an eigenvalue problem as small as possible [160]. The explicit working equations are equivalent to those of the ADC up to the third order (and can be expected to coincide also for the higher order). However, the unitary transformation can be expected to have computational advantages, compared to the ADC, because it only requires the expansion at a given order of some closed-form equations and the calculation of the contributions of that order to the exact ground state.

3.9 MOLECULAR ORBITAL ANALYSIS OF X-RAY SPECTRA

3.9.1 One-Center Models

The instrumental development in the whole family of X-ray spectroscopies during the 1960s and 1970s paved the way for a MO analysis of the spectra, as the resolution element grew sharper than the characteristic energy differences between orbitals. A corresponding application to energy bands in periodic solids could also be discerned. This has had a decisive implication on the utility of X-ray spectra for electronic structure investigations and formed the basis for many close collaboration projects between spectroscopists and quantum chemists or solid-state physicists. As previously discussed in Section 3.6, even asymmetric broadening or fine structure owing to vibronic coupling could be achieved. In this respect X-ray-generated spectra could evidently still not compete with the corresponding spectra in the optical/UV region; however, a decisive advantage of the former is that the MO-derived cross sections, or intensities, allowed unique information on localization properties of the MOs and the associated densities. This owes to the fact that the initial or final hole state in the X-ray process is strictly localized to one atom. Inspecting the transition element, it is realized that the one-center contributions will dominate, implying that, in addition to mapping of local densities, atomic selection rules become operative. The effective selection rules have become a most useful feature in X-ray spectroscopy, especially in X-ray emission, where local atomic decomposition of ground-state occupied MOs or metallic bands can be mapped out.

The canonical MOs can be used to calculate charge and bond order matrices and to perform population analysis. The population numbers can in turn be used in the intensity analysis by means of the one-center models.

The intensity of X-ray emission (XES) and photoelectron emission (XPS) to give a final state with a hole in the ϕ orbital is written as [6, 7]

$$I_{if}(\text{XES}) \propto k_{if}|\langle\phi_{\text{core}}|\hat{t}_1|\phi_p\rangle|^2 \tag{3.48}$$

$$I_{if}(\text{XPS}) \propto k_{if}|\langle\phi_\epsilon|\hat{t}_1|\phi_p\rangle|^2 \tag{3.49}$$

where $\hat{t}_1 = \sum_q \hat{t}_{1q}, \hat{t}_{1q} = x, y, z$ dipole operators; k_{if} is roughly constant over the spectrum energy range and relates to the cube of the transition energy in XES.

Performing an expansion in atomic orbitals $\{\chi_i\}$ and employing the one-center approximation, that is, neglecting cross-atomic terms, the XES intensity is governed by [6]

$$I_{if}(\text{XES}, A) \propto k_{if} \sum_{k=1\pm 2i} W_{lk}|\langle\chi_l|\hat{t}_1|\chi_k\rangle|^2(c_k^p)^2 \tag{3.50}$$

Here l, k denote angular symmetry of the core and valence atomic orbitals localized to center A and i is an integer (as follows from the selection rules for dipole transitions); c_k^p is the expansion coefficient of χ_k in ϕ_p and W_{ik} an angular integration factor. For K_α emission the expression is further simplified to

$$I_{if}(\text{XES}, A) \propto k_{if} \sum_{k\epsilon 2p} (c_k^p)^2 \tag{3.51}$$

while the application of the one-center model to L_β emission leads to the expression

$$I \propto k_{if}\left[|\langle R_{2p}|\hat{t}_1|R_{3s}\rangle|^2 \times 5\sum_{k\epsilon 3s} c_k^2 + |\langle R_{2p}|\hat{t}_1|R_{3d} > |^2 \times 2\sum_{k\epsilon 3d} c_k^2\right] \tag{3.52}$$

The coefficients 5 and 2 result from integration of the angular parts of the atomic wavefunctions. If the radial nl wavefunctions are treated as equal, this expression is further simplified to

$$I \propto k_{if}\left[5\sum_{k\epsilon 3s} c_k^2 + 2\sum_{k\epsilon 3d} c_k^2\right] \tag{3.53}$$

A decomposition for the photoelectron intensity is possible if appropriate assumptions concerning the outgoing electron functions are made. The intensity is then governed by an incoherent sum of local contributions for the relevant ϕ_p MO [89]:

$$I_{if}(\text{XPS}) \propto k_{if} \sum_{A\lambda} \sigma_{A\lambda}(c_{A\lambda}^p)^2 \tag{3.54}$$

where $\sigma_{A\lambda}$ denotes the relative atomic cross section for symmetry λ on atom A. While XES spectra show intensity for bands corresponding to 2p-containing MOs preferentially to the 2*s*-containing ones for first-row elements, the reverse holds for valence XPS spectra.

The XES one-center expression can be mirrored to X-ray absorption (and electron energy loss in the high-energy regime), which then gives information on the localization of the (ground-state) unoccupied orbital in the core excited state. This of course does not have the same strong implications for electronic structure analysis as in the case of XES. Also the XPS expression can be mirrored, namely to *bremsstrahlung*-isochromat (BIS) intensities [171], referring to (ground-state) unoccupied orbitals in the anionic system.

In the case of Auger corresponding expressions can be obtained, but these are more complicated as Auger is a two-eletron process. Although presented with caution some 30 years ago, these expressions have been extensively applied for analysis of molecular Auger spectra. The Auger one-center expressions, as first proposed by Siegbahn [172], are obtained by considering the two-electron Coulomb J and exchange K two-electron integrals present in the transition elements $I_{pq} = 3\sum_{l,m}|J_{lm} - K_{lm}|^2$, $I_{pq} = \sum_{l,m}|J_{lm} + K_{lm}|^2$ and $I_{pq} = 2\sum_{l,m}|J_{lm}|^2$, corresponding to Auger decay to triplet and singlet two-hole states and singlet closed-shell states, respectively:

$$J = \left\langle \phi_p \phi_q \left| \frac{1}{r_{12}} \right| \phi_{\text{core}} \phi_\epsilon \right\rangle$$

$$K = \left\langle \phi_q \phi_p \left| \frac{1}{r_{12}} \right| \phi_{\text{core}} \phi_\epsilon \right\rangle$$

They are decomposed by a one-center expansion as

$$J_{lm} = \sum_{l'm'} \sum_{l''m''} c^{p*}_{nl'm'} c^{q*}_{nl''m''} \delta(m' + m'', m) \times \sum_k d^k(l'm', 00) d^k(lm, l''m'') R^k_{nl'nl'} \tag{3.55}$$

The summation goes over possible atomic l, m channels with c^p and c^q denoting expansion coefficients for p and q MOs pertaining to the atomic orbitals centered on the site of the $\o_{\text{core}}$ ionization. The d^k's are atomic coupling coefficients and the R^k's the atomic two-electron radial integrals involving the Auger (atomic) continuum function, the core orbital, and the atomic valence orbitals; $R^k_{nl'nl'} = R^k[nl'nl', \chi_{\text{core}}\chi_\epsilon]$. This expression gives the intensities as an incoherent sum of atomic cross sections. Corresponding expressions for open-shell molecules have been derived by Ågren [173].

The one-center expressions given above have been routinely used to analyze the different molecular core electron spectroscopies, and they comply with the notion that different core hole derived spectra "map" the same set of final states differently according to the localization character of wavefunctions and orbitals, and they thus supply local information of symmetries and densities. Although these local selection rules were stated several decades ago, they have survived as a most applicable tool in X-ray analysis over the years. This clearly owes to their simplicity and interpretative power and also from that they are straightforwardly obtained from MO theory. In the cases of X-ray absorption and emission they can quite easily be checked against full MO calculations of the intensities, while in the case of Auger spectra [174–177] such checks call for a considerable undertaking in expressing the full two-electron transition elements with a high-energy continuum orbital in the molecular potential. Accordingly only few such applications have been made [178–180].

3.9.2 Multicenter Descriptions of Continuum Orbitals

As previously discussed, according to the Fermi golden rule, the intensity of processes like photoemission and Auger decay is expressed by a transition matrix element between initial and final states of the dipole and, respectively, the Coulomb operator. In both cases the final state belongs to the electronic continuum and we already observed that an L^2 representation lacks a number of relevant properties of a continuum wavefunction. Nevertheless, it was also observed that the transition moment, due to the presence of the initial bound wavefunction, implies an integration essentially over the molecular space and then even an L^2 representation of the final state may provide information on the transition process. We consider now a numerical technique that allows us to compute the intensity for a transition to the electronic continuum from the results of L^2 calculations that have the advantage, in comparison with the simple atomic one-center model, to supply a correct multicenter description of the continuum orbital.

3.9.3 Stieltjes Imaging

The Stieltjes imaging (SI) method provides a mathematically well-defined approach to obtain a property density, as, for instance, the oscillator strength density, in the electronic continuum from the results of a discrete L^2 basis set calculation. As such it may follow the application of a separate initial and final states method for the calculation of a discrete spectrum as well as that of a response function or propagator approach for obtaining a spectral transition intensity. It is a moment method that, following the pioneering work of Langhoff [181], has been largely employed for the calculation of photoionization cross sections and, later, it has been extended to the calculations of other discrete–continuum transitions [178, 182, 183].

The SI approach relies on the fact that a discretized spectrum deriving from an L^2 calculation and covering a large energy range, including the continuum, provides a reasonable approximation of, at least, the lowest order spectral moments S provided the extension of the basis set is sufficiently good:

$$S(k) = \int_0^\infty \frac{df(\epsilon)}{d\epsilon}(\epsilon)^k \, d\epsilon \simeq \sum_{j=1}^{n} f_j(\epsilon_j)^k \tag{3.56}$$

where

$$\frac{df(\epsilon)}{d\epsilon} = \sum_{i}^{\text{discrete}} g_i \delta(\epsilon - \epsilon_i) + g(\epsilon)$$

is the true oscillator strength density, with both discrete g_i and continuous $g(\epsilon)$ contribution, as a function of the excitation energy ϵ, while $[\epsilon_j, f_j]$ is the discretized spectrum obtained by an L^2 calculation.

A short description of the method, as applied to photoionization spectroscopy, may be sketched starting from the expression of the cross section $\sigma(\omega)$ that is proportional to the imaginary part of the polarizability $\alpha(\omega)$ for real values of the photon energy ω:

$$\sigma(\omega) = \frac{4\pi\omega}{c} \mathrm{Im}[\alpha(\omega)] = \frac{2\pi^2}{c} g(\omega) \tag{3.57}$$

The Kramers–Heisenberg expression of the polarizability as a function of a complex frequency z,

$$\alpha(z) = \int_0^{\infty} \frac{df(\epsilon)}{\epsilon^2 - z^2} \tag{3.58}$$

shows that the polarizability is a Stieltjes integral as far as $df(\epsilon)/d\epsilon \geq 0$ in the integration range, which is certainly true for excitations from the ground state. According to the Stieltjes approach, this kind of integral can be convergently approximated, also on the real axis, as

$$\alpha(z) = \int_0^{\infty} \frac{dF^{(n)}(\epsilon)}{\epsilon^2 - z^2} + R_n(z^2) \tag{3.59}$$

where $F^{(n)}(\epsilon)$ is an approximate cumulative oscillator strength function

$$F^{(n)}(\epsilon) = \begin{cases} 0 & 0 < \epsilon < \epsilon_1^{(n)} \\ \sum_{j=1}^{l} f_j^{(n)} & \epsilon_l^{(n)} < \epsilon < \epsilon_{l+1}^{(n)} \\ \sum_{j=1}^{n} f_j^{(n)} & \epsilon_n^{(n)} < \epsilon \end{cases} \tag{3.60}$$

depending on the pseudospectrum $\left\{\epsilon_j^{(n)}, f_j^{(n)}\right\}$ defined by the following $2n$ equations for the negative even spectral moments:

$$S(-2k) = \sum_{j=1}^{n} \frac{f_j^{(n)}}{\left[\epsilon_j^{(n)}\right]^{2k}} \qquad k = 1, 2, \ldots, 2n \tag{3.61}$$

An approximate oscillator strength density $dF^{(n)}(\epsilon)/d\epsilon$ that will converge to the correct one on increasing the order n of the pseudospectrum can be obtained by differentiating, according to the Stieltjes approach, the approximate cumulative function $F^{(n)}(\epsilon)$,

$$g^{(n)}(\epsilon) = \begin{cases} 0 & 0 < \epsilon < \epsilon_1^{(n)} \\ \frac{1}{2}\frac{f_{i+1}^{(n)} + f_i^{(n)}}{\epsilon_{i+1}^{(n)} - \epsilon_i^{(n)}} & \epsilon_i^{(n)} < \epsilon < \epsilon_{i+1}^{(n)} \\ 0 & \epsilon_n^{(n)} < \epsilon \end{cases} \tag{3.62}$$

which defines, through Eq. 3.57, a convergent approximation for the cross section $\sigma(\omega)$.

While the "variational" spectrum $\{\epsilon_i, f_i\}$ is strongly basis set dependent in the continuous portion of the spectrum, the pseudospectrum $\left\{\epsilon_j^{(n)}, f_j^{(n)}\right\}$ is virtually independent of the basis set. The moment problem expressed by Eq. 3.61 can be solved by different numerical techniques, including a linearization by means of a Padé approximant for the polarizability or a continued fraction approximation of a Stieltjes integral different from the polarizability [184], so that not only the even but also the odd negative moments are employed.

By the reproduction of the lowest order spectral moments, the Stieltjes imaging approach provides "optimized" pseudospectra for a representation of the oscillator strength density in the continuum. This point can be further appreciated noticing that the pseudo–excitation energies and oscillator strengths correspond to abscissas and weights of a generalized Gaussian quadrature of order n for the integral in Eq. 3.58, taking $df(\epsilon)/d\epsilon$ as a weight function. In fact, adopting the variable $x = 1/\epsilon$, the integral representing the polarizability can be approximated as

$$\alpha(\omega) = \int_0^\infty \frac{df(\epsilon)}{\epsilon^2 - \omega^2} = \int_0^\infty \frac{df(x)}{1 - \omega^2 x^2} = \sum_{j=1}^{n} \frac{1}{1 - \omega^2 x_j^2} w_j \tag{3.63}$$

where points and weights $[x_j, w_j]$ are determined by the equations

$$s(k) = \int_0^\infty x^k \, df(x) = \sum_{j=1}^{n} (x_j)^k w_j = \int_0^\infty \frac{1}{\epsilon^{k+2}} df(\epsilon) = S(-k-2) \tag{3.64}$$

The points and weights of the Gaussian quadrature solve the moment equations for $k = 0, 1, \ldots, 2n-1$ exactly as the pseudospectrum of order n and the following relationships are valid: $\epsilon_j^{(n)} \leftrightarrow 1/x_j$, $f_j^{(n)} \leftrightarrow \epsilon_j^{(n)^2} w_j$.

The SI method, in the way illustrated so far, has been much employed in calculations of molecular valence photoionization cross sections [5], for both one- and two-photon excitation, in a limited range of a few tens of electron volts above the first ionization threshold. The procedure is, however, quite versatile, and it can be applied to any discretized spectrum given by "energies" and "strengths" as long as these last quantities are positive in order to get a correct continuous intensity. As such it has been employed for the calculation of Auger and resonant Auger spectra [178–180] as well as of shake-off spectra [182].

In fact, using the Fermi golden rule,

$$\Gamma_\epsilon = 2\pi |\langle \Phi_d | H - E_d | \Phi_\epsilon \rangle|^2 \tag{3.65}$$

the following SI procedure can be adopted to estimate the linewidth Γ_{E_d}, proportional to the intensity of an Auger spectrum, of a resonance due to the presence of a discrete state $|\Phi_d >$ embedded and interacting with the degenerate continuum states $|\Phi_\epsilon >$ of a set of open channels. The positive quantity Γ_ϵ in Eq. 3.65 can be computed for the discrete set of continuum energies ϵ_j obtained by L^2 calculations where, for instance, the discrete state is a core-ionized (excited) state and the continuum channels represent the Auger (resonant Auger) decay channels. The SI procedure applied to the "spectrum" $\{\epsilon_j, \Gamma_{\epsilon j}\}$ will produce a set of Stieltjes derivative values at energies generally different from E_d that are, however, rather smooth versus the energy and can then be interpolated at the energy E_d to get the desired resonance linewidth. This method has been effectively applied, in the static exchange approximation, to the calculation of molecular Auger spectra [178–180].

While the mentioned constraint of using positive-definite quantities for the "strengths" can pragmatically be circumvented by a simple modification of the numerical procedure, a more serious limit of the SI technique is its intrinsic limited "energy resolution". The discretized spectrum obtained by an L^2 calculation is characterized by a certain "resolution" in the representation of the continuum that is basis set dependent and also different in the different energy regions. It will in general improve by increasing the dimension of the basis set, but the easily reached redundancy of an L^2 basis puts limits to such improvements. What the SI procedure provides by the introduction of a set of optimized pseudospectra is a more uniform and basis set–independent representation of the continuum. It cannot, however, improve the energy resolution because, by using a limited number of the lowest spectral moments, the dimension of the pseudospectra is much smaller than the dimension of the input discretized spectrum. Thus SI will generally tend to smooth out any narrow structure eventually present in the spectrum, which thus will not be adequately represented by the L^2 calculation.

3.10 ANGULAR DISTRIBUTIONS IN PHOTOEMISSION SPECTRA

As discussed in previous sections, the expression of the intensity for any ionization process always involves a continuum orbital, which may describe the photoelectron in XPS or the secondary emitted electron in Auger decay. In the one-center model the problem is overcome by approximating the molecular continuum orbital by an atomic continuum orbital and finally recurring to available or more easily computable atomic transition moments and two-electron integrals. We have also discussed how the multicenter character of the continuum orbitals can be correctly described by molecular quantum chemical calculations projected on multicenter L^2 basis sets and how a moment technique, like SI, is able to convert the L^2 results in a continuous spectral density and then an intensity for the transition in the continuum. This approach is valid for obtaining an integrated cross section, that is, an intensity corresponding to an electron emission, from a randomly oriented molecule, in any direction with respect to the laboratory frame, related, for instance, to the propagation or polarization direction of the X-ray photon. However, a standard L^2 calculation cannot provide information for computing a differential cross section describing the intensity of a process where the electron is emitted at a specified energy and in a specified direction. This is because the infinitely degenerate character of the molecular continuum orbital is not represented by a projection on a finite basis set and the radial asymptotic behavior of the molecular continuum orbitals is not represented by a projection on an L^2 basis set. For the calculations of properties that are strictly dependent from this characteristics of continuum, like the photoelectron angular distribution in photoemission, but also the cross section branching ratio in resonant Auger decay, as will be discussed in the following, different computational approaches are necessary that are able to describe molecular scattering. A very small number of such approaches has been proposed and we will only consider those that have points of contact with the quantum chemical methods, in particular by keeping the projection on discrete basis sets.

3.10.1 *K*-Matrix Technique

The electronic states above the ionization threshold are characterized by a continuous energy index and by an infinite energy degeneracy physically related to the different propagation direction of the emitted electron. The proper way of describing such states is provided by scattering theory by means of the reactance matrix, related to the scattering matrix. Following the pioneering work of Fano [185], a picture close to the configuration interaction method familiar to quantum chemists for the description of bound electronic states can be adopted where the reactance or K matrix represents the interaction between unperturbed continuum channel states over the electronic Hamiltonian. In order to introduce the model we will here consider a one-electron problem for the electronic continuum of a molecule where the unperturbed states ϕ_ϵ^α are atomiclike continuum orbitals, with a continuous energy index ϵ and a discrete index α that may correspond to the couple of atomic quantum numbers l, m by projection on spherical harmonics that interact by the anisotropy of the molecular Hamiltonian. It should be observed, however, that the K-matrix approach can also be formulated for many-electron wavefunctions or for solving linear response equations, as in the case of RPA [5], as well as for representing the interaction among vibrational continuum channels of a molecule [186].

The one-electron Hamiltonian describing the motion of the emitted electron in a photoionization process is partitioned as

$$\hat{h} = \hat{h}^0 + \hat{V} \tag{3.66}$$

where

$$\hat{h}^0 = \sum_{\alpha=(l,m)}^{\infty} Y_\alpha\rangle\langle Y_\alpha|\hat{h}|Y_\alpha\rangle\langle Y_\alpha \tag{3.67}$$

with the spherical harmonics Y_α centered, for instance, on the molecular center or, in the case of a core ionization that is always well localized, on the specific ionized atom. The index α will then distinguish the unperturbed ionization channels that will be coupled by the anisotropic molecular potential $\hat{V}$. In general, $\hat{h}^0$ will have both discrete and continuous eigenvalues, with the continuum normalized per unit energy interval

$$\hat{h}^0\phi_\epsilon^\alpha = \epsilon\phi_\epsilon^\alpha \qquad \langle\phi_{\epsilon_1}^\alpha|\phi_{\epsilon_2}^\beta\rangle = \delta(\epsilon_1 - \epsilon_2)\delta_{\alpha\beta} \tag{3.68}$$

A continuum eigenstate of the total Hamiltonian $\hat{h}$ with eigenvalue E and degeneracy index β will then be written as a superposition of the unperturbed eigenstates

$$\left|\psi_E^\beta\right\rangle = \sum_\alpha \sum\!\!\!\!\!\!\int d\epsilon\, \mathbf{C}_{\epsilon,E}^{\alpha,\beta}\left|\phi_\epsilon^\alpha\right\rangle \tag{3.69}$$

where the special symbol means summation on the discrete spectrum and integral on the continuous spectrum of channel α, respectively.

Adopting the expression

$$\mathbf{C}_{\epsilon,E}^{\alpha,\beta} = N_E^\alpha\delta(\epsilon - E) + \mathcal{P}\frac{1}{(E-\epsilon)}\sum_\beta N_E^\beta K_{\epsilon,E}^{\alpha,\beta} \tag{3.70}$$

for the coefficients $\mathbf{C}$, where $\mathbf{K}$ is the reactance matrix, N a normalization constant, and $\mathcal{P}$ means that the principal part must be taken on integration over the continuous energy, the eigenvalue equation for $\hat{h}$, projected on the unperturbed eigenstates,

$$\langle\phi_{E'}^\delta|\hat{h}|\psi_E\rangle = E\langle\phi_{E'}^\delta|\psi_E\rangle \tag{3.71}$$

is converted into a set of equations for the K matrix

$$\sum_\alpha \mathcal{P}\sum\!\!\!\!\!\!\int d\epsilon\left[\delta_{\gamma\alpha}\delta(\epsilon-\epsilon') - \frac{V_{\epsilon',\epsilon}^{\gamma,\alpha}}{E-\epsilon}\right]K_{\epsilon,E}^{\alpha,\beta} = V_{\epsilon',E}^{\gamma,\beta} \tag{3.72}$$

where $V_{\epsilon',\epsilon}^{\gamma,\alpha} = \langle\phi_{\epsilon'}^\gamma|\hat{V}|\phi_\epsilon^\alpha\rangle(1-\delta_{\alpha\gamma})$.

The asymptotic form for $r \to \infty$ of the α-channel continuum orbital is

$$\langle\bar{r}|\phi_\epsilon^\alpha\rangle \to \left(\frac{2}{\pi k}\right)^{1/2}\frac{1}{r}\sin[\theta_\alpha(kr) + \Delta_{\alpha\epsilon}]Y_\alpha(\hat{r}) \tag{3.73}$$

with, in the case of ionization of a neutral system,

$$\theta_\alpha(kr) = kr + \frac{1}{k}\ln(2kr) - l\frac{\pi}{2} + \sigma_{el} \tag{3.74}$$

where $\epsilon = \frac{1}{2}k^2$, $\sigma_{\epsilon l}$ is the Coulomb phase shift, and Δ_α is the additional phase shift due to the short-range potential in $\hat{h}^0$. The asymptotic form of the perturbed continuum orbital will then be easily obtained as

$$\left\langle \bar{r} \middle| \psi_E^\gamma \right\rangle \rightarrow \left(\frac{2}{\pi k}\right)^{1/2} \frac{1}{r} \left\{ \sin[\theta_\gamma(kr) + \Delta_{\gamma E}] Y_\gamma(\hat{r}) - \pi \sum_\alpha \mathbf{K}_{E,E}^{\gamma,\alpha} \cos[\theta_\alpha(kr) + \Delta_{\alpha E}] Y_\alpha(\hat{r}) \right\} \tag{3.75}$$

The index γ distinguishes the n linearly independent degenerate eigenstates ψ of $\hat{h}$ that can be obtained from n unperturbed orbitals ϕ,

$$\left|\psi_E^\gamma\right\rangle = \left|\phi_E^\gamma\right\rangle + \sum_\alpha \mathcal{P} \sum\!\!\!\!\!\!\int d\epsilon \frac{\mathbf{K}_{\epsilon,E}^{\alpha,\gamma}}{E-\epsilon} \left|\phi_\epsilon^\alpha\right\rangle \tag{3.76}$$

Such continuum orbitals correspond to stationary wave conditions and are "*K*-matrix normalized,"

$$\left\langle \psi_E^\gamma \middle| \psi_{E'}^\alpha \right\rangle = \delta(E-E') \left(\delta_{\gamma\alpha} + \pi^2 \sum_\beta \mathbf{K}_{E,E}^{\beta,\gamma *} \mathbf{K}_{E,E}^{\beta,\alpha} \right) \tag{3.77}$$

where $\mathbf{K}(E)_{\beta,\alpha} = \mathbf{K}_{E,E}^{\beta,\alpha}$ is the so called *K* matrix on the energy shell. Continuum orbitals normalized per unit energy range and with the incoming-wave boundary conditions appropriate to describe a photoionization process are obtained as

$$\left|\Psi_E^{\alpha(-)}\right\rangle = \sum_\beta [1 - i\pi \mathbf{K}(E)]_{\alpha,\beta}^{-1} \left|\psi_E^\beta\right\rangle \tag{3.78}$$

The *K*-matrix approach is then able to provide continuum orbitals in the anisotropic field of a molecular ion at any energy *E* above the ionization threshold, with the expected infinite degeneracy (index α) and the correct asymptotic behavior for building up an ionized state with the photoelectron propagating in a defined direction with respect to the photon propagation vector or the photon polarization vector. This is relevant in the calculation of the differential cross section or, in other words, in computing the angular dependence of the photoionization process [5]. A standard solution of the eigenvalue problem for $\hat{h}$ by projection on a discrete L^2 basis set would not be able to provide neither the energy continuity of the eigenvalues above the ionization threshold nor their degeneracy. However, we will show in the following that a solution of the eigenvalue problem by the *K*-matrix approach is feasible, with an appropriate choice of the basis set, even in the case of projection of the *K*-matrix equations 3.72 on a discrete L^2 basis set. This is justified by the smooth variation of the interaction matrix $\mathbf{V}_{\epsilon,\epsilon'}^{\alpha,\beta}$ with the continuous energy indices and then by the possibility of an interpolation at any value of *E* inside the energy grid provided by the discrete basis set.

If on the energy grid $[\epsilon_1, \epsilon_2, \ldots, \epsilon_N]$, given by the eigenvalues of $\hat{h}^0$ for a channel α, a column of the matrix $\mathbf{V}$ has the values $V_{\cdot,\epsilon_i}^{\cdot,\alpha} = v_i$, its value at an energy

ϵ, with $\epsilon_j \leq \epsilon \leq \epsilon_{j+1}$, can be written, by interpolation with a polynomial of order n, as

$$\mathbf{V}^{\cdot,\alpha}_{\cdot,\epsilon} \simeq \sum_{r,s=0}^{n} \mathrm{w}^j_{r,s} x^s \tilde{\upsilon}_r \qquad x = \epsilon - \frac{\epsilon_j + \epsilon_{j+1}}{2} \tag{3.79}$$

where $\tilde{\upsilon}_i$ $(i = 0, \ldots, n)$ is the set of $n + 1$ υ values relative to the set x_i $(i = 0, \ldots, n)$ defined by the $n + 1$ grid points ϵ_k closest to ϵ. The interpolation matrix w^j is easily obtained by the conditions

$$\mathbf{V}^{\cdot,\alpha}_{\cdot,\epsilon_i} = \sum_{r,s=0}^{n} \mathrm{w}^j_{r,s} x_i^s \tilde{\upsilon}_r = \upsilon_i \tag{3.80}$$

as

$$\mathrm{w}^j = \begin{pmatrix} 1 & 1 & \ldots & 1 \\ x_0 & x_1 & \ldots & x_n \\ x_0^2 & x_1^2 & \ldots & x_n^2 \\ \vdots & \vdots & & \vdots \\ x_0^n & x_1^n & \ldots & x_n^n \end{pmatrix}^{-1} \tag{3.81}$$

By writing explicitly the sum over the discrete (below the ionization threshold) levels and the integral over the energy continuum, Eqs. 3.72 take the form

$$\mathbf{K}^{\beta,\gamma}_{\epsilon_j,E} - \sum_{\alpha} \left[\sum_{m}^{\text{disc.}} \frac{\mathbf{V}^{\beta,\alpha}_{\epsilon_j,m} \mathbf{K}^{\alpha,\gamma}_{m,E}}{E - \epsilon^{\alpha}_m} + \sum_{i}^{\text{cont.}} \mathcal{P} \int_{\epsilon_i}^{\epsilon_{i+1}} d\epsilon \frac{\mathbf{V}^{\beta,\alpha}_{\epsilon_j,\epsilon} \mathbf{K}^{\alpha,\gamma}_{\epsilon,E}}{E - \epsilon} \right] \tag{3.82}$$

and the energy integral will be approximated, by polynomial interpolation for both **V** and **K**, as

$$\mathcal{P} \int_{\epsilon_i}^{\epsilon_{i+1}} d\epsilon \frac{\mathbf{V}^{\beta,\alpha}_{\epsilon_j,\epsilon}}{E - \epsilon} \mathbf{K}^{\alpha,\gamma}_{\epsilon,E} = \sum_{r,s=0}^{n} \sum_{t,u=0}^{n} \mathrm{w}^i_{r,s} \mathrm{w}^i_{t,u} \tilde{\upsilon}_{j,r} \tilde{k}_{t,E}$$

$$\mathcal{P} \int_{x_i}^{x_i+1} dx \frac{x^{s+u}}{E - \epsilon_i + \epsilon_{j+1}/2 - x} \qquad (s + u \leq n) \tag{3.83}$$

Equations 3.72 can then be written in matrix form as

$$\sum_{\alpha} \sum_{L,I} \left[\delta_{IJ} \delta_{\alpha\beta} - \mathbf{V}^{\beta\alpha}_{JL} \mathbf{D}^{\alpha\alpha}_{LI} \right] \mathbf{K}^{\alpha\gamma}_{IE} = \mathbf{V}^{\beta\gamma}_{JE} \tag{3.84}$$

where the matrix **D** has the structure shown in Figure 3.2 with diagonal contributions

$$D^{\alpha\alpha}_{LI} = \delta_{LI} \left(\frac{1}{E - \epsilon_{L\alpha}} \right)$$

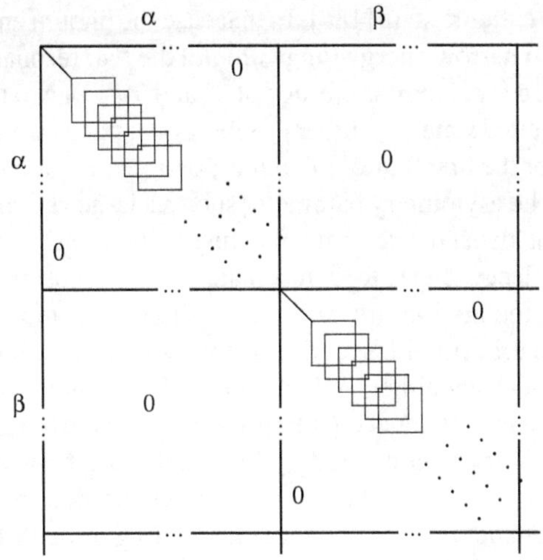

Figure 3.2 Pattern of matrix **D**.

for the discrete part of each channel spectrum and block diagonal terms, in the continuum, deriving from the sum-of-squares matrices of dimension *n*,

$$\mathrm{d}^i_{rt} = \sum_{s,u=0}^{n} \mathbf{w}^i_{r,s}\mathbf{w}^i_{t,u}\ \mathcal{P}\int_{x_i}^{x_{i+1}} dx \frac{x^{s+u}}{E-\epsilon_i+\epsilon_{j+1}/2-x} \tag{3.85}$$

The matrix **D** is employed in the calculation of any quantity depending on the continuum orbital Ψ^α_E as, for instance, the transition dipole moment

$$\left\langle \Psi_0 | \hat{t} | \Psi^\alpha_E \right\rangle = \mathbf{t}^\alpha_E + \sum_\beta \sum_{LI} \mathbf{t}^\beta_L \mathbf{D}^{\beta,\beta}_{L,I} \mathbf{K}^{\beta,\alpha}_{I,E} \tag{3.86}$$

where $\mathbf{t}^\alpha_j = \langle \Psi_0 | \hat{t} | \phi^\alpha_j \rangle$ is also assumed to be a smooth function of energy in the continuum.

The scattering K-matrix method has been applied to the calculation of the asymmetry parameter (β), describing the angular distribution of the photoelectron, as a function of the photon energy in photoemission, mostly from the valence shell. Photoelectron angular distribution for ionization of the core K shell is, in a first approximation, described considering that, due to the essentially *s* atomic character of the core orbital also in a molecular environment, the asymmetry parameter is roughly constant and equal to 2. This is not true, however, when the photon energy is tuned close to the ionization threshold or to a resonance due to a discrete (quasi-stationary) state interacting with the photoemission continua. Resonant photoemission corresponds, in the X-ray region, to participator Auger decay, where the core excited state decays to final states with a single hole in the valence shell; see Figure 3.1. An example is given by the decay of the $O_{1s} \rightarrow 4a_1$ and $O_{1s} \rightarrow 2b_2$ core excited states of

H_2O to the $X(1b_1^{-1})$ cationic state [186]. In that case the photon energy was tuned at a number of values in a narrow energy range around the two resonances and the Auger spectra were collected for electrons emitted at 0° and 90°. Due to the bound character of the $O_{1s} \rightarrow 2b_2$ excited state, a number of vibrational bands are well resolved in the Auger spectra and for the first four of them it is possible to evaluate β as a function of the photon energy. The asymmetry parameter shows a large variation with the photon energy and a strong dependence from the final vibrational state. In the case of participator Auger decay close to a resonance, where both direct and resonant contributions to photoemission are relevant, a single-step model for the decay is needed and the K matrix provides a convenient method to build up such a model. Adopting the Born–Oppenheimer and FC approximations the main ingredient of the model, namely the interchannel coupling V and the dipole transition moment t, are given by an electronic term modulated by FC amplitudes between ground or core-excited state and final states in the continuum. In the present case the vibronic $O_{1s} \rightarrow 2b_2$ states play the role of resonances interacting through V with the different continuum channel identified by a vibronic $1b_1^{-1}$ ionic state times a set of partial waves to represent the photoelectron. Because the asymmetry parameter β depends on the ratio of linear combinations of partial wave transition amplitudes, some of which may show a resonant behavior across the resonance while others, due to no coupling, for symmetry reasons, have only the smooth direct contribution, its variation with the photon detuning can be expected to be rather large. The solution of the K-matrix equations allow to take into account, at the same time, both vibrational/lifetime interference and interference between direct and resonant photoemission. The modulation of V by FC amplitudes that may have different sign for different vibrational states, even for the same final electronic state, due to the different nodal structure of the vibrational functions, makes the coupling rather complex and introduces a variety of dependence of β from the photon energy across the resonances that agrees with the experimental observation. Other properties of the resonant Auger decay that require a one-step model and a scattering method for an appropriate description are the branching ratios, that is, the energy dependence of relative intensities for the participator decay in different final ionic states. As well as the asymmetry parameter β, these relative intensities are expressed by a ratio of quantities depending on the photon energy, which in this case are cross sections where the contribution of the direct photoemission and its interference with the resonant photoemission may lead to deviations of the Auger lineshape from a perfect Lorentzian profile. Although such deviation may be relatively small from the point of view of the bandshape, its effect on the relative intensities is magnified by the ratio. A clear example is offered by the strong dependence on the photon energy and the clear asymmetry around the resonance energy, shown by the branching ratios for the valence shell ionization near the $C_{1s} \rightarrow \pi^*$ core-excited state of CO [187].

An alternative way, based on a pure L^2 formulation, of computing degenerate continumm orbitals at a given energy E for atomic and molecular systems, was proposed by Froese Fischer and Idrees [188], based on an extension of the Rayleigh–Ritz–Galerkin method for bound states. This is a variational approach where the standard eigenvalue problem for a Hamiltonian projected on a not orthonormal basis set

$$(\mathbf{H} - E\mathbf{S})\mathbf{v} = 0 \tag{3.87}$$

is considered in the continuum, where E is a fixed quantity, and then a nontrivial solution of Eq. 3.87 does not exist in general. An approximate solution is offered by the eigenvector corresponding to the eigenvalue of the minimum modulus that, in other words, minimizes the residual

$$\sum_i |\langle \chi_i | H - E | \sum_j \upsilon_j \chi_j \rangle|^2 \tag{3.88}$$

with the normalization constraint $\sum_i |\upsilon_i|^2 = 1$. It is easily shown that this corresponds to the eigenvalue equation

$$(\mathbf{H} - E\mathbf{S})^{\dagger}(\mathbf{H} - E\mathbf{S})\mathbf{v} = \alpha\mathbf{v} \tag{3.89}$$

for the minimum eigenvalue α. This approach has been implemented by using B splines for the radial part of the basis functions χ and spherical harmonics for the angular part and the OCE (one-center expansion) approximation for molecules. More recently [189] a multicentric basis set was employed by adding to a large OCE basis set a limited number of B spline functions centered on the off-center nuclei and defined inside nonoverlapping spheres. According to the B spline approach the eigenvalue problem is projected in a sphere with center (origin of the OCE) on the center of mass of the molecule and radius ($R_{\max}$) large enough to reach the asymptotic region where the numerical continuum orbital can be matched to a known analytical asymptotic expression. By avoiding any specific boundary condition for the B splines at the border $R_{\max}$, the basis set is flexible enough for describing the correct oscillating behavior of the continuum orbital at any energy E limited in its highest value only by the structure of the B splines. If the basis set includes functions with a set of spherical harmonics that can describe n partial waves, the eigenvalue problem in Eq. 3.89 will admit n eigenvalues much closer to zero than the other ones, corresponding to n solutions degenerate at the specified value of E. It has been observed that the lowest eigenvalues and corresponding eigenvectors of the matrix $\mathbf{H} - E\mathbf{S}$ supply a sufficiently accurate approximation of the more computationally costly eiegenvalue problem in Eq. 3.89 and the $(\mathbf{H} - E\,\mathbf{S})\mathbf{v} = a\mathbf{v}$ equation is the one practically solved by inverse iteration. This approach has been applied by Decleva and co-workers mostly for valence photoionization, but also for calculations of core photoionization cross sections, branching ratios, and asymmetry parameters versus photon energy [190]. The continuum orbitals obtained by solving, with the described method, the Kohn–Sham equation for density functionals having the correct asymptotic behavior have been used to calculate transition moments from the bound occupied orbitals and finally independent channel photoionization cross sections and asymmetry parameters. Recently a more accurate and multichannel method has been adopted by this group [191], by solving the time-dependent DFT equation, either iteratively or directly, where the continuum orbitals computed by the minimum eigenvalue approach are employed to impose to the perturbed orbitals the correct outgoing boundary conditions.

3.11 X-RAY ABSORPTION SPECTRA

3.11.1 Near-Edge X-Ray Absorption Fine-Structure Spectra

The use of experimental techniques to generate NEXAFS spectra (near-edge X-ray absorption fine-structure spectra) shows significant capabilities to gain chemical insight; this goes for simple molecules as well as for polymers, liquids, solid-state systems, and surface adsorbates. For example, the notions of "bond length with the ruler" and the "building block principle" have been coined and utilized extensively in NEXAFS analysis [192], the former correlating the energy of continuum resonances (shape resonances) with the interatomic distance pertaining to the molecular orbital housing the excited electron, the latter predicting that the NEXAFS spectrum associated with a particular atom exhibits patterns associated with the particular type of bond of that atom. Furthermore, using linearly polarized X-ray sources, NEXAFS provides unique capabilities for orientational probing of surface-adsorbed species [192], while circularly polarized X rays allow to investigate natural dichroism, which is an important fingerprint in biomolecules. By measuring the angular distribution of fragments emitted along repulsive potential energy curves of core-excited linear molecules in the gas phase, symmetry-resolved spectra are collected, thus getting information on the molecular orientation at the excitation time if the dissociation is fast enough in comparison with molecular rotation [193]. Analysis relying on the building block principle assumes that the total spectrum is decomposed into its constituents (building blocks) by comparing with spectra from analogous molecules which differ by one or two functional groups. The simplest building block is the diatom, the bonding of which determines the position of both discrete and continuum resonances (often being of π^* and σ^* type for organic π-electron systems [194, 195], and in the most basic version of the building block principle the NEXAFS spectrum is given as a superposition of diatomic spectra [192]. Since the experimental spectra are often due to a superposition of X-ray absorption bands from chemically shifted species pertaining to a given kind of atom, a building block decomposition can sometimes be difficult to carry through in practice. An alternative way to proceed is to assemble larger building blocks to the composite molecule, the building blocks can here be in the form of a free molecule or a functional group in an environment where the building block unit presumably is little perturbed [196]. The orientational probing and the structure-to-property relationships rely on a proper assignment of states in the NEXAFS spectra, and the support from simulations for this purpose is indispensable in many cases. The theoretical investigations on these particular aspects of molecular photoabsorption have referred to the one-particle picture either by ab initio $\mathcal{L}^2$ moment theory methods, described above, employing an

MO picture [197], or by semiempirical MSX_α potential barrier and partial-wave expansions for the photoelectron function [198]. While spectral moments in $\mathcal{L}^2$ methods have been generated both by RPA and STEX approximation in the optical and UV regions, mosly STEX has been employed in the X-ray wavelength region [11], due to the above-mentioned shortcoming of RPA for core excitations. The applications of both RPA and STEX have been widened to larger species though an atomic orbital driven "direct" algorithm based on SCF wavefunctions [11].

The STEX technique, and to some lesser extent RPA, has been instrumental in interpreting the available bulk of NEXAFS spectra of molecules in different phases, and there are comparatively few cases where more sophisticated ab initio computational techniques have been called for in order to make proper assignments. The most well-known of these cases is O_2, which presents a very unusual and complex NEXAFS spectrum. In fact, according to the "building block" model, other diatomics, and in some way other molecules show typically a NEXAFS spectrum that for each excitation cite exposes a set of sharp peaks due to strong π^* excitations and weak Rydberg series below the threshold, followed by broad shape resonances above the threshold due to transitions to antibonding orbitals of σ symmetry [199]. The O_2 spectrum, apart for an isolated sharp π^* peak, is characterized, instead, by two large features that are complex sequences of well-resolved peaks of different intensity, covering, totally, an energy range of about 5 eV below the threshold. This is due to the open-shell character of the O_2 ground state giving origin to two exchange splitted (quartet and doublet) core ionization thresholds and to the exceptional circumstance that the two σ^* core excitations are located below the threshold and then possibly mixing with the Rydberg series and, finally, to the different, dissociative and, respectively, bound character of such core excitations, which potentially leads to a very complex vibrational structure of the bands. All this makes up for a NEXAFS spectrum which probably is the most complicated and, consequently, challenging one for both experimental and theory. A long-lasting controversy has emerged regarding the amount of energy splitting of the two triplet states deriving from the coupling of the excited σ^* orbital to the quartet or doublet core ionic states, with estimated values in the range of 0.4–2.75 eV. An early investigation [200] based on limited CI calculations for the two σ^* states and for the Rydberg series in the static-exchange approximation with the Rydberg orbitals explicitly orthogonalized to the σ^* orbital, predicted an exchange splitting of 2.75 eV of the σ^* excitations with the quartet-derived state higher in energy than the doublet-derived state, that is, in opposite order to that predicted for the core ionic states. Later [201], by more extended first-order CI calculations, the same authors showed the presence of a mixing between a σ^* state and a Rydberg state sharing the same (quartet) ion core and reduced the estimated exchange splitting value to 1.64 eV, still keeping, anyway, the view that the two σ^* excited states are mainly responsible for the spectral intensity distribution. More recently [202], rather extensive multireference CI (MRCI) calculations, in the vertical approximation both for the ground and the core-excited states, using the CIPSI (configuration interaction by perturbation with multiconfigurational zero-order wavefunction selected by iterative process) method with aimed selection [203, 204], pointed out that the valence-Rydberg mixing is much more extended and covers almost all the energy range of the complex features below the threshold. By an ab initio estimation of the participator Auger intensity for the decay to the highest occupied molecular orbital (HOMO) and (HOMO-1) hole states [202], the constant-ionic-state (CIS) spectra could be interpreted, confirming that the σ^* character is not limited to the first main features of the absorption spectrum, but it is spread over a larger range, something that reduces the value of an interpretation of the O_2 NEXAFS spectrum as simply formed by two exchange-split components of the σ^* resonance. This point of view was confirmed by following extensive MRCI calculations [205] at different interatomic distances in order to evaluate the potential curves of a number of low-energy core-excited states of σ symmetry. The calculations showed a relevant σ^* Rydberg mixing and a diabatization of the adiabatic potential curves pointed out that the coupling between bound Rydberg and dissociative σ^* diabatic states is very different at the different crossing points. By a qualitative description of the molecular dynamics, ultrafast dissociation was predicted to occur more easily on the lowest σ^* diabatic potential curve in agreement with the experimental observation of atomic peaks only in the lower energy region of the absorption spectrum. The calculated potential curves of six core-excited states associated with the promotion of the O_{1s} electron to three virtual orbitals, σ^*,$3s$,$3p$, with either doublet or quartet ion core, were also employed for a wavepacket simulation of the nuclear dynamics in the core photoabsorption and Auger decay of O_2 [206]. It was found that, due to the multiple curve crossings, the Born–Oppenheimer approximation breaks down and a fully diabatic picture fails to reproduce the spectral shape of the low-energy region of the NEXAFS spectrum. A mixed adiabatic/diabatic picture which classifies crossing points according to the strength of the electronic coupling turned out to be more effective in simulating, by the wavepacket technique for the nuclear dynamics, the overall spectral profile. These results could probably be further improved by the inclusion of nonadiabatic coupling in the model; however, a detailed theoretical assignment of the vibronic peaks of the σ symmetry component of the O_2 NEXAFS spectrum is still an open and difficult problem.

3.11.2 Multiple-Scattering X_α Method

The multiple-scattering X_α (MSX_α) method goes back to the early work of Slater and Johnsson [207] and has become the workhorse for EXAFS calculations and to some extent also for NEXAFS [208–210]. This owes much to the simplicity of the method but also to its interpretative power. Moreover, the fact that it covers both discrete and continuum parts, high up in energy, makes it appealing for routine analysis with high computational throughput. The MSX_α method is based on two assumptions; MS, where the excited electron is considered to be multiply scattered in muffin tin formed potentials, and X_α, where the nonlocal exchange interaction is approximated to be a local exchange interaction regulated by an "X factor." Using multiple-scattering wave theory the local solutions from the various muffin tin potentials can be joined into a continuous electronic wavefunction. Moreover, the calculation of the manyfold of two-electron integrals of ab initio methods is replaced by expressing the Coulomb and exchange potentials directly in terms of the total charge density; for the latter potential this is made possible by the X_α approximation. It is quite evident that modern density functional theory, both technically and philosophically, derives from the early work on the MSX_α method by Slater and Johnsson [207]. The application of the MSX_α method has been extensively described in the book by Stöhr [192], giving numerous illustrating examples.

It is clear that the limits of the MSX_α theory are set by its two main assumptions, scattering in muffin tin potentials and the local exchange approximation. Nowadays, much is known about the latter due to intensive work within DFT on density gradient corrections and, in general, nonlocal corrections to the functionals. Concerning the muffin tin approximation, it is quite clear that it works better far from the ionization edge, where the escaping electron "sees" less details of the molecular potential than close to the edge. Below the edge, the muffin tin approximation evidently falls short of the molecular orbital calculations assuming full nonisotropic shape of the molecular potentials, but it has been fairly successful also closely above the edge, in the so-called shape resonance region. In fact, the latter expression was coined using the MS concept, where the form of the ex atomic potential "shapes" the centrifugal barrier of the potential of the core-ionized atom, thereby creating a barrier where the scattered electron is temporarily trapped. The full molecular potential is typically partitioned into three regions: The first constitutes the inner atomic spheres, the second region is the area complementing the spheres, and the third lies outside a large sphere encompassing the two first regions. The resonance calculated by MS theory is thus often discussed in terms of atomic channels with particular angular momentum labels but also by asymptotic labels in the third region referring to a corresponding wavefunction that is expanded out of the center of the molecule [192]. The division of the potential in this way also provides a possibility to refer resonances to the positions of the spheres. The bond length with the ruler idea derives from a particle in the box argument that the energy position of the shape resonance above the IP is directly related to the width of the potential well, that is, roughly on the distance between the two atoms giving origin to the resonance. For low-Z species with σ and π symmetries one often finds that the shape resonance is of σ^* character, while the π^* resonances often are strongly bound in the lower discrete part of the spectrum.

3.11.3 Extended-Edge X-Ray Absorption Fine-Structure Spectra

As noted above, MSX_α has had many important applications for describing X-ray absorption spectra far beyond the edge and for derivation of intermolecular distances from the oscillating structures there observed. Using similar assumptions as those behind the MSX_α method, an effective equation for the extended-edge X-ray absorption fine-structure (EXAFS) signal can be derived containing "geometric" terms accounting for the finite inelastic mean free path of the photoelectron, for backscattering characteristics of the neighbors, and a sinusoidal distance and phase shift dependence. These terms allow to make an inverse analysis and obtain spectra property to structure information in terms of interatomic bond distances. This goes in particular for systems, like inorganic crystals, with periodic symmetries but also for aperiodic systems if the nearest neighbors have close-lying metallic elements. An example of the latter is the Photosynthesis II system, where EXAFS gives unique information on the central manganese cluster [211]. There are several considerations involved in the EXAFS analysis; perhaps the main one is the neglect, or inclusion, of multiple-scattering contributions to the EXAFS signals, something that is quite strongly dependent on the system under study. Much of the EXAFS research has been inspired by pioneering work of Rehr and Stern, [212–214] with particular contributions in computations by Dehmer and Dill [198, 215, 216] and Natoli [217–219].

Along with the development of synchrotron radiation spectroscopy more advanced variants of EXAFS techniques have been outlined and tested, for instance, "anisotropic EXAFS measured in the Raman mode." This means that polarization-dependent extended X-ray absorption fine structure will show angular information on randomly oriented molecules and amorphous systems. The physical background of such a possibility is based on a polarization–frequency selection of a partially oriented subset from the randomly oriented molecules and on backscattering of the photoelectrons on the surrounding atoms [220, 221]. One can here also mention novel possibilities for "ultrafast EXAFS monitoring of fragments of photodissociation" involving pump–probe techniques that give unique possibilities to monitor the geometry of a dissociative molecule making use of short-probe X-ray pulses synchronized with the pump optical pulse which excites the molecule in a dissociative state. Such a technique has the extra advantage compared with the standard EXAFS method in that information about bond angles for disordered samples can be obtained. This is possible since the polarized pump optical field excites molecules only with a certain space alignment or orientation [221].

3.11.4 X-Ray Circular Dichroism

The use of linearly polarized X rays is nowadays central in the structural analysis of adsorbates [192] due to the simple dipole selection rule for excitations from the K shell that makes the NEXAFS spectra a direct probe of the orientation of an adsorbed molecule with clearly "oriented" excited orbitals of σ^* and π^* character. For this orientational selectivity, together with its chemical selectivity, NEXAFS is one of the most powerful spectroscopies for studying adsorbate interfaces. It was for some time argued if by the absorption of circularly polarized X-ray it was possible to obtain a new fingerprint of organic chiral molecules; in other words, if the natural circular dichroism (CD) effect, known for a very long time in the optical frequency region, could be observable also in the X-ray region. Because X-ray absorption has an atomic character and, of course, CD is not present for the excitation of a spherically symmetric atomic orbital, it was easy to predict a small X-ray CD effect in molecules, essentially deriving from the small distortion of a core orbital by the effect of the molecular (chiral) potential. Nevertheless, considering that "chiral centers" are rather common in large organic molecules of biological interest, it seemed worthwhile to investigate the possibility of measuring such an effect.

From a theoretical point of view, the leading term contributing to CD in randomly oriented molecules is due to the interference of electric and magnetic dipole transition amplitudes, with the rotatory (R) strength taken as a measure of CD. It is expressed from dipole and angular momentum integrals between initial and final states according to the expressions

$$R_{cv}^{L} = \frac{1}{2c}\left\langle \Psi_0 \middle| \mathbf{r} \middle| \Psi_f \right\rangle \cdot \left\langle \Psi_0 \middle| \mathbf{r} \times \nabla \middle| \Psi_f \right\rangle \tag{3.90}$$

$$R_{cv}^{V} = \frac{1}{2c\omega_f}\left\langle \Psi_0 \middle| \nabla \middle| \Psi_f \right\rangle \cdot \left\langle \Psi_0 \middle| \mathbf{r} \times \nabla \middle| \Psi_f \right\rangle \tag{3.91}$$

where ω_f is the excitation energy and L (length) or V (velocity) specifies the gauge adopted. The anisotropy ratio g that is a measure of the CD with respect to the averaged (left and right) absorption intensity [222] is then expressed as

$$g_f = \frac{2\omega_f}{c} \frac{\left\langle \Psi_0 \middle| \mathbf{r} \times \nabla \middle| \Psi_f \right\rangle \cdot \left\langle \Psi_f \middle| \mathbf{r} \middle| \Psi_0 \right\rangle}{\left\langle \Psi_0 \middle| \nabla \middle| \Psi_f \right\rangle \cdot \left\langle \Psi_f \middle| \mathbf{r} \middle| \Psi_0 \right\rangle} \tag{3.92}$$

A couple of theoretical works [223–225] based on RPA and, respectively, STEX approximation for the final core-excited state of methyloxirane, a relatively small molecule frequently used as a benchmark in CD studies, provided a first prediction of X-ray CD at the C K-edge with an anisotropy ratio g of the order of 10^{-3}. From an experimental point of view, the extension of CD spectroscopy to the X-ray region had to await the development of specific insertion devices in modern high-brilliance synchrotron sources. Only recently the first CD measurements of a K-edge absorption spectrum in the gas phase were reported [226] (methyloxirane), showing good agreement with the theoretical prediction of the relaxed STEX calculations. Similar calculations also predicted measurable natural X-ray CD for a number of amino acids [225, 227] at the C, N, and O K-edge; such predictions were substantially confirmed by following experimental measurements on sublimated films of phenylalanine and serine [228]. More recently, the complex polarization approach, briefly reviewed in the following, has been implemented for X-ray circular dichroism and applied to some fullerenes and biomolecules [229–231], thus generalizing the previous STEX and RPA technologies for this application.

3.12 RESONANT X-RAY SCATTERING SPECTRA

The analysis above of the various X-ray spectroscopies assumes a decoupling of the excitation and emission events. An incoming X-ray photon with frequency ω is absorbed and the molecule is core excited to the state $|c\rangle$, and due to Coulomb interaction and vacuum fluctuations the core-excited state decays by emitting Auger electrons and X-ray photons with the energy E to the final state $|f\rangle$. However, unifying these events into a one-step scattering picture can provide a generalization where all ingoing and outgoing scattering channels are treated simultaneously [142]. The development of synchrotron radiation sources have greatly promoted research and analysis of such resonant elastic or inelastic X-ray processes in the one-step picture. The process, also called X-ray Raman scattering, is exceedingly rich in physical interpretations and has over the year been applied to a variety of compounds—atoms, free molecules, liquids, surface adsorbates, and solids. The short lifetime of the intermediate core-excited states presents new physics compared to the optical/UV region, in particular interference effects that are manifested in many different ways. The process is subject to strong selection of participating energy levels for systems containing elements of symmetry [232].

Computationally, the X-ray Raman process has been addressed by both time-dependent and time-independent methods. In the latter case the properties of resonant X-ray spectra (RXS) are guided by the double-differential cross section [233, 234]

$$\sigma(E, \omega) = \sum_f |F|^2 \Phi(\omega - E - \omega_{fo}, \gamma) \tag{3.93}$$

which is the convolution of the spectral distribution Φ of the incident radiation with the RXS cross section $\sigma_o(E, \omega)$ for a monochromatic incident light beam. Here γ is the spectral width of Φ; the index f is dropped in the scattering amplitude: $F_f \to F$; the cross section is written here without the multiplication factor, and atomic units are used. The radiative and nonradiative RXS amplitudes have the same structure near the resonant region [233],

$$F = \sum_c \frac{\langle f|\mathcal{Q}|c\rangle\langle c|\mathcal{D}|o\rangle}{E - \omega_{cf} + \imath\Gamma} \tag{3.94}$$

Here $\omega_{cf} = E_c - E_f$, E_c is the energy of the core-excited state, and Γ is its lifetime broadening. The lifetime broadening of the final state Γ_f is often small and is neglected here for simplicity. The operator $\mathcal{D}$ describes the interaction of the target with the incident X-ray photon. In the case of nonradiative RXS, $\mathcal{Q}$ is the Coulomb operator, while $\mathcal{Q} = \mathcal{D}'^*$ when the emitted particle is the final X-ray photon [233]. The formula above constitutes the key approach to the time-independent resonant Raman cross section as obtained through second-order light-matter perturbation theory. It expresses the Kramers–Heisenberg formula which relates the cross section as a summation of matrix elements involving the full manifold of excited states. In the X-ray region the Kramers–Heisenberg formula is commonly evaluated in a frozen orbital picture and truncated to a few states, in contrast to the UV region where modern response technologies can be employed for a complete sum-over-states evaluation including relaxation. The time-independent methods offer, in principle, the possibility to address large systems with a large number of degrees of freedom.

Common for the time-dependent methods, whether applied in the X-ray or optical regions, is that they offer a conceptual as well as intuitive understanding of the phenomena involved and also make it possible to design new experiments in ways that would be difficult using purely time-independent models. Thus with the use of time-dependent methods it has been possible to highlight a number of phenomena associated with X-ray resonant Raman scattering, like interference, detuning, collapse, dynamic symmetry breaking, Doppler effects, and localization [235–247]. However, this has to be balanced against the fact that time-dependent methods are inherently limited to a few degrees of freedom, something that makes it necessary either to precompute reaction coordinates along one, or a few, directions of molecular motion or limit the study to few-atomic systems if all coordinates are to be considered. It is thus with the free molecules (in fact diatomic molecules) that the most rich and fundamental results have been obtained.

The time-dependent representation for the RXS cross section and scattering amplitude can be obtained from the half-Fourier transform of the denominator on the right-hand side of Eq. 3.94 [248, 249],

$$F = F(\infty) \quad F(\tau) = -\imath \int_0^\tau dt\, e^{\imath(E + E_f + \imath\Gamma)t} \langle f|\phi(t)\rangle \tag{3.95}$$

The wavepacket reads

$$\phi(t) = Qe^{-\imath H_c t}\mathcal{D}|o\rangle \tag{3.96}$$

One assumes here that the molecular Hamiltonian H is the same for all electronic states j, still the notation H_j with index j is useful to identify the electronic shell in which the wavepacket evolves. One can then also apply directly the general theory to nuclear degrees of freedom with the nuclear Hamiltonian depending on the electronic state j. A corresponding time-dependent representation for the RXS cross section (3.93) can be obtained by a Fourier transform of the spectral function,

$$\Phi(\omega, \gamma) = \frac{1}{\pi}\Re e \int_0^\infty dt\, \varphi(t, \gamma)e^{\imath\omega t} \qquad \varphi(t, \gamma) = \int_{-\infty}^{\infty} d\omega\, \Phi(\omega, \gamma)e^{-\imath\omega t} \tag{3.97}$$

Since the Φ-function is real, its Fourier transform has the property $\varphi^*(t, \gamma) = \varphi(-t, \gamma)$; $\varphi(t, \gamma) = \delta(t)$ and $\varphi(t, \gamma) = \text{const}$ correspond to the cases of having white and monochromatic incident light beams, respectively. If the spectral function Φ is approximated by a Gaussian, one has

$$\Phi(\omega, \gamma) = \frac{1}{\gamma\sqrt{\pi}}\exp\left(-\frac{\omega^2}{\gamma^2}\right) \qquad \varphi(t, \gamma) = \exp\left(-\frac{t^2\gamma^2}{2}\right) \tag{3.98}$$

The time-dependent representation for the RXS cross section [249, 250] is obtained by substituting Eqs. 3.95 and 3.97 in Eq. 3.93, giving the dynamical representation for the RXS cross section

$$\sigma(E,\omega) = \frac{1}{\pi}\Re e \int_0^\infty d\tau\, \sigma(\tau)\, \varphi(\tau,\gamma) e^{\imath(\omega - E + E_o)\tau} \tag{3.99}$$

in terms of the autocorrelation function

$$\sigma(\tau) = \langle \psi_E(0) | \psi_E(\tau) \rangle \tag{3.100}$$

where

$$\psi_E(\tau) = e^{-\imath H_f \tau} \psi_E(0) \qquad \psi_E(0) = \int_0^\infty dt\, e^{(\imath E - \Gamma)t} \psi(t) \tag{3.101}$$

The wavepacket $\psi_E(\tau)$ with the initial value $\psi_E(0)$ is the solution of the nonstationary Schrödinger equation with the final-state Hamiltonian H_f, whereas the wavepacket

$$\psi(t) = e^{\imath H_f t} Q e^{-\imath H_c t} \mathcal{D} |o\rangle \tag{3.102}$$

gives different computational strategies in terms of one- or two-step dynamics. A full time-dependent formulation together with wavepacket algorithms thus provides insight and reveals particular time scales and dynamics of the scattering subprocesses that involve nuclear degrees of freedom and channel interference effects. Through the introduction of the quantum concept of a *duration time,* one can thus analyze and predict many effects or processes associated with such time-dependent wavepacket formulations, like "restoration of symmetry selection rules by detuned excitation," meaning that selection rules are dynamically restored; "collapse of vibrational structure," where the short-time dynamics of the RIXS process compresses molecular—Franck–Condon—signatures; "control of molecular dissociation," where the actual detuning and speeding up of the scattering process regulates the amount of dissociation signatures in the spectra; and "appearance of atomic holes," an interesting effect where the interference between the molecular and dissociation channels can suppress spectral peaks and even create a spectral hole. These are a few of many examples of physically interesting phenomena that can be well covered by time-dependent wavepacket calculations and analysis of the X-ray scattering process.

3.13 SUMMARY

A variety of computational methods for X-ray analysis and interpretation have been reviewed, with applications covering a representative cross section of chemical and physical systems and effects. It was indicated that theory and modeling have been indispensable for interpreting and assigning spectra throughout the development of X-ray spectroscopy. From the outset of the present state of the art one can predict development of theory and calculations of X-ray spectra along many lines of research. Several of these rest on clever implementations of electronic structure and scattering theory. One can here foresee a development in terms of response and moment theories and of scattering matrix approaches, including non-Born–Oppenheimer effects. One can predict development of the theories more in conjunction with, or as direct generalizations of, the bound-state electronic structure methods and even the very computer codes, rather than by a development on their own. Applications will probably include further studies of vibronic interaction and fine structures on a fundamental basis, including the coupling with the continuum, anticipating studies of threshold effects, angular distributions, vibrationally enhanced resonances, post-collision interaction effects, and various resonant phenomena in general. We will certainly see much further development of computational science with application to resonant X-ray spectroscopies. An outlook on this development that refers to the coming X-ray free-electron lasers is given in the next, and last, section.

It is clear that the presumed theoretical efforts are actualized and motivated by the ever on-going spectroscopic improvements, in particular by the development of synchrotron radiation facilities with energy and polarization variable excitation sources. First- and second-row diatomics have provided very useful testbeds for the alleged studies. For larger molecules, the number of degrees of freedom for both nuclear and electronic motions, the large number of interacting channels, and the breakdown of the Born–Oppenheimer approximation are facts that will "smear out" fine structures assigned to electronuclear interactions or resonance effects. A line of development refers to a discrete-state electronic structure analysis by means of either advanced many-body methods or clever simplifications thereof. The local character of X-ray transitions and the entailed local effective selection rules are instrumental in this analysis. Chemical information on symmetry, delocalization, hybridization, and bonding can be obtained from the spectra as described in this review. In conjunction with computations the spectra supply information on the conformational geometries and force fields. As progressively larger and more complex systems become tractable, systematic trends can be unraveled in cluster and oligomer sequences approaching polymers, biomolecules, or models of solids or liquids. Thus intermolecular effects will find more interest and the overlap with chemistry, surface science, and solid-state physics will become larger as ab initio calculations on spectra of adsorbates, crystals, and molecular complexes proceed.

Along with the development of multiscale modeling and linear scaling technologies, cluster approaches become promising in the study of extended systems because of the local probe character of X-ray spectra. On the way to ever-larger systems at the mesoscale more approximate—coarse-grain—methods will have to be developed. On the ab initio level approaches based on localized orbitals seem to be promising in that respect or, in the case of polymers, localized Wannier functions. In the case of large systems (or polymers) the size consistency of a method becomes important as is the case with Green's function methods which will automatically fulfill the size consistency requirement and can be easily transformed to an exciton-like representation in the case of periodic systems. Molecular methods can thus complement solid-state methods as well as atomic theory methods in the field of computational X-ray spectroscopy.

We have reviewed different models and methods for computing X-ray spectroscopy in this chapter. These have mostly been derived by rational principles using electronic structure theory. Such theory as implemented in computer codes has been instrumental throughout the years in exploring the physical–chemical origin of observation and in defining the merits and range of validity of models and computational methods. For example, the complete neglect of differential overlap (CNDO) method was decisive in establishing the charge potential model for ESCA shifts in the 1960s [46]. This underlines the fact that the scientific value, the purport and content, of an experimental technology is closely connected to the quality of the computational methods and the models used for its interpretation. That picture will surely be maintained as we now turn to a most exciting phase of X-ray spectroscopy—the use of the X-ray free-electron laser.

3.14 OUTLOOK: X-RAY FREE-ELECTRON LASER

X-ray free-electron lasers (XFELs) are now under construction all over the world. They are generally expected to bring a paradigm shift to the natural sciences, and, in particular, a revival of X-ray spectroscopy, the oldest of our tools to investigate elementary composition and electronic and geometric structure of matter. With ultrashort, femtosecond, pulses, with full space and time coherence at wavelengths matching atomic dimensions, they will open a broad avenue of scientific issues of fundamental and applied character. We can foresee emerging new disciplines like X-ray femtochemistry, dynamic X-ray Raman spectroscopy, and diffractional scattering at an ultrafast time scale, which all will help to solve essential problems in the materials and life sciences, for instance, in protein biology with the possibility to study single proteins and membrane proteins or in materials science with the possibility to follow femtosecond dynamics at atomic dimensions. The X-ray free-electron laser facilitates a "big science" that will be utilized by all science disciplines and will serve as a common meeting place for all their researchers and students. The utilization of improved time and length coherence and other intriguing aspects of the coming X-ray free-electron lasers call for a fundamental theoretical basis and a concomitant development of theory and modeling. New physics will be unraveled and a range of novel application areas will be opened with wide ramifications in basically all branches of natural science. Outstanding problems in research will be addressed associated with the coming XFEL facilities with applications in the areas of molecular and material science, structural chemistry, and biology. New possibilities will be explored to obtain structural and temporal information at multiscale dimensions on man-made, inorganic or biological materials through the use of resonant, inelastic and elastic, X-ray scattering with high power, of the kind that will be offered by the XFEL. Research efforts in theory and modeling can enable new strategies and discoveries in these areas as well as realize many of the great expectations these expensive facilities bring about. All efforts will be pursued in close collaboration with many of the experimental research groups that are, or will be, active in the XFEL field. Our view is that it will be of great strategic importance to meet these large experimental XFEL ventures with a concomitant—but in funding evidently much smaller—effort in theory and simulations. This view rests on the proven success history has shown in finding fruitful collaboration between theory and experiment in the X-ray sciences in the past, some of which was reviewed in this chapter.

Theoretical modeling of the chemical–physical processes which occur at extreme conditions such as in XFELs will have a most important role in a coming strong-field X-ray science. Already now we evidence extensive simulations of the dynamics of the Coulomb explosion induced by high-field ionization of matter [251, 252]. Despite the high intensities, the XFEL–matter interaction is of nonrelativistic and perturbative nature because the so-called Keldysh parameter is always larger than 1 in the X-ray region (laser period is far too small for field or tunnel ionization). Numerous phenomena to be discovered during forthcoming experiments will require new theory building and extensive modeling. This goes for processes like time-resolved X-ray pump–probe spectroscopy, X-ray nonlinear spectroscopy with multi-X-ray photon absorption, and nonlinear X-ray pulse propagation. While nonlinearity in the interaction between matter and electromagnetic fields has become a very important and much attended research field in the optical region, with a broad scope of technical applications, little is known about the theoretical or practical aspects of nonlinearity in X-ray response and spectroscopy. With the enormous intensity of soon-to-come XFEL sources this situation will certainly change and open a very fruitful research field related to nonlinear X-ray spectroscopy and nonlinear X-ray pulse propagation. New "effects" and "processes" will turn up as objects for research when we turn to the X-ray region, for instance, the possible use of the extremely confocal properties of two-X-ray photon absorption in medical therapy. Computations will thus be used to investigate nonlinear effects induced by a strong, coherent, X-ray radiation and in studies of nonlinearity, such as ultra sharp three-dimensional imaging; "X-ray scissors" for bond ruptures; "X-ray tweezers" for ultrasmall confinement; and confocal two-X-ray photon absorption for manipulation at ultrasmall, even atomic, scales. The change of penetration and avoidance of self-absorption imply completely new, unresearched, possibilities for X-ray fluorescence marking and imaging. Simulations of multiphoton absorption are here important from several points of view, for example, to model the systems which can be transparent when the intensity exceeds some critical value. We find that the novel X-ray polarization propagator softwares are excellently suited for simulations of the cross sections of multiphoton absorption in high-intensity X-ray fields.

Understanding nonlinear processes at the smallest accessible spatiotemporal scale will be at the frontier of modern research. In this respect, studies of nonlinear propagation of the XFEL pulses is the mainstream of strong X-ray field physics, where one can expect new and unexpected results. The optical properties of an ensemble of atoms can be strongly altered by exposing the atoms to intense XFEL fields which induce quantum-interference effects by resonant couplings. Stimulated X-ray Raman scattering is such a coherent effect where the interference of the pump and Stokes fields results in a pulse reshaping. For example, first-principles simulations of the strong XFEL pulses propagating through a resonant medium of atomic argon shows rather unexpected dynamics of the pulse shape and of its spectrum. A motivation for studies of strong XFEL pulse propagation is its potential application for the pulse compression into the attosecond region and the active operation with the spectral tunability of the XFEL pulse. Sub-femtosecond pulse compression is a main goal of strong-field X-ray physics, because the attosecond X-ray pulses allow to map the dynamics of the fast electronic subsystem. The pump X-ray pulse generates strong Stokes fields and changes drastically the spectral band through a dynamical Stark effect.

The modeling will be of different kinds, including analytical analysis of various processes and numerical simulations. The latter will include quantum chemical modeling of transition energies, transition dipole moments and nonlinear susceptibilities, multi-X-ray photon absorption, potential surfaces, vibronic coupling, and relativistic effects. Results of these simulations will be used for calculations of linear and nonlinear interaction with XFEL pulses. One can highlight three categories in coming X-ray computational analysis.

3.14.1 Semiclassical Wave Propagation

Accurate descriptions of the XFEL pulse propagation and accounting for the build-up of new coherent X-ray fields require detailed modeling, going beyond simple gain estimates relying on Einstein coefficients and radiation transfer arguments. Such refined modeling is expected to reproduce explicitly the dynamical aspects of the XFEL pulse propagation, including wave coherence properties with an accurate description of refraction and saturation on the subfemtosecond inverse-linewidth time scale. Transient phenomena in laser–matter interactions involving powerful and ultrashort XFEL pulses can be modeled using semiclassical approaches. In this context the electromagnetic wave is described by Maxwell's equations and is coupled to the Bloch model for matter via an expression for the polarization. This part of numerical modeling can be split into two independent codes. The first one implements a strict numerical solution of the coupled Bloch and Maxwell's equations for many-level system. The second code uses the slowly varying amplitude and phase approximation. This approximation used for the solution of coupled paraxial and Bloch equations allows to reduce the computational costs. In fact, this approximation nicely adapts for the pulse propagation in the X-ray region.

3.14.2 X-Ray Polarization Propagator

As reviewed in this chapter the traditional use of X-ray spectroscopy can be traced to the localized nature of the core electron involved in an X-ray transition, which implies effective selection rules, valuable for mapping the local electron structure, and a chemical shift that carries conformational information. We pointed out that from a theoretical point of view the core electron localization is a complicating factor that inflicts large relaxation of the valence electron cloud in a semistationary state that is embedded in an electronic continuum. Treatments of relaxation effects have favored the in-scope-restricted, state-specific methods whereas polarization propagator methods that otherwise form a universal approach to determine spectroscopic properties in the optical and ultraviolet regions have been disfavored. As discussed, the propagator-based formalism has several formal and practical advantages in that it explicitly optimizes the ground-state wavefunction (or density) only, it ensures orthogonality among states, it preserves gauge operator invariance, sum rules, and general size consistency, and it is applicable to all standard electronic structure methods (wavefunction and density based). The resonant-convergent first-order polarization propagator makes it possible to directly calculate the absorption cross section at a particular frequency without explicitly addressing the excited-state spectrum and is open-ended toward extensions to properties and spectra in the X-ray region in general, for instance, X-ray nonlinear spectroscopies such as multiphoton X-ray absorption. It is therefore highly consequential that their applicability now has been extended to the family of X-ray spectroscopies as recently accomplished elsewhere [18–21]. It is our belief that future calculations of X-ray spectra, especially with respect to XFELs, will make extensive use of the X-ray polarization propagator method.

3.14.3 Multiscale/Multiphysics Modeling

The development in recent years of theoretical methods and computer technologies has made it practical to consider rational design of new functional materials based on theoretical predictions. There are now several levels of approaches available for modeling of material properties at different time/length scales relevant in X-ray analysis: electronic, atomistic, mesoscopic, and macroscopic. An aim will be to apply the multiscale modeling concept of seamlessly integrated software as a tool for design of materials with functionality of relevance to the experimental research in the network. We can recognize several natural main levels of approaches which will be used in successive manners for the modeling of materials properties and light–matter interaction at different length/time scales and with relevance to X-ray spectroscopy: electronic structure methods, such as density functional theory implemented for linear scaling and response properties; quantum mechanics/molecular mechanics (QM/MM) methods, with the active region treated within a full quantum mechanical calculation while the electrostatic potential and interaction from the surrounding remainder of the system are determined using a more expedient classical force-field calculation; quantum mechanics/wave mechanics (QMWM) technique, which gives a systematic approach to model macroscopic properties probed by light. The QMWM approach is based on a quantum mechanical account of the many-level electron-nuclear medium coupled to numerical solutions of the density matrix and Maxwell's equations. Larger grains can be turned into continuum leading to the application of special classes of models, for instance, polarizable continuum models, where the effect of an environment or a solvent surrounding is described by the (optical and static) dielectric constants: mesoscale modeling. Many materials properties originate in processes far beyond the nanometer–nanosecond scales currently accessible by conventional molecular dynamic simulation methods. Studies of conformational rearrangements or phase transitions in polymeric materials are typical examples of those. The mesoscale models range from coarse-grain models, free-energy calculations, lattice-Boltzmann, kinetic Monte Carlo, dissipative particle dynamics, and homology modeling.

ACKNOWLEDGMENT

The authors would like to acknowledge the contribution of the "Ministero della Istruzione della Università e della Ricerca" of the Republic of Italy.

REFERENCES

1. H. A. Bethe, E. E. Salpeter, *Quantum Mechanics of One and Two Electron Atoms*, Academic, New York, 1957.
2. I. Cacelli, V. Carravetta, R. Moccia, *Chem. Phys.* **1984**, *90*, 313.
3. H. Bachau, E. Cormier, P. Decleva, J. E. Hansen, F. Martin, *Rep. Prog. Phys.* **2001**, *64*, 1815.
4. J. B. Foresman, M. Head-Gordon, J. A. Pople, M. J. Frisch, *J. Phys. Chem.* **1992**, 135.
5. I. Cacelli, V. Carravetta, A. Rizzo, R. Moccia, *Phys. Rep.* **1991**, *205*, 283.
6. R. Manne, *J. Chem. Phys.* **1970**, *52*, 5733.
7. U. Gelius, *J. Electron. Spectrosc. Rel. Phenom.* **1974**, *5*, 985.
8. W. Hunt, W. Goddard III, *Chem. Phys. Lett.* **1969**, *3*, 414.
9. H. Ågren, V. Carravetta, L. Pettersson, O. Vahtras, *Phys. Rev. B* **1996**, *53*, 16074.
10. H. Ågren, V. Carravetta, O. Vahtras, L. Pettersson, *Chem. Phys. Lett.* **1994**, *222*, 75.
11. H. Ågren, V. Carravetta, O. Vahtras, L. Pettersson, *Theor. Chem. Acc.* **1997**, *97*, 14.
12. N. Kosugi, H. Kuroda, *Chem. Phys. Lett.* **1980**, *74*, 490.
13. N. Kosugi, *Theor. Chim. Acta* **1987**, *72*, 149.
14. T. Dunning, V. McKoy, *J. Chem. Phys.* **1967**, *47*, 1735.
15. P. Jørgensen, J. Olsen, H. J. A. Jensen, *J. Chem. Phys.* **1988**, *74*, 265.
16. M. Feyereisen, J. Nichols, J. Oddershede, J. Simons, *J. Chem. Phys.* **1992**, *96*, 2978.
17. H. Koch, H. Ågren, P. Jørgensen, T. Helgaker, H. J. A. Jensen, *Chem. Phys.* **1993**, *172*, 13.
18. U. Ekström, P. Norman, *Phys. Rev. A* **2006**, *74*, 042722.
19. U. Ekström, P. Norman, V. Carravetta, H. Ågren, *Phys. Rev. Lett.* **2006**, *97*, 143001.
20. P. Norman, D. Bishop, H. Jensen, J. Oddershede, *J. Chem. Phys.* **2001**, *115*, 10323.
21. P. Norman, D. Bishop, H. Jensen, J. Oddershede, *J. Chem. Phys.* **2005**, *123*, 194103.
22. V. Schmidt, *J. Phys.* **1987**, *C9*, 401.
23. C. Liegener, *J. Phys B: At. Mol. Phys.* **1983**, *16*, 4281.
24. P. Bagus, *Phys. Rev.* **1965**, *139*, 619.
25. J. Almlöf, P. Bagus, B. Liu, D. MacLean, U. Wahlgren, M. Yoshimine, Molecule-Alchemy program package, IBM Research Laboratory, 1972. See also IBM Research Report RJ-1077 (1972).

26. S. Svensson, H. Ågren, U. I. Wahlgren, *Chem. Phys. Lett.* **1976**, *38*, 1.
27. H. Jensen, H. Ågren, *Chem. Phys. Lett.* **1984**, *110*, 140.
28. H. Jensen, H. Ågren, *Chem. Phys.* **1986**, *104*, 229.
29. H. Jensen, P. Jørgensen, H. Ågren, *J. Chem. Phys.* **1987**, *87*, 451.
30. D. P. Chong, S. R. Langhoff, *J. Chem. Phys.* **1990**, *93*, 570.
31. D. Chong, *Chem. Phys. Lett.* **1995**, *232*, 486.
32. D. Chong, *J. Chem. Phys.* **1995**, *103*, 1842.
33. C. Bureau, D. Chong, K. Endo, J. Delhalle, G. Lecayon, A. Le Moel, *Nucl. Instrum. Methods Phys. Res. B* **1997**, *269*, 1.
34. C. Bureau, *Chem. Phys. Lett.* **1997**, *269*, 378.
35. M. Stener, A. Lisini, P. Decleva, *Chem. Phys.* **1995**, *191*, 141.
36. C. Hu, D. Chong, *Chem. Phys. Lett.* **1996**, *262*, 729.
37. C. Hu, D. Chong, *Chem. Phys. Lett.* **1996**, *249*, 491.
38. L. Triguero, L. Pettersson, H. Ågren, *Phys. Rev. B* **1998**, *58*, 8097.
39. L. Triguero, L. Pettersson, H. Ågren, *J. Electron Spectrosc. Rel. Phenom.* **1999**, *104*, 195.
40. J. C. Slater, K. H. Johnson, *Phys. Rev. B* **1972**, *5*, 844.
41. M. Nyberg, Y. Luo, L. Triguero, L. Pettersson, H. Ågren, *Phys. Rev. B* **1999**, *60*, 7956.
42. J. Perdew, A. Zunger, *Phys. Rev. B* **1981**, *23*, 5048.
43. G. Tu, V. Carravetta, O. Vahtras, H. Ågren, *J. Chem. Phys.* **2007**, *127*, 174110.
44. E. Sokolowski, C. Nordling, K. Siegbahn, *Phys. Rev.* **1958**, *110*, 776.
45. K. Siegbahn, C. Nordling, A. Fahlman, R. Nordberg, K. Hamrin, J. Hedman, G. Johansson, T. Bergmark, S. Karlsson, I. Lindgren, B. Lindberg, *ESCA—Atomic, Molecular and Solid State Structure Studied by Means of Electron Spectroscopy*, North-Holland, Amsterdam, 1966.
46. K. Siegbahn, C. Nordling, G. Johansson, J. Hedman, P. F. Hedén, K. Hamrin, U. Gelius, T. Bergmark, L. O. Werme, R. Manne, Y. Baer, *ESCA Applied to Free Molecules*, North-Holland, Amsterdam, 1969.
47. B. Lindberg, S. Svensson, P. Malmquist, E. Basilier, U. Gelius, K. Siegbahn, *Chem. Phys. Lett.* **1976**, *40*, 175.
48. W. L. Jolly, D. N. Hendrickson, *J. Am. Chem. Soc.* **1970**, *92*, 1863.
49. W. L. Jolly, in *Electron Spectroscopy*, D. A. Shirley, Ed., North-Holland, Amsterdam, 1972.
50. R. L. Martin, D. A. Shirley, *J. Am. Chem. Soc.* **1974**, *96*, 5299.
51. D. Nordfors, H. Ågren, N. Mårtensson, *J. Electron Spectrosc. Rel. Phenom.* **1990**, *53*, 129.
52. H. Basch, *Chem. Phys. Lett.* **1970**, *5*, 337.
53. M. Barber, P. Seift, D. Cunningham, M. Fraser, *Chem. Commun.* **1970**, *2*, 1938.
54. F. Holsboer, W. Beck, H. Bortunic, *Chem. Phys. Lett.* **1973**, *18*, 217.
55. R. Manne, in *Inner-Shell and X-Ray Physics of Atoms and Solids*, Fabian et al., Eds., Plenum, New York, 1981.
56. H. Siegbahn, O. Goscinski, *Phys. Scripta* **1976**, *13*, 225.
57. U. Gelius, *J. Electron Spectrosc. Rel. Phenom.* **1974**, *9*, 133.
58. V. Carravetta, H. Ågren, A. Cesar, *Chem. Phys. Lett.* **1988**, *148*, 210.
59. O. Goscinski, B. T. Pickup, T. Åberg, *J. Phys. B* **1975**, *8*, 11.
60. L. Hedin, A. Johansson, *J. Phys. B* **1969**, *2*, 1336.

61. H. Siegbahn, R. Medeiros, O. Goscinski, *J. Electron Spectrosc. Rel. Phenom.* **1976**, *8*, 149.
62. J. Pireaux, S. Svensson, E. Basilier, P. Malmqvist, U. Gelius, R. Caudano, K. Siegbahn, *Phys. Rev. A* **1976**, *14*, 2133.
63. H. Ågren, *Chem. Phys. Lett.* **1981**, *83*, 149.
64. H. Siegbahn, M. Lundholm, S. Holmberg, M. Arbman, *Phys. Scripta* **1983**, *27*, 431.
65. H. Siegbahn, M. Lundholm, M. Arbman, S. Holmberg, *Phys. Scripta* **1984**, *30*, 35.
66. M. Lundholm, H. Siegbahn, S. Holmberg, M. Arbman, *J. Electron Spectrosc. Rel. Phenom.* **1986**, *40*, 163.
67. S. Holmberg, R. Moberg, O. Bohman, H. Siegbahn, *J. Phys.* **1987**, *48*(C9), 951.
68. C. Duke, W. Salaneck, T. Fabish, J. Ritsko, J. Thomas, *A. Paton, Phys. Rev.* **1978**, *B18*, 5717.
69. H. Ågren, H. Siegbahn, *J. Chem. Phys.* **1984**, *81*, 488.
70. M. Arbman, H. Siegbahn, L. Petterson, P. Siegbahn, *Mol. Phys.* **1985**, *54*, 1149.
71. H. Ågren, C. Medina-Llanos, K. Mikkelsen, *Chem. Phys.* **1987**, *115*, 43.
72. C. Medina-Llanos, H. Ågren, *Phys. Rev.* **1988**, *B38*, 11785.
73. H. Ågren, R. Arneberg, *Phys. Scripta* **1983**, *28*, 80.
74. K. Mikkelsen, E. Dalgaard, P. Swanstrøm, *J. Chem. Phys.* **1988**, *89*, 3086.
75. C. Medina-Llanos, H. Ågren, K. Mikkelsen, H. Jensen, *J. Chem. Phys.* **1989**, *90*, 6422.
76. H. Ågren, V. Carravetta, *Mol. Phys.* **1985**, *55*, 901.
77. B. Johansson, N. Mårtensson, *Phys. Rev. B* **1980**, *27*, 4427.
78. D. B. Beach, C. J. Eyermann, S. P. Smit, S. F. Xiang, W. L. Jolly, *J. Am. Chem. Soc.* **1984**, *106*, 536.
79. W. L. Jolly, in *Electron Spectroscopy: Theory, Techniques and Applications*, Vol. 1, C. Brundle, A. Baker, Eds., Academic, New York, 1977, Chapter 3, p. 119.
80. W. L. Jolly, C. Gin, D. B. Adams, *Chem. Phys. Lett.* **1977**, *46*, 220.
81. T. H. Lee, W. L. Jolly, A. A. Bakke, R. Weiss, J. G. Vergade, *J. Am. Chem. Soc.* **1980**, *102*, 2631.
82. D. W. Davis, J. W. Rabalais, *J. Am. Chem. Soc.* **1974**, *96*, 5305.
83. V. Carravetta, H. Ågren, *Mol. Phys.* **1985**, *55*, 201.
84. H. Ågren, G. Karlström, *J. Chem. Phys.* **1983**, *79*, 587.
85. D. Nordfors, H. Å gren, *J. Electron Spectrosc. Rel. Phenom.* **1991**, *56*, 1.
86. J. Gasteiger, M. G. Hutchings, *J. Am. Chem. Soc.* **1984**, *106*, 6489.
87. M. Randic, M. Barysz, J. Nowakowski, S. Nikolic, N. Trinajstic, *J. Mol. Struct. (Theochem)* **1989**, *185*, 95.
88. L. Werme, J. Nordgren, H. Ågren, C. Nordling, K. Siegbahn, *Z. Phys.* **1975**, *272*, 131.
89. U. Gelius, S. Svensson, H. Siegbahn, E. Basilier, Å. Faxälv, K. Siegbahn, *Chem. Phys. Lett.* **1974**, *28*, 1.
90. D. Clark, *J. Müller, Chem. Phys.* **1997**, *23*, 429.
91. H. Ågren, L. Selander, J. Nordgren, C. Nordling, K. Siegbahn, J. Müller, *Chem. Phys.* **1979**, *37*, 161.
92. J. Müller, H. Ågren, *Molecular Ions*, J. Berkowitz, K.-O. Groenewald, Eds., Plenum, New York, 1983.
93. O. Goscinski, J. Müller, E. Poulain, H. Siegbahn, *Chem. Phys.* **1978**, *55*, 407.

94. J. Nordgren, L. Selander, L. Pettersson, C. Nordling, K. Siegbahn, H. Ågren, *J. Chem. Phys.* **1982**, *76*, 3928.

95. H. Ågren, J. Müller, *J. Electron Spectrosc. Rel. Phenom.* **1980**, *19*, 285.

96. P. Bagus, F. Schaeffer, *Chem. Phys. Lett.* **1981**, *82*, 505.

97. J. Müller, H. Ågren, O. Goscinski, *Chem. Phys.* **1979**, *38*, 349.

98. H. Ågren, J. Nordgren, *Theor. Chim. Acta* **1981**, *58*, 111.

99. D. Dill, S. Wallace, J. Siegel, J. Dehmer, *Phys. Rev. Lett.* **1978**, *41*, 1230.

100. R. Arneberg, J. Müller, R. Manne, *Chem. Phys.* **1982**, *64*, 249.

101. H. Ågren, R. Arneberg, J. Müller, R. Manne, *Chem. Phys.* **1984**, *83*, 53.

102. W. Domcke, L. Cederbaum, L. Köppel, W. von Niessen, *Mol. Phys.* **1977**, *34*, 1759.

103. F. Gel'mukhanov, L. Mazalov, N. Shklyaeva, *Sov. Phys. JETP* **1975**, *42*, 1001.

104. F. Gel'mukhanov, L. Mazalov, A. Kontratenko, *Chem. Phys. Lett.* **1977**, *46*, 133.

105. F. Kaspar, W. Domcke, L. Cederbaum, *Chem. Phys. Lett.* **1979**, *44*, 33.

106. N. Correia, A. Flores, H. Ågren, K. Helenelund, L. Asplund, U. Gelius, *J. Chem. Phys.* **1985**, *83*, 2035.

107. L. Ungier, T. Thomas, *J. Chem. Phys.* **1985**, *82*, 3146.

108. T. Carroll, T. Thomas, *J. Chem. Phys.* **1988**, *89*, 5983.

109. A. Cesar, H. Ågren, V. Carravetta, *Phys. Rev. A* **1989**, *40*, 187.

110. T. Åberg, *Phys. Scripta.* **1980**, *21*, 495.

111. T. Sharp, H. Rosentock, *J. Chem. Phys.* **1961**, *41*, 3453.

112. E. Doktorov, I. Malkin, V. Manko, *J. Mol. Spectrosc.* **1975**, *56*, 1.

113. P. Malmquist,UUIP 1058, University of Uppsala, Sweden, 1982.

114. A. Palma, L. Sandoval, *Int. J. Quant. Chem.* **1988**, *22*, 503.

115. A. Cesar, H. Ågren, *Int. J. Quant. Chem.* **1992**, *42*, 365.

116. L. Cederbaum, W. Domcke, *J. Chem. Phys.* **1974**, *60*, 2878.

117. L. Cederbaum, W. Domcke, *J. Chem. Phys.* **1976**, *64*, 603.

118. L. S. Cederbaum, H. Köppel, W. von Niessen, *J. Chem. Phys.* **1977**, *34*, 1759.

119. L. S. Cederbaum, W. Domcke, *Adv. Chem. Phys.* **1977**, *36*, 205.

120. T. Darko, I. H. Hillier, J. Kendrick, *Mol. Phys.* **1976**, *32*, 33.

121. R. L. Martin, D. A. Shirley, *J. Chem. Phys.* **1975**, *64*, 3685.

122. H. J. Aa. Jensen, H. Ågren, J. Olsen, in *Modern Techniques in Computational Chemistry: MOTECC-90*, E. Clementi, Ed., Escom, Leiden, 1990, p. 435.

123. G. Angonoa, O. Walter, J. Schirmer, *J. Chem. Phys.* **1987**, *87*, 6789.

124. B. O. Roos, *Chem. Phys. Lett.* **1971**, *15*, 153.

125. B. Roos, P. Taylor, A. Heiberg, *Chem. Phys.* **1980**, *48*, 157.

126. B. Roos, G. Karlström, P. Å. Malmquist, A. Sadlej, P. Widmark, in *Modern Techniques in Computational Chemistry: MOTECC-90,* E. Clementi, Ed., Escom, Leiden, 1990, p. 533.

127. H. Ågren, F. Flores-Riveros, H. Jensen, *Phys. Rev. A* **1986**, *34*, 4606.

128. A. Lisini, G. Fronzoni, P. Decleva, *J. Phys B: At. Mol. Phys.* **1988**, *22*, 3653.

129. G. Fronzoni, G. D. Alti, P. Decleva, A. Lisini, *Chem. Phys.* **1995**, *195*, 171.

130. P. Siegbahn, *J. Chem. Phys.* **1981**, *75*, 2314.

131. R. Buenker, S. Peyerimhoff, *Theor. Chim. Acta* **1974**, *35*, 33.

132. A. Lisini, P. Decleva, *Int. J. Quant. Chem.* **1995**, *55*, 281.
133. W. Rodwell, M. Guest, T. Darko, I. Hillier, J. Kendrick, *Chem. Phys.* **1977**, *77*, 467.
134. M. Guest, W. Rodwell, T. Darko, I. Hillier, J. Kendrick, *J. Chem. Phys.* **1977**, *66*, 5447.
135. A. Barth, R. Buenker, S. Peyerimhoff, W. Butscher, *Chem. Phys.* **1980**, *46*, 146.
136. R. Buenker, S. Peyerimhoff, *Theor. Chim. Acta* **1974**, *39*, 217.
137. L. Cederbaum, W. Domcke, J. Schirmer, *Phys. Rev. A* **1980**, *22*, 206.
138. H. Ågren, A. Cesar, C. Liegener, *Adv. Quant. Chem.* **1992**, *23*, 1.
139. H. Ågren, H. Siegbahn, *Chem. Phys. Lett.* **1980**, *69*, 424.
140. H. Ågren, H. Siegbahn, *Chem. Phys. Lett.* **1980**, *72*, 498.
141. H. Ågren, *J. Chem. Phys.* **1981**, *75*, 1267.
142. T. Åberg, G. Howat, Theory of the Auger effect, in *Handbuch der Physik*, Vol. 1, S. Flügge, W. Melhorn, Eds., Springer, Berlin, 1982.
143. U. Fano, *Phys. Rev.* **1961**, *124*, 1866.
144. C. Liegener, *Chem. Phys. Lett.* **1984**, *106*, 201.
145. D. Jennison, *J. Vac. Sci. Technol.* **1982**, *20*, 548.
146. L. Cederbaum, W. Domcke, *Adv. Chem. Phys.* **1977**, *36*, 205.
147. L. Cederbaum, W. Domcke, J. Schirmer, W. von Niessen, *Adv. Chem. Phys.* **1986**, *65*, 115.
148. J. Oddershede, *Adv. Quant. Chem.* **1978**, *11*, 275.
149. J. Oddershede, *Adv. Chem. Phys.* **1987**, *69*, 201.
150. Y. Öhrn, G. Born, *Adv. Quant. Chem.* **1981**, *13*, 1.
151. P. Jørgensen, J. Simons, in *Second-Quantization Based Methods in Quantum Chemistry*, Academic, New York, 1981.
152. M. Herman, K. Freed, D. Yeager, *Adv. Chem. Phys.* **1981**, *48*, 1.
153. N. Fukuda, F. Iwamoto, K. Sawada, *Phys. Rev. A* **1964**, *135*, 932.
154. P. Ring, P. Schuck, in *The Nuclear Many-Body Problem*, Springer, Berlin, 1980.
155. E. Economou, in *Green's Functions in Quantum Physics*, Springer, Berlin, 1983.
156. C. Liegener, *Chem. Phys. Lett.* **1982**, *90*, 188.
157. J. Schirmer, A. Barth, *Z. Phys.* **1984**, *A317*, 267.
158. J. Ortiz, *J. Chem. Phys.* **1984**, *81*, 5873.
159. F. Tarantelli, A. Tarantelli, A. Sgamelotti, J. Schirmer, L. Cederbaum, *J. Chem. Phys.* **1985**, *83*, 4683.
160. A. Tarantelli, L. Cederbaum, *Phys. Rev. A* **1989**, *39*, 1639.
161. A. Tarantelli, L. Cederbaum, *Phys. Rev. A* **1989**, *39*, 1656.
162. R. Graham, D. Yeager, *J. Chem. Phys.* **1991**, *94*, 2884.
163. J. Almlöf, K. Faegri, Jr., K. Korsell, *J. Comput. Chem.* **1982**, *3*, 385.
164. B. Levy, G. Berthier, *Int. J. Quant. Chem.* **1968**, *2*, 307.
165. K. Faegri, R. Manne, *Mol. Phys.* **1976**, *31*, 1037.
166. R. Manne, K. Faegri, *Mol. Phys.* **1977**, *33*, 53.
167. C. Petrongolo, R. J. Buenker, S. O. Peyerimhoff, *J. Chem. Phys.* **1982**, *76*, 3655.
168. C. Liegener, *Chem. Phys.* **1983**, *76*, 397.
169. C. Liegener, *J. Chem. Phys.* **1983**, *79*, 2924.
170. C. Liegener, *Phys. Rev. A* **1983**, *28*, 256.

171. H. Ågren, N. Mårtensson, R. Manne, *Chem. Phys.* **1985**, *99*, 357.

172. H. Siegbahn, L. Asplund, P. Kelfve, *Chem. Phys. Lett.* **1975**, *35*, 330.

173. H. Ågren, *J. Chem. Phys.* **1981**, *75*, 1267.

174. R. Fink, *J. Electron Spectrosc. Rel. Phenom.* **1995**, *76*, 295.

175. R. Fink, *J. Chem. Phys.* **1997**, *106*, 4038.

176. R. F. Fink, S. L. Sorensen, A. N. de Brito, A. Ausmees, S. Svensson, *J. Chem. Phys.* **2000**, *112*, 6666.

177. R. F. Fink, M. N. Piancastelli, A. N. Grum-Grzhimailo, K. Ueda, *J. Chem. Phys.* **2009**, *130*, 14306.

178. V. Carravetta, H. Ågren, *Phys. Rev. A* **1987**, *35*, 1022.

179. V. Carravetta, H. Ågren, O. Vahtras, H. J. A. Jensen, *J. Chem. Phys.* **2000**, *113*, 7790.

180. V. Carravetta, H. Ågren, *Chem. Phys. Lett.* **2002**, *354*, 100.

181. P. W. Langhoff, in *Theory and Application of Moment Methods in Many-Fermion Systems*, B. Dalton, S. Grimes, J. Vary, S. Williams, Eds., Plenum, New York, 1980, p. 191.

182. H. Ågren, V. Carravetta, A. Cesar, *Chem. Phys. Lett.* **1987**, *139*, 145.

183. I. Cacelli, V. Carravetta, R. Moccia, A. Rizzo, *J. Chem. Phys.* **1988**, *89*, 7301.

184. P. Langhoff, C. Corcoran, J. Sims, F. Weinhold, R. Glover, *Phys. Rev. A* **1976**, *14*, 1042.

185. U. Fano, J. W. Cooper, *Rev. Mod. Phys.* **1968**, *40*, 441.

186. I. Hjelte, A. D. Fanis, V. Carravetta, N. Saito, M. Kitajima, H. Tanaka, H. Yoshida, A. Hiraya, I. Koyano, L. Karlsson, S. Svensson, K. Ueda, M. Piancastelli, *J. Chem. Phys.* **2005**, *122*, 84306.

187. V. Carravetta, F. Gel'mukhanov, H. Agren, S. Sundin, S. Osborne, A. deBrito, O. Bjorneholm, A. Ausmees, S. Svensson, *Phys. Rev. A* **1997**, *56*, 4665.

188. C. F. Fischer, M. Idrees, *Computer Phys.* **1989**, *3*, 53.

189. M. Brosolo, P. Decleva, *Chem. Phys.* **1992**, *159*, 185.

190. M. Stener, D. Toffoli, G. Fronzoni, P. Decleva, *Theor. Chem. Acc.* **2007**, *117*, 943.

191. M. Stener, G. Fronzoni, P. Decleva, *J. Chem. Phys.* **2005**, *122*, 234301.

192. J. Stöhr, *NEXAFS Spectroscopy*, Springer, Berlin, 1992.

193. J. Adachi, N. Kosugi, A. Yagishita, *J. Phys B: At. Mol. Phys.* **2005**, *38*, 127.

194. H. Ågren, V. Carravetta, O. Vahtras, *Chem. Phys.* **1995**, *195*, 47.

195. V. Carravetta, H. Ågren, L. Pettersson, O. Vahtras, *J. Chem. Phys.* **1995**, *102*, 5589.

196. N. Hellgren, J. Guo, C. Såthe, A. Agui, J. Nordgren, Y. Luo, H. Ågren, J. Sundgren, *Ann. Phys. (Leipzig)* **2001**, *79*, 4348.

197. J. Sheehy, T. Gil, C. Winstead, R. Farren, P. Langhoff, *J. Chem. Phys.* **1989**, *91*, 1796.

198. J. Dehmer, D. Dill, in *Electron-Molecule and Photon-Molecule Collisions*, T. Rescigno, V. McKoy, B. Schneider, Eds., Plenum, New York, 1979.

199. N. Kosugi, *Chem. Phys.* **2003**, *289*, 117.

200. N. Kosugi, E. Shigemasa, A. Yagishita, *Chem. Phys. Lett.* **1992**, *190*, 481.

201. A. Yagishita, E. Shigemasa, N. Kosugi, *Phys. Rev. Lett.* **1994**, *72*, 3961.

202. M. N. Piancastelli, A. Kivimäki, V. Carravetta, I. Cacelli, R. Cimiraglia, C. Angeli, M. C. H. Wang, M. de Simone, G. Turri, K. C. Prince, *Phys. Rev. Lett.* **2002**, *88*, 243002.

203. B. Huron, J.-P. Malrieu, P. Rancurel, *J. Chem. Phys.* **1973**, *58*, 5745.

204. C. Angeli, M. Persico, *Theor. Chim. Acc.* **1997**, *98*, 117.
205. I. Hjelte, O. Bjrneholm, V. Carravetta, C. Angeli, R. Cimiraglia, K. Wiesner, S. Svensson, M. N. Piancastelli, *J. Chem. Phys.* **2005**, *123*, 64314.
206. R. Feifel, Y. Velkov, V. Carravetta, C. Angeli, R. Cimiraglia, F. G. P. Salek, S. L. Sorensen, M. N. Piancastelli, A. D. Fanis, M. K. K. Okada, T. Tanaka, H. Tanaka, K. Ueda, *J. Chem. Phys.* **2008**, *128*, 64304.
207. J. Slater, K. Johnsson, *Phys. Rev. B* **1972**, *5*, 844.
208. S. Zabinsky, J. Rehr, A. Ankudinov, R. Albers, M. Eller, *Phys. Rev. B* **1995**, *52*, 2995.
209. A. Ankudinov, B. Ravel, J. Rehr, S. Conradson, *Phys. Rev. B* **1998**, *58*, 7565.
210. J. Rehr, R. Albers, *Rev. Mod. Phys.* **2000**, *72*, 621.
211. D. MacLachlan, B. Hallahan, S. Ruffle, J. N. M. Evans, R. Strange, S. Hasnain, *Biochem. J.* **1992**, *285*, 569.
212. D. E. Sayers, E. Stern, F. Lytle, *Phys. Rev. Lett.* **1971**, *27*, 1204.
213. E. Stern, *Phys. Rev. B* **1974**, *10*, 3027.
214. E. Stern, D. E. Sayers, F. Lytle, *Phys. Rev. B* **1975**, *11*, 4836.
215. J. Dehmer, D. Dill, *Phys. Rev. Lett.* **1975**, *35*, 213.
216. J. Dehmer, D. Dill, *J. Chem. Phys.* **1976**, *65*, 5327.
217. F. Kutzler, C. Natoli, S. D. D. K. Misemer, K. Hodgson, *J. Chem. Phys.* **1980**, *73*, 3274.
218. M. Ruiz-Lopez, M. Loos, J. Goulon, M. Benfatto, C. Natoli, *Chem. Phys.* **1988**, *121*, 419.
219. A. Filipponi, A. D. Cicco, C. Natoli, *Phys. Rev. B* **1995**, *52*, 15122.
220. F. G. H. Ågren, *Phys. Rev. B* **1994**, *15*, 11121.
221. F. Gel'mukhanov, O. Plashkevych, H. Ågren, *J. Phys B: At. Mol. Phys.* **2001**, *34*, 869.
222. A. E. Hansen, T. D. Bouman, *Adv. Chem. Phys.* **1980**, *44*, 545.
223. L. Alagna, S. D. Fonzo, T. Prosperi, S. Turchini, P. Lazzeretti, M. Malagoli, R. Zanasi, C. Natoli, P. Stephens, *Chem. Phys. Lett.* **1994**, *223*, 402.
224. V. Carravetta, O. Plashkevych, H. Ågren, *J. Chem. Phys.* **1998**, *109*, 1456.
225. L. Yang, V. Carravetta, O. Vahtras, O. Plashkevych, H. Ågren, *J. Synchrotron Rad.* **1999**, *6*, 708.
226. S. Turchini, N. Zema, S. Zennaro, L. Alagna, B. Stewart, R. D. Peacock, T. Prosperi, *J. Am. Chem. Soc.* **2004**, *126*, 4532.
227. O. Plachkevytch, V. Carravetta, O. Vahtras, H. Agren, *Chem. Phys.* **1998**, *232*, 49.
228. M. Tanaka, K. Nakagawa, A. Agui, K. Fujii, A. Yokoya, *Phys. Scripta* **2005**, 873.
229. A. Jiemchooroj, U. Ekström, P. Norman, *J. Chem. Phys.* **2007**, *127*, 165104.
230. A. Jiemchooroj, P. Norman, *J. Chem. Phys.* **2008**, *128*, 234304.
231. S. Villaume, P. Norman, *Chirality* **2009**, *21*, 31.
232. Y. Luo, H. Ågren, F. Gel'mukhanov, *J. Phys. B: At. Mol. Phys.* **1994**, *27*, 4169.
233. T. Åberg, B. Crasemann, in *Resonant Anomalous X-Ray Scattering. Theory and Applications*, G. Materlik, C. J. Sparks, K. Fischer, Eds., North-Holland, Amsterdam, 1994, p. 431.
234. F. Gel'mukhanov, H. Ågren, *Phys. Rev. A* **1994**, *49*, 4378.
235. F. Gel'mukhanov, T. Privalov, H. Ågren, *Phys. Rev. A* **1997**, *56*, 256.
236. S. Sundin, F. Gel'mukhanov, H. Ågren, S. J. Osborne, A. Kikas, O. Björneholm, A. Ausmees, S. Svensson, *Phys. Rev. Lett.* **1997**, *79*, 1451.

237. A. Cesar, F. Gel'mukhanov, Y. Luo, H. Ågren, P. Skytt, P. Glans, J.-H. Guo, K. Gunnelin, J. Nordgren, *J. Chem. Phys.* **1997**, *106*, 3439.

238. R. Feifel, F. Burmeister, P. Sałek, M. Piancastelli, M. Bässler, S. Sörensen, C. Miron, H. Wang, I. Hjeltje, O. Björneholm, N. de Brito, F. Gel'mukhanov, H. Ågren, S. Svensson, *Phys. Rev. Lett.* **2000**, *85*, 3133.

239. P. Skytt, P. Glans, J.-H. Guo, K. Gunnelin, S. C. J. Nordgren, F. Gel'mukhanov, A. Cesar, H. Ågren, *Phys. Rev. Lett.* **1996**, *77*, 5035.

240. F. Gel'mukhanov, P. Sałek, T. Privalov, H. Ågren, *Phys. Rev. A* **1999**, *59*, 380.

241. P. Skytt, P. Glans, K. Gunnelin, J.-H. G. J. Nordgren, Y. Luo, H. Ågren, *Phys. Rev. A* **1997**, *55*, 134.

242. O. Bjöarneholm, S. Sundin, S. Svensson, R. Marinho, A. N. de Brito, F. Gel'mukhanov, H. Ågren, *Phys. Rev. Lett.* **1997**, *79*, 3150.

243. P. Sałek, F. Gel'mukhanov, H. Ågren, *Phys. Rev. A* **1999**, *59*, 1147.

244. F. Gel'mukhanov, V. Kimberg, H. Ågren, *Chem. Phys.* **2004**, *299*, 358.

245. O. Björneholm, M. Bässler, A. Ausmees, I. Hjelte, R. Feifel, H. Wang, C. Miron, M. Piancastelli, S. Svensson, S. S. F. Gel'mukhanov, H. Ågren, *Phys. Rev. Lett.* **2000**, *84*, 2826.

246. F. Gel'mukhanov, H. Ågren, *J. Phys. B: At. Mol. Phys.* **1996**, *29*, 2751.

247. S. Svensson, A. Ausmees, S. J. Osborne, G. Bray, F. H. Ågren, A. Naves de Brito, O. Sairanen, A. Kivimäki, E. Nömmiste, H. Aksela, S. Aksela, *Phys. Rev. Lett.* **1994**, *72*, 3021.

248. T.-Y. Wu, T. Ohmura, *Quantum Theory of Scattering*, Prentice-Hall, New York, 1962.

249. F. Gel'mukhanov, H. Ågren, *Phys. Rev. A* **1996**, *54*, 379.

250. F. Gel'mukhanov, L. Mazalov, A. Kondratenko, *Chem. Phys. Lett.* **1977**, *46*, 133.

251. R. Neutze, W. Wouts, D. van der Spoel, J. Hajdu, *Nature* **2000**, *406*, 752.

252. S. Hau-Riege, R. A. London, A. Szoke, *Phys. Rev. E* **2004**, *69*, 05190.

4

MAGNETIC RESONANCE SPECTROSCOPY: SINGLET AND DOUBLET ELECTRONIC STATES

ALFONSO PEDONE
Scuola Normale Superiore, Pisa, Italy

ORLANDO CRESCENZI
Dipartimento di Chimica "Paolo Corradini," Università di Napoli Federico II, Naples, Italy

Computational Strategies for Spectroscopy: From Small Molecules to Nano Systems, First Edition. Edited by Vincenzo Barone.

4.1 INTRODUCTION

Magnetic resonance spectroscopy techniques provide information on many structural and dynamic aspects of chemical systems: This versatility has justified widespread applications in research and technology. At present, techniques such as nuclear magnetic resonance (NMR) and electron paramagnetic resonance (EPR) represent crucial tools in essentially all areas of chemistry, including not only organic and inorganic chemistry but also structural biology of proteins, DNA, RNA, and polysaccharides.

The interaction between computational chemistry and experimental magnetic resonance spectroscopies is increasing at a fast pace in recent years. This parallels a more general trend of successful synergic interactions between experimentalists and computational chemists based on the capability of quantum mechanical methods to provide reliable estimates for a large number of spectroscopic parameters.

As a matter of fact, the vast majority of experimental studies focuses on a relatively well-defined set of parameters. Taking as an example the important case of NMR spectroscopy of organic molecules, the characterization is usually based on measurements of proton and carbon chemical shifts in solution, homonuclear (and possibly heteronuclear) coupling constants, and proton–proton nuclear Overhauser enhancements [or the corresponding rotating-frame effects (ROEs)]. This set of data is certainly reductive if compared with the information content potentially accessible by NMR measurements; however, it does represent a reasonable balance of such factors as operator and instrument time, apparatus availability, costs, amounts of material required, completeness of information, and ease of interpretation.

At any rate, even within these widely employed NMR protocols, computational approaches can play an important role. In particular, the choice among alternative structural hypotheses can often be guided by the correspondence between measured and computed spectroscopic parameters. Instances where this approach has led to disprove a seemingly reliable structural assignment are not uncommon. As a matter of fact, it is reasonable to foresee that this kind of validation will become even more widespread in the near future; rather than any difficulty in the actual calculation of magnetic resonance parameters, stumbling blocks along this direction may be represented by issues of flexibility, protonation microstates, and conformational heterogeneity, which may entail a sizable contribution to the measured values of the magnetic parameters of specific systems and therefore must be carefully addressed in order to obtain computational estimates directly comparable to the experimental counterparts.

From another viewpoint, the computation of NMR parameters for organic/biochemical systems is usually feasible but rarely trivial: The size of the systems imposes obvious limitations on the theory levels that can be employed, with a clear bias toward methods rooted in the density functional theory (DFT) [1, 2]. The issue of structural flexibility has already been hinted at and may well represent the main bottleneck in the generalization of computational applications: In favorable cases, grid searches and relaxed potential energy scans can be used, but when the complexity of the problem increases, the computational effort involved rules out these simple approaches, and other conformational search techniques must be brought into play; these are discussed in detail in other chapters. Of course, for these screening phases one may want to employ lower theory levels than for the actual computation of spectroscopic parameters. In particular, the exploration would be much facilitated by the availability of reliable force fields. Unfortunately, in some areas of organic chemistry this is still not the case: The development of a tailored force field is in principle possible but in general not attractive. Alternatively, less expensive quantum mechanical (QM) methods may represent a viable solution. Needless to say, the adoption of low theory levels during the conformational screening phase implies that the energy cutoff for acceptance of structures that will be passed over to the subsequent refinement steps must be relaxed accordingly.

A more subtle correction related to a dynamical effect is that due to vibrational averaging of spectroscopic parameters. In this case, given the nonclassical nature of nuclear vibrational motions, simple classical simulations would miss important features of the equilibrium distribution of geometric parameters, and therefore a proper quantum mechanical averaging procedure needs to be employed. Recent algorithmic improvements [3–6] allow for an efficient computation of such vibrational averages, which have been shown to significantly influence spectroscopic parameters in a number of important cases. However, at room temperature and in the presence of a bath (e.g., the solvent) quantum effects are often smeared out and can be approximately described in terms of classical dynamics.

From a complementary point of view, dynamical processes that occur on a time scale comparable to that of the spectroscopic transition have a direct influence on signal lineshape, and their description must rely on specific theoretical formulations (e.g., the stochastic Liouville equation [7]). Several examples of the potentialities of such approaches have been provided in the field of EPR [8]. The topic is specifically addressed in chapter 12 of this book and it will not be discussed here.

4.2 NMR AND EPR SPIN HAMILTONIANS

In magnetic resonance spectroscopy, the observed electromagnetic signals are related to transitions between different electronic and/or nuclear spin states; since these states are typically degenerate, an external static magnetic field ($\vec{B}$) is employed to relieve degeneration. In quantum mechanics, the state of a molecule is described by a spatial electronic wavefunction Ψ, an electron spin state defined by the spin quantum number m_S, and a nuclear spin state defined by a set of quantum numbers $m_1, \ldots, m_N$. A generic transition between two spin states (a and b) can be represented by the following symbolism:

$$\Phi^a = |\Psi,\, m_S^a,\, m_1^a, \ldots, m_N^a\rangle \rightarrow \Phi^b = |\Psi, m_S^b, m_1^b, \ldots, m_N^b\rangle \tag{4.1}$$

The underlying assumption, of course, is that the electronic wavefunction is only marginally affected by the transition. Therefore, the corresponding transition energy is

$$\Delta E^{ab} = E(\Phi^b) - E(\Phi^a) = E(m_s^b, m_1^b, \ldots, m_N^b) - E(m_S^a, m_1^a, \ldots, m_N^a) \tag{4.2}$$

The energies of each spin state can be expressed as the expectation values of the corresponding *spin Hamiltonian* $\hat{H}_s$:

$$E(\Phi^a) = \langle \Phi^a | \hat{H}_S | \Phi^a \rangle = \langle m_S^a, m_1^a, \ldots, m_N^a | \hat{H}_S | m_S^a, m_1^a, \ldots, m_N^a \rangle = E(m_S^a, m_1^a, \ldots, m_N^a) \tag{4.3}$$

The concept of a spin Hamiltonian is thus central to this discussion, in which it plays a twofold role: From an experimental viewpoint, "effective" spin Hamiltonians are used to convey a description of the experimental spectral behaviors in terms of numerical values of the magnetic parameters; thus, the structural and dynamical information on the system under examination is summarized and encoded into these empirical parameters. From the viewpoint of computational spectroscopy, the spin Hamiltonian is first of all decomposed into a set of individual operators corresponding to specific physical effects. Once suitable theoretical/computational descriptions are established for these operators a viable link is obtained between computed and observed spectral parameters. In the case of NMR spectroscopy, a general formulation of the spin Hamiltonian is the following:

$$\hat{H}_S(\mathrm{NMR}) = -\sum_{C=1}^{N} \hbar\gamma_C \vec{B} \cdot (\overleftrightarrow{1} - \overleftrightarrow{\sigma}_C) \cdot \vec{I}_C + \frac{\hbar^2}{2} \sum_{C=1}^{N} \sum_{\substack{D=1 \\ D \neq C}}^{N} \gamma_C \gamma_D \vec{I}_C \cdot (\overleftrightarrow{D}_{CD} + \overleftrightarrow{K}_{CD}) \cdot \hat{I}_D \tag{4.4}$$

where γ_c are nuclear magnetogyric ratios; $\vec{I}_C$ nuclear spin operators (related to the nuclear magnetic dipole moments, $\vec{\mu}_C = \hbar\gamma_C \vec{I}_C$); $\overleftrightarrow{\sigma}_C$ magnetic shielding tensors (accounting for the shielding effect of the electrons); $\overleftrightarrow{D}_{CD}$ dipolar coupling tensors (which describe the direct couplings of the nuclear magnetic dipole moments); and $\overleftrightarrow{K}_{CD}$ reduced indirect nuclear spin–spin coupling tensors (which describe the indirect couplings of the nuclear dipoles caused by the surrounding electrons).

In lieu of the reduced nuclear spin–spin coupling tensor, the indirect spin–spin coupling tensor $\overleftrightarrow{J}_{CD} = h\gamma_C\gamma_D\overleftrightarrow{K}_{CD}$ is more usually employed in the NMR spin Hamiltonian.

Equation (4.4) provides an adequate description for most practical applications of NMR spectroscopy. However, the introduction of additional terms may be required in specific situations; thus, for example, in paramagnetic molecules the coupling between electron and nuclear spin magnetic dipoles may give rise to additional spectral features (i.e., the Knight shift [9, 10]), which can be described by introducing the following additional term:

$$\hat{H}(S,I) = \sum_{C=1}^{N} \vec{S} \cdot \overleftrightarrow{A}_C \cdot \vec{I}_C \tag{4.5}$$

where $\overleftrightarrow{A}_C$ is the hyperfine coupling tensor. Electron spin–orbit effects can also affect NMR spectra; however, instead of introducing additional terms into the spin Hamiltonian, these are more conveniently accounted for by a modified spatial wavefunction Ψ_L (where L denotes effects due to the angular orbital moment). In essence, the operators representing orbital–spin interaction are incorporated within the spatial Hamiltonian, and the resulting Schrödinger equation is solved for Ψ_L.

In molecules containing quadrupolar nuclei (nuclei with high spin magnetic dipoles, $|\vec{I}_C| > 1/2$) the different projections of nuclear spin are energetically split by the presence of the various electrostatic charges (both electronic and nuclear) in their vicinity. The following term accounts for such interaction:

$$\hat{H}_Q(I,I) = \sum_{\substack{C=1 \\ |I_C|\geq 1}}^{N} \frac{eQ_C}{6I_C(2I_C-1)\hbar} \vec{I}_C \cdot e\overleftrightarrow{q} \cdot \vec{I}_C \tag{4.6}$$

where the (traceless) tensor $e\overleftrightarrow{q}$ describes the electric field gradient; a component $eq_{\alpha\beta}$ (with $\alpha, \beta = x, y, z$) is the gradient of the α component of an electric field (E_β) in the β direction. Two parameters, the quadrupole coupling constant C_Q and the asymmetry parameter η_Q, are defined from the principal values of the electric field gradient tensor when it is expressed in its principal axis frame (PAF):

$$C_Q = \frac{eQq_{ZZ}^{\mathrm{PAF}}}{h} \tag{4.7}$$

$$\eta_Q = \frac{q_{xx}^{\mathrm{PAF}} - q_{yy}^{\mathrm{PAF}}}{q_{zz}^{\mathrm{PAF}}} \tag{4.8}$$

If now we shift attention toward the description of EPR spectra, the central features which must be accounted for by the spin Hamiltonian are the interactions of the electron spin (S) of a free radical with an external magnetic field (B) and with a generic magnetic nucleus of spin I:

$$H_S(\text{EPR}) = \mu_B \vec{S} \cdot \overleftrightarrow{g} \cdot \vec{B} + \frac{1}{\hbar\gamma_I} \vec{S} \cdot \overleftrightarrow{A} \cdot \vec{\mu}_I \tag{4.9}$$

where the first term is the Zeeman interaction between the electron spin and the external magnetic field through the Bohr magneton $\mu_B = e\hbar/2m_e c$ and the g tensor is defined as follows: $\overleftrightarrow{g} = g_e \overleftrightarrow{1} + \Delta \overleftrightarrow{g}^{\text{corr}} = g_e \overleftrightarrow{1} + \Delta \overleftrightarrow{g}_{\text{RM}} + \Delta \overleftrightarrow{g}_G + \Delta \overleftrightarrow{g}_{\text{OZ/SOC}}$, where $\Delta \overleftrightarrow{g}^{\text{corr}}$ is the correction to the free-electron value ($g_e = 2.0022319$) due to several terms including the relativistic mass ($\Delta \overleftrightarrow{g}_{\text{RM}}$), the gauge first-order corrections ($\Delta \overleftrightarrow{g}_C$), and a term arising from the coupling of the orbital Zeeman (OZ) and the spin–orbit coupling (SOC) operator [11, 12]. The second term in Eq. 4.9 describes the hyperfine interaction between S and the nuclear spin I through the hyperfine coupling tensor $\overleftrightarrow{A}$. In cases of high–spin paramagnetism ($S > \frac{1}{2}$) the EPR effective spin Hamiltonian must be augmented with the zero-field splitting term:

$$\hat{H}(S, S) = \sum_{i=1,\, j \neq i}^{Nue} \vec{S}_i \cdot \overleftrightarrow{D} \cdot \vec{S}_j \tag{4.10}$$

which arises from the magnetic dipolar interactions between the multiple unpaired electrons in the system.

4.3 CALCULATION OF SPIN HAMILTONIAN PARAMETERS

4.3.1 Shielding Constants and Indirect Spin–Spin Coupling Constants

The shielding constants and the indirect spin–spin coupling constants can be evaluated ab initio as the derivatives of the electronic energy with respect to the magnetic induction $\vec{B}$ and the nuclear spin $\vec{I}_K$:

$$\overleftrightarrow{\sigma}_K = \frac{1}{\gamma_K \hbar} \left(\frac{\partial^2 E}{\partial \vec{B}\, \partial \vec{I}_K} \right)_{B=0, I_K=0} \tag{4.11}$$

$$\overleftrightarrow{K}_{KL} = \frac{1}{\gamma_K \gamma_L \hbar^2} \left(\frac{\partial^2 E}{\partial \vec{I}_K\, \partial \vec{I}_L} \right)_{I_K=0, I_L=0} \tag{4.12}$$

In order to evaluate the derivatives, second-order response theory is employed within either a relativistic or a nonrelativistic formulation.

The expressions for closed-shell exact states were first developed by Ramsey in 1953 [13, 14]:

$$\overleftrightarrow{\sigma}_K = \langle 0|\hat{h}_{BK}^{\text{dia}}|0\rangle - 2 \sum_{n_S \neq 0} \frac{\langle 0|\hat{h}_B^{\text{orb}}|n_S\rangle \langle n_S|(\hat{h}_K^{\text{pso}})^T|0\rangle}{E_{n_S} - E_0} \tag{4.13}$$

$$\overleftrightarrow{K}_{KL} = \langle 0|\hat{h}^{\text{dso}}_{KL}|0\rangle - 2\sum_{n_S\neq 0}\frac{\langle 0|\hat{h}^{\text{pso}}_{K}|n_S\rangle\langle n_S|(\hat{h}^{\text{pso}}_{L})^T|0\rangle}{E_{n_S}-E_0}$$
$$-2\sum_{n_T}\frac{\langle 0|\hat{h}^{\text{sd}}_{K}+\hat{h}^{\text{fc}}_{K}|n_T\rangle\langle n_T|(\hat{h}^{\text{sd}}_{L})^T+(\hat{h}^{\text{fc}}_{L})^T|0\rangle}{E_{n_T}-E_0} \tag{4.14}$$

where $|n_S\rangle$ and $|n_T\rangle$ denotes singlet and triplet excited states, respectively. Both expressions contain a diamagnetic part corresponding to an expectation value of the unperturbed state and a paramagnetic part which represents the relaxation of the wavefunction in response to the external perturbations.

In Ramsey expressions, the diamagnetic electronic ($\hat{h}^{\text{dia}}_{BK}$) and spin–orbit ($\hat{h}^{\text{dso}}_{KL}$) operators are given by

$$\hat{h}^{\text{dia}}_{BK} = \frac{\alpha^2}{2}\sum_i \frac{(\vec{r}_{iO}\cdot\vec{r}_{iK})\vec{1}-\vec{r}_{iK}\vec{r}^{\,T}_{iO}}{r^3_{iK}} \tag{4.15}$$

$$\hat{h}^{\text{dso}}_{KL} = \frac{\alpha^4}{2}\sum_i \frac{(\vec{r}_{iK}\cdot\vec{r}_{iL})\vec{1}-\vec{r}_{iK}\vec{r}^{\,T}_{iL}}{r^3_{iK}r^3_{iL}} \tag{4.16}$$

The orbitalic paramagnetic operator ($\hat{h}^{\text{orb}}_{B}$) couples the external field to the orbital motion of the electron by means of the orbital angular momentum operator:

$$\hat{h}^{\text{orb}}_{B} = \frac{1}{2}\sum_i \hat{l}_{iO} \tag{4.17}$$

$$\hat{l}_{iO} = -i\vec{r}_{io}\times\vec{\nabla}_i \tag{4.18}$$

The paramagnetic spin–orbit ($\hat{h}^{\text{pso}}_{K}$) operator, also called the orbital hyperfine operator,

$$\hat{h}^{\text{pso}}_{K} = \alpha^2\sum_i \frac{\vec{l}_{iK}}{\vec{r}^{\,3}_{iK}} \tag{4.19}$$

couples the nuclear magnetic moments to the orbital motion of the electrons whereas the spin dipole ($\hat{h}^{\text{sd}}_{K}$) operator

$$\hat{h}^{\text{sd}}_{K} = \alpha^2\sum_i \frac{3\vec{r}^{\,T}_{iK}\vec{s}_i\vec{r}_{iK}-\vec{r}^{\,2}_{iK}\vec{s}_i}{r^5_{iK}} \tag{4.20}$$

and the Fermi contact ($\hat{h}^{\text{fc}}_{K}$) operator

$$\hat{h}^{\text{fc}}_{K} = -\frac{8\pi\alpha^2}{3}\sum_i \delta(\vec{r}_{iK})\vec{s}_i \tag{4.21}$$

couple the nuclear magnetic moments to the spin of the electron. In the above equations, $\alpha \approx 1/137$ is the fine-structure constant, $\vec{r}_{iO}$ and $\vec{r}_{iK}$ are the positions of the electron i relative to the origin of the vector potential (vide infra) and to the nucleus K, $\delta(\vec{r}_{iK})$ is the Dirac delta function, and $\vec{s}_i$ is the spin of electron i.

4.3.2 Gauge Origin Problem

A problem arising when a magnetic field is present in the Hamiltonian, such as in the calculation of magnetic properties, is the gauge origin problem. The vector potential representing the external magnetic field induction $\vec{B}$ is

$$\vec{A}_O(r) = \tfrac{1}{2}\vec{B} \times (\vec{r} - \vec{O}) \tag{4.22}$$

And therefore the Hamiltonian in Eq. 4.4 is not uniquely defined since we may choose the position of the gauge origin $\vec{O}$ freely and still satisfy the requirement that $\vec{B} = \vec{\nabla} \times \vec{A}_O(\vec{r})$. An exact wavefunction will of course give origin-independent results, as will an HF wavefunction if a complete basis set is employed. However, for practical basis sets the gauge error depends on the distance between the wavefunction and the gauge origin, and some methods try to minimize the error by selecting separate gauges for each (localized) molecular orbital. Two such methods are known as individual gauge for localized orbitals (IGLO) [15] and localized orbital/local origin (LORG) [16].

The implementation that completely eliminates the gauge dependence is known as the gauge including/invariant atomic orbitals (GIAO) [17] and makes the basis functions explicitly dependent on the magnetic field by inclusion of a complex phase factor referring to the position of the basis function (usually the nucleus). The effect is that matrix elements involving GIAOs only contain a difference in vector potentials, thereby removing the reference to an absolute gauge origin.

4.3.3 Field Gradient Calculations

The operators for the electronic and nuclear field gradient contributions at the nuclear center $\vec{R}_X$ are given by

$$q_{\text{el}}^{\alpha\beta}(\vec{r}_i; \vec{R}_X) = -\frac{3(\vec{r}_i - \vec{R}_X)_\alpha (\vec{r}_i - \vec{R}_X)_\beta - \left|(\vec{r}_i - \vec{R}_X)\right|^2 \delta_{\alpha\beta}}{\left|\vec{r}_i - \vec{R}_X\right|^5} \tag{4.23}$$

$$q_{\text{nucl}}^{\alpha\beta}(\vec{r}_i; \vec{R}_X) = \sum_{Y \neq X} Z_Y \frac{3(\vec{R}_Y - \vec{R}_X)_\alpha (\vec{R}_Y - \vec{R}_X)_\beta - \left|(\vec{R}_Y - \vec{R}_X)\right|^2 \delta_{\alpha\beta}}{\left|\vec{R}_Y - \vec{R}_X\right|^5} \tag{4.24}$$

which are inversely proportional to the third power of the distance. In the non-relativistic case the matrix element of $q_{\text{el}}^{\alpha\beta}$ is directly obtained from the expectation value, while the nuclear contribution is an additive constant for a given molecular geometry or lattice structure:

$$q^{\alpha\beta}(\vec{R}_X) = \sum_i \langle \varphi_i(\vec{r})|\hat{q}^{\alpha\beta}_{\mathrm{el}}(\vec{r};\vec{R}_X)|\varphi_i(\vec{r})\rangle + q^{\alpha\beta}_{\mathrm{nucl}}(\vec{R}_X) \tag{4.25}$$

where $\varphi_i(\vec{r})$ are the molecular orbitals.

4.3.4 Calculation of *g*-Tensor and Hyperfine Coupling Constants

The hyperfine coupling tensor $\overleftrightarrow{A}(N)$, which is defined for each nucleus N, can be decomposed into two terms:

$$\overleftrightarrow{A}(N) = a_N \overleftrightarrow{1} + \overleftrightarrow{A}_{\mathrm{dip}}(N) \tag{4.26}$$

The first term (a_N), usually referred to as the Fermi contact interaction, is an isotropic contribution, also known as the hyperfine coupling constant (HCC), and is related to the spin density (ρ_N) at the corresponding nucleus N by

$$a_N = \frac{4\pi}{3}\mu_B\mu_N g_e g_N \langle S_Z\rangle^{-1}\rho_N^{\alpha-\beta} \tag{4.27}$$

$$\rho_N^{\alpha-\beta} = \sum_{\mu,\nu} P^{\alpha-\beta}_{\mu,\nu}\langle\phi_\mu(\vec{r})|\delta(\vec{r}-\vec{r}_N)|\phi_\nu(\vec{r})\rangle \tag{4.28}$$

where $P^{\alpha-\beta}_{\mu,\nu}$ is the difference between the density matrices for electrons with α and β spins, that is, the spin density matrix, while $\langle S_z\rangle$ is the expectation value of the z component of the total electronic spin. The second contribution [$\overleftrightarrow{A}_{\mathrm{dip}}(N)$] is anisotropic and can be derived from the classical expression of interacting dipoles,

$$A^{kl}_{\mathrm{dip}}(N) = \tfrac{1}{2}\mu_B\mu_N g_e g_N\langle S_Z\rangle^{-1}\sum_{\mu,\nu} P^{\alpha-\beta}_{\mu,\nu}\langle\phi_\mu(\vec{r})|r_N^{-5}(r_N^2\delta_{kl}-3r_{N,k}r_{N,l})|\phi_\nu(\vec{r})\rangle \tag{4.29}$$

where $\vec{r}_N = \vec{r} - \vec{R}_N$.

4.4 CALCULATION OF NMR PARAMETERS IN PARAMAGNETIC SPECIES

Nuclear magnetic resonance spectroscopy of paramagnetic (PNMR) species is a valuable experimental technique able to provide unique information on the molecular electronic structure, geometry, and reactivity of radicals such as coordination compounds, metalloproteins, and organic free radicals used as spin labels and spin probes [18]. In the analysis of PNMR spectra, experimentalists usually decompose the chemical shift into three contributions: the "reference" (or "orbital") shift δ^{orb}, the Fermi contact shift δ^{FC}, and the pseudocontact shift δ^{PC}:

$$\delta = \delta^{orb} + \delta^{FC} + \delta^{PC} \tag{4.30}$$

The orbital term is analogous to the usual NMR chemical shift in diamagnetic systems, and to a good approximation it assumes the same value as in an equivalent diamagnetic systems. In a computational context, the isotropic orbital chemical shift is defined as

$$\delta^{orb} = \sigma_{iso}^{ref} - \sigma_{iso}^{orb} \tag{4.31}$$

where σ is the isotropic part of the shielding tensor and σ^{ref} is the shielding constant of the observed nucleus in a diamagnetic reference compound.

The contact shift δ^{FC} accounts for the Fermi contact interaction between the nuclear magnetic moment and the spin density at the location of the nucleus. In the simplest case it is given by

$$\delta^{FC} = \frac{2\pi}{\gamma_I} g_e \mu_B A \frac{S(S+1)}{3kT} \tag{4.32}$$

where A is the isotropic hyperfine coupling constant (in frequency units) and kT represents the thermal energy.

Finally, the pseudocontact shift δ^{PC} reflects the dipolar interaction between the magnetic moments of the radical center and the nucleus:

$$\delta^{PC} = \frac{\mu_B^2 S(S+1)}{3kT} \frac{(3\cos^2\Omega - 1)}{R^3} F(g) \tag{4.33}$$

In a reference system in which the origin is placed at the radical center, Ω represents the angle between the principal symmetry axis and the direction to the nucleus of interest; R is the distance between the induced magnetic moment and the nucleus, and $F(g)$ is an algebraic function of the g-tensor components which accounts for the combined effect of various relaxation times involved.

It follows from these physical interpretations that the Fermi contact shift δ^{FC} reports on the spin density distribution, while long-range structural constraints can be deduced from analysis of the pseudocontact shift δ^{PC}.

The increasing importance recently gained by paramagnetic NMR spectroscopy has led not only to the rapid development of appropriate experimental techniques but also to the implementation and development of the formalisms required for accurate nuclear shielding calculations [19].

4.4.1 First-Principles Calculations of Shielding Tensor in Paramagnetic Systems

In the closed-shell case, which has been the object of the preceding discussion, the NMR shielding tensor can be defined at the level of a single molecule; however, the situation is different in the paramagnetic case. The NMR shielding is now in essence a statistical property which must be computed by averaging over thermally accessible excited states. The general expression for the shielding tensor in open-shell species has been derived by Moon and Patchkovskii [20] and will not be detailed here.

The final expression, obtained in the limit of fast electron spin relaxation and nuclear relaxation times that are sufficiently long to allow for direct observation of the NMR transition, is the following:

$$\sigma_{ab} = \langle E^{(a,b)} \rangle_0 - \frac{1}{kT} \langle E^{(0,b)} E^{(a,0)} \rangle_0 \tag{4.34}$$

with

$$\langle E \rangle_0 = \frac{\sum_k E_k e^{-E_k^{(0,0)}/kT}}{\sum_k e^{-E_k^{(0,0)}/kT}} \tag{4.35}$$

$$E_k^{(a,0)} = \left(\frac{\partial E_k}{\partial B_a} \right)_{\vec{\mu}=\vec{B}=0} \qquad E_k^{(0,b)} = \left(\frac{\partial E_k}{\partial \mu_b} \right)_{\vec{\mu}=\vec{B}=0} \qquad E_k^{a,b} = \left(\frac{\partial^2 E_k}{\partial B_a \partial \mu_B} \right)_{\vec{\mu}=\vec{B}=0} \tag{4.36}$$

The first term of Eq. 4.34 is easily recognized as the Boltzmann average of the orbital NMR shielding tensor; the second term accounts for the interaction of the nuclear magnetic moment with the average spin density, induced by orbital– and electron spin–Zeeman interactions.

In practical calculations of paramagnetic shielding it is thus necessary to evaluate the energies and wavefunctions for all thermally accessible electronic states in the absence of magnetic fields. For each state, $E_k^{(a,b)}$, $E_k^{(a,0)}$, and $E_k^{(0,b)}$ are related to the orbital NMR shielding tensor, the EPR $\overleftrightarrow{g}$ tensor, and the hyperfine $\overleftrightarrow{A}$ tensor, respectively; a Boltzmann averaging is further required to determine the paramagnetic NMR shielding tensor. The fact that the overall effect derives from separate contributions implies that each one of them can be treated independently, even by adopting different levels of theory.

In the simplest instance, namely, a doubly degenerate electronic ground state ($S = \frac{1}{2}$) with no thermally accessible excited states, the energy levels can be discussed in terms of the following effective spin Hamiltonian:

$$\hat{H} = -\vec{B} \cdot \left(\overleftrightarrow{1} - \overleftrightarrow{\sigma}^{\text{orb}} \right) \vec{\mu}_I + \mu_B \vec{B} \cdot \overleftrightarrow{g} \cdot \vec{S} + \frac{1}{\hbar \gamma_I} \vec{\mu}_I \cdot \overleftrightarrow{A} \cdot \vec{S} \tag{4.37}$$

The orbital shielding tensor $\overleftrightarrow{\sigma}^{\text{orb}}$, the electronic paramagnetic resonance $\overleftrightarrow{g}$ tensor, and the hyperfine coupling tensor $\overleftrightarrow{A}$ have been introduced previously.

In the absence of the nuclear magnetic moment ($\vec{\mu}_I = 0$) the energy levels of the Hamiltonian are given by

$$E_m = m \frac{B}{2c} \left(\vec{\xi} \cdot g \cdot g^T \cdot \vec{\xi} \right)^{1/2} \tag{4.38}$$

where m varies between $-S$ and S in integer numbers ($-\frac{1}{2}$, $+\frac{1}{2}$ in this case) and $\vec{\xi} = \vec{B}/B$ specifies the direction of the applied magnetic field. Differentiation with respect to the magnetic field strength yields

$$E_m^{(0,a)} = m\mu_B(g_{ax}^2 + g_{ay}^2 + g_{az}^2)^{1/2} \quad \text{with} \quad a = x,\, y,\, z \tag{4.39}$$

Subsequently, by taking the expectation value of the hyperfine contribution to the effective spin Hamiltonian, for the eigenfunctions of the g-only Hamiltonian (not reported here) and differentiating with respect to the nuclear magnetic moment, the first-order hyperfine contributions are obtained:

$$E_m^{(0,b)} = \frac{m}{\hbar\gamma_I} \frac{(A_{bx}g_{ax} + A_{by}g_{ay} + A_{bz}g_{az})}{\sqrt{g_{ax}^2 + g_{by}^2 + g_{cz}^2}} \tag{4.40}$$

The last term which is required to evaluate the average in Eq. 4.34 is a mixed derivative with respect to the magnetic field strength $\vec{B}$ and nuclear magnetic moment $\vec{\mu}_I$, namely the orbital shielding tensor:

$$E_m^{(a,b)} = \sigma_{ab}^{\text{orb}} \tag{4.41}$$

Combining all the contributions gives

$$\overleftrightarrow{\sigma} = \overleftrightarrow{\sigma}^{\text{orb}} - \frac{\mu_B}{4kT\hbar\gamma_I} \overleftrightarrow{g} \cdot \overleftrightarrow{A}^{\text{T}} \tag{4.42}$$

where the superscript T denotes a matrix transpose. Furthermore, by considering the isotropic and traceless parts of the $\overleftrightarrow{g}$ and $\overleftrightarrow{A}$ tensors separately, the PNMR shielding tensor is given by

$$\overleftrightarrow{\sigma} = \overleftrightarrow{\sigma}^{\text{orb}} - \frac{\mu_B}{4kT\hbar\gamma_I}\left(g_{\text{iso}}A_{\text{iso}}\overleftrightarrow{1} + A_{\text{iso}}\Delta\overleftrightarrow{g} + g_{\text{iso}}\overleftrightarrow{A}_{\text{dip}}^{T} + \Delta\overleftrightarrow{g} \cdot \overleftrightarrow{A}_{\text{dip}}^{T}\right) \tag{4.43}$$

The first term in parentheses corresponds to the contact shift; the second contribution represents the anisotropic part of the contact shift and is traceless, as is the $g_{\text{iso}}\overleftrightarrow{A}_{\text{dip}}^{T}$ term. The interaction between the anisotropic parts of the $\overleftrightarrow{g}$ and $\overleftrightarrow{A}$ tensors, accounted for by the last term, can instead give rise to an isotropic part, the pseudocontact shift.

4.5 ELECTRON-CORRELATED METHODS TO COMPUTE MAGNETIC RESONANCE SPECTROSCOPIC PARAMETERS

The last decade has seen a remarkable development of approaches for including electron correlation effects in NMR and EPR parameter calculations. These can be classified as to whether they are based on a single Slater determinant or a multiconfigurational (MC) ansatz and to whether they use variational methods [such as configuration interaction (CI)] or nonvariational approaches (based on perturbation or CC theory) to treat electron correlation. Several studies have demonstrated that MC approaches are well suited for the description of large static correlation effects whereas single-reference approaches are better suited for treating dynamical correlation effects. In the area of NMR chemical shift calculations of closed-shell molecules, static correlation effects are less important, and as a consequence single-reference approaches are more commonly used.

The least expensive and conceptually simplest correlation treatment that can be applied to medium-size molecules is the second-order Møller–Plesset perturbation theory (MP2), which is the most popular single-reference approach for the low-level treatment of electron correlation [21]. Higher order MP perturbation theory such as MP3 and MP4 are typically less useful; in particular, results coming from MP3 level are inferior to MP2 because of the characteristically oscillatory convergence of perturbation theory. MP4, MP5, and MP6 offer some improvements, but the high computational costs required prevent their routine application [22].

For a more accurate treatment of electron correlation, coupled-cluster (CC) approaches [23] enter into play. While the full CC singles, doubles, triples (CCSDT) model [24] and augmented CCSDT approaches that feature corrections for quadruple and even higher excitations [25, 26] are currently too expensive, CC singles and doubles (CCSD) approximation [27] and CCSD augmented by a perturbative treatment of triple excitations, CCSD(T) [28], are more feasible and rather widely adopted.

The use of variational CI methods [29] met with some popularity in the past. however, their performance is typically inferior with respect to the corresponding CC approaches (e.g., CCSD in comparison with CISD). therefore, they are at present much less used.

MC approaches [30] involve the optimization of molecular orbitals within a restricted subspace of electronic occupations; provided such "active space" is appropriately chosen, they allow for an accurate description of static electron correlation effects. Dynamical correlation effects can also be introduced either at the perturbation theory level [complete active space with second-order perturbation theory (CASPT2), and multireference Møller–Plesset (MR-MP2) methods] [31] or via configuration interaction (MR-CI).

From a practical point of view, the most important methods among the aforementioned post-HF approaches are probably MP2, CCSD(T), and MCSCF.

Several review articles [32–34] have discussed the applications of electron-correlated calculations of NMR chemical shifts to problems in inorganic and organic chemistry, also in the light of comparisons with experimental results. In order to give a taste of the accuracy that can be reached in reproducing the NMR shielding constant of different elements, in Table 4.1 we have reported the shielding constants for HF, H_2O, NH_3, CH_4, CO, and F_2 taken from the work of Gauss and colleagues [35].

Their results show the oscillatory behavior of the convergence of the shielding constant when perturbation theory is used, while CCSD(T) agrees satisfactorily with experiment in all cases. MCSCF provides accurate results for simple cases such as HF

Table 4.1 Comparison of Nuclear Magnetic Shielding Constants (ppm) Calculated at HF-SCF, MP2, MP3, MP4, CCSD, CCSD(T), and MCSCF Levels of Theory Using GIAO Ansatz[a]

Molecule	Nucleus	HF-SCF	MP2	MP3	MP4	CCSD	CCSD(T)	MCSCF	Experiment[b]
HF	^{19}F	413.6	424.2	417.8	419.9	418.1	418.6	419.6	420.0 ± 1.0
	^{1}H	28.4	28.9	29.1	29.2	29.1	29.2	28.5	28.9 ± 0.01
H_2O	^{17}O	328.1	346.1	336.7	339.7	336.9	337.9	335.3	357.6 ± 17.2
	^{1}H	30.7	30.7	30.9	30.9	30.9	30.9	30.2	
NH_3	^{15}N	262.3	276.5	270.1	271.8	269.7	270.7	—	273.3 ± 0.3
	^{1}H	31.7	31.4	31.6	31.6	31.6	31.6	—	
CH_4	^{13}C	194.8	201.0	198.8	199.5	198.7	199.1	198.2	198.4 ± 0.9
	^{1}H	31.7	31.4	31.5	31.5	31.5	31.5	31.3	
CO		−25.5	10.6	−4.2	12.7	0.8	5.6	8.2	3.29 ± 0.9
	^{17}O	−87.7	−46.5	−68.3	−44.0	−56.0	−52.9	−38.9	-38.7 ± 17.2
F_2	^{19}F	−167.9	−170.0	−176.9	−180.7	−171.1	−186.5	−136.6	-196.0 ± 1.0

[a]Results from ref. 35.

[b]Experimental equilibrium values from ref. 36.

and H_2O but is somewhat less satisfactory for the other molecules. Thus, for example, the choice of the active space is particularly critical in the description of F_2 [35]. On the other hand, in those cases where static correlation is important (e.g., for ozone), MCSCF methods prove appropriate [35].

Switching the attention to the performance of electron-correlated post-HF calculations of spin–spin coupling constants, in Table 4.2 we show a comparison of the results reported by Helgaker et al. [32] for the second-row hydrides HF, H_2O, NH_3, and CH_4. In general, the accuracy of the results parallels strictly the quality of the treatment of correlation effects, even though basis set convergence may be an issue. In particular, the most accurate results are obtained by equation-of-motion (EOM)-CCSD, second-order polarization propagator approximation CCSD [SOPPA(CCSD)], and complete active space self-consistent field (CASSCF) calculations. The differences between the calculations reflect mainly the difference in the Fermi contact term. However, for accurate calculations, even the typically small noncontact contributions can significantly contribute to the overall agreement with experimental results.

Overall, in the literature systematic explorations of the performance of post-HF methods in the computation of EPR parameters are less well represented than for the NMR case; however, as a rule of thumb, trends that characterize the performance of NMR chemical shielding calculations can often be extrapolated to the corresponding g-tensor computations; likewise, many observations relative to spin–spin couplings will typically hold for hyperfine coupling constants as well (since both effects are dominated by Fermi contact interactions).

4.6 DFT ROUTE TO MAGNETIC RESONANCE SPECTROSCOPIC PARAMETERS

Methods rooted in density functional theory are particularly appealing in connection with the calculation of magnetic resonance spectroscopic parameters, since they couple remarkable accuracy with a favorable scaling with the number of active electrons [37–40]. In this sense, the so-called hybrid functionals deserve special mention, in which some Hartree–Fock exchange is mixed with its local counterpart. In the past, MP2 used to outperform DFT approaches for applications to organic molecules [38]; however, the steady development of new functionals is changing the situation. Already the PBE0 model [41], a hybrid functional with no adjustable parameters, performs very well in NMR chemical shift computation for common organic nuclei, irrespective of their hybridization state [42]. Both in NMR and in EPR [43], the isotropic shifts of hydrogen atoms are reproduced reasonably well by several functionals.

In many practical applications, the demand on computational performance is strongly relieved, since one is only interested in relative values of the parameters, for example, to chemical shifts as opposed to absolute shielding values. The procedure of subtracting the parameter value computed for the test site from a corresponding reference value involves cancellation of a large portion of the systematic errors. In the same vein, it is usually advantageous to introduce one or more "secondary" references which are chosen based on chemical similarity to the sites of interest. Thus, for

Table 4.2 Comparison of Spin–Spin Coupling Constants and Fermi Contact Contributions (Hz) in HF, H_2O, NH_3, and CH_4 Molecules[a]

Molecule		HF-SCF	MP2	CCSDPPA	SOPPA(CCSD)	EOM-CCSD	CASSCF	Experiment
HF	$^1J(FH)$	654.1	570.01	524.3		513.4	542.60	500 ± 20
	FC	467.5	390.71	329.4		338.2	359.84	
H_2O	$^1J(OH)$	−95.44	−74.73		−81.555	−74.90	−83.934	-80.6 ± 0.1
	FC	−84.74	−64.61		−69.092	−65.45	−72.083	
	$^2J(HH)$	−22.44	−18.33		−8.581	−10.81	−9.602	-7.11 ± 0.03
	FC	−23.23	−19.69		−11.866	−11.12	−12.702	
NH_3	$^1J(NH)$	52.78	42.81			41.81	42.25	43.6
	FC	50.99	41.02			40.19	40.10	
	$^2J(HH)$	−23.55	−19.90			−12.09	−9.77	−10.0
	FC	−23.60	−20.05			−11.72	−11.21	
CH_4	$^1J(CH)$	157.31	130.63	123.865		115.36	116.65	120.78 ± 0.05
	FC	155.90	129.27	122.124		113.84	114.82	
	$^2J(HH)$	−27.16	−21.04	−14.308		−15.80	−13.22	12.564 ± 0.04
	FC	−27.69	−20.53	−14.701		−15.51	−13.80	

[a]Results from ref. 32.

example, acetamide represents a convenient choice for computations aimed at reproducing amide proton chemical shifts in peptides [44]; and the C-1 of chlorobenzene is an optimal reference for chlorine-substituted aromatic carbons [45]. Overall, it is fair to state that DFT calculations have proved capable of predicting the NMR spectra of organic molecules with an accuracy sufficient to allow for fruitful comparison with experimental spectra [e.g., 46].

To provide a hint of how the choice of the functional can influence the value of the computed parameters in a slightly different context, Table 4.3 displays a comparison of the performance of different DFT functionals in reproducing ^{29}Si chemical shifts of a number of sites in silica-based materials; experimentally, NMR parameters of this kind are often employed to obtain structural information. NMR interactions in these materials are essentially dominated by bonding in the first few coordination spheres; therefore, a molecular cluster approach can be adopted in the modeling, in conjunction with atomic basis sets. In particular, separate explorations showed that convergence with respect to cluster size is reached when three complete atomic shells are included around the Si center (shell-3 cluster).

Thus, in the case at hand, the best agreement with experiment is actually provided by HF calculations; moreover, the differences among the functionals explored are not huge.

In a similar vein, Table 4.4 shows a comparison of scalar coupling constants computed with different hybrid and long-range corrected functionals for the geminal ^{29}Si–O–^{29}Si pairs within the siliceous zeolite Sigma-2, modeled by a suitable cluster, and a large basis set (cc-pV5Z on the ^{29}Si coupling partners and 6-31++G* elsewhere).

A rather similar general situation is encountered for EPR: The most refined post-Hartree–Fock (HF) models employing very large basis sets are no doubt able to deliver very accurate results for small rigid systems in vacuo; however, only the DFT methods, coupled to proper modeling of environmental and vibrational averaging effects, are at present capable of providing reliable results also for large systems in condensed phases [43, 48]. Unfortunately, this optimistic view is not completely general, and there exist situations when DFT calculations fail no matter which functional and basis set is adopted. Further progress in the setup of better DFT functionals may well relieve these issues in the long run. For the time being, a viable computational approach is offered by "composite" approaches of the ONIOM (our own *N*-layered *i*ntegrated molecular *o*rbital and molecular *m*echanics) *ilk* [49]. For example, the nitrogen hyperfine coupling constant in nitroxides is underestimated by DFT calculations but can be reproduced accurately at the quadratic configuration interaction (QCISD) level; in this case, the composite approach amounts to performing the computationally expensive QCISD calculation on a "small" model, whose geometry is frozen to that of the corresponding fragment in the "big" system, and introducing the effect of the rest of the molecule at the DFT level; that is, $a_N = a_{N,\mathrm{big}}^{\mathrm{DFT}} + a_{N,\mathrm{small}}^{\mathrm{QCISD}} - a_{N,\mathrm{small}}^{\mathrm{DFT}}$ [50]. We will note in passing that the peculiar local nature of a number of magnetic resonance parameters (including the hyperfine coupling constant of the preceding equation) implies a strong sensitivity to features of the wavefunction sampled at a single position in space (i.e., at the nucleus). In turn,

Table 4.3 Calculated ^{29}Si Isotropic Chemical Shift (ppm) for Shell-3 Cluster Models of Various SiO_2 Polymorphs Using Different DFT Methods in Conjunction with 6-311+G(2df,p) Basis Set[a]

Species/Site	HF	B3LYP	PBE0	M052X	CAM-B3LYP	Experiment
Cristobalite						
Si	−109.1 (0.6)	−108.9 (0.4)	−108.9 (0.4)	−109.3 (0.8)	−109.3 (0.8)	−108.5
Coesite						
Si1	−113.6 (0.3)	−115.0 (1.1)	−114.9 (1.0)	−114.0 (0.1)	−114.5 (0.6)	−113.9
Si2	−108.7 (0.6)	−109.8 (1.7)	−109.8 (1.7)	−109.0 (0.9)	−109.4 (1.3)	−108.1
Sigma-2						
Si1	−115.2 (0.6)	−117.8 (2.0)	−115.5 (0.3)	−115.9 (0.1)	−117.3 (1.5)	−115.8
Si2	−113.2 (0.4)	−114.7 (1.1)	−112.3 (1.3)	−114.1 (0.5)	−114.7 (1.1)	−113.6
Si3	−119.0 (0.7)	−121.2 (1.5)	−118.8 (1.1)	−119.3 (0.4)	−121.0 (1.3)	−119.7
Si4	−108.9 (0.4)	−110.4 (1.9)	−108.3 (0.2)	−109.6 (1.1)	−110.3 (1.8)	−108.5
<Dd>	0.5	1.4	0.9	0.6	1.2	

Note: The errors between experimental and calculated data are reported in parentheses.
[a]Results from ref. 47.

Table 4.4 Calculated $^2J(^{29}Si–O–^{29}Si)$ Coupling Constants for Cluster Model of Zeolite Sigma-2[a]

Coupling Partners	B3LYP	CAM-B3LYP	PBE	PBE0	LC-wPBE	HSEh1PBE	WB97XD	M06	Experiment
Si1-Si4	6.54	6.65	6.22	5.13	7.48	6.19	7.23	5.90	6.3
Si1-Si3	22.29	22.95	22.72	21.78	23.64	21.66	23.57	23.12	23.5
Si2-Si3	16.16	16.50	13.85	15.72	17.17	15.62	17.19	17.24	16.5
Si2-Si4	11.20	11.71	11.85	10.92	12.66	8.75	12.43	12.17	10.0
χ^2[b]	3.1	3.3	11.1	5.8	8.9	5.7	6.1	5.6	—

[a]Results from ref. 48.

[b]$\chi^2 = \sum_{i=1}^{N} [(J_{i,\text{calc}} - J_{i,\text{obs}})/\sigma]^2$.

this translates into stringent requirements on the basis sets adopted. Rather than using very large basis sets, the addition of very tight functions to standard basis sets has proven adequate and computationally convenient: developments in this area are reviewed by Improta and Barone [43]. In a more general perspective, the approach of using very accurate basis sets only at specific sites of interest, while basis sets of lesser quality are adopted in further regions, is a paradigmatic application of the "locally dense basis set" scheme [51–53], which is by no means restricted to the computation of magnetic resonance parameters. As hinted before, the referencing procedure which is implied in the computation of chemical shifts starting from absolute nuclear magnetic shieldings takes care of most systematic errors: For applications based on distributed gauge origins (e.g., within the usual GIAO ansatz) and targeted to typical organic/biological molecules, basis sets of medium size, like 6-311+G(d,p) or 6-311+G(2d,p), have been widely adopted in chemical shift calculations without significant degradation of the accuracy. Larger basis sets are typically required if coupling constants are desired as well [54, 55].

4.7 VIBRATIONAL CORRECTIONS TO NMR AND EPR PROPERTIES

The NMR parameters of an effective spin Hamiltonian obtained from experiment must be regarded as averaged parameters for a vibrating and rotating molecule. Therefore, a direct comparison between theoretical results and experimental observations implies that proper account is given to nuclear motions.

Nuclear shielding and indirect spin–spin coupling constants are strongly dependent on molecular geometry. As a consequence, rovibrational averaging effects may change NMR properties by more than 10%. The dependence of the NMR parameters on the variations of the geometry in the neighborhood of the equilibrium must be known.

From a computational viewpoint, vibrational averaging is an expensive procedure since it requires the calculation of the properties at a potentially large number of nuclear configurations for polyatomic molecules.

Both time-independent and time-dependent approaches are viable to perform such computations. Within the time-independent approach, the most widely used method for computing the rovibrational contributions to NMR properties is by means of second-order perturbation theory. To first order, the vibrationally averaged value of a property Ω is expressed as

$$\langle\Omega\rangle_n = \Omega_e + \sum_i A_i\left(n_i + \frac{1}{2}\right) \tag{4.44}$$

where Ω_e is the value at the equilibrium geometry and

$$A_i = \frac{\beta_{ii}}{\omega_i} - \sum_j \frac{\alpha_j F_{iij}}{\omega_i \omega_j^2} \tag{4.45}$$

α_j and β_{ii} being the first and second derivatives of the property with respect to the ith normal mode while F_{iij} is the third energy derivative. The first term or the right-hand side (RHS) of Eq. 4.45 refers to the harmonic contribution while the second one accounts for anharmonic corrections. By assuming $n_i = 0$, the zero-point vibrational energy contributions to the molecular properties can be evaluated, which accounts for the motion of the nuclear framework at 0 K and typically represents at least 90% of the vibrational corrections. Calculation of the vibrational contributions in perturbation theory is thus reduced to the calculation of a series of geometric derivatives of the molecular energy and properties.

The main limitation in calculating vibrational contributions to properties in polyatomic molecules by means of static (time-independent) calculations is the type of potential energy surface that can be explored. In fact, perturbation theory can only account for vibrational motion occurring near the minimum of single-well potential energy surfaces. This poses severe limitations to a complete and reliable computation of vibrationally averaged properties and vibrationally resolved spectra, especially when the vibrational modes involve a complex conformational rearrangement and/or coupling with solvent motions. A possible alternative route is represented by time-dependent approaches based on classical or quantum treatment of the nuclear dynamics.

Within this approach, the spectroscopic parameters can either be computed "on the fly" or, more frequently, by accurate single-point calculations on different static cluster configurations extracted from the molecular dynamic (MD) trajectory [56]. A disadvantage of the approach is that the lack of a representative "unperturbed" reference geometry does not allow to isolate the different normal-mode contributions to the computed molecular property.

4.8 ENVIRONMENTAL EFFECTS

Several experimental evidences have demonstrated that the magnetic properties of a given molecule may depend strongly on the chemical environment, which influences the system under investigation in several ways: (i) it induces structural modifications (indirect effect); and (ii) for a given structure it modifies the electron density distribution directly due to bulk effects (e.g., its polarity) and specific interactions (e.g., hydrogen bonds).

For these reasons, environmental effects can hardly be neglected in order to get realistic spectroscopic parameters directly comparable to the experimental ones. A suitable theoretical treatment accounting for both specific and bulk solvent effects on the magnetic properties is therefore required.

Bulk solvent effects can be accounted for by at least two different routes. The most direct procedure consists of including in the calculation a number of explicit solvent molecules large enough to reproduce the macroscopic properties of the bulk (such as the dielectric constant). However, this approach can easily lead to high computational costs, unless the number of explicit solvent molecules is kept within reasonable limits, possibly by adopting simplified models to account for those portions of the solvent that are not explicitly described. As a consequence, implicit models [in particular the so-called polarizable continuum model (PCM)] emerged in the last two decades as the most effective tools to treat bulk solvent effects for both ground- and excited-state properties [57].

The basic idea of all continuum models is the partitioning of the solution into two subunits: the "solute," described at the quantum mechanical (QM) level (which, as hinted above, may well be constituted by a cluster of individual molecules, e.g., a solute–solvent cluster), and the "solvent," a continuum medium representing a statistical average over all solvent degrees of freedom at thermal equilibrium.

The PCM method is one of the best known of such models. In essence, it involves the generation of a solvent cavity from spheres centered at each atom in the "solute"; the polarization of the solvent is represented by means of virtual point charges mapped onto the cavity surface and proportional to the derivative of the solute electrostatic potential at each point, calculated from the molecular wavefunction. The point charges are then included into the one-electron Hamiltonian, and therefore they induce a polarization of the solute. An iterative procedure is performed until the wavefunction and the point charges are self-consistent.

For NMR and EPR parameters, the PCM description affects the computation at several levels. First, the reaction field alters both the equilibrium geometry and the electronic distribution of the solute. Second, inclusion of the PCM operator introduces additional terms in the GIAO differentiation. Almost all the existing continuum-based approaches have been used to reproduce solvent effects on magnetic properties and their performances have also been critically compared [58–60].

These studies showed that the agreement with experiment is almost quantitative for aprotic solvents, but there is a noticeable underestimation of solvent shift for protic solvents. In the latter cases, an integrated discrete–continuum scheme which includes a limited number of explicit solvent molecules of the cybotactic region (treated at the QM level) is usually sufficient to restore the agreement between computation and experiments [43, 56, 61–64].

It should be noted that in principle changes in the solute vibrational motion can also be taken into account within the PCM procedure [43], however, this level of detail is not usually pursued.

4.9 CHEMICAL SHIFT ANISOTROPY AND LINESHAPE OF POWDER PATTERNS

As already shown, the total NMR Hamiltonian, from which the spin energy levels are obtained, is the sum of terms representing the Zeeman ($\hat{H}_0$), chemical shift ($\hat{H}_{CS}$), dipole–dipole coupling ($\hat{H}_{dd}$), and quadrupolar ($\hat{H}_Q$) interactions for nuclei with spins greater than $^1/_2$.

In solids and viscous liquids, the latter three terms are anisotropic and are each described by separate interaction tensors describing how the Hamiltonian (and thus the energy levels) varies with molecular orientation.

The chemical shielding is a rank-2 Cartesian tensor represented by a 3 × 3 matrix:

$$\sigma = \begin{pmatrix} \sigma_{xx} & \sigma_{xy} & \sigma_{xz} \\ \sigma_{yx} & \sigma_{yy} & \sigma_{yz} \\ \sigma_{zx} & \sigma_{zy} & \sigma_{zz} \end{pmatrix} \tag{4.46}$$

This is usually expressed in the principal-axis frame (PAF), so that the tensor (σ^{PAF}) is diagonal with principal values $\sigma_{\alpha\alpha}^{\mathrm{PAF}}$. These values are usually replaced by the isotropic shielding (σ_{iso}), the anisotropy Δ, and the asymmetry parameter η as follows:

$$\begin{aligned} \sigma_{\mathrm{iso}} &= \tfrac{1}{3}\left(\sigma_{xx}^{\mathrm{PAF}} + \sigma_{yy}^{\mathrm{PAF}} + \sigma_{zz}^{\mathrm{PAF}}\right) \\ \Delta &= \sigma_{zz}^{\mathrm{PAF}} - \sigma_{\mathrm{iso}} \\ \eta &= \left(\sigma_{xx}^{\mathrm{PAF}} - \sigma_{yy}^{\mathrm{PAF}}\right)/\sigma_{zz}^{\mathrm{PAF}} \end{aligned} \tag{4.47}$$

The chemical shift frequency can then be expressed in terms of these components as

$$\omega_{\mathrm{CS}}(\theta,\phi) = -\omega_0\sigma_{\mathrm{iso}} - \tfrac{1}{2}\omega_0\Delta\{3\cos^2\theta - 1 + \eta\sin^2\theta\cos 2\phi\} \tag{4.48}$$

where the first term of the RHS of Eq. 4.48 is the isotropic chemical shift frequency relative to the bare nucleus and the second term reflects the orientation dependence of the chemical shift frequency. This is expressed by means of the polar angles θ and ϕ, which define the orientation of the magnetic field in the PAF.

The total spectral frequency in absolute units is given by the Larmor frequency plus the chemical shift contribution as

$$\omega = \omega_0 + \omega_{\mathrm{CS}}(\theta,\phi) \tag{4.49}$$

However, in NMR experiments, the absolute frequencies are measured with respect to a specific line in the spectrum of the reference substance. This is referred as chemical shift δ, which is defined as $\delta = -(\sigma - \sigma_{\mathrm{ref}})$. The isotropic chemical shift δ_{iso}, the anisotropy Δ_{CS}, and the chemical shift asymmetry (η_{CS}) are defined in a similar manner to Eq (4.47).

The observed chemical shift is thus

$$\delta = \delta_{\mathrm{iso}} + \tfrac{1}{2}\Delta_{\mathrm{CS}}\{3\cos^2\theta - 1 + \eta_{\mathrm{CS}}\sin^2\theta\cos 2\phi\} \tag{4.50}$$

The transition frequencies expected for quadrupolar nuclei are also dependent on the polar angles defining the orientation of the magnetic field with respect to the PAF of the electric field gradient. It is out of the scope of this chapter to report all the relationships regarding quadrupolar nuclei and the interested reader is remanded to general books on solid-state NMR spectroscopy [65].

The orientation dependence of each nuclear spin interaction means that, for powder samples, the NMR spectrum of a given nucleus consists of a broad powder pattern for each distinct chemical site for that nucleus. The powder pattern can be considered as being made up of an infinite number of sharp lines, one from each different molecular orientation present in the sample.

The powder pattern is commonly computed in the time domain by calculating the free induction decay (FID), which is subsequently Fourier transformed to obtain the frequency spectrum.

In the case of time-independent interactions, such as chemical shielding, heteronuclear dipolar coupling, and quadrupolar coupling, the FID is given by

$$\mathrm{FID}(t) = \frac{1}{8\pi^2}\int_0^{2\pi}\int_0^{\pi} \exp\left(i\sum\nolimits_A \omega_A(\theta,\phi)t\right)\sin\theta\, d\,\theta\, d\phi \tag{4.51}$$

where $\omega_A(\theta,\phi)$ is the contribution to the spectrum from interaction A when the applied magnetic field is oriented by the polar angles (θ,ϕ) with respect to the PAF of the interaction tensor.

4.10 CASE STUDIES

4.10.1 EPR and PNMR Calculations of Aminoxyl Radicals

Aminoxyl radicals are among the most thoroughly studied radicals from both experimental and computational points of view. These are characterized by a long-lived spin-unpaired electronic ground state and by molecular properties strongly sensitive to the chemical surroundings. Thanks to the ongoing development in EPR and ENDOR spectroscopy, these electronic features have led to widespread application of nitroxide derivatives as spin labels in biology, biochemistry, and biophysics to monitor the structure and motion of biological molecules and membranes as well as nanostructures [66, 67].

High-field EPR spectroscopy provides quite rich information consisting essentially of the nitrogen hyperfine ($\overleftrightarrow{A}$) and gyromagnetic ($\overleftrightarrow{g}$) tensors. However, interpretation of these experiments in structural terms strongly benefits from quantum chemical calculations able to dissect the overall observables in terms of the interplay of several subtle effects.

The ab initio computation of nuclear hyperfine tensors of small free-radical systems has a long history [43, 47, 68–83]. Our group recently validated a general computational approach rooted in DFT to the analysis of spin-probing and spin-labeling experiments by providing accurate description of thermodynamic and spectroscopic properties of several aliphatic nitroxides as proxyl and tempo [56]. The performances of the model for a typical problem were tuned not only by the choice of the right density functional and basis set but also by a proper account of stereoelectronic, vibrational, and environmental effects [56].

In the following paragraphs we will discuss in some detail the influence of the surrounding medium on the EPR and PNMR parameters of the 2,2,6,6-tetramethyl-piperidyl-1-oxyl (TEMPO) radical (Figure 4.1) in terms of polarity and hydrogen-bonding

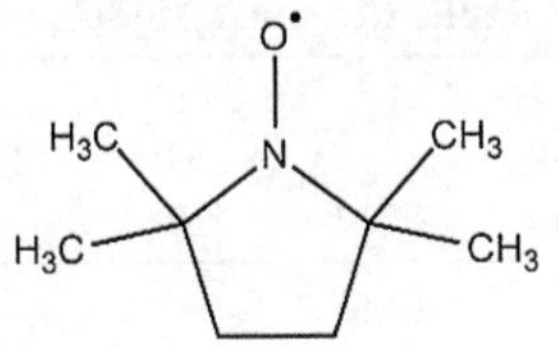

(2,2,5,5-Tetramethylpyrrolidin-1-yl)oxidanyl
PROXYL

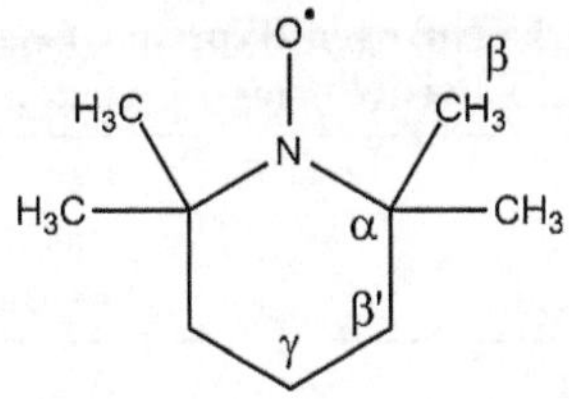

(2,2,6,6-Tetramethylpiperidin-1-yl)oxidanyl
TEMPO

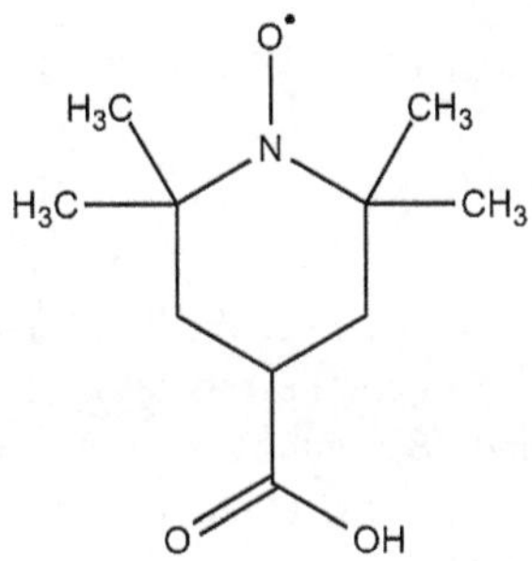

(4-Carboxy-2,2,6,6-tetramethylpiperidin-1-yl)oxidanyl
4-carboxy-TEMPO

Figure 4.1 Cyclic nitroxide radicals studied in this work.

power for several solvents with different dielectric constants. Since the presence of a hydrogen acceptor or donor site makes a nitroxide radical sensitive to the environment pH due to the different a_N values characterizing the protonated and unprotonated forms, we will analyze the titration curve of 4-carboxy-TEMPO (Figure 4.1) in aqueous solution.

4.10.1.1 Nitrogen Hyperfine Coupling Constants The nitrogen isotropic hyperfine coupling constant depends on the polarity of the medium in which the nitroxide is embedded. In fact, two main resonance structures can be written for the nitroxide group (Figure 4.2). Both the solvent polarity and its H-bonding ability favor the pseudoionic structure II, thus increasing the electron density on oxygen and the spin density on nitrogen, which in turn leads to larger a_N values.

The a_N values of TEMPO in different solvents obtained at the B3LYP/N07D level are reported in Table 4.5. At the B3LYP level, the nitrogen hyperfine coupling

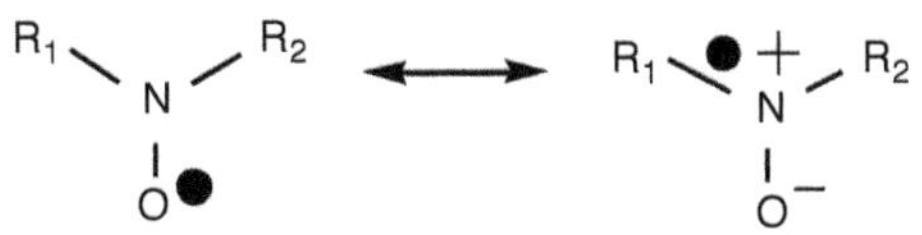

I **II**

Figure 4.2 Main resonance structures of nitroxide radicals.

Table 4.5 Nitrogen Hyperfine Coupling (a_N) Constants Computed in Vacuum and in Different Solvents Using Implicit, Explicit, and Mixed Explicit/Implicit Models

Experiment	Gas Phase	Gas Phase + 1S	Gas Phase + 2S	PCM	PCM + 1S	PCM + 2S
TEMPO 15.28 (cyclohexane)[a]	14.34	—	—	14.65	—	—
TEMPO 15.40 (toluene)[a]	14.34	—	—	14.72	—	—
TEMPO 16.15 (methanol)[a]	14.34	14.9	—	15.22	15.28	—
TEMPO 16,91(water)[b]	14.34	—	15.37	15.26	—	15.91

Note: Geometric structure optimized at the B3LYP/N07D level of theory.
[a]Ref. 84.
[b]Ref. 7.

constant in the gasphase is 14.34 G, whereas the PBE0/N07D level yields a value of 14.95 G, much closer to the experimental estimate of about 15 G. However, solvent shifts delivered by the B3LYP and PBE0 functionals in connection with PCM are very close, so that we will stay, in the present context, at the B3LYP level.

Table 4.5 shows that the solvent shifts computed in aprotic solvents (cyclohexane and toluene) by PCM are in quite good agreement with experimental data. On the other hand, a_N values obtained with PCM in methanol and water do not reproduce the experimental data ($\Delta a_N \approx 1.0$ G in methanol and $\Delta a_N \approx 1.3$ G in water). This means that not only the polarity of the solvent but also the H bonds critically influence the value of a_N. In methanol and water bulk effects alone cannot reproduce the experimental constants. In these solvents it is indeed necessary to take into account the formation of long-living complex adducts between solvent molecules and the solute due to strong specific interactions such as hydrogen bonds. In detail, two different clusters were employed for TEMPO, for methanol and water, respectively (Figure 4.3). The methanol cluster consists of solute and one solvent molecule;

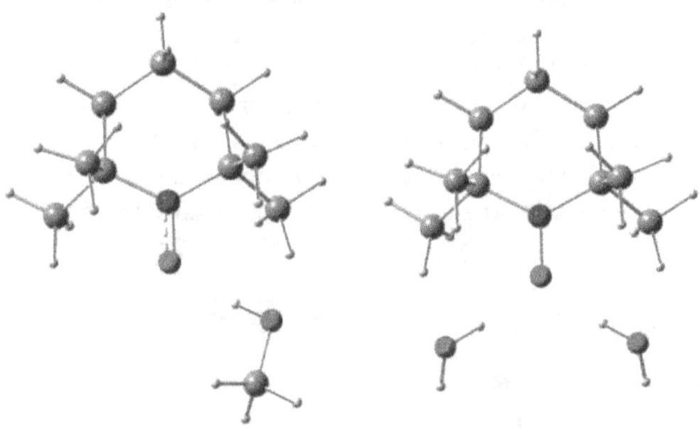

Figure 4.3 Adducts of TEMPO radical with methanol (left) and water (right).

the water cluster needs another solvent molecule, as will be demonstrated in the next section. When the mixed explicit/PCM approach is used, the calculated values in methanol and water are much closer to the experimental data.

4.10.1.2 ^{1}H and ^{13}C PNMR Parameters The first calculations of PNMR parameters of organic free radicals were carried out by Rinckevicius et al. [19], who studied simple nitroxide radicals, and Rastrelli et al. [55], who extended the calculations to transition metal complexes. These calculations evidenced the dominant effect of the Fermi contact contribution on the total chemical shift and the negligible effect of the pseudo contact shift in vacuum and without taking into account vibrational averaging effects.

In Table 4.6 the calculated hyperfine coupling constants of ^{13}C and ^{1}H of the TEMPO radical (see Figure 4.1) have been calculated using the B3LYP functional coupled with the N07D basis set, which has been shown to well reproduce EPR parameters and geometric properties of organic radicals [87]. The effects of solvents with increasing dielectric constants have also been taken into account using the polarizable continuum method for protic solvents like methanol and water; one and two explicit molecules were included in the calculations in a mixed explicit/PCM approach.

Small solvent shifts are calculated for C(α) and C(β) while no solvent effect is encountered for C(β') and C(γ) atoms. The table also reports the relative average error with respect to the experimental data. Unfortunately, it is not clear which solvent was used in those works, but by looking at the value of the nitrogen hyperfine coupling constants (of 16.3 G), it is believed to be methanol. The relative average errors decrease with increasing solvent polarity with the better agreement found for methanol when the mixed explicit/implicit approach was used. On the contrary, the hyperfine coupling constants of ^{1}H are scarcely affected by the solvent polarity, and average relative errors are typically lower than 10% is encountered.

Table 4.6 Calculated Hyperfine Coupling Constants of ^{13}C and ^{1}H Atoms of TEMPO Radical at B3LYP/N07D Level of Theory in Different Solvents

		PCM		Explicit + PCM^{a}		
	Vacuum	Cyclohexane	Toluene	Methanol	Water	Experiment^{b}
C(α)	−2.90	−3.10	−3.10	−3.62	−3.98	−3.6
C(β)	4.27	4.40	4.40	4.9	5.05	4.9
C(β')	0.49	0.58	0.57	0.58	0.56	0.82
C(γ)	−0.23	−0.23	−0.23	−0.24	−0.24	−0.32
$<\varepsilon_{rel}>$	25.2	20.4	20.7	13.7	17.6	—
H(β)	−0.20	−0.19	−0.19	−0.17	−0.17	−0.22
H(β')	−0.35	−0.35	−0.35	−0.38	−0.39	−0.39
H(γ)	0.15	0.16	0.16	0.18	0.18	+0.18
$<\varepsilon_{rel}>$	13.3	12.9	12.9	9.6	8.7	

[a] One explicit methanol and two water explicit molecules embedded in the PCM.
[b] Experimental data from by refs. 85 and 86 for ^{13}C and ^{1}H atoms, respectively.

Table 4.7 Calculated Orbital Shifts (δ^{orb})[a] of ^{13}C and ^{1}H Atoms of TEMPO Radical at B3LYP/N07D Level of Theory in Different Solvents

		PCM		Explicit + PCM[b]	
	Vacuum	Cyclohexane	Toluene	Methanol	Water
C(α)	16.9	17.4	17.5	18.9	19.3
C(β)	−3.1	−3.2	−3.2	−3.4	−3.4
C(β′)	2.2	2.1	2.0	1.7	3.0
C(γ)	−11.2	−11.4	−11.4	−11.7	−11.5
H(β)	−6.6	−6.7	−6.6	−6.7	−6.5
H(β′)	−6.2	−6.1	−6.1	−6.1	−6.1
H(γ)	−6.2	−6.2	−6.2	−6.2	−6.2

[a]The orbital shift was calculated by using the formula $\delta^{orb} = \sigma_{iso}^{ref} - \sigma_{iso}^{orb}$. The signals from the corresponding nuclei in the 2,2,6,6-tetramethylpiperidine precursor of TEMPO were used as the ^{13}C chemical shift reference, apart from the C(γ) for which the methyl carbon C(β) signal was used. For ^{1}H the reference was benzene. The $\sigma_{iso}^{ref}(^{13}C)$ are (in ppm): C(α): 141.2, C(β): 160.8, C(β′):153.7 while $\sigma_{iso}^{ref}(^{1}H) = 23.9$.
[b]One explicit methanol and two water explicit molecules embedded in the PCM.

The effect of solvent polarity on the orbital shift is presented in Table 4.7, which shows a negligible effect on both ^{13}C and ^{1}H. The solvent effect on the total chemical shift (see Table 4.8) is therefore dominated by the Fermi contant shift and follows the HCC trends. The results were obtained using the methanol solvent with one explicit solvent molecule being the closer to the experimental counterparts (average errors less than 12 and 8% for ^{13}C and ^{1}H).

In order to take vibrational averaging effects into proper account, we resorted to classical treatment with a purposely tailored force field using a new accurate force field obtained by extending the ff99SB force field [88] with a reliable

Table 4.8 Calculated Total Chemical Shifts of ^{13}C and ^{1}H Atoms of TEMPO Radical at B3LYP/N07D Level of Theory in Different Solvents

		PCM		Explicit + PCM[a]		
	Vacuum	Cyclohexane	Toluene	Methanol	Water	Experiment[b]
C(α)	−833	−891	−891	−1042	−1147	−1061
C(β)	1249	1287	1284	1433	1477	1462
C(β′)	146	172	169	172	167	249
C(γ)	−79	−79	−79	−82	−82	−95
	23.7	19.0	19.3	12.1	14.0	—
H(β)	−21.4	−20.7	−20.6	−19.2	−19.1	−21
H(β′)	−32.0	−31.9	−31.9	−34.1	−34.9	−31.8
H(γ)	4.8	5.6	5.6	7.2	7.1	6.7
	10.0	6.2	6.2	7.6	8.1	—

[a]One explicit methanol and two water explicit molecules embedded in the PCM.
[b]Experimental data from refs. 85 and 86 for ^{13}C and ^{1}H atoms, respectively.

parameters set able to provide structural, vibrational, and energetic properties of nitroxide systems very close to those calculated at the QM level of theory [61, 89]. However, averaging along the MD simulation has little effect on the EPR parameters of the TEMPO radical and no appreciable improvements are obtained.

4.10.2 A Priori Simulation of CW-EPR Spectrum of Double Spin-Labeled Peptides in Different Solvents

As already mentioned in the previous paragraph, aminoaxyl radicals are usually employed as spin probes of the dynamics and the changes of the structural conformations of proteins and complex biological systems.

In the specific case of proteins, peptides are well-recognized models for studying the stability and folding of helical regions, and EPR techniques have been used to dissect the role of different environments for a long time.

In the past decades, continuous-wave (CW) and pulse [double quantum coherence (DQC) and PELDOR] ESR spectra of double-spin-labeled systems have been reported [90, 91]. The high sensitivity provided by DQC and PELDOR spectra [92] allows reliable determination of distances (1.6–6.0 nm) between labels in frozen solution but cannot be used for distances shorter than 1.6 nm because of the large electron dipolar interaction and the presence of relevant scalar electron exchange interactions prevents the irradiation of a single electron spin, which is a prerequisite for their application [92].

In the latter case, the liquid-solution CW-ESR spectrum is very informative because its shape depends on several structural and dynamic parameters characterizing the double-labeled peptide. For this reason, recent theoretical studies have been focused on the development of effective and flexible computational approaches for the complete a priori simulation of ESR spectra of complex systems in solutions [93].

Since the CW-ESR spectra provides structural information and dynamics at different time scales, proper account of fast and slow motion of the labeled molecules is required for correct reproduction of the spectra. While the fast motion can be derived from a fast-motional perturbative model, in the slow-motion regime the effects on the spin relaxation processes exerted by the molecular motions requires a more sophisticated theoretical approach. The calculation of rotational diffusion in solution can be tackled by solving the stochastic Liouville equation (SLE) or by long-time-scale molecular dynamics simulations [94–96].

The approach recently proposed by Polimeno and Barone to simulate CW-ESR spectra [93] is composed of several steps. First, state-of-the-art QM calculations provide the structural and local magnetic properties of the investigated molecular system. Second, dissipative parameters such as rotational diffusion tensors are calculated by using stochastic Liouville equation. Third, in the case of multiple-label systems, electron exchange and dipolar interactions are computed.

A detailed discussion of the aforementioned approach is out-side the scope of the present chapter and the interested reader is remanded to Chapter 12.

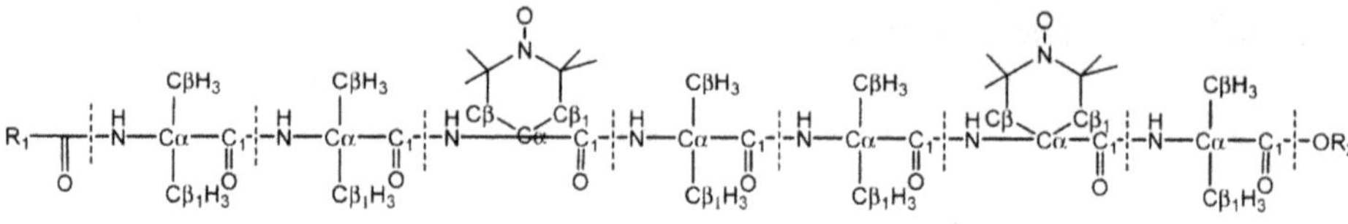

Figure 4.4 Chemical structure of Fmoc-(Aib-Aib-TOAC)$_2$-Aib-OMe (heptapeptide 1); R_1 = 9-fluorenylmethoxy and R_2 = Me.

This approach has been successfully applied to investigate the CW-ESR spectra of the double-spin-labeled, terminally protected heptapeptide Fmoc-(Aib-Aib-TOAC)$_2$-Aib-OMe (see Figure 4.4) in different solvents and at several temperatures [97]. (Fmoc, fluorenyl-9-methoxycarbonyl; Aib, α-aminoisobutyric acid; TOAC, 2,2,6,6-tetramethylpiperidine-1-oxyl-4-amino, 4-carboxylic acid; OMe, methoxy).

This is characterized by the presence of two TOAC nitroxide free radicals at relative position i, $i + 3$, which together with Aib represent two of the strongest helicogenic, C^{α}-tetrasubstituted, α-amino acids. In their work, Carlotto et al. [97] compared the experimental CW-ESR spectra with the theoretical counterpart pertaining to the 3_{10} and α-helix minima obtained by QM computations and unraveled the solvent-driven equilibria between the two conformations.

The 3_{10}-helix crystal structure experimentally determined by X-ray diffractometric analysis in the solid state was very well reproduced after PBE0/6-31G(d) geometry optimizations in vacuo.

Starting from the structure optimized in the gas phase, two different energy minima (see Figure 4.5) corresponding to 3_{10} and α-helical arrangements of the backbone were obtained in aqueous solution (treated at the PCM level).

The inclusion of dispersion interaction terms of the type introduced by Grimme et al. [98] revealed that the α helix was 3.0 kcal mol^{-1} more stable than the 3_{10} helix so that the transition between the two conformations was expected to be solvent dependent.

The results obtained by Carlotto et al. [97] confirmed that the Aib changes the conformation of the heptapeptide from the 3_{10} to the α helix as a function of increasing polarity and hydrogen bond donor of the solvent: the α helix in protic solvents and at low temperature but the 3_{10} helix in aprotic solvents.

Such an equilibrium was further supported comparing the a priori simulated CW-ESR spectra of the heptapeptide in different solvents (acetonitrile, methanol, toluene, and chloroform) and at different temperatures without any adjustable parameters except the relative percentage of 3_{10} and α helices. Figure 4.6 shows the theoretical and experimental spectra in methanol and toluene solutions in the range 270–350 K. The simulations in methanol which consider that only the α helix is present closely reproduce the experimental spectra at all temperatures. Conversely, in the toluene solution, the experimental spectra were well reproduced at high temperatures (350, 340, 330, and 320 K) using comparable percentages of the α helix (60%) and 3_{10}-helix (38%) structures and 2% of monoradical impurity. At low temperatures (below 310 K) the experimental spectra are correctly reproduced by

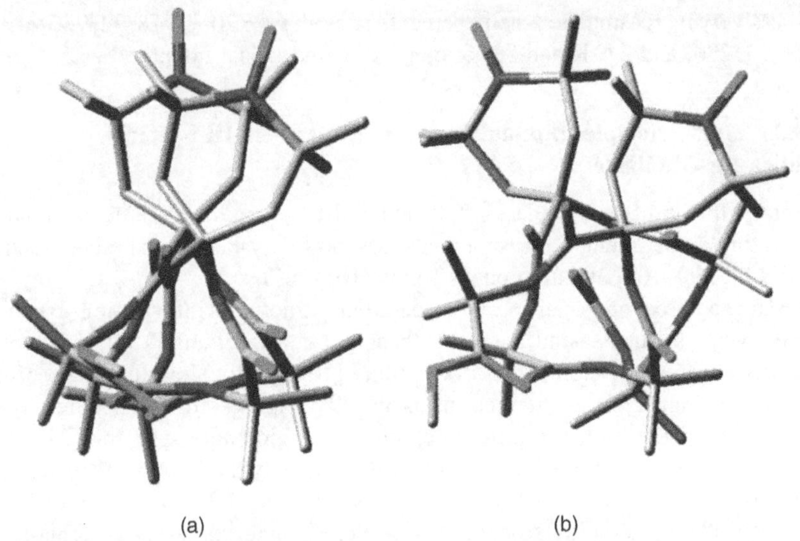

Figure 4.5 Optimized structures of (*a*) 3_{10} helix and (*b*) α helix secondary structures of heptapeptide **1**.

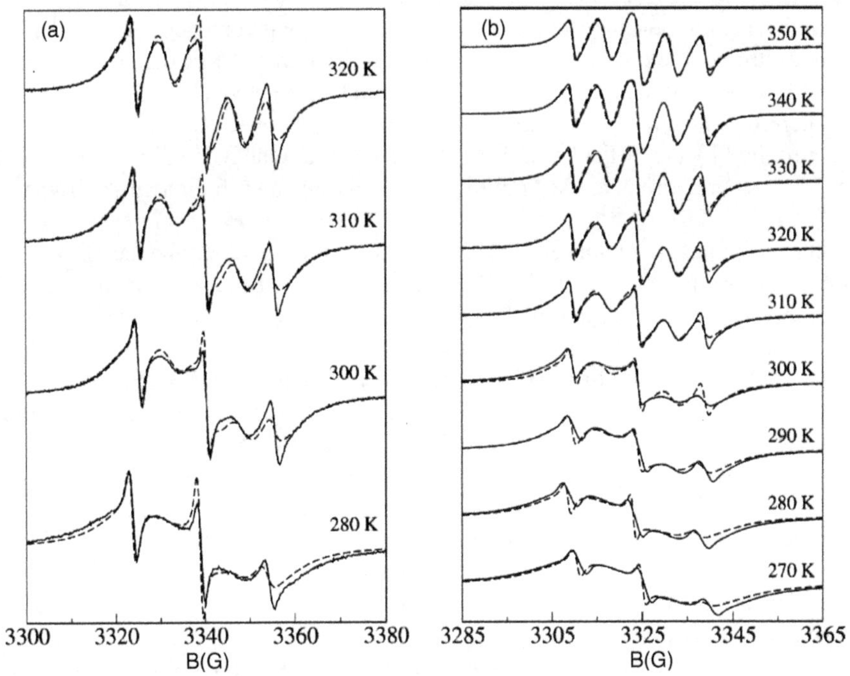

Figure 4.6 Experimental (solid line) and theoretical (dashed lines) CW-ESR spectra of heptapeptide 1 in (*a*) methanol and (*b*) toluene at different temperatures.

progressively increasing the α-helix percentage, with $p_\alpha = 70, 75, 78, 92, 98\%$ at 310, 300, 290, 280, and 270 K and 2% of constant monoradical impurity percentage.

4.10.3 First-Principles Simulations of Solid-State NMR Spectra of Silica-Based Glasses

Currently, the ability to simulate solid-state NMR spectra has become essential for interpreting experimental measurements for both crystalline materials such as zeolites [47, 99–102] and amorphous solids [103–107].

NMR spectroscopy can provide detailed structural information, such as connectivity, distance, bond angles with neighbouring atoms [108], and atomic scale disorder among nonframework cations [109, 110]. Most lineshape broadening due to anisotropic interactions is usually suppressed by specific experiments. Among the high-resolution techniques, magic-angle spinning [111–114] has been extensively applied in silicate glasses to study the largely misunderstood medium-range order of spin-$\frac{1}{2}$ nuclei such as ^{29}Si and ^{31}P while for nuclei with half-integer nuclear spin [so-called half-integer quadrupolar nuclei, here ^{17}O ($I=\frac{5}{2}$) and ^{43}Ca ($I=\frac{7}{2}$)], dynamic angle spinning (DAS), double rotation (DOR), or multiquantum magic-angle spinning (MQMAS) have been more recently introduced [115, 116].

In crystalline silicates the isotropic linewidths are usually sufficiently narrow so that it is possible to resolve crystallographically distinct sites [117]. In contrast, such a resolution cannot be achieved in glasses as a result of the inhomogeneous broadening of the isotropic line owing to the continuous distribution of NMR parameters which arises from a continuous distribution of structural parameters [118, 119].

Therefore, accuracy of the analysis of one-dimensional (1D) NMR spectra is often limited by the difficulty of fitting broad spectra to heavily overlapping NMR lines, which required an assumption of a Gaussian distribution in NMR parameters [119, 120].

In a recent paper, Pedone et al. [105] reproduced for the first time the 1D and 2D NMR solid-state spectra of the $CaSiO_3$ glass by using a combined classical molecular dynamics simulation for generating the glass structure and periodic DFT calculations to obtain NMR parameters, which in turn are introduced to a home-made code named fpNMR to simulate the spectra [105]. In Figure 4.7 the simulated ^{29}Si MAS NMR spectra is compared to the experimental spectrum reported by Zhang et al. [118] while the simulated (MAS, 90°) 2D spectrum of ^{29}Si is reported in Figure 4.8. In both cases agreement between experimental and simulated data is evident. The results of the simulations reported in Figure 4.7 show that each Q^n signal (n is the number of bridging oxygen atoms connected to silicon) can hardly be approximated to a Gaussian function; moreover, the overlap of the signals in the MAS spectrum reproduces very well the experimental data reported by Zhang et al. [118].

The simulated ^{17}O MAS NMR spectra of the $CaSiO_3$ glass at 9.4 and 14.1 T are given in Figure 4.9 and compared to experiment [110]. Even in this case, very good agreement is found in both the position and relative intensity of peaks. The Si–O–Ca

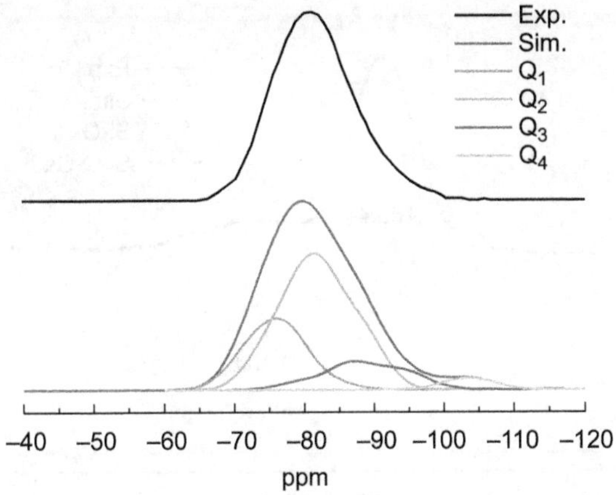

Figure 4.7 Calculated (red) ^{29}Si MAS NMR spectra at 9.4 T of CaSiO3 glass compared to experiment (black) from ref [118]. The separated isotropic line shapes for each Q^n are also reported.

(a narrow component around 109 and 114 ppm at 9.4 T) and the Si–O–Si peaks (a broad component around 30–50 ppm at 9.4 T) are well resolved.

The 2D 3QMAS NMR provides improved resolution among oxygen clusters over 1D MAS NMR, as shown in Figure 4.10, where a marked separation between the populations of nonbridging oxygens (oxygens bonded to one silicon, NBO) and bridging oxygens (oxygens bonded to two silicons, BO) is achieved.

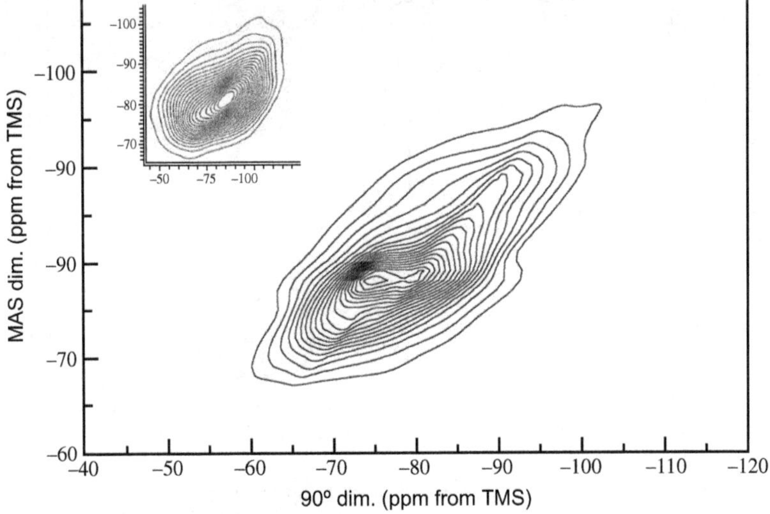

Figure 4.8 KDE simulated ^{29}Si (MAS, 90°) 2D spectrum of the $CaSiO_3$ glass. The experimental spectrum [118] is shown in the inset for comparison.

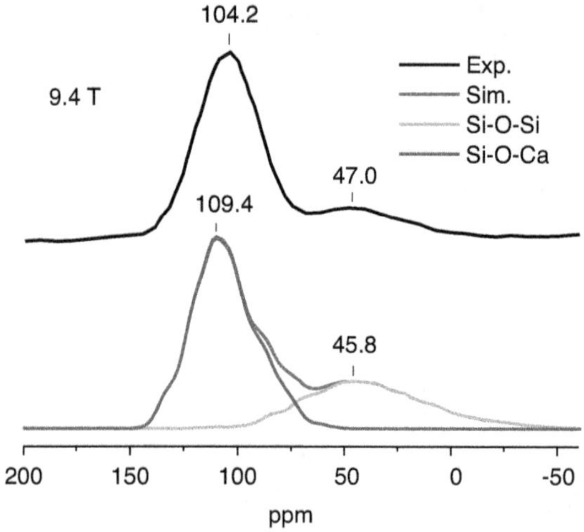

Figure 4.9 Simulated and experimental [110[17]] O MAS spectra of $CaSiO_3$ glass at 9.4 T.

The excellent agreement found in reproducing the experimental spectra demonstrates that DFT calculations accurately predict the δ_{iso} of ^{29}Si and ^{17}O and quadrupolar parameters C_Q and η_Q of ^{17}O. It is thus possible to investigate the relationships between NMR parameters and the local structure around such nuclei.

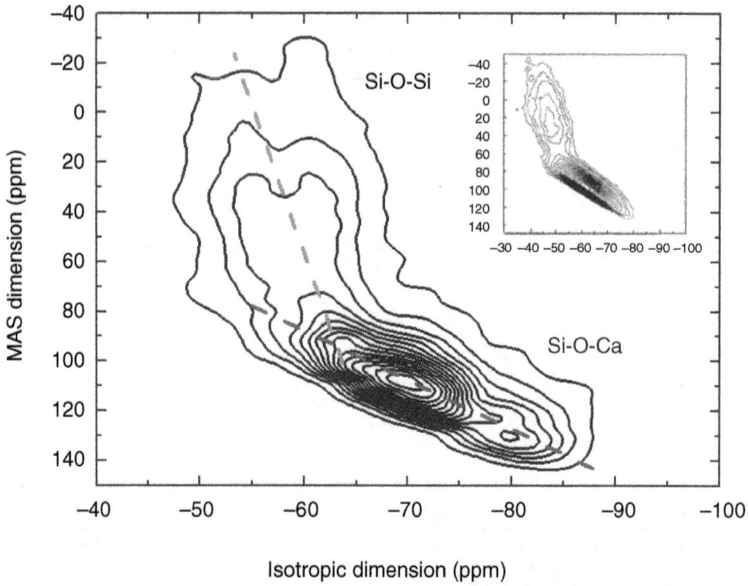

Figure 4.10 Simulated ^{17}O 3QMAS NMR spectrum (9.4 T) for $CaSiO_3$ glass. The inset reports the experimental spectrum collected by Lee et al. [109].

Such attempts have been done in several papers and it is outside the scope of this chapter to review them here [105, 108, 121, 122].

4.10.4 Computational NMR Applications in Structural Biology

Structural biology represents another specific but extremely relevant field of application of NMR. A rough measure of the role of NMR in structural biology can be gained from a statistical analysis of the Protein Data Bank (PDB), the databank of experimentally determined 3D biomolecular structures [123, 124]: About 15% of the entries originate from NMR determinations (also, structures of small peptides are not always deposited in the PDB). While the number of structures determined is increasing at an exponential rate, the balance of experimental techniques (essentially, X-ray diffraction and NMR) has apparently reached a steady value.

A specifically attractive feature of NMR in this context is the fact that the biomacromolecule is studied in solution (or even in micelles, membranes, etc.), that is, in a condition that mimics closely the biological environment. Since crystal contacts are absent, the structure is less likely to be distorted away from the natural conformation. This is especially relevant for floppy molecules, for example, flexible peptides, and for the less rigid regions of large macromolecules. Also, unfolded or partially folded states can be studied by NMR. An important specific asset of NMR is the sensitivity to dynamical processes affecting the biomacromolecule on a variety of time scales. The main limitation of NMR is related to system molecular weight. However, recent developments are striving to expand this limit [125–127]. In favorable cases, a system that would be too big for direct structural determination by NMR can be partitioned into smaller parts, for example, domains [128]. Also, a complete structure determination is not always required, and one is often interested in binding studies or in the characterization of dynamical aspects. In these instances, size limitations are much less of an issue.

In terms of the constituting nuclei, and if one neglects for a moment the possible presence of inorganic cofactors, biomacromolecules can be regarded as giant organic molecules. However, the emphasis of the NMR experiments is quite different in the two fields. In organic chemistry, the focus is typically on the determination of a molecular connectivity; conformational investigations are often of lesser interest, unless they bear upon this primary aim. By contrast, in structural biology applications, the molecular connectivity is known from the outset, and a characterization of the 3D structure is specifically sought. By far the most important NMR parameter for 3D structure calculation is represented by ^{1}H–^{1}H nuclear overhauser effects (NOEs), which arise as a consequence of the dipole–dipole, through-space interaction between neighboring protons [129]. Under some simplifying approximations, the NOE between two protons has a known, simple dependence from their distance. Thus, from a set of measured homonuclear NOEs, a corresponding set of internuclear distances can be reconstructed, and these can be used as restraints to define the 3D structure of the molecule. At least in a semiquantitative sense, the presence of an NOE between two protons indicates spatial proximity and as such provides very direct structural information. Additional structural information is encoded in the scalar couplings (both homonucelar and heteronuclear). The best known and more widely used example is represented by scalar three-bond coupling constants (^3J): Their dependence from the intervening dihedral angle is typically approximated by Karplus-type relationships and can be exploited to provide additional bounds that the final 3D structure should satisfy.

Homonuclear NOEs and scalar J couplings have been for a long time the main experimental parameters entering in an NMR structure determination of nonparamagnetic samples. Additional parameters (including temperature coefficients of amide protons and H/D exchange rates) have typically played a secondary role. More recently, residual dipolar couplings [130–132] have emerged as carriers of valuable structural information, and techniques have been devised to induce a preferred orientation of the protein with respect to the external field (e.g., the use of bicelles [133], or of directionally strained, biocompatible gels with suitable pore size [134, 135].

Chemical shifts are almost invariably assigned in the first stages of a structure calculation [136]. Moreover, their sensitivity to the structural context is strong, and an NMR spectroscopist can often recognize at first sight the signature of an unfolded state with respect to a folded structure in the 1D proton spectrum of a protein. However, the detailed dependence of chemical shifts from biomacromolecular structure is a complex one [137], and most of the encoded information is at present essentially discarded. Database searches and approximate methods are sometimes adopted to gain information on specific structural features that influence a selected subset of chemical shifts in a predictable way. Thus, for example, characteristic trends in the chemical shift of a specific subset of nuclei (^{1}Hα, ^{13}Cα, ^{13}Cβ, and 13C′), in the form of deviations from the corresponding random-coil values, are routinely used to produce a "chemical shift index" [138] that reliably reports on the secondary structure [139]. These methods are fast and can even provide restraints during the structure determination procedure; more sophisticated routes to extract the information encoded in the chemical are also actively developed [140].

However, when the local structural situation is unusual, or more in general when high precision is necessary, these approaches become insufficient. A large number of successful applications to small-to-medium-size organic molecules have shown that DFT methods, possibly coupled with hierarchical treatments (e.g., QM/MM, with the MM charges included in the QM Hamiltonian) and continuum solvent models, can provide accurate values of chemical shifts at reasonable computational costs [46, 56]. Extension to biomacromolecules may at first sight appear straightforward; however, the sheer size of the biochemical systems combined with a typically high structural flexibility and with more specific issues concerning the coexistence of several protonation microstates implies a need for careful consideration [141, 142]. Overall, there is good reason to expect that chemical shift computations by quantum mechanical methods will play an increasingly important role in structural biology.

Admittedly, the inclusion of ab initio chemical shift calculations within the standard process of NMR structure determination is probably not an immediate goal; however, NMR parameters computed from 3D protein structures can find an immediate use to discriminate among different possibilities in cases of structural ambiguities and/or resonance assignment problems. The usual practice of presenting NMR-derived structure in the form of "bundles" allows for a direct exploration of the variation of computed chemical shifts among the individual structures; in a similar vein, the structural dependence of the chemical shifts can be characterized by calculations on individual frames extracted from molecular dynamics trajectories (obtained either at the MM level or by the more sophisticated "mixed" protocols discussed elsewhere in this book). A problem for which ab initio computations are especially promising is represented by the positioning of NMR-silent cofactors, (e.g., Ca^{2+} in a calcium-binding protein). In these cases, structural constraints that can pinpoint the metal in the 3D protein structure are quite difficult to obtain, and site geometry has been more a result of a priori knowledge of the geometry of metal coordination than an experimental outcome [143]. However, the chemical shifts of nuclei neighboring the metal ion are quite sensitive to the detailed geometry of the site. The results from ab initio chemical shift computations should therefore represent an important structural validation [44].

4.11 CONCLUDING REMARKS

The aim of this chapter has been to provide an overview of current computational approaches in some specific areas of magnetic resonance spectroscopy. Apart from theoretical considerations, the focus on singlet and doublet electronic states reflects the fact that in these specific fields computational spectroscopy has nowadays reached the stage of a mature technique. Thus, for example, comparisons between measured and computed values of chemical shifts are becoming an important tool for experimental studies in structural organic chemistry. Likewise, hyperfine coupling constants are routinely calculated to gain insight into the behavior of spin probes. Even more demanding computational applications, for example, those involving the estimation of spin–spin coupling constants, are on their way to enter into routine use even by nonspecialists. In this spirit, the theoretical presentation has been kept at a reasonably accessible level, and a number of "case studies" have been provided in order to illustrate the potentiality of the techniques introduced. This plan should contribute to further promote the introduction of computational approaches within standard experimental studies, which, in this as in many other fields of research, allows for enhanced understanding of phenomena, faster and more efficient characterization protocols, and innovative chemical results.

REFERENCES

1. R. G. Parr, W. Yang, *Density-Functional Theory of Atoms and Molecules*, Oxford University Press, New York, 1989.
2. W. Koch, M. C. Holthausen, *A Chemist's Guide to Density Functional Theory*, Wiley-VCH, Weinheim, 2001.

3. V. Barone, *J. Comp. Chem.* **2005**, *26*, 384.
4. P. Carbonniere, T. Lucca, C. Pouchan, N. Rega, V. Barone, *J. Comp. Chem.* **2005**, *26*, 384.
5. V. Barone, G. Festa, A. Grandi, N. Rega, N. Sanna, *Chem. Phys. Lett.* **2004**, *388*, 279.
6. V. Barone, *J. Phys. Chem. A* **2004**, *108*, 4146.
7. A. Polimeno, J. H. Freed, *Adv. Chem. Phys.* **1992**, *83*, 89.
8. V. Barone, A. Polimeno, *Phys. Chem. Chem. Phys.* **2006**, *8*, 4609.
9. J. Owen, M. Browne, W. D. Knight, C. Kittel, *Phys. Rev.* **1956**, *102*, 1501.
10. W. D. knight, *Philos. Mag. B* **1999**, *79*, 1231.
11. O. L. Makina, J. Vaara, B. Schimmelpfenning, M. Munzarova, V. G. Malkin, M. Kaupp, *J. Am. Chem. Soc.* **2000**, *122*, 9206.
12. M. Kaupp, C. Remenyi, J. Vaara, O. L. Malkina, V. G. Malkin, *J. Am. Chem. Soc.* **2002**, *124*, 2709.
13. N. F. Ramsey, *Phys. Rev.* **1950**, *78*, 699.
14. N. F. Ramsey, *Phys. Rev.* **1953**, *91*, 303.
15. M. Schindler, W. Kutzelnigg, *J. Chem. Phys.* **1982**, *76*, 1919.
16. A. E. Hansen, T. D. Bouman, *J. Chem. Phys.* **1985**, *82*, 5035.
17. R. Ditchfield, *Mol. Phys.* **1974**, *27*, 789.
18. I. Bertini, C. Luchinat, G. Parigi, *Solution NMR of Paramagnetic Molecules*, Elsevier, Amsterdam, 2001.
19. Z. Rinkevicius, J. Vaara, L. Telyantnyk, O. Vahtras, *J. Chem. Phys.* **2003**, *118*, 2550.
20. S. Moon, S. Patchkovskii, Eds., in *Calculation of NMR and EPR Parameters*, M. Kaupp, M. Bühl, V. Malkin, Wiley-VCH, Weinheim, 2004, pp. 325–338.
21. C. Møller, M. S. Plesset, *Phys. Rev.* **1934**, *46*, 618.
22. D. Cremer, Y. He, *J. Phys. Chem.* **1996**, *100*, 6173.
23. J. Gauss, in *Encyclopedia of Computational Chemistry*, P. v. R. Schleyer, P. R. Schreiner, N. L. Allinger, T. Clark, J. Gasteiger, P. Kollman, H. F. I. Schaefer, Eds., Wiley, Chirchester, 1998, p. 615.
24. G. E. Scuseria, H. F. Schaefer, *Chem. Phys. Lett.* **1988**, *152*, 282.
25. M. Kàllay, P. R. Surjan, *J. Chem. Phys.* **2000**, *113*, 1359.
26. M. Nooijen, R. J. Bartlett, *Chem. Phys. Lett.* **2000**, *326*, 255.
27. G. D. Purvis, R. J. Bartlett, *J. Chem. Phys.* **1982**, *76*, 1910.
28. K. Raghavachari, G. W. Trucks, J. A. Pople, M. Head-Gordon, *Chem. Phys. Lett.* **1989**, *157*, 479.
29. I. Shavitt, in *The Electronic Structure of Atoms and Molecules*, H. F. Schaefer, Ed., Addison-Wesley, Reading, MAs, 1972, p. 189.
30. H.-J. Werner, in *Ab Initio Methods in Quantum Chemistry*, Vol. *25*, K. P. Lawley, Ed., Wiley, New York, 1987, p. 1.
31. B. O. Roos, K. Andersson, in *Modern Electronic Structure Theory*, D. R. Yarkony, Ed., World Scientific, Singapore, 1995, p. 55.
32. T. Helgaker, M. Jaszunski, K. Ruud, *Chem. Rev.* **1999**, *99*, 293.
33. J. Gauss, J. F. Stanton, *Adv. Chem. Phys.* **2002**, *123*, 355.

34. M. Bühl, in *Encyclopedia of Computational Chemistry*, P. v. R. Schleyer, P. R. Schreiner, N. L. Allinger, T. Clark, J. Gasteiger, P. Kollman, H. F. I. Schaefer, Eds., Wiley, Chichester, 1998, p. 1835.

35. J. Gauss, J. F. Stanton, in *Calculation of NMR and EPR Parameters*, M. Haupp, M. Bühl, V. G. Malkin, Eds., Wiley-VCH, Weinheim, 2004, p. 123.

36. J. Gauss, J. F. Stanton, *J. Chem. Phys.* **1996**, *104*, 2574.

37. C. Adamo, A. di Matteo, V. Barone, *Adv. Quant. Chem.* **1999**, *36*, 45–76.

38. J. R. Cheeseman, G. W. Trucks, T. A. Keith, M. J. Frisch, *J. Chem. Phys.* **1996**, *104*, 5497.

39. V. G. Malkin, O. L. Malkina, M. E. Casida, D. R. Salahub, *J. Am. Chem. Soc.* **1994**, *116*, 5898.

40. G. Rauhut, S. Puyear, K. Wolinski, P. Pulay, *J. Phys. Chem.* **1996**, *100*, 6310.

41. C. Adamo, V. Barone, *J. Chem. Phys.* **1999**, *110*, 6158.

42. C. Adamo, V. Barone, *Chem. Phys. Lett.* **1998**, *298*, 113.

43. R. Improta, V. Barone, *Chem. Rev.* **2004**, *104*, 1231.

44. C. Benzi, M. Cossi, V. Barone, *Phys. Chem. Chem. Phys.* **2004**, *6*, 2557.

45. O. Crescenzi, G. Correale, A. Bolognese, V. Piscopo, M. Parrilli, V. Barone, *Org. Biomol. Chem.* **2004**, *2*, 1577.

46. A. Bagno, G. Saielli, *Theor. Chem. Acc.* **2007**, *117*, 603.

47. A. Pedone, M. Pavone, M. C. Menziani, V. Barone, *J. Chem. Theory Comput.* **2008**, *4*, 2130.

48. A. Pedone, M. Biczysko, V. Barone, *Chem. Phys. Chem.* **2010**, *11*, 1812.

49. S. Dapprich, I. Komaromi, K. S. Byum, K. Morokuma, M. J. Frisch, *J. Mol. Struct. THEOCHEM* **1999**, *1*, 461.

50. G. A. Saracino, A. Tedeschi, G. D'Errico, R. Improta, L. Franco, M. Ruzzi, C. Corvaia, V. Barone, *J. Phys. Chem. A* **2002**, *106*, 10700.

51. D. B. Chesnut, K. D. Moore, *J. Comp. Chem.* **1989**, *10*, 648.

52. D. B. Chesnut, B. E. Rusiloski, K. D. Moore, D. A. Egolf, *J. Comp. Chem.* **1993**, *14*, 1364.

53. D. B. Chesnut, E. F. C. Byrd, *Chem. Phys.* **1996**, *213*, 153.

54. A. Bagno, *Chem. Eur. J.* **2001**, *7*, 1652.

55. A. Bagno, F. Rastrelli, G. Saielli, *J. Phys. Chem. A* **2003**, *107*, 9964.

56. V. Barone, P. Cimino, O. Crescenzi, M. Pavone, *J. Mol. Struct. THEOCHEM* **2007**, *811*, 323.

57. J. Tomasi, B. Mennucci, R. Cammi, *Chem. Rev.* **2005**, *105*, 2999.

58. A. Pedone, M. Pavone, M. C. Menziani, V. Barone, *J. Chem. Theory Comput.* **2008**, *4*, 2130.

59. C. J. Cramer, D. G. Truhlar, *Chem. Rev.* **1999**, *99*, 2161.

60. A. Klamt, B. Mennucci, J. Tomasi, V. Barone, C. Curutchet, M. Orozco, F. J. Luque, *Acc. Chem. Res.* **2009**, *42*, 489.

61. E. Stendardo, A. Pedone, P. Cimino, M. C. Menziani, O. Crescienzi, V. Barone, *Phys. Chem. Chem. Phys.* **2010**, *12*, 11697.

62. A. Pedone, J. Bloino, S. Monti, G. Prampolini, V. Barone, *Phys. Chem. Chem. Phys.* **2010**, *12*, 1000.

63. A. Pedone, V. Barone, *Phys. Chem. Chem. Phys.* **2010**, *12*, 2722.

64. R. Improta, G. Scalmani, V. Barone, *Chem. Phys. Lett.* **2001**, *336*, 249.
65. M. J. Duer, *Solid State NMR Spectroscopy: Principles and Applications*, Blackwell Science, Oxford, 2002.
66. N. Kocherginsky, H. M. Swartz, *Nitroxide Spin Labels*, CRC Press, New York, 1995.
67. A. H. Buchaklian, C. S. Klug, *Biochemistry* **2005**, *44*, 5503.
68. V. Barone, *J. Chem. Phys.* **1994**, *101*, 6834.
69. V. Barone, *J. Chem. Phys.* **1994**, *101*, 10666.
70. V. Barone, *Theor. Chem. Acc.* **1995**, *91*, 113.
71. V. Barone, in *Advances in Density Functional Theory*, Vol. 1, D. P. Chong, Ed., World Scientific, Singapore, 1995, p. 287.
72. S. M. Mattar, *Chem. Phys. Lett.* **1998**, *287*, 608.
73. S. M. Mattar, A. D. Stephens, *Chem. Phys. Lett.* **2000**, *319*, 601.
74. A. V. Arbuznikov, M. Kaupp, V. G. Malkin, R. Reviakine, O. L. Malkina, *Phys. Chem. Chem. Phys.* **2002**, *4*, 5467.
75. S. M. Mattar, *J. Phys. Chem. B* **2004**, *108*, 9449.
76. J. Neugerbauer, M. J. Louwerse, P. Belanzoni, T. A. Wesolowski, E. J. Baerends, *J. Chem. Phys.* **2005**, *109*, 445.
77. J. C. Schoneborn, F. Neese, W. Thiel, *J. Am. Chem. Soc.* **2005**, *127*, 5840.
78. A. V. Astashkin, F. Neese, A. M. Raitsimiring, J. J. A. Cooney, E. Bultman, J. H. Enemark, F. J. Neese, *J. Am. Chem. Soc.* **2005**, *127*, 16713.
79. S. M. Mattar, *J. Phys. Chem. A* **2007**, *111*, 251.
80. D. Feller, E. R. Davidson, *J. Chem. Phys.* **1988**, *88*, 5770.
81. B. Engels, L. A. Eriksson, S Lunell, in *Advances in Quantum Chemistry*, Academic, San Diego, 1996.
82. S. A. Perera, L. M. Salemi, R. J. Bartlett, *J. Chem. Phys.* **1997**, *106*, 4061.
83. A. R. Al Derzi, S. Fan, R. J. Bartlett, *J. Phys. Chem. A* **2003**, *107*, 6656.
84. H. G. Aurich, K. Hahn, K. Stork, W. Weiss, *Tetrahedron* **1977**, *33*, 969.
85. G. F. Hatch, R. W. Kreilick, *J. Chem. Phys.* **1972**, *57*, 3696.
86. R. W. Kreilick, *J. Chem. Phys.* **1967**, *46*, 4260.
87. V. Barone, P. Cimino, E. Stendardo, *J. Chem. Theor. Comp.* **2008**, *4*, 751.
88. V. Hornak, R. Abel, A. Okur, B. Strockbine, A. Roitberg, C. Simmerling, *Prot. Struct. Func. Bioinform.* **2006**, *65*, 712.
89. P. Cimino, A. Pedone, E. Stendardo, V. Barone, *Phys. Chem. Chem. Phys.* **2010**, *12*, 3741.
90. J. J. Steinhoff, *Front. Biosc.* **2002**, *7*, 97.
91. S. S. Eaton, G. R. Eaton, in *Biological Magnetic Resonance*, Vol. 19 L. Berliner, S. S. Eaton, G. R. Eaton, Eds., Kluwer Academic/Plenum, New York, 2000, pp. 2–28.
92. Y. D. Tsvetkov, in *Biological Magnetic Resonance*, Vol. 21, L. Berlinerand C. J. Bender, Eds., Kluwer Academic/Plenum, New York, 2004, pp. 385–433.
93. A. Polimeno, V. Barone, *Phys. Chem. Chem. Phys.* **2006**, *8*, 4609.
94. D. Sezer, J. H. Freed, B. Roux, *J. Chem. Phys.* **2008**, *128*, 165106.
95. D. Sezer, J. H. Freed, B. Roux, *J. Am. Chem. Soc.* **2009**, *131*, 2597.
96. D. Sezer, J. H. Freed, B. Roux, *J. Phys. Chem. B* **2008**, *112*, 5755.

97. S. Carlotto, P. Cimino, M. Zerbetto, L. Franco, C. Corvaja, M. Crisma, F. Formaggio, C. Toniolo, A. Polimeno, V. Barone, *J. Am. Chem. Soc.* **2007**, *129*, 11248.

98. S. Grimme, J. Antony, T. Schwabe, C. Mück-Lichtenfeld, *Org. Biomol. Chem.* **2007**, *5*, 741.

99. J. Tossell, P. Lazzeretti, *Chem. Phys.* **1987**, *112*, 205.

100. L. M. Bull, B. Bussemer, T. Anupòld, A. Reinhold, A. Samoson, J. Sauer, A. K. Cheetham, R. Dupree, *J. Am. Chem. Soc.* **2000**, *122*, 4948.

101. D. H. Brouwer, G. D. Enright, *J. Am. Chem. Soc.* **2008**, *130*, 3095.

102. J. Tossell, P. Lazzeretti, *Phys. Chem. Miner.* **1988**, *15*, 564.

103. M. Profeta, F. Mauri, C. J. Pickard, *J. Am. Chem. Soc.* **2003**, *125*, 541.

104. M. Profeta, M. Benoit, F. Mauri, C. J. Pickard, *J. Am. Chem. Soc.* **2004**, *126*, 12628.

105. A. Pedone, T. Charpentier, M. C. Menziani, *Phys. Chem. Chem. Phys.* **2010**, *12*, 6054.

106. F. Tielens, C. Gervais, J. F. Lambert, F. Mauri, D. Costa, *Chem. Mater.* **2008**, *20*, 3336.

107. G. Ferlat, T. Charpentier, A. P. Seitsonen, A. Takada, M. Lazzeri, L. Cormier, G. Calas, F. Mauri, *Phys. Rev. Lett.* **2008**, *101*, 065504.

108. T. M. Clarck, P. J. Grandinetti, P. Florian, J. Stebbins, *Phys. Rev. B* **2004**, *70*, 064202.

109. S. K. Lee, J. F. Stebbins, *J. Phys. Chem. B* **2003**, *107*, 3141.

110. S. K. Lee, J. F. Stebbins, *Geochim. Cosmochim. Acta* **2006**, *70*, 4275.

111. E. Dupree, R. F. Pettifer, *Nature (London)* **1984**, *308*, 523.

112. L. Linati, G. Lusvardi, G. Malavasi, L. Menabue, M. C. Menziani, P. Mustarelli, A. Pedone, U. Segre, *J. Non-Cryst. Solids* **2008**, *354*, 84.

113. H. Maekawa, T. Maekawa, K. Kawamura, T. Yokokawa, *J. Non-Cryst. Solids* **1991**, *127*, 53.

114. U. Voigt, H. Lammert, H. Eckert, A. Heuer, *Phys. Rev. B* **2005**, *72*, 64207.

115. L. Frydman, J. S. Harwod, *J. Am. Chem. Soc.* **1995**, *117*, 5367.

116. A. Medek, S. Harwood, L. Frydman, *J. Am. Chem. Soc.* **1995**, *117*, 12779.

117. M. R. Hansen, H. J. Jakobsen, J. Skibsted, *Inorg. Chem.* **2003**, *42*, 2368.

118. P. Zhang, P. J. Grandinetti, J. F. Stebbins, *J. Phys. Chem. B* **1997**, *101*, 4004.

119. J. Mahler, A. Sebald, *Solids State Nucl. Magn. Reson.* **1995**, *5*, 63.

120. H. Eckert, *Prog. NMR Spectrosc.* **1992**, *24*, 159.

121. T. M. Clarck, P. J. Grandinetti, *Solid State NMR* **2005**, *27*, 233.

122. L. L. Hench, H. R. Stanley, A. E. Clark, M. Hall, J. Wilson, Dental applications of bioglass implant, in *Bioceramics*, Butterworth Heinmann, Oxford, **1991**.

123. H. M. Berman, J. Westbrook, Z. Feng, G. Gilliland, T. N. Bhat, H. Weissig, I. N. Shindyalov, P. E. Bourne, *Nucleic Acids Res.* **2000**, *28*, 235.

124. A. E. Ferentz, G. Wagner, *Quart. Rev. Biophys.* **2000**, *33*, 29.

125. K. Pervushin, R. Riek, G. Wider, K. Wüthrich, *Proc. Natl. Acad. Sci. USA* **1997**, *94*, 12366.

126. C. Fernàndez, G. Wider, *Curr. Opin. Struct. Biol.* **2003**, *13*, 570.

127. G. Zhu, X. J. Yao, *Prog. Nucl. Magn. Reson. Spectrosc.* **2008**, *52*, 49.

128. P. B. Card, K. H. Gardner, *Methods Enzymol.* **2005**, *394*, 3.

129. K. Wüthrich, *NMR of Proteins and Nucleic Acids*, Wiley, New York, 1986.

130. M. Blackledge, *Prog. Nucl. Mag. Res. Sp.* **2005**, *46*, 23.

131. R. S. Lipsitz, N. Tjandra, *Annu. Rev. Biophys. Biomol. Struct.* **2004**, *33*, 387.

132. A. Bax, A. Grishaev, *Curr. Opin. Struct. Biol.* **2005**, *15*, 563.

133. N. Tjandra, A. Bax, *Science* **1997**, *278*, 1111.

134. H.-J. Sass, S. J. Stahl, P. T. Wingfield, S. Grzesiek, *J. Biomol. NMR* **2000**, *18*, 303.

135. R. Tycko, F. J. Blanco, Y. Ishii, *J. Am. Chem. Soc.* **2000**, *122*, 9340.

136. J. Cavanagh, W. J. Fairbrother, A. G. I. Palmer, M. Rance, N. Skelton, *Protein NMR Spectroscopy: Principles & Practice*, Academic, New York, 2006.

137. C. A. Hunter, M. J. Packer, C. Zonta, *Prog. Nucl. Mag. Res. Sp.* **2005**, *47*, 27.

138. D. S. Wishart, B. D. Sykes, *J. Biomol. NMR* **1994**, *4*, 171.

139. S. P. Mielke, V. V. Krishnan, *Prog. Nucl. Mag. Res. Sp.* **2009**, *54*, 141.

140. D. A. Case, in *Calculation of NMR and EPR Parameters. Theory and Applications*, M. Kaupp, M. Buhl, V. G. Malkin, Eds., Wiley-VCH Verlag, Weinheim, 2004.

141. L. Szilágyi, *Prog. Nucl. Magn. Reson. Spectrosc.* **1995**, *27*, 325.

142. D. A. Case, *Curr. Opin. Struct. Biol.* **1998**, *8*, 624.

143. J. R. Calhoun, W. Liu, K. Spiegel, M. Dal Peraro, M. L. Klein, K. G. Valentine, A. J. Wand, W. F. DeFrado, *Structure* **2008**, *16*, 210.

5

APPLICATION OF COMPUTATIONAL SPECTROSCOPY TO SILICON NANOCRYSTALS: TIGHT-BINDING APPROACH

FABIO TRANI

Scuola Normale Superiore, Pisa, Italy

5.1 INTRODUCTION

Fluorescent nanoparticles have important applications in many fields, from biological fluorescence imaging, to sensor technology, from optoelectronics to photovoltaics. Multicolor quantum dots are entering as a valid substitute of organic dyes in cancer therapies. It has been shown that they can be used as powerful diagnostic biomarkers [1–3]. A considerable obstacle for most semiconductor quantum dots is the biodegradability of by-products and toxicity-related issues that can lead to serious unpleasant drawbacks and must be overcome by a suitable coverage of the nanocrystal core. Thus, the synthesis of bright, stable, biocompatible water-soluble silicon nanocrystals has opened the route to realistic biomedical applications of functionalized

Computational Strategies for Spectroscopy: From Small Molecules to Nano Systems, First Edition. Edited by Vincenzo Barone.

silicon nanoparticles as cellular probes [4–6]. Silicon nanocrystals can serve as ideal cellular probes, whose sizes should be not so small to fully cause degradation before its use, but surely they have to be as small as possible to avoid releasing of residual toxic elements in the human body. Silicon nanocrystals can be technologically functionalized to be degraded and renally excreted from the human body after their in vivo functions [7, 8]. Huge advances have been done in their fabrication. Recent chemical synthesis techniques allow for an excellent control of nanoparticle size and shape [4,6]. The fabrication of nanocrystals has thus become reproducible, with easy control of their fluorescent properties.

The chemical environment has emerged as a key agent that deeply modifies the quantum dot electronic structure and fluorescent properties. Therefore, apart from their enormous interest as biomarkers, silicon nanocrystals still represent a challenge from the point of view of the scientific understanding of their emission properties. In fact, at variance with crystalline silicon, whose optical activity is negligible, due to the indirect band gap that makes the electron–hole radiative recombination time extremely long (milliseconds), nanostructured silicon shows a strong optical emission that for years has been attributed to the quantum confinement (QC) effect. QC is due to the exciton confinement in a nanometer-sized space region, causing a blue shift and a quantum yield enhancement of the emission spectra. But defect-related emission bands can play an active role in the photoluminescence (PL) of oxidized silicon nanocrystals [9,10]. As a matter of fact, it is very difficult to distinguish and understand the origin of the PL in silicon nanocrystals (Si-nc). Recent studies have shown that measurements under high magnetic fields can give valuable information about the origin of the PL in Si-nc [11]. They have confirmed that, as known from the literature, defect-related emission at the Si–SiO_2 interface is red shifted and less intense than PL due to quantum confinement effects. But nonetheless oxidized Si-nc in solution could even emit in the blue visible spectral range [12]. In this framework, computational spectroscopic tools are recognized as an indispensable tool to understand the nanocrystal chemistry and address the research in this field. The roles of the solvent and the interface are extremely important for the emission properties of Si-nc and their potential technological applications, and the theory can make notable steps forward in the development of new techniques. Computational tools based on multilevel approaches, such as ONIOM [13], and the inclusion of solvent effects for instance through a polarizable continuum model (PCM) [14, 15], are expected to give a boost to the simulation of *realistic* nanocrystals, made of hundreds to thousands of atoms and dispersed in aqueous solutions where the presence of a solvent on the nanocrystal spectroscopy is taken into account. The calculation of infrared spectroscopy helps to reveal the presence of Si–O bonds that deeply influence the nanocrystal emission properties [16]. Yet, the functionalization of nanoparticles with huge organic molecules, DNA fragments, and organic dyes is another aspect whose technological applications are extremely interesting, and from the computational point of view little work has been done [17, 18]. New integrated computational approaches are currently being developed to simulate the effects of the functionalization on the optical properties. They are based on a multilevel approach, where the nanocrystal core is described using a tight-binding model, while the biological fragment attached at the surface is approached within density functional theory. In the following the tight-binding scheme, which is at the basis of such a multilevel scheme, is described in detail.

5.2 TIGHT-BINDING SCHEME

The semiempirical tight-binding approach is a computationally light, well-established tool to study semiconductor nanocrystals [19]. The starting point of the method is the expansion of the nanocrystal wavefunctions into a localized basis set of atomic orbitals,

$$\Psi_i = \sum_{\mu}^{M} c_{\mu i}\phi_{\mu} = \sum_{A}^{N}\sum_{\mu \in A} c_{\mu i}\phi_{\mu} \tag{5.1}$$

Here, the A's label the atoms composing the structure, the μ's indicate the atomic orbitals, the i's are for the molecular orbitals. There are M atomic orbitals and N atoms. In the last expression, the full sum has been divided into a sum over the atoms composing the structure and, for each atom, a sum over the atomic orbitals that are localized on it. In the tight-binding approach, the Hamiltonian H and the overlap S are defined for each couple of atoms:

$$H_{\mu\nu} = \int \phi_{\mu} H \phi_{\nu}\, d\tau \tag{5.2}$$

$$S_{\mu\nu} = \int \phi_{\mu}\phi_{\nu}\, d\tau \tag{5.3}$$

From H and S, the generalized eigenvalue problem is set, and it gives the molecular orbital energy levels

$$\sum_{\nu}^{M} [H_{\mu\nu} - \epsilon_i S_{\mu\nu}] c_{\nu i} = 0 \tag{5.4}$$

With the use of symmetries, the Hamiltonian and overlap matrices can be reduced to a minimum number of independent parameters [20] that are calculated once forever for a prototype system or averaging over a wide class of molecules.

Most of the semiempirical tight-binding methods for nanostructures are based on the parametrization of bulk systems. It consists of an iterative fitting procedure, performed on the tight-binding parameters, to match the bulk silicon band structure calculated using the most advanced techniques [21]. The as-calculated parameters are then applied to the study of the electronic properties of silicon nanostructures. When the nanostructures are well passivated, the surface is expected to play a minor role, and the main electronic and optical properties are determined by the nanocrystal core, which locally behaves as in the bulk [22]. Since the parametrization is done on bulk silicon, the tight-binding results have a better quality upon increasing the nanocrystal size. In fact, for huge nanocrystals, relaxation effects are negligible, correlation effects are not relevant, and the energy levels of delocalized states tend to the bulk values. As a matter of fact, it is extremely important to use a good parametrization. For years, very poor tight-binding parametrizations have been used, since it is difficult to well reproduce the conduction bands with a minimal valence basis set. Only recently very good parametrizations have been proposed for silicon and III–V semiconductors, which accurately reproduce the band structures in most of the first Brillouin zone [21, 23, 24]. Such parametrizations include three-center interaction terms and they are based on an orthogonal basis set. In this case the starting basis set is orthogonal, the overlap is just the unit matrix, and the problem simplifies to a standard (not generalized) eigenvalue problem. It is important to note that the use of an orthogonal basis set is not an approximation (based on the neglect of the overlap), but it is an exact procedure, based on the Lowdin orthogonalization scheme. In his early paper, Slater [20] showed that the Lowdin orbitals, obtained by a particular orthogonalization of the atomic orbitals, transform with the same symmetry properties of the atomic orbitals under the crystal point group.

The tight-binding approach can be used for total energy calculations, too. In this case, the parameters have to be tabulated as a function of the interatomic distance in order to allow atomic relaxation and perform geometric optimization. The parameters can be fitted to the band structures of bulk systems subject to an external pressure [24, 25]. An alternative, bottom-up, approach consists of parameterizing the tight-binding method on small molecules. The most common example is the density functional tight-binding (DFTB) approach, where the Hamiltonian and the overlap are calculated for each couple of elements starting from an isolated diatomic molecule, while a repulsive term is added to the total energy in order to match the geometric configurations and formation energies for a large set of molecules [26–29]. Within DFTB, the parameters are explicitly calculated from Slater-type atomic orbitals. A very important feature of this approach is the self-consistent charge (SCC), an additional term in the Hamiltonian matrix that takes into account the charge displacement and is extremely important to describe noncovalent bonds. The charge transfer among the atoms can be calculated starting from the Mulliken atomic charge, defined as

$$q_A = \sum_{\mu \in A} \sum_{\nu}^{M} P_{\mu\nu} S_{\mu\nu} \tag{5.5}$$

where the density matrix is

$$P_{\mu\nu} = \sum_{i} c_{\mu i} c_{\nu i} \tag{5.6}$$

and the sum is done over the occupied molecular orbitals. The charge fluctuation is the difference with the valence charge of the neutral atom,

$$\Delta q_A = q_A - q_A^0 \tag{5.7}$$

The SCC Hamiltonian contains an additional term H_1, which is proportional to the Coulomb interaction on each atom due to the charge variation on a second atom [27],

$$H^1_{\mu\nu} = \frac{1}{2} S_{\mu\nu} \sum_C^N (\gamma_{AC} + \gamma_{BC})\, \Delta q_C \qquad \mu \in A \qquad \nu \in B \tag{5.8}$$

The SCC Hamiltonian is thus $H_0 + H_1$ where H_0 is the non-SCC Hamiltonian defined in Eq. 5.2.

The γ's are calculated for all the couples of elements. The onsite parameters γ_{AA} are related to the atomic chemical hardness, while the offsite terms reproduce the electrostatic potential generated by a point charge: $\gamma_{AB} \sim 1/|R_A - R_B|$. The energy levels ϵ_i's are calculated self-consistently, starting from the diagonalization of the H^0 term, iteratively updating the charge transfer term H^1 in the Hamiltonian and solving the generalized eigenvalue problem in Eq. 5.4, until the convergence of the eigenvalues is obtained. A generalization of the method to include the spin polarization has also been proposed [28].

DFTB is usually employed to perform total energy calculations and structural optimizations. The total energy is written as

$$E = \sum_{\mu\nu} P_{\mu\nu} H^0_{\mu\nu} + \frac{1}{2} \sum_{AB} \gamma_{AB}\, \Delta q_A\, \Delta q_B + \frac{1}{2} \sum_{AB}^N E^{\text{rep}}_{AB} \tag{5.9}$$

The repulsive potential is calculated for each couple of elements as a function of the interatomic distance. The contributions E^{rep}_{AB} are fitted to minimize the total energy difference with respect to a density functional theory approach for a set of prototype systems [28].

DFTB is an excellent tool for structural optimization and total energy calculations, and it is very useful in molecular dynamics calculations involving a large number of atoms. An extension of the method to time-dependent density functional theory has given promising results [30, 31].

5.3 OPTICAL SPECTROSCOPY

The optical properties are usually calculated within the tight-binding approach using a one-particle scheme. A derivation of the theory, within a localized basis set framework, was given in the early review [32]. That paper describes how to derive the microscopic expressions for the spectroscopic observables for impurities in nonmetals, but it is easily generalizable to semiconductor nanocrystals. From the time-dependent perturbation theory of the interaction of a system with a radiation field, the absorption and emission spectra, the radiative electron–hole recombination time, and the dielectric properties can be calculated starting from a microscopic quantum mechanical point of view. The key ingredients are the transition dipole matrix elements in the tight-binding basis set, from which all the spectroscopic observables are calculated. The tight-binding approach, which is based on a semiempirical parametrization of the Hamiltonian and overlap matrices, requires a further recipe to evaluate the dipole matrix elements. We will describe here a very successful approximation that has been tested to give extremely convincing results for both bulk and nanostructured semiconductors.

The oscillator strength is a dimensionless variable that represents a measurement of the intensity of a given transition and its contribution to the absorption/emission spectra. It is defined as (here and in the following, atomic units are used) [32][1]

$$f_{ia} = \tfrac{2}{3}\,\omega_{ia}|\mathbf{r}_{ia}|^2 \tag{5.10}$$

Here, ω_{ia} is the transition energy and $\mathbf{r}_{ia}$ is the dipole matrix element between the ith occupied and ath unoccupied level. A large oscillator strength corresponds to a fast electron–hole radiative recombination. A fast radiative recombination can be preferred over nonradiative reconversion pathways, leading to a strong emission. The oscillator strength for the lowest energy transition, between the highest occupied molecular orbital (HOMO) and the lowest unoccupied molecular orbital (LUMO), can give an indication of the photoluminescence quantum yield in nanocrystals. A large oscillator strength also means an intense absorption at a given transition energy.

For the calculation of the transition-dipole matrix elements, further approximations are required, because of the lack of an explicit knowledge of the atomic orbitals in the tight-binding scheme. We reasonably assume that the atomic orbitals are strongly localized at the atomic sites and behave as delta functions,

$$\mathbf{r}_{ia} = \sum_{\mu\nu} c_{\mu i} c_{\nu a} \int \phi_\mu \mathbf{r} \phi_\nu \, d\tau \sim \sum_A q_A^{ia} \mathbf{R}_A \tag{5.11}$$

where the last sum is done over the atomic sites and the Mulliken transition charges have been introduced as [30]

$$q_A^{ia} = \frac{1}{2} \sum_{\mu \in A} \sum_{\nu} [c_{\mu i} c_{\nu a} S_{\mu\nu} + c_{\nu i} c_{\mu a} S_{\nu\mu}] \tag{5.12}$$

For an orthogonal tight-binding scheme, the overlap is the unit matrix, and the transition charges are simply

$$q_A^{ia} = \sum_{\mu \in A} c_{\mu i} c_{\mu a} \tag{5.13}$$

[1] A sum into the x, y, z Cartesian components is considered in the square modulus expression, so the final expression is a spatial average.

The absorption cross section for an isolated nanocrystal is obtained from the oscillator strengths as

$$\sigma_{\mathrm{abs}}(\omega) = 2\pi^2 \sum_{ia} f_{ia} S(\omega - \epsilon_{ia}) \tag{5.14}$$

where a broadening function S has been introduced. According to the experimental conditions, a Gaussian or Lorentzian broadening can be chosen with a proper width at half height. For spin-unpolarized calculations, there is a factor of 2 coming from the spin degeneracy. The present approximation has been successfully adopted to describe the optical properties of bulk Si [23] and, more recently, to get reliable absorption spectra and screening properties of silicon nanocrystals upon changing their size and shape [33–36].

We report a representative picture to show the tight-binding results for silicon nanocrystals. Figure 5.1 shows the absorption cross section for a set of silicon nanocrystals upon increasing their size. It can be seen that the absorption spectra move from a multipeak structure that is typical of molecules to a broad, continuous curve that is typical of bulk systems. This is due to the increase in the number of transitions, which makes a nanocrystal as a molecular system that is in the middle between a small molecule and a bulk system. An interesting feature emerges from the analysis of the first transition, defined as the transition between the HOMO and LUMO energy levels.

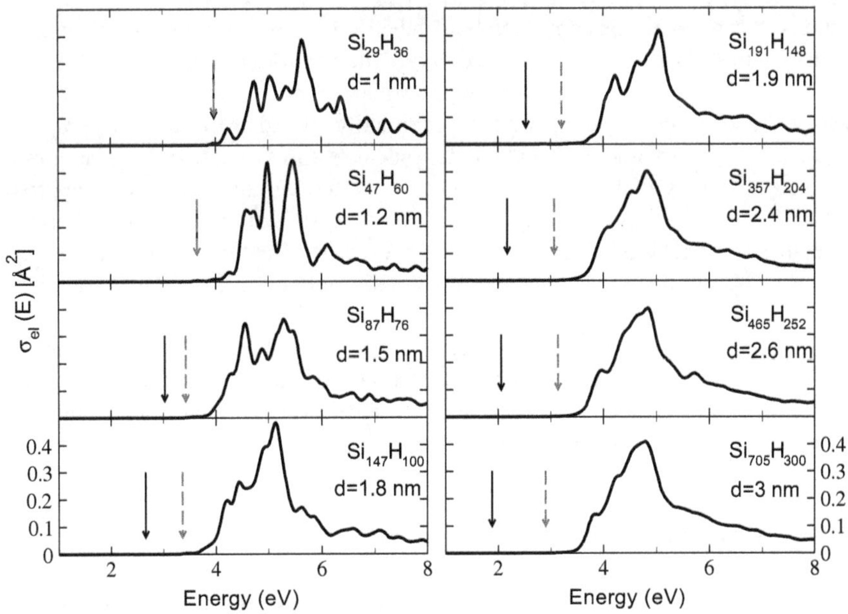

Figure 5.1 Absorption spectra of silicon nanocrystals upon increasing the diameter. Solid lines corresponds to the HOMO–LUMO gap, dashed lines corresponds to the absorption threshold. The absorption cross section has been divided by the number of atomic orbitals in order to compare nanocrystals with different volumes.

As is well known, bulk silicon is characterized by an indirect gap; therefore the electron–hole recombination is forbidden within an electronic picture. The transition becomes allowed when phonon-assisted transitions are considered, but it has a very small transition rate. This makes silicon optically inefficient compared to direct-gap semiconductors. At the nanoscale, the electron–hole recombination is allowed, and the HOMO–LUMO transition strength is large for small nanocrystals. This can be understood looking at Figure 5.1, where the first transition energy (black solid arrow) is compared to the absorption threshold energy (red dashed arrow). For small nanocrystals, both the energies are essentially coincident, which means a strong emission at the HOMO–LUMO energy. Upon increasing the nanocrystal size, the HOMO–LUMO energy becomes smaller and smaller with respect to the absorption threshold. Indeed, the HOMO–LUMO energy tends to the bulk limit of the indirect gap ($\sim$1.2eV), while the absorption threshold tends to the direct gap of silicon, which is much higher in energy ($\sim$3.1eV). Therefore, upon increasing the size, the HOMO–LUMO transition becomes ineffective, it has very small strength, and the phonon-assisted contributions assume a nonnegligible weight.

For the emission spectra, there is a huge difference among molecules, nanostructures, and extended systems. For molecules, the one-particle approximation gives unreasonable results, and the HOMO–LUMO gap is way too high compared to the optical gap coming from photoluminescence experiments. Dynamic correlation effects have a key role for small structures, and the electron–hole confinement gives nonnegligible exciton energies. The situation is very different in the case of extended systems. The one-particle approximation at the DFT level or using the semiempirical (tight-binding) approach for many ordinary structures often gives reasonable band structures when compared to experiments or more refined calculations. This is at the basis of the so-called scissor operator approximation, based on the fact that applying a rigid shift of the conduction bands, independent of the k point, leads to an almost complete matching of the band structures calculated within local density approximation (LDA) or using more refined tools. Nanocrystals are in the middle between molecules and extended systems. Small nanocrystals actually behave as molecules. The relaxation in the excited states is extremely important, leading to a significant Stokes shift, which makes the emission gap different from the absorption threshold. But, upon increasing the nanocrystal size, the Stokes shift becomes negligible, and the one-particle approximation is extremely accurate for the emission spectra calculations [37]. This motivation is at the basis of the wide success of tight-binding methods in the description of nanocrystal properties.

5.4 TIME-DEPENDENT FORMULATION

The one-particle approximation that has been described above does not take into account the local polarization due to the charge transfer from one atom to the other, the so-called local field effects. There are at least two ways to take into account the charge polarization into a tight-binding formulation. The first approach comes from solid-state physics. It is based on the calculation of the screening matrix, which represents the electron screening due to the presence of a test charge added to the system. It is calculated in linear response theory from the inversion of the dielectric matrix. The expressions for the optical observables are similar to the one-particle description, but with a new definition of the oscillator strengths, involving the real space dielectric matrix. A detailed description of the method, based on an early solid-state formulation [38], and its applications to semiconductor nanocrystals can be found in the literature [34–36].

In semiconductor nanostructures, local field effects are mostly due to a surface charge polarization contribution, which is essentially a macroscopic classical term, that is very important in optical properties calculations. For instance, surface polarization is responsible for a strong optical anisotropy of elongated nanocrystals [36]. We can say that the one-particle contribution represents the optical properties of an isolated, stand-alone nanocrystal, the intrinsic properties due to the delocalized states and the quantum confinement effects. One-particle contributions do not take into account the influence of the external environment into the optical properties, such as the macroscopic polarization of the surface bonds. On the contrary, the methods beyond one-particle calculations, based on the inversion of the dielectric matrix or, as we will see below, a time-dependent tight-binding formulation, take into account more properly the influence of the external environment, in particular the charge transfer within the nanocrystals and at the surface.

Figure 5.2 reports the absorption cross section of a small silicon nanocrystal. It is clear that the tight-binding approach with inclusion of local field effects (calculated by inversion of the dielectric matrix) compares very well to the formulation with a classical model of the surface polarization, based on the Clausius Mossotti equa-

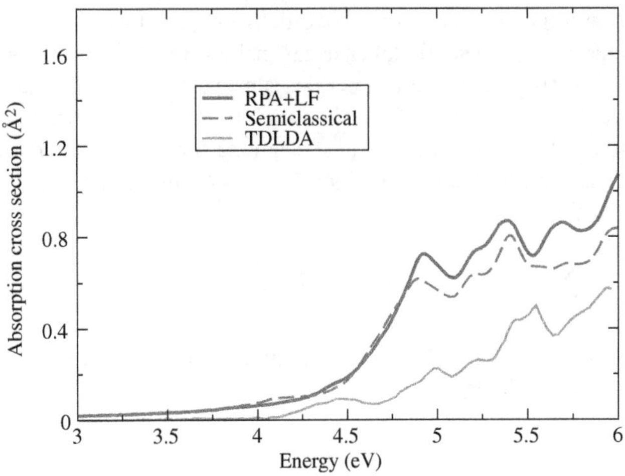

Figure 5.2 Absorption cross section of $Si_{35}H_{36}$ calculated using (1) tight-binding approach with local field effects (solid thick line), (2) the tight-binding energy levels with a classical model for the surface polarization contribution (dashed line) and (3) a time-dependent local density approximation (TDLDA) within density functional theory (solid thin line). TDLDA results from ref. 39.

tion [34]. In the meantime, the agreement with a time-dependent density functional theory is very nice, showing that a tight-binding approach can give reliable results of absorption spectra already at a very small size.

An alternative approach is based on the time-dependent density functional theory [40]. From the linear response theory, it can be shown that proper treatment of the excited states can be obtained from the solutions of a non-Hermitian eigenvalue problem [41],

$$\begin{pmatrix} A & B \\ B & A \end{pmatrix}\begin{pmatrix} X \\ Y \end{pmatrix} = \Omega \begin{pmatrix} 1 & 0 \\ 0 & -1 \end{pmatrix}\begin{pmatrix} X \\ Y \end{pmatrix} \tag{5.15}$$

with

$$A_{ia\sigma,jb\tau} = \omega_{ia}\delta_{\sigma\tau}\delta_{ij}\delta_{ab} + K_{ia\sigma,jb\tau} \tag{5.16}$$

$$B_{ia\sigma,jb\tau} = K_{ia\sigma,bj\tau} \tag{5.17}$$

where $\omega_{ia} = \epsilon_a - \epsilon_i$ and σ, τ are spin variables. The solution of the non-Hermitian eigenvalue problem leads to the excitation energies Ω. The coupling matrix K represents the interaction among the different transitions and depends on the calculation method. Within a restricted tight-binding formulation, the coupling matrix can be approximated as [30]

$$K_{ia\sigma,jb\tau} = \sum_{AB} q_A^{ia} q_B^{jb} [\gamma_{AB} + (2\delta_{\sigma\tau} - 1)m_{AB}] \tag{5.18}$$

where the transition charges and the above-defined γ coefficients are used, m is the spin magnetization and is usually taken as an onsite value, and from this formulation both the singlet and triplet excitation energies can be calculated. This has been named the γ approximation to the TDDFT, and it takes into account the correlation effects to the excited states due to the charge transfer among the atoms. After some straightforward algebra, the non-Hermitian eigenvalue problem can be transformed into a standard eigenvalue problem

$$\sum_{jb} [\omega_{ia}^2 \delta_{\sigma\tau}\delta_{ij}\delta_{ab} + 2\sqrt{\omega_{ia}} K_{ia\sigma,jb\tau}\sqrt{\omega_{jb}}] Z_{jb}^I = \Omega_I^2 Z_{ia}^I \tag{5.19}$$

From the solution of the time-dependent equations, the oscillator strength for the Ith singlet excitation is calculated as[2]

$$f_I = \frac{2}{3}\omega_I \left| \sum_{ia} \mathbf{r}_{ia} \sqrt{\frac{\omega_{ia}}{\omega_I}} (Z_{ia\uparrow}^I + Z_{ia\downarrow}^I) \right|^2 \tag{5.20}$$

[2] See note 1.

The one-particle approximation is recovered when the coupling matrix is neglected, and only the diagonal part of the Hamiltonian is retained in the time-dependent approach. The time-dependent tight-binding approach has been applied to silicon nanocrystals, with very promising results [42, 43].

5.5 CONCLUSIONS

We have exposed a tight-binding approach for the description of silicon nanocrystals. Tight binding is a very powerful tool because it is computationally very light, allowing for the study of thousands of atom structures, at the same time giving a quantum mechanical description of the system, with a time-dependent formulation of the excited states. As it was shown above, tight binding is very effective and accurate in the description of nanocrystal core and bulklike electronic properties. Multilevel schemes for the study of the nanocrystal functionalization, where the core is described at a tight-binding level, while the organic molecule is described at a density functional level, are expected to become an extremely powerful tool in the next years.

REFERENCES

1. I. L. Medintz, H. T. Uyeda, E.R. Goldman, H. Mattoussi, *Nature Mater.* **2005**, *4*, 435–446.
2. M. V. Yezhelyev, A. Al-Hajj, C. Morris, A. I. Marcus, T. Liu, M. Lewis, C. Cohen, P. Zrazhevskiy, J. W. Simons, A. Rogatko, S. Nie, X. Gao, R. M. O'Regan, *Adv. Mater.* **2007**, *19*, 3146–3151.
3. P. Zrazhevskiy, X. Gao, *Nano Today* **2009**, *4*, 414–428.
4. J.-H. Warner, A. Hoshino, K. Yamamoto, R. D. Tilley, *Angew. Chem. Int. Ed.* **2005**, *44*, 4550–4554.
5. Y. He, Z.-H. Kang, Q.-S. Li, C. H. A. Tsang, C.-H. Fan, S.-T. Lee, *Angew. Chem. Int. Ed.* **2009**, *48*, 128.
6. D. Jurbergs, E. Rogojina, L. Mangolini, U. Kortshagen, *Appl. Phys. Lett.* **2006**, *88*.
7. V. S.-Y. Lin, *Nature Mater.* **2009**, *8*, 252–253.
8. J.-H. Park, L. Gu, G. von Maltzahn, E. Ruoslahti, S. N. Bhatia, M. J. Sailor, *Nature Mater.* **2009**, *8*, 331–336.
9. M. V. Wolkin, J. Jorne, P. M. Fauchet, G. Allan, C. Delerue, *Phys. Rev. Lett.* **1999**, *82*, 197–200.
10. N. Daldosso, M. Luppi, S. Ossicini, E. Degoli, R. Magri, G. Dalba, P. Fornasini, R. Grisenti, F. Rocca, L. Pavesi, S. Boninelli, F. Priolo, C. Spinella, F. Iacona, *Phys. Rev. B* **2003**, *68*, 085327.
11. S. Godefroo, M. Hayne, M. Jivanescu, A. Stesmans, M. Zacharias, O. I. Lebedev, G. V. Tendeloo, V. V. Moshchalkov, *Nature Nanotechnol.* **2008**, *3*, 174–178.
12. S.-W. Lin, D.-H. Chen, *Small* **2009**, *5*, 72–76.
13. T. Vreven, K. S. Byun, I. Komaromi, S. Dapprich, J. A. Montgomery, K. Morokuma, M. J. Frisch, *J. Chem. Theory Comput.* **2006**, *2*, 815–826.

14. J. Tomasi, B. Mennucci, R. Cammi, *Chem. Rev.* **2005**, *105*, 2999–3094.
15. M. Cossi, V. Barone, B. Mennucci, J. Tomasi, *Chem. Phys. Lett.* **1998**, *286*, 253–260.
16. M. Rosso-Vasic, E. Spruijt, B. van Lagen, L. De Cola, H. Zuilhof, *Small* **2008**, *4*, 1835–1841.
17. F. A. Reboredo, G. Galli, *J. Phys. Chem. B* **2005**, *109*, 1072–1078.
18. Q. S. Li, R. Q. Zhang, S. T. Lee, T. A. Niehaus, T. Frauenheim, *J. Chem. Phys.* **2008**, *128*, 244714.
19. M. Lannoo, C. Delerue, G. Allan, *J. Lumin.* **1996**, *70*, 170–184.
20. J. C. Slater, G. F. Koster, *Phys. Rev.* **1954**, *94*, 1498–1524.
21. Y. M. Niquet, C. Delerue, G. Allan, M. Lannoo, *Phys. Rev. B* **2000**, *62*, 5109–5116.
22. F. Trani, D. Ninno, G. Cantele, G. Iadonisi, K. Hameeuw, E. Degoli, S. Ossicini, *Phys. Rev. B* **2006**, *73*, 245430.
23. C. Tserbak, H. M. Polatoglou, G. Theodorou, *Phys. Rev. B* **1993**, *47*, 7104–7124.
24. J.-M. Jancu, R. Scholz, F. Beltram, F. Bassani, *Phys. Rev. B* **1998**, *57*, 6493–6507.
25. P. B. Allen, J. Q. Broughton, A. K. McMahan, *Phys. Rev. B* **1986**, *34*, 859–862.
26. D. Porezag, T. Frauenheim, T. Köhler, G. Seifert, R. Kaschner, *Phys. Rev. B* **1995**, *51*, 12947–12957.
27. M. Elstner, D. Porezag, G. Jungnickel, J. Elsner, M. Haugk, T. Frauenheim, S. Suhai, G. Seifert, *Phys. Rev. B* **1998**, *58*, 7260–7268.
28. G. Zheng, M. Lundberg, J. Jakowski, T. Vreven, M. J. Frisch, K. Morokuma, *Int. J. Quant. Chem.* **2009**, *109*, 1841–1854.
29. P. Koskinen, V. Mäkinen, *Comput. Mater. Sci.* **2009**, *47*, 237–253.
30. T. A. Niehaus, S. Suhai, F. Della Sala, P. Lugli, M. Elstner, G. Seifert, T. Frauenheim, *Phys. Rev. B* **2001**, *63*, 085108.
31. T. Frauenheim, G. Seifert, M. Elstner, T. Niehaus, C. Kohler, M. Amkreutz, M. Sternberg, Z. Hajnal, A. Di Carlo, S. Suhai, *J. Phys. Condens. Matter* **2002**, *14*, 3015.
32. D. L. Dexter, *Solid State Phys.* **1958**, *6*, 353–411.
33. F. Trani, G. Cantele, D. Ninno, G. Iadonisi, *Phys. Rev. B* **2005**, *72*, 075423.
34. F. Trani, D. Ninno, G. Iadonisi, *Phys. Rev. B* **2007**, *75*, 033312.
35. F. Trani, D. Ninno, G. Iadonisi, *Phys. Rev. B* **2007**, *76*, 085326.
36. F. Trani, *Surf. Sci.* **2007**, *601*, 2702.
37. E. Degoli, G. Cantele, E. Luppi, R. Magri, D. Ninno, O. Bisi, S. Ossicini, *Phys. Rev. B* **2004**, *69*, 155411.
38. W. Hanke, L. J. Sham, *Phys. Rev. B* **1975**, *12*, 4501.
39. L. X. Benedict, A. Puzder, A. J. Williamson, J. C. Grossman, G. Galli, J. E. Klepeis, J. Y. Raty, O. Pankratov, *Phys. Rev. B* **2003**, *68*, 085310.
40. M. A. L. Marques, C. Ullrich, F. Nogueira, A. Rubio, K. Burke, E. K. U. Gross, *Time-Dependent Density Functional Theory*, Vol. 706 of Lecture Notes in Physics, Springer, Berlin, 2006.
41. R. Bauernschmitt, R. Ahlrichs, *Chem. Phys. Lett.* **1996**, *256*, 454–464.
42. Q. S. Li, R. Q. Zhang, S. T. Lee, T. A. Niehaus, T. Frauenheim, *Appl. Phys. Lett.* **2008**, *92*, 053107.
43. Y. Wang, R. Zhang, T. Frauenheim, T. A. Niehaus, *J. Phys. Chem. C* **2009**, *113*, 12935.

PART IIA

EFFECTS RELATED TO NUCLEAR MOTIONS: TIME-INDEPENDENT MODELS

6

COMPUTATIONAL APPROACH TO ROTATIONAL SPECTROSCOPY

CRISTINA PUZZARINI

Dipartimento di Chimica "G. Ciamician," Università degli Studi di Bologna, Bologna, Italy

Computational Strategies for Spectroscopy: From Small Molecules to Nano Systems, First Edition.
Edited by Vincenzo Barone.

An overview of the theoretical background and computational requirements needed for the accurate evaluation of the spectroscopic parameters of relevance to rotational spectroscopy is given. The accuracy obtainable from state-of-the-art quantum chemical calculations is mainly discussed by means of significant examples, which also allows us to stress the importance of the interplay of theory and experiment in the field of rotational spectroscopy.

6.1 INTRODUCTION

Rotational spectroscopy is by definition a high-resolution spectroscopy as it requires molecules to be detected in the gas phase. Consequently, in the last decades (dating from the mid-twentieth century) rotational spectroscopy turned out to be a powerful tool for investigating the structure and dynamics of molecules [1–3]. Work in the millimeter and submillimeter wavelength range is usually limited to the study of small- to medium-sized molecules, while measurements in the centimeter-wave region allow to investigate larger molecules, even of biological relevance. This distinction is mainly due to the extent of rotational constants, and it will be thus clarified by the summary provided in the next section. Furthermore, rotational spectroscopy turned out to be a particularly useful technique for the detection of new chemical species [e.g., 4–15].

The relevance of rotational spectroscopy is therefore related to molecular structure and molecular properties investigations. Concerning the former, among the methods available for experimental structure determination, rotational spectroscopy is the method of choice when aiming at high accuracy [1]. This is because rotational constants can be obtained with great accuracy and they are strongly related, as it will be clear from the following sections, to the molecular structure. However, such a determination remains a formidable task for polyatomic molecules as it necessitates the investigation of more and more isotopically substituted species and the proper consideration of vibrational effects. The determination of (hyper)fine parameters, such as quadrupole coupling, spin–spin coupling, and spin–rotation constants, is one of the aims of high-resolution rotational spectroscopy as well. These parameters are relevant not only from a spectroscopic point of view, but also from a physical and/or chemical viewpoint, as they might provide detailed information on the chemical bond, structure, and so on (for further insights see, for example, Gordy and Cook [1]). In addition, hyperfine structures are so characteristic that their analysis may help in assigning rotational spectra of unknown species [e.g., 16]. Another field of relevance for rotational spectroscopy is astrochemistry/astrophysics. In fact, molecules found in space are usually first detected in the laboratory by rotational spectroscopy, and then the corresponding determination of precise rotational constants and centrifugal distortion parameters enables the prediction of line positions suitable for radioastronomical detections [e.g., 17–20].

The aim of the present chapter is to provide a resume on the role of quantum chemistry in the field of rotational spectroscopy. Therefore, how the spectroscopic parameters of relevance to rotational spectroscopy can be evaluated by means of quantum chemical calculations will be presented and some emphasis will be given to the computational requirements. It will then be pointed out how quantum chemistry can be used for guiding, supporting, and/or challenging the experimental determinations. As a sort of conclusion the importance of the interplay between theory and experiment in rotational spectroscopy will be demonstrated. By means of significant examples, profits from such an interplay will be shown.

6.2 ROTATIONAL SPECTROSCOPY

In the present section all basic information required for understanding rotational spectroscopy is provided. We give as already assumed the Born–Oppenheimer approximation [21], which allows the separation of nuclear and electronic motion, as well as the separation of the various nuclear motions themselves (vibrational, rotational, translational). We therefore focus only on the quantum mechanics elements related to the rotational motion.

Since rotational spectroscopy is only briefly summarized here, we refer interested readers to textbooks that treat the subject in more detail [1–3].

6.2.1 Symmetry Classification and Angular Momentum

Classification of molecules according to their "rotational" symmetry is a key point in rotational spectroscopy as the expression of the rotational Hamiltonian and the solution of the corresponding eigenvalue equation vary noticeably upon symmetry.

The usual classification of molecules in rotational spectroscopy is the following:

(a) Diatomic or linear molecules
(b) Symmetric-top molecules
(c) Asymmetric-top molecules
(d) Spherical-top molecule

This classification is surely related to the molecular symmetry of the molecule, but it is mostly based on the relative values of principal moments of inertia I_x, I_y, and I_z [1] (where x, y, and z are the principal axes of a *molecule-fixed* coordinate system). As a result, the classification given above can be explained as follows:

(a) Linear (and diatomic) molecules have $I_z = 0$ and $I_x = I_y \equiv I$. Thus, there is only one nonnull moment of inertia.

(b) Symmetric-top molecules still have $I_x = I_y$, but I_z is nonnull. As concerns molecular symmetry, those molecules that have a molecular symmetry axis (z) which is at least a C_3 axis belong to this category. There is a further distinction in prolate (American football-like) top and oblate (pancake-like) top that will be addressed later in the text.

(c) Asymmetric-top molecules have three nonnull moments of inertia: $I_x \neq I_y \neq I_z$. The lack of "rotational" symmetry does not necessarily mean lack of molecular symmetry, as, for example, molecules belonging to the C_{2v} group are asymmetric-top rotors.

(d) Spherical-top molecules have three equivalent moments of inertia: $I_x = I_y = I_z$. From a rotational spectroscopy viewpoint, this class of molecules has a limited interest, since (as will be clear later) they only present perturbation-allowed rotational spectra.

As the rotational Hamiltonian only contains a kinetic energy term, which is expressed in terms of the components of the angular momentum $\hat{\mathbf{J}}$ [1, 2], we first recall the expressions of the eigenvalues for the latter.

Let us start from the commutation relations of $\hat{\mathbf{J}}$ in the *space-fixed* system, where X, Y, and Z denote the corresponding axes. As is well known, $\hat{J}^2 (= \hat{J}_X^2 + \hat{J}_Y^2 + \hat{J}_Z^2)$ commutes with its components, for example,

$$\left[\hat{J}^2, \hat{J}_Z\right] = \hat{J}^2 \hat{J}_Z - \hat{J}_Z \hat{J}^2 = 0 \tag{6.1}$$

but the components of $\hat{\mathbf{J}}$ do not commute among themselves. Therefore, only $\hat{J}^2$ and one projection of $\hat{\mathbf{J}}$, typically chosen as the Z component, have common eigenfunctions, which are usually designated as $|J, M\rangle$. From the commutator rules, the following expressions for the nonvanishing matrix elements of $\hat{J}^2$ and $\hat{J}_Z$ are obtained [1, 2]:

$$\begin{aligned} \langle J,\ M|\hat{J}^2|J,\ M\rangle &= \hbar^2 J(J+1) \\ \langle J,\ M|\hat{J}_Z|J,\ M\rangle &= \hbar M \end{aligned} \tag{6.2}$$

where the quantum number J is a nonnegative integer and $M = J, J-1, J-2, \ldots, -J$.

The corresponding expressions in a *molecule-fixed* coordinate system are also required as for symmetric-top rotors the component of angular momentum about the z axis, $\hat{J}_z$, is a constant of motion and thus commutes with $\hat{J}^2$, which is also a constant of motion [1–3]. In addition to the fact that $\hat{J}_z$ and $\hat{J}_Z$ commute with $\hat{J}^2$, they also commute with each other [1, 2] and have a common set of eigenfunctions, $|J, K, M\rangle$. The eigenvalues of $\hat{J}_z$, which can be found from the commutation rules of the angular momentum operators expressed in the *molecule-fixed* coordinate system [2], are

$$\langle J, K, M|\hat{J}_z|J, K, M\rangle = \hbar K \tag{6.3}$$

where, in analogy to M, $K = J, J-1, J-2, \ldots, -J$.

6.2.2 Hamiltonian for Rigid Rotor

According to the classification given above, we now introduce the rotational Hamiltonian and the corresponding eigenvalues for the various types of molecules. The rigid-rotor approximation is considered in the present section.

6.2.2.1 Diatomic and Linear Molecules The rotational Hamiltonian for a linear (as well as diatomic) molecule is given by the expression

$$\hat{H}_{\text{rot}} = \frac{1}{2}\left(\frac{\hat{\mathbf{J}}^2}{I}\right) = \frac{1}{2}\left(\frac{\hat{J}_x^2}{I_x} + \frac{\hat{J}_y^2}{I_y}\right) \tag{6.4}$$

Making use of the eigenvalues previously introduced, the following expression for the eigenvalues of the rotational Hamiltonian can be derived:

$$E_J = \frac{\hbar^2}{2I}J(J+1) \tag{6.5}$$

This expression can be further simplified by introducing the so-called rotational constant

$$B = \frac{\hbar^2}{2I} \tag{6.6}$$

which leads to:[1]

$$E_J = BJ(J+1) \tag{6.7}$$

6.2.2.2 Symmetric-Top Molecules The rotational Hamiltonian for a symmetric top is given as

$$\hat{H}_{\text{rot}} = \frac{\hat{J}^2}{2I_x} + \left(\frac{1}{2I_z} - \frac{1}{2I_x}\right)\hat{J}_z^2 \tag{6.8}$$

[1] Here and in the next section, rotational constants are given in energy units.

Making use of the eigenvalues previously introduced, the following expression for the eigenvalues of the rotational Hamiltonian in Eq. 6.8 is then derived:

$$E_{J,K} = \frac{\hbar^2}{2}\left[\frac{J(J+1)}{I_x} + \left(\frac{1}{I_z} - \frac{1}{I_x}\right)K^2\right] \tag{6.9}$$

As for linear molecules, this expression can be further simplified by making use of the rotational constants:

$$B^\alpha = \frac{\hbar^2}{2I_\alpha} \tag{6.10}$$

where α stands for x, y, or z, which are commonly denoted in rotational spectroscopy as a, b, and c. It has to be noted that the correspondence between the two sets is not unambiguous; in fact, the particular choice defines a "representation". We refer interested readers to Bunker and Jensen [22]. The a, b, c notation is particularly important as it leads to the usual definition of the rotational constants $A(=\hbar^2/2I_a)$, $B(=\hbar^2/2I_b)$, and $C(=\hbar^2/2I_c)$. Furthermore, the convention $A \geq B \geq C$ applies.

As mentioned above, for symmetric-top molecules we have two cases:

(a) The prolate top, for which $A > B = C$ and thus

$$E_{J,K} = BJ(J+1) + (A-B)K^2 \tag{6.11}$$

(b) The oblate top, from which $A = B > C$ and thus

$$E_{J,K} = BJ(J+1) + (C-B)K^2 \tag{6.12}$$

6.2.2.3 Asymmetric-Top Molecules The determination of the matrix elements of the rotational Hamiltonian for asymmetric-top molecules,

$$\hat{H}_{\text{rot}} = \frac{1}{2}\left(\frac{\hat{J}_x^2}{I_x} + \frac{\hat{J}_y^2}{I_y} + \frac{\hat{J}_z^2}{I_z}\right) \tag{6.13}$$

is more involved, as the energy levels of an asymmetric rotor cannot in general be expressed in closed form [1]. In fact, unlike the symmetric- and linear-rotor Hamiltonians, this Hamiltonian is such that the Schrödinger equation cannot be solved directly. Thus, a closed general expression for the asymmetric-rotor wavefunctions is not possible. However, they may be represented by a linear combination of symmetric-rotor functions, and the eigenvalues can be obtained by the corresponding diagonalization of the Hamiltonian matrix in that basis. We refer the interested reader to the specialized literature on this topic [1–3].

6.2.2.4 Spherical-Top Molecules The rotational energy expression of symmetric-top molecules is the same as that of a linear molecule. However, it has to be noted that, even if the energy

$$E_{J,K} = BJ(J+1) \tag{6.14}$$

is independent of K, each level is $(2J + 1)$ fold-degenerate not only in M but also in K.

6.2.3 Centrifugal Distortion and Vibration–Rotation Interactions

In the previous section the rigid-rotor approximation has been applied, while in the following we account for the nonrigidity of the molecules, which means that the nuclear positions are no longer fixed at their equilibrium values. We first address the effect of the rotation itself on the energy levels (centrifugal distortion), and then the effect of molecular vibrations on the spectroscopic parameters is presented.

The phenomenological Hamiltonian for a semirigid rotor with centrifugal distortion included can be written in the form [1–3]

$$\hat{H}_{\text{rot}} = \frac{1}{2}\sum_{\alpha,\beta} \mu^{e}_{\alpha\beta}\hat{J}_{\alpha}\hat{J}_{\beta} + \frac{1}{4}\sum_{\alpha,\beta\gamma,\delta} \tau_{\alpha\beta\gamma\delta}\hat{J}_{\alpha}\hat{J}_{\beta}\hat{J}_{\gamma}\hat{J}_{\delta} + \sum_{\alpha,\beta\gamma,\delta,\varepsilon,\eta} \tau_{\alpha\beta\gamma\delta\varepsilon\eta}\hat{J}_{\alpha}\hat{J}_{\beta}\hat{J}_{\gamma}\hat{J}_{\delta}\hat{J}_{\varepsilon}\hat{J}_{\eta} + \cdots \tag{6.15}$$

where $\mu^{e}_{\alpha\beta}$ are the elements of the equilibrium inverse moment of inertia tensor. As in the previous section, $\hat{J}_{\alpha}$ is the αth component of the total angular momentum, and the sum over $\alpha, \beta, \gamma, \delta, \varepsilon, \eta$ runs over the inertial axes. While the first term on the right-hand side represents the usual rigid-rotor Hamiltonian, the second and third are those that introduce the centrifugal distortion contributions. The $\tau_{\alpha\beta\gamma\delta}$ and $\tau_{\alpha\beta\gamma\delta\varepsilon\eta}$ are the quartic and sextic centrifugal-distortion constants, respectively. For their expressions, we refer interested readers to the specialistic literature [1–3].

Centrifugal distortion effects can be conveniently treated by means of perturbation theory:

$$\hat{H}_{\text{rot}} = \hat{H}^{0} + \hat{H}' \tag{6.16}$$

where $\hat{H}^{0}$ is the rotational Hamiltonian for a rigid rotor and $\hat{H}'$ is the perturbation operator corresponding to the second and third terms on the right-hand side of Eq. 6.15. Once again, we avoid a detailed discussion of the perturbative treatment of centrifugal distortion effects which can be found in Watson [23]. Here, we briefly recall the corresponding energy expressions. Before proceeding, it should be noted that in the regular expressions the quartic D's and sextic H's centrifugal distortion constants, which are combinations of the $\tau_{\alpha\beta\gamma\delta}$'s and $\tau_{\alpha\beta\gamma\delta\varepsilon\eta}$'s [1, 23–27], respectively, are actually used.

(a) In the case of a *linear molecule*, the expression of $\hat{H}'_{\text{dist}}$ is

$$\hat{H}' = -D_J\hat{J}^4 \tag{6.17}$$

and gives rise to an energy correction of the form

$$E'_{\text{dist}} = -D_J J^2 (J+1)^2 \tag{6.18}$$

(b) For *symmetric-top molecules* $\hat{H}'$ is given by

$$\hat{H}'_{\text{dist}} = -D_J \hat{J}^4 - D_{JK} \hat{J}^2 \hat{J}_z^2 - D_K \hat{J}_z^4 \tag{6.19}$$

and the corresponding energy correction is thus

$$E'_{\text{dist}} = -\left[D_J J^2 (J+1)^2 + D_{JK} J(J+1) K^2 + D_K K^4\right] \tag{6.20}$$

(c) Once again, for asymmetric-top molecules the situation is rather complicated and the inclusion of centrifugal distortion effects proceeds through the introduction of the so-called reduced Hamiltonian. We therefore refer interested readers to Watson's original papers [23–27] as well as to the summaries given in the literature [1, 2].

Effects of vibrations on rotational motion can be conveniently taken into account by means of vibrational perturbation theory. As for centrifugal distortion, we here only recall the relevant issues and refer the reader to the specialistic literature [1, 2, 27–30].

The starting point for the vibrational perturbation theory is the semirigid Hamiltonian due to Watson [28, 29],

$$\hat{H} = \frac{1}{2} \sum_{\alpha,\beta} (\hat{J}_\alpha - \hat{\pi}_\alpha) \mu_{\alpha\beta} (\hat{J}_\beta - \hat{\pi}_\beta) + \frac{1}{2} \sum_r \omega_r \hat{p}_r^2 + V(\mathrm{q}) - \frac{1}{8} \sum_\alpha \mu_{\alpha\alpha} \tag{6.21}$$

which is expressed in the representation of the dimensionless normal coordinates. Here $\hat{\pi}_\alpha$ is the αth component of vibrational angular momentum, $V(\mathrm{q})$ is the potential in terms of the dimensionless normal coordinates, and the final term in Eq. 6.21, the so-called Watson term (denoted $\hat{U}$), is due to the use of a normal-coordinate representation and leads to a nearly constant shift in the spectrum that has negligible spectroscopic importance [30].

The spectrum of the Watson Hamiltonian gives the rovibrational energy levels of the molecule under consideration. The resolution proceeds through the application of perturbation theory (Rayleigh–Schrödinger perturbation theory), which allows the partitioning of the Watson Hamiltonian into the rigid-rotor harmonic oscillator Hamiltonian $\hat{H}^0$ and a perturbation

$$\hat{H} = \hat{H}^0 + \hat{H}' \tag{6.22}$$

where

$$\hat{H}^0 = \sum_\alpha B^\alpha \hat{J}_\alpha^2 + \frac{1}{2} \sum_r \omega_r (\hat{p}_r^2 + q_r^2) \tag{6.23}$$

In the field of rotational spectroscopy, the relevant terms in these perturbative treatments are those that multiply $\hat{J}_{\alpha}^{2}$, as these are the effective rotational constants which contain contributions beyond the rigid-rotator harmonic oscillator approximation. To first order there are no corrections to the simple rigid-rotor rotational constant (equilibrium structure). In second order, there are three contributions that expressed by means of the usual contact transformation method give the second-order result [31]

$$B_i^{\alpha} = B_e^{\alpha} - \sum_r \alpha_r^{\alpha}\left(\upsilon_r + \frac{1}{2}\right) \tag{6.24}$$

with superscript $\alpha = a,\ b,\ c$ and where the sum is taken over all fundamental vibrational modes r. The corresponding vibration–rotation interaction constants, α_r^{α}, are given by

$$\alpha_r^{\alpha} = -2B_e^{\alpha^2}\left[\sum_{\beta}\frac{3(\alpha_r^{\alpha\beta})^2}{4I_{\beta}^{e}} + \sum_{s}\frac{(\zeta_{rs}^{\alpha})^2(3\omega_r^2+\omega_s^2)}{\omega_r(\omega_r^2-\omega_s^2)} + \frac{1}{2}\sum_{s}\frac{\phi_{rrs}a_s^{\alpha\alpha}}{\omega_s^{3/2}}\right] \tag{6.25}$$

with $\alpha_r^{\alpha\beta}$ the derivative of the moment of inertia with respect to normal coordinates [i.e., $(\partial I_{\alpha\beta}/\partial q_r)_e$], ξ_{rs}^{α} the elements of the antisymmetric Coriolis zeta matrix (for a definition, see refs. [1–3]), ω_r the harmonic frequency (associated to the rth normal coordinate), and ϕ_{rrs} the opportune cubic force constant.

6.2.4 Hyperfine Interactions

The fine and hyperfine structure in rotational spectra is due to interactions of the molecular electric and/or magnetic fields with the nuclear moments. The most important of these interactions is the one between the molecular electric field gradient and the electric quadrupole moments of certain nuclei. As far as magnetic interactions are concerned, the end-over-end rotation of a molecule generates a weak magnetic field that interacts with the nuclear magnetic moments to produce a slight magnetic splitting or shift of the lines. In addition to these two interactions, spin–spin interactions between different nuclear spins may arise.

The overall Hamiltonian can be written as a sum of different contributions,

$$\hat{H} = \hat{H}_{\mathrm{rot}} + \hat{H}_{\mathrm{NQC}} + \hat{H}_{\mathrm{SR}} + \hat{H}_{\mathrm{SS}} \tag{6.26}$$

where $\hat{H}_{\mathrm{rot}}$ accounts for the pure rotational part as seen in the previous sections. The additional terms, that is, $\hat{H}_{\mathrm{NQC}}$, $\hat{H}_{\mathrm{SR}}$, and $\hat{H}_{\mathrm{SS}}$, account for nuclear quadrupole-coupling, spin–rotation, and spin–spin interactions, respectively.

For nuclei with a quadrupole moment, the interaction of the latter (defined for nucleus K as eQ_K) with the electric field gradient at that nucleus, q_J^K, is given by [32]:

$$\hat{H}_{\mathrm{NQC}} = \frac{1}{2}\sum_K \frac{eQ_K q_J^K}{I_K(2I_K-1)J(2J-1)}\left[3(\hat{\mathbf{I}}_K\cdot\hat{\mathbf{J}})^2 + \frac{3}{2}(\hat{\mathbf{I}}_K\cdot\hat{\mathbf{J}}) - \hat{\mathbf{I}}_K^2\cdot\hat{\mathbf{J}}^2\right] \tag{6.27}$$

where I_K denotes the nuclear spin and the sum runs over all the K nuclei with $I_K \geq 1$ (i.e., with a quadrupole moment); q_J^K is the expectation value of the space-fixed component V_{ZZ}^K of the electric field gradient tensor at the same nucleus averaged over the rotational motion:

$$q_J^K = \frac{2}{(J+1)(2J+3)} \left[V_{aa}^K \langle \hat{J}_a^2 \rangle + V_{bb}^K \langle \hat{J}_b^2 \rangle + V_{cc}^K \langle \hat{J}_c^2 \rangle \right] \tag{6.28}$$

The elements of nuclear quadrupole-coupling tensor for the nucleus K are then defined as

$$\chi_{\alpha\beta}^K = eQ_K V_{\alpha\beta}^K \tag{6.29}$$

where α, β refer to the inertial axes a, b, and/or c. If we consider for simplicity the case of a linear rotor (for symmetry there is only one nuclear quadrupole-coupling constant, χ), the corresponding correction to rotational energy is then given by

$$E_{\text{NQC}} = -\chi \frac{\frac{3}{4}D(D+1) - I_K(I_K+1)J(J+1)}{2(2J-1)(2J+3)I_K(2I_K-1)} \tag{6.30}$$

with $D = F(F+1) - J(J+1) - I_K(I_K+1)$. Therefore, the overall effect is that the rotational energy levels are split into various sublevels according to the values that can be assumed by the quantum number F.

To describe the interaction between the nuclear magnetic dipole and the effective magnetic field of a rotating molecule, Flygare derived a formulation in terms of a second-rank tensor $\boldsymbol{C}$ coupled with the rotational and nuclear spin momenta [33]:

$$\hat{H}_{\text{SR}} = \sum_K \hat{\mathbf{I}}_K \cdot \boldsymbol{C}^K \cdot \hat{\mathbf{J}} \tag{6.31}$$

where the sum runs over the K nuclei of the molecule with $I_K > 0$. To illustrate, let us consider once again a linear molecule (due to symmetry, there is just one spin–rotation constant, C). In such a case, the hyperfine energy levels are then given by

$$E_{\text{SR}} = \frac{C}{2}[F(F+1) - I_K(I_K+1) - J(J+1)] \tag{6.32}$$

The direct (dipolar) spin–spin interaction between two nuclear magnetic moments $\hat{\mathbf{I}}_K$ and $\hat{\mathbf{I}}_L$ is described by the Hamiltonian [1, 34, 35]

$$\hat{H}_{\text{SS}}^{KL} = \hat{\mathbf{I}}_K \cdot D^{KL} \cdot \hat{\mathbf{I}}_L \tag{6.33}$$

In addition to the dipolar coupling, there are also the so-called indirect contributions to the spin–spin coupling constant, which, while important in nuclear magnetic resonance (NMR) spectroscopy, are usually negligible in rotational spectroscopy [e.g., 36].

6.2.5 Selection Rules and Intensity of Transitions

Given the expressions for rotational energy levels, the subsequent step is defining the selection rules that tell us which are the rotational transitions that take place. The key point is the interaction between the molecular electric dipole components (fixed in the rotating body) and the electric components (fixed in space) of the radiation field. Without going into detail, from the nonvanishing matrix elements of transition dipole moment, the selection rules governing rotational transitions are derived to be [1, 2]

$$\Delta J = 0, \ \pm 1, \qquad \Delta M = 0, \ \pm 1 \tag{6.34}$$

Additional selection rules apply for symmetric-top rotors,

$$\Delta K = 0 \tag{6.35}$$

and asymmetric-top rotors,

$$\Delta K_a = 0, \ \pm 1 \qquad \Delta K_c = 0, \ \pm 1 \tag{6.36}$$

where K_a and K_c represent the quantum numbers of the limiting prolate and oblate symmetric-top rotors, respectively.

With the selection rules given above, the rotational frequencies[2] for

(a) a linear rotor are [1]

$$\nu = 2B(J+1) - 4D(J+1)^3 + H(J+1)^3\left[(J+2)^3 - J^3\right] + \cdots \tag{6.37}$$

(b) a symmetric-top molecule are [1]

$$\begin{aligned}\nu = {}& 2B(J+1) - 4D_J(J+1)^3 - 2D_{JK}(J+1)K^2 + H_J(J+1)^3\left[(J+2)^3 - J^3\right] \\ & + 4H_{JK}(J+1)^3K^2 + 2H_{KJ}(J+1)K^4 + \cdots \end{aligned} \tag{6.38}$$

Although the most fundamental selection rule for rotational spectroscopy is that the molecule should have a nonvanishing permanent dipole moment, we note that molecules without a permanent dipole moment can have perturbation-allowed rotational spectrum [22]. For spherical tops, for example, centrifugal distortion effects can produce a small permanent dipole moment that allows the observation of the rotational spectrum [1, 37].

[2] In the following, rotational constants and spectroscopic parameters are given in frequency units.

Analogously, the selection rules for hyperfine transitions in rotational absorption spectra can be derived:

$$\Delta F = 0, \ \pm 1 \qquad \Delta I_K = 0 \tag{6.39}$$

where F is the quantum number arising from the coupling scheme $\mathbf{F} = \mathbf{J} + \mathbf{I}_K$. For example, let us consider a linear molecule with only one interacting nucleus with spin–rotation constant C. Then, for a generic rotational transition $J \rightarrow J+1$ we obtain three hyperfine components:

$$\nu_{F+1 \leftarrow F} = \nu_0 - C(J+1) + C(F+1) \tag{6.40}$$

$$\nu_{F \leftarrow F} = \nu_0 - C(J+1) \tag{6.41}$$

$$\nu_{F-1 \leftarrow F} = \nu_0 - C(J+1) - CF \tag{6.42}$$

where ν_0 is the unperturbed frequency. A graphical representation is provided by Figure 6.1.

Concerning the intensity of rotational transitions, it should be briefly recalled that the formula for absorption coefficients involves the transition dipole moment. An important quantity often measured is the absorption coefficient at the resonant frequency ν_0, called the peak absorption coefficient $\alpha_{\max}$, which results to be proportional to the transition moment $(|\langle m|\mu|n\rangle|)$ as well as to frequency:

$$\alpha_{\max} \propto \nu_0^2 |\langle m|\mu|n\rangle|^2 \tag{6.43}$$

Once again, we refer the reader to specialistic literature for a deeper insight [e.g., 1].

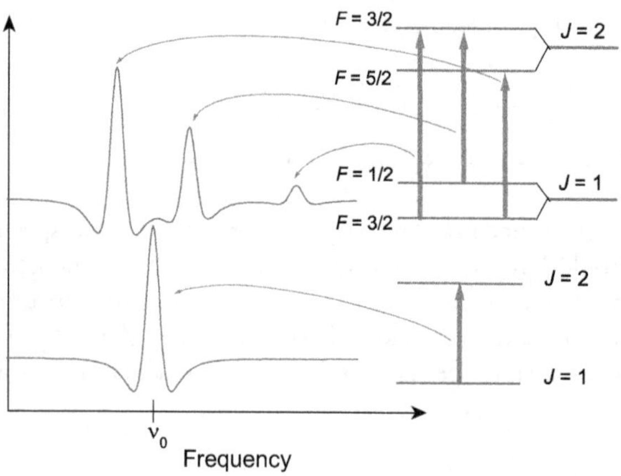

Figure 6.1 Effect of spin–rotation interaction on the $J = 2 \leftarrow 1$ rotational transition of a diatomic molecule with one nuclear spin ($I_K = \frac{1}{2}$, with $C < 0$). The spectrum at the bottom is the one without spin–rotation interaction, while the one at the top illustrates the splittings due to spin–rotation interaction.

6.3 QUANTUM CHEMICAL PREDICTION OF ROTATIONAL SPECTRA

On the basis of what has been summarized in the previous sections, it is clear that the information required for predicting rotational spectra are accurate estimates of:

(a) Rotational parameters
(b) Type of transitions observable (and their intensity)
(c) Fine and hyperfine parameters

As will be made clear in the followings sections, accurate equilibrium structure as well as harmonic and anharmonic force field computations are necessary in order to fulfill point (a), dipole moment evaluations for point (b), and, finally, accurate electric field gradient, spin–rotation, and spin–spin tensor calculations for point (c). Additionally, vibrational corrections for properties related to points (a) to (c) are often required.

6.3.1 Spectroscopic Parameters

In this section we provide the connection between the spectroscopic parameters that are required for predicting rotational spectra and the quantities determinable by quantum chemistry. This is also shown in Table 6.1.

Let us follow the classification given above:

(a) *Rotational Parameters* They include rotational as well as centrifugal distortion constants. The former are inversely proportional to the moments of inertia, which are related to the molecular structure:

$$\mathbf{I} = \sum_K M_K (R_K^2 1 - \mathbf{R}_K \mathbf{R}_K^{\mathrm{T}}) \tag{6.44}$$

where the nuclear coordinates $\mathbf{R}_K$ are obtained from geometric optimizations. The corresponding rotational constants are thus those at the equilibrium (B_e^α of Eq. 6.24). The vibrational ground-state rotational constants can be subsequently obtained by adding the corresponding vibrational corrections, as given in Eq. 6.24. As clear from Eq. 6.25, the latter require anharmonic (cubic) force field calculations.

As concerns centrifugal distortion constants, we briefly note that they require force field calculations as well as the derivatives of the inverse inertia tensor with respect to the normal coordinates ($\mu_{\alpha\beta}^r$). As is clear from the expression

$$\tau_{\alpha\beta\gamma\delta} = -\frac{1}{2}\sum_r \mu_{\alpha\beta}^r (\omega_r)^{-1} \mu_{\gamma\delta}^r \tag{6.45}$$

the harmonic force field (i.e., harmonic frequencies ω_r) is needed for quartic centrifugal distortion constants, while the determination of the sextic ones require the additional evaluation of the cubic force field [27].

Table 6.1 Computational Requirements for Determination of Relevant Parameters in Rotational Spectroscopy

Parameter	Symbol(s)	Computational Task(s)
Rotational constant	A, B, C	Molecular geometry, geometry optimization
Quartic centrifugal distortion constants	$\tau_{\alpha\beta\gamma\delta}$, D_J, D_K, D_{JK}, ...	Harmonic force field, i.e., harmonic frequencies and normal coordinates
Sextic centrifugal distortion constants	$\tau_{\alpha\beta\gamma\delta\varepsilon\eta}$, H_J, H_K, H_{JK}, ...	In addition; cubic force field
Dipole moment	μ	First derivative of energy with respect to external electric field
Nuclear quadrupole coupling	$\chi^K_{\alpha\beta}$, q^K, ...	Electric field gradient (first derivative of energy with respect to nuclear quadrupole moment)
Nuclear spin–rotation tensor	C^K	Second derivative of energy with respect to rotational angular momentum and nuclear spin
Spin–spin (dipolar) interaction	D^{KL}	Molecular geometry
Vibrational corrections	ΔB_{vib}, $\Delta\mu_{\text{vib}}$, $\Delta D^{KL}_{\text{vib}}$, ...	Harmonic and cubic force fields, derivative of properties with respect to normal coordinates

(b) *Dipole Moment* The dipole moment can be determined by computing first derivatives of the energy; more precisely, it is given by the derivative of the energy with respect to the components of an external electric field evaluated at zero field strength:

$$\mu_\alpha = -\left(\frac{dE}{d\varepsilon_\alpha}\right)_{\varepsilon=0} \tag{6.46}$$

where the negative sign is a convention.

(c) *Hyperfine Parameters* From a quantum chemical point of view, the quantity required for the determination of the nuclear quadrupole-coupling tensor is the electric field gradient at the quadrupolar nucleus. This is a first-order property which can be computed as either the first derivative of the energy with respect to the nuclear quadrupole moment or the expectation value of the corresponding (one-electron) operator

$$\hat{V}^K = Z_K \frac{\mathbf{1}(\mathbf{r}-\mathbf{R}_K)^2 - 3(\mathbf{r}-\mathbf{R}_K)(\mathbf{r}-\mathbf{R}_K)^{\mathrm{T}}}{|\mathbf{r}-\mathbf{R}_K|^5} \tag{6.47}$$

with $\mathbf{R}_K$ and $\mathbf{r}$ denoting the position of the Kth nucleus and the electron, respectively.

The nuclear spin–rotation tensors are second-order properties and can be obtained by means of either analytic derivative theory or linear response theory [38]. Without going into detail, we refer interested readers to the literature [e.g., 38, 39]. We only briefly note that the electronic part of the nuclear spin–rotation tensor can be computed as a second derivative with respect to the nuclear spin and the rotational angular momentum as perturbations,

$$\mathbf{C}_K^{\text{el}} = \left(\frac{\partial^2 E^{\text{el}}}{\partial \mathbf{I}_K \partial \mathbf{J}} \right)_{\mathbf{I}_K, \mathbf{J}=0} \tag{6.48}$$

while the nuclear part is determined solely by the molecular geometry,

$$\mathbf{C}_K^{\text{nucl}} = -\alpha^2 \mu_N g_K \sum_{L \neq K} Z_L \frac{(\mathbf{R}_L - \mathbf{R}_K) \cdot (\mathbf{R}_L - \mathbf{R}_K)\mathbf{1} - (\mathbf{R}_L - \mathbf{R}_K)(\mathbf{R}_L - \mathbf{R}_K)}{|\mathbf{R}_L - \mathbf{R}_K|^3} \mathbf{I}^{-1} \tag{6.49}$$

where α is the dimensionless fine constant, g_K is the nuclear g value of the Kth nucleus, and μ_N is the Bohr magneton.

The spin–spin interaction finally does not involve an electronic contribution and can be easily computed from the molecular structure [34, 35] determined within a geometry optimization

$$\boldsymbol{D}_{ij}^{KL} = -\frac{\hbar \mu_0}{8\pi^2} \gamma_K \gamma_L \frac{3(\boldsymbol{R}_{KL})_i (\boldsymbol{R}_{KL})_j - \delta_{ij} \boldsymbol{R}_{KL}^2}{R_{KL}^5} \tag{6.50}$$

where μ_0 is the permeability of the vacuum and γ_K and γ_L are the gyromagnetic ratios for the K and L nuclei, respectively.

6.3.1.1 Vibrational Corrections In order to compare theoretically calculated spectroscopic parameters to experiment, one must consider the effect of molecular vibrations. This is because the properties alluded to above depend upon the structure of the molecule and therefore must be averaged over the vibrational motion of the system under consideration. Force field evaluations in conjunction with vibrational perturbation theory allow the estimation of zero-point vibrational corrections to molecular properties.

The procedure briefly recalled here is based on the perturbative approach described in Auer at al. [40]. The key point is the expansion of the expectation value of the property X under consideration over the vibrational wavefunction in a Taylor series around the equilibrium geometry with respect to normal-coordinate displacements

$$\langle X \rangle = X_{\text{eq}} + \sum_r \left(\frac{\partial X}{\partial Q_r} \right)_{Q=0} \langle Q_r \rangle + \frac{1}{2} \sum_{r,s} \left(\frac{\partial^2 X}{\partial Q_r \partial Q_s} \right)_{Q=0} \langle Q_r Q_s \rangle + \cdots \tag{6.51}$$

where the expansion is truncated after the quadratic term. The expectation values over Q_r and Q_rQ_s are evaluated using perturbation theory [31] and the corresponding expressions, in lowest order, are

$$\langle Q_r \rangle = -\frac{1}{4\omega_r^{3/2}} \sum_s \phi_{rss} \tag{6.52}$$

and

$$\langle Q_r Q_s \rangle = \delta_{rs} \frac{1}{2\omega_r} \tag{6.53}$$

There are other approaches available in the literature [41, 42]. Concerning the properties of interest in the present chapter, the perturbational approach by Ruud et al. [41, 42] based on an expansion around an effective geometry instead of the equilibrium geometry [43] has to be mentioned. Nonperturbative schemes [44–46] are also available, but their application is usually restricted to small systems.

6.3.2 Quantum Chemical Calculations

In the present section all computational requirements for accurately evaluating the quantities listed in the previous section are provided with some detail and, in particular, with emphasis on the levels of theory needed. To fulfil the accuracy requirements, the key point is the employment of the coupled-cluster (CC) level of theory [47]. As well known, the CC singles-and-doubles (CCSD) approximation augmented by a perturbative treatment of triple excitations (CCSD(T)) [48] provides a very good compromise between accuracy and computational cost. As going beyond the CCSD(T) level might be important, the full CC singles–doubles–triples (CCSDT) [49–51] and the CC singles–doubles–triples–quadruples (CCSDTQ) [52] models can also be considered.

6.3.2.1 Equilibrium Structure Accurate equilibrium structures are required in order to get accurate equilibrium rotational constants.

To reach high accuracy and to account simultaneously for basis set effects as well as higher excitations and core correlation effects, the equilibrium geometry can be obtained by making use of composite schemes. In these schemes, the various contributions are evaluated separately at the highest possible level and then combined in order to obtain the best theoretical estimate. This additivity can be exploited at either a gradient level[3] [53, 54] or a geometric parameter level.

[3] CFour (Coupled Cluster techniques for Computational Chemistry), a quantum chemical program package by J. F. Stanton, J. Gauss, M. E. Harding, and P. G. Szalay, with contributions from A. A. Auer, R. J. Bartlett, U. Benedikt, C. Berger, D. E. Bernholdt, Y. J. Bomble, O. Christiansen, M. Heckert, O. Heun, C. Huber, T.-C. Jagau, D. Jonsson, J. Jusélius, K. Klein, W. J. Lauderdale, D. Matthews, T. Metzroth, D. P. O'Neill, D. R. Price, E. Prochnow, K. Ruud, F. Schiffmann, S. Stopkowicz, M. E. Varner, J. Vázquez, J. D. Watts, and F. Wang, and the integral packages MOLECULE (J. Almlöf and P. R. Taylor), PROPS (P. R. Taylor), ABACUS (T. Helgaker, H. J. Aa. Jensen, P. Jørgensen, and J. Olsen), and ECP routines by A. V. Mitin and C. van Wüllen. For the current version, see http://www.cfour.de.

The latter only requires the appropriate geometry optimizations to be carried out, as described by Puzzarini [55] for the extrapolation to the complete basis set limit and Puzzarini and Barone [56, 57] for the inclusion of all contributions. This approach requires the assumption that the convergence behavior of the geometric parameters is the same as that for energy as well as the additivity of contributions applies. The first approximation has been verified by Puzzarini [55]. Within the so-called geometry scheme [55], for example, to take into account the effects of core–valence (CV) electron correlation, geometry optimizations are carried out in conjunction with a core–valence correlation-consistent basis set correlating all electrons as well as in the frozen-core approximation (only valence electrons correlated). Then, the corresponding correction to geometric parameters is given by

$$\Delta r(\mathrm{CV}) = r(\mathrm{all}) - r(\mathrm{valence}) \tag{6.54}$$

where r(all) and r(valence) are the geometries optimized at the CCSD(T) level correlating all and only valence electrons, respectively, in the same basis set.

As far as the first scheme (the so-called gradient scheme) is concerned, it is the nuclear gradient that comprises the various contributions. Let us consider this in more detail. To perform the extrapolation to the complete basis set (CBS) limit, the CBS gradient is given by [53]

$$\frac{dE_{\mathrm{CBS}}}{dx} = \frac{dE^{\infty}(\mathrm{HF-SCF})}{dx} + \frac{d\Delta E^{\infty}(\mathrm{CCSD(T)})}{dx} \tag{6.55}$$

where dE^{∞}(HF-SCF)/dx and $d\Delta E^{\infty}$(CCSD(T))/dx are the nuclear gradients obtained using an exponential extrapolation for the Hartree-Fock self-consistent field (HF-SCF) energy [58] and the n^{-3} extrapolation scheme for the CCSD(T) correlation contribution [59]. The formula given above assumes that a hierarchy of bases has been employed. Usually, Dunning's hierarchy of correlation-consistent valence cc-pVnZ bases [60, 61] is employed, with n denoting the cardinal number of the corresponding basis set. Core correlation effects are considered by adding the corresponding correction, $d\Delta E$(core)/dx, to Eq. 6.55:

$$\frac{dE_{\mathrm{CBS+core}}}{dx} = \frac{dE^{\infty}(\mathrm{HF-SCF})}{dx} + \frac{d\,\Delta E^{\infty}[\mathrm{CCSD(T)}]}{dx} + \frac{d\,\Delta E(\mathrm{core})}{dx} \tag{6.56}$$

with the core correlation energy contribution as the difference of all-electron and frozen-core CCSD(T) calculations using the same core–valence basis set [62, 63]. In a similar manner, corrections due to a full treatment of triples, $d\Delta E$(full – T)/dx, and quadruples, $d\Delta E$(full – Q)/dx, can be accounted for and added to Eq. 6.56:[4]

[4] MRCC, a generalized CC/CI program by M. Kallay, see http://www.mrcc.hu.

$$\frac{dE_{\text{CBS+core+fT}}}{dx} = \frac{dE^{\infty}(\text{HF}-\text{SCF})}{dx} + \frac{d\,\Delta E^{\infty}[\text{CCSD(T)}]}{dx} + \frac{d\,\Delta E(\text{core})}{dx} + \frac{d\,\Delta E(\text{full}-\text{T})}{dx} \quad (6.57)$$

and

$$\frac{dE_{\text{CBS+core+fT+fQ}}}{dx} = \frac{dE^{\infty}(\text{HF}-\text{SCF})}{dx} + \frac{d\,\Delta E^{\infty}[\text{CCSD(T)}]}{dx} + \frac{d\,\Delta E(\text{core})}{dx} + \frac{d\,\Delta E(\text{full}-\text{T})}{dx} + \frac{d\,\Delta E(\text{full}-\text{Q})}{dx} \quad (6.58)$$

The corresponding differences between CCSDT and CCSD(T) and between CCSDTQ and CCSDT are obtained in frozen-core calculations employing small- to medium-sized basis sets. Additional contributions due to relativistic effects can also be considered and the corresponding corrections, obtained in analogy to the previous ones as differences, are added at the gradient level in order to obtain the best theoretical estimate.

6.3.2.2 Anharmonic Force Field Anharmonic force field calculations require the evaluation of the derivatives of the energy with respect to the coordinate system chosen: second derivatives for the harmonic part, third derivatives for the cubic part, fourth derivatives for the quartic part, and so on. The required derivatives of the energy with respect to the nuclear coordinates can be computed using either numerical or analytic techniques. The former approach is clearly of limited accuracy, while the analytic approach has no accuracy problems and is also computationally advantageous. Therefore, when available, the latter is the approach of choice. This is usually the case for the harmonic part of the force field. As analytic schemes for higher derivatives are not yet available for correlated methods, one has to rely on numerical techniques.

In the view of applying second-order vibrational perturbation theory, the best option for the coordinate system is the normal-coordinate representation. The cubic and (semidiagonal) quartic force constants are then derived by numerical differentiation of analytically evaluated second derivatives along the normal coordinates [64–66]:

$$\phi_{rst} = \frac{\phi_{st}^{+} - \phi_{st}^{-}}{2\Delta_r} \quad (6.59)$$

and

$$\phi_{rrst} = \frac{\phi_{st}^{+} + \phi_{st}^{-} - 2\phi_{st}}{\Delta_r^2} \quad (6.60)$$

with ϕ_{st} the quadratic force constant in normal-coordinate representation, ϕ^+ and ϕ^- the corresponding force constants at the displaced geometries, and Δ_r the displacement along the rth normal coordinate.

The usual procedure is to evaluate the force field for the main isotopic species. For the other isotopologues, the force fields are then obtained by a suitable transformation from the original representation into the normal-coordinate representation of the considered isotopic species. Subsequently, spectroscopic constants are determined by means of the vibrational second-order perturbation theory [31].

As concerns the accuracy, for most of the applications in the field of rotational spectroscopy the CCSD(T) level of theory and even the the second-order Møller–Plesset perturbation theory (MP2) [67] in conjunction with medium-sized basis sets (triple-, quadruple-zeta quality) are suitable.

6.3.2.3 Electric and Magnetic Properties

Dipole Moment An extensive benchmark study concerning the quantum chemical determination of dipole moments has been, for example, reported in Bak et al. [68]. The conclusions that can be drawn from such investigation are the following. First of all, it has to be noted that the use of basis sets augmented by diffuse functions is strongly recommended [69, 70]. This is a well-known recommendation and is due to the fact that the dipole operator samples the outer valence region in a molecule. Once additional diffuse functions are included, reasonable results are already obtained at the MP2/aug-cc-pVTZ or CCSD(T)/aug-cc-pVTZ levels. HF-SCF calculations typically overestimate the magnitude of the dipole moment, while, on the other hand, electron correlation effects beyond MP2 are often not substantial except in challenging cases. More precisely, we only briefly note that, whereas the Hartree–Fock model is found to be typically in error by 0.1–0.2 D, the introduction of correlation at the MP2 and CCSD levels greatly reduces the errors (to ~0.05 and ~0.03 D, respectively), and the CCSD(T) errors are found to be small, typically <0.01 D.

For a rigorous comparison of computed dipole moments with experiment, it is noted that vibrational effects should be accounted for as described above.

Nuclear Quadrupole Coupling As evident from Eq. 6.47, the electric field gradient operator exhibits a cubic dependence on $1/|\mathbf{r} - \mathbf{R}_K|$, which means that the inner valence and core region at the corresponding nucleus are those of relevance to this property. Accordingly, additional tight basis functions are needed for the accurate prediction of electric field gradients and, consequently, of nuclear quadrupole-coupling constants [69, 71, 72]. To discuss the computational requirements for these properties, we can refer to the benchmark study carried out in Halkier et al. [69], where the focus was on both the basis set convergence and electron correlation treatment. The results can be summarized as follows. The importance of steep functions in the basis set is demonstrated; in fact, the values obtained with the cc-pVnZ series of basis sets do not converge to the proper limit, while straightforward convergence to the correct limit is obtained when using the cc-pCVnZ sets. Concerning the principal number n of the basis sets, accurate calculations should be carried out with basis sets of at least quadruple-zeta quality. Electron correlation effects are also pronounced; in fact, the deviations of the HF-SCF results from the full-configuration interaction (FCI) limit are quite large, MP2 is usually not able to recover the large correlation contribution, and CC calculations are therefore needed for the accuracy required in the field of rotational spectroscopy; in fact, CCSD(T) results are in excellent agreement with FCI.

Another important issue in the quantum chemical calculations of electric field gradients when heavier nuclei are involved is the consideration of relativistic effects [e.g., 73–75]. Without going into details, we note the increasing importance of relativistic effects when going to heavier elements. In fact, taking halogens as an example, while relativistic effects are negligible for the fluorine electric field gradient, they amount to about 1% for chlorine, to about 7% for bromine, and to about 18% for iodine. This means that already for second- and third-row elements relativity is for this property of relevance and should be accounted for when aiming at quantitative predictions. While for the heavier elements four-component calculations [76] (such as Dirac-HF or corresponding electron correlation treatments starting from the Dirac-HF ansatz) are mandatory, for lighter elements either perturbative schemes (e.g., second-order direct perturbation theory (DPT2) [77]) or quasi-relativistic approaches based on the Douglas–Kroll–Hess transformation [78] are sufficient.

Vibrational corrections for electric field gradients can be computed as explained above, but in most cases they are not very pronounced.

Spin–Rotation Interaction As explained above, the nuclear spin–rotation tensor consists of an electronic and a nuclear part, with the former computed as the second derivative of the electronic energy with respect to the rotational angular momentum and the appropriate nuclear spin as perturbations [38]. This is efficiently done using analytic second-derivative techniques [65], but, while calculations carried out using standard basis functions suffer from a slow basis set convergence, the use of perturbation-dependent basis functions significantly accelerates the basis set convergence [38].

As concerns the computational requirements, we note that electron correlation effects are important and that the CCSD(T) approach appears to be the method of choice for the corresponding calculations [e.g., 79]. With respect to the basis set convergence, even if a significant speed-up is already obtained by using perturbation-dependent basis functions, it should be noted that the basis set requirements are rather demanding: In general, basis sets of at least quadruple-zeta quality are required to obtain converged values. Also important are in some cases consideration of additional diffuse functions [38, 39] and the inclusion of vibrational effects [79, 80].

Dipolar Spin–Spin Coupling As mentioned above, the determination of the dipolar coupling constants only requires knowledge of the molecular geometry (see Eq. 6.50). Quantum chemical investigations, however, are necessary to provide the vibrational corrections which can be evaluated using the perturbative techniques previously described. In general, we note that the agreement between theory and experiment is consistently improved when vibrational effects are accounted for [81]. In view of accurately predicting experimental data, another aspect should be mentioned.

Let us consider in the following the case of a linear molecule. Rotational spectroscopy actually provides an effective spin–spin coupling constant, D_{eff}^{KL}, which also includes the (traceless) anisotropic part ΔJ^{KL} of the indirect spin–spin coupling

$$D_{\text{eff}}^{KL} = D^{KL} + \tfrac{1}{3}\Delta J^{KL} \tag{6.61}$$

From an experimental point of view, there is no possibility to distinguish between these two contributions, and only quantum chemical calculations can provide a means to separate them. Techniques for the calculation of indirect spin–spin couplings have been developed [82–85] and can also be used for evaluation of ΔJ^{KL}. Nevertheless, the indirect part of the spin–spin coupling is usually significantly smaller than the dipolar one (especially when there are not heavy elements involved) and in most cases can thus be neglected, though there is some evidence that this is not always the case [e.g., 86].

6.3.3 Simulation of Rotational Spectra

Once quantum chemistry has provided all the information required, that is, rotational and centrifugal distortion constants and, if the case, hyperfine parameters as well as line intensities, a graphical simulation of the rotational spectrum can be performed. The latter requires the knowledge of the experimental technique involved. For example, if the frequency modulation with second harmonic detection is performed, then the second derivative of the natural spectrum is obtained (as seen in Figure 6.2). The graphical representation of the computed spectrum can then be

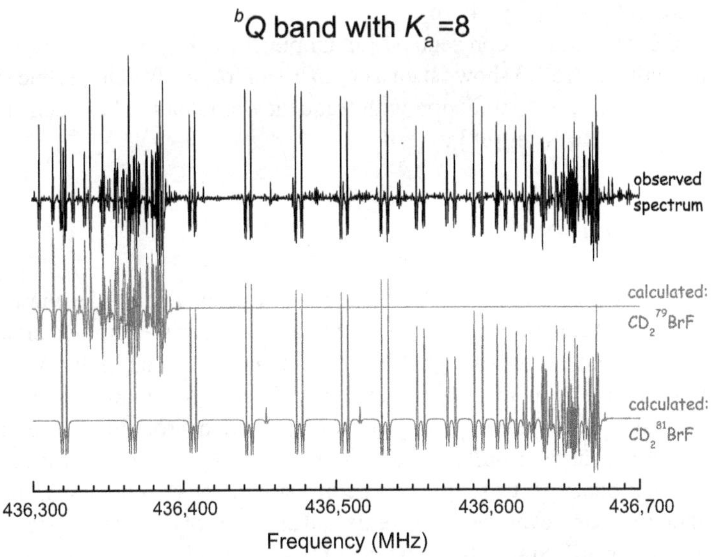

Figure 6.2 Comparison of recorded and calculated spectra for portion of band head at ~436 GHz (^{b}Q band with $K_a = 8$) of both $CD_2{}^{79}BrF$ and $CD_2{}^{81}BrF$ [87].

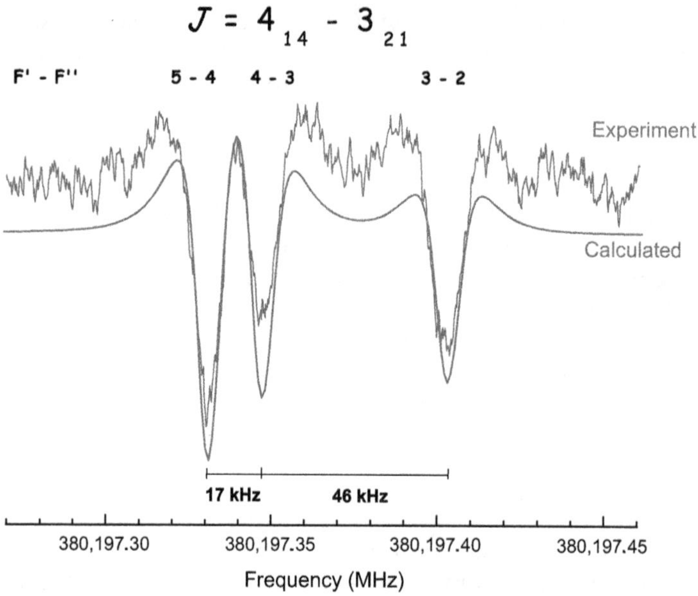

Figure 6.3 Comparison of recorded and calculated hyperfine structure of the $J = 4_{1,4} \leftarrow 3_{2,1}$ transition of $H_2^{16}O$ [88].

directly compared to the experimental one and, in this way, it can be of great help in the assignment and fitting procedures. This is especially true when complicated spectra are involved (see, e.g., Figure 6.2) and/or the spectroscopic parameters involved are unknown [e.g., 4, 16].

Figure 6.2 provides a representative example for a Doppler-limited rotational spectrum, while Figure 6.3 shows such a comparison for resolved hyperfine structure (Lamb-dip technique in conjunction with frequency modulation has been employed for recording the spectrum [88]).

6.4 APPLICATIONS

In this section applications will be presented in particular in the view of how quantum chemical calculations can be used to assist experimental investigations in the field of rotational spectroscopy. First, it will be shown that high-level calculations can provide reliable and accurate values for the corresponding spectroscopic and/or hyperfine parameters. Going a step further, as rotational spectroscopy provides highly accurate results, in particular for small molecules in the gas phase, data from rotational spectroscopy are ideally suited for benchmarking purposes. On the other hand, however, measurements and analyses of rotational spectra are not often straightforward. State-of-the-art quantum chemical computations can therefore be used to guide experimental investigations and in particular to assist in the determination of the spectroscopic parameters of interest. Quantum chemistry in this way allows to verify ("benchmark") results from rotational spectroscopy. As a sort of conclusion, it will be demonstrated how fruitful the interplay of theory and experiment can be in the field of rotational spectroscopy. For example, it will be shown how it is possible to derive equilibrium geometric parameters from a joint experimental–theoretical approach.

6.4.1 Benchmarking Quantum Chemistry with Rotational Spectroscopy

Quantum chemistry has nowadays reached such an advanced level that highly accurate results can be achieved for energies and properties of small- to medium-sized molecules. As stressed in the previous section, the requirements for these high-level calculations are efficient treatment of electron correlation via coupled-cluster theory, basis set extrapolation techniques, incorporation of core correlation, and relativistic as well as vibrational effects together with the use of suitable additivity schemes. Nevertheless, despite all the progress made so far, it is still essential to benchmark the results from quantum chemical calculations, and, as pointed out above, rotational spectroscopy offers such an opportunity.

Among the various spectroscopic parameters, rotational constants seem to be appropriate for such a comparison, as they can be accurately determined from experiment. From a computational point of view, the composite scheme

$$\text{fc-CCSD(T)/cc-pV}\infty\text{Z} + \Delta\text{core/cc-pCVQZ} + \Delta\text{T/cc-pVTZ} + \Delta\text{Q/cc-pVDZ}$$

presented in the previous section can be used to get an accurate estimate for equilibrium rotational constants. The corresponding accuracy can be estimated from a statistical analysis of the performance of quantum chemical results with respect to the corresponding experimental data. This has been recently carried out for the rotational constants of 16 closed-shell molecules (HF, N_2, CO, F_2, HCN, HNC, CO_2, H_2O, NH_3, CH_4, HCCH, HOF, HNO, HNNH, CH_2CH_2, and H_2CO) for which a total of 97 isotopologues were considered [89]. The results can be summarized as follows. Table 6.2 and Figure 6.4 provide a detailed comparison of the performance of the

Table 6.2 Statistical Analysis of Relative Errors (%) in Computed Rotational Constants with Respect to B_e Values Derived from Experimental B_0 Values by Subtracting Computed Vibrational and Electronic Contributions

Computational Approach	$\bar{\Delta}$	$\bar{\Delta}_{abs}$	Δ_{std}	Δ_{max}
fc-CCSD(T)/cc-pVQZ	−0.379	0.434	0.298	1.264
fc-CCSD(T)/cc-pV5Z	−0.298	0.320	0.180	0.726
fc-CCSD(T)/cc-pV6Z	−0.277	0.296	0.158	0.643
fc-CCSD(T)/cc-pV ∞ Z	−0.253	0.274	0.141	0.556
fc-CCSD(T)/cc-pV6Z + core	0.041	0.057	0.100	0.534
fc-CCSD(T)/cc-pV6Z + core + ΔT	0.071	0.087	0.110	0.584
fc-CCSD(T)/cc-pV6Z + core + ΔT + ΔQ	−0.027	0.063	0.072	0.325
fc-CCSD(T)/cc-pV∞Z + core + ΔT + ΔQ	−0.003	0.041	0.071	0.262

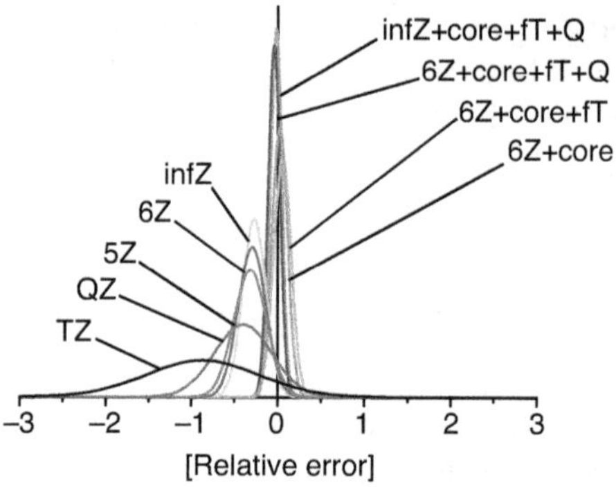

Figure 6.4 Normal distribution $\rho(R)$ of the errors in the calculated rotational constants in comparison with B_e values derived from experimental B_0 values and computed vibrational corrections [89].

different computational levels, as the calculated rotational constants, B_e, are compared to the values derived from experiment by subtracting vibrational and electronic corrections. From Figure 6.4, it is evident that improvements in the basis set narrow the error distribution. While for the cc-pVTZ basis a rather broad distribution is obtained, the cc-pV6Z distribution is already rather sharp and only a slight further improvement is seen when applying basis set extrapolation techniques. However, the center of the error distributions for frozen-core CCSD(T) results is somewhat off the y axis, thus indicating a systematic deficiency. The error distributions are shifted toward the y axis only when core correlation effects are included. While consideration of a more rigorous treatment of triple excitations at the CCSDT level has only marginal effects (and actually worsens the agreement), inclusion of quadruple excitations at the CCSDTQ level leads to a further gain in accuracy and to a substantially narrower distribution function.

If we compare theoretical and experimental ground-state rotational constants (instead of the equilibrium ones), we note that for the best theoretical estimates (which include vibrational corrections) the agreement with the corresponding experimental B_0 values is on average better than 0.1% The corresponding mean absolute error is 0.04% and the standard deviation is 0.07%. For prediction of ground-state rotational constants, the importance of vibrational corrections for high accuracy needs to be stressed; in fact, a narrow error distribution is only obtained when they are included. In fact, larger deviations between experiment and theory are observed for all computational results that do not consider vibrational corrections, the mean absolute error and the standard deviation being 0.7 and 0.6%, respectively [89]. Finally, we refer interested readers to Puzzarini et al. [89] for a more detailed discussion.

6.4.2 "Benchmarking" Rotational Spectroscopy with Quantum Chemistry

Measurements and analyses of rotational spectra can be not at all straightforward. In such cases high-level quantum chemical computations are needed to guide the

Table 6.3 Comparison of Experimental and Computed Diagonal Elements of Spin–Rotation Tensors (kHz) of CF_2 and CCl_2 [90]

	C_{aa}	C_{bb}	C_{cc}
CF2			
Best theoretical estimate[a]	380.86	33.14	13.83
Experiment (2000)[b]	783.(19)	108.(6)	−24.(6)
Experiment (revised, 2005)[c]	379.(11)	31.(4)	11.(4)
CCl_2			
Best theoretical estimate[a]	56.83	2.69	1.25
Experiment[d]	58.2(9)	2.8(2)	1.4(3)

[a]From ref. 90.
[b]From ref. 91.
[c]From ref. 93.
[d]From ref. 92.

investigation and in particular to assist the determination of the spectroscopic parameters of interest. In this way, it can be said that quantum chemistry allows to verify ("benchmark") results from rotational spectroscopy. To this end, the determination of the hyperfine parameters of dihalogen carbenes provides a significant example of the fact that quantum chemistry can be used for "benchmarking" results from experiments. More precisely, Puzzarini et al. [90] used quantum chemical calculations to verify the experimental values for the halogen spin–rotation constants of CF_2 [91] and CCl_2 [92]. For this purpose, sequences of correlation-consistent basis sets were used in conjunction with the CCSD(T) method. Theoretical best estimates were then obtained via extrapolation to the complete basis set limit and consideration of core correlation corrections as well as effects due to additional diffuse functions. Zero-point vibrational corrections were also incorporated in order to have a meaningful comparison to experiment. While the comparison with the experimental results revealed a good agreement in the case of CCl_2, a severe disagreement was noted in the case of CF_2 (see Table 6.3). The good agreement obtained in the case of CCl_2 as well as the high level of theory employed in the calculations supported the reliability of the computed values for CF_2 and the conclusion that the corresponding experimental values were erroneous. Such a suggestion was actually confirmed by a reanalysis of the experimental data which revealed a misassignment of the hyperfine components and led to spin–rotation parameters in good agreement with theory [93].

The general conclusion that can be drawn from the example reported above is that, when reliable literature data are not available, the verification of the assignments is possible only via comparison of the derived experimental parameters to the corresponding computed values, as different assignments may lead to satisfactory interpretations of the experimental results. As a sort of summary for this paragraph, it can be pointed out that quantum chemical calculations are essential to verify the "correctness" of the analysis of the experimental data as well as to provide reliable input values for the analysis of the experimental hyperfine structure.

6.4.3 Interplay between Experiment and Theory

Based on all the considerations given above, it is evident that rotational spectroscopy can benefit by a fruitful interplay of theory and experiment. The power of such an interplay will be demonstrated in the following by means of a few examples.

6.4.3.1 Investigation of "Unknown" Rotational Spectra In view of the high accuracy reachable by quantum chemistry in predicting rotational constants, computed values can be used to guide experimental investigation, to support the recording as well as the assignment of the rotational spectra, either when detection of new molecules is involved or when no experimental data are available. To this purpose, a few representative examples will be presented in the following.

A prominent example in this context is the recent detection of oxadisulfane (HSOH) via rotational spectroscopy [4]. The successful identification of HSOH among the products of the pyrolysis of $(t\text{-Bu})_2SO$ was possible due to accurate predictions of the spectroscopic parameters of HSOH. In fact previous searches for HSOH without such predictions were unsuccessful [4]. As outlined by Winnewisser et al. [4], quantum chemical calculations were used to predict the HSOH rotational–torsional spectrum: The equilibrium rotational constants were obtained at the CCSD(T)/cc-pCVQZ level of theory and then augmented by vibrational corrections at the CCSD(T)/cc-pVTZ level. Dipole moment components were also computed in order to predict the type of rotational transitions detectable and their intensity.

There are other examples in the literature that show how quantum chemical predictions of rotational constants may support and guide experimental detections of new molecules. On this topic, the combined theoretical and experimental investigation of the previously unknown carbon chain molecule HC_5P carried out by Bizzocchi et al. [94] deserves to be mentioned. Phosphabutadiyne, HC_5P, was generated through a gas-phase, high-temperature reaction between PCl_3 and toluene; then, scanning the rotational spectrum in selected frequency intervals in the 80–90-GHz frequency region, the authors were able to identify a sequence of weak lines whose pattern was consistent with that of a linear molecule characterized by a value of B_0 very close to the predicted one [94].

Another significative example is also provided by the recent assignment of the rotational spectra of various isotopic species of *trans*-1-chloro-2-fluoroethylene, *trans*-CHCl=CHF [16]. During such investigation, despite the presumably accurate predictions for the rotational spectra of the deuterated species, based on an empirical scaling of the quantum chemically computed values via the experimental data available for the main isotopologue [95], some problems were encountered in the assignment of the corresponding spectra. To be more specific, the spectrum of $CH^{35}Cl{=}CDF$ could not be initially assigned, while the spectrum for $CD^{35}Cl{=}CHF$ could be assigned, though the determined spectroscopic parameters did not agree with the scaled values. This was unexpected, as the scaling procedure, though empirical in nature, usually improves the quantum chemically predicted constants and in this way facilitates detection and analysis of the corresponding spectra. However, when repeating the scaling using the experimental constants determined for $CD^{35}Cl{=}CHF$, no problem

Table 6.4 Experimental and Theoretical Rotational Constants (MHz) for Various Isotopologues of *trans*-1-Chloro-2-fluoroethylene [16]

	Experiment	Theory	Experiment	Theory
	$CH^{35}Cl{=}CHF$		$CH^{37}Cl{=}CHF$	
A_0	53655.7296(13)	53740.67	53612.9224(17)	53698.13
B_0	2476.60705(40)	2479.28	2415.96632(38)	2418.56
C_0	2366.41041(45)	2369.04	2310.90479(41)	2313.47
	$CD^{35}Cl{=}CHF$		$CD^{37}Cl{=}CHF$	
A_0	42810.6833(22)	42879.45	42766.8580(24)	42835.82
B_0	2476.0507(34)	2478.81	2415.2413(42)	2417.95
C_0	2339.6221(43)	2342.32	2285.1521(42)	2287.77
	$CH^{35}Cl{=}CDF$		$CH^{37}Cl{=}CDF$	
A_0	43786.717(19)	43836.35	43763.8174(32)	43813.62
B_0	2466.640(20)	2469.31	2405.669(85)	2408.27
C_0	2334.144(20)	2336.68	2279.400(85)	2281.88
	$CD^{35}Cl{=}CDF$		$CD^{37}Cl{=}CDF$	
A_0	36119.5308(18)	36163.81	36093.02672(63)	36137.48
B_0	2466.1772(27)	2468.84	2405.054(7)	2407.66
C_0	2307.5064(12)	2310.04	2253.834(7)	2256.30

occurred and the spectra for all the other deuterated species could be detected and properly assigned. In addition, good agreement between the experimental and the scaled spectroscopic parameters was obtained (discrepancies lower than 0.05%). The failure of the applied scaling procedure cast some doubts on the results reported in Cazzoli et al. [95] for the two main isotopic species. Consequently, the rotational spectra of $CH^{35}Cl{=}CHF$ and $CH^{37}Cl{=}CFH$ were also reconsidered [16]. The new set of spectroscopic parameters obtained for these two isotopologues, unlike the previous set, was in good agreement with both the theoretical results and the data obtained for the deuterated species. We therefore conclude that accurate theoretical predictions played in their subsequent successful assignments a crucial role in elucidating the misassignments made in Cazzoli et al. [95] and in their subsequent successful assignments [16]. Table 6.4 summarizes the results obtained in Cazzoli et al. [16] for rotational constants and shows the excellent agreement that can be reached between theory and experiment even for a medium-sized molecule containing a second-row atom. Figure 6.5 shows the comparison between experiment and theory for *trans*-$CH^{37}Cl{=}CHF$: a portion of the $\Delta J = 0$, $K_a = 6$ ($\sim$665 GHz) band is depicted. In the inset a detailed portion of the band head is shown.

A very significant example that allows to stress the predictive capabilities of quantum chemistry in the field of high-resolution rotational spectroscopy is provided by recent analysis and assignment of the hyperfine structure of the rotational spectrum for $H_2{}^{17}O$ [80]. Without going into details, we commit to Figure 6.6 the conclusions. To summarize, the analysis of the recorded hyperfine structure in the rotational spectrum of $H_2{}^{17}O$ was hampered by the great complexity of the hyperfine structure itself. In the present case, the hyperfine structure resulted to be particularly crowded

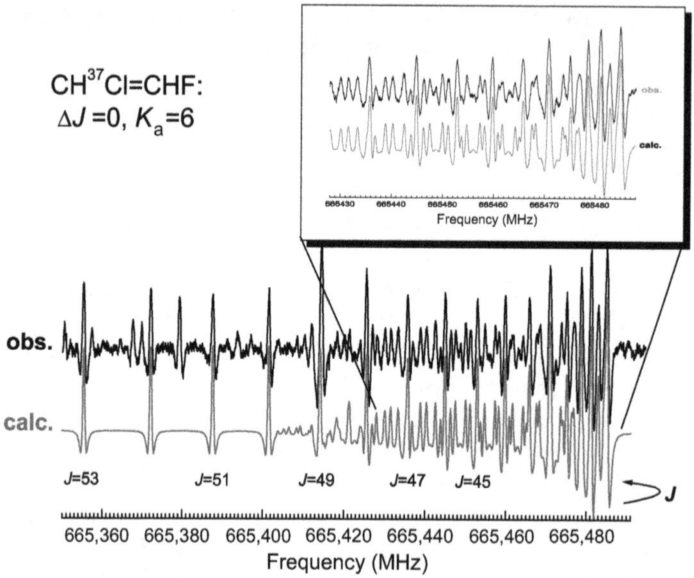

Figure 6.5 Comparison between experiment and theory for *trans*-CH^{37}Cl=CHF: a portion of the $\Delta J = 0$, $K_a = 6$ (~665-GHz) band [16].

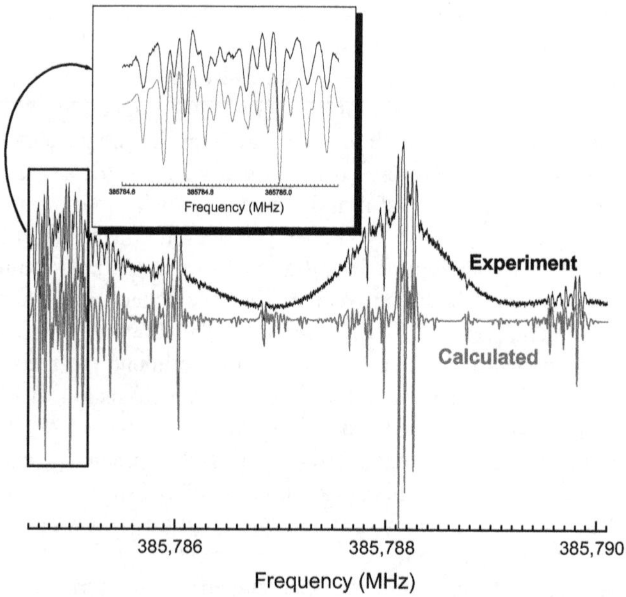

Figure 6.6 The $J = 4_{1,4} \leftarrow 3_{2,1}$ rotational transition (ortho) of $H_2{}^{17}O$ recorded at $P = 0.5$ mTorr (mod. depth = 12 kHz). Calculated spectrum is also depicted [80].

because of the large number of parameters involved, and such a complicated hyperfine structure turned out to be hardly interpretable without the proper hints for them provided by quantum chemistry. The quantitative agreement between experiment and theory is revealed by Figure 6.6 itself. We refer interested readers to Puzzarini et al. [80] as well as to what is reported in the following.

6.4.3.2 Molecular Structure Determination When possible, the best choice for reporting structural information is providing the true equilibrium structure defined as the corresponding minimum on the Born–Oppenheimer potential energy surface. Among the methods available for experimental structure determination, rotational spectroscopy is the method of choice when aiming at high accuracy [1]. However, it remains a formidable task to determine accurate equilibrium structures for polyatomic molecules because of the increasing number of structural parameters (which consequently necessitates the investigation of more and more isotopically substituted species of the molecule) and because of inadequacies in the consideration of vibrational effects. As concerns the latter, quantum chemical calculations can provide the required vibrational corrections. The combination of experimental ground-state rotational constants with calculated vibrational corrections

$$B_e^\alpha = B_0^\alpha(\text{exp}) + \frac{1}{2}\sum_r \alpha_r^\alpha(\text{theo}) \tag{6.62}$$

has turned out to be a very powerful approach to accurately determining "experimental" structures for polyatomic molecules [96–107]. Due to the computational contribution, these structures are usually referred to as "empirical," "mixed experimental/theoretical," or "semiexperimental." Thirty years ago, in a study of methane, Pulay, Meyer, and Boggs corrected experimental rotational constants for effects of vibration–rotation interaction and obtained a structure [96] which is still in excellent agreement with a very recent recommendation [108], attesting to the power of this procedure. Efforts of the same types were subsequently made by many others. Let us mention the relevant contributions by Allen et al. [97–99], Botschwina et al. [100, 101], Demaison [109], Craig [110, 111], Groner [112, 113], Gauss [114], Stanton [115, 117], and their collaborators. The accuracy of such a joint experimental and theoretical procedure for the determination of equilibrium structures has recently been investigated by Pawlowski et al. [105], who concluded that errors in the determined empirical bond lengths are typically below 0.001 Å for first-row elements, as long as an electron-correlated approach is used in the calculation of the vibrational corrections. It has also been shown that this approach can be successfully applied to polyatomic molecules containing either second-row atoms or heavier nuclei [87, 118–120].

As concerns the procedure, the empirical equilibrium structure is obtained by a least-squares fit of the molecular structural parameters to the mixed experimental/theoretical equilibrium moment of inertia I_e^i which are obtained in a straightforward manner from the corresponding equilibrium rotational constants B_e^i [see Eq. (6.10)]. Therefore, the fit requires at least as many independent rotational constants as there are structural degrees of freedom. Nevertheless, the use of more constants is preferred, as in this way the consistency of the available experimental data can be checked. In the fitting procedure the same weight can be used for all considered moments of inertia as the experimental rotational constants are much more accurate than the theoretical vibration–rotation interaction constants. In fact, the errors in empirical equilibrium geometries arise mainly from errors in the calculated vibration-rotation interaction constants [105]. How the errors in the vibration-rotation interaction constants propagate to errors in the equilibrium geometries has deeply been investigated by Pawlowski et al. [105]. For details, the reader is referred to this reference. An additional issue that should be stressed is the need to obtain a well-conditioned system of equations, which means that the set of rotational constants should be balanced; that is, it should contain information on isotopic substitutions for almost all nuclei. Otherwise, the determined structural parameters are not precise and may even be unreliable. Finally, it is important to point out that, while the individual α_r^α constants may be affected by resonances in the Coriolis contribution, their sum

$$B_e^{\alpha 2}\left[\sum_{\beta,r}\frac{3(\alpha_r^{\alpha\beta})^2}{4I_\beta^e}-\frac{1}{2}\sum_{r,\,s}\frac{(\zeta_{rs}^{\alpha})^2(\omega_r-\omega_s)^2}{\omega_r\omega_s(\omega_r+\omega_s)}+\frac{1}{2}\sum_{r,s}\frac{\phi_{rrs}a_s^{\alpha\alpha}}{\omega_s^{3/2}}\right] \tag{6.63}$$

is unaffected.

As already mentioned, the semiexperimental approach described above has been found to provide reliable and accurate results also for molecules containing heavy elements. To this purpose, as an example of the procedure described above we give the recent determination of the equilibrium structure of bromofluoromethane, CH_2FBr, by means of the investigation of the rotational spectra of various isotopic species [87]. To be more specific, based on the experimental ground-state rotational constants for eight isotopic species and quantum chemical calculations of the anharmonic force field, the empirical equilibrium r_e structure was determined (Figure 6.7). As evident from Table 6.5, the semiexperimental equilibrium geometry is accurate and in good

Table 6.5 Comparison of Different Experimental Structures of Bromofluoromethane to Theoretical Best Estimate [87]

	Best Estimate[a]	Empirical r_e	r_0	r_s	$r_m^{(1)}$	$r_m^{(1L)}$
C-F	1.3578	1.35757(13)	1.3714(34)		1.3641(19)	1.3674(15)
C-Br	1.9247	1.92854(12)	1.9287(33)	1.928(13)	1.9274(10)	1.9286(8)
C-H	1.0833	1.08333(4)	1.0782(12)	1.0864(20)	1.0854(4)	1.0699(37)
∠FCBr	110.14	110.151(32)	110.06(85)		110.36(16)	110.24(20)
∠HCF	109.50	109.552(10)	109.27(27)		109.13(7)	109.28(7)
∠HCBr	107.29	107.233(8)	107.69(23)	107.3(15)	107.36(5)	107.19(5)
∠HCH	113.05	113.057(7)	112.83(18)	113.43(22)	113.47(8)	113.60(7)

Note: Distances are in angstrom and angles in degree.

[a]CCSD(T)/cc-pV∞ + Δcore (CCSD(T)/cc-pCVQZ) + Δfull-T (cc-pVTZ) + Δrel (DPT2: MP2/cc-pVQZ). For details, see ref. 87.

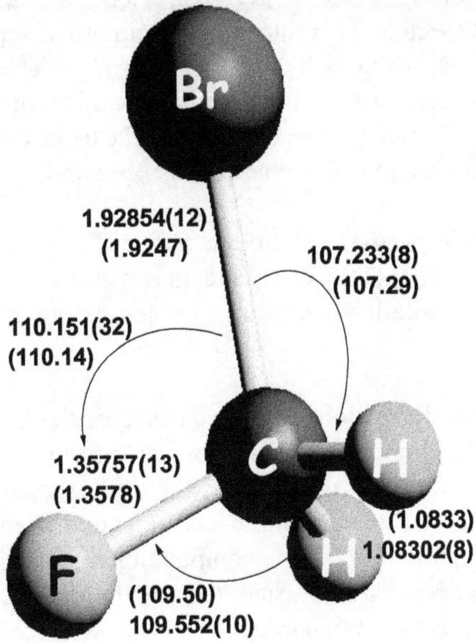

Figure 6.7 Equilibrium structure of bromofluoromethane: comparison of empirical structure and theoretical best estimate (given in parentheses) [87].

agreement with the high-level pure ab initio structure, while the r_0, a partial r_s, and the $r_m^{(1)}$ and $r_m^{(1L)}$ structures, obtained from the analysis of the experimental data alone, do not provide good description of the equilibrium structure.

Another interesting example is provided by the equilibrium structure of substituted diacetylenes. The semiexperimental approach allows us to compare their structures as obtained at the same level of accuracy. Such a comparison is reported in Thorwirth et al. [121] where the authors compared and discussed the molecular geometries of diacetylene itself and various substituted diacetylenes/acetylenes. These are summarized in Table 6.6. The main conclusion that can be drawn is that the HCCC fragment remains almost unchanged by substitution of one hydrogen of HCCCCH. From Table 6.6 the effects of conjugation (i.e., π-electron delocalization)

Table 6.6 Equilibrium Structure of HCCC Fragment in Diacetylenes (Bond Distances in Angstrom)

	Reference	C–H	C≡C	C–C
r_e^{emp} HC_4–H	121	1.0615	1.2085	1.3727
r_e^{emp} HC_4–F	122	1.0614	1.2080	1.3731
r_e^{emp} HC_4–CH_3	123	1.0613(3)	1.2085(6)	1.3734(14)
r_e^{emp} HC_4–CN	124	1.60220(3)	1.20948(5)	1.36519(12)

are also evident: The formal C–C single-bond length is shorter and the C–C triple-bond longer than expected. In Table 6.6 the empirical equilibrium structure of cyanobutadiyne, HC_5N, derived by Bizzocchi et al. is also reported [124]. This example allows us to note how conjugation is affected by the presence of the cyano group: The single C–C bond becomes longer and the triple C–C bond shorter. On the whole, a lower delocalization of π-electrons is observed.

6.4.3.3 Molecular Properties There are different molecular properties that are of interest to rotational spectroscopy and that can consequently be determined from the analysis of rotational spectra. These are either electric or magnetic properties.

Dipole Moment The dipole moment, which has intrinsic importance as a molecular property, has a twofold importance as it provides information on line intensities as well as on the type of transitions observable. Rotational spectroscopy by means of the Stark effect can accurately determine it, and quantum chemistry can support the corresponding investigation. The basic computational requirements for the accurate evaluations of this property are, as usual, the employment of correlated methods in conjunction with large basis sets, and we refer the reader to the brief overview given in the section concerning the quantum chemical calculations as well as to the references there cited [68–70].

In the present discussion it has to be noted that the availability of a theoretically justified formula for the extrapolation to the complete basis set limit [70] allows us to improve even more the accuracy reachable by quantum chemistry in estimating this property. Another issue that should be taken into account in predicting molecular properties is the zero-point vibrational effect, as quantum chemical results are equilibrium values, whereas experimental data usually refer to the vibrational ground state.

As an instructive example, here we briefly discuss the recent determination of the dipole moment components of bromofluoromethane [125]. A thorough investigation was carried out at the coupled-cluster level accounting for basis set truncation error, core correlation corrections, and vibrational effects. The best theoretical estimate was then employed for predicting the Stark spectra, which was largely complicated by the hyperfine structure due to the bromine atom (see Figure 6.8). In Table 6.7 the experimental and theoretical results are compared: An impressive good agreement between the best estimated vibrationally averaged and experimental vibrational ground-state dipole moment components has to be noted, the discrepancies being on the order of 2% for μ_a and 0.2% for μ_b.

Nuclear Quadrupole Coupling Equation 6.29 allows us to mention an important application related to this property. In fact, the interplay of accurate experimental investigations and quantum chemical evaluations of electric field gradients provides the unique opportunity to determine nuclear quadrupole moments. Actually, this procedure, based either on atomic or molecular quadrupole-coupling constants, is the major source for this type of nuclear data [126–128].

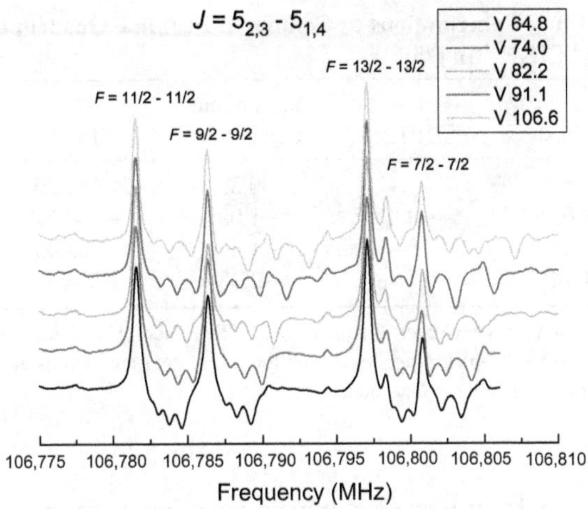

Figure 6.8 Stark spectra of the $F = \frac{9}{2}$ and $F = \frac{11}{2}$ hyperfine components of the $J = 5_{2,3} \leftarrow 5_{1,4}$ rotational transition recorded with different electric field applied (64.8, 74.0, 82.2, 91.1, and 106.6 V) [125].

Despite the importance of the nuclear quadrupole coupling in affecting the rotational spectra, the literature concerning the theoretical predictions of its related constants is rather limited. A significant example is provided by the recent investigation performed on the hyperfine structure in the rotational spectra of bromofluoromethane [75]. The experimental determination was supported by quantum chemical calculations of the nuclear quadrupole-coupling and spin–rotation tensors of ^{79}Br and ^{81}Br, performed at the CCSD(T) level in conjunction with core-valence correlation–consistent bases. Zero-point vibrational (ZPV) corrections were computed at the MP2 level in conjunction with the cc-pCVTZ basis set, whereas relativistic effects on the electric field gradient at the bromine

Table 6.7 Theoretical and Experimental Dipole Moment Components (D) of $CH_2{}^{79}BrF$ [125]

	μ_a	μ_b	μ_{tot}
CCSD(T)/aug-cc-pVQZ	−0.341	−1.696	−1.730
CCSD(T)/aug-cc-pV5Z	−0.346	−1.700	−1.735
CBS	−0.350	−1.702	−1.738
CBS + CV^a	−0.355	−1.710	−1.746
CBS + CV + ZPV^b	−0.339	−1.701	−1.734
Experiment^c	0.3466(11)	1.704(26)	1.739(26)

^{a}Values extrapolated to the complete basis set limit (CBS) plus core correlation corrections (CV).
^{b}Zero-point vibrational corrections (ZPV) added to the CBS + CV values.
^{c}Absolute value. Reported uncertainties are 3 times the standard deviation of the fit.

Table 6.8 Individual Contributions to Computed Bromine Quadrupole-Coupling Tensors (MHz) of $CH_2F^{79}Br$ [75]

	Nonrelativistic	DPT2	Vibrational Correction	Total	Experiment
χ_{aa}	421.84	27.47	0.39	449.70	443.431(8)
χ_{bb}	−138.68	−9.06	1.08	−146.67	−144.932(19)
χ_{cc}	−283.17	−18.41	−1.47	−303.05	−298.499(19)
χ_{ab}	−264.74	−17.48	−2.18	−284.40	278.56(19)[a]

Note: The nonrelativistic values have been obtained at the CCSD(T)/cc-pCVQZ level, the relativistic DPT2 corrections at the CCSD(T)/cc-pVQZ-unc level, and the vibrational corrections at the MP2/cc-pCVTZ level. Total values are compared to experiment.
[a]Absolute value.

nucleus and thus on the bromine quadrupole-coupling tensor were obtained at the CC level using DPT2 [74]. The comparison between experimental and theoretical values of the bromine quadrupole-coupling tensor for the main isotopic species is reported in Table 6.8. At first sight, one may note an overall good agreement, the discrepancies ranging from 1.4 to 2.1%. Table 6.8 also gathers the various contribution to the final best theoretical values for the bromine quadrupole-coupling tensors. It is relevant to note that relativistic effects are rather large (i.e., on the order of 6%) and thus essential to obtain a satisfactory agreement between theory and experiment. Based on the procedure mentioned above, a redetermination of the bromine quadrupole moment (307.5(10) mbarn) has been recently carried out [129]. This updated value leads in the case of CH_2BrF to final values for the elements of the bromine quadrupole-coupling tensor that deviate by less than 0.4% from the corresponding experimental numbers.

Spin–Rotation Interaction Among the magnetic parameters, spin–rotation constants have a particular relevance as, in addition to the major part they have in the hyperfine structure of rotational spectra, their measurement plays an important role for the experimental determination of absolute nuclear magnetic shielding constants [130] and, thus, for establishing corresponding absolute NMR scales [36, 131–134]. In fact, accordingly to Ramsey's formulation [135], the NMR shielding tensor σ can be decomposed into two contributions: the paramagnetic contribution (σ^p) and the diamagnetic one (σ^d), where the former can be determined from the spin–rotation tensor [33]. Additionally, while σ^d, which depends on the ground state of the molecule, can be accurately calculated by means of standard quantum chemical methods, σ^p is more difficult to calculate accurately. Therefore, when establishing absolute magnetic shielding scale, highly precise experimental determinations of spin–rotation constants are employed for evaluating σ^p. We refer interested readers to Puzzarini et al. [80] for a significant example concerning ^{17}O.

The investigation of the hyperfine structure of the rotational spectrum of $H^{13}CN$ by means of the Lamb-dip technique [136] provides a good example of how theory may also supply missing experimental data. In detail, it turned out to be essential to fix in

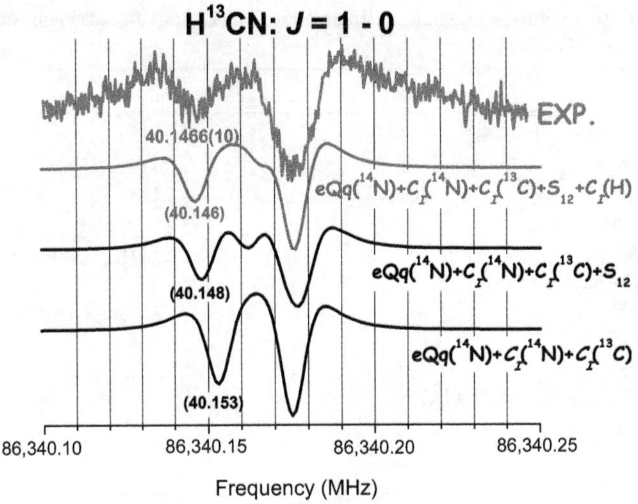

Figure 6.9 A portion of the $J=1 \leftarrow 0$ transition (the $\Delta F_N = 2 \leftarrow 1$ components) is shown. In addition to the experimental spectrum (top), the contributions of each hyperfine parameter to the hyperfine pattern are also depicted (from the bottom): only $eQq(^{14}\mathrm{N}) + C_I(^{14}\mathrm{N}) + C_I(^{13}\mathrm{C})$; $eQ_q(^{14}\mathrm{N}) + C_I(^{14}\mathrm{N}) + C_I(^{13}\mathrm{C}) + S_{12}$; $eQ_q(^{14}\mathrm{N}) + C_I(^{14}\mathrm{N}) + C_I(^{13}\mathrm{C}) + S_{12} + C_I(\mathrm{H})$ [136].

the fitting procedure the spin–rotation constant of hydrogen at the theoretical value [137] for two reasons. First, the inclusion or not of $C_I(\mathrm{H})$ affects the derived value of the spin–spin interaction constant $S_{\mathrm{H}-^{13}\mathrm{C}}$ (S_{12} in Figure 6.9), while all the other parameters are completely unaffected. In fact, only when $C_I(\mathrm{H})$ is included in the fit is a reliable value (i.e., in agreement with theoretical predictions) for $S_{\mathrm{H}-^{13}\mathrm{C}}$ obtained. The second reason is graphically explained in Figure 6.9: $C_I(\mathrm{H})$ is required to correctly reproduce the hyperfine structure of the $J=1 \leftarrow 0$ rotational transition.

Importance of Vibrational Corrections Of great relevance in this context are the results collected in Table 6.9 for the oxygen spin–rotation tensor of $H_2{}^{17}O$ [80]. In fact, the vibrational corrections turned out to be unusually large, that is, ~28% for the C_{aa} component, ~11% for C_{bb}, and ~6% for C_{cc}. It is thus clear that vibrational corrections cannot be neglected even for qualitative predictions in such a case. For example, only proper consideration of vibrational effects leads to the correct prediction that the C_{aa} and C_{bb} components of the oxygen spin–rotation tensor are nearly degenerate for the vibrational ground state of $H_2{}^{17}O$. Concerning the comparison of the perturbative (VPT2) and variational (discrete variable representation, DVR, with quasi-analytic treatment of kinetic energy, DVR-QAK) approaches for the treatment of vibrational corrections, we note that both schemes provide the same results within 2–7%. Therefore, it can be concluded that the perturbative treatment is suitable for evaluating vibrational corrections even in cases where they are rather large. For further discussions, we refer readers to Puzzarini et al. [80].

Table 6.9 Equilibrium Values and Vibrational Corrections for Oxygen Spin–Rotation Tensor (kHz) of $H_2{}^{17}O$ [80]

Basis Set	C_{aa}	C_{bb}	C_{cc}
Equilibrium Values[a]			
aug-cc-pCVTZ	−22.14	−24.67	−17.19
aug-cc-pCVQZ	−22.15	−24.99	−17.36
aug-cc-pCV5Z	−22.24	−25.15	−17.45
aug-cc-pCV6Z	−22.25	−25.20	−17.48
Vibrational Corrections (VPT2)			
aug-cc-pCVTZ	−5.88	−2.73	−0.97
aug-cc-pCVQZ	−5.90	−2.72	−0.98
aug-cc-pCV5Z	−5.90	−2.73	−0.98
Vibrational Corrections (DVR-QAK)			
aug-cc-pCVTZ	−6.33	−2.78	−1.01
aug-cc-pCVQZ	−6.35	−2.78	−1.02
aug-cc-pCV5Z	−6.35	−2.78	−1.02

[a]Computed at the semiexperimental equilibrium geometry.

6.5 PERSPECTIVES: OPEN-SHELL SPECIES

The treatments and examples provided so far only concern closed-shell molecules. The situation appears more involved when open-shell species are considered. In fact, in addition to the nuclear rotational angular momentum $\hat{\mathbf{R}}$, the electronic angular momentum $\hat{\mathbf{L}}$ (quantum number: Λ) and the electronic spin angular momentum $\hat{\mathbf{S}}$ (quantum number: Σ) contribute to the total angular momentum $\hat{\mathbf{J}}$:

$$\hat{\mathbf{J}} = \hat{\mathbf{R}} + \hat{\mathbf{L}} + \hat{\mathbf{S}} \tag{6.64}$$

while in the treatments of the rotational problems for closed-shell species, it is essentially assumed that $\hat{\mathbf{J}} = \hat{\mathbf{R}}$. As a consequence, further interactions must be accounted for in the rotational Hamiltonian; more specifically, these are the spin–orbit, electron spin–rotation and electron spin–nuclear spin interactions. We report below only a brief account, and we refer interested readers to the specialized literatures [e.g., 1–3].

Spin-Orbit Coupling Associated with the resultant electron spin angular moment is a magnetic moment which can interact with the magnetic field generated by the circulating charge due to a nonnull orbital angular momentum, thus leading to the so-called spin-orbit interaction. This is described by the Hamiltonian

$$\hat{H}_{\mathrm{SO}} = A\hat{\mathbf{L}} \cdot \hat{\mathbf{S}} \tag{6.65}$$

where A is the spin–orbit coupling tensor.

Spin-Rotation Interaction Another type of interaction involves the coupling between the rotational angular momentum and the electronic spin angular momentum. Such an interaction is qualitatively due to the fact that electrons do not follow the molecular frame exactly and consequently rotationally dependent magnetic fields are generated leading to the following term in the Hamiltonian

$$\hat{H}_{\text{fs}} = \gamma \hat{\mathbf{R}} \cdot \hat{\mathbf{S}} \tag{6.66}$$

where γ is the electron spin–rotation coupling tensor.

Spin–Spin Interaction In states where $S \geq 1$, in addition to the spin–rotation term, spin–spin coupling terms arise, but they are less important in the field of rotational spectroscopy.

6.5.1 Hund's Coupling Cases

The importance of the various interactions mentioned above is then reflected in the resulting energy-level pattern. Hund provided a classification according to the magnitude of the corresponding splittings with respect to rotational spacing. Such a classification then led to the so-called Hund's coupling cases, which are summarized in Table 6.10 [2].

As evident from Table 6.10, there is an initial classification into states for which the various Λ components are well separated in energy with respect to BJ and cases in which the interaction is so weak that the separations are small and the states nearly degenerate. Then, there is a further subclassification according to the extent of the spin–orbit coupling; once again this criterion is with respect to BJ. For further details and analyses we refer the reader to the literature [e.g., 1, 2]. We only report below some details concerning the rotational spectrum of molecules that can be well described by the Hund's case *b*, as this is by far relevant in rotational spectroscopy.

Table 6.10 Summary of Hund's Coupling Cases[a] [2]

Coupling of $\hat{\mathbf{L}}$ to Axis	Spin–Orbit Coupling[b]
Strong: states with different $\lvert\Lambda\rvert$ well separated	*Case a:* Strong. States with different $\lvert\Omega\rvert$ well separated
Strong: states with different $\lvert\Lambda\rvert$ well separated	*Case b:* Weak. States with different $\lvert\Omega\rvert$ close
Weaker: coupling weaker than spin–orbit coupling	*Case c:* Strong. States with different $\lvert\Omega\rvert$ well separated
Weak: states with different $\lvert\Lambda\rvert$ close	*Case d:* Weak. States with different $\lvert\Omega\rvert$ close

[a] The criteria of well separated and close are defined by $\Delta E \gg BJ$ and $\Delta E \ll BJ$, respectively.
[b] $\Omega = \Lambda + \Sigma$.

6.5.2 Rotational Spectrum for Hund's Case *b* Limit

As an example, let us consider the case of a doublet nondegenerate electronic ground state (such as $^2\Sigma$, ^{2}B, ...). In such a case, there are is no strong spin–orbit coupling. Therefore, this type of molecule approximates closely Hund's case *b* [138].

The rotational Hamiltonian consists of three contributions:

$$\hat{H} = \hat{H}_{\text{rot}} + \hat{H}_{\text{fs}} + \hat{H}_{\text{hfs}} \tag{6.67}$$

where $\hat{H}_{\text{rot}}$ accounts for the pure rotational part as seen in the first section of this chapter (Eq. 6.15), and $\hat{H}_{\text{fs}}$ is the Hamiltonian accounting for the electron spin–rotation interaction, previously described by Eq. 6.66. The hyperfine-structure Hamiltonian can be separated in two parts:

$$\hat{H}_{\text{hfs}} = \hat{H}_{\text{esns}} + \hat{H}_{\text{nsr}} \tag{6.68}$$

where the first term describes the isotropic and anisotropic coupling of electron spin with nuclear spins, while the second term describes the nuclear–spin interaction, which is the same as in Eq. 6.31. If one or more quadrupolar nuclei are present, then the Hamiltonian of Eq. 6.67 also contains the nuclear quadrupole-coupling Hamiltonian (Eq. 6.27). Furthermore, $\hat{H}_{\text{hfs}}$ may also include the $\hat{H}_{\text{ss}}$ of Eq. 6.33, which accounts for the dipolar nuclear spin–nuclear spin interaction. From a quantum chemical point of view, the quantities that have not been introduced so far are related to the $\hat{H}_{\text{esns}}$ and the $\hat{H}_{\text{fs}}$ terms. These are briefly introduced in the following.

Concerning the $\hat{H}_{\text{esns}}$ Hamiltonian, the quantities that need to be computed are the isotropic and anisotropic spin–spin coupling tensor. The isotropic contribution, the so-called Fermi contact term, is related to the spin density at the nucleus K under consideration [139],

$$A_0^{(K)} = \frac{8\pi}{3}\frac{g_e}{g_0} g_K \beta_K \sum_{\mu\nu} P_{\mu\nu}^{\alpha-\beta} \langle \phi_\beta | \delta(r_{nK}) | \phi_\nu \rangle \tag{6.69}$$

where g_0 and g_e are the g values of the electron in the radical and the free electron, respectively, g_K is the nuclear g factor, and β_K the nuclear magneton; $P^{\alpha-\beta}$ is the difference between the density matrices for electrons with α and β spins, that is, the spin density matrix. The anisotropic contribution, denoted as dipolar hyperfine coupling, can be derived from the classical expression of interacting dipoles [140]

$$A_{ij}^{(K)} = \frac{8\pi}{3}\frac{g_e}{g_0} g_K \beta_K \sum_{\mu\nu} P_{\mu\nu}^{\alpha-\beta} \langle \phi_\mu | r_{nK}^{-5} (r_{nK}^2 \delta_{ij} - 3 r_{nK,i} r_{nK,j}) | \phi_\nu \rangle \tag{6.70}$$

Therefore, the essential quantities to be calculated are the spin density at the K nucleus and the dipole–dipole coupling terms, respectively. The computational requirements for them have been largely investigated in the literature and we refer interested readers there [141–143]. We only briefly recall that the CCSD(T) level of theory in conjunction with medium to large basis sets, including tight functions, is able to provide quantitative predictions [e.g., 141, 143, 144].

The $\hat{H}_{fs}$ Hamiltonian requires the evaluation of the electron spin–rotation coupling tensor. Curl [145] derived an approximate relation between the latter and the electronic g tensor:

$$\gamma_{\alpha\alpha} = -2B_e^{\alpha}\Delta g_{\alpha\alpha} \tag{6.71}$$

where α runs over the inertial axes and $\Delta g_{\alpha\alpha}$ is the transversal component of the g-shift (i.e., different between the g tensor and the free–electron g value: $\Delta g_{ij} = g_{ij} - g_e\delta_{ij}$). Computationally, the electronic g tensor is a second-order property and can be calculated as

$$\mathbf{g} = \frac{1}{\mu_B}\left(\frac{\partial^2 E}{\partial \mathbf{B}\partial \mathbf{S}}\right)_{\mathbf{B},\mathbf{S}=0} \tag{6.72}$$

where **B** denotes an external magnetic field and **S** the electronic spin. Then, in view of its computation, the g tensor itself can be decomposed in various contributions [e.g., 146–148]. Calculations of the electronic g tensor are more involved; as a consequence, only recently the first CC implementation appeared [148]. For details concerning the computational requirements we refer interested readers to Gauss et al. [148] and references therein.

Once the rotational energies are computed, the rotational spectrum is obtained from the selection rules for electric dipole transitions:

$$\Delta N = \pm 1 \qquad \Delta J = 0, \ \pm 1$$

To illustrate the predictive capabilities of quantum chemistry in the field of rotational spectroscopy of open-shell species, in Table 6.11 the comparison between experimental and calculated isotropic and anisotropic hyperfine coupling constants of NH_2 is reported. The computed values have been obtained at the CCSD(T) level employing the aug-cc-pCVQZ basis set augmented by additional tight functions on H [143]. Corrections due to full-treatment of triples and quadruples have also been

Table 6.11 Isotropic and Anisotropic Hyperfine Coupling Constants (MHz) of NH_2[a]

	Experiment[b]	Theory
a_F (H)	−67.182(108)	−68.19
T_{aa}(H)	18.217(174)	19.02
T_{bb}(H)	−13.05(27)	−14.11
a_F (N)	28.061(52)	27.44
T_{aa}(N)	−43.035(91)	−42.40
T_{bb}(N)	−44.630(118)	−43.10

[a]The symbol convention of the experimental paper has been employed: a_F and $T_{\alpha\alpha}$ stand for the isotropic and anisotropic hyperfine coupling constants, respectively.
[b]From ref. 149.

accounted for (for details see ref. [143]). We only note that a nearly quantitative agreement can be obtained even for these more challenging parameters.

6.6 CONCLUSIONS

It has been shown that quantum chemical calculations are nowadays able to provide very accurate predictions for the molecular and spectroscopic properties related to rotational spectroscopy and the computational requirements have been discussed in some detail. To sum up, it has been shown that, for predicting rotational spectra, the information required are accurate estimates of:

(a) Rotational parameters
(b) Type of transitions observable (and their intensity)
(c) Fine and hyperfine parameters

The corresponding required calculations have been presented; that is, to fulfil point (a) accurate equilibrium structures as well as harmonic and anharmonic force field computations are necessary, for point (b) accurate dipole moment evaluations are needed, and for point (c) electric field gradient, spin–rotation and spin–spin tensor calculations are required. Furthermore, for meaningful predictions and/or comparisons to experiment, the vibrational corrections related to points (a) to (c) are also required.

Then, through a few selected examples, the predictive capabilities of quantum chemical computations have also been pointed out: They take a fundamental role in the field of high-resolution spectroscopy by guiding, supporting, and/or challenging the experimental determinations.

ACKNOWLEDGMENTS

The author acknowledges fruitful collaborations as well as discussions with the following colleagues: A. Baldacci (Venezia, Italy), V. Barone (Pisa, Italy), L. Bizzocchi (Bologna, Italy), G. Cazzoli (Bologna, Italy), S. Coriani (Trieste, Italy), C. Degli Esposti (Bologna, Italy), L. Dore (Bologna, Italy), A. Gambi (Udine, Italy), J. Gauss (Mainz, Germany), M. E. Harding (Mainz, Germany, and Austin, Texas), M. Heckert (Mainz, Germany), T. Metzroth (Mainz, Germany), A. Rizzo (Pisa, Italy), S. Stopkowicz (Mainz, Germany), J. F. Stanton (Austin, Texas), F. Temps (Kiel, Germany), and J. Vazquez (Austin, Texas). This work has been supported by RFO funds (University of Bologna).

REFERENCES

1. W. Gordy, R. L. Cook, in *Microwave Molecular Spectra*, 3rd ed., A. Weissberger, Ed., Wiley, New York, 1984.
2. H. W. Kroto, *Molecular Rotation Spectra*, Wiley, London, 1975.

3. C. H. Townes, A. L. Schawlov, *Microwave Spectroscopy*, McGraw-Hill, New York, 1955.
4. G. Winnewisser, F. Lewen, S. Thorwirth, M. Behnke, J. Hahn, J. Gauss, E. Herbst, *Chem. Eur. J.* **2003**, *9*, 5501.
5. L. Sari, M. C. McCarthy, H. F. Schaefer, III, P. Thaddeus, *J. Am. Chem. Soc.* **2003**, *125*, 11409.
6. A. J. Apponi, M. C. McCarthy, C. A. Gottlieb, P. Thaddeus, *J. Chem. Phys.* **1999**, *111*, 3911.
7. M. C. McCarthy, A. J. Apponi, P. Thaddeus, *J. Chem. Phys.* **1999**, *111*, 7175.
8. M. C. McCarthy, S. Thorwirth, C. A. Gottlieb, P. Thaddeus, *J. Am. Chem. Soc.* **2004**, *126*, 4096.
9. M. J. Travers, M. C. McCarthy, C. A. Gottlieb, P. Thaddeus, *Astrophys. J.* **1997**, *483*, L135.
10. S. Thorwirth, J. Gauss, M. C. McCarthy, F. Shindo, P. Thaddeus, *Chem. Common* **2008**, 5292.
11. L. Bizzocchi, C. Degli Esposti, P. Botschwina, *J. Chem. Phys.* **2003**, *119*, 170.
12. M. C. McCarthy, A. J. Apponi, V. D. Gordon, C. A. Gottlieb, P. Thaddeus, T. D. Crawford, J. F. Stanton, *J. Chem. Phys.* **1999**, *111*, 6750.
13. K. Suma, Y. Sumiyoshi, Y. Endo, *Science* **2005** *308*, 1885.
14. K. Suma, Y. Sumiyoshi, Y. Endo, *J. Am. Chem. Soc.* **2005**, *127*, 14998.
15. M. Koerber, O. Baum, J. Hahn, T. F. Giesen, S. Schlemmer, J. Gauss, *Inorg. Chem.* **2009**, *48*, 2269.
16. G. Cazzoli, C. Puzzarini, A. Gambi, J. Gauss, *J. Chem. Phys.* **2006**, *125*, 054313.
17. R. J. McMahon, M. C. McCarthy, C. A. Gottlieb, J. B. Dudek, J. F. Stanton, P. Thaddeus, *Astrophys. J.* **2003**, *590*, L61.
18. S. Brünken, M. C. McCarthy, P. Thaddeus, P. D. Godfrey, R. D. Brown, *Astron. Astrophys.* **2006**, *459*, 317.
19. M. C. McCarthy, C. A. Gottlieb, H. Gupta, P. Thaddeus, *Astrophys. J.* **2006**, *652*, L141.
20. S. Brünken, H. Gupta, C. A. Gottlieb, M. C. McCarthy, P. Thaddeus, *Astrophys. J.* **2007**, *664*, L43.
21. M. Born, J. R. Oppenheimer, *Ann. Phys.* **1927**, *84*, 457.
22. P. R. Bunker, P. Jensen, *Molecular Symmetry and Spectroscopy*, NRC Reserach Press, Ottawa, 1998, p. 211.
23. J. K. G. Watson, in *Vibrational Spectra and Structure*, Vol. 6, J. During, Ed., Elsevier, Amsterdam, 1997, p. 1.
24. J. K. G. Watson, *J. Chem. Phys.* **1967**, *46*, 1935.
25. J. K. G. Watson, *J. Chem. Phys.* **1968**, *48*, 181.
26. J. K. G. Watson, *J. Chem. Phys.* **1968**, *48*, 4517.
27. M. R. Aliev, J. K. G. Watson, in *Molecular Spectroscopy: Modern Research*, Vol. III, K. N. Rao, Ed., Academic, New York, 1985, pp. 1–67.
28. J. K. G. Watson, *Mol. Phys.* **1968**, *15*, 479.
29. J. K. G. Watson, *Mol. Phys.* **1970**, *19*, 465.
30. H. Meyer, *Ann. Rev.* **2002**, *53*, 141.
31. I. A. Mills, in *Modern Spectroscopy: Modern Research*, K. N. Rao, C. W. Matthews, Eds., Academic, New York, 1972, p. 115.

32. E. A. C. Lucken, *Nuclear Quadrupole Coupling Constants*, Academic, London/New York, 1969.
33. W. H. Flygare, *J. Chem. Phys.* **1964**, *41*, 793.
34. A. Abragam, *Principles of Nuclear Magnetism*, Oxford University Press, New York, 1961, p. 97.
35. K. Schmidt-Rohr, H.W. Spiess, *Multidimensional Solid State NMR and Polymers*, Academic, New York, 1994, p. 19.
36. D. L. Bryce, R. E. Wasylishen, *Acc. Chem. Res.* **2003**, *36*, 327.
37. D. Papousek, M. R. Aliev, *Molecular Vibrational-Rotational Spectra*, Elsevier, New York, 1982.
38. J. Gauss, K. Ruud, T. Helgaker, *J. Chem. Phys.* **1996**, *105*, 2804.
39. J. Gauss, D. Sundholm, *Mol. Phys.* **1997**, *91*, 449.
40. A. A. Auer, J. Gauss, J. F. Stanton, *J. Chem. Phys.* **2003**, *118*, 10407.
41. K. Ruud, *Calculation of NMR and EPR Parameters: Theory and Application*, M. Kaupp, M. Buhl, V. G. Malkin, Eds., Wiley, Weinheim, 2004, Chapter 10.
42. K. Ruud, P.-O. Åstrand, P. R. Taylor, *J. Chem. Phys.* **2000**, *112*, 2668.
43. P.-O. Åstrand, K. Ruud, P. R. Taylor, *J. Chem. Phys.* **2000**, *112*, 2655.
44. C. E. Dykstra, D. J. Malik, *J. Chem. Phys.* **1987**, *87*, 2806.
45. D. Sundholm, J. Gauss, A. Schäfer, *J. Chem. Phys.* **1996**, *105*, 11051.
46. J. Vázquez, M. E. Harding, J. F. Stanton, J. Gauss, *J. Chem. Theory Comput.* **2011**, *7*, 1428.
47. R. J. Bartlett, *Modern Electronic Structure Theory*, D. R. Yarkony, Ed., World Scientific, Singapore, 1995, p. 1047 ff.
48. K. Raghavachari, G. W. Trucks, J. A. Pople, M. Head-Gordon, *Chem. Phys. Lett.* **1989**, *157*, 479.
49. J. Noga, R. J. Bartlett, *J. Chem. Phys.* **1987**, *86*, 7041.
50. G. E. Scuseria, H. F. Schaefer III, *Chem. Phys. Lett.* **1988**, *152*, 382.
51. J. D. Watts, R. J. Bartlett, *J. Chem. Phys.* **1993**, *93*, 6104.
52. M. Kallay, P. R. Surján, *J. Chem. Phys.* **2001**, *115*, 2945.
53. M. Heckert, M. Kállay, D. P. Tew, W. Klopper, J. Gauss, *J. Chem. Phys.* **2006**, *125*, 044108.
54. M. Heckert, M. Kállay, J. Gauss, *Mol. Phys.* **2005**, *103*, 2109.
55. C. Puzzarini, *J. Phys. Chem.* **2009**, *113A*, 14530.
56. C. Puzzarini, V. Barone, *Phys. Chem. Chem. Phys.* **2009**, *11*, 11463.
57. C. Puzzarini, *Int. J. Quant. Chem.* **2010**, *110*, 2483.
58. D. Feller, *J. Chem. Phys.* **1993**, *98*, 7059.
59. T. Helgaker, W. Klopper, H. Koch, J. Noga, *J. Chem. Phys.* **1997**, *106*, 9639.
60. T. H. Dunning, Jr., *J. Chem. Phys.* **1989**, *90*, 1007.
61. A. K. Wilson, T. van Mourik, T. H. Dunning, Jr., *J. Mol. Struct. (THEOCHEM)* **1996**, *388*, 339.
62. D. E. Woon, T. H. Dunning, Jr., *J. Chem. Phys.* **1995**, *103*, 4572.
63. K. Peterson, T. H. Dunning, Jr., *J. Chem. Phys.* **2002**, *117*, 10548.
64. W. Schneider, W. Thiel, *Chem. Phys. Lett.* **1989**, *157*, 367.

65. J. F. Stanton, J. Gauss, *Int. Rev. Phys. Chem.* **2000**, *19*, 61.
66. J. F. Stanton, C. L. Lopreore, J. Gauss, *J. Chem. Phys.* **1998**, *108*, 7190.
67. C. Møller, M. S. Plesset, *Phys. Rev.* **1934**, *46*, 618.
68. K. L. Bak, J. Gauss, T. Helgaker, P. Jørgensen, J. Olsen, *Chem. Phys. Lett.* **2000**, *319*, 563.
69. A. Halkier, H. Larsen, J. Olsen, P. Jørgensen, J. Gauss, *J. Chem. Phys.* **1999**, *110*, 734.
70. A. Halkier, W. Klopper, T. Helgaker, P. Jørgensen, *J. Chem. Phys.* **1999**, *111*, 4424.
71. A. Halkier, H. Koch, O. Christiansen, P. Jørgensen, T. Helgaker, *J. Chem. Phys.* **1997**, *107*, 849.
72. L. Olsen, O. Christiansen, L. Hemmingsen, S. P. A. Sauer, K. V. Mikkelsen, *J. Chem. Phys.* **2002**, *116*, 1424.
73. M. Pernpointner, M. Seth, P. Schwerdtfeger, *J. Chem. Phys.* **1998**, *108*, 6722.
74. S. Stopkowicz, J. Gauss, *J. Chem. Phys.* **2008**, *129*, 164119.
75. G. Cazzoli, C. Puzzarini, S. Stopkowicz, J. Gauss, *Mol. Phys.* **2008**, *106*, 1181.
76. K. G. Dyall, K. Fægri, *Relativistic Quantum Chemistry*, Oxford University Press, New York, 2007.
77. W. Kutzelnigg, in *Relativistic Electronic Structure Theory, Part 1: Fundamentals*, P. Schwerdtfeger, Ed., Elsevier, Amsterdam, 2002, p. 664 ff.
78. A. Wolf, M. Reiher, B. A. Hess, in *Relativistic Electronic Structure Theory, Part 1: Fundamentals*, P. Schwerdtfeger, Ed., Elsevier, Amsterdam, 2002, p. 622 ff.
79. C. Puzzarini, J. F. Stanton, J. Gauss, *Int. Rev. Phys. Chem.* **2010**, *29*, 273.
80. C. Puzzarini, G. Cazzoli, M. E. Harding, J. Vázquez, J. Gauss, *J. Chem. Phys.* **2009**, *131*, 234304.
81. C. Puzzarini, T. Metzroth, J. Gauss, to be published.
82. T. Enevoldsen, J. Oddershede, S. P. A. Sauer, *Theor. Chem. Acc.* **1998**, *100*, 275.
83. O. Vahtras, H. Ågren, P. Jørgensen, H. J. A. Jensen, S. B. Padkjaer, T. Helgaker, *J. Chem. Phys.* **1992**, *96*, 6120.
84. S. A. Perera, M. Nooijen, R. J. Bartlett, *J. Chem. Phys.* **1996**, *104*, 3290.
85. A. A. Auer, J. Gauss, *J. Chem. Phys.* **2001**, *115*, 1619.
86. D. L. Bryce, R. E. Wasylishen, J. Autschbach, T. Ziegler, *J. Am. Chem. Soc.* **2004**, *124*, 4894.
87. C. Puzzarini, G. Cazzoli, A. Baldacci, A. Baldan, C. Michauk, J. Gauss, *J. Chem. Phys.* **2007**, *127*, 164302.
88. G. Cazzoli, C. Puzzarini, M. E. Harding, J. Gauss, *Chem. Phys. Lett.* **2009**, *473*, 21.
89. C. Puzzarini, M. Heckert, J. Gauss, *J. Chem. Phys.* **2008**, *128*, 194108.
90. C. Puzzarini, S. Coriani, A. Rizzo, J. Gauss, *Chem. Phys. Lett.* **2005**, *409*, 118.
91. N. Hansen, H. Mäder, F. Temps, *Chem. Phys. Lett.* **2000**, *327*, 97.
92. N. Hansen, H. Mäder, F. Temps, *Phys. Chem. Chem. Phys.* **2001**, *3*, 50.
93. N. Hansen, H. Mäder, F. Temps, *Chem. Phys. Lett.* **2005**, *409*, 149.
94. L. Bizzocchi, C. Degli Esposti, P. Botschwina, *J. Chem. Phys.* **2003**, *119*, 170.
95. G. Cazzoli, C. Puzzarini, A. Gambi, *J. Chem. Phys.* **2004**, *120*, 6495.
96. P. Pulay, W. Meyer, J. E. Boggs, *J. Chem. Phys.* **1978**, *68*, 5077.
97. A. L. L. East, C. S. Johnson, W. D. Allen, *J. Chem. Phys.* **1993**, *98*, 1299.
98. A. L. L. East, W. D. Allen, S. J. Klippenstein, *J. Chem. Phys.* **1995**, *102*, 850.

99. W. D. Allen, A. L. L. East, A. G. Csaszar, in *Structures and Conformations of Nonrigid Molecules*, J. Laane, M. Dakkouri, B. van der Veken, H. Oberhammer, Eds., Kluwer, Dordrecht, 1993, p. 343.
100. M. C. McCarthy, C. A. Gottlieb, P. Thaddeus, M. Horn, P. Botschwina, *J. Chem. Phys.* **1995**, *103*, 7820.
101. C. Bartel, P. Botschwina, H. Bürger, A. Guarnieri, Ä. Heyl, A. Huckauf, D. Lentz, T. Merzliak, E. B. Mkdami, *Angew. Chem.* **1998**, *110*, 3036.
102. A. A. Auer, J. Gauss, *Phys. Chem. Chem. Phys.* **2001**, *3*, 3001.
103. J. Gauss, J. F. Stanton, *J. Phys. Chem. A* **2000**, *104*, 1319.
104. J. Gauss, J. F. Stanton, *J. Phys. Chem. A* **2000**, *104*, 2865.
105. F. Pawlowski, P. Jørgensen, J. Olsen, F. Hegelund, T. Helgaker, J. Gauss, K. L. Bak, J. F. Stanton, *J. Chem. Phys.* **2002**, *116*, 6482.
106. R. W. Larsen, F. Pawlowski, F. Hegelund, P. Jørgensen, J. Gauss, B. Nelander, *Phys. Chem. Chem. Phys.* **2003**, *5*, 5031.
107. P. G. Szalay, L. S. Thøgersen, J. Olsen, M. Kállay, J. Gauss, *J. Phys. Chem. A* **2004**, *108*, 3030.
108. J. F. Stanton, *Mol. Phys.* **1999**, *97*, 841.
109. J. Demaison, *Mol. Phys.* **2007**, *105*, 3109.
110. N. C. Craig, P. Groner, D. C. McKean, *J. Phys. Chem. A* **2006**, *110*, 7461.
111. N. C. Craig, D. Feller, P. Groner, H. Y. Hsin, D. C. McKean, D. J. Nemchick, *J. Phys. Chem. A* **2007**, *111*, 2498.
112. P. Groner, R. D. Warren, *J. Mol. Struct.* **2001**, *599*, 323.
113. P. Groner, S. G. Kukolich, *J. Mol. Struct.* **2006**, *780–781*, 178.
114. S. Thorwirth, M.E. Harding, D. Muders, J. Gauss, *J. Mol. Spectrosc.* **2008**, *251*, 220.
115. R. J. McMahon, R. J. Halter, R. L. Fimmen, R. J. Wilson, S. A. Peebles, R. L. Kuczkowski, J. F. Stanton, *J. Am. Chem. Soc.* **2000**, *122*, 939.
116. R. J. Halter, R. L. Fimmen, R. J. McMahon, S. A. Peebles, R. L. Kuczkowski, J. F. Stanton, *J. Am. Chem. Soc.* **2001**, *123*, 12353.
117. J. Vázquez, J. F. Stanton, *J. Phys. Chem. A* **2002**, *106*, 4429.
118. O. Baum, S. Esser, N. Gierse, S. Brünken, F. Lewen, J. Hahn, J. Gauss, S. Schlemmer, T. F. Giesen, *J. Mol. Struct.* **2006**, *795*, 256.
119. J. F. Stanton, J. Gauss, O. Christiansen, *J. Chem. Phys.* **2001**, *114*, 2993.
120. C. Puzzarini, G. Cazzoli, A. Gambi, J. Gauss, *J. Chem. Phys.* **2006**, *125*, 054307.
121. S. Thorwirth, M. E. Harding, D. Muders, J. Gauss, *J. Mol. Spectrosc.* **2008**, *251*, 220.
122. P. Botschwina, C. Puzzarini, *J. Mol. Spectrosc.* **2001**, *208*, 292.
123. G. Cazzoli, L. Cludi, M. Contento, C. Puzzarini, *J. Mol. Spectrosc.* **2008**, *251*, 229.
124. L. Bizzocchi, C. Degli Esposti, P. Botschwina, *J. Mol. Spectrosc.* **2004**, *225*, 145.
125. G. Cazzoli, C. Puzzarini, A. Baldacci, A. Baldan, *J. Mol. Spectrosc.* **2007**, *241*, 112.
126. P. Pyykkö, *Z. Naturforsch.* **1992**, *47a*, 189.
127. P. Pyykkö, *Mol. Phys.* **2001**, *99*, 1617.
128. P. Pyykkö, *Mol. Phys.* **2008**, *106*, 1965.
129. J. Gauss, M. E. Harding, S. Stopkowicz, C. Puzzarini, to be published.
130. W. H. Flygare, *Chem. Rev.* **1974**, *74*, 653.

131. C. J. Jameson, *Chem. Rev.* **1991**, *91*, 1375.
132. R. E. Wasylishen, D. L. Bryce, *J. Chem. Phys.* **2002**, *117*, 10061.
133. M. Gee, R. E. Wasylishen, A. Laaksonen, *J. Phys. Chem. A* **1999**, *103*, 10805.
134. D. Sundholm, J. Gauss, A. Schäfer, *J. Chem. Phys.* **1996**, *105*, 11051.
135. N. F. Ramsey, *Phys. Rev.* **1950**, *78*, 699.
136. G. Cazzoli, C. Puzzarini, *J. Mol. Spectrosc.* **2005**, *233*, 280.
137. G. Cazzoli, C. Puzzarini, J. Gauss, *Astrophys. J. Suppl.* **2005**, *159*, 181.
138. E. L. Hill, J. H. Van Vleck, *Phys. Rev.* **1928**, *32*, 250.
139. E. Fermi, *Z. Phys.* **1930**, *60*, 320.
140. A. Frosch H. M. Foley, *Phys. Rev.* **1952**, *88*, 1337.
141. A. R. Al Derzi, S. Fau, R. J. Bartlett, *J. Phys. Chem.* **2003**, *107A*, 6656.
142. R. Improta, V. Barone, *Chem. Rev.* **2004**, *104*, 1231.
143. C. Puzzarini, V. Barone, *J. Chem. Phys.* **2010**, *133*, 184301.
144. C. Puzzarini, V. Barone, *J. Chem. Phys.* **2008**, *129*, 084306.
145. R. F. Curl, *Mol. Phys.* **1965**, *9*, 585.
146. M. Engström, B. Minaev, O. Vahtras, H. Ågren, *Chem. Phys.* **1998**, *237*, 149.
147. F. Neese, *Mol. Phys.* **2007**, *105*, 2507.
148. J. Gauss, M. Kállay, F. Neese, *J. Phys. Chem.* **2009**, *113A*, 11541.
149. M. Tonooka, S. Yamamoto, K. Kobayashi, S. Saito, *J. Chem. Phys.* **1997**, *106*, 2563.

7

TIME-INDEPENDENT APPROACH TO VIBRATIONAL SPECTROSCOPIES

CHIARA CAPPELLI

Dipartimento di Chimica e Chimica Industriale, Università di Pisa, Pisa, Italy

MALGORZATA BICZYSKO

Scuola Normale Superiore, Pisa, Italy and Dipartimento di Chimica "Paolo Corradini," Università di Napoli Federico II, Naples, Italy

Computational Strategies for Spectroscopy: From Small Molecules to Nano Systems, First Edition.
Edited by Vincenzo Barone.

In this chapter, the time-independent approach to the calculation of vibrational spectra by means of quantum mechanical methodologies is presented. The general aspects of the modeling of vibrational spectra for isolated systems are summarized by first resorting to the double-harmonic approximation and by further discussing the inclusion of anharmonic effects through perturbation theory. The complications arising as a result of the presence of a surrounding solvating environment are evidenced and the continuum solvation approach to the matter is presented, from both the classical and quantum mechanical points of view. Few key examples are discussed in order to illustrate the overall performance of the methodologies and to give a general feeling of the state-of-the art of the field to date.

7.1 INTRODUCTION

Vibrational spectra depend on the molecular composition and bond topology so that fine details extracted from the large wealth of information enclosed in the spectrum permit us to characterize the molecule in terms of conformation, chemical linkage, and mutual interactions among atoms and atomic charges modulated by the intrinsic temperature and environmental effects. For such reasons, vibrational spectroscopies employing natural (IR, Raman) or, more recently, polarized [vibrational circular dichroism (VCD), Raman optical activity (ROA)] radiation are among the most powerful techniques for characterizing the structure and dynamical behavior of molecular systems over a wide range of dimensions and lifetimes. However, the proper assignment of spectra, which is necessary to fully understand the physical–chemical properties of the system under study, is often not straightforward, especially for unstable species or nonstandard bonding patterns. In the last years, computation of harmonic force fields by quantum mechanical (QM) approaches has proven to be of ever-increasing aid in this connection, in particular due to the development of more reliable models with good scaling properties and the corresponding user-friendly computer codes. Traditionally, computations of vibrational properties were set within the so-called double-harmonic approximation (vide infra) but, more recently, the quest for an improved precision of the computed parameters, especially for unstable species, is requiring effective models beyond the harmonic approximation. For small molecules, converged rovibrational levels can be obtained by fully variational methods. However, for large molecules, some level of approximation is unavoidable, concerning both the form of the potential and the rovibrational treatment. Both time-dependent (which are analyzed in Chapter 11 of this book) and time-independent routes can be pursued to face this problem. The most successful time-independent approaches are at present based on truncated two- or three-mode potentials and self-consistent (VSCF) and/or second-order perturbative (VPT2) vibrational treatments. In particular, second-order perturbation theory, which provides closed expressions for most of the spectroscopic parameters required for the analysis of the experimental frequencies, still appears very effective for the study of medium-sized polyatomic molecules. Intensities at the same level of approximation are also being developed for the different spectroscopic techniques mentioned above. While these kinds of computations can be performed quite effectively also for large systems and allow for a selection of a reduced number of active modes, the generation of the underlying energy derivatives becomes unfeasible for refined post-Hartree–Fock approaches. Here, the Kohn–Sham (KS) route to density functional theory (DFT) has revolutionized the scenario, providing sufficiently accurate results at reasonable computational cost and, especially, allowing for the development of effective linear scaling approaches. Since the focus of this chapter is on systems of medium-to-large size, in the following we will make reference mainly to DFT methods, whereas more sophisticated approaches will be considered for benchmarking purposes only on small systems. Additionally we will start the discussion with isolated systems, which together with their interest in issues related, for instance, to atmospheric chemistry or to astrophysics represent also the starting point for the investigation of the role of intrinsic and environmental effects in determining overall experimental observables. Once the vibrational problem for isolated species is solved in a satisfactory way, proper account of environmental effects comes into play for the condensed phases, where most of the experiments are performed. In the case of solvents not forming too strong hydrogen bonds, that is, for which the electrostatic component of the solute–solvent interaction constitutes the dominating term, continuum solvation models have been proven to be particularly effective, so that the extension to solution of the computation of essentially all the quantities available for isolated molecules has been proposed in the literature, gaining reliability and computational time comparable to the corresponding calculations in vacuo. In the following sections we will try to provide a general overview of the topics mentioned above, with the aim of giving a feeling of the status and perspectives of the time-independent approach to vibrational spectroscopy without any ambition of completeness, but, rather, with the hope of providing a contribution toward a more critical use of the available computational tools also accessible to nonspecialists.

7.2 GENERAL ASPECTS OF MODELING VIBRATIONAL SPECTRA: ISOLATED SYSTEMS

7.2.1 Vibrational Frequencies

As mentioned in the introduction, most of the QM calculations of vibrational frequencies are performed within the double-harmonic approximation, that is, the truncation of the expansion of the potential energy as a function of the nuclear coordinates to the quadratic term (mechanical harmonic approximation) and the consideration of the linear term only in the expansion of the dipole moment as a function of the nuclear coordinates (electric harmonic approximation). In such a framework, the QM calculation of vibrational frequencies can be reduced to the evaluation of the components of the Hessian matrix followed by diagonalization of the corresponding mass-weighted matrix [1]. Let us start by writing out the energy of a system around its equilibrium geometry $\mathbf{r}_0$ by a Taylor expansion truncated up to the second order:

$$E(\mathbf{r}) = E(\mathbf{r}_0) + \Delta\mathrm{r}^+ \mathbf{H}\,\Delta\mathbf{r} \tag{7.1}$$

where $\mathbf{H}$ is the Hessian matrix (i.e., the matrix containing second derivatives of the energy with respect to the coordinates evaluated at $\mathbf{r}_0$), $\Delta\mathbf{r}$ is the column vector containing displacements of coordinates with respect to their equilibrium value, and $\Delta\mathbf{r}^+$ is the corresponding row (transposed) vector. In the particular case of Cartesian coordinates (x), to which the second derivative matrix $\mathbf{F}$ is associated, the kinetic energy is diagonal, and the diagonalization of the $\mathbf{F}$ matrix after mass weighting leads to a set of decoupled harmonic oscillators in terms of the so-called normal modes $\mathbf{Q}$:

$$\mathbf{Q} = \mathbf{L}^+\mathbf{M}^{1/2}\,\Delta\mathbf{x} \tag{7.2}$$

where, by convention, all the components of $\mathbf{Q}$ vanish at the reference geometry, $\mathbf{M}$ is the diagonal matrix collecting atomic masses, and $\mathbf{L}$ is the matrix of columnwise eigenvectors of the mass-weighted Cartesian force constant matrix $\mathbf{M}^{1/2}\mathbf{F}\mathbf{M}^{1/2}$. The second-derivative matrix $\mathbf{\Phi}$ over normal modes

$$\mathbf{\Phi} = \mathbf{L}^+\mathbf{M}^{-1/2}\mathbf{F}\mathbf{M}^{-1/2}\mathbf{L} \tag{7.3}$$

is diagonal when evaluated at the equilibrium geometry with eigenvalues λ proportional to the squares of harmonic vibrational frequencies ω. It should be stressed that there are several ways to describe vibrational motions in molecular systems: Vibrations can be naturally described in terms of internal coordinates or local modes, which, however, for large systems leads to complicated expansions of the kinetic energy operator. So, even if the adoption of internal coordinates would allow for a better analysis of the vibrational frequencies in terms of chemical groups, substituents, and so on, the evaluation of energy derivatives by QM methods is naturally performed in Cartesian coordinates and leads to simpler equations for normal modes. Transformation to internal coordinates can possibly be performed for interpretative purposes but faces against issues related to redundancies, nonlinear transformations, and so on. Such a subject is covered in most textbooks on vibrational spectroscopy [1, 2].

Moving to consider the actual ways vibrational frequencies are computed in QM codes and following Frisch et al. [3], and limiting the analysis to Hartree–Fock (HF) or DFT wavefunctions, we will start the discussion by assuming the SCF energy to be written as [4]

$$E = \mathrm{tr}(\mathbf{hR}) + \tfrac{1}{2}\mathrm{tr}[\mathbf{RG}(\mathbf{R})] + V \tag{7.4}$$

The analytical calculations of energy second derivatives with respect to nuclear coordinates can be obtained by formal double derivation of the previous equation as

$$E^{xy} = \mathrm{tr}(\mathbf{h}^{xy}\mathbf{R}) + \tfrac{1}{2}\mathrm{tr}(\mathbf{RG}^{xy}(\mathbf{R})) + \mathrm{tr}(\mathbf{R}^y\mathbf{F}^{(x)}) - \mathrm{tr}(\mathbf{S}^{xy}\mathbf{W}) - \mathrm{tr}(\mathbf{S}^x\mathbf{W}^y) + V^{xy} \tag{7.5}$$

where **R** is the density matrix, $\mathbf{h}^{xy}$ collects one-electron integral second derivatives, $\mathbf{G}^{xy}(\mathbf{R})$ is the contraction of **R** with two-electron integral second derivatives, $\mathbf{R}^y$ is the density matrix derivative, $\mathbf{F}^{(x)}$ is the Fock matrix formed with derivative integrals [i.e., $\mathbf{F}^{(x)} = \mathbf{h}^x + \mathbf{G}^x(\mathbf{R})$], $\mathbf{S}^x$ and $\mathbf{S}^{xy}$ overlap first and second derivatives, $\mathbf{W} = \mathbf{RFR}$, and $\mathbf{W}^y = \mathbf{R}^y\mathbf{FR} + \mathbf{RF}^y\mathbf{R} + \mathbf{RFR}^y$. The V^{xy} are nuclear repulsion energy second derivatives and $\mathbf{F}^y = \mathbf{F}^{(y)} + \mathbf{G}(\mathbf{R}^y)$.

Besides the evaluation of the Fock operator, density matrix, and integral derivatives, the analytical evaluation of energy second derivatives is achieved once the density matrix first derivatives with respect to geometric perturbations are known. The evaluation of the latter terms is the bottleneck of computational procedures because of its cost in terms of CPU time and disk storage. In the common practice, such a quantity is obtained by resorting to the first-order coupled perturbed HF (or KS) technique [4], which conceptually starts from the HF (or KS) equations, expands all the matrices in terms of the perturbation, and, by collecting all the terms at the same order, yields sets of equations which are usually solved iteratively.

The evaluation of vibrational frequencies beyond the harmonic approximation, which requires the evaluation of high-order (third and fourth) energy derivatives will be treated in a further section.

7.2.2 Intensities (IR, Raman, and Their Chiral Counterparts)

7.2.2.1 IR Intensities Once we have defined normal frequencies and modes through the diagonalization of the energy second-derivative matrix, at least another quantity is needed for the description of vibrational IR spectra: the vibrational intensity, which is related to the transition probability between vibrational states as induced by the radiation.

For infrared-active modes, in terms of the molar absorption coefficient (or extinction coefficient) a, the absorption spectrum of a molecule in vacuo is given by

$$a = \frac{8\pi^3 N_A}{3hc} \nu \sum_k p_g D_{gk} f(\nu_{gk}, \nu) \tag{7.6}$$

with N_A being Avogadro's number, c the velocity of light in vacuo, ν_{gk} the frequency of the excitation $g \rightarrow k$, and p_g the fractional population of g. The dipole strength of the excitation $g \rightarrow k$, namely, D_{gk}, is given as

$$D_{gk} = |\langle \psi_g | \mu | \psi_k \rangle|^2 \tag{7.7}$$

and $f(\nu_{gk}, \nu)$ the normalized line shape function as

$$\int_0^{\infty} f(\nu_{gk}, \nu) d\nu = 1 \tag{7.8}$$

Within the double (electric and mechanical) harmonic approximation, for the fundamental transition in the kth normal mode

$$D(0 \rightarrow 1)_k = \frac{\hbar}{4\pi \nu_k} \left(\frac{\partial \mu}{\partial Q_k} \right)^2 \tag{7.9}$$

From the macroscopic point of view the quantity usually measured is the absorbance Abs relative to a band. In terms of wavenumbers $\tilde{\nu}$, (Abs) is defined by means of the well-known Lambert–Beer law as

$$\text{Abs}(\tilde{\nu}) = a(\tilde{\nu}) C \ell \tag{7.10}$$

where C is the molar concentration of the species under study and, ℓ is the length of the cell which contains the absorbing sample.

The total intensity of the transition is obtained by integrating a over the entire range of frequencies it spans (the band). This quantity can be expressed by means of the integrated absorption coefficient A:

$$A = \ln 10 \int a(\tilde{\nu})\, d\tilde{\nu} \tag{7.11}$$

and then, by assuming the double-harmonic approximation to be valid, the following expression for A^{gas} can be derived:

$$A = \frac{\pi N_A}{3c^2} \left(\frac{\partial \mu}{\partial Q_k} \right)^2 \tag{7.12}$$

The starting point to compute the $\partial\mu/\partial Q$ term is the expression of the tth component of the molecule dipole moment μ as a function of the density matrix **R** (HF or DFT calculations are considered):

$$\mu_t = -\text{tr}[\mathbf{R}\mathbf{m}_t] + \mu_{N,t} \tag{7.13}$$

where $\mathbf{m}_t$ collects the integrals of the tth Cartesian component of the dipole moment operator and $\mu_{N,t}$ indicates the nuclear contribution.

The first derivative of μ with respect to the normal coordinate Q_i is obtained by differentiating Eq. 7.13:

$$\frac{\partial \mu_t}{\partial Q_i} = -\text{tr}\left[\mathbf{R}^i \mathbf{m}_t + \mathbf{R}\mathbf{m}_t^i\right] + \mu_{N,t}^i \tag{7.14}$$

In the previous equation $\mathbf{R}^i$ is the density matrix derivative with respect to Q_i. It can be calculated through a coupled perturbed Hartree–Fock (CPHF) procedure. Notice that the presence of the $\mathbf{m}_t^i$ term, that is, of the dipole moment matrix derivative in Eq. 7.14, is due to the dependence of m_t on basis functions, which in turn depend on nuclear coordinates.

7.2.2.2 Raman Activities Moving to Raman activities, the starting quantity is the electric moment μ induced in the system by a monochromatic radiation (usually in the visible or ultraviolet range):

$$\mu = \alpha \mathbf{F} \tag{7.15}$$

with **F** being the radiation electric field and α being the molecular polarizability tensor.

In particular, if a molecular vibration causes a change in the molecular polarizability, the scattered radiation will contain the sum (and difference) of the incident frequency and the molecular vibrational frequency: This is the *vibrational Raman scattering*, which we will consider in the following.

The radiant intensity $I^{sc}(\mathbf{k})$ of the vibrational Raman scattering along the direction of the wave vector **k** is, for isolated molecules [5],

$$I^{sc}(\mathbf{k}) = Ik^4 \left| \sum_{r \neq 0} \frac{\langle m|\mu \cdot \mathbf{e}|r\rangle \langle r|\mu \cdot \mathbf{e}'|0\rangle}{\hbar(\omega_{r0} - \omega_{\text{in}})} + \frac{\langle m|\mu \cdot \mathbf{e}'|r\rangle \langle r|\mu \cdot \mathbf{e}|0\rangle}{\hbar(\omega_{r0} + \omega_{\text{sc}})} \right|^2 \tag{7.16}$$

where I is the intensity of the incident radiation, μ is the dipole moment operator, **e** and $\mathbf{e}'$ are the unit vectors of the scattered and incident radiation electric field. $\hbar\omega_{ro}$ is the energy difference between the vibronic states r and 0 (the ground state) and $\omega_{\text{in}} = kc$ and $\omega_{\text{sc}} = k'c$ are the frequencies of the incident and scattered radiation. Equation 7.16 implicitly assumes that only the ground state is populated, and for this reason it can describe only the Stokes scattering.

In the Born–Oppenheimer approximation each state $|r\rangle$ can be factorized into an electronic and a nuclear part. In particular, we will assume that the final state $|m\rangle$ is the product of the ground electronic state and an excited vibrational state.

Following Placzek's approach [e.g., 5–8], for the nonresonant Raman scattering of a randomly oriented ensemble within the double-harmonic approximation, and by assuming $\omega_{\text{sc}} = \omega_{\text{in}}$, it is possible to obtain an expression of I^{sc} (**k**) as a function of derivatives of the frequency-dependent electronic polarizability with respect to normal modes Q_i:

$$I^{sc}(\mathbf{k}) = \frac{h}{4\pi\omega_{10}} Ik^4 \frac{S}{45} \tag{7.17}$$

where $\hbar\omega_{10}$ is the energy difference between the first and ground vibrational states. If the scattered radiation is collected perpendicularly to the linearly polarized incident radiation, the scattering factor S is defined as

$$S = 45\alpha'^2 + 7\gamma'^2 \tag{7.18}$$

In Eq. 7.18, α' is the derivative of the isotropic part of the polarizability with respect to Q_i:

$$\alpha' = \frac{1}{3}\frac{\partial}{\partial Q_i}(\alpha_{xx} + \alpha_{yy} + \alpha_{zz}) \tag{7.19}$$

and γ'^2 has the expression

$$\gamma'^2 = \frac{1}{2}\left\{\left(\frac{\partial\alpha_{xx}}{\partial Q_i} - \frac{\partial\alpha_{yy}}{\partial Q_i}\right)^2 + \left(\frac{\partial\alpha_{yy}}{\partial Q_i} - \frac{\partial\alpha_{zz}}{\partial Q_i}\right)^2 + \left(\frac{\partial\alpha_{zz}}{\partial Q_i} - \frac{\partial\alpha_{xx}}{\partial Q_i}\right)^2 + 6\left[\left(\frac{\partial\alpha_{xy}}{\partial Q_i}\right)^2 + \left(\frac{\partial\alpha_{yz}}{\partial Q_i}\right)^2 + \left(\frac{\partial\alpha_{zx}}{\partial Q_i}\right)^2\right]\right\} \tag{7.20}$$

The α_{ij} terms are the elements of the polarizability tensor evaluated at frequency ω.

Until very recently, only numerical algorithms for the evaluation of geometric polarizability derivatives were available: However, in recent years, analytical algorithms have been proposed in the literature for both HF [9] and DFT wavefunctions [10].

*7.2.2.3 **Vibrational Circular Dichroism*** The differential response of a chiral sample to left and right circularly polarized light can be represented by the quantity $\Delta\epsilon$, defined as [11]

$$\Delta\epsilon = \epsilon_L - \epsilon_R \tag{7.21}$$

where $\epsilon_{L,R}$ are the molar absorption coefficients for left and right circularly polarized light, respectively. Ignoring for the moment solvent effects, the differential molar absorption coefficient at frequency ν, $\Delta\epsilon\ (\nu)$, is

$$\Delta\epsilon(\nu) = 4\gamma\nu\sum_i R_i f(\nu_i, \nu) \tag{7.22}$$

where ν_i is the frequency of the ith transition, γ a numerical coefficient given, for example, by Stephens [11], $f(\nu_i, \nu)$ the normalized lineshape function, and

$$R_i = \mathrm{Im}\left[\langle 0|\mu_{\mathrm{el}}|1\rangle_i \cdot \langle 0|\mu_{\mathrm{mag}}|1\rangle_i\right] \tag{7.23}$$

the rotational strength. In Eq. 7.23, μ_{el} and μ_{mag} are the electric and magnetic dipole moment operators, respectively, and $|0\rangle$, $|1\rangle$ are vibrational states. The transition moments in Eq. 7.23 may be written as [11]

$$\langle 0|(\mu_{\mathrm{el}})_\beta|1\rangle_i = \left(\frac{\hbar}{4\pi\nu_i}\right)^{1/2}\sum_{\lambda\alpha} S_{\lambda\alpha,i}P^{\lambda}_{\alpha\beta} \tag{7.24}$$

$$\langle 0|(\mu_{\text{mag}})_\beta|1\rangle_i = -(4\hbar^3\pi\nu_i)^{1/2}\sum_{\gamma\alpha} S_{\lambda\alpha,i}M^\lambda_{\alpha\beta} \tag{7.25}$$

where $P^\lambda_{\alpha\beta}$ and $M^\lambda_{\alpha\beta}$ are the atomic polar tensors (APTs) and atomic axial tensors (AATs), respectively, and the $S_{\lambda\alpha,i}$ matrix converts Cartesian displacement coordinates $X_{\lambda\alpha}$ of the nucleus λ to normal coordinates Q_i:

$$X_{\lambda\alpha} = \sum_i S_{\lambda\alpha,i}Q_i \tag{7.26}$$

The ν_i and $S_{\lambda\alpha,i}$ are obtained simultaneously by diagonalization of the mass-weighted Cartesian force field (the Hessian). The APTs and AATs of nucleus λ are defined by [11]

$$P^\lambda_{\alpha\beta} = \left(\frac{\partial\langle\psi_g|(\mu_{el})_\beta|\psi_g\rangle}{\partial X_{\lambda\alpha}}\right)_{r_0} \tag{7.27}$$

$$= 2\left\langle\left(\frac{\partial\psi_g}{\partial X_{\lambda\alpha}}\right)_{r_0}\Big|(\mu_{\text{el}})_\beta\Big|\psi_g\right\rangle$$

and

$$M^\lambda_{\alpha\beta} = I^\lambda_{\alpha\beta} + J^\lambda_{\alpha\beta} \tag{7.28}$$

with

$$I^\lambda_{\alpha\beta} = \left\langle\left(\frac{\partial\psi_g}{\partial X_{\lambda\alpha}}\right)_{r_0}\Big|\left(\frac{\partial\psi_g}{\partial B_\beta}\right)_{B_\beta=0}\right\rangle \tag{7.29}$$

$$J^\lambda_{\alpha\beta} = \frac{i}{4\hbar c}\sum_\lambda \epsilon_{\alpha\beta\gamma}r^0_{\lambda\gamma}Z_\lambda e \tag{7.30}$$

Here $Z_\lambda e$ and $\mathbf{r}^0_\lambda$ are the charge and position of nucleus λ at the equilibrium geometry $\mathbf{r}^0$, ψ_G is the wavefunction of the ground electronic state g, and $\partial\psi_g/\partial X_{\lambda\alpha}$ and $\partial\psi_g/\partial B_\beta$ are the derivatives of the wavefunction of g with respect to nuclear displacement and magnetic field, respectively.

Due to their definition, AATs depend on the gauge origin. In the practice, the use of an exact wavefunction would give origin-independent results as well as a HF wavefunction over a complete basis set. Obviously, only finite basis sets can be employed, and thus a methodology to assure gauge independence is required. The use of gauge including/invariant atomic orbitals (GIAOs) [12] eliminates the gauge dependence by using basis functions χ_μ (**B**) explicitly dependent on the magnetic field by inclusion of a complex phase factor referring to the position of the basis function (usually the nucleus):

$$\chi_\mu(\mathbf{B}) = \chi_\mu(0)\exp\left[-\frac{i}{2c}(\mathbf{B}\times\mathbf{r}_\mu)\cdot r\right] \tag{7.31}$$

where $\mathbf{r}_\mu$ is the position vector of basis function χ_μ with respect to the gauge origin and $\chi_\mu(0)$ denotes the usual field-independent basis function.

The calculation of AATs requires the evaluation of the derivative of the wavefunction with respect to the magnetic field $\partial\psi_G/\partial B_\beta$ (compare Eq. 7.29). Within the GIAO framework, it is possible to exploit a CPHF procedure to obtain the required derivative: In this case the perturbation to consider is the (imaginary) external magnetic field. The CPHF equations involved refer to a basis set dependence on the perturbation, where the derivative of the basis function becomes

$$\chi_\mu^1 = \frac{\partial\chi_\mu}{\partial B_\beta} = -\frac{i}{2c}(r_\mu\times r)_\beta\chi_\mu(0) \tag{7.32}$$

7.2.2.4 Vibrational Raman Optical Activity As VCD is the chiral analogue to IR absorption, vibrational Raman optical activity (VROA) is the chiral analogue to Raman scattering. The quantity which is measured in VROA is the differential scattering intensity between right and left circularly polarized light $I_k^R - I_l^L$ [13, 14],

$$\Delta_k = I_k^R - I_k^L \tag{7.33}$$

where $I_k^{L,R}$ are the scattered intensities with linear k polarization for right (R) and left (L) circularly polarized incident light and k denotes the Cartesian component. In the case of a sample in the gas phase, the differential scattering intensities between right and left circularly polarized light for the polarized right-angle scattering of Raman optical activity is given by [15]

$$\Delta_z(90) = 6\beta(G')^2 - 2\beta(A)^2 \tag{7.34}$$

where

$$\beta(G')^2 = \tfrac{1}{2}(3\alpha_{ki}^{\upsilon}G'^{\upsilon}_{ki} - \alpha_{kk}^{\upsilon}G'^{\upsilon}_{ii}) \tag{7.35}$$

$$\beta(A)^2 = \tfrac{1}{2}\omega_{\text{rad}}\alpha_{ki}^{\upsilon}\epsilon_{kjl}A_{jli}^{\upsilon} \tag{7.36}$$

ω_{rad} is the radiation angular frequency, ϵ_{kjl} is the unit third-rank antisymmetric tensor, and the other quantities are defined as

$$\alpha^{\upsilon}_{ki} G'^{\upsilon}_{ki} = < 0|\alpha_{ki}|1 >< 1|G'_{ki}|0 > = \frac{1}{2\omega}\left(\frac{\partial \alpha_{ki}}{\partial Q}\right)_0 \left(\frac{\partial G'_{ki}}{\partial Q}\right)_0 \tag{7.37}$$

$$\alpha^{\upsilon}_{ki} \epsilon_{kjl} A^{\upsilon}_{jli} = < 0|\alpha_{ki}|1 >< 1|\epsilon_{kjl} A_{jli}|0 > = \frac{1}{2\omega}\left(\frac{\partial \alpha_{ki}}{\partial Q}\right)_0 \epsilon_{kjl} \left(\frac{\partial A_{jli}}{\partial Q}\right)_0 \tag{7.38}$$

In the equations above, the Placzek approximation [7] is assumed. The quantities reported in Eqs. 7.37–7.38 are the electric dipole–electric dipole polarizability α, the imaginary part of the electric dipole–magnetic dipole polarizability $\mathbf{G}'$, and the real part of the electric dipole–electric quadrupole polarizability $\mathbf{A}$ [15]; ω is the frequency associated with the vibrational transition and Q is the corresponding normal coordinate. The subscript 0 indicates that the quantities are calculated at the equilibrium geometry. We note that, as already mentioned for VCD, in this case some quantities, that is, the mixed electric dipole–magnetic dipole and the electric dipole–electric quadrupole polarizabilities, individually depend on a choice of origin. However, such an origin dependence cancels out in the anisotropic invariants. Nevertheless, in approximate calculations using finite basis sets, this cancellation of the origin dependence can only be achieved through the use of gauge-independent basis sets, such as London atomic orbitals [16] or GIAO [12].

The ab initio calculation of VROA [17] was originally done through numerical differentiation; however, recently analytical algorithms have been proposed and implemented [18], so that the limit of the affordable systems has been enormously enhanced.

7.2.3 Illustrative Examples: Harmonic Computations

A good agreement between computed and experimental vibrational properties is a mandatory starting point for reliable spectral assignments. Although a direct quantitative comparison with experiments often requires going beyond the harmonic approximation (see Section 7.2.4), in many cases the effect of anharmonicity can be accounted for in an approximate manner. For example, harmonic frequencies, which are relatively simple to compute, can be scaled by a (uniform or frequency-dependent) factor leading to reasonable agreement with experimental data [19–22]. Additionally, for very large systems even computation of vibrational properties within a harmonic framework might not be a trivial task and effective approaches allowing us to compute only the relevant part of harmonic vibrational spectra [23, 24] and/or mixed quantum mechanical/molecular mechanical (QM/MM) computations [25, 26] might greatly improve understanding of experimental outcomes for complex systems.

For small molecules it is possible to perform accurate computations with post-Hartre–Fock approaches, and in this respect harmonic frequencies computed at the CCSD(T) (coupled clusters with single, double, and perturbative inclusion of triple excitation [27]) level, with basis sets of at least triple-ζ quality reach an overall accuracy of 15–20 cm^{-1} for closed-shell systems [e.g., 28, 29]. For radicals, the situation is not so well assessed, but some recent investigations confirm that analogous accuracy can be reached [e.g., 30–34]. However, the unfavorable scaling of the CCSD(T) model with the number of active electrons limits its applicability to very small systems only. In addition, a simple reduction of the computational cost by combining correlated QM methods with small basis sets is not to be recommended due to the quite unpredictable accuracy of the results. Thus, the extension of computational studies to large systems requires cheaper and at the same time reliable electronic structure models.

It is widely recognized that vibrational properties can be effectively computed with KS-DFT approaches [22, 35, 36], which over the years has become the method of choice to assist spectroscopic measurements of medium-to-large molecular systems. In this respect, hybrid functionals, in particular B3LYP [37], provide satisfactory results when coupled to basis sets of at least double-ζ plus polarization quality supplemented by diffuse *sp* functions [38–41]. Recently, several new functionals have been proposed in order to correct the well-known DFT failures in the description of dispersion interactions or electronic excited states of charge transfer character. However, in current developments the major effort has focused on energetic properties, that is, on the correct prediction of equilibrium structures and on excited-state computations, thus only slightly overlooking the correct description of vibrational properties. On these grounds, here we would like to shed some light on such issues, analyzing the performances of several standard and last-generation density functionals in predicting vibrational properties within the harmonic approximation. Taking into account that studies of larger systems require computational models coupling reliability with favorable scaling with the number of electrons, we will highlight the results delivered by the N07D [42–45] basis set with functionals from the B3LYP family, which has been shown to provide an accuracy comparable to the most advanced computational approaches at a reduced computational cost [46–49].

It seems appropriate to start our analysis by discussing the accuracy of harmonic frequencies with reference to the recently introduced benchmark set F38 [50], built up to cover a broad range of frequencies for small molecules, and applied to assess the accuracy of harmonic vibrational frequencies for several density functional [50–52] methods. The F38 reference data set is based on the best experimental estimates of harmonic frequencies [53, 54], with the single exception of the umbrella mode of the NH_3, which is taken from a CCSD(T)/cc-pVQZ calculation [55]. The results presented in Table 7.1 point out the good overall accuracy of harmonic frequencies computed by the B3LYP, TPSSh, B97-1, B98, and B2PLYP DFT models, characterized by mean unsigned errors (MUEs) within 25–35 cm^{-1}. Furthermore, at the B3LYP level the extension of the basis set beyond N07D has only negligible effects

Table 7.1 MUEs, Maximum Negative (MIN) and Positive (MAX) Deviations of Harmonic Vibrational Frequencies (cm^{-1}) Computed by Several DFT Models for Molecules from F38 Benchmark Set and Compared to F38 Reference Data

Method	Basis Set	MUE	MIN	MAX	References
BLYP	MG3S[a]	67	−200	38	51
B3LYP	N07D	33	−85	106	49
	aug-cc-pVTZ	33	−72	133	49
	MG3S[a]	31	−69	128	51
CAM-B3LYP	N07D	52	−110	129	49
B2PLYP	N07D	23	−56	100	49
	aug-cc-pVTZ	18	−47	99	49
PBE	MG3S[a]	56	−170	75	51
PBEh	MG3S[a]	46	—	—	50
PBE0	N07D	50	−86	140	49
LC-ωPBE	N07D	74	−135	192	49
TPSS	MG3S[a]	39	−145	84	51
TPSSh	MG3S[a]	28	—	—	50
B97-1	MG3S[a]	33	−81	61	49
B97-3	MG3S[a]	46	—	—	50
B98	MG3S[a]	32	—	—	50
ωB97	N07D	62	−93	159	49
ωB97X	N07D	60	−97	152	49
M05	MG3S[a]	54	—	—	50
M05-2X	MG3S[a]	70	−72	262	52
M06-L	MG3S[a]	39	−117	85	51
M06	N07D	57	−124	134	49
M06-2X	N07D	66	−77	163	49
M06-2X	MG3S[a]	56	−55	244	52
M06-HF	MG3S[a]	95	—	—	50
M08-SO	MG3S[a]	52	−92	207	52
M08-HX	MG3S[a]	63	−67	227	52
VSXC	MG3S[a]	24	−106	57	51
HSE06	N07D	50	−86	138	49
BMK	MG3S[a]	52	—	—	50
HF-LYP	MG3S[a]	233	—	—	50

Note: Benchmark harmonic frequency values as compiled in refs. 50 and 52 based on data from refs. 53–55.
[a] From ref. 56.

on the accuracy of vibrational properties. Among all tested DFT methods, B2PLYP yields the most accurate harmonic frequencies, especially if coupled to large basis sets. However, although B2PLYP/N07D results are better than those obtained with other DFT functionals, the MP2 contribution to B2PLYP leads to computed harmonic frequencies not fully converged with respect to the basis set.

Other DFT functionals do not provide vibrational frequencies with the accuracy required for a univocal assignment of experimental spectra. In particular, functionals from the Minnesota family [50–52] show both large negative and positive deviations from experiments, the latter more than 200 cm^{-1} on average. Additionally, it can be noted that long-range corrected functionals show larger MUEs than their uncorrected counterparts, thus indicating that local properties should not be overlooked during the development of long-range/dispersion corrections.

Together with vibrational frequencies, the interpretation of experimental spectra also requires reliable results for vibrational intensities. It has been widely recognized that the prediction of intensities is more difficult than the computation of band positions [57–61]. Since IR and Raman band intensities depend on dipole moment and polarizability derivatives, respectively (see Eqs. 7.12 and 7.18), their reliable estimate requires an accurate description of the electronic charge density and its changes along normal modes. These conditions imply, in turn, that IR intensities are sensitive to electron correlation effects and basis set extension. On the other side, experimental IR intensities are difficult to be determined due to a number of effects such as band overlap, resonances, intensity sharing, and influence of rotational fine structure or instrumental limits [59], which usually lead to inaccuracies of about 10% [60, 61]. In the case of IR intensities computed within the double-harmonic approximation, it has been shown by Schaefer et al. [59, 60] that increasing the level of electron correlation leads to converged values and that quantitative IR intensity predictions can be obtained at the CCSD(T) level in conjunction with basis sets of at least aug-cc-pVTZ quality [59]. However, for large molecular systems it is necessary to assess the accuracy of less expensive approaches. Schlegel et al. [57] examined DFT IR intensities computed with various commonly used local, gradient-corrected and hybrid density functionals in conjunction with basis sets of different quality. It has been concluded that the best agreement with the post-Hartree–Fock results [of quadratic configuration interaction singles and doubles (QCISD) and CCSD(T) quality] is achieved by hybrid functionals (B3LYP, B3P86, B3PW91), showing also that starting from the standard 6-31G(d) basis set the addition of *p*-type polarization functions to hydrogens and first sets of diffuse functions on other atoms is crucial to improve the results, whereas further basis set extension shows less pronounced effects. There are very few reports on the QM prediction of Raman intensities, which are related to the square of polarizability derivatives (see Eqs. 7.18, 7.19, and 7.20). However, also in this case it has been shown that the effects due to electron correlation are less significant than those originating from the truncation of the basis set [58], and in particular DFT approaches provide intensities comparable to their MP2 counterparts at a significantly lower computational cost [58]. In addition, still considering basis set effects, it has been shown that the computation of Raman intensities requires larger basis sets than infrared intensities. Nevertheless, it is possible to develop purposely tailored basis sets which, despite their medium size, are flexible enough to correctly describe the tail region of the electron density. In this respect, results of accuracy similar to those obtained with the much larger aug-cc-pVTZ basis set can be obtained with the pVTZ basis set developed by Sadlej [62], which is optimized for electric properties [58] or with the recently introduced "spectroscopic" aug-N07D basis set [42]. Moreover, it has been shown recently that long-range corrected functionals, such as LC-ωPBE [63] and CAM-B3LYP [64], yield good results on Raman intensities [65] without worsening the accuracy of their uncorrected counterparts as far as infrared intensities are concerned.

Additional issues are related to the prediction of VCD rotational strengths or VROA activities, which require the evaluation of a magnetic component and invariably a methodology to assure for the gauge origin independence. It should be noted that, by definition, chiral molecules have no symmetry, so that they are difficult to study with post-Hartree–Fock approaches. It is thus particularly relevant to notice that reliable VCD spectra can be computed by DFT models [17, 66–68], although basis sets giving minimal basis set errors and density functionals as close as possible to exact are required. In the practice, it has been shown that basis sets such as TZ2P and cc-pVTZ give results close to the basis set limit, while smaller basis sets (such as, e.g., 6-31G*) are substantially less accurate [67]. Additionally, similar to other vibrational properties, the best results have been obtained with hybrid density functionals, in particular with B3LYP or B3PW91. Again, at least to the best of our knowledge, little is known on the performance of recently developed functionals in the prediction of VCD or VROA spectra. For this reason, let us analyze (*R*)-1-methylphenyloxirane (RMPO), a molecule for which good agreement with experiment has been obtained by computations at the B3LYP/aug-cc-pVTZ level [69]. Figure 7.1 compares the VCD spectra computed within the DFT/N07D scheme for several functionals with B3LYP/aug-cc-pVTZ results. First, we observe that the spectrum obtained at the B3LYP level with the N07D basis is in good agreement with B3LYP/aug-cc-pVTZ results, thus confirming the applicability of the N07D basis to VCD. Next, by comparing spectral shapes computed with different functionals, it comes out clear that B97-1 shows intensities and band position patterns having the best agreement with the B3LYP reference, whereas other tested functionals turned out to be less reliable.

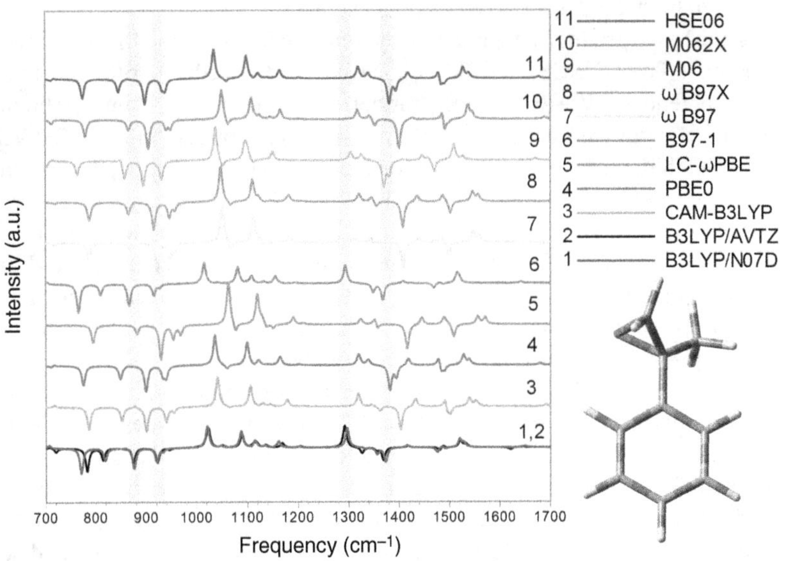

Figure 7.1 Performance of different DFT approaches in predicting the VCD spectra of (*R*)-1-methylphenyloxirane [70].

Although further benchmark tests are required to assess the accuracy of recently developed functionals with respect to all kinds of vibrational properties computed within the harmonic approximation, it seems reasonable to conclude this section remarking that, based on the results obtained to date, only the B2PLYP model in some aspects outperforms the standard B3LYP functional. Additionally, for the cases where long-range and/or dispersion effects need to be described, a viable solution to compute vibrational properties is to go through CAM-B3LYP or B3LYP-D.

7.2.4 Going Beyond the Double-Harmonic Approximation

As mentioned above, harmonic frequency calculations allow for the determination of vibrational spectra and molecular properties in a simple manner. This approach assumes an ideal case of a perfectly symmetric harmonic representation of the true shape of the potential energy surface (PES), but in reality the PES is always "anharmonic," leading to bond breaking, and in many cases proper account of anharmonic contributions becomes mandatory. In particular, the harmonic approximation is not able to predict either overtones or combination bands, which can have an impact on the interpretation of experimental spectra. In this respect, for small molecules converged rovibrational levels can be obtained through fully variational methods [71–75] based on the complete Hamiltonian and accurately represented full-dimensional anharmonic potential energy surfaces. However, even if a blackbox-type algorithm for the variational computation of energy levels and wavefunctions using a (ro)vibrational Hamiltonian expressed in an arbitrarily chosen body-fixed frame and in any set of internal coordinates of full or reduced vibrational dimensionality has been recently presented [75], the computational cost of such a treatment for large molecules makes some approximations unavoidable, with regard to both the shape of the potential and the rovibrational treatment. Here we can distinguish two groups of approaches: the first one being set within the vibrational self consistent field (VSCF) [76–80] framework and based on truncated two- to four-mode potentials. VSCF has been further extended [81] to include vibrational Møller–Plesset perturbation theory (VMP, known also as correlation-corrected VSCF, cc-VSCF) [82, 83], vibrational configuration interaction (VCI) [77, 84], or vibrational coupled cluster (VCC) [85]. This methodology is generally well suited to describe a large variety of systems, including large-amplitude and floppy motions, but requires extensive sampling of the PES, which leads to relatively large computational costs. To overcome this difficulty, effective procedures to construct potential energy surfaces [86, 87], including approaches dedicated specifically to vibrational structure calculations [88–92], are under constant development.

In this chapter we will focus instead on another approach, the vibrational second-order perturbation theory (VPT2) [93–98], which provides closed expressions for most of the spectroscopic parameters required for the analysis of the experimental frequencies. VPT2 has been shown to be very effective for the study of semirigid polyatomic molecules of medium size [38–41, 99, 100] at relatively low computational cost. This level of theory requires quadratic, cubic, and semi-diagonal quartic force constants. Starting from analytical Hessians, the needed cubic (ϕ_{rst}) and quartic constants (ϕ_{rstt}) can be computed by one-dimensional numerical differentiation using structures displaced from the equilibrium geometry by small increments $\pm\delta Q$ along normal modes:

$$\phi_{rst} = \sqrt{\omega_r\omega_s\omega_t}\Phi_{rst} \tag{7.39}$$

$$\Phi_{rst} = \frac{1}{3}\left(\frac{\Phi_{st}(\delta Q_r) - \Phi_{st}(-\delta Q_r)}{2\delta Q_r} + \frac{\Phi_{tr}(\delta Q_s) - \Phi_{tr}(-\delta Q_s)}{2\delta Q_s} + \frac{\Phi_{rs}(\partial Q_t) - \Phi_{rs}(-\delta Q_t)}{2\delta Q_t}\right) \tag{7.40}$$

$$\Phi_{rstt} = \frac{\Phi_{rs}(\delta Q_t) + \Phi_{rs}(-\delta Q_t) - 2\Phi_{rs}(0)}{\delta Q_t^2} \tag{7.41}$$

$$\Phi_{rrtt} = \frac{1}{2}\left(\frac{\Phi_{rr}(\delta Q_t) - \Phi_{rr}(-\delta Q_t) - 2\Phi_{rr}(0)}{2\delta Q_t^2} + \frac{\Phi_{tt}(\delta Q_r) - \Phi_{tt}(-\delta Q_r) - 2\Phi_{tt}(0)}{2\delta Q_r^2}\right) \tag{7.42}$$

Within the VPT2 theory, the vibrational Hamiltonian $\mathbf{H}_{\text{vib}}$ can be expressed in wavenumber units as a sum of the zero-order harmonic term $\mathbf{H}^0_{\text{vib}}$, the first-order term $\mathbf{H}^1_{\text{vib}}$ accounting for the cubic components of the potential, and the second-order term $\mathbf{H}^2_{\text{vib}}$, which includes all quartic components of the potential and the kinetic energy corrections arising from the vibrational angular momentum j_α:

$$\begin{aligned}\mathbf{H}_{\text{vib}} &= \mathbf{H}^0_{\text{vib}} + \mathbf{H}^1_{\text{vib}} + \mathbf{H}^2_{\text{vib}} \\ &= \frac{1}{2}\sum_r \omega_r(p_r^2 + q_r^2) \\ &\quad + \frac{1}{6}\sum_{rst}\phi_{rst}q_rq_sq_t + \frac{1}{24}\sum_{rstu}\phi_{rstu}q_rq_sq_tq_u + \sum_\alpha B^0_\alpha\, j_\alpha^2\end{aligned} \tag{7.43}$$

where the q_r are reduced normal coordinates, p_r their conjugated momenta, and ω harmonic frequencies and α denotes the rotational axis and B^0_α the corresponding equilibrium rotational constant. The vibrorotational Hamiltonian $\mathbf{H}_{\text{vibrot}}$ is then defined by adding to Eq. 7.43 rotational energy terms related to the centrifugal distortion and the Coriolis coupling between vibrational and rotational (J_α) angular momenta:

$$\begin{aligned}\mathbf{H}_{\text{vibrot}} &= \mathbf{H}_{\text{vib}} \\ &\quad + \sum_\alpha B^e_\alpha J_\alpha^2 + \sum_{\alpha,\beta}\sum_r B^\alpha_r q_r J_\alpha J_\beta \\ &\quad + \frac{1}{2}\sum_{\alpha,\beta}\sum_{r,s} B^\alpha_{rs} q_r q_s J_\alpha J_\beta - 2\sum_\alpha B_\alpha J_\alpha\, j_\alpha\end{aligned} \tag{7.44}$$

where $B_r^{\alpha\beta}$ and $B_{rs}^{\alpha\beta}$ are rotational constant derivatives with respect to normal coordinates.

For a given set of vibrational states, the harmonic problem is solved first in order to identify all terms $|\mathbf{v}_0\rangle$ of Eq. 7.44 which generate resonant or nearly resonant interactions. Then the vibrational wavefunctions $|\mathbf{v}_1\rangle$ are obtained by second-order perturbation theory on the basis of all nonresonant energy terms, while the resonant energy terms are treated separately via variational computations. Such a treatment assures to exclude nearly degenerate terms from the perturbative treatment. It is worth remarking that, in general, the choice of appropriate thresholds and the number of intervening states involve some arbitrariness, thus requiring some experience and spectroscopic sensitivity. However, in the VPT2 implementation [98], whose results will be shown in this chapter, the criteria proposed by Martin et al. [101] are exploited and this appropriately setup automatic procedure has been shown to provide accurate results at least for fundamental bands [39, 100]. Additionally, these well-tested schemes allow for reliable VPT2 computations even for macromolecular systems with a large number of normal modes, where anharmonic resonances are unavoidable.

In the perturbative model, the vibrational energy (in wavenumbers) of asymmetric tops is given by

$$E_n = \xi_0 + \sum_i \omega_i\left(n_i + \frac{1}{2}\right) + \sum_i \sum_{j<i} \xi_{ij}\left(n_i + \frac{1}{2}\right)\left(n_j + \frac{1}{2}\right) \tag{7.45}$$

where the ω's are the harmonic wavenumbers and the ξ's are the anharmonic contributions given by the equations [98, 102]

$$A\xi_0 = \sum_{i=1}^{f} \frac{\Phi_{iiii}}{\lambda_i} - \frac{7}{9}\sum_{i=1}^{f} \frac{\Phi_{iii}^2}{\lambda_i^2} + 3\sum_{i=1}^{f}\sum_{j\neq i=1}^{f} \frac{\Phi_{ijj}^2}{\lambda_j(4\lambda_j - \lambda_i)} \tag{7.46}$$

$$-16\sum_{i=1}^{f}\sum_{j>i}^{f}\sum_{k>j}^{f} \frac{\Phi_{ijk}^2}{\Delta_{ijk}} - 16\sum_{\alpha=x,y,z} \beta_\alpha^0\left[1 + 2\sum_{i=1}^{f}\sum_{j>i}^{f}\left(\zeta_{ij}^\alpha\right)^2\right] \tag{7.47}$$

$$B_i\xi_{ii} = \Phi_{iiii} - \frac{5\Phi_{iii}^2}{3\lambda_i} - \sum_{j\neq i=1}^{f} \frac{(8\lambda_i - 3\lambda_j)\Phi_{iij}^2}{\lambda_j(4\lambda_i - \lambda_j)} \tag{7.48}$$

$$C_{ij}\xi_{ij} = \Phi_{iijj} - \frac{2\Phi_{iij}^2}{4\lambda_i - \lambda_j} - \frac{2\Phi_{ijj}^2}{4\lambda_j - \lambda_i} - \frac{\Phi_{iii} - \Phi_{ijj}}{\lambda_i} - \frac{\Phi_{jjj} - \Phi_{iij}}{\lambda_j} \tag{7.49}$$

$$+\sum_{k\neq i\neq j=1}^{f}\left[\frac{2(\lambda_i + \lambda_j - \lambda_k)\Phi_{ijk}^2}{\Delta_{ijk}} - \frac{\Phi_{iik}\Phi_{jjk}}{\lambda_k}\right] + 4(\lambda_j + \lambda_i)\sum_{\alpha=x,y,z} \beta_\alpha^0\left(\zeta_{ij}^\alpha\right)^2 \tag{7.50}$$

where $A = 64hc$, $B_i = 16hc\lambda_i$, $C_{ij} = 4hc(\lambda_i\lambda_j)^{1/2}$, $\lambda_i^2 = 2\pi c\omega_i$

$$\Delta_{ijk} = \lambda_i^2 + \lambda_j^2 + \lambda_k^2 - 2(\lambda_i\lambda_j + \lambda_i\lambda_k + \lambda_j\lambda_k) \tag{7.51}$$

and β_α^0 and ζ_{ij}^α are the rotational and Coriolis coupling constants, respectively. Fundamental bands ν_i, overtones $[2\nu_i]$, and combination bands $[\nu_i + \nu_j]$ are given by

$$\nu_i = \omega_i + 2\xi_{ii} + \frac{1}{2}\sum_{j\neq i}\xi_{ij} \tag{7.52}$$

$$[2\nu_i] = 2\omega_i + 6\xi_{ii} + \sum_{j\neq i}\xi_{ij} = 2\nu_i + 2\xi_{ii} \tag{7.53}$$

$$[\nu_i + \nu_j] = \omega_i + \omega_j + 2\xi_{ii} + 2\xi_{jj} + 2\xi_{ij} + \frac{1}{2}\sum_{l\neq i,j}(\xi_{il} + \xi_{jl} = \nu_i + \nu_j + \xi_{ij}) \tag{7.54}$$

Let us now denote the nth derivative of a property Ω with respect to the dimensionless reduced normal coordinates $q_i, q_j, q_k, \ldots$ calculated at the equilibrium geometry by Ω_{ijk}:

$$\Omega_{ijk\ldots} = \left(\frac{\partial^n\Omega}{\partial q_i\partial q_j\partial q_k\ldots}\right)_e$$

At the same level of perturbation theory the vibrationally averaged value of Ω is expressed as

$$\langle\Omega\rangle_n = \Omega_e + \sum_i A_i(n_i + \tfrac{1}{2}) \tag{7.55}$$

where Ω_e is the value at the equilibrium geometry and

$$A_i = \frac{\Omega_{ii}}{\omega_i} - \sum_j \frac{\Omega_j\Phi_{iij}}{\omega_j\omega_i^2} \tag{7.56}$$

In the following, the first term of the right-hand side of Eq. 7.56 will be referred to as harmonic and the second one as anharmonic.

The expression for IR intensities is more complicated [103] since they correspond to transitions between two different vibrational states. The pioneering work of Handy and co-workers [97] gave a general expression for the VPT2 intensities and the first implementation in the Spectro code. However, quite recently more compact expressions have been derived by Vazquez and Stanton [104], which are rewritten below due to a few typographic errors in the original references. Using the notation $\mathbf{P}^\alpha$ and $\mathbf{P}^\alpha_{ijk\ldots}$

to refer to any Cartesian coordinate α of the dipole moment and its derivatives, the α Cartesian component of the transition dipole moment for the fundamental $i(\langle\mathbf{0}|\mathbf{P}^{\alpha}|\mathbf{0}+v_i\rangle)$ corrected for anharmonicity is given by the relation,

$$
\begin{aligned}
\langle\mathbf{P}^{\alpha}\rangle_{0,i} = & \frac{1}{\sqrt{2}}\mathbf{P}^{\alpha} + \frac{1}{4\sqrt{2}}\sum_{j}\mathbf{P}^{\alpha}_{ijj} \\
& - \frac{1}{8\sqrt{2}}\sum_{jk}\Phi_{ijk}\mathbf{P}^{\alpha}_{ik}\left\{\left(\frac{1}{\omega_i+\omega_j+\omega_k} - \frac{1}{\omega_i-\omega_j-\omega_k}\right) + \frac{2}{\omega_k}\Phi_{kjj}\mathbf{P}^{\alpha}_{ik}\right\} \\
& + \frac{1}{16\sqrt{2}}\sum_{jkl}\left\{\Phi_{ikl}\Phi_{jkl}\mathbf{P}^{\alpha}_{j}\left[\frac{4\omega_j(\omega_l+\omega_k)(1-\delta_{ij})(1-\delta_{ik})(1-\delta_{il})}{(\omega_i^2-\omega_j^2)[\omega_i^2-(\omega_l+\omega_k)^2]}\right.\right. \\
& \qquad - \frac{(\omega_l+\omega_k)\left[3\omega_i^2-(\omega_l+\omega_k)^2\right]\delta_{ij}(1+\delta_{ik})(1-\delta_{il})}{\omega_i\left[\omega_i^2-(\omega_l+\omega_k)^2\right]^2} \\
& \qquad \left. - \frac{4\omega_j(7\omega_i\omega_k+3\omega_k\omega_j+3\omega_k^2+4\omega_i\omega_j+4\omega_i^2)(1-\delta_{ij})(1-\delta_{ik})\delta_{il}}{\omega_k(2\omega_i+\omega_k)\left(\omega_i^2-\omega_j^2\right)(\omega_i+\omega_j+\omega_k)}\right] \\
& \quad + \Phi_{ijk}\Phi_{llk}\mathbf{P}^{\alpha}_{j}\left[\frac{\delta_{ij}}{\omega_i\omega_k}\left(1+\frac{2}{9}\delta_{ik}\delta_{il}\right) - \frac{4\omega_j(1-\delta_{ij})(1-\delta_{ik})(1-\delta_{il})}{\omega_k\left(\omega_i^2-\omega_j^2\right)}\right. \\
& \qquad \left.\left. - \frac{4\omega_j\delta_{ik}(1-\delta_{ij})}{\omega_i\left(\omega_i^2-\omega_j^2\right)}\left(1+\frac{2}{3}\delta_{il}\right)\right]\right\} \\
& - \frac{1}{8\sqrt{2}}\sum_{jk}\Phi_{ijkk}\mathbf{P}^{\alpha}_{j}\left(\frac{1}{\omega_i+\omega_j} - \frac{1-\delta_{ij}}{\omega_i-\omega_j}\right) \\
& + \frac{1}{2\sqrt{2}}\sum_{jk\neq i}\mathbf{P}^{\alpha}_{k}\left\{\sum_{\gamma}B^{\gamma}_{e}\zeta^{\gamma}_{ij}\zeta^{\gamma}_{jk}\left[\frac{\sqrt{\omega_i\omega_k}}{\omega_j}\left(\frac{1}{\omega_i+\omega_k} + \frac{1-\delta_{ik}}{\omega_i-\omega_k}\right)\right.\right. \\
& \qquad \left.\left. - \frac{\omega_j}{\sqrt{\omega_i\omega_k}}\left(\frac{1}{\omega_i+\omega_k} - \frac{1-\delta_{ik}}{\omega_i-\omega_k}\right)\right]\right\}
\end{aligned}
\tag{7.57}
$$

where we have exploited the fact that B_e is diagonal, so that

$$
\sum_{\gamma\varsigma}B^{\gamma\varsigma}_{e}\zeta^{\gamma}_{ij}\zeta^{\varsigma}_{jk} = \sum_{\gamma}B^{\gamma}_{e}\zeta^{\gamma}_{ij}\zeta^{\gamma}_{jk} \tag{7.58}
$$

For combination and overtone transitions, the α Cartesian component of the transition dipole moments, respectively $\langle \mathbf{0}|\mathbf{P}^{\alpha}|\mathbf{0}+\upsilon_i+\upsilon_j\rangle$ and $\langle \mathbf{0}|\mathbf{P}^{\alpha}|\mathbf{0}+2\upsilon_i\rangle$, are given by

$$\langle \mathbf{P}^{\alpha}\rangle_{0,i(1+\delta_{ij})+j(1-\delta_{ij})} = \sum_k \left\{ \left[1 + \frac{(1-\sqrt{2})\delta_{ij}}{\sqrt{2}} \right] \times \left[\frac{1}{2}\mathbf{P}^{\alpha}_{ij} + \frac{1}{4}\Phi_{ijk}\mathbf{P}^{\alpha}_{k} \left(\frac{1}{\omega_i+\omega_j-\omega_k} - \frac{1}{\omega_i+\omega_j+\omega_k} \right) \right] \right\} \tag{7.59}$$

It should be noted that divergences need to be considered not only for frequencies but also for intensities whenever $\omega_i = 2\omega_j$ or $\omega_i = \omega_j + \omega_k$. However, as already discussed, in these cases it is possible to remove potentially divergent terms and treat them variationally in a successive step [98, 105].

7.2.5 Illustrative Examples on Calculation of Vibrational Spectra Beyond Harmonic Approximation

In this section we will discuss issues related to the computation of vibrational properties beyond the harmonic approximation within the VPT2 approach [93, 98]. It is widely recognized that VPT2 computations coupled with semidiagonal quartic force fields evaluated at the CCSD(T) (coupled clusters with single, double, and perturbative inclusion of triple excitations [106]) level in conjunction with basis sets of at least triple-ζ quality usually provide results with an accuracy of the order of 10–15 cm^{-1} for the fundamental transitions [29, 95, 107–117]. However, as already stated in Section 7.2.3, computations at the CCSD(T) level are still limited to small systems, so that the extension of accurate computational studies to large systems requires cheaper and at the same time reliable electronic structure methods. In this respect, the density functional theory stands as a valuable route and several VPT2 computations based on DFT anharmonic force fields have been reported for small- and medium-sized semirigid molecules [38–40, 101]. Moreover, as far as anharmonic vibrational computations are concerned, hybrid [e.g., CCSD(T)/DFT] approaches can also be applied, mainly by means of two possible routes. In the simplest one, harmonic frequencies computed at the highest level of theory are a posteriori corrected with anharmonic contributions ($\Delta\nu$) derived from VPT2 computations performed at a lower level, that is, $\nu_{\text{Higher/Lower}} = \omega_{\text{Higher}} + \Delta\nu_{\text{Lower}}$. Such an approximation, in particular within the CCSD(T)/DFT scheme, has already been validated for several closed- and open-shell systems [e.g., 30, 34, 118–120]. The second route introduces the harmonic frequencies evaluated at the highest level directly into the VPT2 computations along with the third- and fourth-order force constants obtained at the low level of theory. For the latter case, an automatic procedure which compares normal modes computed by the two levels of theory and replaces harmonic data accordingly is particularly effective and facilitates the application of a hybrid route for large systems, for which the ordering of several closely lying vibrations might be exchanged [49]. It should be noted that such a procedure can significantly improve the quality of the results in difficult cases, that is, when there are large discrepancies between harmonic frequencies computed at the two levels of theory or when Fermi resonances are present. Considering DFT and hybrid computations, it is important to dissect the overall accuracy of the vibrational frequencies into harmonic and anharmonic contributions. Let us take as an example H_2CN, FCS, and NH_2 radicals, for which the accuracy of VPT2 frequencies delivered by CCSD(T) computations has been confirmed by comparison with experimental data [33, 121]. First, DFT and hybrid results delivered by several recently developed density functionals are discussed. For this purpose, Table 7.2 compares mean absolute errors (MAEs), with respect to CCSD(T) computations, over all normal modes of H_2CN, N_2H, and FCS, for which a good agreement with both experiment and the benchmark CCSD(T) studies has been obtained at the B3LYP/N07D level. It has to be noted that an accuracy only slightly lower than that of B3LYP, with discrepancies on the order of 15–25 cm^{-1} with respect to CCSD(T), has been achieved for anharmonic frequencies computed by several functionals, except LC-ωPBE, and the ones from ωB97(X) and M06 families, which show MAEs in range 30–60 cm^{-1}. Additionally, larger discrepancies are observed for the highest frequencies, as depicted in Figure 7.2. Additional insight can also be derived from Figure 7.3, which shows differences in the anharmonic contribution with respect to CCSD(T) results as a function of the frequency for all normal modes of the selected radicals. It is clear that, among the tested functionals, those originating from B3LYP and PBE0, together with the recently introduced B2PLYP2 and HSE06, yield accurate anharmonic corrections. In contrast, recently developed functionals belonging to the M06 and ωB97 families

Table 7.2 MUEs of Anharmonic Vibrational Frequencies (cm^{-1}) Computed by Several DFT/N07D and Hybrid CCSD(T) + DFT Models for H_2CN, FCS, and NH_2 Radicals as Compared to Experimental Data[a]

Method	DFT	Hybrid CCSD(T) + DFT
B3LYP	16	10
B3LYP-D	19	11
CAM-B3LYP	24	12
PBE0	24	12
LC-ωPBE	59	18
M06	49	48
M06-2X	43	24
ωB97	33	13
ωB97X	30	10
ωB97XD	21	10
HSE06	23	13
B2PLYP	21	9

[a] Experimental data: FCS [122], NH_2 [123, 124], and H_2CN [125–127]. Note that the value for ν_5 has been excluded due to difficulties in the assignment of the experimental band as described in ref. 33.

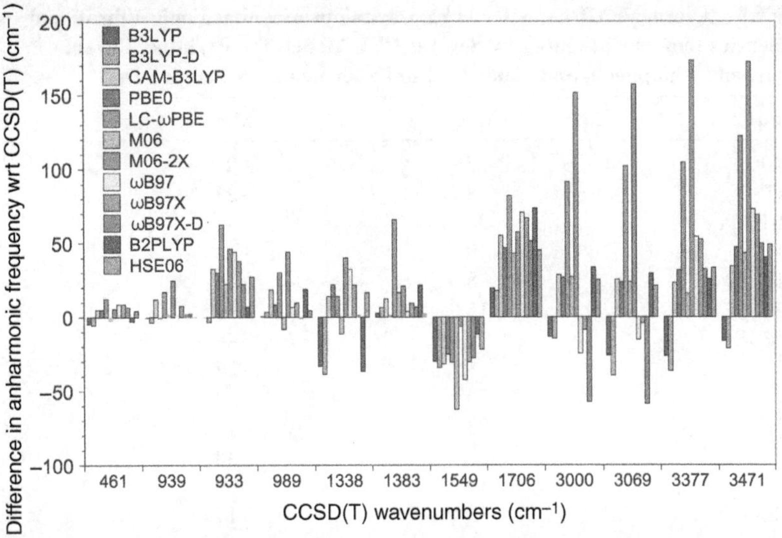

Figure 7.2 Performance of different DFT approaches in prediction of anharmonic frequencies. Relative discrepancies from the values computed at CCSD(T) level for all normal modes of H_2CN, NH_2, and FCS, listed according to increasing wavenumbers. Values from ref. 34.

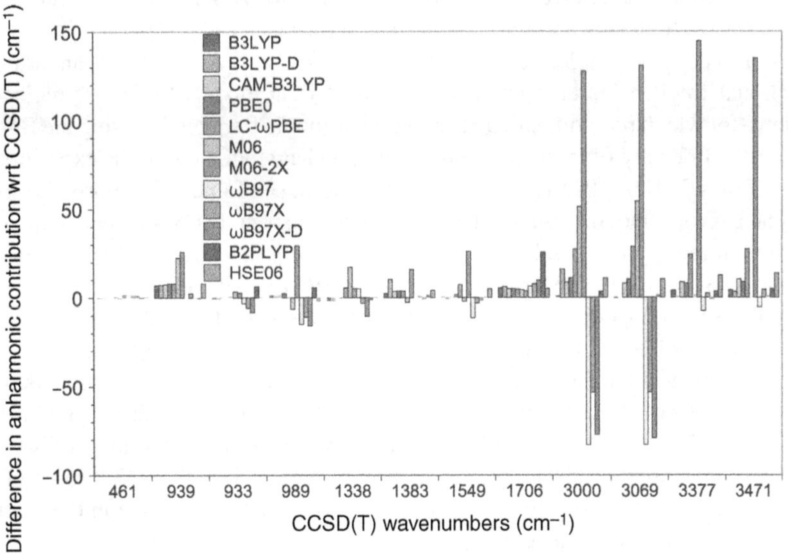

Figure 7.3 Performance of different DFT approaches in predicting anharmonic contributions. Relative discrepancies from the values computed at CCSD(T) level for all normal modes of H_2CN, NH_2, and FCS, which are listed according to increasing wavenumbers. Values from ref. 34.

Table 7.3 Averaged MUEs, MIN MAX Deviations of Anharmonic Vibrational Frequencies (cm^{-1}) Computed by Several DFT Models for Pyridine, Furan, Pyrrole, and Thiophene, and Compared to Experimental Reference Data[a]

Method	Mue	Min	MAX
B3LYP	7	−21	17
B3LYP-D	8	−27	13
CAM-B3LYP	20	—	51
B2PLYP	10	−13	34
B2PLYP-D	8	−18	27
PBE0	15	−10	47
LC-ωPBE	39	—	90
B97-1	8	−24	13
ωB97	25	−20	76
ωB97X	23	−14	66
ωB97XD	21	−48	48
M06	19	−59	52
M06-2X	33	−14	157
HSE06	15	−9	48

[a] Experimental data: pyridine [131], furan and pyrrole [132], and thiophene [133].

yield significantly less accurate results, far off the accuracy required for spectroscopic studies. It is also noteworthy that the performance of the M06 functionals is significantly worsened by the addition of Hartree–Fock exchange (2X). It has also been checked that large errors in anharmonic corrections obtained at the M06/N07D and ωB97/N07D levels are not removed by larger basis sets [34].

As a second example, let us consider the overall accuracy of anharmonic vibrational frequencies computed by the DFT/N07D models for larger molecular systems, for which previous calculations of anharmonic frequencies using the B3LYP or the B97-1 density functionals yielded very good agreement with the experimental results [38–40, 100, 101, 128, 129]. For this purpose, Table 7.3 reports the mean unsigned errors and maximum deviations with respect to experimental data for pyridine, furan, pyrrole, and thiophene. These results show clearly that functionals based on a Becke exchange term, including the B2PLYP method, provide anharmonic frequencies in good agreement with experiments, consistent with the accuracy of the harmonic contributions and anharmonic corrections discussed before. Qualitatively correct frequencies are also predicted by the PBE0 and HSE06 functionals, both showing MUEs of about 15 cm^{-1}. All the other DFT models considered yield MUEs in the range 20–40 cm^{-1}. Overall, the results presented in this section, in line with what has been already discussed for harmonic frequencies, show that most of the recently developed density functionals are significantly less accurate in the calculation of vibrational frequencies [34, 49, 130].

In addition to the analysis of frequencies, anharmonicity also affects spectral shapes. Let us first show the VCD spectra of (*R*)-1-methylphenyloxirane. In Figure 7.4 anharmonic corrections to frequencies are combined with harmonic computations of band intensities, and the results show that even a mixed

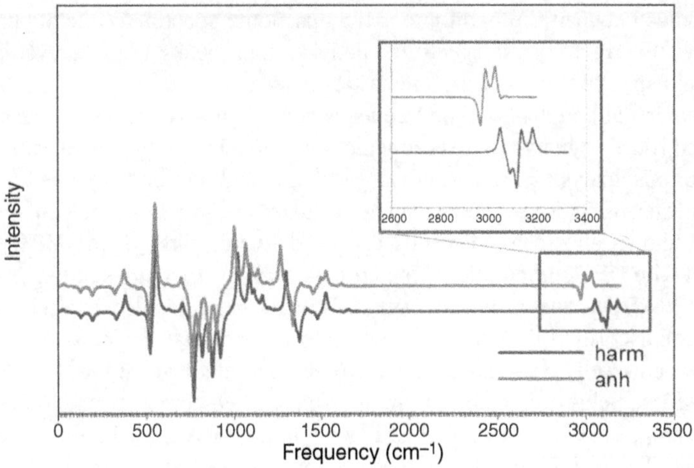

Figure 7.4 Harmonic and anharmonic VCD spectra for (*R*)-1-methylphenyloxirane [70].

harmonic/anharmonic treatment may cause nonnegligible changes in the spectrum appearance. However, the full treatment of anharmonicity requires to include anharmonic contributions in the intensities, so that the influence of overtones and combination bands to the spectrum shape can be included. This is exemplified in Figure 7.5, showing the VPT2-corrected IR spectrum of furan, a medium-sized molecule. The 1200–1800 cm^{-1} region is of particular interest, showing noticeable

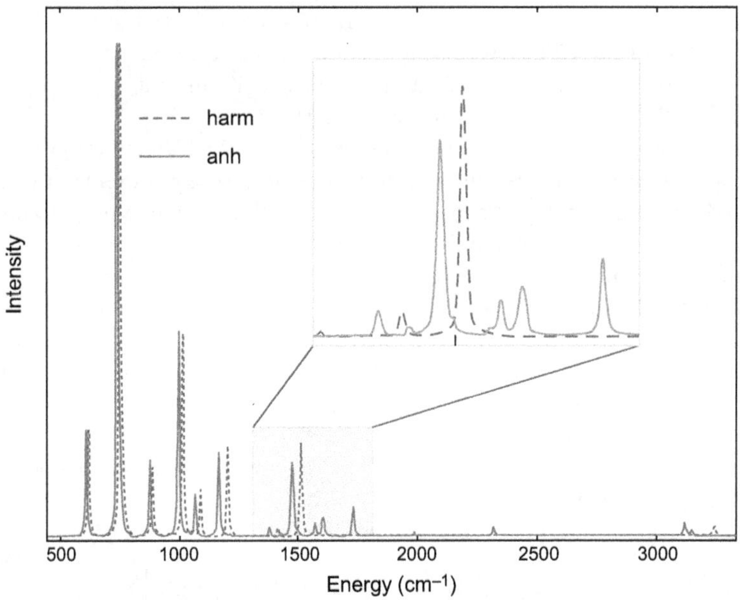

Figure 7.5 IR harmonic (dashed line) and anharmonic (solid line) absorption spectra of furan in the gas phase, spanning a range of 500–3500 cm^{-1}, calculated at the B3LYP/N07Ddiff level.

discrepancies between the harmonic and anharmonic spectra. While a single main peak is visible in the first spectrum, at least four peaks are observable in the anharmonic spectrum, which presents a far richer structure.

In summary, although harmonic frequency computations offer a valuable and well-established route to analyzing experimental vibrational data, in general, for a deeper and ubiquitous analysis of experimental results, it shall be recommended to include anharmonic terms, which at present can be evaluated even for large molecular systems. There, as shown above, good results can be delivered through the VPT2 scheme combined with DFT computations. In particular, in view of studies of larger systems, the B3LYP/N07D model provides results of remarkable reliability at a relatively low computational cost. In addition, it seems safe to state that the most cost-effective approach is currently to add anharmonic corrections calculated at the B3LYP level to harmonic force fields obtained using more sophisticated computational models, such as, for example, CCSD(T) or B2PLYP with large basis sets. In this respect, the B2PLYP/AVTZ//B3LYP/N07D approach combines accurate harmonic frequencies with the possibility of taking into account vibrational effects beyond the harmonic approximation even for quite large systems of biological and/or technological interest.

7.2.6 From One to Two Dimensions: 2D IR on a Time-Independent Perspective

In recent years, the field of vibrational spectroscopy has been further widened through the development of two-dimensional (2D) techniques [134–138], which permit to follow what happens to the vibrations in a molecule after an initial excitation and provides a 2D spectrum that reveals information on dephasing of the individual vibration and coupling between different vibrations. Thus, in principle, 2D IR signals can be useful for the determination of structural parameters as far as reliable and accurate theoretical methods for the interpretation of experimental spectra in terms of structural motifs and patterns are available [134].

A popular approach to interpret bidimensional spectra is the vibrational exciton Hamiltonian [139], which assumes vibrational modes to be represented in terms of local modes not interacting with the remaining vibrational degrees of freedom of the system, so that the exciton Hamiltonian can be expressed in terms of local-mode basis states. Within this framework, the key ingredients to construct the vibrational exciton Hamiltonian are vibrational frequencies of any pair of local modes (site energies) and the coupling among them. The latter can in turn be modeled with different approaches, the simplest being the so-called transition dipole coupling (TDC) model [140–142], which solves the problem in terms of the electrostatic coupling between point dipoles.

In such an approach the coupling constant k_{ij} between two local modes i and j is given by (in units of mdyn/Å amu) [141, 142]:

$$k_{ij} = \frac{0.1}{\epsilon} \frac{\left(\frac{\partial \mu_i}{\partial q_i}\right)\left(\frac{\partial \mu_j}{\partial q_j}\right) - 3\left[\left(\frac{\partial \mu_i}{\partial q_i}\right)\mathbf{n}_{ij}\right]\left[\left(\frac{\partial \mu_j}{\partial q_j}\right)\mathbf{n}_{ij}\right]}{\mathbf{R}_{ij}^3} \tag{7.60}$$

where $\partial\mu_k/\partial q_k$ are dipole partial derivatives with respect to local mode k (i.e., the vibrational transition dipole moment, in D/Å amu$^{-1/2}$), $\mathbf{R}_{ij}$ the distance (Å) between local oscillators i and j, and $\mathbf{n}_{ij}$ unit vector along that direction. The dielectric constant ϵ is assumed to be unity in the gas phase. Once k_{ij} is obtained, the elements of the exciton Hamiltonian, that is, the site energies ϵ_i and the coupling β_{ij} between oscillators i and j, are

$$\beta = \frac{\hbar k_{ij}}{2(k_i k_j)^{1/4}} \qquad \epsilon_i = \hbar k_i^{1/2} \tag{7.61}$$

where k_k are local-mode force constants.

Because of its simplicity, the TDC model has been extensively used to evaluate vibrational couplings, but there exist many examples proving that it is not quantitatively reliable [143].

Despite this, refinements of the TDC model have been suggested by including higher order multipole contributions [144, 145]. A more accurate approach, especially when coupled to ab initio methods, is the Hessian matrix reconstruction (HMR) analysis [146, 147], which, starting from ab initio geometry optimization and vibrational analysis, transforms the information into frequencies of well-defined local modes and intermode vibrational couplings, so that once again the corresponding exciton Hamiltonian is constructed.

By considering two interacting normal vibrations (Q_j and Q_j) to be related to two local modes (q_1, q_2) (the normal modes of a smaller part of the system) by a proper unitary transformation [147], it follows that

$$\begin{pmatrix} Q_1 \\ Q_2 \end{pmatrix} \simeq \mathrm{U}\begin{pmatrix} q_1 \\ q_2 \end{pmatrix} = \begin{pmatrix} \cos\theta & \sin\theta \\ -\sin\theta & \cos\theta \end{pmatrix}\begin{pmatrix} q_1 \\ q_2 \end{pmatrix} \tag{7.62}$$

with the mixing angle θ being

$$\theta = \tfrac{1}{2}\arctan\left(\frac{2k_{ij}}{k_i - k_j}\right) \tag{7.63}$$

By means of the same transformation, both local-mode vibrational frequencies (or site energies ϵ_i) and the intersite coupling β can be obtained:

$$\begin{pmatrix} \epsilon_i & \beta \\ \beta & \epsilon_j \end{pmatrix} = \mathrm{U}^{-1}\begin{pmatrix} \Omega_i & 0 \\ 0 & \Omega_j \end{pmatrix}\mathrm{U} \tag{7.64}$$

where Ω_i is the vibrational frequency corresponding to the normal mode Q_i. Using the same transformation, the force constant of each local mode, k_i, as well as the coupling, k_{ij}, are obtained from the normal-mode force constants. Once again β_{ij} and ϵ_i are obtained through Eq. 7.61.

7.3 MODELING VIBRATIONAL SPECTRA OF SYSTEMS IN CONDENSED PHASE

The development of computational approaches to vibrational spectra of systems in the condensed phase requires the analysis of the physics of the "solvated" system, thus introducing in the model ways for treating the interaction between the system and its surrounding. This is a general requirement for a procedure to be successful in the description (and further prediction) of spectra; it is in fact well known that all spectral features (frequencies, intensities, and bandshape) are hugely influenced by the interactions between the system and the environment.

There are basically two ways of computationally treating solvated systems: The first is the use of discrete approaches in which both the target system and the environment are treated at the atomistic level. The alternative is represented in the so-called focused approaches, where only a limited portion of the system (the solute) is treated at the atomistic level, whereas all the rest ("the solvent," or more generally "the environment") lacks part of its explicit description and is viewed as a continuum. Many intermediate approaches are possible, however, in which the focused part is progressively extended so as to include the target molecule and a small part of the surrounding environment, leaving the rest to be treated as a continuum [148, 149].

From a computational point of view, for a given theoretical level, purely continuum solvation approaches are by far less expensive than explicit models. Such a feature is especially relevant if one is compelled to exploit QM techniques, such as those usually required to study problems in the field of computational spectroscopy. In this case, in fact, the modern implementations of continuum solvation models are able to keep the cost of the calculation for solvated systems basically the same as the corresponding calculation for the isolated system [149].

7.3.1 Continuum Solvation Models: Efficient Way to Account for Solvent Effects in Vibrational Spectroscopy

It has currently been demonstrated in the literature that, due to the progress of computational quantum chemistry, a realistic description of vibrational frequencies for polyatomic molecules in solution is now feasible.

Due to the huge number of numerical applications to chemical systems of the calculation of harmonic frequencies within a continuum solvation framework, in this contribution we will only focus on the methodological developments, the presentation of selected applications only having the role of illustrating the concepts.

The extension of continuum solvation modes to evaluate vibrational frequencies of molecular systems in solution was pioneered by Rivail and co-workers in the 1980s [150] by exploiting a semiempirical QM molecular model coupled with a continuum description of the medium. Further extension to ab initio QM methods, including the treatment of electron correlation effects and electrical and mechanical anharmonicities, was then proposed [151–153] in the framework of the polarizable continuum model (PCM). Wang et al. [154] used an ab initio self consistent reaction field (SCRF) Onsager model to compute vibrational frequencies at different levels of the ab initio QM molecular theory, and the G-COSMO model was applied by Stefanovich and Truong to the evaluation of vibrational frequencies at the DFT level [155]. The multipole self consistent reaction field (SCRF) model, developed by the group of Rivail, was also extended to the calculation of frequency shifts at the HF, MP2, and DFT levels, including nonequilibrium effects [156]. Since these pioneering studies, the most relevant contributions in this field are those related to the development of the polarizable continuum model family of methods [149], which have been extended to the treatment of solvent-induced vibrational frequency shifts and vibrational intensities in a unified and coherent formulation [149, 157]. Thus, models to treat several vibrational spectroscopies, such as IR [158], Raman [159], IR linear dichroism [160], VCD [161], VROA [162], and 2D IR [163], accounting also for mechanical anharmonic effects [164], have been proposed and tested by including local field effects in the formulation as well as an incomplete solute–solvent regime (nonequilibrium) and, when necessary, by extending the model to the treatment of specific solute–solvent (or solute–solute) interactions.

Before commenting on the relatively recent progress in such a field, it is worth mentioning the theory of vibrations of molecules in a continuum environment, originally formulated from a classical point of view. Although the role of classical theories is nowadays quite marginal, understanding this constitutes the basis of knowledge in understanding the QM treatment of the subject, which will be described in Section 7.3.1.2.

7.3.1.1 Classical Approaches The state-of-the-art of classical models for the description of solvent-induced frequency shifts has been briefly reviewed by Rao et al. [165]. Such classical methodologies have been mostly based on continuum solvation models, for which an analytical expression can be derived. The most important contributions in this field are those due to West and Edwards [166], Bauer and Magat [167], Kirkwood [168], Buckingham [169, 170], Pullin [171], and Linder [172], all based on a description of the solvated systems in terms of the Onsager model [173], which represents the solvated solute as a polarizable point dipole in a spherical cavity immersed in a continuum, infinite, homogeneous, and isotropic dielectric medium. It is especially important to cite in this context the work of Bauer and Magat [167] in which the solvent-induced shift Δv on the vibrational frequency v of a given normal mode is

$$\frac{\Delta v}{v} = C \frac{\epsilon_0 - 1}{2\epsilon_0 + 1} \tag{7.65}$$

where C is a constant depending on the solute [174] and ϵ_0 the solution (solvent) static dielectric permittivity.

Investigation in the literature in this field shows that special efforts were done in the past to investigate the relation between IR intensities in the gas (A^{gas}) and liquid (A^{sol}) phases in the case of pure liquids [175–177] and systems in solution [169, 178–181]. Almost all the classical models for solvent effects on IR intensities, such as the ones due to Buckingham [169, 170], Mecke [182], Polo and Wilson [176], Mirone [181], and Warner and Wolfsberg [183], are based on the Onsager continuum model.

The classical approaches cited above focus on the ratio between integrated intensities in gas and solution phases, $f = A^{\text{sol}}/A^{\text{gas}}$, for which the following relation holds [157, 184]:

$$f = \left(\frac{\mathbf{E}^{\text{sol}}}{\mathbf{E}^{\text{gas}}}\right)^2 \tag{7.66}$$

where $\mathbf{E}^{\text{sol}}$ and $\mathbf{E}^{\text{gas}}$ are the vibrating electric fields acting on the molecule in the liquid and gas phases. Actually, $\mathbf{E}^{\text{sol}}$ is defined as the microscopic *local electric field* acting on the molecule, which is different from the macroscopic Maxwell field $\mathbf{E}^M$ acting inside the liquid. In the Onsager approach, the local field is written as a function of the Maxwell field and the electric dipole moment μ of the molecule, so that $\mathbf{E}^{\text{sol}}$ is expressed as the sum of two terms: The one depending on $\mathbf{E}^M$ is called the "cavity field" and the other, which is related to the dipole moment, is the "reaction field":

$$\mathbf{E}^{\text{sol}} = \frac{3\epsilon_0}{2\epsilon_0 + 1}\mathbf{E}^M + \frac{2\epsilon_0 - 2}{(2\epsilon_0 + 1)r^3}\mu \tag{7.67}$$

where ϵ_0 is again the static dielectric permittivity of the liquid (the solution).

The electric dipole moment μ in Eq. 7.67 can be in turn written as

$$\mu = \mu_{\text{perm}} + \alpha\mathbf{E}^{\text{sol}} \tag{7.68}$$

where μ_{perm} is the permanent dipole moment of the isolated molecule and the $\alpha\mathbf{E}^{\text{sol}}$ term is the field-induced dipole moment. By assuming the reorientation time of the molecules to be larger than the vibration period of the radiation field, it is possible to assume that only the induced moment contributes to the vibrating electric field at the absorption frequency. With this assumption and by using the Lorenz–Lorentz equation, it is possible to derive an expression for $\mathbf{E}^{\text{sol}}$ as a function of ϵ, n, and $\mathbf{E}^M$. In addition, within the IR range of frequencies it is reasonable to assume the dielectric constant of the solution, ϵ, to be equal to the square of the solution refractive index n_s^2. With this assumption and by considering that, in order to have the same probing intensity I both in solution and in vacuo, it has to be $(\mathbf{E}^M/\mathbf{E}^{\text{gas}})^2 = 1/n_s$ [184], the Polo–Wilson equation for pure liquids ($n = n_s$) [176] is obtained,

$$f = \frac{1}{n}\left(\frac{n^2 + 2}{3}\right)^2 \tag{7.69}$$

as well as the Mallard–Straley [179] and Person [180] equation for solutions,

$$f = \frac{1}{n_s}\left[\frac{n^2 + 2}{(n^2/n_s^2) + 2}\right]^2 \tag{7.70}$$

Slightly different is Buckingham's approach [169, 170], which assumes the solution to be composed of small (with respect to the radiation wavelength) solvent macroscopic spheres comprising a single solute molecule and surrounded by pure solvent; each sphere is independent of the others (i.e., the solution is dilute). The ratio between the integrated absorption in solution and in the gas phase can be written as

$$f \propto \left(\frac{\partial \overline{\mathbf{M}}^{\text{sol}} / \partial Q}{\partial \mu^{\text{gas}} / \partial Q} \right)^2 \tag{7.71}$$

where $\overline{\mathbf{M}}^{\text{sol}}$ is the dipole of the sphere averaged over all solvent configurations and μ^{gas} is the dipole moment of the isolated molecule. It is possible to show that [169, 170]

$$\frac{\partial \overline{\mathbf{M}}^{\text{sol}}}{\partial Q} = \left[\frac{9n_s^2}{(n_s^2+2)(2n_s^2+1)} \right]^2 \frac{\partial \bar{\mu}^{\text{sol}}}{\partial Q} \tag{7.72}$$

where $\bar{\mu}^{\text{sol}}$ is the dipole moment of the solute molecule in a sphere that is very small relative to the macroscopic sphere. The factor in brackets arises from the oscillating dipole induced in the solvent portion between the microscopic and the macroscopic spheres. This part of the solvent interacts with the solute as a continuum.

By expanding μ^{sol} as a function of the dipole of the isolated molecule and the polarizability α of the molecule, it is possible to obtain an expression for $\partial \bar{\mu}^{\text{sol}} / \partial Q$ as a function of ϵ, the solute refractive index n, the solution refractive index n_s, and α [169, 170]. Note that the Buckingham approach accounts for nonequilibrium solvent effects (see below), described in terms of the optical dielectric constant ϵ_{opt}.

Similar to IR, classical theories have also been proposed in the literature for Raman intensities in solution [181, 185–191]. The starting point is again the definition of the "local field" $\mathbf{E}^{\text{sol}}$ acting on the molecule. In all cases the local field factor is defined as $f = S^{\text{sc}}_{\text{sol}} / S^{\text{sc}}_{\text{vac}}$, with S^{sc} being the scattering intensity.

The need for a local field correction in Raman spectra [157] was first suggested by Woodward and George [192] without any attempt to gain any quantitative evaluation of the effect [184]. Starting once again from the Onsager theory, Pivovarov derived an expression for the ratio between polarizability derivatives in solution and in vacuo (i.e., Raman intensities) [187, 188]:

$$f_P = \frac{\partial \alpha_{\text{eff}} / \partial Q}{\partial \alpha / \partial Q} = \frac{3n_s^2}{(2n_s^2+1)[1-(2n_s^2-2)/(2n_s^2+1)(\alpha/r^3)]^2} \tag{7.73}$$

where α_{eff} is the effective polarizability of the molecule in the cavity (see below for a discussion), α the polarizability of the isolated molecule, n_s the refractive index of the medium, and r the radius of the (spherical) cavity.

Still starting from Onsager's theory, Mirone et al. [181, 189, 191] reported the following expression for the ratio between Raman intensities in solution and in vacuo:

$$f_M = \left[\frac{3\epsilon_{\text{opt}}}{(2\epsilon_{\text{opt}} + 10)\left[1 - (2\epsilon_{\text{opt}} - 2)/(2\epsilon_{\text{opt}} + 1)(\alpha/r^3)\right]} \right]^4 \tag{7.74}$$

with $\epsilon_{\text{opt}} = n_s^2$. By further assuming the ratio α/r^3 to be approximated by using the Lorenz–Lorentz formula, eq. 7.74 becomes

$$f_M = \left[\frac{n^2 + 2}{(n^2/\epsilon_{\text{opt}}) + 2} \right]^4 \tag{7.75}$$

If specialized to pure liquids, the latter expression reduces to the one proposed by Eckhardt and Wagner [190].

A comparison between the QM PCM formulation of IR and Raman intensities (vide infra) and classical equations can be found elsewhere [158, 159].

To end this discussion on classical approaches to vibrational intensities, it is worth noting that, at least to the best of our knowledge, equations similar to the ones reported above have not been developed for VCD and VROA. This is reasonably due to the fact that they are recently developed techniques, so that the availability of QM techniques to be exploited in the comparison between experimental and calculated values has overcome the necessity of developing simple classical methodologies.

7.3.1.2 Quantum Mechanical Models The QM evaluation of vibrational frequencies and intensities for systems in the condensed phase by means of QM methods has been made feasible in recent years by the development of derivative techniques for QM wavefunctions [193] with the further inclusion of solvation effects (see Tomasi et al. [149] for a review paper on the state-of-the-art on this subject).

In fact, the strategy which is commonly followed in the QM calculation of the quantities entering the description of vibrational spectra of systems in the condensed phase is to start from the theory developed for isolated systems and to supplement that theory with solvent peculiarities. In the framework of continuum solvation methods, this implies the development of reliable and computationally affordable algorithms for the evaluation of (free) energy first and second derivatives with respect to nuclear coordinates and/or external electric or magnetic fields (if required), no matter what specific vibrational spectroscopy is to be considered.

Besides technicalities, for which the interested reader is addressed to the relevant literature [149], from a conceptual point of view, if one takes as a reference the calculation in vacuo, the presence of the solvent introduces several complications. In fact, besides the "direct" effect of the solvent on the solute electronic distribution (which implies changes in the solute properties, that is, dipole moment, polarizability, and higher order responses), it should be taken into account that "indirect" solvent effects exist, that is, the solvent reaction field perturbs the molecular PES. This implies that the molecular geometry of the solute (the PES minima) and vibrational frequencies (the PES curvature around minima in the harmonic approximation) are affected by the presence of a solvating environment. Also, the dynamics of the solvent molecules around the solute (the so-called "nonequilibrium effect") should be considered to gain a realistic picture of the system and, depending on the nature of the solute–solvent system, specific (solute–solvent or solute–solute) interactions can be present, which can largely affect the calculated properties [157, 184]. Last, it should be considered that in the framework of continuum solvation models the electric field acting on the molecule in the cavity is different from the Maxwell field in the dielectric: however, the response of the molecule to the external perturbation depends on the field locally acting on it ("local field" effects) [158–162, 194, 195]. This last effect modifies the solute response to external electric and magnetic fields (the radiation), that is, vibrational intensities. In order to develop a reliable continuum model for vibrational properties of systems in the condensed phase, all such effects should be accurately modeled [157, 184, 196].

Reaction Field Effects Exactly as for isolated molecules, the evaluation of vibrational frequencies and normal modes for solvated molecules requires the evaluation of energy derivatives with respect to nuclear coordinates calculated at the equilibrium nuclear configuration. Within the continuum solvation approach, the energetic quantity to be differentiated is the free energy [174]. The QM analogues for "vibrational intensities" depend on the spectroscopy under study, but in any case derivative methods are needed

Also, because such derivatives are to be evaluated at the equilibrium geometry, a key point is its determination on the solvated PES, which still requires a viable method to calculate free-energy gradients. The problem of the formulation of free-energy derivatives with respect to different perturbations (electric, magnetic, nuclear and mixed) within continuum solvation models has been treated in the past [149]. It is beyond the scope of this conceptual presentation to go into the details of such issues. It is sufficient here to recall some peculiarities which arise in the particular case of derivatives with respect to nuclear (normal) coordinates which determine the quantities of interest in vibrational spectra. In the case of the PCM family of continuum solvation models, in which a molecule-shaped cavity is used (instead of, e.g., spherical cavities), the analytical expressions of free-energy first and second derivatives include terms dependent on the cavity terms due to the fact that the cavity is assumed to "follow" the nuclear distortion along an optimization path or during the vibration. Although this peculiarity can improve the quality of the results because it is a more physical choice, it can cause numerical inaccuracy and instability. Such instability can cause artificial effects in the computed result, which can be severe in case further numerical differentiation is to be performed, such as the case of the evaluation of anharmonic contributions.

Since the first implementations of analytical free-energy second derivatives in the late 1990s, an enormous quantity of applications have appeared in the literature. It would be impossible, and beyond the scope of this chapter, to review such a field, starting from the simplest application, which is probably the calculation of harmonic frequencies to evaluate zero-point and thermal contributions to free energies, and ending with the use of calculated frequency data to interpret vibrational spectra. It is instead worth mentioning the evaluation of anharmonic effects within the framework of continuum solvation methodologies, which has received only little attention so far and has only very recently been implemented in a reliable and computationally viable way [164]. In this framework, continuum solvation methods are particularly attractive, due to their already stated reliability coupled to computational costs fully comparable with those of the corresponding computations in the gas phase.

The evaluation of anharmonic force fields for solvated systems follows what has already been reported in Section 7.2.4 for isolated molecules. The key issue for an effective extension of the methodology to the condensed phase is related to the handling of the cavity containing the solute molecule. When the derivatives of the (free) energy in solution with respect to the atomic positions need to be computed, the cavity surface must be a continuous and smooth function of the same atomic positions. The importance of this issue has been recognized in recent years [197–199], mainly because smooth energy derivatives are needed in the study of solvent effects on the equilibrium structure of molecules. A robust and reliable method fulfilling these characteristics originally proposed by York and Karplus [200] has been extended to second derivatives and the corresponding fully analytical expression for the second derivatives of the PCM contribution to the energy [201, 202]. In the same framework, mechanical anharmonic effects for PCM have been recently tested [164] (see below for a brief discussion on the performances of the method).

To end this short section on reaction field effects, we remark on another implication of the calculation of free-energy second derivatives, that is, that the assessment of a complete equilibrium scheme or the account for vibrational and/or electronic nonequilibrium solvent effects [203, 204] should be done (see below for more details on this subject).

The Local Field By analogy to what has been reported in Section 7.3.1.1, also in the case of quantum mechanical formulations of continuum solvation models, there is the need to consider the formal noncoincidence between the radiation electric field (static $\mathbf{E}^{\mathrm{loc}}$ and optical $\mathbf{E}_{\omega}^{\mathrm{loc}}$) acting on the molecule in the cavity and the corresponding Maxwell fields in the medium, E and E_{ω} [157–159, 161, 162, 184].

However, the response of the molecule to the external perturbation clearly depends on the field locally acting on it. The so-called local field effect is normally taken into account by means of the classical approaches already discussed (Section 7.3.1.1), where macroscopic quantities are related to the microscopic electric response of the liquid constituents in the gas phase by using a simple multiplicative factor. In particular, it is assumed that [205]

$$\mathbf{E}_{\omega}^{\mathrm{loc}} = \frac{n^2+2}{3}\mathbf{E}_{\omega} \qquad \mathbf{E}^{\mathrm{loc}} = \frac{\epsilon+2}{3}\mathbf{E} \tag{7.76}$$

A more general framework to treat local field effects in linear and nonlinear optical processes in solution has been pioneered, among others [206], by Wortmann and

Bishop [207] using a classical Onsager reaction field model, but such a model has not been extended to vibrational spectra.

Still within a continuum solvation approach, a unified treatment of the "local field" problem has been developed in the PCM for (hyper)polarizabilities [194, 195, 208] and extended to several optical [209, 210] and spectroscopic properties, including IR, Raman, VCD, and VROA spectra [157–159, 161, 162, 196].

The basis of the PCM approach to the local field relies on the same assumption as in the classical approaches; that is, it is assumed that the "effective" field experienced by a molecule in the cavity is the sum of the reaction and cavity fields. The former is related to the dielectric polarization induced from the solute charge distribution, whereas the cavity field depends on the dielectric polarization induced by the applied field once the cavity has been created.

In PCM, reaction field effects on molecular properties and spectroscopies that arise from the cavity field effects are accounted for by introducing the concept of effective molecular response properties, which directly relate the response of the molecular solutes to the Maxwell field in the liquid, both static **E** and dynamic $\mathbf{E}_\omega$ [158, 194, 208].

By analogy to Onsager's theory, it is assumed that the response of the molecule to an external probing field can be expressed in terms of an "external dipole moment" μ^*, which is the sum of the molecular dipole moment and the dipole moment $\tilde{\mu}$ arising from the molecule-induced dielectric polarization. The details of the formulation of the latter term can be found elsewhere [158, 208]. Suffice it to recall that such a term is defined starting from an additional charge distribution spread on the cavity surface (the external charge), which represents the component of the solvent polarization that is induced by the external field *oscillating at the frequency* ω of the radiation and is computed through the optical dielectric constant of the medium.

All effective properties in solution, which account for cavity field effects, are formulated in terms of $\tilde{\mu}$. The IR intensity is [158]

$$A^{\text{sol}} = \frac{\pi N_A}{3 n_s c^2}\left(\frac{\partial(\mu+\tilde{\mu})}{\partial Q_i}\right)^2 \tag{7.77}$$

For Raman intensities

$$I^{\text{sc}}_{\text{sol}}(\mathbf{k}) = \frac{h}{4\pi\omega_{10}} I k^4 \frac{S_{\text{sol}}}{45} \tag{7.78}$$

where

$$S_{\text{sol}} = 45\alpha'^2_{\text{sol}} + 7\gamma'^2_{\text{sol}} \tag{7.79}$$

and

$$\alpha'_{\text{sol}} = \frac{1}{3}\frac{\partial}{\partial Q_i}(\alpha^*_{xx} + \alpha^*_{yy} + \alpha^*_{zz}) \tag{7.80}$$

$$\gamma'^2_{\text{sol}} = \frac{1}{2}\left\{\left(\frac{\partial\alpha^*_{xx}}{\partial Q_i} - \frac{\partial\alpha^*_{yy}}{\partial Q_i}\right)^2 + \left(\frac{\partial\alpha^*_{yy}}{\partial Q_i} - \frac{\partial\alpha^*_{zz}}{\partial Q_i}\right)^2 + \left(\frac{\partial\alpha^*_{zz}}{\partial Q_i} - \frac{\partial\alpha^*_{xx}}{\partial Q_i}\right)^2 \right.$$

$$\left. +6\left[\left(\frac{\partial\alpha^*_{xy}}{\partial Q_i}\right)^2 + \left(\frac{\partial\alpha^*_{yz}}{\partial Q_i}\right)^2 + \left(\frac{\partial\alpha^*_{xx}}{\partial Q_i}\right)^2\right]\right\} \tag{7.81}$$

In Eqs. 7.80 and 7.81, α^* is the dynamic effective polarizability [159].

Within the same framework, the VCD and VROA intensities can be obtained as

$$R_i^{\text{sol}} = \text{Im}\langle 0|(\mu_{\text{el}} + \tilde{\mu}_{\text{el}})|1\rangle_i \cdot \langle 0|\mu_{\text{mag}}|1\rangle_i \tag{7.82}$$

and

$$\langle 0|(\mu_{\text{el}} + \tilde{\mu}_{\text{el}})_\beta|1\rangle_i = \left(\frac{\hbar}{4\pi\nu_i}\right)^{1/2} \sum_{\lambda\alpha} S^{\text{sol}}_{\lambda\alpha,i}\left[(P^\lambda_{\alpha\beta})^{\text{sol}} + \tilde{P}^\lambda_{\alpha\beta}\right] \tag{7.83}$$

where $\tilde{P}^\lambda_{\alpha\beta}$ accounts for the "cavity field" terms and is obtained as

$$\tilde{P}^\lambda_{\alpha\beta} = \left(\frac{\partial\left\langle\psi_g^{\text{sol}}|(\tilde{\mu}_{\text{el}})_\beta|\psi_g^{\text{sol}}\right\rangle}{\partial X_{\lambda\alpha}}\right)_{r_0^{\text{sol}}} \tag{7.84}$$

$$= 2\left\langle\left(\frac{\partial\psi_g^{\text{sol}}}{\partial X_{\lambda\alpha}}\right)_{r_0^{\text{sol}}}|(\tilde{\mu}_{\text{el}})_\beta|\psi_g^{\text{sol}}\right\rangle \tag{7.85}$$

By using GIAOs, the magnetic contribution to R_i^{sol} can be obtained:

$$\langle 0|(\mu_{\text{mag}})_\beta|1\rangle_i = -(4\hbar^3\pi\nu_i)^{1/2}\sum_{\lambda\alpha} S^{\text{sol}}_{\lambda\alpha,i}(M^\lambda_{\alpha\beta})^{\text{sol}} \tag{7.86}$$

Notice that, similar to the electric contribution, μ_{mag} should also be reformulated to account for the cavity field effects. However, by assuming the response of the solvent to magnetic perturbations to be described only in terms of its magnetic permittivity (which is usually close to unity), it is reasonable to consider that the magnetic analogous to the electric cavity field gives minor contributions to R_i^{sol}. For this reason, to the best of our knowledge, such a matter has not been treated in the literature so far.

Moving to VROA, by analogy with what was already discussed, the relevant quantities become

$$(I_z^R - I_z^L)^{\mathrm{sol}}(90) = 6\tilde{\beta}(G')^2 - 2\tilde{\beta}(A)^2 \tag{7.87}$$

$$\tilde{\beta}(G')^2 = \frac{3\tilde{\alpha}^{\upsilon}_{ki}\tilde{G}'^{\upsilon}_{ki} - \tilde{\alpha}^{\upsilon}_{kk}\tilde{G}'^{\upsilon}_{ii}}{2} \tag{7.88}$$

$$\tilde{\beta}(A)^2 = \tfrac{1}{2}\omega_{\mathrm{rad}}\tilde{\alpha}^{\upsilon}_{ki}\epsilon_{kjl}\tilde{A}^{\upsilon}_{jli} \tag{7.89}$$

$$\tilde{\alpha}^{\upsilon}_{ki}\tilde{G}'^{\mu}_{ki} =< 0_{\mathrm{sol}}|\tilde{\alpha}_{ki}|1_{\mathrm{sol}} >< 1_{\mathrm{sol}}|\tilde{G}'_{ki}|0_{\mathrm{sol}} >= \frac{1}{2\omega_{\mathrm{sol}}}\left(\frac{\partial\tilde{\alpha}_{ki}}{\partial Q_{\mathrm{sol}}}\right)_0\left(\frac{\partial\tilde{G}'_{ki}}{\partial Q_{\mathrm{sol}}}\right)_0 \tag{7.90}$$

$$\tilde{\alpha}^{\upsilon}_{ki}\epsilon_{kjl}\tilde{A}^{\upsilon}_{jli} =< 0_{\mathrm{sol}}|\tilde{\alpha}_{ki}|1_{\mathrm{sol}} >< 1_{\mathrm{sol}}|\epsilon_{kjl}\tilde{A}_{jli}|0_{\mathrm{sol}} >= \frac{1}{2\omega_{\mathrm{sol}}}\left(\frac{\partial\tilde{\alpha}_{ki}}{\partial Q_{\mathrm{sol}}}\right)_0\epsilon_{kjl}\left(\frac{\partial\tilde{A}_{jli}}{\partial Q_{\mathrm{sol}}}\right)_0 \tag{7.91}$$

In these equations, α^* is the effective Raman electric dipole–electric dipole polarizability [159], $\tilde{\mathbf{G}}'$ is the imaginary part of the effective electric dipole–magnetic dipole polarizability, and $\tilde{\mathbf{A}}$ is the real part of the effective electric dipole–electric quadrupole polarizability, all containing both reaction field and cavity field effects. The definitions of the quantities above are

$$\tilde{\alpha}_{ki} = 2\sum_{r\neq l}\frac{\omega_{rl}\,\mathrm{Re}[< l|\mu_k + \tilde{\mu}_k|r >< r|\mu_i + \tilde{\mu}_i|l >]}{\omega_{rl}^2 - \omega_{\mathrm{rad}}^2} \tag{7.92}$$

$$\tilde{G}'_{ki} = -2\omega_{\mathrm{rad}}\sum_{r\neq l}\frac{\mathrm{Im}[< l|\mu_k + \tilde{\mu}_k|r >< r|m_i|l >]}{\omega_{rl}^2 - \omega_{\mathrm{rad}}^2} \tag{7.93}$$

$$\tilde{A}_{kil} = 2\sum_{r\neq l}\frac{\omega_{rl}\mathrm{Re}[< l|\mu_k + \tilde{\mu}_k|r >< r|\Theta_{il}|l >]}{\omega_{rl}^2 - \omega_{\mathrm{rad}}^2} \tag{7.94}$$

By resorting to the linear response framework [211, 212], the tensors in Eqs. 7.92–7.94 may be expressed as response properties by a proper choice of operators. More specifically,

$$\alpha^{*}_{ki} = - \langle\langle \tilde{\mu}_k; \quad \tilde{\mu}_i \rangle\rangle \, \omega_{\mathrm{rad}} \tag{7.95}$$

$$\tilde{G'}_{ki} = - \langle\langle \tilde{\mu}_k; \quad m_i \rangle\rangle \, \omega_{\mathrm{rad}} \tag{7.96}$$

$$\tilde{A}_{kil} = - \langle\langle \tilde{\mu}_k; \quad \Theta_{il} \rangle\rangle \, \omega_{\mathrm{rad}} \tag{7.97}$$

Solvation Regime When dealing with solvent effects, it is to be kept in mind that absorption and emission are time-dependent processes. In this case (as ours), the time dependency of the spectroscopic phenomenon is not explicitly accounted for in the computational methodology, that is, time-independent methods are exploited. Moreover, the motions associated to the degrees of freedom of the solvent molecules involve different time scales. In the particular case of vibrational spectroscopy, typical vibration times being on the order of 10^{-14}–10^{-12}s, it is clear that the orientational component of the solvent polarization cannot instantaneously readjust to follow the oscillating "solute," so that a nonequilibrium solute–solvent system should in principle be considered.

According to the standard approach in the field [149, 174], also in the case of vibrational spectroscopies, the solvent polarization can be formally viewed as composed of diverse contributions each related to the various degrees of freedom of the solvent molecules. Usually such contributions are grouped in two terms [149, 213]: The "slow" polarization accounts for the motions which are slower than those involved in the physical phenomenon under examination (the molecular vibration in our case), whereas the "fast" polarization includes the faster contributions. Then, it is further assumed that only the slow motions are instantaneously equilibrated to the momentary molecule charge distribution. The fast terms instead do not readjust, so that a nonequilibrium solvent–solute system arises.

In the case of vibrations of solvated molecules such a two-term partition has been proposed in the literature [203]. In this case the slow term includes the contributions arising from translation and rotation of the molecule as a whole, whereas the fast term accounts for the electronic and vibrational molecular motions. As a result of a shift from an equilibrated solute–solvent system, the fast polarization changes so as to equilibrate with the new solute charge distribution, whereas the slow term remains fixed to the value corresponding to the solute charge distribution of the initial state; in this situation a nonequilibrium solute–solvent regime occurs.

After a pioneering exploration of such a field due to the group of Rivail [156], such a conceptual scheme has been proposed and implemented for PCM to treat nonequilibrium effects on IR frequencies and intensities [203], Raman activities [204], and VCD rotational strengths [161].

In the particular case of Raman effects the dynamic aspects are to be described in terms of two time scales. One is connected to the vibrational motions of the nuclei [204], the other to the oscillation of the radiation electric field, that is, to oscillations in the solute electronic density [159]. In the presence of a solvent medium, both time scales originate nonequilibrium effects in the solvent response, being in general much faster than the time scale of the solvent inertial response, so that both should be (and have been) taken into account in the formulation of the methodology.

Specific Solute–Solute and Solute–Solvent Effects A well-known disadvantage of continuum solvation methodologies is their inadequacy to treat strong interacting solute–solvent systems. This does not mean that specific interactions are totally discarded in such methodologies but that only a portion of them, namely the mean electrostatic contribution, is considered, whereas the other terms, connected to the directionality of the interaction, are missed.

Often, in computational spectroscopy the observed phenomena are dominated by the so-called first-solvation-shell effects, such as hydrogen bonding. In such cases the continuum approach still remains a valid methodology for accounting for a part of the solute–solvent interaction, especially due to its low cost, especially in comparison with the generally higher cost of the calculation of the spectroscopic observable. However, if quantitative description of the experimental findings is to be obtained, methods for going beyond the pure continuum approach are to be developed. The usual approach consists of the redefinition of the solute as composed of the target chromophore plus the least possible number of explicit solvent molecules, further immersed in the continuum dielectric, which is able to take into account the bulk effect of the solvent. Obviously, there is in this case a kind of ambiguity in determining which part of the system constitutes the solute and which one the solvent, that is, where the solute stops and the continuum begins, so that such a choice is far from being straightforward but is also not only strongly dependent on the particular spectroscopy under study but, in the particular case of vibrations, is also in principle dependent upon the vibrational mode under investigation.

Some Illustrative Applications The aim of this short section is to illustrate the concepts discussed so far by resorting to some examples taken from the literature. This means that this section is far from being a complete overview of the field of the application of continuum solvation models to calculate vibrational spectra.

Let us start by discussing the role of mechanical anharmonic effects, which have been recently tested for PCM by Cappelli et al. [164]. There, formaldehyde and some simple mono- and dipeptide prototypes were considered, chosen for comparison with a previous study [214] on the calculation of amide mode anharmonicity in vacuo. As an example of the performance of the method, Figure 7.6 reports the correlation between calculated and experimental anharmonicities for a selection of mono- and dipeptides in aqueous solution. The quality of the results is very good, the only notable issue being the description of the amide A–amide A diagonal anharmonicity of formamide, which is however well known to dimerize under the conditions of the experimental measurements. Such an effect is discarded in the calculations.

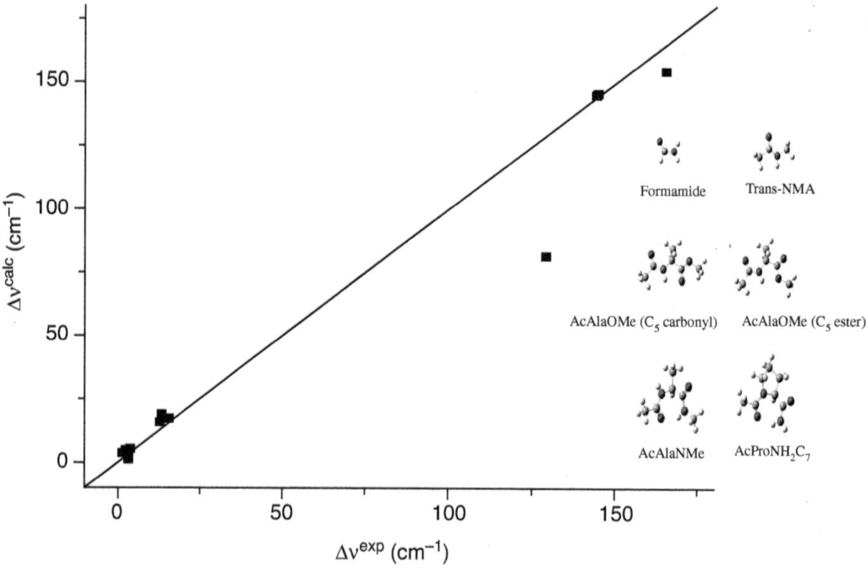

Figure 7.6 Correlation between experimental and calculated anharmonicities for some simple dipeptides in aqueous solution; values in cm^{-1}. Experimental values taken from ref. 214 and references therein: $R^2 = 0.96$.

Moving to the QM formulation of local field effects, let us show some examples illustrating the performance of the method with respect to classical theories and the reliability in predicting/describing experimental findings.

Table 7.4 reports a comparison between PCM and classical local (reaction + cavity) field factors for IR intensities of a series of simple aldehydes in aqueous solution [158]. Here $f_\mu + \bar{\mu}$ is obtained as the ratio between the calculated PCM IR intensity (with the account of both reaction and cavity field) and the corresponding value for the isolated molecule. PCM factors are generally different from classical formulations, and the difference is not limited to the same compound, but there is also a discrepancy in the observed trend in passing from one species to another. The largest difference between PCM and classical data is shown by the MSP equation, as reasonably expected due to the fact that MSP does not take into account any dependence on the static dielectric constant of the solvent. Such a dependence is instead present in the reaction field term of the PCM calculated data.

Table 7.5 focuses on the cavity field term only and compares PCM and the classical Onsager factor. The latter is a fairly good approximation of PCM when the molecular cavity is almost spherical [i.e., in the case of HCHO and $(CH_3)_3CCHO$]. In the other cases the difference is larger, the maximum deviation occurring for HC≡CCHO, whose molecular cavity, more or less of a cylindrical shape, is badly approximated by a sphere.

The same analysis applied to Raman activities (see Table 7.6) shows again that PCM factors are quite different from classical ones (Mirone's theory is reported; see

Table 7.4 IR Intensities: PCM ($f_{\mu+\tilde{\mu}}$) and Semiclassical f Values for Selection of Aldehydes in Aqueous Solution

	f_H	f_B	f_{MSP}	$f_{\mu+\tilde{\mu}}$
HCHO	1.667	1.442	1.210	1.434
CH_3CHO	1.567	1.355	1.188	1.586
CH_3CH_2CHO	1.640	1.418	1.205	1.609
$(CH_3)_3CCHO$	1.673	1.447	1.212	1.791

Note: From Hirota [178]:

$$f_H = \left[\frac{(n^2+2)(2\epsilon+1)}{3(n^2+2\epsilon)}\right]^2$$

From Buckingham [169, 170]:

$$f_B = \left[\frac{9\epsilon_{\text{opt}}}{(\epsilon_{\text{opt}}+2)(2\epsilon_{\text{opt}}+1)}\right]^2\left[\frac{(n^2+2)(2\epsilon+1)}{3(n^2+2\epsilon)}\right]^2$$

From Mallard-Straley [179], Person [180]:

$$f_{\text{MSP}} = \frac{1}{\sqrt{\epsilon_{\text{opt}}}}\left[\frac{n^2+2}{(n^2/\epsilon_{\text{opt}})+2}\right]^2$$

In all the previous equations n is the refractive index of the pure solute (i.e. of the aldehyde under study), whereas ϵ and ϵ_{opt} are the static or optical dielectric permittivities of the solvent, which is water in this case.

Eq. 7.75). In particular, the range of variation of PCM results (1.67–2.92) is smaller than the classical one (2.57–5.02), such a discrepancy one again being ascribed to the different level of description of the solute and of the cavity between PCM and the semiclassical model. In particular, in the case of CCl_4, the assumption of a spherical cavity is reasonable (the PCM cavity is not so different from a sphere) so that the differences are mainly due to the approximation of the molecule to a polarizable

Table 7.5 IR Intensities: Classical (f_{Ons}) and PCM ($f_{\tilde{\mu}}$) Cavity Field Factors for Selection of Aldehydes in Aqueous Solution

	f_{Ons}	$f_{\tilde{\mu}}$
HCHO		1.287
CH_3CHO		1.257
CH_3CH_2CHO	1.370	1.250
$(CH_3)_3CCHO$		1.285
HC≡CCHO		1.235

Note:

$$f_{Ons} = \left(\frac{3\epsilon_{\text{opt}}}{2\epsilon_{\text{opt}}+1}\right)^2$$

with ϵ_{opt} being the optical dielectric permittivity of the solvent, i.e. water in the present case.

Table 7.6 Raman Activities: f_{PCM} and f_M (see Eq. 7.75) of Various Solutes (CCl_4, $CHCl_3$, CH_3CN and H_2O) in Various Solvents

	CCl_4		$CHCl_3$		CH_3CN		H_2O	
Solvents	f_{PCM}	f_M	f_{PCM}	f_M	f_{PCM}	f_M	f_{PCM}	f_M
Carbon tetrachloride	2.51	3.67	2.55	3.62	2.26	3.24	2.08	3.15
Carbon disulfide	2.77	5.02	2.92	4.60	2.74	4.04	2.42	3.98
Acetonitrile	2.34	2.88	2.10	2.85	2.16	2.61	1.67	2.57

dipole, whereas in the other cases both assumptions cause the discrepancies. As a further discussion, it is also worth noting that, similar to all classical theories, the Mirone factor (Eq. 7.75) only depends on a given solute–solvent couple; that is, it is the same for all of the bands of the spectrum. On the contrary, in the case of PCM a different factor is obtained for each band (see Table 7.7), and thus, relative scattering factors change in passing not only from a medium to another but also from a band to another.

Since their formulation, no extended testing of vibrational nonequilibrium effects has been performed so far. The few studies present in the literature seem to suggest substantial effects in the evaluation of absolute infrared absorption intensities in polar solvents [213] but negligible effects in the case of Raman activities [204]. Vibrational harmonic frequencies seem instead to be even less sensitive. An example of this behaviour is shown in Table 7.8.

To end this brief section, a few examples illustrate the overall performance of the PCM in reproducing solvent effects on vibrational spectroscopy. Figure 7.7 shows the correlation between absolute calculated and experimental IR and Raman intensities for two push–pull molecules in solvents of different polarity (more details can be found in Corni et al. [216]. The quality of the results is good, especially by considering that the reproduction of absolute intensities is a delicate task. Even more representative of the performances which can be obtained from the continuum solvation approach in reproducing vibrational spectra is the case shown in Figure 7.8. There, calculated IR and VCD spectra of (*s*)-nacetylproline amide in the gas phase and aqueous solution in the range of the amide-I mode are reported, in comparison with experimental data from Oh et al. [217].

Table 7.7 Raman Activities: f_{PCM} for the Various Bands of H_2O in Different Solvents

	f_{PCM}			f_M
Solvents	δ	ν_s	ν_{as}	All
Carbon tetrachloride	1.53	2.08	1.89	3.15
Carbon disulfide	1.65	2.42	2.15	3.98
Acetonitrile	1.39	1.67	1.57	2.57

Note: For the sake of comparison f_M is also shown; δ is the HOH bending mode, ν_s is the symmetric O–H stretching mode, and ν_{as} is the asymmetric one.

Table 7.8 Comparison between Equilibrium (eq) and Nonequilibrium (neq) B3LYP/6-31G(d) Frequency ($\Delta\nu$, cm^{-1}) and Intensity ($\Delta\epsilon$, km/mol) Shifts with Respect to Gas Phase

	DMK			MEK			SBMK			TBMK		
	eq	neq	exp	eq	neq	exp	eq	neq	exp	eq	neq	exp
					$\Delta\nu$							
1,2-Dichloroethane	29	24	27	27	22	27	27	22	24	26	22	22
Acetonitrile	33	25	26	31	23	27	31	24	24	30	23	22
					$\Delta\epsilon$							
1,2-Dichloroethane	92	49	46 ± 10	99	55	55 ± 7	114	64	64 ± 7	112	61	68 ± 6
Acetonitrile	113	49	55 ± 7	121	54	55 ± 7	152	76	68 ± 6	137	59	65 ± 7

Note: Abbreviations: Dimethyl ketone (DMK), methyl ethyl ketone (MEK), *sec*-buthyl methyl ketone (SBMK), and *tert*-butyl methyl ketone (TBMK). Experimental data from ref. 215.

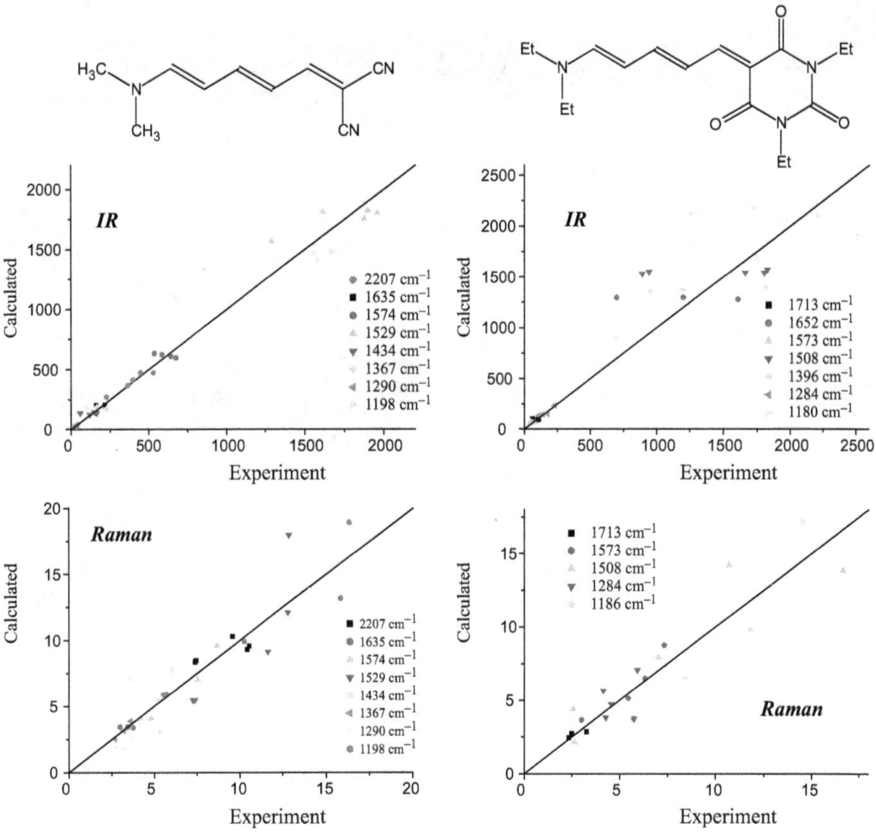

Figure 7.7 Correlation between experimental and PCM calculated IR intensities and Raman activities (with reaction, cavity field, and nonequilibrium effects) for selected modes of the two push–pull molecules depicted in various solvents. Values from ref. 216. Units are km/mol for IR intensities and $10^{-4}\,\mathrm{cm}^2\,\mathrm{g}^{-1/2}$ for Raman activities.

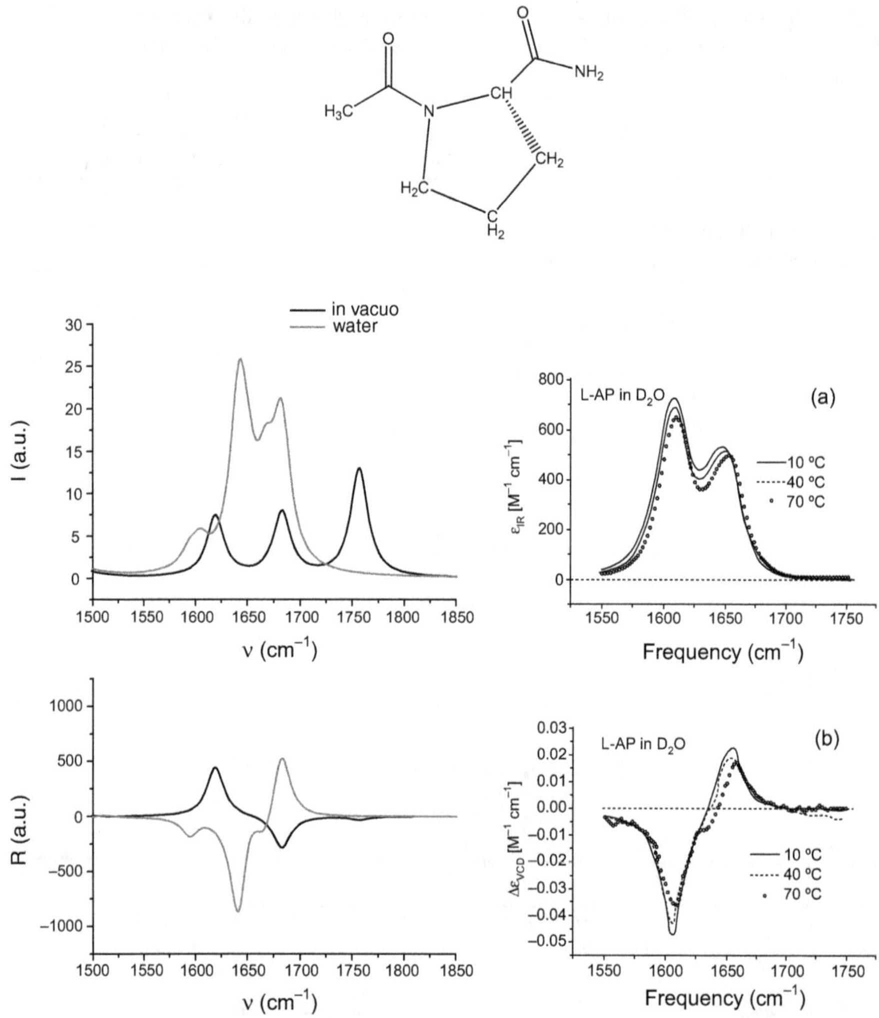

Figure 7.8 Comparison between calculated and experimental IR and VCD spectra of (*s*)-*N*-methyl-acetylproline amide. Both reaction and cavity field effects are considered in the PCM calculated values. Data from ref. 218.

The results of the calculation show that the double band in the IR spectrum cannot be reproduced by discarding solvent effects. The differences between vacuum and solution are even more dramatic in the case of VCD, where the correct $-/+$ pattern is reproduced only if solvation effects are taken into account, whereas a wrong spectrum is obtained in the other case. Note that the spectra shown in Figure 7.8 result from the combination of reaction field effects not only in the calculation of the spectroscopic quantities (cavity field is also included in intensities) but also in the assessment in the relative populations of the various conformers, which change markedly moving from vacuo to aqueous solution. In particular, in this case the combination of conformational (indirect) and direct solvent effects (on the spectroscopic observable) is what leads to a good reproduction of the experimental observations.

7.4 CONCLUSIONS AND PERSPECTIVES

In this chapter an overview of the theoretical foundations of time-independent approaches to vibrational spectroscopies has been reported, and their performances have been assessed through some illustrative examples. Obviously the latter are far from being exhaustive, due on the one hand to the large literature available (e.g., in the case of the calculation of IR frequencies) and on the other hand to the lack of extensive studies (e.g., regarding most of the issues discussed for solvation effects on intensities). However, the examples we have chosen should give a general view of the state-of-the art of the field up to date. In particular, it should be quite clear that, at least for semirigid systems, results of good interpretative valence can be obtained by time-independent computations, both in the gas phase and in solution.

Besides their relatively low computational cost, a noticeable advantage of time-independent over time-dependent approaches stems from the fact that they work directly in the frequency domain, thus giving access to a detailed analysis of vibrational contributions. Moreover, time-independent computations permit a reduced dimensionality treatment of the vibrational problem, which paves the was toward a better understanding of spectroscopic properties of macromolecular systems. Finally, anharmonic corrections to both frequencies and intensities can actually be evaluated even for medium-to-large molecular systems through VPT2 approaches in the gas phase and in solution. It should also be stressed that the approaches presented in this chapter permit us to compare directly experimental and theoretical spectra as well as to evaluate and dissect both dynamic and environmental effects which determine spectral properties. Overall, the computational strategies here presented, together with their integration into a computational chemistry package, allow for straightforward but at the same time detailed and accurate computational studies of various kinds of vibrational spectra, even for quite large systems of biological and/or technological interest, also by taking into account environmental effects.

However, the field is far from being established, and further developments are required, especially in order to extend the time-independent treatment to large macromolecular systems in the condensed phase or to allow for reliable studies in case some part of the vibrational problem is related to highly anharmonic PESs involving large-amplitude motions or solvent librations. There, use of effective QM/MM schemes combined with implicit solvent methods would probably represent a viable choice to reduce the computational cost of otherwise prohibitively expensive anharmonic frequency analysis. Another field where the potentialities of time-independent approaches have not yet been fully assessed is that of ultrafast vibrational spectroscopies, whose interest has enormously increased in the last years, due to their widely recognized role in the study of both the structure and dynamics of condensed phases and biomolecules.

Additional issues to be considered are related to the possibility of restricting the anharmonic treatment only to a small part of the total system, directly related to the spectroscopic observable, that is, the most intense bands in the IR spectrum, or to selected large-amplitude motions. There, time-independent computations might benefit from the application of simplified problem-tailored approaches, which would permit us to study a spectroscopically relevant part of a molecular subsystem by highly accurate electronic structure methods and vibrational approaches and at the same time to treat the remaining part at a more approximate level without losing the benefits of a unified and comprehensive picture. With regard to this issue, it is worth noting that hybrid DFT functionals, in particular the B3LYP family, perform well for a variety of molecular systems and vibrational properties, so that they represent the methods of choice for large biological systems, in particular if coupled to purposely developed "spectroscopic" basis sets. Eventually, more sophisticated post-Hartree–Fock methods can be employed to improve absolute frequencies of specific vibrational modes, while DFT-based approaches can still be used to obtain anharmonic corrections or solvent shifts.

REFERENCES

1. E.B. Wilson, J. C. Decius, P.C. Cross, *Molecular Vibrations: The Theory of Infrared and Raman Vibrational Spectra*, McGraw-Hill, New York, 1955.
2. S. Califano, *Vibrational State*, Wiley, London, 1976.
3. M. J. Frisch, M. Head-Gordon, J. A. Pople, *Chem. Phys.* **1990**, *141*, 189.
4. R. McWeeny, *Methods of Molecular Quantum Mechanic*, Academic, London, 1992.
5. D. P. Craig, T. Thirunamachandran, *Molecular Quantum Electrodynamics: An Introduction to Radiation Molecule Interaction*, Dover Publications, Mineola, NY, 1984.
6. D. A. Long, *The Raman Effect*, Wiley, Chichester, 2002.
7. G. Placzek, in *Handbuch der Radiologie*, Vol. 6, E. Marx, Ed., Akademische Verlagsgesellschaft, Leipzig, 1934, p. 205.
8. A. C. Albrecht, *J. Chem. Phys.* **1961**, *34*, 1476.
9. O. Quinet, B. Champagne, *J. Chem. Phys.* **2001**, *115*, 6293.
10. D. Rappoporta, F. Furche, *J. Chem. Phys.* **2007**, *126*, 201104.
11. P. Stephens, *J. Phys. Chem.* **1985**, *89*, 748.
12. R. Ditchfield, *Mol. Phys.* **1974**, *27*, 789.
13. L. Barron, A. D. Buckingham, *Mol. Phys.* **1971**, *20*, 1111.
14. L. D. Barron, *Molecular Light Scattering and Optical Activity*, Cambridge University Press, Cambridge, 1982.
15. T. Helgaker, K. Ruud, K. Bak, P. Jørgensen, J. Olsen, *Faraday Discuss.* **1991**, *99*, 165.
16. F. London, *J. Phys. Radium* **1937**. *8*, 397.
17. K. Ruud, A. Thorvaldsen, *Chirality* **2009**, *21*, E54.

18. V. Liegeois, K. Ruud, B. Champagne, *J. Chem. Phys.* **2007**, *127*, 204105.
19. G. Rauhut, P. Pulay, *J. Phys. Chem.* **1995**, *99*, 3093.
20. P. Sinha, S. E. Boesch, C. Gu, R. A. Wheeler, A. K. Wilson, *J. Phys. Chem. A* **2004**, *108*, 9213.
21. P. Andersson, P. Uvdal, *J. Phys. Chem. A* **2005**, *109*, 2937.
22. J. Merrick, D. Moran, L. Radom, *J. Phys. Chem. A* **2007**, *111*, 11683.
23. C. Herrmann, J. Neugebauer, M. Reiher, *New J. Chem.* **2007**, *31*, 818.
24. A. Ghysels, V. Van Speybroeck, E. Pauwels, S. Catak, B. R. Brooks, D. Van Neck, M. Waroquier, *J. Comp. Chem.* **2010**, *31*, 994.
25. V. Barone, M. Biczysko, G. Brancato, Extending the range of computational spectroscopy by QM/MM approaches: Time-dependent and time-independent routes, in *Advances in Quantum Chemistry, Vol. 59: Combining Quantum Mechanics and Molecular Mechanics, Some Recent Progresses in QM/MM*, J. Sabin, S. Canuto, Eds., Academic, New York, 2010, pp. 17–56.
26. Q. Cui, M. Karplus, *J. Chem. Phys.* **2000**, *112*, 1133.
27. K. Raghavachari, G. W. Trucks, J. A. Pople, M. Head-Gordon, *Chem. Phys. Lett.* **1989**, *157*, 479.
28. D. Begue, P. Carbonniere, V. Barone, C. Pouchan, *Chem. Phys. Lett.* **2005**, *416*, 206.
29. D. P. Tew, W. Klopper, M. Heckert, J. Gauss, *J. Phys. Chem. A* **2007**, *111*, 11242.
30. C. Puzzarini, V. Barone, *J. Chem. Phys.* **2008**, *129*, 084306/ 1–7.
31. C. Puzzarini, V. Barone, *Phys. Chem. Chem. Phys.* **2008**, *10*, 6991.
32. C. Puzzarini, V. Barone, *Chem. Phys. Lett.* **2008**, *462*, 49.
33. C. Puzzarini, V. Barone, *Chem. Phys. Lett.* **2009**, *467*, 276.
34. C. Puzzarini, M. Biczysko, V. Barone, *J. Chem. Theory Comput.* **2010**, *6*, 828.
35. B. G. Johnson, P. M. W. Gill, J. A. Pople, *J. Chem. Phys.* **1993**, *98*, 5612.
36. M. Wong, *Chem. Phys. Lett.* **1996**, *256*, 391.
37. A. D. Becke, *J. Chem. Phys.* **1993**, *98*, 5648.
38. V. Barone, *Chem. Phys. Lett.* **2004**, *383*, 528.
39. V. Barone, *J. Phys. Chem. A* **2004** *108*, 4146.
40. R. Burcl, N. C. Handy, S. Carter, *Spectrochim. Acta A* **2003**, *59*, 1881.
41. A. D. Boese, J. Martin, *J. Phys. Chem. A* **2004**, *108*, 3085.
42. Double and triple-ζ basis sets of N07 family, are available for download, available http://idea.sns.it, accessed April 17, 2010.
43. V. Barone, P. Cimino, E. Stendardo, *J. Chem. Theory Comput.* **2008**, *4*, 751.
44. V. Barone, P. Cimino, *Chem. Phys. Lett.* **2008**, *454*, 139.
45. V. Barone, P. Cimino, *J. Chem. Theory Comput.* **2009**, *5*, 192.
46. V. Barone, J. Bloino, M. Biczysko, *Phys. Chem. Chem. Phys.* **2010**, *12*, 1092.
47. V. Barone, M. Biczysko, P. Cimino, Interplay of stereo electronic vibrational and environmental effects in tuning physico-chemical properties of carbon centered radicals, in *Carbon-Centered Free Radicals and Radical Cations*, M. D. E. Forbes, Ed., Willey, Hoboken, NJ, 2010, pp. 105–139.
48. J. Bloino, M. Biczysko, F. Santoro, V. Barone, *J. Chem. Theory Comput.* **2010**, *6*, 1256.
49. M. Biczysko, P. Panek, G. Scalmani, J. Bloino, V. Barone, *J. Chem. Theory Comput.* **2010**, *6*, 2115.

50. Y. Zhao, D. G. Truhlar, *Theor. Chim. Acta* **2008**, *120*, 215.
51. Y. Zhao, D. G. Truhlar, *J. Chem. Phys.* **2006**, *126*, 194101.
52. Y. Zhao, D. Truhlar, *J. Chem. Theory Comput.* **2008**, *4*, 1849.
53. J. M. L. Martin, J. El-Yazal, J.-P. Francois, *Mol. Phys.* **1995**, *86*, 1437.
54. *NIST Chemistry Webbook*, available: http://webbook.nist.gov.chemistry, accesed December 9, 2006.
55. J. M. L. Martin, *J. Chem. Phys.* **1994**, *100*, 8186.
56. B. J. Lynch, D. G. Truhlar, *J. Phys. Chem. A* **2003**, *107*, 3898.
57. M. D. Halls, H. B. Schlegel, *J. Chem. Phys.* **1998**, *109*, 10587.
58. M. D. Halls, H. B. Schlegel, *J. Chem. Phys.* **1999**, *111*, 8819.
59. B. Galabov, Y. Yamaguchi, R. B. Remington, H. F. Schaefer III, *J. Phys. Chem. A* **2002**, *106*, 819.
60. J.R. Thomas, B.J. DeLeeuw, G. Vacek, T.D. Crawford, Y. Yamaguchi, H. F. Schaeffer III, *J. Chem. Phys.* **1993**, *99*, 403.
61. B. Galabov, T. Dudev, *Vibrational Intensities*, Elsevier Science, Amsterdam, 1996.
62. A. Sadlej, *Theor. Chim. Acta* **1992**, *79*, 123.
63. D. Jacquemin, E. Perpète, G. Scalmani, M. J. Frisch, R. Kobayashi, C. Adamo, *J. Chem. Phys.* **2007**, *126*, 144105/ 1–12.
64. T. Yanai, D. P. Tew, N. C. Handy, *Chem. Phys. Lett.* **2004**, *393*, 51.
65. C. A. Jimenez-Hoyos, B. G. Janesko, G. E. Scuseria, *Phys. Chem. Chem. Phys.* **2008**, *10*, 6621.
66. F. J. Devlin, P. J. Stephens, J. R. Cheeseman, M. J. Frish, *J. Phys. Chem. A* **1997**, *101*, 6322.
67. P. J. Stephens, F. J. Devlin, J.-J. Pan, *Chirality* **2008**, *20*, 643.
68. L. D. Barron, A. D. Buckingham, *Chem. Phys. Lett.* **2010**, *492*, 199.
69. P. Fristrup, P. R. Lassen, C. Johannessen, D. Tanner, P.-O. Norrby, K. J. Jalkanen, L. Hemmingsen, *J. Phys. Chem. A* **2006**, *110*, 9123.
70. V. Barone, M. Biczysko, J. Bloino, to be published.
71. S. Carter, S. Culik, J. M. Bowman, *J. Chem. Phys.* **1997**, *107*, 10458.
72. J. Koput, S. Carter, N. C. Handy, *J. Chem. Phys.* **2001**, *115*, 8345.
73. P. Cassam-Chenai, J. Lievin, *Int. J. Quan. Chem.* **2003**, *93*, 245.
74. P. Jensen, P. R. Bunker, Eds., *Computational Molecular Spectroscopy*, Wiley, Chichester, 2000.
75. E. Matyus, G. Czako, A. Csazar, *J. Chem. Phys.* **2009**, *130*, 134112.
76. J. M. Bowman, *Science* **2000**, *290*, 724.
77. J. M. Bowman, S. Carter, X. Huang, *Int. Rev. Phys. Chem.* **2003**, *22*, 533.
78. S. Carter, N. C. Handy, *J. Chem. Phys.* **2000**, *113*, 987.
79. J. Chaban, J. Jung, R. Gerber, *J. Chem. Phys.* **1999**, *111*, 1823.
80. G. Rauhut, T. Hrenar, *Chem. Phys.* **2008**, *346*, 160.
81. O. Christiansen, *Phys. Chem. Chem. Phys.* **2007**, *9*, 2942.
82. L. S. Norris, M. A. Ratner, A. E. Roitberg, R. B. Gerber, *J. Chem. Phys.* **1996**, *105*, 11261.
83. O. Christiansen, *J. Chem. Phys.* **2003**, *119*, 5773.
84. S. Carter, A. R. Sharma, J. M. Bowman, P. Rosmus, R. Tarroni, *J. Chem. Phys.* **2009**, *131*, 224106.
85. O. Christiansen, *J. Chem. Phys.* **2004**, *120*, 2149.

86. D. A. R. S. Latino, R. P. S. Fartaria, F. F. M. Freitas, J. Aires-De-Sousa, F. M. S. Silva Fernandes, *Int. J. Quant. Chem.* **2010**, *110*, 432.
87. O. Godsi, C. R. Evenhuis, M. A. Collins, *J. Chem. Phys.* **2006**, *125*, 104105.
88. S. Hirata, K. Yagi, *Chem. Phys. Lett.* **2008**, *464*, 123.
89. B. J. Braams, J. M. Bowman, *Int. Rev. Phys. Chem.* **2009**, *28*, 577.
90. M. Sparta, I. Hlyvik, D. Toffoli, O. Christiansen, *J. Phys. Chem. A* **2009**, *113*, 8712.
91. L. Pele, R. Gerber, *J. Chem. Phys.* **2008**, *128*, 165105.
92. S. Carter, N. C. Handy, *Chem. Phys. Lett.* **2002**, *352*, 1.
93. I. M. Mills, *Molecular Spectroscopy: Modern Research*, Academic, New York, 1972.
94. R. D. Amos, N. C. Handy, W. H. Green, D. Jayatilaka, A. Willets, P. Palmieri, *J. Chem. Phys.* **1991**, *95*, 8323.
95. K. Ruud, P. R. Taylor, T. Helgaker, *J. Chem. Phys.* **2003**, *119*, 1951.
96. K. Ruud, P. O. Astrand, P. R. Taylor, *J. Chem. Phys.* **2000**, *112*, 2668.
97. J. F. Gaw, A. Willetts, W. Green, N. C. Handy, in *Advances in Molecular Vibrations and Collision Dynamics*, J. M. Bowman, Ed., JAI Press, Greenwich, CT, 1990, pp. 169–185.
98. V. Barone, *J. Chem. Phys.* **2005**, *122*, 014108.
99. E. Miani, E. Cane, P. Palmieri, A. Trombetti, N. C. Handy, *J. Chem. Phys.* **2000**, *112*, 248.
100. V. Barone, G. Festa, A. Grandi, N. Rega, N. Sanna, *Chem. Phys. Lett.* **2004**, *388*, 279.
101. J. M. L. Martin, T. J. Lee, P. R. Taylor, J.-P. Francois, *J. Chem. Phys.* **1995**, *103*, 2589.
102. W. H. Miller, R. Hernandez, N. C. Handy, D. Jayatilaka, A. Willets, *Chem. Phys. Lett.* **1990**, *172*, 62.
103. A. Willets, N. C. Handy, W. H. Green, D. Jayatilaka, *J. Phys. Chem.* **1990**, *94*, 5608.
104. J. Vazquez, J. F. Stanton, *Mol. Phys.* **2006**, *104*, 377.
105. J. Vazquez, J. F. Stanton, *Mol. Phys.* **2007**, *105*, 101.
106. K. Raghavachari, G. W. Trucks, J. A. Pople, M. Head-Gordon, *Chem. Phys. Lett.* **1989**, *157*, 479.
107. A. L. L. East, W. D. Allen, S. J. Klippenstein, *J. Chem. Phys.* **1995**, *102*, 8506.
108. J. M. L. Martin, T. J. Lee, P. R. Taylor, J.-P. Francois, *J. Chem. Phys.* **1995**, *103*, 2589.
109. C. E. Dateo, T. J. Lee, *Spectrochim. Acta A* **1997**, *53*, 1065.
110. J. Breidung, W. Thiel, *Theor. Chem. Acc.* **1998**, *100*, 183.
111. J. F. Stanton, C. L. Lopreore, J. Gauss, *J. Chem. Phys.* **1998**, *108*, 7190.
112. J. F. Stanton, J. Gauss, *J. Chem. Phys.* **1998**, *108*, 9218.
113. J. Breidung, W. Thiel, J. Gaus, J. F. Stanton, *J. Chem. Phys.* **1999**, *110*, 3687.
114. C. Puzzarini, *J. Chem. Phys.* **2005**, *123*, 024313/ 1–14.
115. L. Bizzocchi, C. Degli Esposti, C. Puzzarini, *Mol. Phys.* **2006**, *104*, 2627.
116. C. Puzzarini, *J. Mol. Struct.* **2007**, *242*, 70.
117. A. Baldacci, P. Stoppa, A. Pietropolli Charmet, S. Giorgianni, G. Cazzoli, C. Puzzarini, C. W. Larsen, *J. Phys. Chem. A* **2007**, *111*, 7090.
118. D. Begue, P. Carbonniere, C. Pouchan, *J. Phys. Chem. A* **2005**, *109*, 4611.
119. P. Carbonniere, T. Lucca, C. Pouchan, N. Rega, V. Barone, *J. Comp. Chem.* **2005**, *26*, 384.
120. C. Puzzarini, V. Barone, *Phys. Chem. Chem. Phys.* **2008**, *10*, 6991.
121. C. Puzzarini, *Theor. Chem. Acc.* **2008**, *120*, 325.

122. N. Caspary, B. Wurfel, A. Thoma, G. Schallmoser, V. Bondybey, *Chem. Phys. Lett.* **1993**, *212*, 329.
123. A. McKellar, M. Vervloet, J. Burkholder, C. Howard, *J. Mol. Struct.* **1990**, *142*, 319.
124. J. Burkholder, C. Howard, A. McKellar, *J. Mol. Struct.* **1988**, *127*, 415.
125. M. Pettersson, J. Lundell, L. Khriachtchev, M. Rasanen, *J. Chem. Phys.* **1998**, *109*, 618.
126. D. Cowles, M. Travers, J. Frueh, G. Ellison, *J. Chem. Phys.* **1991**, *94*, 3517.
127. M. Jacox, *J. Phys. Chem.* **1987**, *91*, 6595.
128. P. O'Maley, *J. Mol. Struct.* **2005**, *755*, 147.
129. J. Bloino, M. Biczysko, O. Crescenzi, V. Barone, *J. Chem. Phys.* **2008**, *128*, 244105.
130. M. Biczysko, P. Panek, V. Barone, *Chem. Phys. Lett.* **2009**, *475*, 105.
131. T. D. Klots, *Spectrochim. Acta A* **1998**, *54*, 1481.
132. A. Mellouki, J. Lievin, M. Herman, *Chem. Phys.* **2001**, *2712*, 239.
133. T. D. Klots, R. D. Chirico, W. V. Steele, *Spectrochim. Acta A* **1994**, *50*, 765.
134. M. Cho, *Chem. Rev.* **2008**, *108*, 1331.
135. J. Jeon, S. Yang, J.-H. Choi, M. Cho, *Acc. Chem. Res.* **2009**, *42*, 1280.
136. Z. Ganim, H. S. Chung, A. W. Smith, L. P. Deflores, K. C. Jones, A. Tokmakoff, *Acc. Chem. Res.* **2008**, *41*, 432.
137. T. I. C. Jansen, J. Knoester, *Acc. Chem. Res.* **2009**, *42*, 1405.
138. Y. S. Kim, R. M. Hochstrasser, *J. Phys. Chem. B* **2009**, *113*, 8231.
139. S. Mukamel, *Principles of Nonlinear Optical Spectroscopy*, Oxford University Press, Oxford, 1995.
140. S. Krimm, Y. Abe, *Proc. Natl. Acad. Sci. U. S. A.* **1972**, *69*, 2788.
141. H. Torii, M. Tasumi, in *Infrared Spectroscopy of Biomolecules*, H. H. Mantsch, D. Chapman, Eds., Wiley, New York, 1996.
142. H. Torii, M. Tasumi, *J. Chem. Phys.* **1992**, *96*, 3379.
143. H. Torii, M. Tasumi, *J. Raman Spectrosc.* **1998**, *29*, 81.
144. P. Hamm, S. Wouterson, *Bull. Chem. Soc. Jpn.* **2002**, *75*, 985.
145. A. Moran, S. Mukamel, *Proc. Nat. Acad. Sci.* **2004**, *101*, 506.
146. S. Ham, S. Cha, J.-H. Choi, M. Cho, *J. Chem. Phys.* **2003**, *119*, 1451.
147. R. D. Gorbunov, D. S. Kosov, G. Stock, *J. Chem. Phys.* **2005**, *122*, 224904.
148. C. J. Cramer, D. G. Truhlar, *Chem. Rev.* **1999**, *99*, 2161.
149. J. Tomasi, B. Mennucci, R. Cammi, *Chem. Rev.* **2005**, *105*, 2999.
150. J.-L. Rivail, D. Rinaldi, in *Intermolecular Forces*, B. Pullman, Ed., Reidel, Dordrecht, 1982.
151. F. J. Olivares del Valle, J. Tomasi, *Chem. Phys.* **1987**, *114*, 231.
152. F. J. Olivares del Valle, M. Aguilar, S. Tolosa, J. C. Contador, J. Tomasi, *Chem. Phys.* **1990**, *143*, 371.
153. M. A. Aguilar, F. J. Olivares del Valle, J. Tomasi, *Chem. Phys.* **1991**, *150*, 151.
154. J. Wang, R. J. Boyd, A. Laaksonen, *J. Chem. Phys.* **1996**, *104*, 7261.
155. E. V. Stefanovich, T. N. Truong, *J. Chem. Phys.* **1996**, *105*, 2961.
156. J.-L. Rivail, D. Rinaldi, V. Dillet, *Mol. Phys.* **1996**, *89*, 1521.

157. C. Cappelli, Continuum solvation approaches to vibrational properties, in *Continuum Solvation Models in Chemical Physics: Theory and Applications*, B. Mennucci, R. Cammi, Eds., Wiley, Chichester, 2007, p. 167.

158. R. Cammi, C. Cappelli, S. Corni, J. Tomasi, *J. Phys. Chem. A* **2000**, *104*, 9874.

159. S. Corni, C. Cappelli, R. Cammi, J. Tomasi, *J. Phys. Chem. A* **2001**, *105*, 8310.

160. C. Cappelli, S. Corni, B. Mennucci, J. Tomasi, R. Cammi, *Int. J. Quant. Chem.* **2005**, *104*, 716.

161. C. Cappelli, S. Corni, B. Mennucci, R. Cammi, J. Tomasi, *J. Phys. Chem. A* **2002**, *106*, 12331.

162. M. Pecul, E. Lamparska, L. Frediani, C. Cappelli, K. Ruud, *J. Phys. Chem. A* **2006**, *110*, 2807.

163. A. Biancardi, C. Cappelli, B. Mennucci, R. Cammi, *J. Phys. Chem. B* **2010**, *114*, 4924.

164. C. Cappelli, S. Monti, G. Scalmani, V. Barone, *J. Chem. Theory Comput.* **2010**, *6*, 1660.

165. C. N. R. Rao, S. Singh, P. Senthilnathan, *Chem. Soc. Rev.* **1976**, *5*, 297.

166. W. West, R. Edwards, *J. Chem. Phys.* **1937**, *5*, 14.

167. E. Bauer, M. Magat, *J. Phys. Radium* **1938**, *9*, 319.

168. J. Kirkwood, *J. Chem. Phys.* **1939**, *7*, 911.

169. A. D. Buckingham, *Proc. Roy. Soc. (London)* **1958**, 169.

170. A. D. Buckingham, *Proc. Roy. Soc. (London)* **1960**, 32.

171. A. D. E. Pullins, *Spectrochim. Acta* **1958**, *13*, 125.

172. B. Linder, *J. Phys. Chem.* **1992**, *96*, 10708.

173. L. Onsager, *J. Am. Chem. Soc.* **1936**, *58*, 1486.

174. J. Tomasi, M. Persico, *Chem. Rev.* **1994**, *94*, 2027.

175. N. Q. Chako, *J. Chem. Phys.* **1934**, *2*, 644.

176. S. R. Polo, M. K. Wilson, *J. Chem. Phys.* **1955**, *23*, 2376.

177. A. A. Clifford, B. Crawford, *J. Phys. Chem.* **1966**, *70*, 1536.

178. E. Hirota, *Bull. Chem. Soc. Japan* **1954**, *27*, 295.

179. W. C. Mallard, J. W. Straley, *J. Chem. Phys.* **1957**, *27*, 877.

180. W. B. Person, *J. Chem. Phys.* **1958**, *28*, 319.

181. P. Mirone, *Spectrochim. Acta* **1966**, *22*, 1897.

182. R. Mecke, *Discuss. Faraday Soc.* **1950**, *9*, 161.

183. J. W. Warner, M. Wolfsberg, *J. Chem. Phys.* **1983**, *78*, 1722.

184. C. Cappelli, Theoretical models for the study of vibrational properties of molecules in solution, thesis, Scuola Normale Superiore, Pisa, 2001.

185. D. G. Rea, *J. Opt. Soc. Am.* **1959**, *49*, 90.

186. J. R. Nestor, E. R. Lippincott, *J. Raman Spectrosc.* **1973**, *1*, 305.

187. V. M. Pivovarov, *Opt. Spec.* **1959**, *6*, 60.

188. V. M. Pivovarov, *Opt. Spec.* **1960**, *9*, 139.

189. P. Mirone, *Chem. Phys. Lett.* **1969**, *4*, 323.

190. G. Eckhardt, W. Wagner, *J. Mol. Spectrosc.* **1966**, *19*, 407.

191. G. Fini, P. Mirone, P. Patella, *J. Mol. Spectrosc.* **1968**, *28*, 144.

192. L. A. Woodward, J. H. B. George, *Proc. Phys. Soc.* **1951**, *64*, 780.

193. Y. Yamaguchi, J. D. Goddard, Y. Osamura, H. Schaefer, *A New Dimension to Quantum Chemistry: Analytic Derivative Methods in Ab Initio Molecular Electronic Structure Theory*, Oxford University Press, New York, 1994.
194. R. Cammi, B. Mennucci, J. Tomasi, *J. Phys. Chem. A* **2000**, *104*, 4690.
195. R. Cammi, B. Mennucci, Macroscopic nonlinear optical properties from cavity models, in *Continuum Solvation Models in Chemical Physics: Theory and Applications*, B. Mennucci, R. Cammi, Eds., Wiley, Chichester, 2007, p. 238.
196. J. Tomasi, R. Cammi, B. Mennucci, C. Cappelli, S. Corni, *Phys. Chem. Chem. Phys.* **2002**, *4*, 5697.
197. H. Li, J. H. Jensen, *J. Comput. Chem.* **2004**, *25*, 1449.
198. P. Su., H. Li, *J. Chem. Phys.* **2009**, *130*, 074109.
199. A. W. Lange, J. M. Herbert, *J. Phys. Chem. Lett.* **2010**, *1*, 556.
200. D. M. York, M. Karplus, *J. Phys. Chem. A* **1999**, *103*, 11060.
201. G. Scalmani, M. J. Frisch, *J. Chem Phys.* **2010**, *132*, 114110.
202. M. J. Frisch et al., Gaussian 09, Revision A.02, Gaussian, Wallingford, CT, 2009.
203. C. Cappelli, S. Corni, R. Cammi, B. Mennucci, J. Tomasi, *J. Chem. Phys.* **2000**, *113*, 11270.
204. C. Cappelli, S. Corni, J. Tomasi, *J. Chem. Phys.* **2001**, *115*, 5531.
205. C. J. F. Bottcher, P. Bordewijk, *Theory of Electric Polarization,* Vol. II. *Dielectric in Time–Dependent Fields*, Elsevier, Amsterdam, 1978.
206. P. Macak, P. Norman, Y. Luo, H. Ågren, *J. Chem. Phys.* **2000**, *112*, 1868.
207. R. Wortmann, D. M. Bishop, *J. Chem. Phys.* **1998**, *108*, 1001.
208. R. Cammi, B. Mennucci, J. Tomasi, *J. Phys. Chem.* **A** **1998**, *102*, 870.
209. C. Cappelli, B. Mennucci, J. Tomasi, R. Cammi, A. Rizzo, G. Rikken, R. Mathevet, C. Rizzo, *J. Chem. Phys.* **2003**, *118*, 10712.
210. C. Cappelli, B. Mennucci, J. Tomasi, R. Cammi, A. Rizzo, *J. Phys. Chem. B* **2005**, *109*, 18706.
211. J. Olsen, P. Jørgensen, Time-dependent response theory with applications to self-consistent field and multi-configurational self-consistent field wave functions, in *Modern Electronic Structure Theory, Part II*, D. R. Yarkony, Ed., World Scientific, Singapore, 1995, p. 857.
212. R. Cammi, L. Frediani, B. Mennucci, K. Ruud, *J. Chem. Phys.* **2003**, *119*, 5818.
213. B. Mennucci, R. Cammi, J. Tomasi, *J. Chem. Phys.* **1998**, *109*, 2798.
214. J. Wang, R. M. Hochstrasser, *J. Phys. Chem. B* **2006**, *110*, 3798.
215. M. I. Redondo, M. V. Garcia, J. Morcillo, *J. Mol. Struct.* **1988**, *175*, 313.
216. S. Corni, C. Cappelli, M. Del Zoppo, J. Tomasi, *J. Phys. Chem. A* **2003**, *107*, 10261.
217. K.-I. Oh, J. Han, K.-K. Lee, S. Hahn, H. Han, M. Cho, *J. Phys. Chem. B* **2006**, *110*, 13335.
218. C. Cappelli, B. Mennucci, *J. Phys. Chem. B* **2008**, *112*, 3441.

8

TIME-INDEPENDENT APPROACHES TO SIMULATE ELECTRONIC SPECTRA LINESHAPES: FROM SMALL MOLECULES TO MACROSYSTEMS

MALGORZATA BICZYSKO AND JULIEN BLOINO

Scuola Normale Superiore, Pisa, Italy and Dipartimento di Chimica "Paolo Corradini," Università di Napoli Federico II, Naples, Italy

FABRIZIO SANTORO

CNR—Consiglio Nazionale delle Ricerche, Istituto di Chimica dei Composti OrganoMetallici, UOS di Pisa, Area della Ricerca del CNR, Pisa, Italy

VINCENZO BARONE

Scuola Normale Superiore di Pisa, Pisa, Italy

Computational Strategies for Spectroscopy: From Small Molecules to Nano Systems, First Edition.
Edited by Vincenzo Barone.

State-of-the-art time-independent approaches to simulate vibronic spectra from first principles, covering representative models available for systems ranging from small molecules with several interacting electronic states to macrosystems, are presented. For the former, the computation of rovibronic spectra beyond the Born–Oppenheimer (BO) and harmonic approximations is feasible, while large systems require simplified yet effective and reliable models. In this respect, prescreening procedures set in the harmonic approximation able to select a priori the most relevant transitions and generalized toward the prediction of the outcome of a variety of one- and two-photon electronic spectroscopies are presented. These approaches allow to compute absorption, emission, and circular dichroism spectra and can be adapted to a broad range of approximations on both the potential energy surfaces and the electronic transition moments (or tensors) to evaluate vibronic transitions within both "vertical" and "adiabatic" frameworks. Furthermore, these methods have been extended to account for the effect of the temperature and can be coupled with continuum models for the simulation of the effects of a surrounding medium on both the position and width of the spectral signal. The discussed cases illustrate the extent of the accuracy achievable through high-quality computational spectroscopy simulations and the amount of specific information deliverable.

8.1 INTRODUCTION

This chapter is devoted to time-independent computational methods for the simulation and analysis of the vibrational structure of electronic spectra. In general, an electronic excitation can involve any kind of electronic states, including neutral or ionic and bound or dissociative ones. In addition, the molecular structure can be very flexible, and the region of the electronic potential energy surfaces (PESs) relevant for the spectral signal can be strongly anharmonic. Finally, the initial and/or final electronic state can be subjected to strong nonadiabatic couplings, such as those triggered by the existence of a conical intersection, whose probability of occurrence increases with the dimension of the system. At the state-of-the-art, unfortunately, no unique rigorous method can be proposed for such a general situation. For small molecular systems, quantum mechanical (QM) approaches for computational spectroscopy are nowadays capable of providing results comparable to the most accurate experimental measurements [1]. Among the most challenging examples, we can mention theoretical spectroscopic studies of systems with several interacting electronic states, based on the evaluation of accurate ab initio post-Hartree–Fock PES, and variational calculations of rovibronic energy levels beyond the BO approximation [2–4]. The high accuracy of the results achievable in these cases clearly demonstrates the potentiality of computational chemistry studies to become key tools for the prediction and understanding of spectroscopic properties of all kinds of molecular systems. Suitable and effective approaches can be selected to face the specific problem under investigation and, within approximations and limitations that will become clear in the following, the lineshape of a variety of different spectroscopic signals for a wide range of systems (mainly quasi-rigid ones and with negligible nonadiabatic couplings) can be nowadays simulated with well-automatized methods. On these premises, it can be foreseen that, together with further developments, the latter will help to boost the computational studies in electronic spectroscopy in the next future. It shall be noticed that, up to very recently, direct relation between experimental and computed spectroscopic results have been often limited to comparison between energy levels (vibrational, electronic, and rovibrational), which as a rule must be extracted from the experimental data by a nontrivial interpretation (involving at the very least band assignment). Alternatively, when computed spectroscopic constants are compared with their experimental counterparts, the latter are generated by applying model Hamiltonians to simulate experimental spectra through trial-and-error procedures. Although very successful over the years, such approaches relied on the predetermined physical picture of the phenomena, which can be a significant limitation in complex cases. In the present chapter, we will present, instead, computational models which are designed to limit as much as possible the introduction of a priori restrictions on the nature of the spectroscopic phenomena, such as, for example, the possible anharmonic resonances or the number of simultaneously excited vibrations. Moreover, as anticipated above, a particular attention will be paid to approaches able to simulate the electronic spectra lineshape, allowing a direct vis-à-vis comparison between experimental and theoretical results in large systems. We will focus on time-independent eigenstate approaches for bound transitions based on a Taylor expansion of a local region of the PES at relevant stationary points. Complementary methods set in the time-dependent framework which allow for the computation of spectral shapes without requiring the knowledge of the molecular eigenstates are discussed in Chapter 10.

The chapter is organized as follows, after a brief summary of the theoretical background underlying the computation of vibronic spectra, a number of possible methods and approaches will be discussed and analyzed with the aim of helping the reader to choose the most suitable (and feasible) one for the particular problem of interest. Specifically, the exposition will be focused on time-independent methods for bound states that allow to obtain the spectral shape by summing over all the possible state-to-state vibronic transitions. These methods rely on the knowledge of all the relevant eigenstates of the system, and they represent the most natural choice for a detailed analysis and assignments of high-resolution spectra. On the other hand, the number of vibronic states of a given electronic state to be taken into account is in general very high and increases steeply with the molecular size since, considering, for example, a Franck–Condon transition, all the final vibrational states are accessible if they overlap with the initial vibrational state. Therefore, despite the existence of effective algorithms for the variational computation of vibronic eigenstates, the computation of all the required eigenstates for a typical electronic spectrum is restricted to small systems if no simplified assumption is made on the PES to support them. In this context, this chapter reviews a variety of models ranging from (electronic) multistate anharmonic models accounting for vibronic and rovibronic couplings, suitable for small molecules, to a hierarchy of models set within (or starting from) the harmonic approximation for the treatment of large systems, with hundreds of normal modes, of direct biological and technological interest. We will deal with both one- and two-photon (nonresonant) processes, highlighting in which limit their vibrational structure may be simulated within a common computational framework. In such a scheme, an analogous treatment can be feasible also for different spectroscopies.

A particular attention will be given to discuss and compare recently developed prescreening techniques which allow for an effective calculation of the spectra of sizable molecules within the harmonic approximation. Thanks to the remarkable level of automatization they have reached, those procedures have been suitable for implementation in almost blackbox procedures, and a number of them is now accessible even to nonspecialists in commercial packages [5–7] and in stand-alone codes [8–13].

Despite these progresses, many computational studies in closely related research fields but more focused on the development and application of accurate and effective methods for the description of electronic states or, for example, of the environmental effects on electronic levels still avoid the direct calculation of the vibronic spectral shape. Actually, the latter is commonly represented by a phenomenological lineshape superimposed on the computed vertical transition, considered coincident with the spectrum maximum. Here, an analysis in terms of the moments of the spectrum will give the possibility to rediscuss the theoretical foundations of such a comparison, highlighting its potentiality and limitations and clarifying in which models the two quantities are theoretically predicted to be coincident.

8.2 GENERAL FRAMEWORK

8.2.1 Born–Oppenheimer Approximation and Nonadiabatic Effects

Consider the molecular Hamiltonian for a system of N_a nuclei and n_e electrons

$$\hat{H} = \hat{T} + U + \hat{H}^{SO} \tag{8.1}$$

where $\hat{T}$ is the kinetic operator, U the potential energy, and $\hat{H}^{SO}$ the spin–orbit coupling. The translational motion of the center of mass (CM) can be exactly separated passing from a laboratory-fixed (LF) to a space-fixed (SF) reference frame, parallel to the LF frame and moving with the CM,

$$\hat{H} = -\frac{1}{2M}\nabla^2_{\mathrm{CM}} + \hat{T} + U + \hat{H}^{SO} \tag{8.2}$$

where $\hat{T} = \hat{T}_e + \hat{T}_N$, $\hat{T}_e$ and $\hat{T}_N$ being the kinetic electronic and nuclear rotational–vibrational operators, respectively (notice that, when the CM of the nuclei is adopted, a mass polarization term appears in $\hat{T}_e$ [14]). Many theoretical explorations and reviews have been dedicated to the general problem of obtaining tractable Hamiltonians ensuring the best separation between rotations and vibrations in systems undergoing large-amplitude internal motions (see, e.g., refs. 15 and 16 and the recent review by Meyer [14]). Such an optimal separation can be obtained by introducing a reference frame rotating with the molecule, referred to a body-fixed (BF) frame and transforming the $3 \times (N_a + n_e - 1)$ coordinates in the three Euler angles describing rotation, $3N_a - 6$ internal coordinates $\boldsymbol{R}$, and $3n_e$ electronic coordinates $\boldsymbol{r}$. Following Sutcliffe [15], one obtains the general BF molecular Hamiltonian

$$\hat{H}^{BF} = \hat{T}_{vr}\left(\hat{\boldsymbol{J}}, \hat{\boldsymbol{j}}, \boldsymbol{R}\right) + \hat{T}_v + \hat{T}_e + U(\boldsymbol{r}, \boldsymbol{R}) + \hat{H}^{SO}(\boldsymbol{r}, \boldsymbol{R}) \tag{8.3}$$

where the rotational Hamiltonian $\hat{T}_{vr}$ depends on the total BF angular momentum $\hat{\boldsymbol{J}}$ and the electronic angular momentum, sum of the total orbital and spin momenta, $\hat{\boldsymbol{j}} = \hat{\boldsymbol{L}} + \hat{\boldsymbol{S}}$. Here $\hat{T}_{vr}$ also includes the residual couplings with the internal coordinates. Defining the electronic Hamiltonian as $\hat{H}_e = \hat{T}_e + U(\boldsymbol{r}, \boldsymbol{R})$, the eigenvalues problem

$$\hat{H}_e \Phi_i(\boldsymbol{r}, \boldsymbol{R}) = V_i(\boldsymbol{R}) \Phi_i(\boldsymbol{r}, \boldsymbol{R}) \tag{8.4}$$

defines the adiabatic electronic wavefunctions $\Phi_i(\boldsymbol{r}, \boldsymbol{R})$ and electronic energy potential $V_i = (\boldsymbol{R})$. We now neglect the spin–orbit coupling $\hat{H}^{SO}$ and adopt the Born–Oppenheimer expansion for the full molecular wavefunction:

$$\Psi(\boldsymbol{r}, \boldsymbol{R}) = \sum_i \Phi_i(\boldsymbol{r}, \boldsymbol{R})\chi_i(\boldsymbol{R}) \tag{8.5}$$

By substituting Eq. 8.5 into Eq. 8.3, multiplying from the left by $\Phi_j(\boldsymbol{r}, \boldsymbol{R})^*$, and integrating over the electronic coordinates, we get a set of coupled differential equations for the nuclear wavefunction [17] $\chi_j(\boldsymbol{R})$,

$$\left[\hat{T}_N + V_j(\boldsymbol{R})\right]\chi_j(\boldsymbol{R}) - \sum_i \Lambda_{ji}\chi_i(\boldsymbol{R}) = E\chi_j(\boldsymbol{R}) \tag{8.6}$$

where the coupling terms Λ_{ji} are described below.

We now limit our discussion to semirigid molecules, where the BF reference frame is obtained by applying the Eckart conditions [18], and the vibrations are expressed in terms of the normal coordinates $\boldsymbol{Q}$. Moreover, for simplicity, we consider a rotationless molecule, so that the vibrational kinetic operator assumes the simple expression $\hat{T}_v = -1/2\nabla^2$. In this framework, we can further specify that

$$\Lambda_{ji} = \langle\Phi_j(\boldsymbol{r}, \boldsymbol{Q})|\nabla|\Phi_i(\boldsymbol{r}, \boldsymbol{Q})\rangle\nabla + \tfrac{1}{2}\langle\Phi_j(\boldsymbol{r}, \boldsymbol{Q})|\nabla^2|\Phi_i(\boldsymbol{r}, \boldsymbol{Q})\rangle \tag{8.7}$$

The coupling terms Λ_{ji} are made of two contributions, the nonadiabatic derivative couplings, $F_{ji}(\boldsymbol{Q}) = \langle\Phi_j(\boldsymbol{r}, \boldsymbol{Q})|\nabla|\Phi_i(\boldsymbol{r}, \boldsymbol{Q})\rangle$, which are vectors in the nuclear space, and the scalar nonadiabatic couplings [second term on the right-hand side (RHS) of Eq. 8.7]. By differentiating Eq. 8.4 with respect to the normal coordinates $\boldsymbol{Q}$, one easily obtains the following expression for the derivative couplings [17]:

$$F_{ji}(\boldsymbol{Q}) = \frac{\langle\Phi_j(\boldsymbol{r}, \boldsymbol{Q})|\nabla\hat{H}_e|\Phi_i(\boldsymbol{r}, \boldsymbol{Q})\rangle}{V_i(\boldsymbol{Q}) - V_j(\boldsymbol{Q})} \tag{8.8}$$

It is straightforward to see that the derivative couplings diverge when the two electronic states, $|\Phi_i\rangle$ and $|\Phi_j\rangle$, become degenerate. The same trend is observed also for the scalar coupling, as shown by the relation

$$\langle\Phi_j(\boldsymbol{r}, \boldsymbol{Q})|\nabla^2|\Phi_i(\boldsymbol{r}, \boldsymbol{Q})\rangle = \nabla F_{ji}(\boldsymbol{Q}) - \langle\Phi_j(\boldsymbol{r}, \boldsymbol{Q})\nabla\big|\nabla\Phi_i(\boldsymbol{r}, \boldsymbol{Q})\rangle \tag{8.9}$$

From the above discussion, it follows that, when the electronic states involved in an electronic transition are well separated in energy (at least in the region of the nuclear coordinates that is relevant for the spectrum), the nonadiabatic coupling terms Λ_{ji} can be neglected, thus obtaining the standard BO adiabatic approximation. In this limit, the calculation of the spectrum can be pursued according to a single (electronic) state adiabatic approach. This means that, even when the frequency region of interest spans several electronic states, the spectra can be computed as a sum of spectra from a single initial electronic state to a single final electronic state. When this is not possible, a multistate approach is necessary. These cases encompass the Jahn–Teller effect, where the electronic degeneracy is imposed by symmetry, cases of accidental degeneracy and cases of symmetry-allowed degeneracy occurring when two states belong to with different irreducible representation (irrep) in a high-symmetry subdomain of the nuclear coordinates but belong to the same irreps when the symmetry is lowered by internal motions. State couplings may alter electronic spectra, even when the interacting adiabatic states are not very close in energy, as in the case of intensity-borrowing mechanisms such as the Herzberg–Teller effect. As we will briefly discuss in the following sections, in these conditions, it is still possible to follow an adiabatic approach and include these couplings through the perturbation theory. Up to now, we have discussed nonadiabatic effects arising from the vibronic coupling of electronic and internal motions. We conclude this description noticing that relevant nonadiabatic effects can also have a more complex rovibronic nature when rotational motions also play a role. This is the case, for example, of the Renner–Teller (RT) effect, which is relevant for molecules that can assume linear configurations, where $C_{\infty v}$ or $D_{\infty v}$ symmetries determine the existence of degenerate states. The nature of this coupling is nicely seen on a triatomic AB_2 molecule, adopting the Jacobi coordinates, with the BB distance written r and its reduced mass as μ_r, the distance between the barycenter of BB and A as R (with the associated mass μ_R) and their angle as γ. In this case, fixing z to be the BB axis, the RT part of the kinetic operator $\hat{T}_r$ in Eq. 8.3 is [19, 20]

$$\hat{T}^{\mathrm{RT}} = -\frac{(\mu_r r^2 - \mu_R R^2)\hat{J}_z \hat{j}_z}{\sin^2 \gamma} z \tag{8.10}$$

which diverges for linear configurations, causing relevant nonadiabatic effects.

The divergence of nonadiabatic couplings in Eq. 8.7 leads to computational problems in full quantum treatments, like those reviewed in this chapter, due to the difficulty to integrate them over the vibrational wavefunctions. Different strategies have been proposed to face these problems. Probably the most traditional solution is to work within the so-called crude-adiabatic approximation, where the electronic wavefunctions at a fixed nuclear configuration $\boldsymbol{Q}_0$ are used in the expansion in Eq. 8.5,

$$\Psi(\boldsymbol{r}, \boldsymbol{Q}) = \sum_i \Phi_i(\boldsymbol{r}, \boldsymbol{Q}_0)\chi_i(\boldsymbol{Q}) \tag{8.11}$$

Such an expansion is, for example, at the base of the Herzberg–Teller theory of vibronic coupling we will present in Section 8.3.1.1. Unfortunately, the convergence of the expansion given in Eq. 8.11 is very slow, making its use unpractical in many cases [21]. A different approach is to define a new set of diabatic states (represented by the row of states $|\Psi^{\mathrm{d}}\rangle$), through an orthogonal transformation of a small subset of the relevant adiabatic states $|\Psi^{\mathrm{ad}}\rangle$:

$$|\Psi^{\mathrm{d}}\rangle = |\Psi^{\mathrm{ad}}\rangle \boldsymbol{T}(\boldsymbol{Q}) \tag{8.12}$$

Strictly diabatic states are defined in order to have vanishing derivative couplings [17],

$$\boldsymbol{F}^{\mathrm{d}} = \boldsymbol{T}^{\mathrm{T}} \boldsymbol{F}^{\mathrm{ad}} \boldsymbol{T} + \boldsymbol{T}^{\mathrm{T}} \nabla \boldsymbol{T} = 0 \tag{8.13}$$

where F^{ad} is the matrix of the derivative couplings in the adiabatic representation. Notice that in general the condition in Eq. 8.13 can be fulfilled exactly only when the whole set of adiabatic states is taken into account or a single nuclear degree of freedom is considered [21]. Therefore, strictly diabatic states cannot be defined in general. Nonetheless, several routes exist to define quasi-diabatic states by finding a transformation in Eq. 8.12 to make the derivative couplings negligible. These methods can be divided depending on the data they use as input to define the diabatic representation: While derivative-based methods work directly on the adiabatic derivative couplings, energy-based methods simply use energetic parameters. Finally, property-based methods work by imposing a predetermined behavior to an easily computable molecular property, suitably chosen in order to represent the "diabatic" character of the electronic states. Many of these methods are based, for instance, on the diagonal or off-diagonal matrix elements of the electric dipole expressed in the adiabatic basis set. An extensive review on this matter can be found in work by Köppel [21], while examples of adiabatic PESs along with their diabatic counterparts obtained through block-diagonalization of the electronic Hamiltonian are shown later in Figure 8.15.

In the diabatic representation we have to solve a set of coupled equations analogous to Eq. 8.6 [22],

$$\left[\hat{T}_v + W_{jj}(\boldsymbol{Q})\right]\chi_j(\boldsymbol{Q}) + \sum_i W_{ji}\chi_i(\boldsymbol{Q}) = E\chi_j(\boldsymbol{Q}) \tag{8.14}$$

In this case, the nonadiabatic couplings have been omitted since they are negligible by definition. At the same time, the diabatic states are not anymore eigenfunctions of the electronic Hamiltonian in Eq. 8.4, and therefore new potential couplings, W_{ji}, arise,

$$W_{ji}(\boldsymbol{Q}) = \langle \Phi_j^d(\boldsymbol{r}, \boldsymbol{Q}) | \hat{H}_e | \Phi_i^d(\boldsymbol{r}, \boldsymbol{Q}) \rangle \tag{8.15}$$

that are nonetheless smooth functions of the nuclear coordinates $\boldsymbol{Q}$. Eq. 8.14 must be specified for the particular system under investigation. It is recast in a time-dependent perspective in Chapter 10.

8.2.2 Time-Independent Expressions for Different Electronic Spectroscopies

Time-independent approaches to the calculation of electronic spectra are characterized by a summation of individual transitions from the initial eigenstates $|\Psi_i\rangle$ weighted by their Boltzmann population ρ_i to the manifold of the final eigenstates $|\Psi_f\rangle$ of the system.

In the following, whenever not stated differently, we neglect rotational states and report the radiative intensity expressions after taking an isotropic average over all the possible molecular orientations. Therefore a molecular state will be regarded as a vibronic (vibro-electronic) state.

8.2.2.1 One-Photon Processes: OPA, OPE, ECD One-photon absorption (OPA), electronic circular dichroism (ECD) and one-photon emission (OPE) spectra can be generally written in the form

$$I = \alpha\omega^{\beta} \sum_i \sum_f \rho_i \left[\boldsymbol{d}_{if}^{A} \cdot \boldsymbol{d}_{if}^{B*} \right] \delta(\omega_f - \omega_i \pm \omega) \tag{8.16}$$

where δ is the Dirac delta function, $\boldsymbol{d}_{if}^{A}$ and $\boldsymbol{d}_{if}^{B}$ are transition dipole moment integrals between the initial and final vibronic states, and the the sign $\pm$ holds for emission/absorption, respectively.

The intensity for one-photon absorption, emission, or electronic circular dichroism is obtained by replacing I, α, β $\boldsymbol{d}_{if}^{A}$, and $\boldsymbol{d}_{if}^{B}$ with the values given below:

$$\text{OPA}: \quad I = \epsilon(\omega) \quad \alpha = \frac{10\pi\mathcal{N}_A}{3\epsilon_0 \ln(10)\hbar c} \quad \beta = 1 \quad \boldsymbol{d}_{if}^{A} = \boldsymbol{d}_{if}^{B} = \boldsymbol{\mu}_{if}$$

$$\text{OPE}: \quad I = \frac{I_{\text{em}}}{N_i} \quad \alpha = \frac{2\mathcal{N}_A}{3\epsilon_0 c^3} \quad \beta = 4 \quad \boldsymbol{d}_{if}^{A} = \boldsymbol{d}_{if}^{B} = \boldsymbol{\mu}_{if}$$

$$\text{ECD}: \quad I = \Delta\epsilon(\omega) \quad \alpha = \frac{40\mathcal{N}_A \pi}{3\epsilon_0 \ln(10)\hbar c^2} \quad \beta = 1 \quad \boldsymbol{d}_{if}^{A} = \boldsymbol{\mu}_{if}, \quad \boldsymbol{d}_{if}^{B} = \Im(\boldsymbol{m}_{if})$$

where $\epsilon(\omega)$ is the molar absorption coefficient for a given angular frequency ω and $\Delta\epsilon(\omega)$ is the difference (referred to as anisotropy) between the molar absorption coefficients ϵ_L and ϵ_R relative to the left (L) and right (R) circularly polarized light, respectively. For one-photon emission, I_{em}/N_i is the energy emitted by 1 mol s^{-1}, where N_i is the number of molecules in the initial state i. Finally, $\mathcal{N}_A$ is the Avogadro constant, c the speed of light, and ϵ_0 the vacuum permittivity; $\boldsymbol{\mu}_{if}$ is the electric transition dipole moment between the vibronic states i and f and $\Im(\boldsymbol{m}_{if})$ is the imaginary part of the magnetic transition dipole moment between the vibronic states i and f, $\boldsymbol{m}_{if}$.

A more detailed description of the calculation of I for OPA, OPE, and ECD is available elsewhere [23] and Chapter 2.

We will use from now on the symbol $\boldsymbol{d}^Z$ to represent indifferently $\boldsymbol{d}^A$ and $\boldsymbol{d}^B$. We define the integral $\boldsymbol{d}_{if}^Z$ and its conjugate as

$$\boldsymbol{d}_{if}^Z = \langle \Psi_i | \boldsymbol{d}^Z | \Psi_f \rangle \tag{8.17}$$

$$\boldsymbol{d}_{if}^{Z*} = \langle \Psi_f | \boldsymbol{d}^{Z*} | \Psi_i \rangle \tag{8.18}$$

where Ψ_i and Ψ_f are the molecular wavefunctions of the vibronic states i and f, respectively.

In the BO approximation, the total molecular state i can be written, in Dirac notation, as $|\Psi_i\rangle = |\Phi_{\underline{i}} \chi_{r(\underline{i})}\rangle$, where hereafter an underlined subscript indicates an electronic state ($\underline{i}$) while a vibrational state r associated to the electronic state $\underline{i}$ is indicated with $r(\underline{i})$. Moreover, the dipole moment $\boldsymbol{d}^Z$ is separated into its electronic and nuclear components, $\boldsymbol{d}_e^Z$ and $\boldsymbol{d}_N^Z$, respectively, and therefore, from Eq. 8.17, we have

$$\langle \Psi_i | \boldsymbol{d}^Z | \Psi_f \rangle = \langle \chi_{r(\underline{i})} \Phi_{\underline{i}} | \boldsymbol{d}_e^Z | \Phi_{\underline{f}} \chi_{s(\underline{f})} \rangle + \langle \chi_{r(\underline{i})} \Phi_{\underline{i}} | \boldsymbol{d}_N^Z | \Phi_{\underline{f}} \chi_{s(\underline{f})} \rangle \tag{8.19}$$

Since we are dealing with electronic transitions ($j \neq i$), the second term on the RHS of the previous equation is null due to the orthogonality of the electronic wavefunctions, so that

$$\langle \Psi_i | \boldsymbol{d}^Z | \Psi_f \rangle = \langle \chi_{r(\underline{i})} \Phi_{\underline{i}} | \boldsymbol{d}_e^Z | \Phi_{\underline{f}} \chi_{s(\underline{f})} \rangle \tag{8.20}$$

or equivalently

$$\langle \Psi_i | \boldsymbol{d}^Z | \Psi_f \rangle = \langle \chi_{r(\underline{i}))} | \boldsymbol{d}_{e,if}^Z | \chi_{s(\underline{f})} \rangle$$

where $\boldsymbol{d}_{e,if}^Z$ is the electronic transition dipole moment $\langle \Phi_{\underline{i}} | \boldsymbol{d}_e^Z | \Phi_{\underline{f}} \rangle$.

In these conditions, Eq. 8.16 can be written as

$$I = \alpha \omega^\beta \sum_{\underline{i}} \sum_{r(\underline{i})} \sum_{\underline{f}} \sum_{s(\underline{f})} \rho_i \left[\langle \chi_{r(\underline{i})} | \boldsymbol{d}_{e,if}^{\mathrm{A}} | \chi_{s(\underline{f})} \rangle \langle \chi_{s(\underline{f})} | \boldsymbol{d}_{e,if}^{B*} | \chi_{r(\underline{i})} \rangle \right] \delta(\omega_f - \omega_i \pm \omega) \tag{8.21}$$

8.2.2.2 ***Two-Photon Absorption and Circular Dichroism*** The BO approximation also makes it possible to work out tractable expressions to investigate the vibrational effect for two-photon processes like, for example, vibrational resonance Raman (RR), two-photon absorption (TPA), and two-photon circular dichroism (TPCD), which are schematically depicted in Figure 8.1 along with the one-photon ones.

Two-Photon Absorption As extensively discussed in Chapter 2, the extinction coefficient in TPA for an excitation by two equal photons is [24]

$$\delta^{\mathrm{TPA}}(\omega) = \frac{(2\pi)^2 \omega^2 \mathcal{N}_A}{30 c^2 (4\pi\epsilon_0)^2} \sum_i \sum_f \rho_i \delta_{if}^{\mathrm{TPA}}(\omega) \delta(\omega_f - \omega_i - 2\omega) \tag{8.22}$$

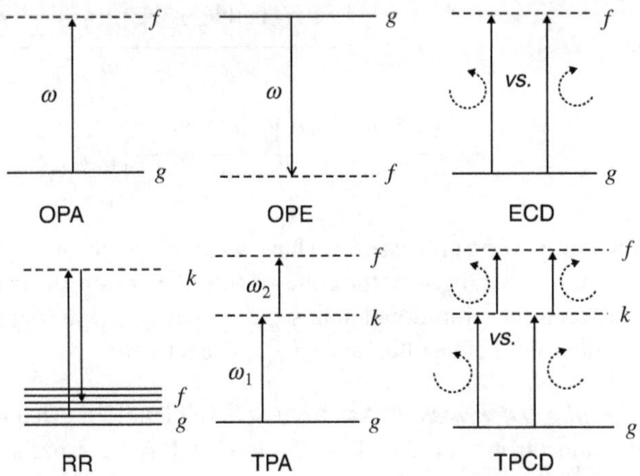

Figure 8.1 Schematic representation of one- and two-photon electronic transitions related to several spectroscopy techniques. (The authors thank Dr. N. Lin for providing a preliminary version of this figure.)

with

$$\delta_{if}^{\mathrm{TPA}} = \sum_{\tau,\eta} \Big[\mathcal{F} S_{if}(\tau,\tau,\omega) S_{if}^{*}(\eta,\eta,\omega) + \mathcal{G} S_{if}(\tau,\eta,\omega) S_{if}^{*}(\tau,\eta,\omega)$$
$$+ \mathcal{H} S_{if}(\tau,\eta,\omega) S_{if}^{*}(\eta,\tau,\omega) \Big] \tag{8.23}$$

$$S_{if}(\tau,\eta,\omega) = \frac{1}{\hbar} \sum_{m} \left[\frac{\langle \Psi_i | \mu(\tau) | \Psi_m \rangle \langle \Psi_m | \mu(\eta) | \Psi_f \rangle}{\omega_m - \omega_i - \omega} + \frac{\langle \Psi_i | \mu(\eta) | \Psi_m \rangle \langle \Psi_m | \mu(\tau) | \Psi_f \rangle}{\omega_m - \omega_i - \omega} \right] \tag{8.24}$$

where τ and η represent any Cartesian coordinate and $\mathcal{F}$, $\mathcal{G}$ and $\mathcal{H}$ assume respectively the values 2, 2, 2 for linearly polarized light and −2, 3, 3 for circularly polarized light. At variance with one-photon processes, the transition tensor elements $S_{if}(\tau, \eta, \omega)$ depend on the exciting frequency and on the properties of the intermediate states $|\Psi_m\rangle$.

Within the BO approximation, the energy of the vibronic state $|\Psi_i\rangle$ is $V_i = \hbar(\omega_{\underline{i}} + \omega_{r(i)})$, where the $\underline{i}$ and $r(\underline{i})$ subscripts indicate the electronic or vibrational contributions, respectively (notice that the rotational energy has been neglected). When the exciting light frequency is far from resonance with the one-photon transition to any intermediate state $|\Psi_m\rangle$, we can recover an expression for TPA which is analogous to one-photon processes. In fact, we can neglect the vibrational energy in the denominator, so that $\omega_m - \omega_i \approx \omega_{\underline{m}} - \omega_{\underline{i}}$, and use the closure relation in the vibrational states to obtain

$$\langle \chi_{r(\underline{i})}|S_{e,if}(\tau,\eta,\omega)|\chi_{s(\underline{f})}\rangle = \frac{1}{\hbar}\langle \chi_{r(\underline{i})}|\sum_{\underline{m}}\left[\frac{\langle \Phi_{\underline{i}}|\mu(\tau)|\Phi_{\underline{m}}\rangle\langle \Phi_{\underline{m}}|\mu(\eta)|\Phi_{\underline{f}}\rangle}{\omega_{\underline{m}}-\omega_{\underline{i}}-\omega} + \frac{\langle \Phi_{\underline{i}}|\mu(\eta)|\Phi_{\underline{m}}\rangle\langle \Phi_{\underline{m}}|\mu(\tau)|\Phi_{\underline{f}}\rangle}{\omega_{\underline{m}}-\omega_{\underline{i}}-\omega}\right]|\chi_{s(\underline{f})}\rangle \tag{8.25}$$

where, similar to one-photon processes, we have written the vibronic TPA transition amplitudes in terms of the integral of the components of electronic transition tensors $S_{e,if}(\tau, \eta, \omega)$ between the vibrational states $|\chi_{r(\underline{i})}\rangle$ and $|\chi_{s(\underline{f})}\rangle$ associated with the initial and final electronic states $|\Phi_{\underline{i}}\rangle$and $|\Phi_{\underline{f}}\rangle$, respectively.

Two-Photon Circular Dichroism As discussed in detail in Chapter 2, TPCD arises from the anisotropy in the TPA of left and right circularly polarized light, $\Delta\delta^{\mathrm{TPCD}} = \delta^{\mathrm{TPA},L} - \delta^{\mathrm{TPA},R}$, which can be computed as [24]

$$\Delta\delta^{\mathrm{TPCD}}(\omega) = \frac{4(2\pi)^2\omega^2 N_A}{15c^3(4\pi\epsilon_0)^2}\sum_i\sum_f \rho_i R_{if}^{\mathrm{TPCD}}(\omega)\delta(\omega_f-\omega_i-2\omega) \tag{8.26}$$

where

$$R_{if}^{\mathrm{TPCD}}(\omega) = -b_1\mathcal{B}_1(\omega) - b_2\mathcal{B}_2(\omega) - b_3\mathcal{B}_3(\omega) \tag{8.27}$$

with

$$\begin{aligned}
\mathcal{B}_1(\omega) &= \frac{1}{\omega^3}\sum_{\tau,\eta} M_{if}(\tau,\eta,\omega)P_{if}^*(\tau,\eta,\omega)\\
\mathcal{B}_2(\omega) &= \frac{1}{2\omega^3}\sum_{\tau,\eta} T_{if}^+(\tau,\eta,\omega)P_{if}^*(\tau,\eta,\omega)\\
\mathcal{B}_3(\omega) &= \frac{1}{\omega^3}\sum_{\tau,\eta} M_{if}(\tau,\tau,\omega)P_{if}^*(\eta,\eta,\omega)
\end{aligned}$$

The second-rank tensor $P_{if}(\omega)$ depends on the velocity dipole operator, while $M_{if}(\omega)$ depends on the velocity dipole operator and on the magnetic dipole operator and finally $T_{if}^+(\omega)$ on the velocity dipole operator and the velocity form of the electric quadrupole operator, respectively. Their mathematical expressions are reported and described in detail in Chapter 2. Once more, like we did for TPA, invoking the BO approximation and integrating over the electronic coordinates, the TPCD intensity between vibronic states can be written in terms of elements of electronic transition tensors $P_{e,if}(\tau, \eta, \omega)$, $M_{e,if}(\tau, \eta, \omega)$, and $T_{e,if}^+(\tau, \eta, \omega)$ between the vibrational states $|\chi_{r(\underline{i})}\rangle$ and $|\chi_{s(\underline{f})}\rangle$ associated with the initial and final electronic states $|\Phi_{\underline{i}}\rangle$ and $|\Phi_{\underline{f}}\rangle$, respectively.

8.2.2.3 Vibrational Resonance Raman Resonance Raman is a scattering phenomenon involving two photons, one incident and the other scattered. Let us consider a monochromatic incident radiation I with an angular frequency ω_I propagating in the direction defined by the unit vector n_0^I and impacting on the sample and let us analyze the scattered radiation (s) of frequency ω_s in the direction n_0^s, where n_0^s and n_0^I define the so-called scattering plane and $\theta = \cos^{-1}(n_0^I \cdot n_0^s)$ is the scattering angle. Let us now introduce two possible polarizations for the incident and scattered light, respectively perpendicular ($\perp$) and parallel ($||$) to the scattered plane. The scattered intensity at $\theta = 90°$ of any polarization ($\perp^s + ||^s$) for an incident light with perpendicular polarization ($\perp^I$) is [25],

$$\mathcal{I}(\pi/2; \perp^s + ||^s, \perp^I) = \frac{\omega_S^4 \mathcal{I}}{16\epsilon_0^2 c_0^4 \pi^2} \frac{45a^2 + 7\gamma^2 + 5\delta^2}{45} \quad (8.28)$$

where $\mathcal{I}$ is the incident field irradiance and, as done in previous sections, an orientational average has been taken considering the molecules freely rotating in space. The terms α, γ, and δ depend on the molecular polarizability tensor and therefore are functions of ω_I and ω_s and of molecular parameters. They are respectively the mean polarizability and the symmetric and antisymmetric anisotropy, and their working expression will be given below. It is expedient to avoid dependence on $\mathcal{I}$ and to define a differential cross section with respect to a scattering solid angle Ω,

$$\dot{\sigma}(\omega_I, \omega_s) = \frac{\partial \sigma}{\partial \Omega} = I(\pi/2; \perp^s + ||^s, \perp^i)/\mathcal{I} \quad (8.29)$$

It is worthy to point out that some authors define $\dot{\sigma}(\omega_I, \omega_s)$ not per unit of incident irradiance $\mathcal{I}$ but per unit of incident photon flux $\mathcal{I}/\hbar\omega_I$ and, additionally, they report the second differential cross section $\ddot{\sigma}(\omega_I, \omega_s) = \partial^2\sigma/\partial\Omega\, \partial\omega_s$. This alternative quantity does not depend on ω_s^4 as in Eq. 8.29 but on $\omega_s^3 \omega_I$.

Let us now introduce the polarizability tensor for a transition $|\Psi_i\rangle \rightarrow |\Psi_f\rangle$. Its ($\tau\eta$)th Cartesian component is

$$\alpha_{fi}(\tau, \eta) = \frac{1}{\hbar} \sum_m \left\{ \frac{\langle \Psi_f | \mu(\tau) | \Psi_m \rangle \langle \Psi_m | \mu(\eta) | \Psi_i \rangle}{\Delta\omega_{mi} - \omega_I - i\gamma_m} + \frac{\langle \Psi_f | \mu(\tau) | \Psi_m \rangle \langle \Psi_m | \mu(\eta) | \Psi_i \rangle}{\Delta\omega_{mi} + \omega_I - i\gamma_m} \right\} \quad (8.30)$$

where ω_I is the incident frequency, $\boldsymbol{\mu}$ is the electric dipole, the sum is taken over all the possible intermediate states $|\Psi_m\rangle$, $\Delta\omega_{mi} = \omega_m - \omega_i$, where $\hbar\omega_i$ and $\hbar\omega_m$ are the energies of the states $|\Psi_i\rangle$ and $|\Psi_m\rangle$, respectively, and finally γ_m is the lifetime of the excited states $|\Psi_m\rangle$.

The following equalities hold [25]:

$$a = \tfrac{1}{3}\left[\alpha_{fi}(x, x) + \alpha_{fi}(y, y) + \alpha_{fi}(z, z)\right] \quad (8.31)$$

$$\begin{aligned} \gamma^2 = \tfrac{1}{2} &\left[\left|\alpha_{fi}(x, x) - \alpha_{fi}(y, y)\right|^2 + \left|\alpha_{fi}(x, x) - \alpha_{fi}(z, z)\right|^2 + \left|\alpha_{fi}(y, y) - \alpha_{fi}(z, z)\right|^2 \right] \\ &+ \tfrac{3}{2}\left[\left|\alpha_{fi}(x, y) + \alpha_{fi}(y, x)\right|^2 + \left|\alpha_{fi}(x, z) + \alpha_{fi}(z, x)\right|^2 + \left|\alpha_{if}(y, z) + \alpha_{if}(z, y)\right|^2 \right] \end{aligned} \quad (8.32)$$

$$\delta^2 = \tfrac{1}{2}\left[\left|\alpha_{fi}(x,y) - \alpha_{fi}(y,x)\right|^2 + \left|\alpha_{fi}(x,z) - \alpha_{fi}(z,x)\right|^2 + \left|\left|\alpha_{if}(y,z) - \alpha_{if}(z,y)\right|^2\right] \tag{8.33}$$

Assuming the validity of the BO approximation, we deal with vibronic states that can be written as a product $|\Psi_i\rangle = |\Phi_i\rangle \otimes |\chi_{r(\underline{i})}\rangle$, where $|\Phi_{\underline{i}}\rangle$ is the electronic state and $|\chi_{r(\underline{i})}\rangle$ the associated vibrational state.

When the incident frequency is close to resonance with the transition energy ω_{mi}, the first resonant term in Eq. 8.30 becomes dominant with respect to the second off-resonant one, which can be safely neglected. Moreover, considering the specific case where both initial and final states belong to the ground electronic state $|\Phi_{\underline{g}}\rangle$ (vibrational resonance Raman) and integrating over the electronic degrees of freedom, we get

$$\alpha_{fi}(\tau,\eta) = \sum_{\underline{m},t(\underline{m})} \frac{\langle\chi_{f(\underline{g})}|\mu_{e,gm}(\tau)|\chi_{t(\underline{m})}\rangle\langle\chi_{t(\underline{m})}|\mu_{e,mg}(\eta)|\chi_{i(\underline{g})}\rangle}{\Delta\omega_{\underline{mg}} + \omega_{t(\underline{m})} - \omega_{i(\underline{g})} - \omega_I - i\gamma_{\underline{m}}} \tag{8.34}$$

where the subscripts i and f label now directly the initial and final vibrational states of the transition, respectively, $\mu_{e,gm}(\tau) = \langle\Phi_{\underline{g}}|\mu(\tau)|\Phi_{\underline{m}}\rangle$ and $\Delta\omega_{\underline{mg}} = \omega_{\underline{m}} - \omega_{\underline{g}}$, and we have further assumed that the lifetime γ_m of the intermediate states $|\Phi_{\underline{m}}, \chi_{t(\underline{m})}\rangle$ does not depend on the vibrational state $|\chi_{t(\underline{m})}\rangle$, so that it is possible to drop the t subscript $(\gamma_{\underline{m}})$. As it can be seen from Eqs. 8.28–8.33, the transition probability depends on the square moduli of the polarizability tensor elements. It is worthy to notice that, when more than one excited electronic state $|\Phi_{\underline{m}}\rangle$ falls in the near-resonance regime, the sum in Eq. 8.34 must be carried over all of those intermediate electronic states, giving rise to possible rise to interferential features. Nonetheless, even in this case, it is possible to compute a polarizability tensor for each intermediate state $|\Phi_{\underline{m}}\rangle$ and then sum them to obtain the total polarizability tensor. Therefore, in the following we will consider a single intermediate state $|\Phi_{\underline{m}}\rangle$ and drop the summation over $\underline{m}$.

8.3 SINGLE-STATE HARMONIC APPROACHES FOR LARGE SYSTEMS

Even if nonadiabatic problems are not rare, a vast amount of experimental studies deals with transitions between nondissociative electronic states which do not show conical intersections and strong nonadiabatic effects. In these cases, even if the frequency range of interest encompasses several final electronic states, the spectrum can be obtained as a sum of the spectra of each of them. It is therefore possible to carry out the complete study of interest with a single final electronic state at a time (single-state approach). This section is focused on such an approach, with the aim to present robust and rather general tools tailored for large systems and grounded in the harmonic approximation that are now available also to nonspecialists. For those cases in which nonadiabatic effects cannot be neglected, the single-state approach may still allow to perform useful preliminary analysis to dissect single-state and multistate effects.

8.3.1 Franck–Condon and Herzberg–Teller Approximations of Electronic Transition Amplitudes

8.3.1.1 Electronic Transition Amplitudes We assume that a single-state approach is feasible, thus setting our discussion in the framework of the BO approximation.

One-Photon Processes To compute the transitions intensities for one-photon spectra following Eq. 8.21, it is necessary to evaluate the electric transition dipole moment $\boldsymbol{\mu}_{e,if}$ and, in case of ECD spectroscopy, also the magnetic one, $\boldsymbol{m}_{e,if}$. Since the transition dipole moment $\boldsymbol{d}^{Z}_{e,if}$, which can represent either $\boldsymbol{\mu}_{e,if}$ or $\Im(\boldsymbol{m}_{e,if})$, is in general an unknown function of the vibrational coordinates, it must be approximated. The most common approximation, stated by Franck [26] and formalized by Condon [27], assumes that the transition takes place in such a short time that the position of the nuclei remains almost unchanged and the transition dipole can be considered as constant. While this approximation is fairly good when the transition is fully allowed and the minima of the potential energy surfaces of the initial and final electronic states are almost vertical to each other, it shows serious limitations when the transition is weakly-allowed or dipole-forbidden. The limitation is even more strongly felt for electronic circular dichroism, where the product $\boldsymbol{\mu}_{if} \cdot \Im(\boldsymbol{m}_{if})$ may be almost negligible even if the corresponding OPA transition is strongly allowed, whenever the electric and magnetic transition moments are nearly orthogonal.

An early extension to the Franck–Condon (FC) principle was proposed by Herzberg and Teller [28] and accounted for a linear variation of the transition dipole moment with respect to the normal coordinates of the initial state $\boldsymbol{Q}_{\underline{i}}$. By extension, it is possible to generalize the approach by developing the transition dipole moment $\boldsymbol{d}^{Z}_{e,if}$ in a Taylor series about the equilibrium geometry of the initial state ($\boldsymbol{Q}^{\text{eq}}_{\underline{i}} = \boldsymbol{0}$),

$$\boldsymbol{d}^{Z}_{e,if}\left(\boldsymbol{Q}_{\underline{i}}\right) \approx \boldsymbol{d}^{Z}_{e,if}\left(\boldsymbol{Q}^{\text{eq}}_{\underline{i}}\right) + \sum_{k=1}^{N}\left(\frac{\partial \boldsymbol{d}^{Z}_{e,if}}{\partial \mathrm{Q}_{\underline{i}}(k)}\right)_0 \mathrm{Q}_{\underline{i}}(k) + \frac{1}{2}\sum_{k=1}^{N}\sum_{l=1}^{N}\left(\frac{\partial^2 \boldsymbol{d}^{Z}_{e,if}}{\partial \mathrm{Q}_{\underline{i}}(k)\partial \mathrm{Q}_{\underline{i}}(l)}\right)_0 \mathrm{Q}_{\underline{i}}(k)\mathrm{Q}_{\underline{i}}(l) + \cdots \tag{8.35}$$

with N the number of normal modes.

The FC approximation corresponds to the first term on the RHS of Eq. 8.35, the Herzberg–Teller (HT) one to the second one, and the acronym FCHT will refer to both terms taken simultaneously. The remaining terms in Eq. 8.35 will not be taken into account in the following discussion.

Using Eq. 8.35, the transition dipole moment integral $\boldsymbol{d}^{Z}_{if}$ is given by the relation

$$\langle\Psi_i|\boldsymbol{d}^{Z}|\Psi_f\rangle \approx \boldsymbol{d}^{Z}_{e,if}\left(\boldsymbol{Q}^{\text{eq}}_{\underline{i}}\right)\langle\chi_{r(\underline{i})}|\chi_{s(\underline{f})}\rangle + \sum_{k=1}^{N}\left(\frac{\partial \boldsymbol{d}^{Z}_{e,if}}{\partial \mathrm{Q}_{\underline{i}}(k)}\right)_0 \langle\chi_{r(\underline{i})}|\mathrm{Q}_{\underline{i}}(k)|\chi_{s(\underline{f})}\rangle \tag{8.36}$$

where, as usual, we have assumed here that the Eckart conditions are met and $|\chi_{r(\underline{i})}\rangle$ and $|\chi_{s(\underline{f})}\rangle$ represent the initial vibrational state r associated with the electronic state $\underline{i}$ and the final vibrational state s associated with the electronic state $\underline{f}$. The overlap integral $\langle\chi_{r(\underline{i})}|\chi_{s(\underline{f})}\rangle$ is referred to as the FC integral.

The linear terms in Eq. 8.36, corresponding to the summation over k on the RHS, introduce an intensity-borrowing effect due to the coupling of the states involved in the electronic transition with other closely lying electronic states, with which they mix upon small displacements along the normal coordinates. The general problem of vibronic coupling for semirigid molecules can be cast in the framework of the diabatic linear coupling vibronic model (see Section 8.5), where the diabatic states are ideally independent of the nuclear coordinates and therefore have constant electronic transition moments. In that framework, the dependence of the adiabatic transition dipole moments on the nuclear coordinates arises naturally from the coupling of the diabatic states. Notwithstanding the possibility to resort to this more general solution, when the interacting states are sufficiently separated in energy and the interaction is weak, it is possible to rely on the perturbative description originally proposed by HT, which allows us to account for the main borrowing effects, keeping the simplicity of an adiabatic single-state description.

The first derivatives in Eq. 8.35 can be written as [29]

$$\left(\frac{\partial \boldsymbol{d}^{Z}_{e,\underline{if}}}{\partial \mathrm{Q}_{\underline{i}}(k)}\right)_0 = \left\langle \left(\frac{\partial \Phi_{\underline{i}}(\boldsymbol{r},\boldsymbol{Q}_{\underline{i}})}{\partial \mathrm{Q}_{\underline{i}}(k)}\right)_0 \middle| \boldsymbol{d}^{Z}_{e} \middle| \Phi_{\underline{f}}\left(\boldsymbol{r},\boldsymbol{Q}_{\underline{i}}\right)\right\rangle + \left\langle \Phi_{\underline{i}}\left(\boldsymbol{r},\boldsymbol{Q}_{\underline{i}}\right) \middle| \boldsymbol{d}^{Z}_{e} \middle| \left(\frac{\partial \Phi_{\underline{f}}(\boldsymbol{r},\boldsymbol{Q}_{\underline{i}})}{\partial \mathrm{Q}_{\underline{i}}(k)}\right)_0\right\rangle \tag{8.37}$$

To compute the derivatives of the electronic wavefunction, we adopt the Herzberg–Teller expression of the electronic Hamiltonian [30],

$$\hat{H}_e\left(\boldsymbol{r},\boldsymbol{Q}_{\underline{i}}\right) = \hat{H}_e\left(\boldsymbol{r},\boldsymbol{Q}^{\mathrm{eq}}_{\underline{i}}\right) + \sum_k \left(\frac{\partial \hat{H}_e(\boldsymbol{r},\boldsymbol{Q}_{\underline{i}})}{\partial \mathrm{Q}_{\underline{i}}(k)}\right)_0 \mathrm{Q}_{\underline{i}}(k) + \cdots \tag{8.38}$$

and use the perturbation theory. More specifically, the unperturbed Hamiltonian is $\hat{H}_e(\boldsymbol{r},\boldsymbol{Q}^{\mathrm{eq}}_{\underline{i}})$ with eigenvalues $V_{\underline{m}}(\boldsymbol{Q}^{\mathrm{eq}}_{\underline{i}})$ and eigenfunctions $\Phi_{\underline{m}}(\boldsymbol{r},\boldsymbol{Q}^{\mathrm{eq}}_{\underline{i}})$ (the same ones appearing in the crude adiabatic approximation of Eq. 8.11), while the remaining terms on the RHS in Eq. 8.38 act like perturbators. Limiting the Taylor expansion in Eq. 8.38 to the first order, the perturbed electronic wavefunction $\underline{m}$ can be written as

$$\Phi_{\underline{m}}\left(\boldsymbol{r},\boldsymbol{Q}_{\underline{i}}\right) = \Phi_{\underline{m}}\left(\boldsymbol{r},\boldsymbol{Q}^{\mathrm{eq}}_{\underline{i}}\right) + \Phi_{\underline{m}}^{(1)}\left(\boldsymbol{r},\boldsymbol{Q}_{\underline{i}}\right) \tag{8.39}$$

with

$$\Phi_{\underline{m}}^{(1)}\left(\boldsymbol{r},\boldsymbol{Q}_{\underline{i}}\right) = -\sum_{\underline{n}\neq\underline{m}} \Phi_{\underline{n}}\left(\boldsymbol{r},\boldsymbol{Q}_{\underline{i}}^{\mathrm{eq}}\right) \times \sum_{k} \frac{\left\langle \Phi_{\underline{n}}(\boldsymbol{r},\boldsymbol{Q}_{\underline{i}}^{\mathrm{eq}}) \middle| \left(\partial \hat{H}_e(\boldsymbol{r},\boldsymbol{Q}_{\underline{i}})/\partial Q_{\underline{i}}(k)\right)_0 \middle| \Phi_{\underline{m}}(\boldsymbol{r},\boldsymbol{Q}_{\underline{i}}^{\mathrm{eq}}) \right\rangle}{V_{\underline{m}}\left(\boldsymbol{Q}_{\underline{i}}^{\mathrm{eq}}\right) - V_{\underline{n}}\left(\boldsymbol{Q}_{\underline{i}}^{\mathrm{eq}}\right)} \mathrm{Q}_{\underline{i}}(k) \quad (8.40)$$

Assuming, for simplicity, that the initial electronic state $\underline{i}$ does not change with the nuclear coordinates at first order, we obtain

$$\left(\frac{\partial \boldsymbol{d}^{Z}_{e,\underline{if}}}{\partial \mathrm{Q}_{\underline{i}}(k)}\right)_0 = -\sum_{\underline{n}\neq\underline{f}}\sum_{k} \frac{\left\langle \Phi_{\underline{i}}(\boldsymbol{r},\boldsymbol{Q}_{\underline{i}}^{\mathrm{eq}}) \middle| \boldsymbol{d}^{Z}_{e} \middle| \Phi_{\underline{n}}(\boldsymbol{r},\boldsymbol{Q}_{\underline{i}}^{\mathrm{eq}}) \right\rangle \left\langle \Phi_{\underline{n}}(\boldsymbol{r},\boldsymbol{Q}_{\underline{i}}^{\mathrm{eq}}) \middle| (\partial \hat{H}_e(\boldsymbol{r},\boldsymbol{Q}_{\underline{i}})/\partial \mathrm{Q}_{\underline{i}}(k))_0 \middle| \Phi_{\underline{f}}(\boldsymbol{r},\boldsymbol{Q}_{\underline{i}}^{\mathrm{eq}}) \right\rangle}{V_{\underline{f}}(\boldsymbol{Q}_{\underline{i}}^{\mathrm{eq}}) - V_{\underline{n}}(\boldsymbol{Q}_{\underline{i}}^{\mathrm{eq}})} \quad (8.41)$$

Moreover, it is easily seen from Eq. 8.41 that the derivatives of the transition dipole moment arise from the vibronic coupling due to the changes in the electronic Hamiltonian with the nuclear coordinates. Additionally, since all the terms in the Hamiltonian in Eq. 8.38 must belong to the total symmetric irreps Γ_A, the derivative $(\partial \hat{H}_e(\boldsymbol{r},\boldsymbol{Q}_{\underline{i}})/\partial \mathrm{Q}_{\underline{i}}(k))$ belongs to the irreps of $\mathrm{Q}_{\underline{i}}(k)$, namely $\boldsymbol{\Gamma}(\mathrm{Q}_{\underline{i}}(k))$. Therefore, symmetry considerations provide the conditions for a nonvanishing derivative of the transition dipole moment polarized along a given Cartesian axis $\tau = x, y, z$,

$$\left(\frac{\partial \boldsymbol{d}^{Z}_{e,\underline{if}}}{\partial \mathrm{Q}_{\underline{i}}(k)}\right)_0 (\tau) \neq 0 \quad \text{if} \begin{cases} \boldsymbol{\Gamma}(\Phi_{\underline{f}}) \otimes \boldsymbol{\Gamma}(\mathrm{Q}_{\underline{i}}(k)) \otimes \boldsymbol{\Gamma}(\Phi_{\underline{n}}) \supset \boldsymbol{\Gamma}_A \\ \boldsymbol{\Gamma}(\Phi_{\underline{i}}) \otimes \boldsymbol{\Gamma}(\tau) \otimes \boldsymbol{\Gamma}(\Phi_{\underline{n}}) \supset \boldsymbol{\Gamma}_A \end{cases} \quad (8.42)$$

On the same foot, it is easily proven that, at the equilibrium geometry of the electronic initial state ($\underline{i}$), the first derivative of the energy of the final state ($\underline{f}$) can only be different from zero along total symmetric coordinates. In fact, from Eq. 8.38, we have

$$\left\langle \Phi_{\underline{f}}\left(\boldsymbol{r},\boldsymbol{Q}_{\underline{i}}^{\mathrm{eq}}\right) \middle| \left(\frac{\partial \hat{H}_e\left(\boldsymbol{r},\boldsymbol{Q}_{\underline{i}}\right)}{\partial \mathrm{Q}_{\underline{i}}(k)}\right)_0 \middle| \Phi_{\underline{f}}\left(\boldsymbol{r},\boldsymbol{Q}_{\underline{i}}^{\mathrm{eq}}\right) \right\rangle \quad (8.43)$$

which is different from zero only for $\boldsymbol{\Gamma}(\mathrm{Q}_{\underline{i}}(k)) \in \boldsymbol{\Gamma}_A$.

Two-Photon Processes To evaluate two-photon absorption and circular dichroism intensities, it is possible to obtain mathematical expressions similar to those of one-photon processes, assuming that the dependence of the transition tensors on the frequency is small in the energy window of interest of the spectrum. In that case, for TPA, we can fix $S_{e,if}(\tau, \eta, \omega)$ in Eq. 8.23 to its value at the vertical excitation energy, ω_V, and introduce FC and HT contributions through the usual linear expansion in normal coordinates [24],

$$S_{e.if}\left(\tau, \eta, \boldsymbol{Q}_{\underline{i}}, \omega_V\right) \approx S_{e,if}\left(\tau, \eta, \boldsymbol{Q}_{\underline{i}}^{\text{eq}}, \omega_V\right) + \sum_k \left(\frac{\partial S_{e,if}\left(\tau, \eta, \boldsymbol{Q}_{\underline{i}}, \omega_V\right)}{\partial \mathrm{Q}_{\underline{i}}(k)}\right)_0 \mathrm{Q}_{\underline{i}}(k) \tag{8.44}$$

An analogous approach leads to an expression of the TPCD intensity in terms of the FC and HT contributions from each of the three electronic tensors. For example, the electronic tensor $P_{e,if}$ is given by the relation [24]

$$P_{e,if}\left(\tau, \eta, \boldsymbol{Q}_{\underline{i}}, \omega_V\right) = \langle \Phi_{\underline{i}} | P_{if}\left(\tau, \eta, \boldsymbol{Q}_{\underline{i}}, \omega_V\right) | \Phi_{\underline{f}} \rangle \approx P_{e,if}\left(\tau, \eta, \boldsymbol{Q}_{\underline{i}}{}^{\text{eq}}\right) + \sum_k \left(\frac{\partial P_{e,if}\left(\tau, \eta, \boldsymbol{Q}_{\underline{i}}, \omega_V\right)}{\partial \mathrm{Q}_{\underline{i}}(k)}\right)_0 \mathrm{Q}_{\underline{i}}(k) \tag{8.45}$$

Vibrational Resonance Raman Regarding the vibrational resonance Raman, the polarizability tensor of Eq. 8.34 can also be expressed by the Taylor expansion given in Eq. 8.35 allowing us to distinguish FC and HT contributions. To simplify the equations, we will adopt a shorter notation for the derivatives of the transition dipole moment,

$$\left(\frac{\partial \boldsymbol{d}_{e,if}^{z}}{\partial \mathbf{Q}_{\underline{i}}(k)}\right)_0 = \dot{\boldsymbol{d}}_{e,if}^{z}(k)$$

The polarizability tensor is then given by the relation

$$\begin{aligned}
\alpha_{fi}(\tau, \eta) = {} & \frac{1}{\hbar} \sum_{\underline{m}, t(\underline{m})} \frac{\langle \chi_{f(g)} | \mu_{e,g\underline{m}}(\tau, \boldsymbol{Q}_{\underline{i}}^{\text{eq}}) | \chi_{t(\underline{m})} \rangle \langle \chi_{t(\underline{m})} | \mu_{e,\underline{m},g}(\eta, \boldsymbol{Q}_{\underline{i}}^{\text{eq}}) | \chi_i \rangle}{\Delta\omega_{\underline{mg}} + \omega_{t(\underline{m})} - \omega_{i(\underline{g})} - \omega_I - i\gamma_{\underline{m}}} \\
& + \frac{1}{\hbar} \sum_{\underline{m}, t(\underline{m})} \frac{\langle \chi_{f(\underline{g})} | \sum_k \dot{\mu}_{e,\underline{g},\underline{m}}(\tau, k) | \chi_{t(\underline{m})} \rangle \langle \chi_{t(\underline{m})} | \mu_{e,\underline{m},\underline{g}}(\eta, \boldsymbol{Q}_{\underline{i}}^{\text{eq}}) | \chi_{i(\underline{g})} \rangle}{\Delta\omega_{\underline{mg}} + \omega_{t(\underline{m})} - \omega_{i(\underline{g})} - \omega_I - i\gamma_{\underline{m}}} \\
& + \frac{1}{\hbar} \sum_{\underline{m}, t, (\underline{m})} \frac{\langle \chi_{f(\underline{g})} | \mu_{e,\underline{gm}}(\tau, \boldsymbol{Q}_{\underline{i}}^{\text{eq}}) | \chi_{t(\underline{m})} \rangle \langle \chi_{t(\underline{m})} | \sum_k \dot{\mu}_{e,\underline{mg}}(\eta, k) | \chi_{i(\underline{g})} \rangle}{\Delta\omega_{\underline{mg}} + \omega_{t(\underline{m})} - \omega_{i(\underline{g})} - \omega_I - i\gamma_{\underline{m}}} \\
& + \frac{1}{\hbar} \sum_{\underline{m}, t(\underline{m})} \frac{\langle \chi_{f(\underline{g})} | \sum_k \dot{\mu}_{e,\underline{gm}}(\tau, k) | \chi_{t(\underline{m})} \rangle \langle \chi_{t(\underline{m})} | \sum_l \dot{\mu}_{e,\underline{mg}}(\eta, l) | \chi_{i(\underline{g})} \rangle}{\Delta\omega_{\underline{mg}} + \omega_{t(\underline{m})} - \omega_{i(\underline{g})} - \omega_I - i\gamma_{\underline{m}}}
\end{aligned} \tag{8.46}$$

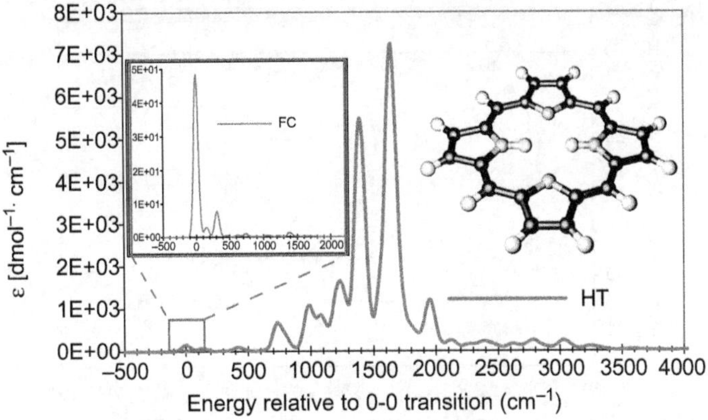

Figure 8.2 Absorption spectra of $S_1 \leftarrow S_0$ electronic transition of porphyrin computed with FC|AH and FCHT|AH approaches, which span an energy range of 0–4000 cm^{-1} with respect to the 0–0 transition.

The first term of Eq. 8.46 is the FC term, while the remaining terms arise from the Herzberg–Teller (HT) borrowing mechanism of electronic transitions. By inspection of Eq. 8.33, one may notice that the antisymmetric anisotropy vanishes for pure FC transitions but is in general different from zero when the HT effect is considered.

8.3.1.2 Practical Example: OPA Spectra of Porphyrin and ECD Spectra of ax-R3MCP The influence of the HT terms[1] will be illustrated with the OPA spectrum of porphyrin and ECD spectrum of the axial-methyl conformer of (*R*)-(+)-3-methyl-cyclopentanone (*ax*-R3MCP).

Free-base porphyrin (H2P) is similar to the major building block (Mg–porphyrin) of chlorophyll but belongs to the D_{2h} point group, at variance with the higher D_{4h} symmetry shown by some metal–porphyrin complexes. For this reason, its two bands Q (dark) and B (intense, also called the Soret band) are composed of two pairs of transitions, namely (Q_x,Q_y) and (B_x,B_y) with components polarized in the x and y directions (considering the molecule in the xy plane and the x the axis which passes through the two central H atoms). However, the Q bands are very weak and a proper calculation of their spectra must take into account the HT effect. Figure 8.2 shows the absorption spectra related to the $S_1(Q_x) \leftarrow S_0$ electronic transition of porphyrin computed with different approximation levels of the transition dipole moment in the Taylor series. The comparison of the FC and FCHT spectral shapes shows the dramatic impact of neglecting the transition dipole derivatives on the spectra. The spectrum intensity is mainly related to the HT term and the progression of normal modes of a_{1g} or b_{1g} symmetry. The a_{1g} modes give rise to progressions which have both FC and HT contributions (polarized along x), while b_{1g} mode progressions result

[1] Calculations were performed with the adiabatic model presented in Secion 8.3.2.1.

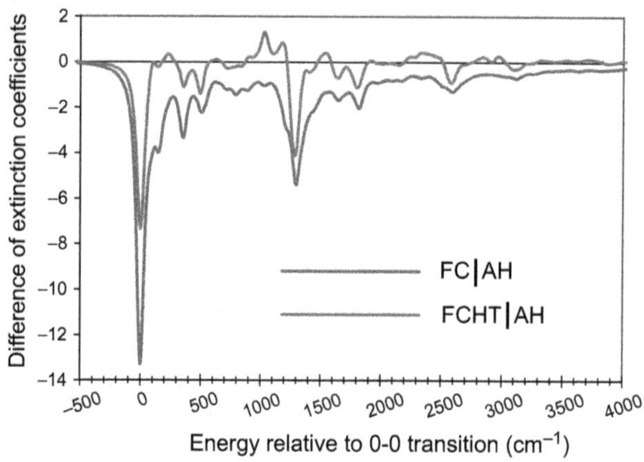

Figure 8.3 ECD spectra of the $S_2 \leftarrow S_0$ electronic transition of *ax*-R3MCP computed with FC|AH (adiabatic Hessian) and FCHT|AH approaches, which span an energy range of 0–4000 cm^{-1} with respect to the 0–0 transition.

only in HT contributions (polarized along y). Symmetry considerations show that the HT terms polarized along x mainly reflect the intensity borrowing from the B_x Soret band, while y-polarized HT terms derive from a mixing with the B_y state (minor contributions may also come from the Q_y state) [31]. Additionally, the FCHT spectrum shows a very rich vibrational structure at variance with its FC counterpart, which is dominated by the 0–0 transition.

The second example is related to the effect of the inclusion of the HT terms on the possible change of the sign of some vibronic lines in the ECD spectra [32, 33]. In fact, as shown in Figure 8.3, the FC and FCHT spectra lineshapes differ significantly where the HT contribution is higher than FC and so controls the sign of the transition intensities. In those cases, the FC approach cannot correctly reproduce the spectrum. It should be noted that those remarkable HT effects cannot be observed for the OPA spectrum of *ax*-R3MCP, showing clearly the higher sensitivity of ECD to the approximation of the transition dipole moment. This phenomenon is related to the dot product of the two different transition dipole moments and, more precisely, to their relative orientation, which stands as an additional factor with respect to OPA that can lead to a breakdown of the FC approximation. For instance, when the mutual orientation of the electric and magnetic dipole moments is close to 90°, small changes might introduce strong HT effects even if the two individual moments change only slightly with the nuclear coordinates. In those cases a sign reversal of the computed rotatory strength can even be observed.

8.3.2 Harmonic Calculation of Transition Intensities

To avoid an overload of indexes, hereafter we will slightly change the notation. Specifically, a bar and a double bar will be used to indicate the initial and final states, respectively[2]. As a result, the BO approximation is noted as $|\bar{\Psi}\rangle = |\bar{\Phi}\bar{\chi}\rangle$ for the initial state and $|\bar{\bar{\Psi}}\rangle = |\bar{\bar{\Phi}}\bar{\bar{\chi}}\rangle$ for the final one. For large systems (hundreds of normal modes), a general solution of the anharmonic vibrational problem cannot be achieved, and hence the harmonic approximation is unavoidable, at least as a starting point for the description of the PES and the vibrational wavefunctions.

With this approximation, the multidimensional vibrational wavefunction χ_v can be written as a product of one-dimensional vibrational wave funcions χ_{vk},

$$|\chi_v\rangle = \prod_{k=1}^{N} |\chi_{vk}\rangle \tag{8.47}$$

For purposes of compactness, the convenient Dirac notation will be adopted from now on, so that $|\chi_v\rangle \equiv |v\rangle$, where v represents the vector of quantum numbers v_k for each vibrational mode k.

Using second quantization, the second term on the RHS of Eq. 8.36, which depends on the normal coordinates $\bar{\boldsymbol{Q}}$, can be reformulated as

$$\langle\bar{v}|\bar{\mathrm{Q}}_k|\bar{\bar{v}}\rangle = \sqrt{\frac{\hbar}{2\bar{\omega}_k}}\langle\bar{v}|\left(\bar{a}_k^{\dagger} + \bar{a}_k\right)|\bar{\bar{v}}\rangle \tag{8.48}$$

$$= \sqrt{\frac{\hbar}{2\bar{\omega}_k}}\left[\sqrt{\bar{v}_k}\langle\bar{v} - \bar{1}_k|\bar{\bar{v}}\rangle + \sqrt{\bar{v}_k + 1}\langle\bar{v} + \bar{1}_k|\bar{\bar{v}}\rangle\right] \tag{8.49}$$

where $\bar{a}_k$ and $\bar{a}_k^{\dagger}$ are the annihilation and creation operators, respectively.

Introducing Eq. 8.49 in Eq. 8.36, we obtain the following Taylor expansion for the transition dipole moment:

$$\langle\bar{\Psi}|\boldsymbol{d}^Z|\bar{\bar{\Psi}}\rangle \approx \boldsymbol{d}^Z_{e,if}(\bar{\boldsymbol{Q}}^{\mathrm{eq}})\langle\bar{v}|\bar{\bar{v}}\rangle$$

$$+\sum_{k=1}^{N}\left(\frac{\partial \boldsymbol{d}^Z_{e,if}}{\partial \bar{\mathrm{Q}}_k}\right)_0 \sqrt{\frac{\hbar}{2\bar{\omega}_k}}\left[\sqrt{\bar{v}_k}\langle\bar{v} - \bar{1}_k|\bar{v}\rangle + \sqrt{\bar{v}_k + 1}\langle\bar{v} + \bar{1}_k|\bar{\bar{v}}\rangle\right] \tag{8.50}$$

It should be noted that analogous expressions depending on the general overlaps $\langle\bar{v}|\bar{\bar{v}}\rangle$ can be obtained for TPA (from Eq. 8.44), TPCD (from Eq. 8.45), and vibrational resonance Raman (from Eq. 8.46). A major issue to calculate the integrals in Eq. 8.50 arises from the fact that the vibrational wavefunctions pertaining to the two electronic states are expressed in a different set of normal coordinates. This problem can be

[2] This notation, different from the usual prime and double-prime symbols, has been chosen to limit the overload of superscripts and subscripts used to describe the quantities in the equations.

overcome by the linear transformation proposed by Duschinsky [34] to express the normal coordinates of one state with respect to the other's,

$$\bar{\boldsymbol{Q}} = \boldsymbol{J}\bar{\bar{\boldsymbol{Q}}} + \boldsymbol{K} \tag{8.51}$$

where $\boldsymbol{J}$ is the so-called Duschinsky matrix and represents the mixing of the normal modes during the transition, $\boldsymbol{K}$ is the shift vector of the normal modes between the initial and final states, and $\bar{\boldsymbol{Q}}$ represents the normal coordinates of the initial state and $\bar{\bar{\boldsymbol{Q}}}$ those of the final one.

This transformation is a good approximation when the molecule does not undergo a noticeable distortion during the transition. An extensive discussion of the transformation between the normal coordinates including a possible distortion of the molecule and the limitations of the Duschinsky transformation, as well as the necessity of minimizing the rotation effects, has been done by several authors [35–38].

The rotation or Duschinsky matrix is given by

$$\boldsymbol{J} = (\bar{\boldsymbol{L}})^{-1}\bar{\bar{\boldsymbol{L}}} \tag{8.52}$$

where $\bar{\boldsymbol{L}}$ and $\bar{\bar{\boldsymbol{L}}}$ are the transformation matrices from mass-weighted Cartesian coordinates to normal coordinates of the initial and final states, respectively, following the relations

$$\bar{\boldsymbol{Q}} = \bar{\boldsymbol{L}}\boldsymbol{M}^{1/2}\bar{X} \quad \bar{\bar{\boldsymbol{Q}}} = \bar{\bar{\boldsymbol{L}}}\boldsymbol{M}^{1/2}\bar{\bar{\boldsymbol{X}}}$$

with $\boldsymbol{M}$ the diagonal matrix of the atomic masses and $\boldsymbol{X}$ the Cartesian coordinates of the nuclear displacements upon the vibrations.

The overlap integrals can then be evaluated analytically [39–45] or recursively [46–49]. While the former method allows straightforward calculations and avoids possible error propagation, it suffers from a quickly growing complexity and a lack of versatility when dealing with medium-to-large systems. As a consequence, the recursive approaches, more suited for a general-purpose application, are usually implemented. Here we report the approach presented by Ruhoff [48] and based on the generating functions of Sharp and Rosenstock [39]:

$$\begin{aligned}\langle \bar{v}|\bar{\bar{v}}\rangle = \frac{1}{\sqrt{2\bar{v}_k}}\Bigg[&\mathrm{B}_k\langle \bar{v}-\bar{1}_k|\bar{\bar{v}}\rangle + \sqrt{2(\bar{v}_k-1)}\mathrm{A}_{kk}\langle \bar{v}-\bar{2}_k|\bar{\bar{v}}\rangle \\ &+\sum_{\substack{l=1\\ l\neq k}}^{N}\sqrt{2\bar{v}_l}\mathrm{A}_{kl}\langle \bar{v}-\bar{1}_k-\bar{1}_l|\bar{\bar{v}}\rangle + \sum_{l=1}^{N}\sqrt{\frac{\bar{\bar{v}}_l}{2}}\mathrm{E}_{lk}\langle \bar{v}-\bar{1}_k|\bar{\bar{v}}-\bar{\bar{1}}_l\rangle\Bigg]\end{aligned} \tag{8.53}$$

and

$$\begin{aligned}\langle \bar{v}|\bar{\bar{v}}\rangle = \frac{1}{\sqrt{2\bar{\bar{v}}_k}}\Bigg[&\mathrm{D}_k\langle \bar{v}|\bar{\bar{v}}-\bar{\bar{1}}_k\rangle + \sqrt{2(\bar{\bar{v}}_k-1)}C_{kk}\langle \bar{v}|\bar{\bar{v}}-\bar{\bar{2}}_k\rangle \\ &+\sum_{\substack{l=1\\ l\neq k}}^{N}\sqrt{2\bar{\bar{v}}_l}\mathrm{C}_{kl}\langle \bar{v}|\bar{\bar{v}}-\bar{\bar{1}}_k-\bar{\bar{1}}_l\rangle + \sum_{l=1}^{N}\sqrt{\frac{\bar{v}_l}{2}}\mathrm{E}_{kl}\langle \bar{v}-\bar{1}_l|\bar{\bar{v}}-\bar{\bar{1}}_k\rangle\Bigg]\end{aligned} \tag{8.54}$$

where the matrices and vectors $\boldsymbol{A}$, $\boldsymbol{B}$, $\boldsymbol{C}$, $\boldsymbol{D}$, and $\boldsymbol{E}$ are the Sharp and Rosenstock matrices. Specifically, calling $\bar{\boldsymbol{\Gamma}}$ and $\bar{\bar{\boldsymbol{\Gamma}}}$ the diagonal matrices of the reduced frequencies $\bar{\omega}_k/\hbar$ and $\bar{\bar{\omega}}_k/\hbar$, respectively, and $\boldsymbol{I}$ the identity matrix, the required quantities to evaluate Eqs. 8.53 and 8.54 are defined as

$$\begin{aligned}
\boldsymbol{A} &= 2\bar{\boldsymbol{\Gamma}}^{1/2}\boldsymbol{J}\boldsymbol{X}^{-1}\boldsymbol{J}^{\mathrm{T}}\bar{\boldsymbol{\Gamma}}^{1/2} - \boldsymbol{I}\\
\boldsymbol{B} &= -2\bar{\boldsymbol{\Gamma}}^{1/2}\left(\boldsymbol{J}\boldsymbol{X}^{-1}\boldsymbol{J}^{\mathrm{T}}\bar{\boldsymbol{\Gamma}} - \boldsymbol{I}\right)\boldsymbol{K}\\
\boldsymbol{C} &= 2\bar{\bar{\boldsymbol{\Gamma}}}^{1/2}\boldsymbol{X}^{-1}\bar{\bar{\boldsymbol{\Gamma}}}^{1/2} - \boldsymbol{I}\\
\boldsymbol{D} &= -2\bar{\bar{\boldsymbol{\Gamma}}}^{1/2}\boldsymbol{X}^{-1}\boldsymbol{J}^{\mathrm{T}}\bar{\boldsymbol{\Gamma}}\boldsymbol{K}\\
\boldsymbol{E} &= 4\bar{\bar{\boldsymbol{\Gamma}}}^{1/2}\boldsymbol{X}^{-1}\boldsymbol{J}^{\mathrm{T}}\bar{\boldsymbol{\Gamma}}^{1/2}
\end{aligned} \tag{8.55}$$

with

$$\boldsymbol{X} = \boldsymbol{J}^{\mathrm{T}}\bar{\boldsymbol{\Gamma}}\boldsymbol{J} + \bar{\bar{\boldsymbol{\Gamma}}} \tag{8.56}$$

We should also mention that, recently, Borrelli and Peluso [50] proposed a perturbative method to handle the normal-mode mixing and calculate the FC integrals.

8.3.2.1 Adiabatic and Vertical Models Two general routes have been proposed to compute "single-state" vibronic spectra within the harmonic approximation. The first model lies on the observation that the most intense transitions are vertical so that a correct description of the PES of the final state about the geometry of the initial state is more suited to the analysis of the region of the spectral maximum and of the broad features of the low-resolution spectrum. In fact, the latter mostly reflects the short-time dynamics of the system after an instantaneous promotion on the final state. In its simplest form, this vertical approach has been named the linear coupling method (LCM) [51] and is also known as the vertical gradient (VG). The VG model assumes that the equilibrium geometries of the initial and final electronic states are displaced, but the Duschinsky effect is null ($\boldsymbol{J} = \boldsymbol{I}$ in Eq. 8.52), and the vibrational frequencies are the same in both electronic states. Thus, the approach does not require the effective computation of the excited state's equilibrium geometry, frequencies, and normal modes and only the energy gradient of the final-state PES needs to be evaluated at the equilibrium geometry of the initial state. Such a task is much less time consuming than the Hessian computations. Consequently, at the cost of not taking into account changes in vibrational frequencies and/or normal modes between the excited and ground states, the VG model makes it feasible to treat larger systems. Another vertical approach, which we will call vertical hessian (VH), describes the final-state PES on the ground of its gradient and Hessian computed at the initial-state equilibrium geometry, thus accounting for the changes in vibrational properties between the electronic states. In the literature such an approach is sometimes referred to as vertical FC or VHFC [52], but here and in the following, for better clarity, we separated the acronym specifying the approximation of the final PES (VG, VH) from that concerning the approximation on the transition moment (FC, HT, FCHT). At variance with vertical approaches, which concentrate on the region of spectral maximum, in "adiabatic" models, both initial and final PESs are built starting from a harmonic analysis at the respective equilibrium geometries. These approaches treat both states on the same foot, that is, at the same level of accuracy, and are particularly suited for the high-resolution description of the spectra close to their origin. Consequently, adiabatic schemes such as Adiabatic Hessian (AH), where both equilibrium geometry and normal modes of the final state are explicitly computed, are well suited to cases when an accurate reproduction of the fine structure of the spectrum is required, in particular in studies where assignment of the excited-state frequencies is needed [53–58]. However, adiabatic approaches are limited by the computational costs of the geometric optimization and frequency calculations in the excited state, which can be prohibitive for large systems. It should be noted that, within the "single-state" model, once the initial- and final-state harmonic PESs are obtained, the machinery to compute the spectra is the same for vertical and adiabatic harmonic approaches. The difference between the models (adiabatic and vertical) can be better appreciated by the way the shift vector and rotation matrix of the linear relation between the normal modes of both states, given by the Duschinsky transformation in Eq. 8.51, are computed. In the case of the adiabatic models, the shift vector is

$$\boldsymbol{K} = \bar{\boldsymbol{L}}^{-1}\boldsymbol{M}^{1/2}\,\Delta\boldsymbol{X} \tag{8.57}$$

where $\Delta\boldsymbol{X} = \bar{\bar{\boldsymbol{X}}}^{\text{eq}} - \bar{\boldsymbol{X}}^{\text{eq}}$ is a vector representing the shift of nuclear Cartesian coordinates between the initial ($\bar{\boldsymbol{X}}$) and final ($\bar{\bar{\boldsymbol{X}}}$) states.

Conversely, the equilibrium geometry of the final state in vertical approaches is not directly computed, and $\boldsymbol{K}$ is derived by extrapolating the equilibrium geometry of the final state. To clarify this point, let us formally expand the harmonic PES of the final state around its (unknown) equilibrium geometry,

$$\bar{\bar{V}}(\bar{\bar{\boldsymbol{Q}}}) = \tfrac{1}{2}\bar{\bar{\boldsymbol{Q}}}^{\text{T}}\bar{\bar{\boldsymbol{\Omega}}}^{2}\bar{\bar{\boldsymbol{Q}}} + E_{\text{ad}} \tag{8.58}$$

where $\bar{\bar{\boldsymbol{\Omega}}}$ is the diagonal matrix of the harmonic frequencies $\bar{\bar{\omega}}$ of the final state and E_{ad} is the energy difference between the final- and initial-state energies in their minima. Notice that E_{ad} is a signed quantity that is positive in absorption and negative in emission. Introducing the Duschinsky relation given in Eq. 8.51, the harmonic potential $\bar{\bar{V}}$ can be expanded into

$$\bar{\bar{V}}(\bar{\boldsymbol{Q}}) = \tfrac{1}{2}\bar{\boldsymbol{Q}}^{\text{T}}\boldsymbol{J}\bar{\bar{\boldsymbol{\Omega}}}^{2}\boldsymbol{J}^{\text{T}}\bar{\boldsymbol{Q}} - \boldsymbol{K}^{\text{T}}\boldsymbol{J}\bar{\bar{\boldsymbol{\Omega}}}^{2}\boldsymbol{J}^{\text{T}}\bar{\boldsymbol{Q}} + \tfrac{1}{2}\boldsymbol{K}^{\text{T}}\boldsymbol{J}\bar{\bar{\boldsymbol{\Omega}}}^{2}\boldsymbol{J}^{\text{T}}\boldsymbol{K} + E_{\text{ad}} \tag{8.59}$$

where we made use of the orthogonality of the Duschinsky matrix.

A direct link of the quantities in Eq. 8.59 to energy derivatives of the final PES at the equilibrium geometry of the initial state can be easily obtained by expanding the $\bar{\bar{V}}$

potential in a Taylor expansion with respect to the normal coordinates of the initial state up to the second order,

$$\bar{\bar{V}}(\bar{Q}) = \bar{\bar{V}}(\bar{Q}^{\mathrm{eq}}) + \left(\frac{\partial \bar{\bar{V}}}{\partial \bar{Q}}\right)^{\mathrm{T}} \bar{Q} + \frac{1}{2}\bar{Q}^{\mathrm{T}} \frac{\partial^2 \bar{\bar{V}}}{\partial \bar{Q}^2} \bar{Q} \tag{8.60}$$

$$= \bar{\bar{V}}(\bar{Q}^{\mathrm{eq}}) + \boldsymbol{g}^{\mathrm{T}}\bar{Q} + \frac{1}{2}\bar{Q}^{\mathrm{T}}\boldsymbol{F}\bar{Q} \tag{8.61}$$

where $\boldsymbol{g}$ is the gradient and $\boldsymbol{F}$ the Hessian of the final-state PES on the base of the initial-state mass-weighted normal coordinate and $\bar{\bar{V}}(\bar{Q}^{\mathrm{eq}})$ is the vertical excitation energy, written E_V in the following. It is straightforward to link those quantities to the gradient ($\bar{\bar{g}}_X$) and Hessian ($\bar{\bar{F}}_X$) calculated at the equilibrium geometry of the initial state with respect to the Cartesian coordinates with the relations

$$\boldsymbol{g} = \bar{\boldsymbol{L}}^{-1}\boldsymbol{M}^{-1/2}\bar{\bar{g}}_X \quad \boldsymbol{F} = \bar{\boldsymbol{L}}^{-1}\boldsymbol{M}^{-1/2}\bar{\bar{\boldsymbol{F}}}_X\boldsymbol{M}^{-1/2}\{\bar{\boldsymbol{L}}^{\mathrm{T}}\}^{-1} \tag{8.62}$$

By analogy between Eqs. 8.59 and 8.61, one obtains

$$\bar{\bar{g}}_X{}^{T}\boldsymbol{M}^{-1/2}\{\bar{\boldsymbol{L}}^{-1}\}^{\mathrm{T}} = -\boldsymbol{K}^{\mathrm{T}}\boldsymbol{J}\bar{\bar{\Omega}}^{2}\boldsymbol{J}^{\mathrm{T}} \tag{8.63}$$

$$\bar{\boldsymbol{L}}^{-1}\boldsymbol{M}^{-1/2}\bar{\bar{\boldsymbol{F}}}_X\boldsymbol{M}^{-1/2}\{\bar{\boldsymbol{L}}^{\mathrm{T}}\}^{-1} = \boldsymbol{J}\bar{\bar{\Omega}}^{2}\boldsymbol{J}^{\mathrm{T}} \tag{8.64}$$

$$E_V = E_{\mathrm{ad}} + \tfrac{1}{2}\boldsymbol{K}^{\mathrm{T}}\boldsymbol{J}\bar{\bar{\Omega}}^{2}\boldsymbol{J}^{\mathrm{T}}\boldsymbol{K} \tag{8.65}$$

From the first line in the set of equations given previously, we can get an expression for the shift vector as a function of the gradient $\bar{\bar{g}}_X$ and the inverse of the Hessian $\bar{\bar{\boldsymbol{F}}}^{-1} = \boldsymbol{J}\bar{\bar{\Omega}}^{-2}\boldsymbol{J}^{\mathrm{T}}$ at the initial-state equilibrium geometry through the relation

$$\boldsymbol{K} = -\bar{\bar{\boldsymbol{F}}}^{-1}\bar{\boldsymbol{L}}^{-1}\boldsymbol{M}^{-1/2}\bar{\bar{g}}_X \tag{8.66}$$

As reported above, in the simplest form VG, the vertical approach assumes that the Hessian matrix is the same in both initial and final states. As a consequence, we have $\bar{\boldsymbol{L}} = \bar{\bar{\boldsymbol{L}}}$ so that $\boldsymbol{J}$ is the identity matrix $\boldsymbol{I}$. Moreover, it is assumed that the final-state frequencies are identical to the initial-state ones. From this set of hypotheses, we get $\bar{\bar{\boldsymbol{F}}}^{-1} = \bar{\bar{\Omega}}^{-2}$ and

$$\boldsymbol{K} = -\bar{\bar{\Omega}}^{-2}\bar{\boldsymbol{L}}^{-1}\boldsymbol{M}^{-1/2}\bar{\bar{g}}_X \tag{8.67}$$

As mentioned before, the shift vector calculated in this case is not directly related to the structural parameters ($\Delta\boldsymbol{X}$), contrary to adiabatic models, but depends on the vibrational frequencies of the final state (but in VG approximation $\bar{\bar{\Omega}} = \bar{\Omega}$) and its gradient. Therefore, in all the practical cases where the approximations behind VG and the harmonic approximation itself are not accurate, the elements of the

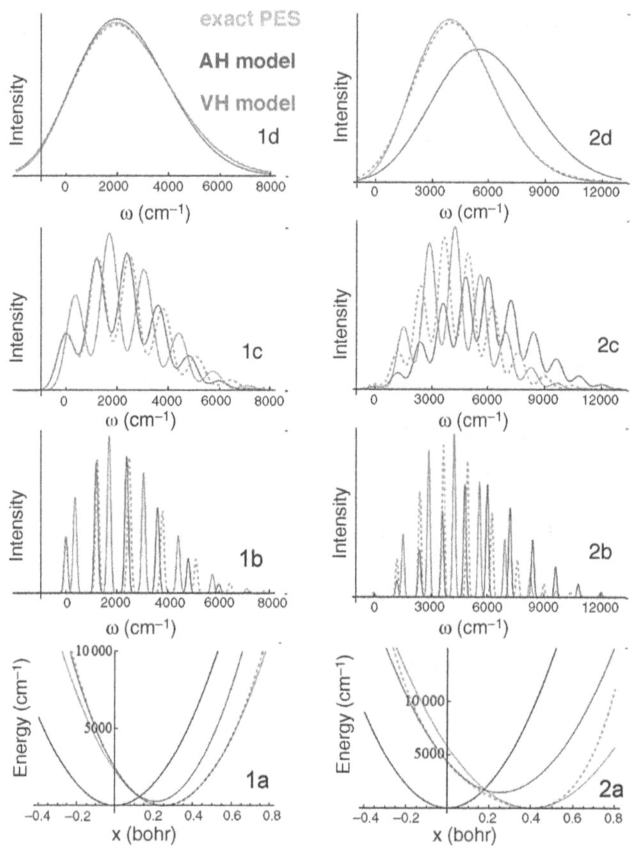

Figure 8.4 FC absorption spectra for model monodimensional system (reduced mass of 6 amu) whose PES is harmonic in the initial state and quartic in the final state and comparison with predictions given by AH and VH harmonic models. The ground-state frequency is $1500\,\text{cm}^{-1}$ while the second derivative of the excited quartic PES corresponds to a frequency of $1350\,\text{cm}^{-1}$ at the vertical position (ground-state geometry) and of $1200\,\text{cm}^{-1}$ at the excited- state equilibrium geometry. Two cases are presented: In the left panels the excited-state geometry is displaced by 0.25 bohr (0.21 according to VH model), while in the right panels it is displaced by 0.45 bohr (0.30 according to VH). Bottom panels (1*a* and 2*a*) report the ground-state PES (black), the excited-state PES (green dashed) and its AH (blue) and VH (red) approximations. With the same colors upper panels report the spectra convoluted with a Gaussian with three different half widths at half maximum [80 (1*b* and 2*b*), 400 (1*c* and 2*c*), and 1200 (1*d*, *a* and 2*d*) cm^{-1}]. (See online version for color figure.)

"vertical" $\boldsymbol{K}$ should be considered "effective" displacements, at variance with the true displacement adopted in adiabatic methods (see the bottom panels of Figure 8.4 discussed above).

As we anticipated above, both adiabatic and vertical models can be coupled to any level of approximation of the transition dipole moments $\boldsymbol{d}^{A}_{e,if}$ and $\boldsymbol{d}^{B}_{e,if}$ in Eq. 8.35

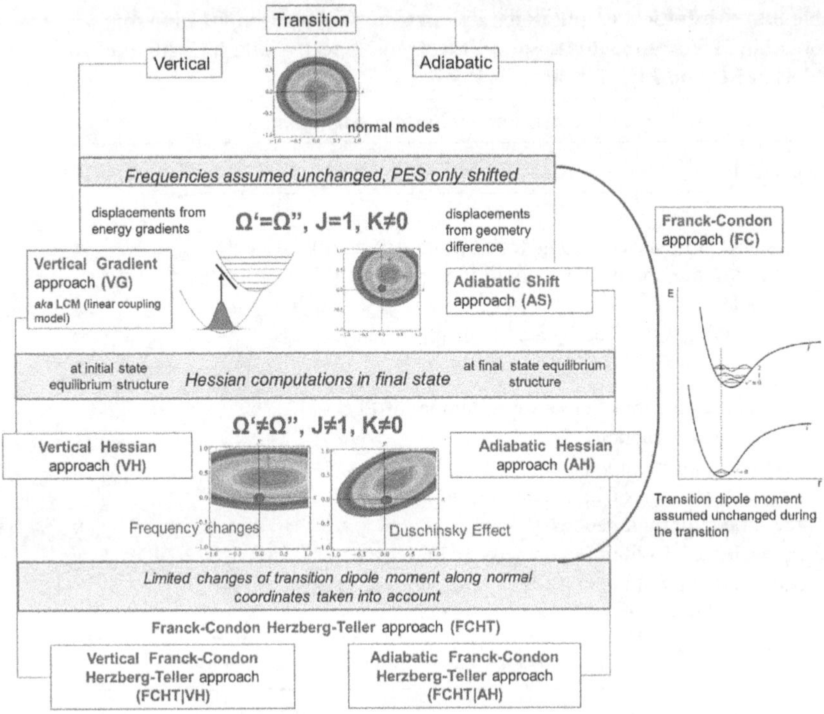

Figure 8.5 Schematic representation of different approaches available to compute vibrationally resolved electronic spectra.

(or tensors if two-photon absorption is considered) to define the physical model applied to the problem under study. Nonetheless, one should be aware that, in a "perturbative" framework, this might lead to mix different perturbative orders [59]. Limiting our study to FC, FCHT, and HT, the corresponding combinations are schematically represented in Figure 8.5. For instance, the acronyms FC|AH, FCHT|AH, and HT|AH are used to refer to the adiabatic models defined by the harmonic representation of the PES of each electronic state calculated at its equilibrium geometry, which differ only by the approximation of the transition dipole moment, respectively FC, FCHT, and HT. As shown in the figure, it is also possible to define a matching adiabatic model to the VG, namely the adiabatic shift (AS). In both cases, the changes in vibrational frequencies and the rotation of normal modes upon the electronic excitation are not taken into account. It is worthy to notice at this point that, while it is possible to combine VG and AS models with the FC and the FCHT approximations, as often done for VG in the literature [51], the effort required, at least at the time-dependent density functional theory (TD-DFT) level, for the numerical differentiation of the transition dipole derivatives neutralizes the computational convenience of those models since, at the same cost, the excited-state normal modes can also be obtained.

Table 8.1 Ab Initio Computations Required to Generate Input Data for Simulation of Vibrationally Resolved Electronic Spectra with VG, AS, and AH Models in FC and FCHT Framework

Potential Energy Surface Approximations				
Computation	VG	AS	VH	AH
Initial state				
Cartesian coordinates of atoms (equilibrium structure)	x	x	x	x
Energy at minimum of PES (equilibrium geometry)	x	x	x	x
Frequencies	x	x	x	x
Normal modes, expressed by atomic displacements	x	x	x	x
Final state				
Cartesian coordinates of atoms at minimum of PES (equilibrium structure)		x		x
Energy at equilibrium geometry of initial state	x		x	
Energy at minimum of PES (equilibrium geometry)		x		x
Forces at equilibrium geometry of initial state	x		x	
Frequencies at equilibrium geometry of initial state			x	
Frequencies at minimum of PES (equilibrium geometry)				x
Normal modes, expressed by atomic displacements				x

Electronic Transition Amplitudes Approximations			
Approximation	FC	HT	FCHT
Electronic transition amplitudes	x		x
Derivatives of electronic transition amplitudes		x	x

From a practical point of view, the diverse combined models presented above can be applied to any kind of spectroscopy described in Section 8.2. As a practical summary, Table 8.1 lists the data required as input by each one.

The choice of the most appropriate approach will depend on the problem under study and the feasibility of the computations to obtain all the necessary information. It is noteworthy that all models require the optimized geometry for the initial state along with the calculation of its Hessian matrix. However, as discussed above, they differ significantly for the data required for the final state, with a possible large impact on the total computational times.

The simplest, VG model requires only the energy gradient of the excited state to be calculated at the geometry of the ground state. Because of these low requirements, the VG model provides the most feasible up-to-date approach for the studies of spectra on a broad energy range and/or for macromolecules. At variance, the AS model requires the determination of the equilibrium structure for the final state, but not the frequencies, so it might be considered as a solution for cases where the main interest is in the spectral features close to the transition origin but no precise frequencies are required. It is interesting to notice that both simplified models constrain the total zero-point vibrational energy to be the same in the initial and final states.

Vertical and adiabatic approaches evaluate differently the transition energy between initial and final states. The vertical transition energy (E_V) is more accurate within the vertical framework, while E_{ad} is more accurate within the adiabatic framework. When the harmonic approximation is exact, the AH and VH approaches are equivalent. In all practical cases, they differ and both have strong and weak points. Generally speaking, we can state that AH is better suited for the analysis of the bands closer to the 0–0 origin and high-resolution spectra, while VH is more accurate to describe the region of the spectrum maximum and the low-resolution spectrum. These features can be illustrated by applying them on a simple monodimensional model system, where the exact final-state PES is a quartic function of the coordinate. Figure 8.4 compares the exact and model PESs (bottom panels) and the resulting exact, VH and AH spectra at different resolutions for two different choices of the parameters. In the calculations reported in the left panels, parameters (given in the caption) are chosen to make the exact potential more similar to the AH approximation than to the VH one, and the opposite has been done in the right column. Analysis of the panels in the left column of Figure 8.4 shows that VH is as good as AH for low-resolution spectra, but its predictions of the positions of the high-resolution bands are considerably off, even those close to the origin 0–0 (set to zero in energy), while AH performs pretty well also at high resolution. At variance, in the case reported in the right column, AH prediction of the band maximum (best analyzed in the low-resolution spectrum) shows a sensible error, while VH prediction is very accurate. Nonetheless, also in this case the high-resolution bands close to the 0–0 transition are better simulated by the AH model. Notice in fact that the 0–0 transition of the exact quartic potential is at zero energy and is very weak. The AH nicely captures this feature while, according to VH, the 0–0 transition is predicted to lie at about 1500 cm^{-1} and to be rather strong. Focusing on absorption spectra, AH and VH models are almost equivalent as far as the computational cost is concerned since the most expensive computation is usually the excited-state Hessian, required in both approaches (AH also requires the excited-state geometry optimization). The AS and VG models are computationally cheaper, but neglecting the difference in the Hessians of the initial and final states introduces further approximations, usually worsening the reproduction of the spectrum in the region of the maximum (AS) or in the region of the 0–0 (VG).

When the harmonic description of the PES fails, an anharmonic description is required, but this is often unpractical for medium-to-large systems. Within the vertical framework, Hazra et al. [52] proposed a partial analysis restricted to the modes for which the harmonic curvature of the PES leads to imaginary frequencies. This separation is based on the hypothesis that the PES can be studied independently along each "anharmonic" mode. Further simplification can be added by assuming that there is no mixing between each anharmonic mode and all the other ones. As a result, the overlap integrals between the initial and final vibronic states are considered as a product of a $(N-n)$-dimensional harmonic overlap integral and the n anharmonic ones, where n represents the number of anharmonic modes. Since it would be out of the scope of this presentation, we will not give further details on this specific case of the FC/FCHT vertical approaches and will assume here that the harmonic approximation stands in all the studied cases.

8.3.2.2 Simplified Vertical and Adiabatic Approaches for Vibrational Resonance Raman It is noteworthy that the most popular approaches for the computation of vibrational RR spectra, namely the gradient approximation and the transform theory (in its most widely adopted version) can be described in the framework of the simplified AS and VG harmonic approaches discussed in the above section. These methodologies focus on fundamental transitions, that is, the transitions between the ground state and a state with one quantum in a single mode, and assume that no Duschinsky rotation ($\boldsymbol{J}=\boldsymbol{I}$ in Eq. 8.52) and no frequency change ($\bar{\bar{\boldsymbol{\Omega}}}=\bar{\boldsymbol{\Omega}}$) take place between the two electronic states $\underline{g}$ and $\underline{m}$ involved in the transition. This model, which can be classified as VG or AS depending on the way the shift vector K is defined, is sometimes known in the context of resonance Raman as the independent-mode displaced harmonic oscillator (IMDHO) model. Furthermore, only the FC term in the polarizability tensor is considered in Eq. 8.46. With these approximations, using Eq. 8.53, one simply obtains that

$$\langle \bar{\mathbf{0}} | \bar{\bar{\mathbf{0}}} + \bar{\bar{\nu}}_r \rangle = -\frac{\Delta_r}{\sqrt{2}} \langle \bar{\mathbf{0}} | \bar{\bar{\mathbf{0}}} + \bar{\bar{\nu}}_r - \bar{\bar{1}}_r \rangle \tag{8.68}$$

where Δ_r is the dimensionless displacement defined as $\Delta_r = \mathrm{K}_r(\bar{\omega}_r/\hbar)^{1/2}$.

Transform Theory Transform theory is based on the well-known relation between the absorption cross section and the polarizability tensor [60, 61]. At $T=0\,\mathrm{K}$,

$$\sigma_{\mathrm{abs}}(\omega_I) \propto \omega_I \Im\left[\alpha_{\bar{0},\bar{0}}(x,x) + \alpha_{\bar{0},\bar{0}}y, y + \alpha_{\bar{0},\bar{0}}(z,z)\right] \tag{8.69}$$

It can be then shown that [60]

$$\alpha(\tau,\eta)^{\bar{0}_r \to \bar{1}_r} = \mu_{e,gm}(\tau, \boldsymbol{Q}_{\underline{g}}^{\mathrm{eq}})\mu_{e,gm}(\eta, \boldsymbol{Q}_{\underline{g}}^{\mathrm{eq}})\frac{\Delta_r}{\sqrt{2}}[\Phi(\omega_I) - \Phi(\omega_I - \omega_r)] \tag{8.70}$$

where $\Phi(\omega_I)$ is related to the normalized spectrum $\bar{\sigma}_{\mathrm{abs}}(\omega_I) = \sigma_{\mathrm{abs}}(\omega_I)/\omega_I$ by

$$\Phi(\omega_I) = i\pi\bar{\sigma}_{\mathrm{abs}}(\omega_I) + \mathcal{P}\int d\omega \frac{\bar{\sigma}_{\mathrm{abs}}(\omega)}{\omega - \omega_I} \tag{8.71}$$

where $\mathcal{P}$ is the principal value of the integral.

Since the transform theory adopts Eq. 8.68, it can be seen as an application of the AS model to vibrational resonance Raman, even if, as clarified below, the displacement in Eq. 8.68 can also be obtained from the VG model.

Gradient Approximation To describe the so-called gradient approximation in the framework adopted in this chapter, let us consider only the FC terms in the polarizability tensor and only the fundamental transitions $0_r \to 1_r$ for each mode r. If $\Delta_r \ll 1$, as noticed by Long [25] and Warshel and Dauber [62], we have $\langle \bar{0}_r | \bar{\bar{0}}_r \rangle \approx \langle \bar{1}_r | \bar{\bar{1}}_r \rangle \approx 1$ and $\langle \bar{0}_r | \bar{\bar{1}}_r \rangle = -\langle \bar{1}_r | \bar{\bar{0}}_r \rangle \approx -\Delta_r/\sqrt{2}$ and all the other integrals equal to zero. By neglecting all the terms beyond the first order in the displacements Δ_r, the polarizability tensor becomes simply

$$\alpha(\tau,\eta)^{\bar{0}_r\to\bar{1}_r} = \mu_{e,gm}(\tau,\boldsymbol{Q}_{\underline{g}}^{\mathrm{eq}})\mu_{e,gm}(\eta,\boldsymbol{Q}_{\underline{g}}^{\mathrm{eq}})\left[\frac{\langle\bar{0}_r|\bar{\bar{0}}_r\rangle\langle\bar{\bar{0}}_r|\bar{1}_r\rangle}{\Delta\omega_{\underline{mg}}-\omega_I-i\gamma_{\underline{m}}}+\frac{\langle\bar{0}_r|\bar{\bar{1}}_r\rangle\rangle\langle\bar{\bar{1}}_r|\bar{1}_r\rangle}{\Delta\bar{\omega}_{\underline{mg}}+\bar{\omega}_I-i\gamma_{\underline{m}}}\right] \tag{8.72}$$

$$\alpha(\tau,\eta)^{\bar{0}_r\to\bar{1}_r} = \mu_{e,gm}(\tau,\boldsymbol{Q}_{\underline{g}}^{\mathrm{eq}})\mu_{e,gm}(\eta,\boldsymbol{Q}_{\underline{g}}^{\mathrm{eq}})\left[\frac{\bar{\omega}_r\Delta_r}{\sqrt{2}(\Delta\omega_{\underline{mg}}-\omega_I-i\gamma_{\underline{m}})(\Delta\omega_{\underline{mg}}-\bar{\omega}_r-\omega_I-i\gamma_{\underline{m}})}\right] \tag{8.73}$$

where $\Delta\omega_{\underline{mg}}$ is the electronic adiabatic energy difference.

When $\Delta\omega_{\underline{mg}}-\omega_I \gg \bar{\omega}_r$, $\alpha(\tau,\eta)^{\bar{0}_r\to\bar{1}_r}$ is proportional to $\Delta_r\bar{\omega}_r$ and, from Eqs. 8.28 and 8.33, it follows that the differential cross section $\sigma^{\bar{0}_r\to\bar{1}_r} \propto (\Delta_r\omega_r)^2$. Notice that, in the limit of validity of the adopted IMDHO model with equal frequencies for the potential energy surfaces ($\boldsymbol{J}=\boldsymbol{I}$ and $\bar{\bar{\boldsymbol{\Omega}}}=\bar{\boldsymbol{\Omega}}$), Δ_r can be obtained avoiding geometry optimization of the excited-state, simply computing the derivative of the excited-state energy with respect to the normal coordinate $\bar{Q}_r$, at the ground-state equilibrium geometry (i.e., following the VG approach). At variance, if Δ_r is obtained by optimization of the excited-state PES, the IDMHO becomes identical to the AS one:

$$\left(\frac{\partial\bar{V}}{\partial\bar{Q}_r}\right)_0 = V_{\underline{m}}(r) = -\omega_r^{3/2}\Delta_r \tag{8.74}$$

from which it comes out that the ratio between the RR intensities of the fundamentals of two modes, r and s, is

$$\frac{I^{\bar{0}_r\to\bar{1}_r}}{I^{\bar{0}_s\to\bar{1}_s}} = \frac{\left[V_{\underline{m}}(r)/V_{\underline{m}}(s)\right]^2\omega_s}{\omega_r} = \frac{\omega_r^2\Delta_r^2}{(\omega_s^2\Delta_s^2)} \tag{8.75}$$

In a series of seminal papers [63, 64], Heller and co-workers showed by a time-dependent approach that in pre-resonance conditions, when only the very short time dynamics on the excited state must be considered, the first equality in Eq. 8.75 still holds, even when the independent-mode model is not valid and the second equality in Eq. 8.75 does not hold. In true resonance cases and staying within the harmonic approximation, Duschinsky mixings and frequency changes must be considered and this can be done by direct application of Eq. 8.46.

8.3.2.3 ***Practical Example: OPA Spectra for*** **ax-*R3MCP*** The differences and similarities between the various physical models described in the previous section will be discussed with the example of the $S_2 \leftarrow S_0$ electronic transition of the *ax*-R3MCP.

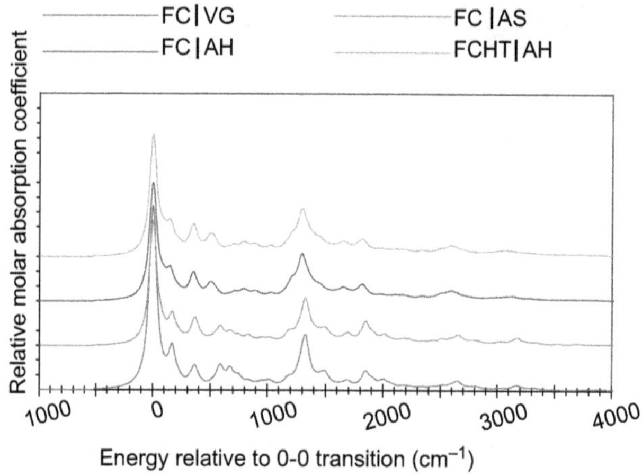

Figure 8.6 Convoluted OPA spectra of $S_2 \leftarrow S_0$ electronic transition of *ax*-R3MCP computed with vertical and adiabatic FC approaches.

Figures 8.6 and 8.7 show the simulated OPA spectra, obtained with various approximations related to the changes between the electronic states upon the electronic transition, and the expansion of the transition dipole moments. First, let us consider the general shape of OPA spectra, convoluted using Lorentzian functions with a half-width at half-maximum (HWHM) of $80\,\text{cm}^{-1}$. In general, spectra convoluted with distribution (Gaussian or Lorentzian) functions of large HWHM are more suitable to compare with the typical low-resolution UV–vis spectra. It is immediately visible that in such a case all approaches yield essentially equivalent spectrum line shapes. In particular, results obtained with the simplest FC|VG approach are in qualitative agreement with the more demanding computations performed within the adiabatic FC|AH framework. This illustrates the interest of the FC|VG approach that can be sufficient to correctly reproduce the general features of the experimental spectra and can often be applied to simulate low-resolution spectra, in particular for larger systems for which more computationally demanding approaches could be unfeasible. However, the situation is different if the fine structure of the spectra is needed, since, in this case, it is crucial to properly describe the final-state normal modes and frequencies. In fact, as clearly shown in Figure 8.7, various approaches yield quite different spectra patterns. Moreover, larger deviations can be expected for molecules (and/or states) showing larger changes in frequencies and/or significant Duschinsky mixings upon excitation. As an example, in work by Lin et al. [32], sensible differences between VG and AH results have been documented for the $n \rightarrow \pi^*$ $(S_1 \leftarrow S_0)$ electronic transition of both axial and equatorial conformers of R3MCP. Such a transition involves the lone pair of the oxygen and the π^* orbital residing on the CO bond, so that the CO stretch is responsible for the most prominent progression of the spectrum and its vibrational frequency strongly decreases upon excitation. In this situation, models like FC|VG, which do not take into account this effect, predict too large spacings between the

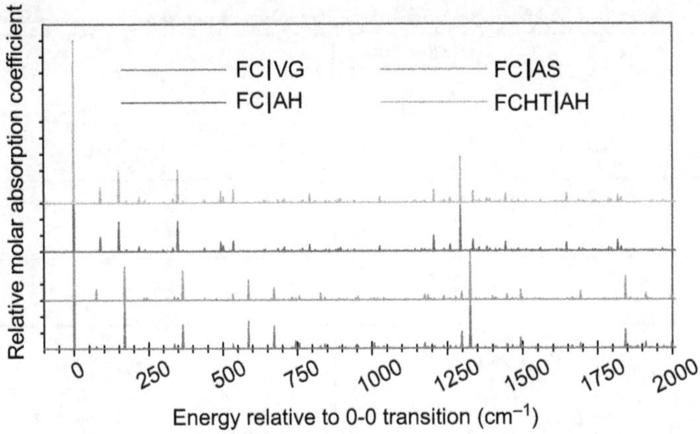

Figure 8.7 Stick OPA spectra of $S_2 \leftarrow S_0$ electronic transition of *ax*-R3MCP computed with vertical and adiabatic FC approaches.

bands, while application of the FC|AH model results in a significant improvement of the agreement with the experiments.

8.3.2.4 Practical Example: One- and Two-Photon Spectra for R3MCP In this section, we report the $S_1 \leftarrow S_0$ and $S_2 \leftarrow S_0$ one- and two-photon absorption, and circular dichroism (CD) spectra of R3MCP, computed according to the FCHT|AH and FCHT|VG models.

These results have been discussed in detail elsewhere [65]. Figure 8.8 shows room temperature spectra obtained by averaging with their Boltzmann weights the spectra of both equatorial and axial conformers. As explained in Section 8.3.1, in the limit of the model we adopt, the lineshapes of one- and two-photon nonresonant spectra are equal at the FC level since they only depend on FC factors. Therefore, differences can only arise from the HT contributions. Figure 8.8 shows that these differences are remarkable for the $S_1 \leftarrow S_0$ transition. Specifically, while OPA and TPA FCHT spectra have similar shapes, with the FC term prevailing in the overall contribution, as reported by Lin et al. [65], they are both rather different from the ECD spectrum. Such differences become drastic for the TPCD case, whose lineshape shows a clear sign inversion (not possible at the FC level) due to the interference among the several different transition amplitudes, which contribute to the spectroscopic signal (see Eq. 8.27). For the $S_2 \leftarrow S_0$ transition, apart from the overall sign of the spectrum (positive by definition for absorption and negative, in this specific example, for both one- and two-photon CD spectra), differences among the four presented spectra are much more moderate than they are in the $S_1 \leftarrow S_0$ case.

As far as the comparison of the predictions of VG and AH models are concerned, they show remarkable differences in some cases, especially for TPCD, suggesting that the more complex a signal is (in the sense that it depends on many transition amplitudes) and the more important the HT effects are, the larger are the differences that can arise between the different models used to approximate the

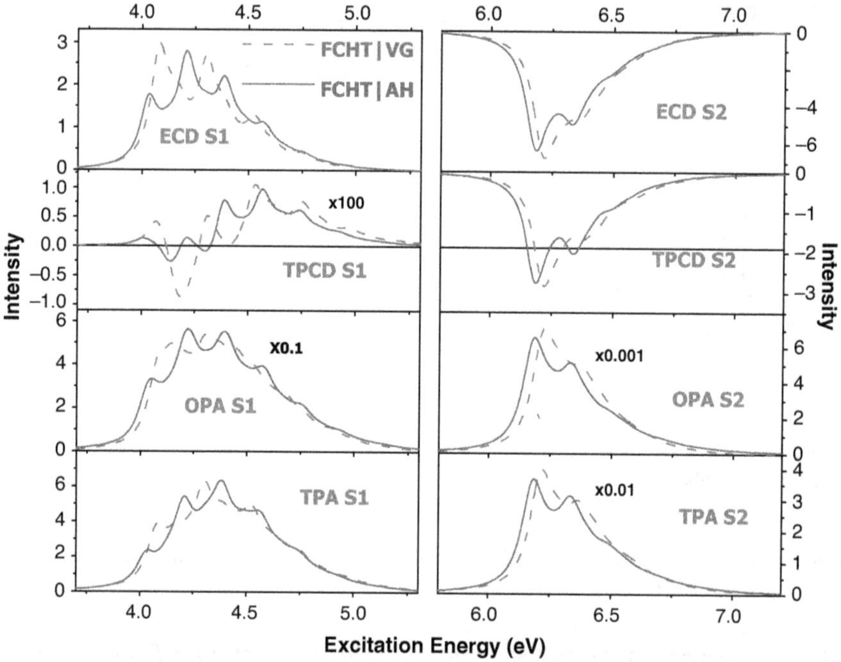

Figure 8.8 ECD and TPCD and OPA and TPA spectra for S_1 and S_2 states of R3MCP computed with FCHT|VG and FCHT|AH models. All spectra are convoluted with a Lorentzian broadening of 0.05 eV and obtained by averaging with Boltzmann weights the spectra of the individual conformers. Units are dm^3 mol^{-1} cm^{-1} for both ECD and OPA and 10^{-30} cm^4 s photon^{-1} mol^{-1} for both TPCD and TPA.

final-state PES. We conclude the discussion on this example by recalling that the first experimental TPCD spectra have been recently reported in literature [66]. The results reported in Figure 8.8, showing that no a priori correlation exists between the signs of ECD and TPCD spectra (as it clearly appears in the $S_1 \leftarrow S_0$ case), highlight the fact that these two spectroscopies can provide complementary information so that their comparison could be beneficial in determining the absolute configuration of chiral molecules.

8.3.3 Analysis of Spectral Moments

A moment analysis of the spectrum can yield interesting general information. For the sake of simplicity, we consider the spectral shape of a one-photon transition, but in the limits reported in Section 8.3.1.1, the conclusion will be valid also for nonresonant two-photon absorption and circular dichroism processes. From now on, we neglect the frequency-dependent prefactor

$$I(\omega) = \sum_i \sum_f \rho_i \left[\boldsymbol{d}_{if}^A \cdot \boldsymbol{d}_{if}^{B*} \right] \delta(\omega_f - \omega_{\underline{i}} \pm \omega) \tag{8.76}$$

The kth moment of the spectrum is defined as [67]

$$\mathcal{M}^{(k)} = \int \hbar\omega^k I(\omega)\, d\hbar\omega \tag{8.77}$$

In the following, we assume that the BO approximation is valid and the Eckart conditions are met. Additionally, for the sake of simplifying the notation, all transitions are supposed to involve a single initial and a single final electronic state. Consequently, the summation will be done over the initial vibrational states (also noted *i*) from now on, with the vibrational wavefunction simply noted as χ. The subscript *if* will also be dropped in the transition dipole moment ($\boldsymbol{d}^Z_{e,if} = \boldsymbol{d}^Z_e$). In these conditions, the Taylor expansion given in Eq. 8.35 can be noted in the condensed form

$$\boldsymbol{d}^Z_e \approx \boldsymbol{d}^Z_{e,0} + \sum_k \dot{\boldsymbol{d}}^Z_{e,k} \bar{Q}_k \tag{8.78}$$

with

$$\boldsymbol{d}^Z_{e,0} \equiv \boldsymbol{d}^Z_{e,if}(\boldsymbol{Q}^{\rm eq}_{\underline{i}}) \qquad \dot{\boldsymbol{d}}^Z_{e,k} \equiv \left(\frac{\partial \boldsymbol{d}^Z_{e,if}}{\partial Q_{\underline{i}}(k)}\right)_0$$

In Chapter 10 it is shown that the spectrum can be recast in the time-domain expression

$$I(\omega) = 2\pi\hbar \int \exp[(\pm i)\omega t] I(t)\, dt \tag{8.79}$$

where $+$ and $-$ stand for absorption and emission, respectively, and from now on in the symbols $\pm$ and $\mp$ the upper sign will be intended for absorption and the lower for emission,

$$I(t) = \sum_i \rho_i \left\langle \chi_i \middle| \boldsymbol{d}^A_e \exp\left(\frac{-i\hat{\bar{\bar{H}}}t}{\hbar}\right) \boldsymbol{d}^{B*}_e \exp\left(\frac{+i\hat{\bar{\bar{H}}}t}{\hbar}\right) \middle| \chi_i \right\rangle \tag{8.80}$$

Following Lax [67], we invert Eq. 8.79:

$$I(t) = \int \exp[(\mp i)\omega t] I(\omega)\, d(\hbar\omega) \tag{8.81}$$

Next, we expand the LHS and the exponential of the integrand in the RHS of Eq. 8.81 in Maclaurin series (about $t = 0$) and equate terms with the same power, thus getting

$$\frac{d^k I(t)}{d(it/\hbar)^k} = (\mp)^k \int \hbar\omega^k I(\hbar\omega)\, d(\omega) = (\mp)^k \mathcal{M}^{(k)} \tag{8.82}$$

Therefore, by deriving $I(t)$ in Eq. 8.80 with respect to the time, we easily get the (unnormalized) moments of the spectrum. Specifically for $\mathcal{M}^{(0)}$, $\mathcal{M}^{(1)}$, and $\mathcal{M}^{(2)}$ we have [67]

$$\mathcal{M}^{(0)} = \sum_i \rho_i \langle \chi_i | \boldsymbol{d}_e^A \boldsymbol{d}_e^{B*} | \chi_i \rangle \tag{8.83}$$

$$\mathcal{M}^{(1)} = (\pm) \sum_i \rho_i \Big\langle \chi_i \Big| \Big[\boldsymbol{d}_e^A \hat{\bar{\bar{H}}} \boldsymbol{d}_e^{B*} - \boldsymbol{d}_e^A \boldsymbol{d}_e^B \hat{\bar{H}} \Big] \Big| \chi_i \Big\rangle \tag{8.84}$$

$$\mathcal{M}^{(2)} = \sum_i \rho_i \Big\langle \chi_i \Big| \Big[\boldsymbol{d}_e^A \hat{\bar{\bar{H}}}^2 \boldsymbol{d}_e^{B*} - 2\boldsymbol{d}_e^A \hat{\bar{\bar{H}}} \boldsymbol{d}_e^{B*} \hat{\bar{\boldsymbol{H}}} + \boldsymbol{d}_e^A \boldsymbol{d}_e^{B*} \hat{\bar{H}}^2 \Big] \Big| \chi_i \Big\rangle \tag{8.85}$$

The 0th moment $\mathcal{M}^{(0)}$ is simply the integral of the spectrum over the frequency domain, and hence the total intensity. It can be straightforwardly evaluated up to the first-order expansion of the electronic dipole moments in the normal coordinates of the initial state (Eq. 8.78) by noting that only terms with even powers of each normal coordinate $\bar{Q}_k$ can give a contribution to the expectation value in Eq. 8.83, due to symmetry rules (see Eq. 8.48), and that

$$\langle \bar{v}_k | \bar{Q}_k^2 | \bar{v}_k \rangle = \frac{\hbar}{2\bar{\omega}_k} (2\bar{v}_k + 1) \tag{8.86}$$

As a result, we obtain

$$\mathcal{M}^{(0)} = \boldsymbol{d}_{e,0}^A \cdot \boldsymbol{d}_{e,o}^{z*} + \sum_k \frac{\hbar}{2\bar{\omega}_k} \dot{\boldsymbol{d}}_{e,k}^A \cdot \dot{\boldsymbol{d}}_{e,k}^{B*} \coth\left(\frac{\beta\hbar\bar{\omega}_k}{2}\right) \tag{8.87}$$

where $\beta = 1/k_B T$, with k_B the Boltzmann constant. It is noteworthy that coth $[\beta\hbar\bar{\omega}_k] = (2\bar{v}_k(\bar{T}) + 1)$, where $\bar{v}_k(\bar{T})$ is the average quantum number of mode k at temperature T in the initial electronic state, so that the 0th moment can also be written as

$$\mathcal{M}^{(0)} = \boldsymbol{d}_{e,0}^A \cdot \boldsymbol{d}_{e,o}^{Z*} + \sum_k \frac{\hbar}{2\bar{\omega}_k} \dot{\boldsymbol{d}}_{e,k}^A \cdot \dot{\boldsymbol{d}}_{e,k}^{B*} [2\bar{v}_k(T) + 1] \tag{8.88}$$

The first moment (normalized through division by $\mathcal{M}^{(0)}$) represents the average frequency, also called the center of gravity of the spectrum. It can be calculated analytically taking advantage of the Taylor expansion given in Eq. 8.61 and recalling that the expectation value of the kinetic operator is

$$\langle \chi_i | T_v | \chi_i \rangle = \frac{1}{2} \sum_k \langle \chi_i | \bar{P}_k^2 | \chi_i \rangle = \frac{\hbar}{4} \sum_k \bar{\omega}_k (2\bar{v}_k + 1) = \frac{1}{2} E_i \tag{8.89}$$

where $\bar{P}_k = -i\hbar\,\partial/\partial\bar{Q}_k$ is the momentum operator associated with $\bar{Q}_k$.

It is useful to write the first moment as a sum of three terms depending on the derivative's order of the transition dipole moments $\boldsymbol{d}_e^A$ and $\boldsymbol{d}_e^B$,

$$\mathcal{M}^{(1)} = \mathcal{M}^{(1)}_{\mathrm{FC}} + \mathcal{M}^{(1)}_{\mathrm{FCHT}} + \mathcal{M}^{(1)}_{\mathrm{HT}} \tag{8.90}$$

$$\mathcal{M}^{(1)}_{\mathrm{FC}} = (\pm)\boldsymbol{d}^A_{e,0}\cdot\boldsymbol{d}^{B*}_{e,0}\sum_i \rho_i\left[\langle\chi_i|E_V+\hat{T}-\hat{\bar{H}}|\chi_i\rangle + \sum_k\left\langle\chi_i\left|\frac{1}{2}\mathrm{F}_{kk}\bar{Q}_k^2\right|\chi_i\right\rangle\right] \tag{8.91}$$

$$\mathcal{M}^{(1)}_{\mathrm{FCHT}} = (\pm)\sum_i\rho_i\sum_k\left\langle\chi_i\left|\left(\boldsymbol{d}^A_{e,0}\cdot\dot{\boldsymbol{d}}^{B*}_{e,k}+\boldsymbol{d}^{B*}_{e,0}\cdot\dot{\boldsymbol{d}}^{A}_{e,k}\right)g_k\bar{Q}_k^2\right|\chi_i\right\rangle \tag{8.92}$$

$$\mathcal{M}^{(\mathrm{T})}_{\mathrm{HT}} = (\pm)\sum_i\rho_i\sum_k\sum_l\sum_m\sum_n\left\langle\chi_i\left|\frac{1}{2}\left(\dot{\boldsymbol{d}}^A_{e,k}\cdot\dot{\boldsymbol{d}}^{B*}_{e,l}\right)\mathrm{F}_{mn}\bar{Q}_k\bar{Q}_l\bar{Q}_m\bar{Q}_n\right|\chi_i\right\rangle \tag{8.93}$$

$$+(\pm)\sum_i\rho_i\sum_k\left\langle\chi_i\left|\left(\dot{\boldsymbol{d}}^A_{e,k}\cdot\dot{\boldsymbol{d}}^{B*}_{e,k}\right)\bar{Q}_k^2(E_V-\hat{\bar{H}})\right|\chi_i\right\rangle \tag{8.94}$$

$$-(\pm)\sum_i\rho_i\sum_k\sum_l\sum_m\left\langle\chi_i\left|\left(\dot{\boldsymbol{d}}^A_{e,k}\cdot\dot{\boldsymbol{d}}^{B*}_{e,l}\right)\bar{Q}_k\frac{\bar{P}_m^2}{2}\bar{Q}_l\right|\chi_i\right\rangle \tag{8.95}$$

Now, similar to Eq. 8.86, the following relations can be straightforwardly obtained:

$$\left\langle\bar{v}\left|\bar{Q}_k^4\right|\bar{v}\right\rangle = \frac{\hbar^2}{4\bar{\omega}_k^2}(6\bar{v}_k^2+6\bar{v}_k+3) \tag{8.96}$$

$$\left\langle\bar{v}\left|\bar{Q}_k\bar{P}_k^2\bar{Q}_k\right|\bar{v}\right\rangle = \frac{\hbar^2}{4}(2\bar{v}_k^2+2\bar{v}_k+3) \tag{8.97}$$

Using the above equations together with Eq. 8.89 for the contributions of the operator $\hat{P}_k^2$ and keeping in mind that only even powers of each normal mode $\bar{Q}_k$ may contribute to the expectation value, it is straightforward to obtain an analytic expression of the three terms on the RHS of Eq. 8.90,

$$\mathcal{M}^{(1)}_{\mathrm{FC}} = (\pm)\dot{\boldsymbol{d}}^A_{e,0}\cdot\dot{\boldsymbol{d}}^{B*}_{e,0}\left\{E_V+\frac{\hbar}{4\bar{\omega}_k}\sum_k(\mathrm{F}_{kk}-\bar{\omega}_k^2)[2\bar{v}_k(T)+1]\right\} \tag{8.98}$$

$$\mathcal{M}^{(1)}_{\mathrm{FCHT}} = (\pm)\sum_k\frac{\hbar}{2\bar{\omega}_k}\left\{\left(\boldsymbol{d}^A_{e,0}\cdot\dot{\boldsymbol{d}}^{B*}_{e,k}+\boldsymbol{d}^{B*}_{e,0}\cdot\dot{\boldsymbol{d}}^{B}_{e,k}\right)g_k[2\bar{v}_k(T)+1]\right\} \tag{8.99}$$

$$\mathcal{M}_{\mathrm{HT}}^{(1)} = (\pm)\sum_k \left\{ \dot{\boldsymbol{d}}_{e,k}^{A} \cdot \dot{\boldsymbol{d}}_{e,k}^{B*} \frac{3\hbar^2 \mathrm{F}_{kk}}{8\bar{\omega}_k^2} \left[2\bar{v}_k(T)^2 + 2\bar{v}_k(T) + 1\right] \right.$$
$$+ \sum_{l\neq k} \left[\dot{\boldsymbol{d}}_{e,k}^{A} \cdot \dot{\boldsymbol{d}}_{e,k}^{B*} \mathrm{F}_{ll} + \dot{\boldsymbol{d}}_{e,k}^{A} \cdot \dot{\boldsymbol{d}}_{e,l}^{B*} \mathrm{F}_{kl}\right] \frac{\hbar^2}{8\bar{\omega}_k \bar{\omega}_l} [2\bar{v}_k(T) + 1][2\bar{v}_l(T) + 1]$$
$$+ \dot{\boldsymbol{d}}_{e,k}^{A} \cdot \dot{\boldsymbol{d}}_{e,k}^{B*} \left(E_V - \frac{\hbar}{2} \sum_l \bar{\omega}_l [2\bar{v}_l(T) + 1] \right) \frac{\hbar}{2\bar{\omega}_k} [2\bar{v}_k(T) + 1]$$
$$+ \dot{\boldsymbol{d}}_{e,k}^{A} \cdot \dot{\boldsymbol{d}}_{e,k}^{B*} \frac{\hbar^2}{4} \left[2\bar{v}_k(T)^2 + 2\bar{v}_k(T) + 3\right]$$
$$\left. + \sum_{l\neq k} \dot{\boldsymbol{d}}_{e,k}^{A} \cdot \dot{\boldsymbol{d}}_{e,k}^{B*} \frac{\hbar^2 \bar{\omega}_l}{4\bar{\omega}_k} [2\bar{v}_k(T) + 1)(2\bar{v}_l(T) + 1] \right\} \tag{8.100}$$

The expression for $\mathcal{M}_{\mathrm{HT}}^{(1)}$ can be further simplified summing the terms in the last three rows and noticing that

$$\bar{v}_r(T)^2 = \bar{v}_r(T)^2 + \tfrac{1}{4}\left[\coth\left(\tfrac{1}{2}\beta\hbar\omega_r\right)^2 - 1\right] \tag{8.101}$$

$$2\bar{v}_r(T)^2 = 2\bar{v}_k(T) + 1 = \coth\left(\tfrac{1}{2}\beta\hbar\omega_r\right)^2 \tag{8.102}$$

Therefore we obtain the final expression

$$\mathcal{M}_{\mathrm{FC}}^{(1)} = (\pm)\boldsymbol{d}_{e,0}^{A} \cdot \boldsymbol{d}_{e,0}^{B*} \left[E_V + \frac{\hbar}{4\bar{\omega}_k} \sum_k (\mathrm{F}_{kk} - \bar{\omega}_k^2) \coth\left(\frac{\beta\hbar\bar{\omega}_k}{2}\right) \right] \tag{8.103}$$

$$\mathcal{M}_{\mathrm{FCHT}}^{(1)} = (\pm)\sum_k \frac{\hbar}{2\bar{\omega}_k} \left[\left(\boldsymbol{d}_{e,0}^{A} \cdot \dot{\boldsymbol{d}}_{e,k}^{B*} + \boldsymbol{d}_{e,0}^{B*} \cdot \dot{\boldsymbol{d}}_{e,0}^{A} \right) g_k \coth\left(\frac{\beta\hbar\bar{\omega}_k}{2}\right) \right] \tag{8.104}$$

$$\mathcal{M}_{\mathrm{HT}}^{(1)} = (\pm)\sum_k \left\{ \dot{\boldsymbol{d}}_{e,k}^{A} \cdot \dot{\boldsymbol{d}}_{e,k}^{B*} \left[\frac{\hbar^2}{2} + E_V \frac{\hbar}{2\bar{\omega}_k} \left(\frac{\beta\hbar\bar{\omega}_k}{2}\right) + \frac{3\hbar^2 \mathrm{F}_{kk}}{8\bar{\omega}_k^2} \coth^2\left(\frac{\beta\hbar\bar{\omega}_k}{2}\right) \right] \right.$$
$$\left. + \sum_{l\neq k} \left[\dot{\boldsymbol{d}}_{e,k}^{A} \cdot \dot{\boldsymbol{d}}_{e,k}^{B*} \mathrm{F}_{ll} + \dot{\boldsymbol{d}}_{e,k}^{A} \cdot \dot{\boldsymbol{d}}_{e,l}^{B*} \mathrm{F}_{kl}\right] \frac{\hbar^2}{8\bar{\omega}_k \bar{\omega}_k} \coth\left(\frac{\beta\hbar\bar{\omega}_k}{2}\right) \coth\left(\frac{\beta\hbar\bar{\omega}_l}{2}\right) \right. \tag{8.105}$$

Equations 8.98 and 8.100 allow for a critical analysis of the relation between the normalized first moment $\mathcal{M}^{(1)}/\mathcal{M}^{(0)}$ and the vertical excitation E_V. The latter is usually compared with the spectrum maximum $E_{\max}$, and the following analysis should clarify why this is possible and in which limits they should coincide theoretically. Equation 8.98 shows that, at the FC level, if the normal modes are simply displaced between the two electronic states (case of the VG approach), $\mathcal{M}^{(1)}/\mathcal{M}^{(0)}$ coincides with E_V. However, $\mathcal{M}^{(1)}/\mathcal{M}^{(0)}$ only coincides with E_{max} when the spectrum is symmetric around the latter. In most cases, the absorption/emission spectrum is asymmetric with respect to E_{max}, showing a wing longer than the other (the blue/red one for absorption/emission) and tends to a symmetric shape only in the limit of infinite broadening or very large displacements. As a consequence, even in the cases where VG and FC approximations hold exactly, a red/blue shift of E_{max} with respect to E_V should be expected for absorption/emission spectra. In the more general harmonic case, when the normal modes of the final states have also different frequencies and are mixed through the Duschinsky effect, at $T = 0$ K, $\mathcal{M}^{(1)}/\mathcal{M}^{(0)}$ is superior to E_V for absorption and inferior for emission, when the trace (Tr) of $\boldsymbol{F}$ (Tr[$\boldsymbol{F}$]) is larger than Tr[$\bar{\boldsymbol{\Omega}}^2$]. In the particular case of no Duschinsky mixing, this leads to the simple rule $\mathcal{M}^{(1)}/\mathcal{M}^{(0)} > E_V$ for absorption when (Tr[$\bar{\bar{\boldsymbol{\Omega}}}$] > Tr[$\bar{\boldsymbol{\Omega}}$]) and the opposite holds for emission. At finite temperature, the differences in low-frequency modes play a more important role in ruling the value $\mathcal{M}^{(1)}/\mathcal{M}^{(0)} - E_V$ because of the term $\coth(\beta\hbar\omega_k/2)$ in Eq. 8.98. Finally, when the HT effect is significant, the relation between $\mathcal{M}^{(1)}/\mathcal{M}^{(0)}$ and E_V depends on more factors, including the Duschinsky mixing, and the two quantities should not be expected to be equal.

The second moment gives the width of the spectrum while the third one gives information on its skewness or asymmetry. A general analytical computation becomes very cumbersome so we will limit the equation to the FC approximation of the second moment, namely $\mathcal{M}_{\mathrm{FC}}^{(2)}$. The second moment in the FC approximation is

$$
\begin{aligned}
\frac{\mathcal{M}_{\mathrm{FC}}^{(2)}}{(\boldsymbol{d}_{e,0}^{A} \cdot \boldsymbol{d}_{e,0}^{B*})^2} &= \sum_i \rho_i \langle \chi_i | (\hat{\bar{\bar{H}}} - \hat{\bar{H}})^2 | \chi_i \rangle \\
&= E_V^2 + E_V \sum_r \frac{\hbar}{2\bar{\omega}_r}(F_{rr} - \bar{\omega}_r^2)[2\bar{\nu}_r(T) + 1] \\
&\quad + \sum_{r,s\neq r} \frac{\hbar^2}{16\bar{\omega}_r\bar{\omega}_s}(F_{rr} - \bar{\omega}_r^2)(F_{ss} - \bar{\omega}_s^2)[2\bar{\nu}_r(T) + 1][2\bar{\nu}_s(T) + 1] \\
&\quad + \sum_r \frac{\hbar}{2\bar{\omega}_r} g_r^2 [2\bar{\nu}_r(T) + 1] + \frac{3\hbar^2}{16\bar{\omega}_r^2}(F_{rr} - \bar{\omega}_r^2)^2 \left[2\bar{\nu}_r(T)^2 + 2(\bar{\nu}_r(T) + 1)\right] \\
&\quad + \sum_{r,s\neq r} \frac{\hbar^2}{4\bar{\omega}_r\bar{\omega}_s} F_{rs}^2 [2\bar{\nu}_r(T) + 1][2\bar{\nu}_s(T) + 1]
\end{aligned}
$$

where the first equality arises from Eq. 8.85 recalling that the commutator $[\hat{\bar{\bar{H}}}, \hat{\bar{H}}]$ does not contribute to the expectation value. On the ground of Eq. 8.98. we can finally obtain the following measure of the width of the spectrum:

$$\sigma = \frac{\left[\mathcal{M}_{\mathrm{FC}}^{(2)} - (\mathcal{M}_{\mathrm{FC}}^{(1)})^2\right]^{1/2}}{\boldsymbol{d}_{e,0}^{A} \cdot \boldsymbol{d}_{e,0}^{B*}}$$

$$= \left[\sum_r \frac{\hbar}{2\bar{\omega}_r} g_r^2 \coth\left(\frac{\beta\hbar\omega_r}{2}\right) + \sum_r \frac{\hbar^2}{8\bar{\omega}_r^2}(F_{rr} - \bar{\omega}_r^2)^2 \coth\left(\frac{\beta\hbar\omega_r}{2}\right)^2 \right.$$
$$\left. + \sum_{r,s} \frac{\hbar^2}{4\bar{\omega}_r\bar{\omega}_s} F_{rs}^2 \coth\left(\frac{\beta\hbar\omega_r}{2}\right) \coth\left(\frac{\beta\hbar\omega_r}{2}\right)\right]^{1/2}$$

From the previous equation, we can conclude that the spectral width increases with the displacements (related to the gradient $\boldsymbol{g}$), the difference between the diagonal force constants in the final state F_{rr} and the squared frequencies in the initial state $\bar{\omega}_r^2$, and finally the Duschinsky mixings (related to the off-diagonal force constants F_{rs})

8.3.4 Solvent Broadening: System/Bath Approach

We present a derivation of the broadening due to the solvent according to a system/bath quantum approach, originally worked out in the field of solid-state physics to treat the effect of electron/phonon couplings in the electronic transitions of electron traps in crystals [67, 68]. This approach has the advantage to treat all the nuclear degrees of freedom of the system solute/medium on the same foot, namely as coupled oscillators. The same type of approach has been adopted by Jortner and co-workers [69] to derive a quantum theory of thermal electron transfer in polar solvents. In that case, the solvent outside the first solvation shell was treated as a dielectric continuum and, in the frame of the polaron theory, the vibrational modes of the outer medium, that is, the polar modes, play the same role as the lattice optical modes of the crystal investigated elsewhere [67, 68]. The total Hamiltonian of the solute (s) and the medium (m) can be formally written as

$$H^{\mathrm{tot}} = H^{(s)}(\boldsymbol{Q}_s) + H^{(m)}(\boldsymbol{Q}_m) + H^{(m,s)}(\boldsymbol{Q}_s, \boldsymbol{Q}_m) \tag{8.106}$$

Invoking the harmonic approximation for the PESs of the initial $|\Phi_i\rangle$ and final $|\Phi_f\rangle$ electronic states, we can write

$$H^{(s)} = |\Phi_i\rangle\langle\Phi_i| \left(\hat{T}^{(s)} + \tfrac{1}{2}\bar{\boldsymbol{Q}}^{(s)^{\mathrm{T}}}\{\bar{\Omega}^{(s)}\}^2\bar{\boldsymbol{Q}}^{(s)}\right)$$
$$+ |\Phi_f\rangle\langle\Phi_f| \left(T^{(s)} + \tfrac{1}{2}\bar{\boldsymbol{Q}}^{(s)^{\mathrm{T}}} F^{(s)}\bar{\boldsymbol{Q}}^{(s)} + g^{(s)^{\mathrm{T}}}\bar{\boldsymbol{Q}}^{(s)}\right) \tag{8.107}$$

$$H^{(m)} = |\Phi_i\rangle\langle\Phi_i| \left(\hat{T}^{(m)} + \tfrac{1}{2}\bar{\boldsymbol{Q}}^{(m)^{\mathrm{T}}}\{\bar{\Omega}^{(m)}\}^2\bar{\boldsymbol{Q}}^{(m)}\right)$$
$$+ |\Phi_f\rangle\langle\Phi_f| \left(T^{(m)} + \tfrac{1}{2}\bar{\boldsymbol{Q}}^{(m)^{\mathrm{T}}} F^{(m)}\bar{\boldsymbol{Q}}^{(m)} + g^{(m)^{\mathrm{T}}}\bar{\boldsymbol{Q}}^{(m)}\right) \tag{8.108}$$

where $\bar{\boldsymbol{Q}}^{(s)}$ and $\bar{\boldsymbol{Q}}^{(m)}$ are respectively the normal coordinates of the solute and medium in the initial electronic state and $\bar{\boldsymbol{\Omega}}^{(s)}, \bar{\boldsymbol{\Omega}}^{(m)}$ the corresponding diagonal matrix of the frequencies. The final-state PES is still assumed to be harmonic, but displacements of the equilibrium positions, changes of the frequencies, and mode mixing (Duschinsky effect) may occur upon the electronic transition within the two subsets of solute and medium modes. These effects are described by a second-order Taylor expansion of the final-state PES along the initial-state coordinates. Therefore, $\boldsymbol{g}^{(s)}$ and $\boldsymbol{F}^{(s)}$ are the gradient and the Hessian matrix (in principle, nondiagonal) of the final-state PES along the solute coordinates and $\boldsymbol{g}^{(m)}$ and $\boldsymbol{F}^{(m)}$, the same quantities along the solvent modes. The solute/medium mode couplings are neglected, so $H^{(m,s)}(\boldsymbol{Q}^{(s)}, \boldsymbol{Q}^{(m)}) = 0$. As a direct consequence of this approximation, the total spectrum can be written as a convolution of the spectra of the independent subsystems [e.g., 70],

$$S(\omega) = \int S^{(s)}(\omega - \omega')S^{(m)}(\omega')\, d\omega' \tag{8.109}$$

$$S^{(s)}(\omega) = \sum_i \rho_i^{(s)} \left| \left\langle \chi_i^{(s)} | \mu_{e,if} | \chi_f^{(s)} \right\rangle \right|^2 \delta\left(\omega_f^{(s)} - \omega_i^{(s)} - \omega\right) \tag{8.110}$$

$$S^{(m)}(\omega) = \sum_i \rho_i^{(m)} \left| \left\langle \chi_i^{(m)} | \chi_f^{(m)} \right\rangle \right|^2 \delta\left(\omega_f^{(m)} - \omega_i^{(m)} - \omega\right) \tag{8.111}$$

where $\rho_i^{(s)}$ and $\rho_i^{(m)}$ are the Boltzmann weights of the solute and medium initial states, $|\chi_i^{(s)}\rangle$ and $|\chi_i^{(m)}\rangle$, respectively. In the above equations, we have skipped the prefactors for the sake of brevity and made the reasonable approximation that the electronic transition dipole moment $\boldsymbol{\mu}_{e,if}$ does not depend on medium coordinates. The separation of the Hamiltonians also implies that the average energies (i.e., the first moment $\mathcal{M}^{(1)}$) and the variances ($\sigma^2 = \mathcal{M}^{(2)} - \mathcal{M}^{(1)2}$) of the two spectra can simply be added to one another in the total spectrum. As shown by Eq. 8.111, the solute spectrum is actually broadened by the one of the solvent. The latter, due to the very dense bath of vibrational modes, is usually approximated as a continuous distribution matching the correct first and second moments.

To compute these moments, and in lack of detailed information, the Duschinsky effect is usually neglected between the solvent modes and, in Eq. 8.108, the medium Hamiltonian for the final state, namely $|\Phi_f\rangle\langle\Phi_f|\bar{\bar{H}}^{(m)}$ can be recast as

$$\hat{\bar{\bar{H}}}^{(m)} = \hat{T}^{(m)} + \tfrac{1}{2}\bar{\boldsymbol{Q}}^{(m)\mathrm{T}}\{\bar{\bar{\Omega}}^{(m)}\}^2\bar{\boldsymbol{Q}}^{(m)} + g^{(m)}\bar{\boldsymbol{Q}}^{(m)\mathrm{T}} + \bar{\bar{E}}^r \tag{8.112}$$

where $\bar{\bar{E}}^r = \sum_k \bar{\bar{E}}_k^r = \frac{1}{2}\sum_k (g_k/\bar{\omega}_k)^2$ is the reorganization energy of the solvent in the final electronic state. Moments of the solvent spectrum can be easily obtained from the results of the previous section. For the Hamiltonian in Eq. 8.112, we have

$$\mathcal{M}^{(1)(m)} = \sum_k \frac{\hbar}{2\bar{\omega}_k}(\bar{\bar{\omega}}_k^2 - \bar{\omega}_k^2)\coth\left(\frac{\beta\hbar\bar{\omega}_k}{2}\right) + \bar{\bar{E}}^r \tag{8.113}$$

$$\{\sigma^{(m)}\}^2 = \frac{\mathcal{M}_{\mathrm{FC}}^{(2)} - (\mathcal{M}_{\mathrm{FC}}^{(1)})^2}{(\boldsymbol{d}_{e,0}^{\mathrm{A}} \cdot \boldsymbol{d}_{e,0}^{\mathrm{B}*})^2}$$

$$= \sum_k \frac{\hbar\bar{\bar{\omega}}_k^2}{\bar{\omega}_k} \bar{\bar{E}}_k^r \coth\left(\frac{\beta\hbar\bar{\omega}_k}{2}\right) + \sum_k \frac{\hbar^2}{8\bar{\omega}_k{}^2}(\bar{\bar{\omega}}_k^2 - \bar{\omega}_k^2)\coth^2\left(\frac{\beta\hbar\bar{\omega}_k}{2}\right) \quad (8.114)$$

Results in Eq. 8.114 are identical to those derived by Markham [68] for an analogous Hamiltonian worked out to describe spectra of electron traps in crystals. In the classical limit, $\beta \gg \hbar\bar{\omega}_k$, we can take a first-order expansion of the hyperbolic cotangent function coth $(\beta\hbar\bar{\omega}_k/2) \approx 2k_BT/\hbar\bar{\omega}_k$, thus obtaining the expressions $\mathcal{M}_c^{(1)(m)}$ and $\{\sigma_c^{(m)}\}^2$:

$$\mathcal{M}_c^{(1)(m)} = \sum_k k_BT\left(\frac{\bar{\bar{\omega}}_k^2}{\bar{\omega}_k^2} - 1\right) + \bar{\bar{E}}^r \quad (8.115)$$

$$\left\{\sigma_c^{(m)}\right\}^2 = 2k_BT\sum_k \frac{\bar{\bar{\omega}}_k^2}{\bar{\omega}_k^2}\bar{\bar{E}}_k^r + \sum_k \frac{(k_BT)^2}{2}\left(\frac{\bar{\bar{\omega}}_k^2}{\bar{\omega}_k^2} - 1\right)^2 \quad (8.116)$$

The above expressions can be further simplified by assuming that the frequencies of the solvent modes are the same in the initial and final electronic states, thus giving

$$\mathcal{M}_c^{(1)(m)} = \bar{\bar{E}}^r \quad (8.117)$$

$$\{\sigma_c^{(m)}\}^2 = 2k_BT\bar{\bar{E}}^r \quad (8.118)$$

O'Rourke [71] and Markham [68] showed that, in this limit (and assuming a single frequency for the solvent modes), the lineshape of the solvent spectrum becomes a Gaussian. It should be highlighted that Marcus obtained the same result as reported in Eq. 8.118 for the broadening due to polar interactions between the solute and the medium [72], on the ground of a particle description of polar media [73, 74], treating the medium at the classical level. In his approach, the nonpolar contributions are neglected and $\bar{\bar{E}}^r$ is the polar contribution to the difference between nonequilibrium (*neq*) and equilibrium (*eq*) Helmotz free energy in the final electronic state at the FC solute geometry. We recall that in the *neq* solvation regime, only the fast, electronic solvent polarization is in equilibrium with the solute final-state charge density, while the *eq* regime is characterized by the full equilibration of the medium with the final state (see Chapter 1). Before concluding this section, it is worth noting that Marcus [75] introduced a second solvent contribution to the broadening, arising from the first coordination shell of the solvent. This can be easily done in our framework by dividing the sum in Eq. 8.114 in two additive contributions of "first" $\{\sigma^{(m,\text{ first})}\}^2$ and "outer" $\{\sigma^{(m,\text{ outer})}\}^2$ solvation shells, respectively. For the first solvation shell, Marcus takes into account the possibility that the frequency along the corresponding solvent modes is not the same in the two electronic states. The expression derived classically by Marcus for the width introduced by the first shell matches the first term on the RHS of Eq. 8.116, while the second term in that equation only arises following a quantum treatment.

Summarizing, the spectral lineshape of the solute in a polar solution can be written as the convolution

$$S(\omega) = \int d\omega_1 \int d\omega_2 \int d\omega_3 \, S^{(s)}(\omega - \omega_1) S^{(r)}(\omega_1 - \omega_2) S^{(m,\text{first})}(\omega_2 - \omega_3) S^{(m,\text{outer})}(\omega_3) \tag{8.119}$$

where $S^{(s)}$ is the solute stick spectrum, $S^{(m,\text{first})}$ and $S^{(m,\text{ outer})}$ are the lineshapes due to first-shell and outer sphere solvent, respectively, and $S^{(r)}$ takes into account all residual broadening causes, like excited-state finite-lifetime and nonpolar solute/solvent interactions. As discussed in detail in Chapter 1, the latter can usually be neglected in polar solvents, while the *neq* and *eq* solvation regimes of the bulk solvent are amenable of a description in terms of the polarizable continuum model (PCM).

8.4 PRESCREENING OF VIBRONIC TRANSITIONS

Because of the redundant calculations induced by the recursive formulas in Eqs. 8.53 and 8.54, it is often more efficient to store the overlap integrals than to recompute them each time they are needed. The convenience to resort to massive storage was more evident in the past when the processor frequency was far lower and led to the elaboration of several effective algorithms. A particularly important issue was the design of a versatile and fast indexing solution to retrieve a given integral in memory. With this in mind, we can cite several major models, among which the binary tree is one of the most prominent. Gruner and Brumer [76] proposed a simple and efficient method using binary trees to find any overlap by associating a change in a given mode k to the left subtree and a change of the associated quantum number v_k to the right subtree. A complete description of the structure of a binary tree would be outside the scope of this chapter, and the interested reader can find a detailed presentation elsewhere [77, 78]. For the sake of completeness, we can mention that several other schemes were presented later, such as the one proposed by Ruhoff and Ratner [79] or Toniolo and Persico [80] using the definition of the recursion formulas to restrict the binary tree to a chosen subset of overlap integrals. More recently, Hazra and Nooijen [81] and then Dierksen and Grimme [82] proposed refined versions of this procedure. Finally, other routes have been explored to index overlap integrals in memory and avoid the memory overload created by the usage of a binary tree, such as the hash table of Schumm et al. [83]. Nowadays, the increased power of the processors and the new possibilities offered by their ever-growing parallelization capabilities make the storage issues less critical since recalculations can sometimes be faster. In this section we will focus on a number of prescreening methods developed to select a priori the relevant transitions for the spectrum.

Indeed, the major difficulty to compute single-state vibronic spectra in bound (nondissociative) systems lies in the large amount of discrete transitions to consider, which increases steeply with the size of the system and the energy window of interest. However, in practice, most transitions have a very low intensity that can be safely neglected. Thus, only a limited finite number of transitions gives an actual contribution to the vibrational structure of the electronic spectrum. A first, preponderant step before considering the generation of any theoretical spectrum is then to define a consistent way of selecting these transitions. Without pretending to be exhaustive, we will briefly review different ways to perform such a task, describing their main features and discussing their shortcomings, when the latter have an impact on the overall calculations. It is noteworthy that alternative solutions to this problem are offered by time-dependent approaches that are complementary to time-independent ones, in the sense that, renouncing to a state-to-state description of the spectrum, they can directly describe the effect of the complete ensemble of excitable stationary states. Methods rooted in the time-dependent framework are described in Chapter 10.

A first, straightforward approach is to define an energy window for the simulation of the spectrum, with its lower and upper bounds chosen with respect to experimental parameters or defined arbitrarily, and use it to select which transitions can take place. The criterion to choose the transitions is applied in "real time" in the sense that the transition must be first considered and its energy calculated before being taken into account or discarded. Thus, in order to avoid an infinite control over the transitions energies, reliable algorithms [84, 85] have been devised. For instance, Kemper et al. [84] proposed a backtracking algorithm to count all possible states for a given energy interval. Contrary to most similar algorithms presented before, such as the one of Beyer and Swinehart [85], their method retained the information on the levels involved in the transition, making easier the calculation of the overlap integrals. It is noteworthy that this procedure was designed for an arbitrary precision of the energy levels, including anharmonicity. The algorithm generates a list of all possible transitions, which can then be computed. However, to avoid a massive storage when many transitions must be taken into account, the criterion is applied on-the-fly in practice.

The originally proposed procedure only took into account transitions from the ground vibrational state of the initial electronic state (temperature of 0 K). Berger and Klessinger [86] worked out a generalization able to take into account also a distribution of initial vibrational states through an interlocked algorithm, where the procedure for the selection of the final states is nested inside the other one treating the initial states. The complete algorithm uses only two thresholds corresponding to the lower and upper energy bounds chosen by the user. For each combination of quantum numbers in the initial electronic state corresponding to a vibrational state with valid energy for the first algorithm, the second backtracking procedure of Beyer and Swinehart is run to select the vibrational final states. Such a routine is run until both backtracking procedures have reached their end, with no more combination of initial and final states to consider in the allowed energy interval.

Because it is simple to implement and intuitive physically, this method, or similar counting algorithms, has been commonly used to compute FC integrals and simulate

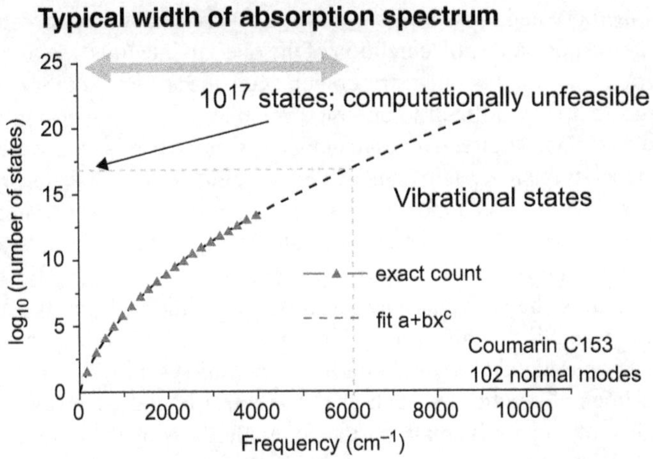

Figure 8.9 Number of vibrational modes to be considered increases steeply with molecule dimension, for example, coumarin C153.

absorption or emission spectra [76, 82, 86, 87]. However, this kind of scheme shows a serious drawback in its poor scaling with the spectrum energy range and the size of the system, that is, the number of vibrational modes. In fact, while the counting methods can give fast results on a narrow energy window around the 0–0 transition, their computational times grow very quickly when a larger energy window of the spectrum is required, as shown in Figure 8.9 for coumarin C153.

Hazra and Nooijen [81] proposed a different approach where the selection criterion is not the energy of the transitions but their probability. In their approach, the overlap integrals are categorized in levels defined by the sum of the quantum numbers in the final state. Hence, starting from the level $L_0(\langle\bar{\mathbf{0}}|\bar{\bar{\mathbf{0}}}\rangle)$, all possible integrals from the level L_1, which can be directly calculated based on the integrals between the vibrational ground states, are treated. Then, from the overlap integrals of L_0 and L_1, those of level L_2 are computed. It is interesting to note that these levels are analogous to the clusters in coupled-cluster electronic theory. Apart from the normalization factors, in fact, the cluster of the final vibrational states belonging to level "n" can be formally generated from the ground vibrational state $|\bar{\bar{0}}\rangle$ by applying the nth order of the power expansion of the excitation operator $\exp[\mathcal{T}] = \Pi_k \exp[\mathcal{T}_k]$, where $\mathcal{T}_k = a_k^\dagger$ and $a_k^\dagger$ are the usual creation operators. Each time an overlap integral is calculated, the corresponding probability of transition is confronted to a threshold and discarded if lower. As a consequence, after having reached a maximum, the number of overlap integrals in higher levels will gradually drop until it vanishes. This approach is independent from the energy bounds of the spectrum so its performance depends solely on the studied system. However, a drawback arises when coupling this approach with the recursion formulas, as done by Hazra and Nooijen. In fact, if a term used on the RHS of Eq. 8.53 or 8.54 is missing because it has been removed due to a value below the probability threshold, the calculation is still performed without it. In other words, a given overlap integral $\langle\bar{v}|\bar{\bar{v}}\rangle$ is computed with the overlap integrals available in memory and those missing, supposed very small, are neglected. Such a choice breaks the normalization condition of the overlap integrals, making it difficult to reliably control the quality of the prescreening and the accuracy of the calculations. This also makes it very difficult to choose a reliable threshold working for a large panel of systems, since its effectiveness cannot be assessed precisely and the impact of the discard depends on the system under study. It should be noted that the problem can be partly overcome by recalculating the missing precursors. However, because only two levels are stored at a time, the recursion calculation of these integrals can be time consuming and reduce the efficiency of the method.

Notwithstanding the above caveats, it must be highlighted that the analysis of the transition probabilities offers a consistent way to simulate a spectrum, which can be easily generalized to a large number, if not all systems, without depending on the experimental conditions, such as the spectral bounds. However, when the energy window of interest is small, on-the-fly methods using those spectral bounds as presented above can be faster, but approximation methods done in "real time" remain difficult to handle from the perspective of the storage as there is no way to know beforehand the number of overlap integrals that will have to be computed and saved.

As a consequence, recent developments [82, 88, 89] have been focused on a priori approaches to evaluate the most important transitions before their actual calculation. General-purpose evaluation methods are fairly recent and were designed to deal with the newly accessible simulations of UV–visible spectra for medium-to-large systems. They are mostly based on the estimates of the probabilities of bulks of transitions at a fraction of the times required to compute the actual overlap integrals. The major impediment then is the difficulty to devise a consistent methodology, which can work on most, or possibly all, systems whatever the approximation of the transition dipole moments chosen. Because of their recentness and the general need to approximate the value of the overlap integrals, several ways have been explored and patterns designed to define the criteria for the a priori prescreenings. We will cite here some typical examples illustrating various approaches based on the approximation of the Duschinsky rotation [82, 87] using sum rules derived in the coherent-states [88] approach (introduced by Doktorov et al. [47] for the calculation of the overlap integrals) or based on the intensities of specific classes (the meaning of this term will be cleared in the following) of transitions for the evaluation of each term on the RHS of Eqs. 8.53 and 8.54 [31, 89].

A first and simple approximation is to neglect the mode mixing and consider a one-to-one relation between the modes of the initial and final states, with the Duschinsky transformation matrix $\boldsymbol{J}$ equal to the identity matrix (notice that VG and AS models belong by definition to this approximation). The interest of this approximation, called parallel-mode approximation, is that the multidimensional FC integrals can be calculated as products of one-dimensional integrals using the relation

$$\langle \bar{v} | \bar{\bar{v}} \rangle = \prod_{k=1}^{N} \langle \bar{v}_k | \bar{\bar{v}}_k \rangle \tag{8.120}$$

The one-dimensional FC integrals can then be straightforwardly calculated using analytic [90, 91] or recursion formulas [92] for monodimensional oscillators.

However, considering the practical applications of such a scheme, a first difficulty arises from the fact that the normal modes are rarely uncoupled and the one-to-one correspondence between those of the initial and final states is not valid in most cases. A first issue, resolved by Ervin et al. [93], is the possibility of rotation of the normal modes during the electronic transitions. It can be simply overcome by first calculating the exact Duschinsky matrix from which the greatest overlap between each mode of the initial and final electronic states is kept and the others, lower in intensity, discarded. In practice, for each column k, the highest value J_{kl}^2 is taken and the corresponding value J_{kl} is set to 1, while the other ones are disregarded and their values set to 0. Finally, the modes are reordered so that the rotation matrix is equivalent to the identity matrix. This procedure allows us to assimilate each mode of one electronic state, with the most similar mode of the other state, in the sense that it corresponds to the largest projection. However, in most cases, the discarded coupling can be strong and such a severe approximation can lead to unpredictable errors when comparing the obtained spectrum with the real system. Moreover, another problem immediately pointed out by Ervin et al. [87] is that the parallel-mode approximation can simply fail in some cases. It can happen that, after applying exactly the procedure described above for each column l, one then finds rows with more than one nonzero element, for example, J_{kl} and J_{ml}, and other ones with none. While Ervin et al. suggested a manual reassignment, such a solution cannot be automated and remains arbitrary. As a result, a treatment of the rotation matrix purely restricted to the parallel-mode approximation level is not sustainable for a general-purpose procedure.

A compromising method between a complete treatment of the mode mixing, using the correct rotation matrix, and the complete neglect done in the parallel-mode approximation was later proposed by the same authors [87]. It is based on a division of the normal modes in two groups, depending on the nature of the normal modes, more precisely if they are uncoupled or coupled, with the former treated with the parallel-mode approximation and the latter treated exactly. Practically, the Duschinsky matrix is treated as a block-diagonal matrix, instead of an identity matrix. However, to retain most of the information on the changes undergone by the system during the transitions, the model Duschinsky matrix must be as similar as possible to the exact one. Hence, a high threshold (about 95%) must be used when selecting coupled and uncoupled modes in a given electronic state based on their projection on the other state. Practically, this means that for a given mode k of the initial state the first n modes (l) of the final state with the highest projection coefficients (J_{kl}^2) so that the condition $\sum_{k=1}^{n} J_{kl}^2 \geq 0.95$ is met are considered to be coupled. As a consequence, in case of strong coupling, the gain of this method with respect to the complete treatment of the Duschinsky matrix can be relatively low. However, the method can be really effective for symmetrical rigid systems where the coupling of the modes is limited. Additionally, a more subtle problem can occur if one wishes to account for a higher level of approximation of the transition dipole moments beyond the FC approximation, such as HT. In this case, the terms outside the blocks are necessary to correctly calculate the transition dipole moment integrals and the block diagonalization can introduce important errors which cannot be easily evaluated beforehand.

Dierksen and Grimme [82] also proposed a method which makes use of the block diagonalization of the Duschinsky matrix but adopts a different approach to proceed. Instead of simply discarding the elements outside the block, they calculate a block-diagonal model rotation matrix by replacing the exact normal modes of the final state by an approximate set. Their procedure requires a threshold on the sum of the elements J_{ik}^2 for each row and column. This threshold is used to choose the blocks in the original Duschinsky matrix and the new transformation matrix $\bar{\bar{L}}$ is generated so that out-of-block elements are canceled. Contrary to Ervin et al. [87], the new Duschinsky matrix obtained in such a way is not used for the actual calculations but as a prescreening. For each block of this matrix, the FC integrals with the corresponding transition energies satisfying the conditions fixed by the bounds of the spectrum are calculated using the recursion formulas of Doktorov et al. [47]. The overlap integrals above a second threshold are kept while the other ones are discarded. The "complete" FC integrals are obtained by multiplying those preserved in each block and compared to a third threshold. If they are above it, the correct overlap integral between the same combinations bands is calculated with the original Duschinsky matrix.

While first designed for FC calculations, an advantage of using the model system is the possibility to adapt straightforwardly the prescreening to the FCHT and HT models. Moreover, with respect to the previously discussed method, this approach improves the reliability of the simulated spectrum by taking into account the correct Duschinsky effect. However, similar to the "coupled" approach of Ervin et al. [87], the efficiency and resulting speed of the method are strongly bound to the mode coupling in the original system. A workaround to limit the computational cost would be to lower the first threshold, but at the expense of the model rotation matrix differing strongly from the original one, with the risk of partially invalidating the prescreening. As a consequence, the method is best suited to rigid and symmetrical systems [56, 94]. Finally, the introduction of additional thresholds may raise difficulties for a full automatization in a blackbox procedure.

Recently, Jankowiak et al. [88] proposed a new approach based on the coherent-state representation used by Doktorov et al. [47] to obtain the recursion formulas needed to compute the FC integrals. These formulas have been shown by Liang and Li [95] to be equivalent to the ones derived by Ruhoff [96] from the analytic approach of Sharp and Rosenstock [39]. The method initially limited to 0 K FC spectra has been generalized to deal with FCHT spectra [97] and finite-temperature effects [98].

Using a generating function similar to the one introduced by Malkin et al. [99], Jankowiak et al. obtained several analytic sum rules. From these sum rules, it is then possible to estimate the contribution of any overlap integral or an entire group of them to the total intensity. Several scenarios were defined to maximize the efficiency of the overall procedure. Due to its analytic definition, the method relies on very little arbitrary parameters to work, so it can be safely used in a blackbox procedure.

In the next section, we will focus on a specific prescreening method, developed by our group, which has been used to produce the spectra presented in this chapter. Since it is based on the introduction of quantities and categorizations whose analysis can help in rationalizing the relevant factors and the most important transitions that determine the spectrum shape, we will present it in greater detail.

8.4.1 *Class*-Based Prescreening Approach

The prescreening method presented here is based on a categorization of the multidimensional vibrational states of the final state in *classes*, which are defined as the number of simultaneously excited modes in a given electronic state. By convention, the state of reference is the final one and we will always refer to it when referring to a given *class*. For instance, *class* 1 (C_1) represents all transitions to final vibrational states with a single excited mode k, $\langle \bar{\mathbf{v}} | \bar{\bar{\mathbf{0}}} + \bar{\bar{v}}_k \rangle$ and *class* 0 contains the overlap integral to the vibrational ground state of the final electronic state, $\langle \bar{\mathbf{v}} | \bar{\bar{\mathbf{0}}} \rangle$. It is interesting to compare this partition in classes with the partition in levels proposed by Hazra and Nooijen [81] by going back to the analogy with the coupled-cluster expansion. Indeed, considering the generation of all possible final states through the excitation operator $\mathcal{T}$, $\exp[\mathcal{T}]|\bar{\bar{\mathbf{0}}}\rangle = \Pi_k \exp\,[\mathcal{T}_k]|\bar{\bar{\mathbf{0}}}\rangle$, while level n corresponds to the transitions generated by the nth-order term in the Taylor expansion of $\exp[\mathcal{T}]$ (i.e., exciting simultaneously all the modes k), *class* n corresponds to the transitions generated expanding at all orders the operators $\exp[\mathcal{T}_k]$ of n modes (chosen in all possible combinations) and at zero order those of the remaining $N - n$ modes. An interesting feature of the selection procedure we present here is its capacity to treat both FC and HT spectra, including temperature effects. This versatility is coupled to the low computational cost of the actual prescreening procedure [31, 70, 89], making it applicable to a wide range of experimental conditions. The overlap integrals in *classes* 1 and 2 are computed up to full convergence. In practice, one sets a chosen limit, but it can be as large as needed to reach full convergence since their calculations are cheap. Characteristic quantities can then be extracted from the ensemble of data gathered at this point. Those values are subsequently used to choose the most relevant transitions to compute in each *class* of higher order, starting from *class* 3. The prescreening gives a priori an estimate of the maximum quantum number which must be considered for each mode. In practice, the selection procedure for a given *class* is done as follows. A first threshold, $C_{1_{\max}}$, is used to define the highest number of quanta the singly excited mode of the final state in *class* C_1 can get. The corresponding overlap integrals ($\langle \bar{\mathbf{v}} | \bar{\bar{\mathbf{0}}} + \bar{\bar{v}}_k \rangle$ with $\bar{\bar{v}}_k [\![1, C_{1_{\max}}]\!]$) are calculated. The FC factors corresponding to the transitions from the ground, and the highest ($|\bar{\mathbf{v}}_{\max}\rangle$) if temperature is taken into account, the vibrational initial states to all possible final vibrational states ($|\bar{\bar{\mathbf{0}}} + \bar{\bar{v}}_k\rangle$), respectively F_{C_1} and $F_{C_1}^{(T)}$, are stored in memory. More explicitly, for each excited mode k in the final state with $\bar{\bar{v}}_k \in [\![1, C_{1_{\max}}]\!]$, the stored quantities are

$$F_{C_1}(k, \bar{\bar{v}}_k) = \left| \frac{1}{\sqrt{2\bar{\bar{v}}_k}} \left[\mathrm{D}_k \langle \bar{\mathbf{0}} | \bar{\bar{\mathbf{0}}} + \bar{\bar{v}}_k - \bar{\bar{1}}_k \rangle + \sqrt{2(\bar{\bar{v}}_k - 1)}\, \mathrm{C}_{kk} \langle \bar{\mathbf{0}} | \bar{\bar{\mathbf{0}}} + \bar{\bar{v}}_k - \bar{\bar{2}}_k \rangle \right] \right|^2 \tag{8.121}$$

$$F_{C_1}^{(T)}(k, \bar{\bar{v}}_k) = \left| \frac{1}{\sqrt{2\bar{\bar{v}}_k}} \left[\mathrm{D}_k \langle \bar{\boldsymbol{v}}_{\max} | \bar{\bar{\mathbf{0}}} + \bar{\bar{v}}_k - \bar{\bar{1}}_k \rangle + \sqrt{2(\bar{\bar{v}}_k - 1)} \mathrm{C}_{kk} \langle \bar{\boldsymbol{v}}_{\max} | \bar{\bar{\mathbf{0}}} + \bar{\bar{v}}_k - \bar{\bar{2}}_k \rangle \right. \right.$$
$$\left. \left. + \sum_{l=1}^{N} \sqrt{\frac{\bar{\bar{v}}_l}{2}} \mathrm{E}_{kl} \langle \bar{\boldsymbol{v}}_{\max} - \bar{1}_l | \bar{\bar{\mathbf{0}}} + \bar{\bar{v}}_k - \bar{\bar{1}}_k \rangle \right] \right|^2 \tag{8.122}$$

where, as it can be realized from analysis of the definitions in Eq. 8.55, the terms D_k and C_{kk} give respectively information on the effect of the shifts in equilibrium positions and the frequencies on the overlap integrals of overtones and more precisely on the vibrational progression of mode k and E_{kl} gives information of the mode mixing of this mode due to the transition. The second quantity is defined only for $T > 0\,\mathrm{K}$; otherwise $F_{C_1}(k, \bar{\bar{v}}_k) = F_{C_1}^{(T)}(k, \bar{\bar{v}}_k)$.

If the HT approximation is considered, an additional quantity is stored, H_{C_1}, which gives an upperbound estimation of the square of the pure HT contribution for a given mode k and the corresponding transition $\langle \bar{\mathbf{0}} | \bar{\bar{\mathbf{0}}} + \bar{\bar{v}}_k \rangle$,

$$H_{C_1}^{(T)}(k, \bar{\bar{v}}_k) = \sum_{l=1}^{N} \sum_{\tau=x,y,z} \left| \left(\frac{\partial d_{e,if}^{A}(\tau)}{\partial \bar{Q}_l} \right)_{\mathbf{0}} \right| \left| \left(\frac{\partial d_{e,if}^{B*}(\tau)}{\partial \bar{Q}_l} \right)_{\mathbf{0}} \right| \times \left| \sqrt{\frac{\hbar}{2\bar{\omega}_l}} \langle \bar{\mathbf{0}} + \bar{1}_l | \bar{\bar{\mathbf{0}}} + \bar{\bar{v}}_k \rangle \right|^2 \tag{8.123}$$

with the summation over each Cartesian coordinate of the transition dipole moments $\boldsymbol{d}_{e,if}^{A}$ and $\boldsymbol{d}_{e,if}^{B}$. Absolute values are taken to get rid of the sign that can be variable in circular dichoism intensities.

As done previously, a second set, $H_{C_1}^{(T)}$, is defined if $T > 0\,\mathrm{K}$. It is given by the relation

$$H_{C_1}^{(T)}(k, \bar{\bar{v}}_k) = \sum_{l=1}^{N} \sum_{\tau=x,y,z} \left| \left(\frac{\partial d_{e,if}^{A}(\tau)}{\partial \bar{Q}_l} \right)_{\mathbf{0}} \right| \left| \left(\frac{\partial d_{e,if}^{B*}(\tau)}{\partial \bar{Q}_l} \right)_{\mathbf{0}} \right|$$
$$\times \left| \sqrt{\frac{\hbar}{2\bar{\omega}_l}} \left[\sqrt{\bar{v}_{l_{\max}}} \left[\langle \bar{\boldsymbol{v}}_{\max} - \bar{1}_l | \bar{\bar{\mathbf{0}}} + \bar{\bar{v}}_k \rangle + \sqrt{\bar{v}_{l_{\max}} + 1} \langle \bar{\boldsymbol{v}}_{\max} + \bar{1}_l | \bar{\bar{\mathbf{0}}} + \bar{\bar{v}}_k \rangle \right] \right] \right| \tag{8.124}$$

where $\bar{v}_{l_{\max}}$ is the quantum number related to the mode l in the initial state $|\bar{\boldsymbol{v}}_{\max}\rangle$.

Next, a last set, F_{C_2}, is extracted from *class* 2 and uses a second threshold $C_{2_{\max}}$. It is used to obtain rough information on the relevance of Duschinsky mixing of the normal modes in determining the FC integrals. It contains all combinations of modes k and l but only considering the cases of an equal number of quanta for both modes ($\bar{\bar{v}}_k = \bar{\bar{v}}_l$ with $\bar{\bar{v}}_k, \bar{\bar{v}}_l \in [\![1, C_{2_{\max}}]\!]$):

$$F_{C_2}(k,l,\bar{\bar{v}}_k = \bar{\bar{v}}_l) = \left|\langle \bar{\mathbf{0}}|\bar{\bar{\mathbf{0}}} + \bar{\bar{v}}_k + \bar{\bar{v}}_l\rangle\right|^2 - \frac{F_{C_1}(k,\bar{\bar{v}}_k) \times F_{C_1}(l,\bar{\bar{v}}_l)}{|\langle \bar{\mathbf{0}}|\bar{\bar{\mathbf{0}}}\rangle|^2} \tag{8.125}$$

Notice that, according to definition, $F_{C_2} = 0$ if the modes k and l are not involved in Duschinsky mixings. Such a definition avoids double weighting the effects due to displacements and frequency changes in the prescreening procedure, since they are already taken into account by F_{C_1}. Similar to F_{C_1} and H_{C_1}, a temperature-related set, $F_{C_1}^{(T)}$ defined in the same way as F_{C_2} is extracted taking into account only the transitions from the highest vibrational state $|\bar{\boldsymbol{v}}_{\max}\rangle$. Consequently, $F_{C_1}^{(T)}$ is obtained through the relation

$$F_{C_2}^{(T)}(k,l,\bar{\bar{v}}_k = \bar{\bar{v}}_l) = \left|\langle \bar{\boldsymbol{v}}_{\max}|\bar{\bar{\mathbf{0}}} + \bar{\bar{v}}_k + \bar{\bar{v}}_l\rangle\right|^2 - \frac{F_{C_1}^{(T)}(k,\bar{\bar{v}}_k) \times F_{C_1}^{(T)}(l,\bar{\bar{v}}_l)}{|\langle \bar{\boldsymbol{v}}_{\max}|\bar{\bar{\mathbf{0}}}\rangle|^2} \tag{8.126}$$

At this point, since all necessary data sets are defined, we can explain in greater detail the reason for the usage of the temperature-dependent ones in the overall prescreening. Indeed, while for $T = 0\,\mathrm{K}$ spectra the F_{C_1} and F_{C_2} tensors carry all the necessary information on the causes of vibrational progressions, namely the geometry displacements, frequency changes, and Duschinsky mixings, and therefore their analysis can allow the selection of relevant transitions for reaching spectrum convergence, additional information is necessary for finite-temperature spectra. In fact, it is clear that thermal excitation in the initial state stands as an additional cause of (or at least may strongly enhance) progressions on the final normal modes. This information is in principle contained also in transitions belonging to higher order classes whose brute-force calculation would strongly increase the computational burden. In this context, the definition and usage of the additional tensors $F_{C_1}^{(T)}$ and $F_{C_2}^{(T)}$ are expedient inasmuch they allow us to take this information into account through cheaply computable quantities.

Once all the necessary elements for the prescreening have been obtained, the independent treatment of *class* 3 and above can be initiated. Indeed, as explained before, the information needed to choose the most relevant transitions in those *classes* is entirely contained in *classes* 1 and 2. Using the data contained in the arrays F_{C_1}, F_{C_2}, H_{C_1} in case of HT calculations, and their temperature-specific counterparts, respectively $F_{C_1}^{(T)}$, $F_{C_2}^{(T)}$, and $H_{C_1}^{(T)}$, if $T > 0\,\mathrm{K}$, the set of maximum quantum numbers ($\bar{\bar{\boldsymbol{v}}}_{\max}$) for a given class is established. The accuracy of the calculations is ruled by a user-defined limit, $N_I^{\max}$, which represents the maximum number of integrals to compute in a given class. A small value of $N_I^{\max}$ speeds up the generation of the spectrum but at the expense of its accuracy, while a high limit will improve the quality of the overall spectrum by requiring more calculations. The selection scheme operates through comparisons of the elements of each data array with a suitable set of thresholds (ϵ_i). While in the following, for the sake of clarity, we treat each threshold for each of the six data array independently ($\epsilon_1,\ldots,\epsilon_6$), it has been observed in a relevant number of tests that they can be bound to be equal, or at east proportional to each other, so that the prescreening algorithm is actually ruled by a single threshold (ϵ). Assuming that the calculations are performed at the HT level of approximation for the transition dipole moments and $T > 0$ K, so that all six previously defined arrays are necessary, we set six thresholds, ϵ_1 (bound to F_{C_1}), ϵ_2 (for F_{C_2}), ϵ_3 (for H_{C_1}), ϵ_4 (for $F_{C_1}^{(T)}$), ϵ_5 (for $F_{C_2}^{(T)}$), and ϵ_6 (for $H_{C_1}^{(T)}$). For each mode k, six maximum quantum numbers $\bar{\bar{v}}_{k_{\max}}(x)$ (where $x \in [\![1,6]\!]$) are set independently from each other following the same procedure. Starting from a sufficiently large value, $\bar{\bar{v}}_k = \max(\mathcal{C}_{1_{\max}}, \mathcal{C}_{2_{\max}})$, the quantum number is decremented until the following conditions are met:

$$
\begin{aligned}
&\bar{\bar{v}}_{k_{\max}}(1) \rightarrow F_{C_1}(k, \bar{\bar{v}}_k) \geq \epsilon_1 \\
&\bar{\bar{v}}_{k_{\max}}(2) \rightarrow F_{C_2}(k, l, \bar{\bar{v}}_k) \geq \epsilon_2 \quad \forall l \neq k \\
&\bar{\bar{v}}_{k_{\max}}(3) \rightarrow H_{C_1}(k, \bar{\bar{v}}_k) \geq \epsilon_3 \\
&\bar{\bar{v}}_{k_{\max}}(4) \rightarrow F_{C_1}^{(T)}(k, \bar{\bar{v}}_k) \geq \epsilon_4 \\
&\bar{\bar{v}}_{k_{\max}}(5) \rightarrow F_{C_2}^{(T)}(k, l, \bar{\bar{v}}_k) \geq \epsilon_5 \quad \forall l \neq k \\
&\bar{\bar{v}}_{k_{\max}}(6) \rightarrow H_{C_1}^{(T)}(k, \bar{\bar{v}}_k) \geq \epsilon_6
\end{aligned}
$$

The actual maximum number of quanta is then chosen as the maximum of these values:

$$\bar{\bar{v}}_{k_{\max}} = \max(\bar{\bar{v}}_{k_{\max}}(1), \bar{\bar{v}}_{k_{\max}}(2), \bar{\bar{v}}_{k_{\max}}(3), \bar{\bar{v}}_{k_{\max}}(4), \bar{\bar{v}}_{k_{\max}}(5), \bar{\bar{v}}_{k_{\max}}(6))$$

Once the set $\bar{\bar{\boldsymbol{v}}}_{\max}$ has been defined, the corresponding number of integrals to calculate is roughly estimated, for a given class $\mathcal{C}_n$, as $N_I = {}_N C_n \times \langle \bar{\bar{\boldsymbol{v}}}_{\max} \rangle^n$, where ${}_N C_n$ represents the number of combinations of the n excited oscillators and $\langle \bar{\bar{\boldsymbol{v}}}_{\max} \rangle$ is the arithmetic mean of the N maximum quantum numbers. If the number of integrals to compute, N_I, is higher than the allowed limit, $N_I^{\max}$, the thresholds ϵ_1 to ϵ_6 are increased and the set of maximum quantum numbers $\bar{\bar{\boldsymbol{v}}}_{\max}$ is reestimated. Once the condition $N_I \leq N_I^{\max}$ is fulfilled, all FC integrals are computed using the correct maximum number of quanta $\bar{\bar{v}}_{k_{\max}}$ for each mode.

The tests are sufficiently fast to allow a rather large number of trials in a very short computational time. Hence, the thresholds ϵ_1 to ϵ_6 can be chosen to be very low (e.g., $\approx 10^{-9}$).

It should be noted that the inclusion of temperature effects raises additional issues bound to the choice of the starting vibrational states in the initial electronic state. An evident way to limit the treatment is to use a threshold on the Boltzmann population of each vibrational state. In practice, this threshold is set with respect to the population of the ground state. Similar to the final states, a division in *classes* is performed among all selected initial states. For each *class*, a set is defined by the initial states sharing the same simultaneously excited modes, so that they differ only by the quantum numbers of these modes, and each set (in previous papers named "mother states" [31, 70]) is treated separately.

A more detailed description of the prescreening algorithm can be found in the literature [31, 70, 89].

8.4.1.1 Generalization of Class-*Based Prescreening Method for Vibrational Resonance Raman* While initially designed for one-photon spectroscopies, the prescreening procedure described above has been trivially generalized also for two-photon spectroscopies [65]. Here we limit our discussion to its possible extensions to cope with vibrational RR. Additionally, we will only consider the case of transitions from the vibrational ground state, $|\bar{\mathbf{0}}\rangle$, so that only Stokes bands are taken into account. Inspection of Eq. 8.46 shows that the FC and HT contributions to vibrational RR can be computed in a time-independent perspective due to the recursive formulas in Eq. 8.54 by fully taking into account the displacements, frequency changes, and Duschinsky mixings of the normal modes. However, such an approach can be very cumbersome from a computational perspective for two main reasons: first, the explicit calculation of the polarizability tensor through the direct summation over the vibrational states of the intermediate electronic state, whose number in sizable molecules is huge (easily exceeding 10^{20}), and, second, the extremely large number of possible final vibrational states belonging to the ground electronic state, which can lead to unfeasible calculations. Because of the inherent complexity bound to a complete treatment of the transitions in vibrational RR, we will limit the present discussion to cases where the final states are fundamental bands or overtones. Indeed, most experimental RR spectra are usually measured in a rather narrow energy window encompassing fundamentals and only low-excited overtones and combination bands, and most of the theoretical calculations are only limited to fundamentals. With this choice, the number of possible final states is drastically reduced and can be easily handled without further limitations. In this respect, it is also important to realize that, while a too restrictive selection of the final states may end up in the absence of some bands in the simulated RR spectrum, an incomplete inclusion of the intermediate states $|\chi_{t(\underline{m})}\rangle$ in Eq. 8.46 may lead to inaccuracies in the predicted bands that are not easily controllable. Therefore, an effective selection of the intermediate states cannot be avoided in order to design a reliable method for sizable molecules. Actually, from Eq. 8.46, it is easily realized that, from a sum-over-state time-independent perspective, the computation of the polarizability tensor for a given final state $|\bar{\bar{\mathbf{v}}}\rangle$ (the initial state is always the ground state $|\bar{\bar{\mathbf{0}}}\rangle$) is equivalent to the computation of two absorption spectra (with possible FC and HT effects) for the electronic transition $|\Phi_g\rangle \rightarrow |\Phi_m\rangle$: the first from the ground vibrational state $|\chi_{i(\underline{g})}\rangle \equiv |\mathbf{0}\rangle$ (spectrum recorded at $T=0$ K) and the second from a selected hot- vibrational state $|\chi_{f(\underline{g})}\rangle \equiv |\mathbf{v}\rangle \neq |\mathbf{0}\rangle$. If more than one intermediate electronic state $|\Phi_m\rangle$ must be considered, the calculations must be repeated for each of them and the polarizability tensors summed before computing the square contributions in Eq. 8.29. These features of the polarizability tensor allow for immediate adoption of the prescreening method described in the previous section for zero- and finite-temperature spectra, thus showing that it represents an effective tool also for RR spectra. Nonetheless, it is necessary to generalize the analytical sum rules (that play the role of $\mathcal{M}^{(0)}$) for the absorption spectrum to check the spectrum convergence. This is done quite trivially in FC cases, noting that the sum of the numerator of the polarizability between the initial $|\chi_{i(\underline{g})}\rangle$ and final $|\chi_{f(\underline{g})}\rangle$ states must be zero, because the two wavefunctions are orthogonal and for a complete set of intermediate states it is $\sum_{t(\underline{m})}|\chi_{t(\underline{m})}\rangle\langle\chi_{t(\underline{m})}| = 1$. In case of HT transitions, suitable sum rules to check the calculation convergence can be obtained by exploiting the same techniques adopted in Section 8.3.3.

8.4.2 Spectra Convergence

By discarding selectively the transitions of low probability to generate the vibronic spectrum, a loss of accuracy may be observed depending on the quality of the approximation. Such a matter is of particular importance when dealing with a priori prescreening methods, since the exact extent of the approximation is not known when the most relevant transitions are chosen. As a consequence, it is necessary to have a consistent way to evaluate the reliability of the overall simulation. This is done by calculating the so-called convergence of the spectrum, which can be done performing an analytical calculation of the 0th moment $\mathcal{M}^{(0)}$ of the given spectrum. For one-photon transitions up to first-order (HT) expansion of the transition dipoles on the nuclear coordinates, this moment has been given in Eq. 8.87. Additionally, the expression for OPA including a second-order expansion has been given by Barone et al. [5], and a generalization to two-photon absorption and emission processes is reported by Lin et al. [65]. Summing the individual contribution of each transition taken into account in the intensity given in Eq. 8.76, the spectrum convergence is obtained by dividing the total intensity obtained from this summation by $\mathcal{M}^{(0)}$. For an accurate simulation, such quantity should converge to the ideal limit 1. The spectrum convergence represents a powerful tool to evaluate the overall quality of a prescreening procedure. It should be noted that this is true only if the overlap integrals are calculated without approximations, which is not the case for the method proposed by Hazra and Nooijen [81] presented at the beginning of Section 8.4. In the following, we will use the spectrum convergence to compare the efficiency of prescreening methods, that is, the quality of the simulation obtained using them with respect to the computational time, and present other means to evaluate the convergence based on the spectral shape. While the discussion will be mainly focused on the *class*-based prescreening method, the general argument remains true for most a priori selection schemes, among which those presented in Section 8.4. The interested reader will find more detailed discussions elsewhere [5, 6].

As a first example, we compare the performance of several prescreening schemes for the computation of the photoelectron spectrum of a large polycyclic aromatic hydrocarbon (PAH) derivative with 462 normal modes. Figure 8.10 compares the spectrum convergence against the required computer time for three a priori selection schemes designed by Dierksen et al. [82], Jankowiak et al. [88], and Santoro et al. [89] with several values of $N_I^{\max}$ tested for the latter. It is immediately visible that all three methods are sufficiently accurate to reach near-complete (larger than

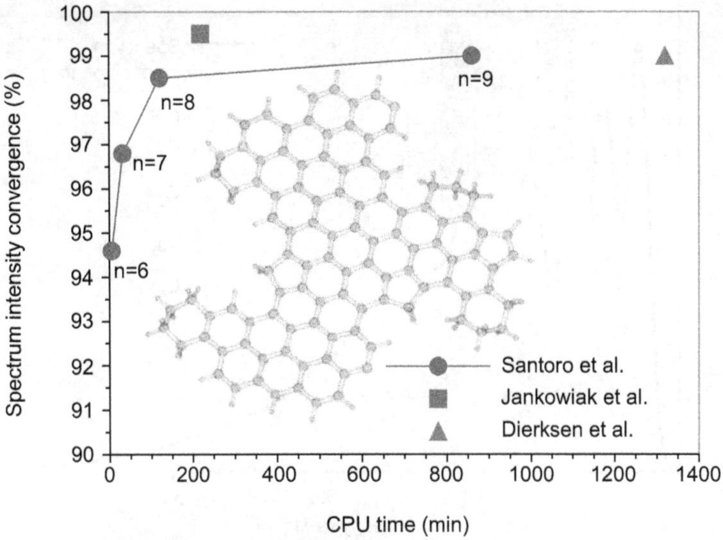

Figure 8.10 Convergence of spectrum calculation for PAH, a macromolecule with 462 normal modes and 10^{37} vibrations in the first 5000 cm^{-1}. Comparison of total spectrum intensity computed with three different prescreening schemes (by Santoro et al. [89], Jankowiak et al. [88], and Dierksen et al. [82]) are reported. For FCclasses, computations with $N_I^{\max}$ set to 10^n ($n=6,7,8,9$) are reported [89].

99%) convergence. Furthermore, it can be noticed that the method presented by Jankowiak et al., which is based on a selection scheme relying directly on the analytic sum rules, performs better in the vicinity of full convergence. The *class*-based approach [89], on the other hand, referred to as FCclasses in the figure, provides results already satisfactory at a very low computational cost. On the other hand, the prescreening method of Dierksen and Grimme shows higher computational times because of the counting algorithm based on the spectral bounds used to define the pool of transitions from which the actual prescreening is performed using their probability. Hence, as mentioned previously, its performance is highly dependent on the upper bound of the spectrum.

Since we will now linger on subtler aspects of the convergence and various ways to appreciate the efficiency of the prescreening, we will put emphasis on the *class*-based method (called FCClasses). First, let us discuss a particular feature of this approach, from which it got named, the categorization of the transitions in *classes*. With respect to our previous discussion in Section 8.4.1, the term *classes* will be used here to refer exclusively to the number of simultaneously excited modes in the final state, independently of the initial state, so that when referring to *class* C_2, for instance, we will consider all transitions to combination states involving two simultaneously excited modes, without any restriction on the vibrational initial states. The spectrum convergence with respect to the *classes* is shown in Figure 8.11 for coumarin 153, which has 102 normal modes. Similarly to PAH, the method is

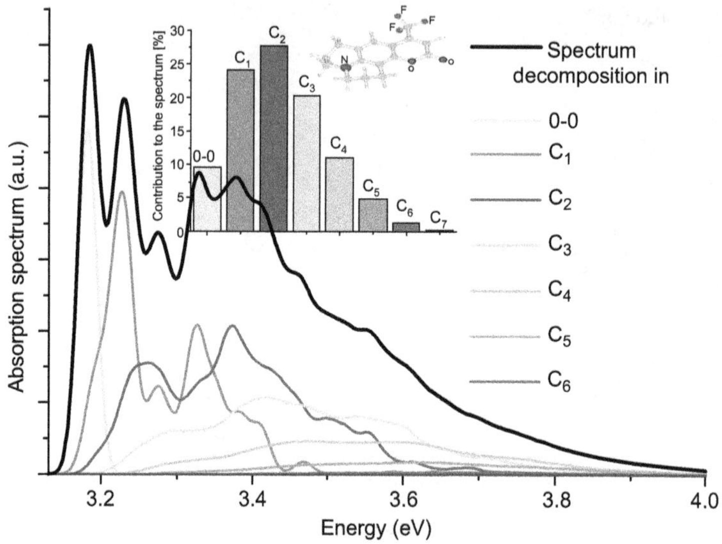

Figure 8.11 Convergence of spectrum calculation for coumarin 153 (102 normal modes) with respect to the *classes*; In the upper panel, contributions of specific *classes* are compared with the total spectrum (see legend).

able to perform very well, recovering up to 99% of the complete spectrum with $N_I^{\max}$ set to 10^8. It is clear that the contribution to the intensity of *classes* higher than C_5 decreases steeply with the order of the *class*, and the difference between the spectrum intensity calculated up to C_7 and up to C_6 is smaller than 1%, confirming the good spectrum convergence with respect to *classes*. It is also shown that contributions of the *classes* become flatter with the increase of the order of the *class* and thus do not influence the spectrum shape.

Other technical aspects of the convergence will be presented through two, rather different cases, the phosphorescence spectrum of a large biomolecule, namely chlorophyll c_2, and the photoelectron spectrum of a nanosystem, adenine adsorbed on a Si_{119} cluster. The results obtained for the latter, a total system with 636 normal modes, are compared to the ones of the isolated adenine molecule in Table 8.2. It is interesting to analyze the spectrum convergence for both systems with respect to the mean number of final vibrational states to treat in each *class* C_n, which is directly related to the number of transitions to compute, and investigate the efficiency of the selection procedure. Table 8.2 lists the binomial coefficients ${}_NC_n$ for the isolated adenine and adenine@Si_{119}, along with the intensity convergence obtained with $N_I^{\max}$ set to the default value of 10^8. It is noteworthy that in both cases, either the isolated molecule with 39 normal modes or the macrosystem with more than 600, almost all the spectrum intensity has been recovered at an equivalent computational cost with a spectrum convergence of about 98%. This is particularly interesting since, for the cluster, the default value of $N_I^{\max}$ is insufficient to keep all the integrals chosen from the initial evaluation of $\bar{\bar{\nu}}_{\max}$ (see Section 8.4.1 for more details), even for the small

Table 8.2 Convergence of Spectra Computations for Adenine and Adenine@Si(100)

	Adenine		Adenine@Si(100)	
Class (n)	${}_N C_n$	Progression	${}_N C_n$	Progression
3	9.14×10^3	84.54%	4.27×10^7	87.31%
4	8.23×10^4	93.57%	6.7×10^9	94.82%
5	5.76×10^5	97.48%	8.54×10^{11}	97.37%
6	3.26×10^6	98.32%	8.98×10^{13}	97.88%
7	1.54×10^7	98.39%	8.08×10^{15}	97.93%

Note: For each *class* C_1 the number of combinations of the n excited oscillators, ${}_N C_n$, and the corresponding spectrum progression are listed. The C_1 and C_2 transitions have been computed with analytical formulas allowing a maximum quantum number $v_i = 30$ and $v_1 = v_2 = 20$ (MaxC1 = 30, MaxC2 = 20), respectively. For the *classes* C_n, with $n \geq 3$, the transitions to be computed have been selected setting the parameter $N_I^{\max}$ to 10^8 (the default value).

class C_3 with only three simultaneously excited modes. Indeed, despite the significant difference in the systems' size, in both cases, the electronic transition is localized on the adenine molecule, and the categorization in classes allows us to take benefit of this feature, automatically selecting the reduced-dimensionality model–system comprising only the relevant modes for the spectrum. This particular case shows the interest of a priori strategies to select only the relevant transitions and discard the less probable ones beforehand so that even large systems do not have much impact on the computational time required to carry out the whole simulation (if not, all the vibrational modes are significantly perturbed by the electronic transition), opening the route, for example, to the explicit simulation of large systems like dyes in a protein environment.

For the phosphorescence spectrum of chlorophyll c_2, a large molecule with 73 atoms and 213 normal modes, both intensity and lineshape convergence with respect to the maximum number of integrals (from 10^2 to 10^{12}) are shown in Figure 8.12. As expected, a very small number of integrals is not sufficient to obtain a good convergence of the spectrum intensity, and less than half of it is recovered for $N_I^{\max} = 10^2$. While improving this result, calculations up to 10^{12} integrals per *class* are still insufficient to reach a full convergence of the spectrum. Moreover, it should be noted that this gain is obtained at the expense of a huge increase in computational times when switching from 10^9 (with a convergence of about 80%) to 10^{12} integrals (with a convergence close to 90%). A more encouraging result lies in the convergence of the lineshape. Indeed, as shown in Figure 8.12, the latter is much faster than for intensity. Such a phenomenon is observed in most cases, especially when dealing with large systems. It can be related to the fact that the eventual loss of convergence usually occurs for rather high-order *classes*. However, as can be seen in the example of coumarin C153 in Figure 8.11, as the order n increases, the *class* contribution to the total spectral intensity reaches a maximum (in general for a very low value of n) and then decreases rather steeply. More importantly, the contribution to the lineshape becomes also flatter and flatter, with a very shallow maximum slowly moving farther

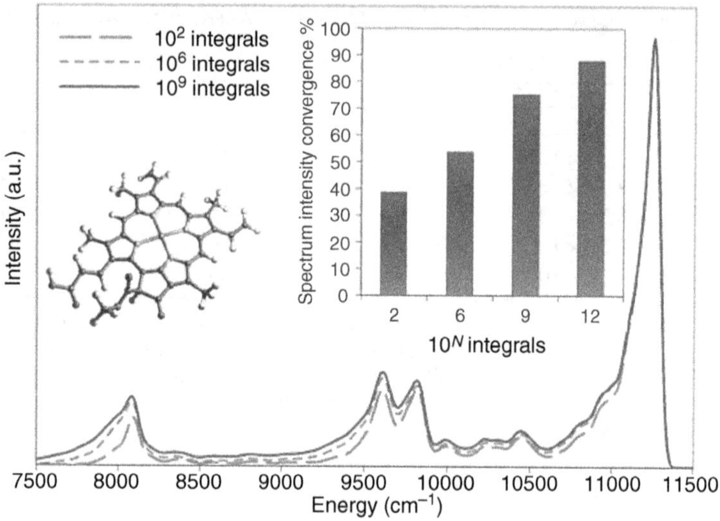

Figure 8.12 Convergence of spectrum calculation for chlorophyll c_2, molecule with 213 normal modes. Comparison of spectrum shape calculated with N_I^{max} set to 10^2 (dashed line), 10^6 (fine-dashed line), and 10^9 (solid line), while the onset shows convergence of spectrum intensity computed with N_I^{max} up to 10^{12}.

from the 0–0 transition. Such a behavior is expected since higher order *classes* collect a very large number of weak transitions to excited states with rather different energies. This expedient property can be used to generate the reliable spectrum at relatively low computational cost with the main error represented by a small drift in the wing of the spectrum farther from the 0–0 transition. In fact, comparison of the spectrum lineshapes calculated with N_I^{max} set to 10^2, 10^6, and 10^9 clearly shows that the main spectral features are well reproduced even if the total spectrum intensity is far from convergence. Incidentally, the spectra calculated with $N_I^{max} = 10^9$ or larger are identical on this scale. Thus, inspection of the spectrum lineshape indicates that the most important transitions have been taken into account and that accurate enough spectra can be computed with N_I^{max} set to 10^9, despite the spectrum intensity being below 90%. However, it should also be stressed that, when good reproduction of the high-energy wing of the spectrum (the one suffering from the largest relative error) is needed, which is of particular interest for the computation of nonradiative transition rates, a careful check of the convergence in that energy region must be performed and purposely tailored methods may be more suited. Before concluding this section it is worthy to highlight that any time-independent method, despite the effectiveness of the adopted prescreening, will necessarily encounter problems when the physics of the system is such that the number of the truly relevant transitions is too large to be computed and cannot be reduced without losing part of the intensity. On the other hand, and technically speaking, this happens when the number of important transitions is really huge, that is, larger than 10^{12}, and it is highly unlikely that one is interested in their detailed analysis. In those cases, the most effective computational route, even for the near future, passes through a combination of time-dependent methods (see Chapter 10) to obtain low-resolution converged spectra and time-independent methods to individuate and analyze the most important stick transitions.

8.5 MULTISTATE AND ANHARMONIC APPROACHES

As discussed in Section 8.2.1, when nonadiabatic couplings cannot be neglected, the BO approximation is not reliable and coupled electronic states must be considered simultaneously with their interactions. For small systems, several full-dimensional approaches based on the vibronic or spin–rovibronic wavefunctions and taking into account simultaneously at least two electronic states have been developed [2, 100–104]. To quote some examples, the full vibronic Hamiltonians have been derived and employed for linear tetra-atomic molecules showing Renner-Teller interactions [103] or CX_3Y-like molecules of C_{3v} symmetry showing Jahn–Teller interactions [104]. In the following, we will present the computational approaches based on the full rovibronic Carter–Handy Hamiltonian [100], developed for triatomic molecules and expressed in internal coordinates, which allows us to take into account up to three interacting electronic states [2, 100, 101].

The complete Carter–Handy Hamiltonian [2, 100] expressed in internal coordinates, $\boldsymbol{R}_n$ and θ, which correspond to the bond lengths (for three-atomic molecules $n = 1, 2$) and angle, respectively, is first separated between electronic and nuclear contributions,

$$\hat{H} = \hat{T}_N + \hat{H}_e \tag{8.127}$$

where $\hat{H}_e$ is the pure electronic Hamiltonian and $\hat{T}_N$ the nuclear Hamiltonian given by the relation

$$\hat{T}_N = \hat{T}_v + \hat{T}_{vr} \tag{8.128}$$

where $\hat{T}_v$ represents the vibrational kinetic energy operator in internal valence coordinates and $\hat{T}_{vr}$ is the rotational kinetic operator that also includes the coupling between rotational ($\hat{\boldsymbol{J}}$), electronic ($\hat{\boldsymbol{L}}$), and spin ($\hat{\boldsymbol{S}}$) angular momenta,

$$\begin{aligned}\hat{T}_v(R_1, R_2, \theta) = &-\frac{1}{4}\left[\left(\frac{1}{\mu_1 R_1^2} + \frac{1}{\mu_2 R_2^2} - \frac{2\cos\theta}{m_B R_1 R_2}\right)\left(\frac{\partial^2}{\partial\theta^2} + \cot\theta\frac{\partial}{\partial\theta}\right)\right] \\ &- \frac{1}{2\mu_1}\frac{\partial^2}{\partial R_1^2} - \frac{1}{2\mu_2}\frac{\partial^2}{\partial R_2^2} - \frac{\cos\theta}{m_B}\frac{\partial^2}{\partial R_1 \partial R_2} \\ &+ \frac{1}{m_B}\left(\frac{1}{R_1}\frac{\partial}{\partial R_2} + \frac{1}{R_2}\frac{\partial}{\partial R_1}\right)\left(\sin\theta\frac{\partial}{\partial\theta} + \cos\theta\right)\end{aligned} \tag{8.129}$$

$$
\begin{aligned}
\hat{T}_{vr}(R_1,R_2,\theta,\hat{\boldsymbol{J}},\hat{\boldsymbol{L}},\hat{\boldsymbol{S}}) = {} & \frac{1}{8\cos^2(\theta/2)}\left\{\frac{1}{\mu_1 R_1^2}+\frac{1}{\mu_2 R_2^2}+\frac{2}{m_B R_1 R_2}\right\}(\hat{J}_z,\hat{L}_z,\hat{S}_z)^2 \\
& + \frac{1}{8\sin^2(\theta/2)}\left\{\frac{1}{\mu_1 R_1^2}+\frac{1}{\mu_2 R_2^2}+\frac{2}{m_B R_1 R_2}\right\}(\hat{J}_z,\hat{L}_z,\hat{S}_z)^2 \\
& + \frac{1}{8}\left\{\frac{1}{\mu_1 R_1^2}+\frac{1}{\mu_2 R_2^2}+\frac{2\cos\theta}{m_B R_1 R_2}\right\}(\hat{J}_y,\hat{L}_y,\hat{S}_y)^2 \\
& - \frac{1}{4\sin\theta}\left\{\frac{1}{\mu_1 R_1^2}+\frac{1}{\mu_2 R_2^2}\right\}[\hat{J}_z+\hat{L}_z+\hat{S}_z,\hat{J}_x+\hat{L}_x+\hat{S}_x]_+ \\
& + \frac{i}{2}\left\{\frac{1}{\mu_1 R_1^2}+\frac{1}{\mu_2 R_2^2}\right\}\left(\frac{\cot\theta}{2}+\frac{\partial}{\partial\theta}\right)(\hat{J}_y,\hat{L}_y,\hat{S}_y) \\
& + \frac{i}{2m_B}\left\{\frac{1}{R_2}\frac{\partial}{\partial R_1}-\frac{1}{R_1}\frac{\partial}{\partial R_2}\right\}(\hat{J}_y,\hat{L}_y,\hat{S}_y) \\
& + A_{SO}\hat{\boldsymbol{L}}\cdot\hat{\boldsymbol{S}}
\end{aligned} \tag{8.130}
$$

where the spin–orbit coupling is represented by the constant or geometry-dependent A_{SO} term. Time-independent variational computations of the eigenstates of these Hamiltonians permit the detailed analysis of the interplay between the different effects due to conical intersections, RT interactions, spin–orbit couplings, and anharmonic resonances. In such a manner, calculations beyond the BO approximation, taking into account all relevant electronic states and considering all possible couplings, allow for the accurate simulation of vibronic spectra even in difficult cases. However, such a complete approach is not feasible, in general, for larger systems. Nonetheless, in many cases, the multielectronic state problem can be satisfactorily addressed for semirigid systems within the so-called multimode vibronic coupling model (MVCM) [22, 107], which is based on a quasi-diabatic representation of the electronic states as described in Section 8.2.1.

In its original formulation, named linear vibronic coupling model (LVCM), the quasi-diabatic Hamiltonian included second-order diagonal and first-order off-diagonal terms. Such an approach has been successfully applied over the years to simulate spectra for systems with coupled electronic states [22]. More recently, the MVCM approach has been extended to include also quadratic terms in the off-diagonal matrix elements, leading to the so-called quadratic vibronic coupling model (QVCM) [108–110]. The QVCM implementation by Nooijen [108] includes the off-diagonal coupling constants, which involve modes of the same symmetry and allows us to treat simultaneously a large number of electronic states. Recently, it has been extended to quartic coupling constants and generalized [111] to one-photon chiral spectroscopies. Stanton et al. [109] introduced the "adiabatic parameterization" approach, along with the description of the diagonal blocks of the potential, which contains up to quartic or quadratic terms, depending on if they are related to totally symmetric or nonsymmetric coordinates, respectively [110, 112]. The MVC model has also been recently extended to take into account, in a nonperturbative manner, the spin–orbit interactions [113] and has been applied to the studies of spectra in molecules with conical intersections and spin–orbit coupling. The main shortcomings of the time-independent (e.g., Lanczos-based) computations based on the multimode vibronic coupling model are related to the steeply increasing size of the multimode expansion, which limits their feasibility up to a few modes. Promising developments include, for instance, the employment of effective Hamiltonians worked out in the Green function formalism [114] or parallel algorithms [115], together with the optimal design of reduced-dimensionality vibronic basis sets. However, the most effective implementations of MVCM for larger systems remain those based on time-dependent methods such as the multiconfigurational time-dependent Hartree (MCTDH) [116], which are described in details in Chapter 10.

Considering the anharmonic effects, their proper treatment is by far more complicated in case of vibrationally resolved electronic transitions than in vibrational spectroscopy since two different electronic states must be treated at the same time. Indeed, equilibrium geometries can be quite different, so that the description of a larger area of the PES is required, with the resulting problems of couplings, limits of polynomial expansions, and so on. Furthermore, the normal-modes of the two states may be sufficiently different to preclude the normal-mode description of the vibronic problem and switching to internal coordinates can be more suitable [117, 118]. Finally, a proper treatment of anharmonicity would require the computation of the full-dimensional PES, a computational effort which is still out of reach for sizable molecular systems. Theoretical models to generate vibrationally resolved electronic spectra may in principle include anharmonicity [52, 86, 119–124] but, at present, a general approach to go beyond harmonic approximation applicable to molecular systems including more than a few atoms is still lacking. However, several schemes have been proposed to improve the accuracy of the simulated spectra by using vibronic models set beyond harmonic approximation, which can be applied to small systems or well-defined local modes with limited dimensionality approaches. An example of such approaches, which are well represented by the works of Berger et al. [86], is based on the description of PES anharmonicity through one-dimensional cuts along all or a set of normal modes [52, 86, 122]. This can be successfully applied for systems with strongly anharmonic potentials (e.g., double well) but weak intermode couplings, in particular for cases where the normal modes of the initial and final electronic states are very similar even if the associated one-dimensional (1D) cuts of the PES vary significantly. Another example is represented by the approach of Hazra et al. [52], set in the vertical framework and mentioned in Section 8.3.2.1, where the anharmonic PES is expanded about the ground-state equilibrium geometry to describe the intrinsic anharmonicity of systems whose PESs show imaginary frequencies, as it happens when the electronic transition causes a lowering of the molecular symmetry. In this approach, all normal modes, except those with imaginary frequencies for which 1D anharmonic treatment is performed, are described at the harmonic level. An alternative route is represented by full-dimensional anharmonic treatments of the vibronic problem in polyatomic molecules. Such approaches have been applied by Luis et al. [120, 121] by means of the second-order perturbation theory of the FC factors initially calculated at the harmonic level. Alternatively, variational approaches set within vibrational self-consistent field (VSCF) or vibrational configuration-interaction (VCI) frameworks [123, 124] have been introduced. In full-dimensional approaches, the anharmonic vibrational wavefunctions of the electronic states involved in the transition are expanded with harmonic oscillator basis functions relative to the normal coordinates of their respective electronic states. Effective computational routes can be designed by adopting the harmonic oscillator wavefunctions of one electronic state as a basis set to describe the anharmonic vibrational states of both electronic states involved in the transition [124, 125]. In this general framework, both perturbative and variational approaches are capable of taking fully into account linear and nonlinear anharmonic effects (up to a desired extent) along with the normal mode and frequency changes during the electronic transition. Although very promising, such models are computationally demanding and, in practice, they have been applied only to relatively small systems. A simplified route to include anharmonic effects has been proposed for semirigid systems based on the assumption that the anharmonic character of the ground electronic state PES is essentially retained during the electronic excitation. In this framework, anharmonic corrections, derived from ground-state data, for example, through second-order perturbative vibrational theory [126], are also used to correct the excited-state frequencies [53] and FC factors [127]. Though approximative, these approaches allow for significant improvements concerning the band positions and can be applied to relatively large systems. In particular, the approach which will be applied in Section 8.6 is based on the derivation of excited-state mode-specific scaling factors starting from the ground-state ones. In turn, the latter can be obtained either by means of perturbative [126] or variational anharmonic frequency calculations or derived through a comparison with easily accessible ground-state experimental data. For each particular normal mode $\bar{Q}_k$, the frequency scaling vector $\bar{\boldsymbol{\alpha}}$ is computed first using the formula $\bar{\alpha}_k = \bar{\nu}_k/\bar{\omega}_k$, where $\bar{\boldsymbol{\nu}}$ is the vector of the anharmonic (or experimental) frequencies and $\bar{\boldsymbol{\omega}}$ its counterpart for the harmonic frequencies. The Duschinsky coefficients J_{lk} are then applied to derive the relation between the initial (k) and final (l) state mode-specific anharmonicity scaling factors:

$$\bar{\bar{\alpha}}_l = \sum_k^N J_{kl}^2 \bar{\alpha}_k \tag{8.131}$$

8.5.1 Multimode Vibronic Coupling Model

While, as discussed above, general solutions to the multistate nonadiabatic problem in Eq. 8.6 are out of reach for large systems, the latter can be suitably tackled for semirigid systems due to MVCM models based on a quasi-diabatic description of the electronic states. Such models allow for effective computations when the harmonic approximation is, at least, a good starting point for the description of the diabatic potentials [22], $W_{ji}(\boldsymbol{Q})$. More specifically, let us consider a harmonic description of the initial electronic state of the transition, and let us assume from now on, for the sake of simplicity, that it coincides with the ground state and is not coupled to other electronic states. We define the dimensionless normal coordinates q according to $\bar{q}_i = (\omega_i/\hbar)^{1/2}\bar{Q}_i$, where ω_i is the frequency associated with mode $\bar{Q}_i$ on the initial state. The nuclear kinetic operator and the ground-state potential can be written as

$$T_N = -\frac{\hbar}{2}\sum_i \omega_i \frac{\partial^2}{\partial \bar{q}_i^2} \tag{8.132}$$

$$W_{00} = \frac{\hbar}{2}\bar{\boldsymbol{q}}^{\mathrm{T}}\Omega\bar{\boldsymbol{q}} \tag{8.133}$$

Assuming a quasi-diabatic representation, the diabatic diagonal excited potentials $W_{jj}(\bar{\boldsymbol{q}})$ and the off-diagonal couplings $W_{ji}(\bar{\boldsymbol{q}})$ are expanded in Taylor series with respect to the coordinates $\bar{\boldsymbol{q}}$ about the ground equilibrium geometry $\bar{\boldsymbol{q}} = \boldsymbol{0}$ (vertical approach):

$$W_{jj} = W_{00}(\boldsymbol{q}) + \bar{\boldsymbol{q}}^{\mathrm{T}}\kappa^{(j)} + \bar{\boldsymbol{q}}^{\mathrm{T}}\gamma^{(j)}\bar{\boldsymbol{q}} + \cdots \tag{8.134}$$

$$W_{ji} = W_{ji}(\boldsymbol{0}) + \bar{\boldsymbol{q}}^{\mathrm{T}}\lambda^{(j,i)} + \cdots \tag{8.135}$$

$$\kappa_n^{(j)} = \left(\frac{\partial W_{jj}}{\partial q_n}\right)_{\boldsymbol{0}} \tag{8.136}$$

$$\gamma_{nm}^{(i,j)} = \frac{1}{2}\left[\left(\frac{\partial^2 W_{jj}}{\partial q_n\,\partial q_m}\right)_{\boldsymbol{0}} - \delta_{nm}\hbar\omega_n\right] \tag{8.137}$$

$$\lambda_n^{(j,i)} = \left(\frac{\partial W_{jj}}{\partial q_n}\right)_{\boldsymbol{0}} \tag{8.138}$$

The way in which the diagonal excited-state potentials are written in Eq. 8.138 emphasizes the fact that their difference with respect to the ground potential is expanded in power of the dimensionless normal coordinates. According to the discussion in Section 8.3.2.1, linear terms in Eq. 8.138 introduce equilibrium displacements and bilinear terms, the so-called Duschinsky effect (see below). In symmetric systems, the value of the parameters in Eq. 8.138 are restricted by the requirement that each term of the total Hamiltonian must belong to the total symmetric irreducible representation (irreps) Γ_A. Therefore, in analogy with the discussion on the Herzberg–Teller effect in Section 8.3.1.1, it can be easily proven that displacements ($\kappa_n^{(j)} \neq \boldsymbol{0}$) are only allowed for totally symmetric modes, and the interstate coupling constants ($\lambda_n^{(j,i)} \neq \boldsymbol{0}$) only when

$$\boldsymbol{\Gamma}_j \otimes \boldsymbol{\Gamma}_n \otimes \boldsymbol{\Gamma}_i \supset \boldsymbol{\Gamma}_A \tag{8.139}$$

where Γ_j and Γ_i are respectively the irreps of diabatic states j and i and Γ_n is the irrep of the coordinate q_n. It is noteworthy that for an electronic two-state system the latter can be simply computed from ab initio adiabatic energy at the ground vertical excitation. Following the previous notation, the adiabatic potentials, V_1 and V_2 are given by the relation

$$[V_1(\boldsymbol{q}) - V_2(\boldsymbol{q})]^2 = [W_{11}(\boldsymbol{q}) - W_{22}(\boldsymbol{q})]^2 + 4\left[\sum \lambda_n^{(1,2)} qn\right]^2 \tag{8.140}$$

where

$$\lambda_n^{(1,2)} = \left[\frac{1}{8}\frac{\partial^2}{\partial q_n^2}|V_1(z) - V_2(z)|^2\right]^{1/2} \tag{8.141}$$

where the last equality holds if the difference between the diagonal diabatic potentials does not depend quadratically on q_n. These approaches have been also adopted for the particular case of Jahn–Teller and pseudo-Jahn–Teller interactions, where the nonadiabatic interaction is imposed by symmetry [128, 129].

The calculation of the molecular eigenstates with the MVCM model, necessary in traditional time-independent methods, can prove to be very cumbersome or even unfeasible. However, time-independent effective solutions, practicable for reduced-dimensionality models (in practice when the number of relevant normal coordinates is less than 10), may be obtained by taking advantage of the Lanczos iterative tridiagonalization of the Hamiltonian matrix [130]. The Lanczos algorithm proves to be very suitable for the computation of low-resolution spectra; however, its effectiveness is better highlighted in a time-dependent framework. In fact, it can be easily realized that Lanczos states are only sequentially coupled, and it is therefore clear that only a limited number of states is necessary to describe short-time dynamics since the latter is the only relevant information for low-resolution spectra (see Chapter 10).

8.6 EXPERIMENTAL AND SIMULATED SPECTRA

In this section, we illustrate and discuss some applications of the computational approaches presented in this chapter to highlight their flexibility, expected accuracy, and interpretative capability and to provide some guidelines for the choice of the most suitable approach for a given problem, taking into proper account their computational feasibility. The chosen examples range from small molecules, whose spectroscopy is ruled by several interacting electronic states, to large systems of direct biological or technological interest, amenable to a more affordable single-state description. In such a manner, we review practical applications of approaches aimed at tackling different challenges from sophisticated models with nonadiabatic and anharmonic effects to simpler yet accurate tools set in the harmonic framework for the study of macromolecules. Concerning strongly nonadiabatic systems, we will focus on triatomic molecules. Notwithstanding their small dimensionality, these systems often have complex electronic spectra due to the nonadiabatic interactions enhanced, for instance, by the degeneracy of ground or low-lying excited electronic states at linear configuration. Such a phenomenon causes the so-called RT interaction [131], whose theoretical foundations have been briefly reviewed in Section 8.5, which earned a significant interest in theoretical spectroscopy. In the present contribution, we illustrate the interpretative capability of state-of-the-art theoretical approaches discussing the calculation of spectra for triatomic molecules showing up to three-state interactions (RT and/or HT) for which full-dimensional vibronic studies beyond the BO and harmonic approximations can be performed [2, 100, 101]. It will be shown how computational spectroscopy tools can be effectively applied in such complex cases. Additionally, we will illustrate how harmonic approaches to the computation of vibrationally resolved electronic spectra [5, 6] stand as general and easy-to-use tools that can be applied to simulate spectra for a large variety of systems ranging from small molecules in the gas phase to macrosystems in condensed phases.

8.6.1 Accuracy and Interpretation

In many practical cases, a detailed analysis of the experimental electronic spectra is quite difficult due, for example, to the often nontrivial identification of the electronic band origin, multimode effects, possible overlaps of several electronic transitions, and nonadiabatic and/or anharmonic effects. Although such complications are challenging also for the theoretical approaches, there are several examples (see Section 8.6.1.1) which clearly show how theoretical simulations can provide a valuable tool with remarkable interpretative potential. When choosing the most proper model for a specific system, it should be realized that it is in general unknown a priori whether nonadiabatic effects exist in the region of the coordinate space relevant for the spectral features. In this respect, particular care needs to be taken while applying approaches, like the vertical ones, which may be operated with a minimal exploration of the final-state PES. On the other side, more extended examinations of PES, such as those necessary to locate the excited-state minima, can bring to light issues that may otherwise remain unobserved. After that, for cases where relevant nonadiabatic effects exist, proper modeling of the spectroscopy of the system under investigation cannot be done without a multi–electronic state treatment. However, one should be aware that the disagreement between experimental and simulated spectra should not be attributed to nonadiabatic effects prior to exploring in detail the possible features of "single-state" vibronic models (e.g., full dimensionality). Beyond the possible nonadiabatic interactions, anharmonicity stands as an additional general factor to be taken into account in the spectroscopy of real systems. While a general route to its full treatment is out of reach for sizable systems, we will illustrate that it can be accounted for in a simplified manner in semirigid systems with nearly "diagonal" normal modes of the reference state, leading, nonetheless, to significant improvement of the simulated spectra accuracy [53]. In the following, we will present a few examples showing the advantages of modern computational spectroscopy approaches with respect to traditional spectroscopy models.

8.6.1.1 Absorption Spectra for Triatomic Systems Showing Up to Three-State Interactions Theoretical spectroscopy models based on full or effective Hamiltonians, which might be applied for cases with the simultaneous presence of RT and SO interactions, have been extensively reviewed [106, 132]. They might be set within variational or perturbative frameworks, which differ conceptually in the description of the possible anharmonic resonances. Effective Hamiltonians derived through perturbation theory require a priori the definition of all possible kinds of interactions [106]. The simplest models lie in Hougen's [133] theory of the Fermi resonances, which is based on the assumption that only terms differing by two quanta in the bending and one quanta in the stretching do interact. In fact, the picture is often more complicated and the relative magnitude between RT and SO interactions can lead to quite different energy-level patterns (see Figure 8.13), which in turn cause various types of interactions. Such a complex situation can be described using the variational approach and the full kinetic spin–rovibronic Hamiltonian, where all possible interactions are directly taken into account and the analysis of the anharmonic resonances is not limited to predefined cases. For triatomic radicals, such

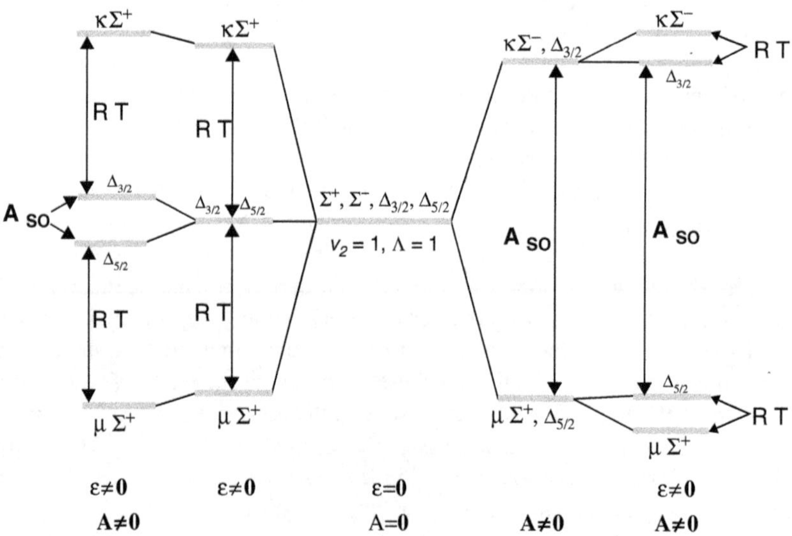

Figure 8.13 Schematic representation of spin–rovibronic energy level patterns caused by large/small RT and SO interactions. As an example splitting of spin–rovibronic energy levels related to the first bending quanta of triatomic molecule in a $^2\Pi$ electronic state ($v_2 = 1, \Lambda = 1$) are presented. The RT interaction leads to a splitting into the $\mu\Sigma^-$, Δ (doubly degenerate), and $\kappa\Sigma^+$ energy levels (the relative position of the two Σ levels correspond to the positive value of the RT parameter); the SO interaction leads to the splitting into $\frac{1}{2}$ and $\frac{3}{2}$ (doubly degenerate) spin components (the relative position corresponds to the positive value of A_{SO}). The simultaneous presence of the RT (ϵ is the RT parameter) and the SO interaction (A_{SO} is the spin–orbit coupling constant) leads to a complete splitting of the energy levels. Energy levels are labeled according to the notation introduced by Hougen [105] and described in detail in ref. 106.

computations can be performed using the Carter–Handy Hamiltonian [100] (given in Eq. 8.127) in conjunction with highly accurate multireference configuration interaction (MRCI) [134] computations for the description of the PES. Details on the variational approach are given elsewhere [2, 4].

Full-dimensional analytical adiabatic ab initio potential energy, SO couplings, and transition dipole moment surfaces may be obtained from the properties computed for selected geometry structures through fitting to a polynomial form in internal coordinates (R_1, R_2, θ) and taken into proper account. The diabatic PES and adiabatic couplings are obtained from MRCI [134] adiabatic PES using block diagonalization of the electronic Hamiltonian [135]. Fully variational computations based on the Carter–Handy Hamiltonian have been successfully performed for several molecules, while the accuracy of the computed energy levels [2, 4, 136, 137] and dipole-allowed transition intensities [4] has been shown from comparison with experimental data. Here, we will discuss the computational results for the HBN, HCP^+, and HBS^+ radicals analyzing the vibronic and RT interactions. Additionally, the $A^2B_2 \leftarrow X^2A_1$ absorption spectrum of the NO_2 molecule, which is complicated by the conical intersection between both electronic states, will be presented [101]. Theoretical analyses of anharmonic resonances in such complex cases are facilitated by examination of the total (e.g., vibronic, spin–rovibronic) wavefunctions, which can be represented by the cuts of vibrational parts corresponding to the two interacting electronic states as shown in Figure 8.14. In simultaneous RT and SO interactions the degeneracy of vibronic levels is removed. To ease the interpretation, we will adopt the labeling of spin–rovibronic energy levels as generally used for RT systems and

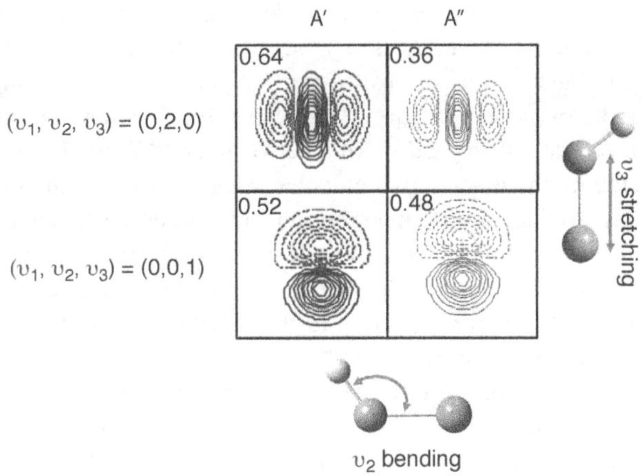

Figure 8.14 Total spin–rovibronic wavefunctions represented by plots of vibrational parts corresponding to two interacting electronic states. Shown are the cuts of the pure vibrational part along two internal coordinates, bending (x axis) and X–Y stretching (y axis), with the R_{HX} kept at the equilibrium value. Positive and negative lobes are ploted with continuous and dashed lines, respectively. The two electronic components (A' and A'') are plotted in adjacent panels with contributions from each electronic state to the total wavefunction shown in the upper right corner of each panel.

described in detail elsewhere [30, 106]. Here, we just note that the energy levels are labeled according to the remaining good quantum number describing the projection of the total spatial angular momentum excluding the spin.[3] It is shown in Figure 8.13 how the degenerate first bending quantum for a general molecule HXY showing RT effects is split due to the RT interaction into states of Σ and Δ symmetry,[4] while the SO coupling causes an additional splitting, leading to four states of different energy. For higher bending quanta, the energy level patterns are even more complex and additional complications might arise from the vibronic interactions with the third electronic state. For the HBN radical in the low-energy region, all levels belong to the doubly degenerate $X^2\Pi$ electronic state and so are affected by RT and SO interactions solely. In fact, all levels with $v_2 > 0$ are split in two components (μ and κ) due to the RT interaction, which is dominant over the SO effect. All levels belonging to the κ (upper) surfaces show also anharmonic interactions, as, for example, a very strong Fermi resonance between the $(v_1,0,1)\Pi$ and $(v_1,2,0)\kappa\Pi$ states presented in Figure 8.15. The potential energy surfaces as well as the crossing between the $X^2\Pi$ and $A^2\Sigma^+$ electronic states, more than 11,000 cm^{-1} above the minimum of the former, are also shown. However, some relevant vibronic interactions are present also at lower energies, with the most noticeable case represented by the interaction between, $A(0,0,2)\Sigma$ and $X\ (0,1,5)\mu\Sigma^+$ states.

To illustrate in more detail cases of anharmonic resonances, we show more suitable examples represented by HCP^+ and HBS^+ radicals along with their deuterated counterparts [138, 139]. Figure 8.16 illustrates an analysis of the interaction magnitude, which is related to the vibrational part of the total rovibronic wavefunction shape perturbation, as schematically represented by the color scale ranging from blue to red for weak to strong interactions, respectively. For example, it can be observed that for both molecular systems the $(0,2,0)\kappa\Pi_{1/2}$ and $(0,2,0)\mu\Pi_{1/2}$ energy levels for the hydrogenated and deuterated species, respectively, are essentially unperturbed, as represented by the two-nodal picture of the wavefunctions. On the other hand, for HBS^+, there is a clear half-to-half mixing between the $(0,2,0)\mu\Pi_{1/2}$ and $(0,0,1)\Pi_{1/2}$ states as depicted by the essentially equivalent shapes of wavefunction for the two energy levels involved. A similar interaction is also observed for HCP^+, but in this case the lower vibronic state shows a significant bending character while the higher one can be attributed to the stretching quanta. On the other hand, for both deuterated species, the vibronic levels are in general further apart on the energy scale; thus only weak interactions take place and all states have a clear bending or stretching character. It should be noted that all interactions depicted in Figure 8.16 are well defined by the standard Fermi interaction, $v_2 + 2v_3 = 2$, so they can also be effectively described by simplified models from Hougen's [133] theory. A different situation is depicted in Figure 8.17 where, for DCP^+, pair interactions can be observed while, for HCP^+, all energy levels belonging to the $v_2 + 2v_3 = 5$ Fermi polyad [except $(0,5,0)\kappa$] interacts.

[3] $\Lambda + l$, where l and Λ are projections of vibrational and orbital angular momenta with respect to the molecule axis, and Σ, Π, Δ energy levels correspond to $\Lambda + l = 1, 2, 3$, respectively, while v_1,v_2, and v_3 correspond to the H–X stretching, bending, and X–Y stretching respectively.

[4] The μ and κ components correspond to the higher and lower potential energy surfaces, respectively.

v	Δv_{so}	Assignment	
0.0	-42.5	$X(0,0,0)\Pi$	
645.8		$X(0,1,0)\mu\Sigma+$	
851.6		$X(0,1,0)\kappa\Sigma-$	
1328.4	-4.0	$X(0,2,0)\mu\Pi$	
1563.7	-15.5	$X(0,2,0)\kappa\Pi$	very strong Fermi resonance between $(v_1,0,1)\Pi$ and $(v_1,2,0)\kappa\Pi$ states
1641.1	-21.2	$X(0,0,1)\Pi$	very strong Fermi resonance between $(v_1,0,1)\Pi$ and $(v_1,2,0)\kappa\Pi$ states
1978.4		$X(0,3,0)\mu\Sigma+$	
2327.6		$X(0,3,0)\kappa\Sigma-$	weak Fermi resonance between $(v_1,1,1)\kappa\Sigma-$ and $(v_1,3,0)\kappa\Sigma^-$ states
2788.4	-42.5	$X(1,0,0)\Pi$	
3155.1	-30.2	$X(0,0,2)\Pi$	member of (v_2+2v_3)=2n Fermi polyad
4752.3	-24.3	$X(0,0,3)\Pi$	member of (v_2+2v_3)=2n Fermi polyad
5495.8	-42.5	$X(2,0,0)\Pi$	
6239.2		$X(0,0,4)\Pi$	member of (v_2+2v_3)=2n Fermi polyad
6389.9		$A(0,0,0)\Sigma$	
7030.4		$A(0,1,0)\Pi$	
7653.5		$A(0,2,0)\Sigma$	
7807.1	-23.2	$X(0,0,5)\Pi$	member of (v_2+2v_3)=2n Fermi polyad
8122.2	-42.4	$X(3,0,0)\Pi$	
8160.3		$A(0,0,1)\Sigma$	
8285.1		$A(0,3,0)\Pi$	
9196.9		$A(1,0,0)\Sigma$	
9263.3		$X(0,0,6)\Pi$	member of (v_2+2v_3)=2n Fermi polyad
9821.4		$A(1,1,0)\Pi$	weak vibronic interaction between $A(1,1,0)\Pi$ and $X(2,6,0)\Pi$ states.
9917.8		$A(0,0,2)\Sigma$	strong vibronic interaction between $A(0,0,2)\Sigma$ and $X(0,1,5)\mu\Sigma^+$ states

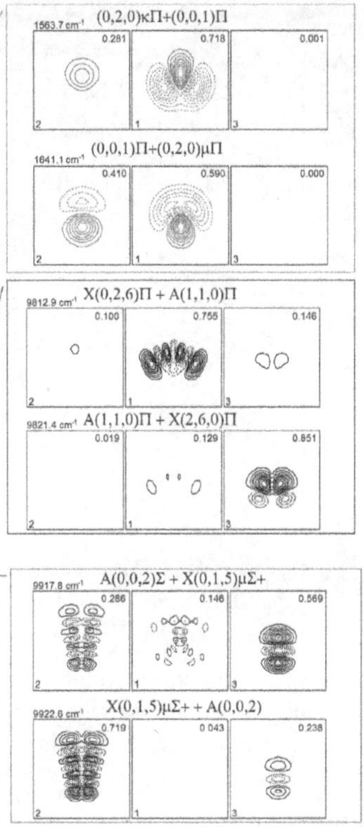

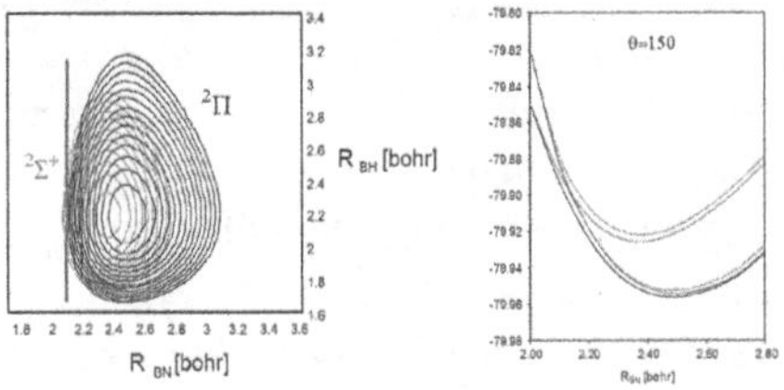

Figure 8.15 HBN radical: potential energy surfaces and selected energy levels of Σ and Π symmetry. Adiabatic (red, black) and diabatic (violet, magenta) PES. 2D cut at linear geometries, crossing seam shown as a green line; 1D cuts along BN stretching coordinate for R_{BH} and θ fixed at 2.1 Å and 150°, respectively. The component of the A″ symmetry of the $^2\Pi$ state shown as a blue line. Assignment based on plots and expansion coefficients of vibrational part of wavefunctions. All values in cm^{-1}; levels showing resonances are marked.

In this latter case, a description based on Hougen's theory with the use of the simple effective Hamiltonian is insufficient, and a good agreement with experiment requires more sophisticated and less limited models. On the other hand, for DCP^+, due to essentially pair interactions, a phenomenological treatment has shown to be suitable to obtain reliable results [139]. Similarly, the conical-intersection effects on the absorption spectrum of NO_2 [101] can be better understood analyzing the vibronic transitions in terms of the individual nonadiabatic states, their main BO components, and the A^2B_2 Fermi polyads. It has been shown that the intensity distribution in different spectral regions depends strongly on the X^2A_1/A^2B_2 electronic interaction and the vibrational resonances in the A^2B_2 state. For the energies below 13,000 cm^{-1}, the spectrum is dominated by weak cold and hot vibrational transitions related to the X^2A_2 electronic state. Above this limit, the absorption structure is dominated by A^2B_2 combination progressions which show a clear pattern up to 16,000 cm^{-1} and can be

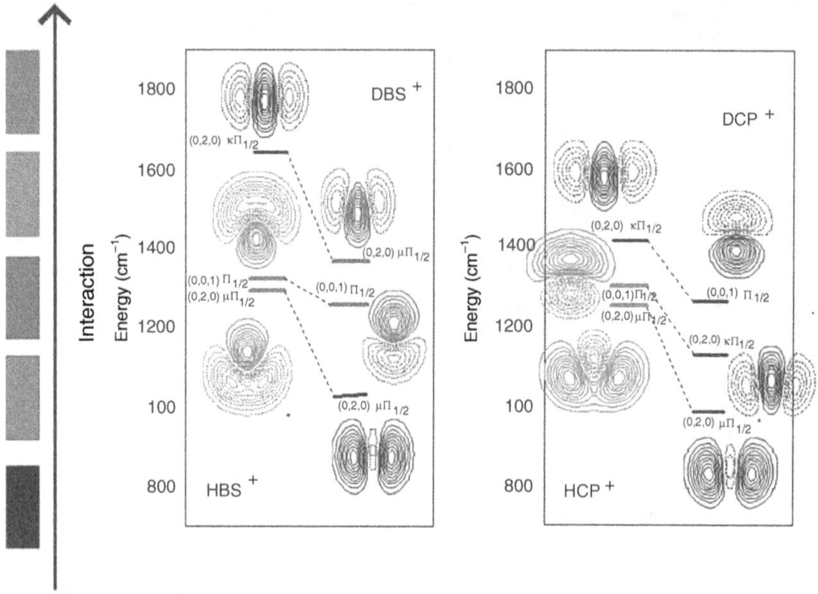

Figure 8.16 Vibronic wavefunction as tool to analyze nonadiabatic interactions: interaction magnitude. Shown are the cuts of the pure vibrational part along two internal coordinates, bending (x axis) and X–Y stretching (y axis). (See online version for color figure.)

analyzed in terms of $2v_1 + v_2 = n$ Fermi polyads. In the higher energy window, vibronic states begin to mix and the bands cannot be assigned in such a simple manner as the A^2B_2 polyads lose their physical meaning. In summary, it can be concluded that simplified models based on the perturbative description of anharmonic resonances might not be sufficient in some cases. Additionally, for molecules showing very large SO splitting, the standard assumption that SO coupling is independent of the molecular geometry may be inadequate. In those cases, reliable computational studies require a further extension of the model taking into account explicitly the geometry dependence of the SO coupling. Such a refined treatment has been performed, for instance, for the $BrCN^+$ radical [140] whose $X^2\Pi$ electronic state is characterized by a very large SO splitting.

8.6.1.2 The $S_1 \leftarrow S_0$ Electronic Transition of Anisole The recently published computational study on the vibrational structure of the absorption spectrum of the $S_1 \leftarrow S_0$ electronic transitions of anisole [53] represents an example of the accuracy achievable when time-independent simulations of vibronic spectra are coupled to good-quality ab initio computations for geometries and force fields in both electronic states. For anisole, methods based on the density functional theory and its time-dependent extension for electronic excited states [B3LYP/6-311 + G(d,p) and TD-B3LYP/6-311 + G(d,p)] have been applied to perform geometry optimizations and harmonic frequency calculations, while the energy of the electronic transition has been refined by EOM-CCSD/CCSD//aug-cc-pVDZ computations. The remarkable

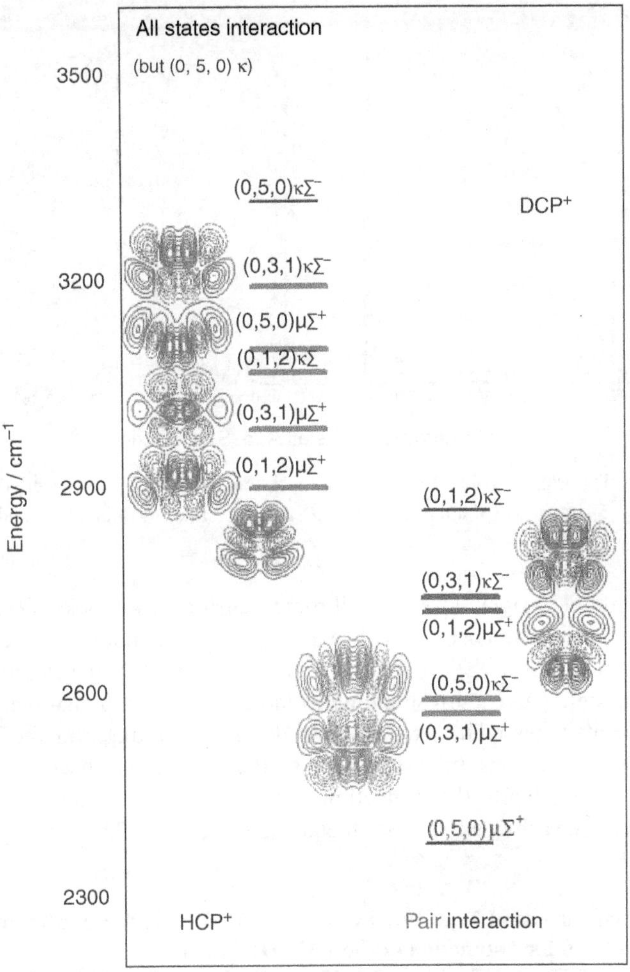

Figure 8.17 Vibronic wavefunction as tool to analyze nonadiabatic interactions: type of interactions. Shown are the cuts of pure vibrational part along two internal coordinates, bending (*x* axis) and X–Y stretching (*y* axis).

overall agreement between theoretical and experimental [141] rotational constants (with an average deviation of about 0.5% for both electronic states) confirms the good quality of the calculated geometry structures. The vibrational frequencies in the first excited electronic state have been corrected according to the ground-state experimental frequencies (EA) or to the calculated perturbative anharmonic frequencies [126] (TA), and the spectrum computations converged up to 99.6% of the total analytical spectrum intensity, that is, the zero-order moment. The simulated vibronic profile convoluted with a Gaussian distribution function (FWHM of $2\,\text{cm}^{-1}$) is compared to the highly accurate experimental spectrum from the resonance-enhanced multiphoton ionization (REMPI) simulation [142] reported in Figure 8.18.

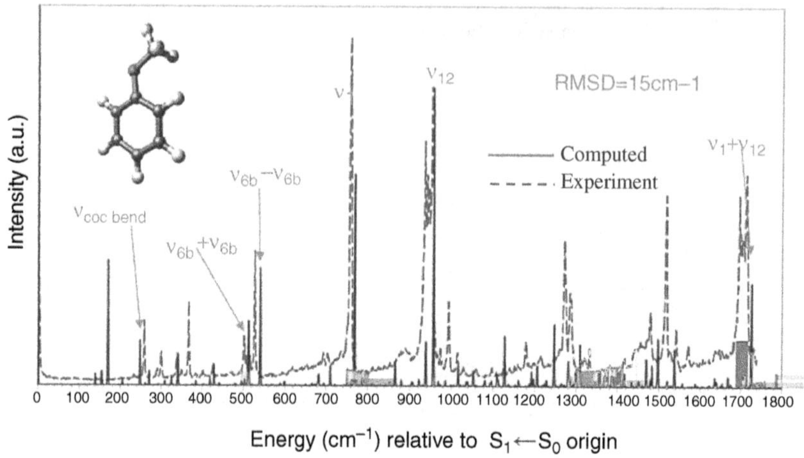

Figure 8.18 Theoretical (blue lines) and experimental REMPI [142] (red dashed lines) spectra of $S_1 \leftarrow S_0$ transition of anisole along with assignment of most intense bands; see ref. 53 for details.

To carry out the assignment of the vibronic transitions, we took advantage of the Duschinsky relation given in Eq. 8.51 to express the normal modes of the excited state as linear combinations of the ground-state ones. Table 8.3 lists the experimental and computed vibronic bands corresponding to the most intense transitions along with their assignment expressed in terms of the modes of S_1 resulting from the Duschinsky rotation of the S_0 modes (please note that the coefficients used for the linear combinations are actually the projection elements, J^2_{kl}). On balance, a very good agreement has been achieved (the root-mean-square deviation between computed and

Table 8.3 Assignments of Most Intense Bands of $S_1 \leftarrow S_0$ Electronic Transitions of Anisole Considered for Estimation of the RMS Deviation

Experiment[a]	$\bar{\nu}_{TA}$	$\bar{\nu}_{EA}$	Assignment[b]
259	249	256	$\{\nu_{\text{COCbendinq}}\}$
501	512	509	$\{-0.61\nu_{6b} - 0.39\nu_{6a}\}$
527	540	538	$\{-0.60\nu_{6a} + 0.38\nu 6b\}$
759	769	772	$\{\nu_1\}$
937[c]	922	922	$\{\nu_{17a}\} + \{\cdots\}$
943[c]	938	941	$\{\nu_{17a}\} + \{\cdots\}$
948[c]	—	—	Combination
954[c]	957	960	$\{-0.70\nu_{12}\}$
1713[d]	1726	1732	$\{\nu_1\} + \{\nu_{12}\}$

Note: All energies are relative to the 0–0 origin. Frequencies are in cm^{-1}.
[a] Experimental data from ref. 142.
[b] In parentheses the fundamental modes of S_1 resulting from Duschinsky rotation between S_0 vibrations.
[c] Average of four bands considered for estimation of root-mean-square (RMS) deviation.
[d] Average of 1696 and 1713 cm^{-1} considered for estimation of RMS deviation.

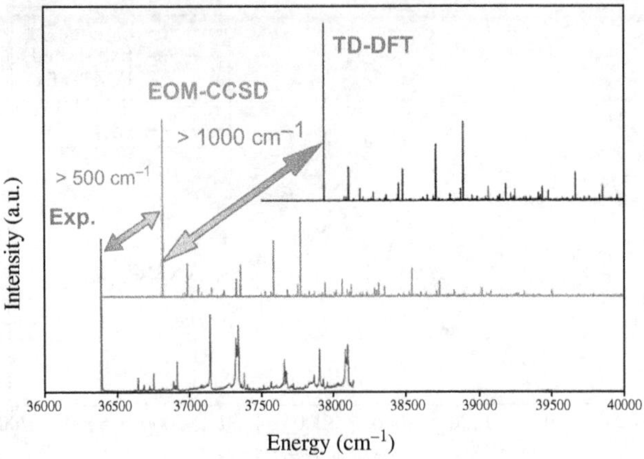

Figure 8.19 Theoretical and experimental REMPI [142] spectra of $S_1 \leftarrow S_0$ transition of anisole expressed as function of absolute energy (cm^{-1}). (See ref. 53 for details.)

experimental bands is 15 cm^{-1}), as detailed by Bloino et al. [53]. In order to correctly reproduce the band intensities and the rich vibrational structure of the REMPI spectra, it has been necessary to account for the changes in structure, vibrational frequencies, and normal modes between the involved electronic states. It is worthwhile highlighting that the remarkable overall agreement, even for the band positions, has only been possible by correcting the vibrational frequencies for anharmonicity.

The discrepancy between the absolute position of the experimental and simulated spectra shown in Figure 8.19 remains the main shortcoming of the purely theoretical approach: To achieve a good match between computed and experimental spectra, the energy of the electronic transition should be computed with an accuracy of $\approx$10 cm^{-1}. The DFT/TD-DFT computations are able to provide quite reasonable estimates of the relative energetics of the electronic states (within 0.2 eV), which can be further refined based on coupled cluster calculations (0.05 eV). But, for the purpose of the assignment of vibronic bands in such highly accurate spectrum, it is compulsory to compare the spectra shifted to the 0–0 origin. In such a way the remarkable agreement between theoretical and experimental spectra allowed for the revision of some experimentally observed vibronic transitions. Specifically, for many bands that had been assigned to S_1 fundamentals, consideration of the relative intensities has suggested instead a different interpretation in terms of combinations or overtones [53]. In conclusion, we have shown that the approach presented here is more reliable than a comparison based purely on computed frequencies, and represents a valuable tool for the interpretation of experimental results.

8.6.1.3 The $A^2B_1 \leftarrow \tilde{X}^2A_1$ Electronic Transition of Phenyl Radical The analysis of the $A^2B_1 \leftarrow \tilde{X}^2A_1$ electronic transition of phenyl radical [143] allows us to present a critical comparison of the interpretative potential of full-dimensional vibronic models with respect to more approximate approaches. Fully converged (>99%) vibronic spectra computed within the FCHT|AH framework, corrected for anharmonicity and

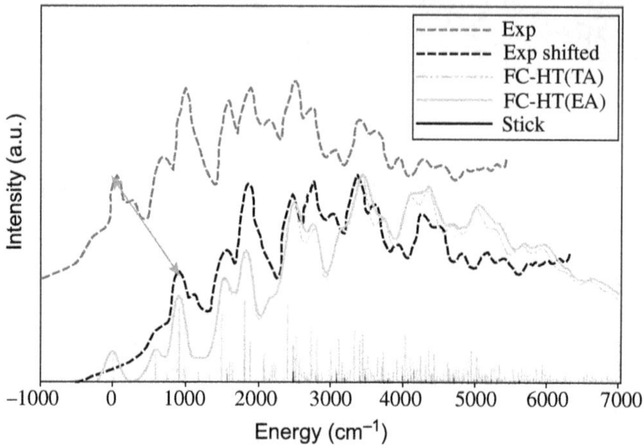

Figure 8.20 Theoretical, convoluted, and stick FCHT|AH spectrum of $A^2B_1 \leftarrow \tilde{X}^2A_1$ electronic transition of phenyl is compared to experimental spectrum and its counterpart arbitrarily shifted along energy axis [144].

convoluted using a Gaussian function with a FWHM of 200 cm^{-1}, are compared to the recently published experimental data [144] in Figure 8.20.

It can be observed that the simulated spectrum reproduces correctly the main features of its experimental counterpart while the most striking difference is a shift between the computed and simulated spectra on an energy scale relative to the 0–0 transition. It is worth recalling that the part of the spectrum close to the origin is very weak and the measurements were taken close to the performance limit of the spectrometer further enhanced by application of a multiple-pass technique, as described in detail elsewhere [144]. Thus, the reason for the discrepancy can be traced back to the weakness of the 0–0 transition. A closer look at the experimental spectrum reveals a weak progression preceding the first intense band, formerly assigned to the electronic transition origin. Actually, comparison with the theoretical spectrum suggests that the transition assigned as 0–0 may be already the result of a transition to an excited vibrational state of A^2B_1. Further support to this hypothesis is given by comparison with the experimental spectrum shifted by 873 cm^{-1} to match the transition origin reported elsewhere [145] and shown in Figure 8.20. It is remarkable that in such a case a very good agreement, also for the band position of the most intense transitions between experiment and simulated spectra, is observed.

It should be noted that a number of theoretical attempts to interpret the electronic spectra of the phenyl radical at the theoretical level have been done before [143], albeit purely based on reduced-dimensionality vibronic models. As an example, Kim et al. [146] computed the theoretical spectrum from analytical FC integrals assuming that only some modes could contribute to the spectrum and disregarding the possibility of mode mixing. The 12 "active" normal modes (11 of a_1 symmetry and 1 of b_1 symmetry) have been chosen "manually" by analyzing the ground- and excited-state normal modes obtained from ab initio computations. The resulting theoretical spectrum for the $A^2B_1 \leftarrow \tilde{X}^2A_1$ electronic transition showed only some weak features, except for the 0–0 origin, at variance with the rich vibronic structure of the experimental one. Such a discrepancy was attributed to the existence of another nonplanar local minimum characterized by a larger geometry relaxation, thus inducing a richer vibronic structure. Alternatively, it was suggested that a symmetry-forbidden transition to a close dark state ($B^2A_1 \leftarrow \tilde{X}^2A_1$) could contribute through vibronic coupling. The latter possibility has been recently investigated by the theoretical simulation of photodetachment spectra of phenide ion [147], which allows us to probe dark states not accessible by optical transitions. A theoretical photoelectron spectrum has been computed with a quadratic vibronic Hamiltonian, taking into account the A^2B_1/B^2A_1 coupling and including 15 modes, and compared to the UV absorption spectrum from Radziszewski [144]. The effect of the coupling to the dark state has been investigated [147] by comparing the vibronic profile obtained considering and neglecting the vibronic coupling with the experimental data without including nonadiabatic effects. The authors concluded that the comparison of coupled and uncoupled results points out a strong impact of nonadiabatic coupling, especially for the high-energy wing of the A^2B_1 band and the entire B^2A_1 band.

However, as discussed above, the implementation of full-dimensional single-state FCHT|AH computations leads to very good agreement between experimental and theoretical spectra. Additionally, the most intense transitions are indeed related to only a few vibrations, all of them of a_1 symmetry, as suggested by Kim et al. [146]. As a consequence, the significant discrepancy observed between simulated spectra from Biczysko et al. [143] and the almost featureless one from Kim et al. [146] must be attributed to the simplified model with a reduced dimensionality. It should also be pointed out that the Duschinsky mixing is not negligible and needs to be considered to account accurately for the spectrum shape, but neglecting the Duschinsky rotation is not the main source of discrepancies, as confirmed by FC|AS computations also reported by Biczysko et al. [143]. Notwithstanding the above conclusions, the possible existence of a vibronic coupling with the dark B^2A_1 electronic state, which could influence the $A^2B_1 \leftarrow \tilde{X}^2A_1$ transition, as postulated elsewhere [146, 147], cannot be excluded. Nevertheless, it seems evident that full-dimensional vibronic models are necessary to correctly reproduce the spectrum shape and should be fully exploited prior to analyzing the possible role of nonadiabatic effects.

8.6.2 Spectra for Larger Systems of Biological or Technological Interest

In this last section, we present some examples of computational spectroscopy studies for complex biological systems or nanomaterials of direct technological interest. Specifically, we will show the results for the UV spectrum of chlorophyll *a* in methanol solution and the investigation of the electronic and optical properties of organic light-emitting diodes (OLEDs). Recent developments in computational approaches, together with the increased computational resources, nowadays allow studies at the QM level of both ground and excited states, even for systems as large as those reported here. Due to this breakthrough, it may be expected that QM computations of optical properties combined with spectroscopic experiments will contribute to shedding further light on the understanding of the molecular mechanism of light harvesting in photosystem II [148] or photophysical properties of OLEDs and light-emitting electrochemical cells (LECs) [149–151].

8.6.2.1 UV Spectra of Chlorophyll The UV spectrum of chlorophyll *a* [152, 153] has been modeled from chlorophyll a_1, a large molecule with 46 atoms and 132 normal modes. For such a system, fully QM simulations of vibronic spectra within the FC|AH or FCHT|AH frameworks might be possible but are computationally demanding, in particular if large energy windows, encompassing several electronic transitions, need to be studied. On the other hand, the computational cost can be significantly reduced resorting to the FC|VG approach that allows a relatively cheap and straightforward computation of low-resolution electronic spectra for large molecules in the gas phase and in solution. As discussed in Section 8.3.2.1, the application of this model is better suited to simulations of the overall features of the spectra, that is, their shape, than the fine details of the latter, so its predictions are more appropriate to analyze experimental UV–vis spectra recorded in condensed phase at room temperature.

In the present example, the electronic QM computations have been performed with the DFT/N07D model while the effect of the methanol solvent has been included by means of the polarizable continuum model, where the solvent is represented by a homogeneous dielectric polarized by the solute, placed within a cavity built as an envelope of spheres centered on the solute atoms [154] (see Chapter 1 for details). The solvent has been described in the nonequilibrium limit where only its fast (electronic) degrees of freedom are equilibrated with the excited-state charge density while the slow (nuclear) degrees of freedom remain equilibrated with the ground state. This assumption is well suited to describe the broad features of the absorption spectrum in solution due to the different time scales of the electronic and nuclear response components of the solvent reaction field [89].

The simulated spectra have been computed in the gas phase and the methanol solvent, and the complete spectrum in the 250–700-nm range was obtained by summation of the contributions from transitions to the first eight singlet excited electronic states. Both computed spectra are compared to the experimental data recorded in the methanol solution [152, 153] in Figure 8.21. It is immediately visible that, while both computed spectra reproduce qualitatively the lineshape of their experimental counterpart, a much better agreement, in particular for the absolute positions of vibronic bands, has been obtained for the one simulated in methanol. In fact, for the latter, a uniform 200-cm^{-1} shift on the energy scale leads to a very good quantitative agreement with experiment, and such a shifted spectrum is also shown. Additionally, it is possible to analyze individual contributions of single-state transitions to the spectrum lineshape. For chlorophyll a_1, the spectrum lineshape is dominated by the contributions from transitions to the S_1, S_3, and S_4 excited electronic states, with the nonnegligible contributions from transitions to S_2 and S_8. Overall it can be concluded that a reliable spectrum lineshape including vibrational and environmental effects can be simulated by the simple FC|VG approach combined with relatively inexpensive studies at the TD-DFT/DFT/N07D//CPCM level.

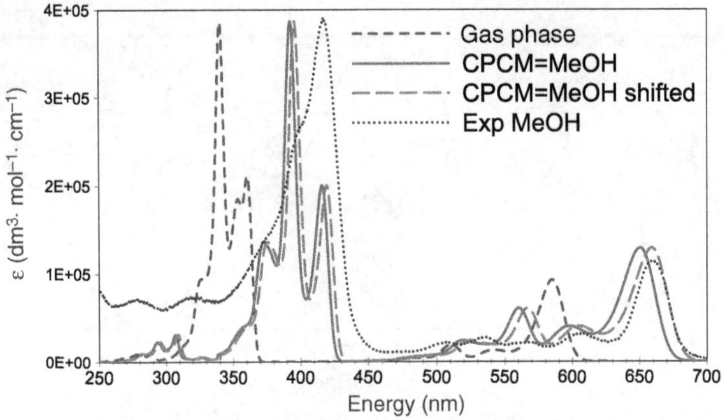

Figure 8.21 Absorption (FC|VG) spectrum of chlorophyll a_1 in 250–700-nm energy range, sum and single contributions of transitions to eight first singlet electronic states, simulated in methanol solution and compared to experimental data obtained in methanol solvent [152, 153].

8.6.2.2 Theoretical Prediction of Emission Color in Phosphorescent Iridium(III) Complexes Theoretical simulations of optical properties may lead to the in silico design of new materials with predetermined emission properties. In this respect, it is particularly important to take into account the vibrational structure of the electronic spectra, a task made possible by the effective theoretical approaches described in this chapter. The improvement from the purely electronic picture is crucial since the electronic spectra bandshape ultimately determines the color perceived by the human eye. Based on this observation, we present the simulation of the vibrationally resolved electronic spectra of two prototype cationic Ir(III) complexes showing high emission quantum efficiencies [155]. Considering the frontiers of modern technology, Ir(III) complexes are attracting widespread interest due to their high stability, emission color tunability, and strong SO coupling, leading to improved quantum efficiency of light-emitting devices. Figure 8.22 compares the theoretical results (with the single empirical adjustment of the uniform blue shift of 0.24 eV applied to both complexes) to experimental data, clearly showing a very good agreement between the experimental and computed bandshapes. The computational/experimental agreement can also be evaluated by comparing the emission color in terms of the CIE color coordinates defined by the International Commission on Illumination (Commission internationale de l'éclairage, CIE), which can be obtained calculating the spectral overlap with the standard CIE red, green, and blue color matching functions [156]. The CIE coordinates (not accessible by definition with mere electronic calculations) computed for both complexes are reported in Figure 8.22 along with their experimental counterparts, showing that the computational study was able to reproduce quantitatively the difference in emitting color between the two Ir(III) complexes, highlighting the predictive capability of applied integrated theoretical approach.

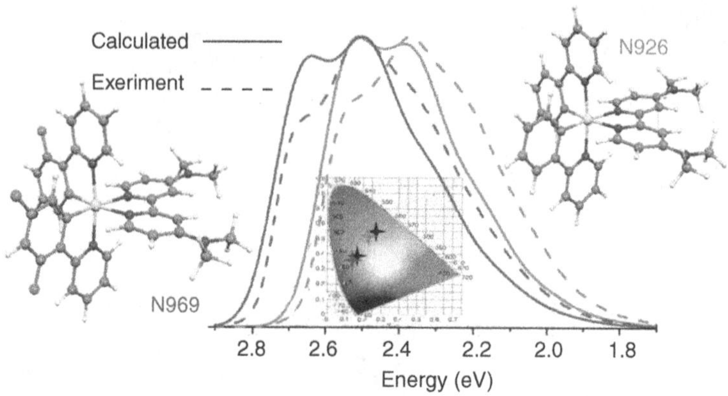

Figure 8.22 Calculated (solid lines) and experimental (dashed lines) emission spectra for N969 (blue) and N926 (red), along with corresponding calculated (blue/red stars) and experimental (black stars) CIE coordinates for N969 and N926.

8.7 CONCLUSIONS

Several theoretical approaches to compute vibrationally resolved electronic spectra have been presented along with their possible applications for the study of a variety of molecular systems. Fully dimensional anharmonic models to compute spectra beyond BO approximation have been discussed using examples of small molecular systems, along with applicability of effective schemes set within "single-state" harmonic farmeworks for large systems. It should be noted that particular computational tools based on the a priori selection schemes, despite being tailored for large systems, can be utilized as well to generate high-quality spectra for small systems when nonadiabatic and anharmonic couplings are negligible. Additionally, it needs to be stressed that the availability of easy-to-use, general, and robust computational tools able to simulate good-quality spectra even for large systems with hundreds of normal modes, whenever harmonic approximation is reliable, paves the way to spectroscopic studies of systems of direct biological and/or technological interest, improving their interpretation and understanding.

ACKNOWLEDGMENTS

This work was supported by the Italian MIUR and IIT (Project Seed HELYOS). The large-scale computer facilities of the M^3-VILLAGE network (http://m3village.sns.it) are kindly acknowledged.

REFERENCES

1. P. Jensen, P. R. Bunker, *Computational Molecular Spectroscopy*, Wiley, Chichester, 2000.
2. S. Carter, N.C. Handy, C. Puzzarini, R. Tarroni, P. Palmieri, *Mol. Phys.* **2000**, *98*, 1697.

3. M. Biczysko, R. Tarroni, S. Carter, *J. Chem. Phys.* **2003**, *119*, 4197.
4. R. Tarroni, S. Carter, *Mol. Phys.* **2004**, *102*, 2167.
5. V. Barone, J. Bloino, M. Biczysko, F. Santoro, *J. Chem. Theory Comput.* **2009**, *5*, 540.
6. J. Bloino, M. Biczysko, F. Santoro, V. Barone, *J. Chem. Theory Comput.* **2010**, *6*, 1256.
7. M. J. Frisch et al., Gaussian 09 Revision A.2, Gaussian, Wallingford, CT, 2009.
8. F. Santoro, *FCclasses*, a Fortran 77 code, 2008 webpage, available: http://village.pi.iccom.cnr.it, accessed July 12, 2011.
9. H.-C. Jankowiak, J. Huk, R. Berger, hotFCHT 2010, webpage: http://fias.uni-frankfurt.de/berger/group/hotFCHT/index.html, accessed May 31, 2010.
10. M. Dierksen, FCFAST 2005 Version 1.0.
11. R. Borrelli, A. Peluso, MolFC, 2010 webpage, available: http://www.theochem.unisa.it/molfc.html, accessed May 31, 2010.
12. K. M. Ervin, PESCAL, 2010 webpage, available: http://wolfweb.unr.edu/~ervin/pes/, accessed May 31, 2010.
13. M. Nooijen, A. Hazra, VIBRON, a program for vibronic coupling and Franck-Condon calculations, with contributions from J. F. Stanton and K. Sattelmeyer, University of Waterloo, Waterloo, Ontario, Canada, 2003, http://www.science.uwaterloo.ca/~nooijen/Links/download.html, accessed June 13, 2010.
14. H. Meyer, *Annu, Rev. Phys. Chem.* **2002**, *53*, 141.
15. B. R. Sutcliffe, in *Current Aspects of Quantum Chemistry*, Vol. 21, R. Carbo, Ed., Elsevier, 1982.
16. M. J. Bramley, W. H. J. Green, N. C. Handy, *Mol. Phys.* **1991**, *73*, 1183.
17. L. S. Cederbaum, Born-Oppenheimer approximation and beyond, in *Conical Intersections, Electronic Structure, Dynamics and Spectroscopy*, W. Domcke, R. Yarkony, H. Köppel, Eds. World Scientific, Singapore 2004, pp. 3–40.
18. C. Eckart, *Phys. Rev.* **1937**, *47*, 552.
19. C. Petrongolo, *J. Chem. Phys.* **1988**, *89*, 1287.
20. F. Santoro, C. Petrongolo, G. Schatz, *J. Phys. Chem. A* **2002**, *106*, 8276.
21. H. Köppel, Diabatic representation: Methods for the construction of diabatic electronic states, in *Conical Intersections, Electronic Structure, Dynamics and Spectroscopy*, W. Domcke, R. Yarkony, H. Köppel, Eds., World Scientific, Singapore, 2004, pp. 175–204.
22. H. Köppel, W. Domcke, L. Cederbaum, The multi-mode vibronic-coupling approach, in *Conical Intersections, Electronic Structure, Dynamics and Spectroscopy*, W. Domcke, R. Yarkony, H. Köppel, Eds., World Scientific, Singapore, 2004, pp. 323–368.
23. IDEA: In-silico developments for emerging applications, available: http://idea.sns.it, accessed January 29, 2010.
24. N. Lin, F. Santoro, Y. Luo, X. Zhao, V. Barone, *J. Phys. Chem. A* **2002**, *113*, 4198.
25. D. A. Long, *The Raman Effect: A Unified Treatment of the Theory of Raman Scattering by Molecules*, Wiley, Chichester, 2002.
26. J. Franck, *Trans. Faraday Soc.* **1926**, *21*, 536.
27. E. U. Condon, *Phys. Rev.* **1928**, *32*, 858.
28. G. Herzberg, E. Teller, *Z. Phys. Chem. B Chem. E* **1933**, *21*, 410.
29. G. Orlandi, W. Siebrand, *J. Chem. Phys.* **1973**, *58*, 4513.

30. G. Herzberg, *Molecular Spectra and Molecular Structure: III Electronic Spectra and Electronic Structure of Polyatomic Molecules*, D. Van Nostrand, New York, 1966.
31. F. Santoro, A. Lami, R. Improta, J. Bloino, V. Barone, *J. Chem. Phys.* **2008**, *128*, 224311/1–17.
32. N. Lin, F. Santoro, X. Zhao, A. Rizzo, V. Barone, *J. Phys. Chem. A* **2008**, *112*, 12401.
33. F. Santoro, V. Barone, *Int. J. Quant. Chem.* **2010**, *110*, 476.
34. F. Duschinsky, *Acta Physicochim. URSS* **1937**, *7*, 551.
35. N. J. D. Lucas, *J. Phys. B At. Mol. Phys.* **1973**, *6*, 155.
36. I. Özkan, *J. Mol. Spectrosc.* **1990**, *139*, 147.
37. C. Zhixing, *Theor. Chem. Acc.* **1989**, *75*, 481.
38. E. Coutsias, C. Seok, K. Dill, *J. Comput. Chem.* **2004**, *25*, 1849.
39. T. E. Sharp, H. M. Rosenstock, *J. Chem. Phys.* **1963**, *41*, 3453.
40. R. Islampour, M. Dehestani, S. H. Lin, *J. Mol. Spectrosc.* **1999**, *194*, 179.
41. A. M. Mebel, M. Hayashi, K. K. Liang, S. H. Lin, *J. Phys. Chem. A* **1999**, *103*, 10674.
42. H. Kikuchi, M. Kubo, N. Watanabe, H. Suzuki, *J. Chem. Phys.* **2003**, *119*, 729.
43. J. Liang, H. Li, *Mol. Phys.* **2005**, *103*, 3337.
44. J.-L. Chang, *J. Chem. Phys.* **2008**, *128*, 174111/1–10.
45. H. Kupka, P. H. Cribb, *J. Chem. Phys.* **1986**, *85*, 1303.
46. L. S. Cederbaum, W. Domcke, *J. Chem. Phys.* **1976**, *64*, 603.
47. E. V. Doktorov, I. A. Malkin, V. I. Man'ko, *J. Mol. Spectrosc.* **1977**, *64*, 302.
48. P. T. Ruhoff, *Chem. Phys.* **1994**, *186*, 355.
49. P.-Å. Malmqvist, N. Forsberg, *Chem. Phys.* **1998**, *228*, 227.
50. R. Borrelli, A. Peluso, *J. Chem. Phys.* **2008**, *129*, 064116.
51. P. Macak, Y. Luo, H. Ågren, *Chem. Phys. Lett.* **2000**, *330*, 447.
52. A. Hazra, H. H. Chang, M. Nooijen, *J. Chem. Phys.* **2004**, *121*, 2125/1–12.
53. J. Bloino, M. Biczysko, O. Crescenzi, V. Barone, *J. Chem. Phys.* **2008**, *128*, 244105/1–15.
54. M. de Groot, W. J. Buma, *Chem. Phys. Lett.* **2007**, *435*, 224.
55. M. de Groot, W. J. Buma, *Chem. Phys. Lett.* **2006**, *420*, 459.
56. M. Dierksen, S. Grimme, *J. Chem. Phys.* **2004**, *120*, 3544/1–11.
57. I. Pugliesi, N. M. Tonge, K. E. Hornsby, M. C. R. Cockett, M. J. Watkins, *Phys. Chem. Chem. Phys.* **2007**, *9*, 5436.
58. N. M. Tonge, E. C. MacMahon, I. Pugliesi, M. C. R. Cockett, *J. Chem. Phys.* **2007**, *126*, 154319/1–11.
59. S. H. Lin, H. Eyring, *Proc. Natl. Acad. Sci. USA* **1974**, *71*, 382.
60. J. Neugebauer, B. A. Hess, *J. Chem. Phys.* **2004**, *120*, 11564.
61. L. Peticolas, T. Rush III, *Comput. Chem.* **1995**, *16*, 1261.
62. A. Warshel, P. Dauber, *J. Chem. Phys.* **1977**, *66*, 5477.
63. E. J. Heller, R. L. Sundberg, D. Tannor, *J. Phys. Chem.* **1982**, *86*, 1822.
64. E. J. Heller, S.-Y. Lee, *J. Chem. Phys.* **1979**, *71*, 4777.
65. N. Lin, F. Santoro, A. Rizzo, Y. Luo, X. Zhao, V. Barone, *J. Phys. Chem. A* **2009**, *113*, 4198.
66. C. Toro, L. De Boni, N. Lin, F. Santoro, A. Rizzo, F. E. Hernandez, *Chem. Eur. J.* **2010**, *16*, 3504.

67. M. Lax, *J. Chem. Phys.* **1952**, *20*, 1752.
68. J. J. Markham, *Rev. Mod. Phys.* **1959**, *31*, 956.
69. N. R. Kestner, J. Logan, J. Jortner, *J. Phys. Chem.* **1974**, *78*, 2148.
70. F. Santoro, A. Lami, R. Improta, V. Barone, *J. Chem. Phys.* **2007**, *126*, 184102/1–11.
71. R. C. O'Rourke, *Phys. Rev.* **1953**, *91*, 65.
72. R. A. Marcus, *J. Chem. Phys.* **1965**, *43*, 1261.
73. R. A. Marcus, *J. Chem. Phys.* **1963**, *38*, 335.
74. R. A. Marcus, *J. Chem. Phys.* **1963**, *39*, 1734.
75. R. A. Marcus, *J. Phys. Chem.* **1989**, *93*, 307.
76. D. Gruner, P. Brumer, *Chem. Phys. Lett.* **1987**, *138*, 310.
77. D. E. Knuth, *Art of Computer Programming,* Vol. 1: *Fundamental Algorithms*, 3rd ed., Addison-Wesley, Reading, MA, 1997.
78. N. Wirth, *Algorithms and Data Structures*, Prentice-Hall, Englewood Cliffs, NJ, 1985.
79. P. T. Ruhoff, M. A. Ratner, *Int. J. Quant. Chem.* **2000**, *77*, 383.
80. A. Toniolo, M. Persico, *J. Comp. Chem.* **2001**, *22*, 968.
81. A. Hazra, M. Nooijen, *Int. J. Quant. Chem.* **2003**, *95*, 643.
82. M. Dierksen, S. Grimme, *J. Chem. Phys.* **2005**, *122*, 244101/1–9.
83. S. Schumm, M. Gerhards, K. Kleinermanns, *J. Phys. Chem. A* **2000**, *104*, 10645.
84. M. J. H. Kemper, J. M. F. Van Dijk, H. M. Buck, *Chem. Phys. Lett.* **1978**, *53*, 121.
85. T. Beyer, D. F. Swinehart, *Commun. Assoc. Comput. Machin.* **1973**, *16*, 379.
86. R. Berger, C. Fischer, M. Klessinger, *J. Phys. Chem. A* **1998**, *102*, 7157.
87. K. M. Ervin, T. M. Ramond, G. E. Davico, R. L. Schwartz, S. M. Casey, W. C. Lineberger, *J. Phys. Chem. A* **2001**, *105*, 10822.
88. H.-C. Jankowiak, J. L. Stuber, R. Berger, *J. Chem. Phys.* **2007**, *127*, 234101/1–23.
89. F. Santoro, R. Improta, A. Lami, J. Bloino, V. Barone, *J. Chem. Phys.* **2007**, *126*, 084509/1–13.
90. E. Hutchisson, *Phys. Rev.* **1930**, *36*, 410.
91. E. Hutchisson, *Phys. Rev.* **1931**, *37*, 45.
92. W. Smith, *J. Phys. B* **1969**, *2*, 1.
93. K. M. Ervin, J. Ho, W. C. Lineberger, *J. Phys. Chem.* **1988**, *92*, 5405.
94. M. Dierksen, S. Grimme, *J. Phys. Chem. A* **2004**, *108*, 10225.
95. J. Liang, H. Li, *Chem. Phys.* **2005**, *314*, 317.
96. P. T. Ruhoff, *Chem. Phys.* **1994**, *186*, 355.
97. S. Coriani, T. Kjœrgaard, P. Jørgensen, K. Ruud, J. Huh, R. Berger, *J. Chem. Theory Comput.* **2010**, *6*, 1028.
98. J. S. Huh, H.-C. Jankowiak, J. L. Stuber, R. Berger, private communication.
99. I. A. Malkin, V. I. Man'ko, D. A. Trifonov, *J. Math. Phys.* **1973**, *14*, 576.
100. S. Carter, N. C. Handy, P. Rosmus, G. Chambaud, *Mol. Phys.* **1990**, *71*, 605.
101. F. Santoro, C. Petrongolo, Lanczos calculation of the X/A nonadiabatic Franck-Condon absorbtion spectrum of NO_2, in *Advances in Quantum Chemistry*, Vol. 36, Academic, New York, 2000, pp. 323–339.
102. P. Jensen, T. E. Odaka, W. P. Kraemer, T. Hirano, P. R. Bunker, *Spectrochim. Acta A* **2002**, *58*, 763.

103. L. Jutier, C. Lèonard, F. Gatti, *J. Chem. Phys.* **2009**, *130*, 134301/1–10.
104. A. V. Marenich, J. E. Boggs, *J. Chem. Phys.* **2005**, *122*, 024308/1–11.
105. J. T. Hougen, *J. Chem. Phys.* **1962**, *36*, 519.
106. J. M. Brown, The Renner-Teller effect: The effective Hamiltonian approach, in *Computational Molecular Spectroscopy*, P. Jensen, P. R. Bunker, Eds., Wiley, Chichester, 2000, pp. 516–537.
107. H. Köppel, W. Domcke, L. S. Cederbaum, *Adv. Chem. Phys.* **1984**, *57*, 59.
108. M. Nooijen, *Int. J. Quant. Chem.* **2003**, *95*, 768.
109. T. Ichino, A. J. Gianola, W. C. Lineberg, J. F. Stanton, *J. Chem. Phys.* **2006**, *125*, 084312.
110. J. F. Stanton, *J. Chem. Phys.* **2007**, *126*, 134309.
111. M. Nooijen, *Int. J. Quant. Chem.* **2006**, *106*, 2489.
112. T. Ichino, J. Gauss, J. F. Stanton, *J. Chem. Phys.* **2009**, *130*, 174105.
113. M. S. Schuurman, D. E. Weinberg, D. Yarkony, *J. Chem. Phys.* **2007**, *127*, 104309/1–12.
114. P. S. Christopher, M. Shapiro, P. Brumer, *J. Chem. Phys.* **2006**, *124*, 184107/1–11.
115. M. S. Schuurman, D. E. Weinberg, D. Yarkony, *Chem. Phys.* **2008**, *347*, 57.
116. H. Beck, A. Jäckle, G. Worth, H.-D. Meyer, *Phys. Rep.* **2000**, *324*, 1.
117. R. Borrelli, A. Peluso, *J. Chem. Phys.* **2006**, *125*, 194308/1–8.
118. A. Peluso, R. Borrelli, A. Capobianco, *J. Phys. Chem. A* **2009**, *113*, 14831.
119. F. Chau, J. M. Dyke, E. P. Lee, D. Wang, *J. Electron Spectrosc. Relat. Phenom.* **1998**, *97*, 33.
120. J. M. Luis, D. M. Bishop, B. Kirtman, *J. Chem. Phys.* **2004**, *120*, 813.
121. J.M. Luis, M. Torrent-Sucarrat, M. Solà, D.M. Bishop, B. Kirtman, *J. Chem.Phys.* **2005**, *122*, 184104/1–13.
122. W. Eisfield, *J. Phys. Chem. A* **2006**, *110*, 3903.
123. J. M. Luis, B. Kirtman, O. Christiansen, *J. Chem. Phys.* **2006**, *125*, 154114.
124. V. Rodriquez-Garcia, K. Yagi, K. Hirao, S. Iwata, S. Hirata, *J. Chem. Phys.* **2006**, *125*, 104109.
125. J. M. Bowman, X. Huang, L. B. Harding, S. Carter, *Mol. Phys.* **2006**, *104*, 33.
126. V. Barone, *J. Chem. Phys.* **2005**, *122*, 014108/1–10.
127. H. Wang, C. Zhu, J.-G. Yu, S. H. Lin, *J. Phys. Chem. A* **2009**, *113*, 14407.
128. H. Köppel, Jahn-Teller and pseudo-Jahn-Teller intersections: Spectroscopy and vibronic dynamics, in *Conical Intersections, Electronic Structure, Dynamics and Spectroscopy*, W. Domcke, R. Yarkony, H. Köppel, Eds., World Scientific, Singapore, 2004, pp. 429–472.
129. I. B. Bersuker, Ed., *The Jahn-Teller Effect: A Bibliographic Review*, IFI/Plenum, New York, 1984.
130. C. Lanczos, *Res. Nat. Bur. Stand.* **1950**, *45*, 225.
131. R. Renner, *Z. Phys.* **1934**, *92*, 172.
132. P. Jensen, G. Osmann, P. R. Bunker, The Renner effect, in *Computational Molecular Spectroscopy*, P. Jensen, P. R. Bunker, Eds., Wiley, Chichester, 2000, pp. 487–515.
133. J. T. Hougen, *J. Chem. Phys.* **1962**, *37*, 403.
134. H. J. Werner, P. Knowles, *J. Chem. Phys.* **1988**, *89*, 5803.
135. T. Pacher, L. S. Cederbaum, H. Köppel, *Adv. Chem. Phys.* **2003**, *84*, 293.

136. R. Tarroni, A. Mitrushenkov, P. Palmieri, S. Carter, *J. Chem. Phys.* **2001**, *115*, 11200.

137. R. Tarroni, S. Carter, *J. Chem. Phys.* **2005**, *123*, 014320.

138. M. Biczysko, R. Tarroni, *Mol. Phys.* **2002**, *100*, 3667.

139. M. Biczysko, R. Tarroni, *Phys. Chem. Chem. Phys.* **2002**, *4*, 708.

140. M. Biczysko, R. Tarroni, *Chem. Phys. Lett.* **2005**, *415*, 223.

141. C. G. Eisenhardt, G. Pietraperzia, M. Becucci, *Phys. Chem. Chem. Phys.* **2001**, *3*, 1407.

142. L. J. H. Hoffmann, S. Marquardt, A. S. Gemechu, H. Baumgärtel, *Phys. Chem. Chem. Phys.* **2006**, *8*, 2360.

143. M. Biczysko, J. Bloino, V. Barone, *Chem. Phys. Lett.* **2009**, *471*, 143.

144. J. G. Radziszewski, *Chem. Phys. Lett.* **1999**, *301*, 565.

145. J. H. Miller, L. Adrews, P. Lund, P. N. Schatz, *J. Chem. Phys.* **1980**, *73*, 4932.

146. G.-S. Kim, A. Mebel, S. Lin, *Chem. Phys. Lett.* **2002**, *361*, 421.

147. V. S. Reddy, T. S. Venkatesan, S. Mahapatra, *J. Chem. Phys.* **2007**, *126*, 074306.

148. S. Vassiliev, D. Bruce, *Photosynth. Res.* **2008**, *97*, 75.

149. C. Adachi, M. A. Baldo, M. E. Thompson, S. Forrest, *J. Appl. Phys* **2001**, *90*, 5048.

150. J. Slinker, D. Bernards, P. Houston, H. Abruna, S. Bernhard, G. Malliaras, *Chem. Commun.* **2003**, 2392.

151. J. Slinker, A. A. Gorodetsky, M. S. Lowry, J. Wang, S. Parker, R. Rohl, S. Bernhard, G. Malliaras, *J. Am. Chem. Soc.* **2004**, *126*, 2763.

152. H. Du, R. A. Fuh, J. Li, A. Corkan, J. S. Lindsey, *Photochem. Photobiol.* **1998**, *68*, 141.

153. H. H. Strain, M. R. Thomas, J. J. Katz, *Biochim. Biophys. Acta* **1963**, *75*, 306.

154. M. Cossi, G. Scalmani, N. Rega, V. Barone, *J. Comp. Chem.* **2003**, *24*, 669.

155. F. De Angelis, F. Santoro, M. K. Nazeruddin, V. Barone, *J. Phys. Chem. B* **2008**, *112*, 13181.

156. M. E. Beck, *Int. J. Quant. Chem.* **2004**, *101*, 683.

PART IIB

EFFECTS RELATED TO NUCLEAR MOTIONS: TIME-DEPENDENT MODELS

9

EFFICIENT METHODS FOR COMPUTATION OF ULTRAFAST TIME- AND FREQUENCY-RESOLVED SPECTROSCOPIC SIGNALS

MAXIM F. GELIN AND WOLFGANG DOMCKE
Department of Chemistry, Technische Universität München, Garching, Germany
DASSIA EGOROVA
Institute of Physical Chemistry, Universität Kiel, Kiel, Germany

Computational Strategies for Spectroscopy: From Small Molecules to Nano Systems, First Edition.
Edited by Vincenzo Barone.

We outline recent developments in the theoretical description of femtosecond time- and frequency-resolved spectroscopy. The focus is on the equation-of-motion phase-matching approach (EOM-PMA), which does not require the evaluation of multitime nonlinear response functions and thus allows the computation of time- and frequency-resolved signals for complex dissipative systems. In comparison with other methods, the EOM-PMA exhibits a considerably improved scaling of the computational effort with the system size. Pulse duration and pulse overlap effects are automatically taken into account, which allows realistic simulations of the spectra. We first address the computation of two-pulse time- and frequency-resolved spectra, such as fluorescence up-conversion, pump–probe (PP), and two-pulse photon echo (PE) spectra. We then outline the three-pulse EOM-PMA for the computation of third-order four-wave-mixing signals such as, for example, homodyne or heterodyne three-pulse photon echo and coherent anti-Stokes–Raman scattering. The methods are illustrated by the calculation of various nonlinear spectra of model systems exhibiting ultrafast dissipative dynamics. We focus on coherent processes in electronic spectroscopy and the capabilities of currently available spectroscopic techniques to provide information on the underlying dynamical mechanisms in molecular systems.

9.1 INTRODUCTION

Nonlinear optical spectroscopy comprises a family of techniques such as fluorescence up-conversion, pump–probe, transient grating, photon echo, coherent anti-Stokes–Raman scattering, which generally are referred to as four-wave-mixing spectroscopies [1]. These techniques differ in the number, ordering, and phase-matching directions of the pulses involved and in the specific informations they deliver on the material system under study. However, all these techniques share one fundamental property: The corresponding signals are uniquely determined by the third-order polarization $P(t)$.

There exist two major theoretical methods for the computation of $P(t)$, the perturbative and the nonperturbative approaches. In the perturbative approach, $P(t)$ is expressed as a triple time integral involving the nonlinear response function [1]

$$P(t) = \int_0^\infty dt_1 \int_0^\infty dt_2 \int_0^\infty dt_3 \, S(t_1, t_2, t_3) E(t - t_3) E(t - t_3 - t_2) E(t - t_3 - t_2 - t_1) \quad (9.1)$$

Here $E(t)$ represents the electric field of the pulses involved and depends parametrically on their carrier frequencies, phases, and mutual delays. The form of Eq. 9.1 is very appealing because the response function $S(t_1, t_2, t_3)$ represents the system dynamics in the absence of external fields. For simple material systems, such as few-level systems or damped harmonic oscillators, $S(t_1, t_2, t_3)$ can be calculated analytically [1]. However, as the material systems become more complex, the evaluation of the response functions necessitates a number of approximations [2–6] or requires extensive numerical calculations [7, 8] and/or (within additional approximations) computer simulations [9, 10]. Performing the triple time integration in Eq. 9.1 for each particular value of the pulse carrier frequencies and delay times is also time consuming.

The conceptual alternative to the perturbative approach is the nonperturbative evaluation of time- and frequency-resolved spectra. The idea of the nonperturbative approach is simple. To study the dynamics of the vast majority of chemically interesting systems, we have to resort to numerical methods and/or simulations. Suppose we wish to calculate the spectroscopic response of such a system. To do so, we have to take into account interaction of the system with the pertinent laser pulses. Since we have to resort to a numerical simulation anyway, it seems logical to incorporate all relevant laser fields into the system Hamiltonian (which thus becomes time dependent) and to numerically calculate the dynamics of the driven system [11]. Since no assumptions are made about the relative timings of the pulses involved, all effects due to pulse overlaps are accounted for automatically. Furthermore, no assumptions have to be made about the weakness of the laser fields. The approach is thus potentially useful for describing strong-field effects, which are crucial if we wish to use laser pulses to manipulate and control the material system dynamics [12, 13]. When dealing with complex multilevel systems (notably with strong electronic and vibrational couplings as well as with bath-induced relaxations), the nonperturbative approach has proven its superiority over perturbative treatments; see, for example, recent applications to two- [14] and three- [15] pulse spectroscopic signals.

In this chapter, a relatively new alternative formalism for the calculation of time- and frequency-resolved spectroscopic signals, the so-called EOM-PMA [15], is outlined. This method shares features both with the perturbative approach [1] and the nonperturbative approach [11]. In the EOM-PMA, one directly computes components of the third- (or higher) order polarization corresponding to a particular phase-matching condition, in contrast to the a posteriori decomposition of the nonperturbatively computed total polarization [11]. Rather than expressing the signals in terms of multitime nonlinear response functions (cf. Eq. 9.1), the signals are obtained as expectation values involving certain auxiliary density matrices. The calculation of spectroscopic signals is thus reduced to the time propagation of a few modified density matrices. The computational cost of these density matrix propagations is comparable to that of the propagation of the field-free density matrix. Since the a posteriori decomposition of the total polarization is avoided, the EOM-PMA is computationally more efficient than the nonperturbative approach.

To make the chapter self-contained and easy to read, we introduce the necessary definitions and starting equations in Section 9.2. Sections 9.3 and 9.4 describe the two- and three-pulse EOM-PMA, respectively. In Section 9.5, we discuss the time- and frequency-gated (TFG) spontaneous emission (SE) and optical two-dimensional (2D) three-pulse (3P) PE spectra for a model system, which accounts for strong electronic and electronic–vibrational coupling, vibrational relaxation, and dephasing. We demonstrate how different spectroscopic techniques complement each other in providing information about the dynamics of the material system.

We focus in this chapter on time- and frequency-resolved electronic spectroscopy. The basic ideas and the formalism are rather straightforwardly transferable to multipulse infrared spectroscopy [16]. The intention of this chapter is to familiarize the reader with the methods which are the most efficient in applications to complex molecular systems. References to alternative approaches can be found in a monograph [1] and recent review articles [3–6, 8, 14, 15]. Aspects of computational efficiency of various methods are discussed throughout the chapter.

9.2 BASIC EQUATIONS

In the context of electronic spectroscopy of molecules, we represent the system Hamiltonian H as the sum of an electronic ground-state Hamiltonian, H_g, and an excited-state Hamiltonian, H_e,

$$H = H_g + H_e \tag{9.2}$$

The latter may represent a number of (intermolecularly coupled) electronic states. In the diabatic representation the Hamiltonians are written as

$$H_g = |g\rangle h_g \langle g| \tag{9.3}$$

$$H_e = \sum_i |i\rangle(h_i + \epsilon_i)\langle i| + \sum_{i \neq j} |i\rangle U_{ij} \langle j| \tag{9.4}$$

Here the bra–ket notation is used to denote the electronic ground state ($|g\rangle$) and the ensemble of excited states ($|i\rangle$). The h_g and h_i represent the corresponding vibrational Hamiltonians. The U_{ij} are electronic coupling matrix elements and ϵ_i are the vertical excitation energies of the excited states $|i\rangle$.

The interaction of the molecular system with N laser pulses is written in the dipole approximation as follows:

$$H_N(t) = -\sum_{a=1}^{N} \left[\exp\{+i\mathbf{k}_a\mathbf{r}\}\upsilon_a^{(+)}(t) + \exp\{-i\mathbf{k}_a\mathbf{r}\}\upsilon_a^{(-)}(t)\right] \tag{9.5}$$

$$\upsilon_a^{(+)}(t) = V\lambda_a E_a(t-\tau_a)\exp\{-i\omega_a t\} \quad \upsilon_a^{(-)}(t) \equiv \left[\upsilon_a^{(+)}(t)\right]^{\dagger} = V\lambda_a E_a(t-\tau_a)\exp\{i\omega_a t\} \tag{9.6}$$

Here λ_a, $\mathbf{k}_a$, ω_a, $E_a(t)$, and τ_a denote the amplitude, wave vector, frequency, dimensionless envelope, and central time of the pulses. The transition dipole moment operator is taken in the form

$$V \equiv X + X^{\dagger} \quad X = \sum_i V_{gi}|g\rangle\langle i| \tag{9.7}$$

V_{gi} being the matrix element of this operator between the $|g\rangle$ and $|i\rangle$ diabatic electronic states; V_{gi} may depend on the vibrational degrees of freedom due to non-Condon effects.

For optical transitions, the relevant pulse carrier frequencies ω_a and the vertical excitation energies ϵ_i are of the order of the energy gap ϵ_e between the minima of the ground-state and excited-state potential energy surfaces. Since ϵ_e is much larger than all other relevant energies, it is convenient to adopt the rotating-wave approximation (RWA) for the system-field interaction Hamiltonian. The RWA amounts to the omission of the counterrotating terms ($\sim\exp\{\pm i(\epsilon_i+\omega_a)t\}$) while retaining the corotating terms ($\sim\exp\{\pm i(\epsilon_i-\omega_a)t\}$) in the material system responses to the applied fields. The use of the RWA is justified if the phase factors $\exp\{\pm i(\epsilon_i+\omega_a)t\}$ are highly oscillatory on the time scale of the system dynamics. Physically, this means that upward transitions should be accompanied by the absorption of a photon, while downward transitions should be accompanied by the emission of a photon. It is well established that the RWA is accurate for electronic spectroscopy with visible or UV pulses. It can also be applied in infrared spectroscopy if there exists a clear separation of high- and low-frequency vibrational modes.

Within the RWA, the interaction Hamiltonian (9.5) can be written as follows:

$$\mathcal{H}_N(t) = -\sum_{a=1}^{N}\left[\exp\{+i\mathbf{k}_a\mathbf{r}\}\upsilon_a^{>}(t)+\exp\{-i\mathbf{k}_a\mathbf{r}\}\upsilon_a^{<}(t)\right] \tag{9.8}$$

$$\upsilon_a^{>}(t) = X^{\dagger}\lambda_a E_a(t-\tau_a)\exp\{-i\omega_a t\} \quad \upsilon_a^{<}(t) \equiv \left[\upsilon_a^{<}(t)\right]^{\dagger} = X\lambda_a E_a(t-\tau_a)\exp\{i\omega_a t\} \tag{9.9}$$

[Hereafter, the symbol $\mathcal{H}_N(t)$ is used to denote the system-field interaction Hamiltonian in the RWA in order to distinguish it with $H_N(t)$ in Eq. 9.5]. It is convenient to reduce all the energies in the excited electronic states and the carrier frequencies of the pulses by the value of ϵ_e, that is, to replace $\epsilon_i \to \epsilon_i-\epsilon_e$ and $\omega_a \to \omega_a-\epsilon_e$. This convention is used in the following.

Within the RWA, the master equation for the reduced density matrix is ($\hbar=1$)

$$\partial_t\rho(t) = -i[H+\mathcal{H}_N(t),\rho(t)]-D\rho(t) \tag{9.10}$$

D being a suitable dissipative operator [17, 18]. For simplicity of notation, D is written as a time-independent operator. Upon the substitution $D\rho(t) \to \int_0^t dt'\,D(t-t')\rho(t')$, all the derived formulas remain true for a general non-Markovian dissipative operator. Strictly speaking, D may depend on the laser fields involved. It can be shown, however, that this effect can be neglected even for rather strong pulses [19]. We assume that at time $t=0$ (before all the pulses are switched on) the material system is in its ground electronic state. Therefore, Eq. 9.10 must be solved with the initial condition

$$\rho(0) = |g\rangle\rho_g\langle g| \qquad \rho_g = Z_g^{-1}\exp\left\{-\frac{h_g}{k_BT}\right\} \tag{9.11}$$

Here ρ_g is the Boltzmann operator, Z_g is the corresponding partition function, k_B is the Boltzmann constant, and T is the temperature. The master equation (9.10) with the initial condition (9.11) is a starting point for all subsequent calculations.

9.3 TWO-PULSE SPECTROSCOPIES

9.3.1 Spontaneous Emission Signal

The material system is promoted by the pump pulse to the excited electronic state from which it can (spontaneously) emit photons. Since the SE is a quantum process, it is described in the formalism of the quantization of the radiation field [1, 20]. The pump process, on the other hand, can be considered semiclassically. We thus start with the master equation (9.10) with $N=1$,

$$\partial_t \rho(t) = -i[H + \mathcal{H}_1(t), \rho(t)] - D\rho(t) \tag{9.12}$$

which describes the dynamics of the material system excited by the pump pulse. We introduce the propagator $G(t, t')$ for Eq. 9.12,

$$\rho(t) = G(t, t')\rho(t') \quad t \geq t' \tag{9.13}$$

The ideal time- and frequency-resolved SE spectrum, that is, the steady-state rate of the change of the number of photons with the frequency ω_S at time t, is given by the expression [20]

$$S_{\mathrm{SE}}(t, \omega_S) \sim \mathrm{Re}\int_0^t dt'\, e^{-i\omega_S(t-t')} C_{\mathrm{SE}}(t, t') \tag{9.14}$$

$$C_{\mathrm{SE}}(t, t') \equiv \mathrm{tr}\{X^{\dagger} G(t, t') X G(t', 0)\rho(0)\} \tag{9.15}$$

Eqations 9.14 and 9.15 are valid for a pump pulse of arbitrary intensity and duration. As is easy to demonstrate with Eq. 9.12, the off-diagonal elements (electronic coherences) of the density matrix, $\langle g|\rho(t)|i\rangle$, are proportional to the phase factor $\exp\{i\mathbf{k}_1\mathbf{r}\}$, while the diagonal elements (electronic populations), $\langle g|\rho(t)|g\rangle$ and $\langle i|\rho(t)|i\rangle$, are independent of this phase. Therefore, the SE spectrum (9.14) is also independent of the phase factor, and we can put $\mathbf{k}_1\mathbf{r}=0$.

The straightforward computation of the SE spectrum according to Eqs. 9.14 and 9.15 requires two time propagations, via $G(t, t')$ and $G(t', 0)$, on a two-dimensional grid (t, t'). Since this procedure is time consuming, we directly calculate the ideal time- and frequency-resolved spectrum $S_{\mathrm{SE}}(t, \omega_S)$ rather than the two-time correlation function $C_{\mathrm{SE}}(t, t')$. This allows us to avoid the time integration over t' in Eq. 9.14. The time dependence of the spectrum $S_{\mathrm{SE}}(t, \omega_S)$ at a fixed SE frequency ω_S is obtained by the propagation of just two auxiliary density matrices.

To this end, let us consider the auxiliary master equation [21]

$$\partial_t \sigma(t) = -i[H + \mathcal{H}_1(t), \sigma(t)] - D\sigma(t) + i\tilde{v}_S^{<}(t)\sigma(t) \tag{9.16}$$

with $\sigma(0) = |g\rangle\rho_g\langle g|$. Equation 9.16 differs from Eq. 9.10 by the presence of the term $i\tilde{v}_S^{<}(t)\sigma(t)$. Explicitly,[1]

$$\tilde{v}_S^{<}(t) = \lambda_S \exp\{i\omega_S t\}X \quad \left[\tilde{v}_S^{<}(t)\right]^{\dagger} \equiv \tilde{v}_S^{<}(t) \tag{9.17}$$

where λ_S is a small parameter. The term $i\tilde{v}_S^{<}(t)\sigma(t)$ is not Hermitian and enters Eq. 9.16 just as a multiplicative operator, rather than through a commutator, so that Eq. 9.16 is not a true Liouville–von Neumann equation. This reflects the fact that the SE is determined by the time correlation function of the polarization, rather than by the polarization itself [22, 23].

Solving Eq. 9.16 perturbatively in λ_S, one obtains the formal solution $\sigma(t) = \sigma_0(t) + \lambda_S\sigma_1(t) + O(\lambda_S^2)$ and the SE spectrum

$$S_{\mathrm{SE}}(t, \omega_S) \sim \mathrm{Im}\,\mathrm{tr}\{\tilde{v}_S^{>}(t)\sigma_1(t)\} \tag{9.18}$$

Noting that $\sigma_0(t) \equiv \rho(t)$ (the latter is the solution of Eq. 9.12), we have the result

$$S_{\mathrm{SE}}(t, \omega_S) = \mathrm{Im}\,A_{\mathrm{SE}}(t, \omega_S) \quad A_{\mathrm{SE}}(t, \omega_S) = \mathrm{tr}\{\tilde{u}_S^{>}(t)[\sigma(t) - \rho(t)]\} + O(\lambda_S^3) \tag{9.19}$$

The simultaneous propagation of $\rho(t)$ and $\sigma(t)$ via Eqs. 9.12 and 9.16 gives a time-dependent cut of the time- and frequency-resolved SE spectrum at a particular frequency ω_S.

Equation 9.19 determines the SE spectrum which is measured under ideal time and frequency resolution [1]. The actual TFG SE spectrum, $S_{\mathrm{SE}}^{\mathrm{TFG}}(t, \omega_S)$, which is measured with a finite time and frequency resolution, is related to the ideal spectrum as follows [20–23]:

$$S_{\mathrm{SE}}^{\mathrm{TFG}}(t, \omega_S) \sim \mathrm{Re}\int_{-\infty}^{\infty} dt' \int_0^{t'} dt''\, E_t(t'-t)E_t(t''-t)\exp\{-(\gamma + i\omega_S)(t'-t'')\}C_{\mathrm{SE}}(t, t'') \tag{9.20}$$

Here $E_t(t)$ is the time gate function and γ determines the width of the frequency filter. It is easy to verify that the TFG SE spectrum can be expressed through the convolution of A_{SE} (t, ω_S) with the corresponding joint TFG function Φ_{SE}:

$$S_{\mathrm{SE}}^{\mathrm{TFG}}(t, \omega_S) \sim \mathrm{Im}\int_{-\infty}^{\infty} d\omega' \int_{-\infty}^{\infty} dt'\, \Phi_{\mathrm{SE}}(t-t', \omega_S - \omega')A_{\mathrm{SE}}(t', \omega') \tag{9.21}$$

[1] The definition of $v_S^{<}$ and $v_S^{>}$ is the same as for $v_a^{<}$ and $v_a^{>}$ (Eq. 9.9) but with no pulse envelope $[E_a(t - \tau_a) = 1]$.

where

$$\Phi_{\mathrm{SE}}(\tau,\Omega) = E_t(-\tau)\Phi(\tau,\Omega;\gamma) \tag{9.22}$$

The joint TFG function is defined through the gate-pulse envelope as follows:

$$\Phi(\tau,\Omega;\gamma) = \int_{-\infty}^{-\tau} d\zeta\, E_t(\zeta)\exp\{(\gamma + i\Omega)(\tau+\zeta)\} \tag{9.23}$$

For the exponential envelope

$$E_t(t) = \exp\{-\Gamma|t|\} \tag{9.24}$$

($1/\Gamma$ being the pulse duration), we have

$$\Phi(\tau,\Omega;\gamma) = \theta(\tau)\frac{\exp\{-\Gamma\tau\}}{\Gamma+\gamma+i\Omega} + \theta(-\tau)\left[\frac{\exp\{(\gamma+i\Omega)\tau\}}{\Gamma+\gamma+i\Omega} + \frac{\exp\{(\gamma+i\Omega)\tau\} - \exp\{\Gamma\tau\}}{\Gamma-\gamma-i\Omega}\right] \tag{9.25}$$

where $\theta(\tau)$ is the Heaviside step function.

Integrating the ideal time- and frequency-resolved SE spectrum (9.14) over the SE frequency, we get the frequency-integrated ideal SE signal

$$S_{\mathrm{SE}}(t) \equiv \int_{-\infty}^{\infty} d\omega_S\, S_{\mathrm{SE}}(t,\omega_S) = \langle X^\dagger X\rho(t)\rangle \tag{9.26}$$

which is solely determined by the density matrix $\rho(t)$ of Eq. 9.12. The experimentally measured frequency-integrated SE signal, which is obtained by integration of the TFG SE spectrum (9.20) over the SE frequency ω_S, is connected with its ideal counterpart (9.26) via a simple convolution:

$$S_{\mathrm{SE}}^{\mathrm{TFG}}(t) \equiv \int_{-\infty}^{\infty} d\omega_S\, S_{\mathrm{SE}}^{\mathrm{TFG}}(t,\omega_S) = \int_{-\infty}^{\infty} dt' E_t^2(t-t') S_{\mathrm{SE}}(t') \tag{9.27}$$

We remark that the TFG function 9.22 does not coincide with its counterpart defined by Mukamel et al. [22]. The reason is that the present definition of the ideal SE spectrum, Eq. 9.14, differs from the definition used elsewhere [22, 23]. The ideal SE spectra may be regarded as Wigner transforms $\int d\tau \exp\{-i\omega_S\tau\}\times C_{\mathrm{SE}}(t+\sigma\tau, t-(1-\sigma)\tau)$ of the fundamental correlation function C_{SE} with $\sigma=0$ (present work) or $\sigma=\frac{1}{2}$ [22, 23]. The two spectra are connected with each other via an appropriate time–frequency convolution. We have taken $\sigma=0$ because this choice arises naturally in our calculations and because Eqs. 9.14 and 9.15 reduce to the corresponding formulas in terms of the optical response functions [1] in the weak pump limit. This choice renders the TFG function (9.22) complex, so that both real and imaginary parts of $A_{SE}(t,\bar{\omega}_S)$ are required to calculate the TFG SE spectrum (9.21). If $\sigma = \frac{1}{2}$, both the ideal spectrum and the TFG functions are automatically real.

9.3.2 Two-Pulse-Induced Third-Order Polarization

The formalism described above can immediately be generalized for the calculation of two-pulse (2P) PP and PE spectra. In this case, the material system (9.3–9.4) is assumed to interact with two classical pulses, a pump pulse ($a=1$) and a probe pulse ($a=2$). The corresponding interaction Hamiltonian is given by Eq. 9.8 with $N=2$. In the EOM-PMA, we wish to evaluate the field-induced polarization $P(t)=\text{tr}\{V\rho(t)\}$ in the leading (linear) order in the probe field ($a=2$), while keeping all orders in the pump field ($a=1$). We start from our basic kinetic equation (9.10) for $N=2$. Solving this equation perturbatively in λ_2, we arrive at the result [21]

$$P(t) = P_{\text{PP}}(t) + P_{\text{PE}}(t) + O(\lambda_2^2) \tag{9.28}$$

$$P_{\text{PP}}(t) = i\exp\{-i\mathbf{k}_2\mathbf{r}\}\int_0^t dt' E_2(t'-\tau_2)e^{i\omega_2 t'}C_{\text{PP}}(t,t') + \text{H.c.} \tag{9.29}$$

$$C_{\text{PP}}(t,t') \equiv \text{tr}\{X^{\dagger}G(t,t')[X,G(t',0)\rho(0)]\} \tag{9.30}$$

$$P_{\text{PE}}(t) = i\exp\{i(2\mathbf{k}_1-\mathbf{k}_2)\mathbf{r}\}\int_0^t dt' E_2(t'-\tau_2)e^{i\omega_2 t'}C_{\text{PE}}(t,t') + \text{H.c.} \tag{9.31}$$

$$C_{\text{PE}}(t,t') \equiv \text{tr}\{XG(t,t')[X,G(t,0)\rho(0)]\} \tag{9.32}$$

[the propagators $G(t,t')$ are defined in Eq. 9.13]. As is well known, the total polarization consists of two contributions (9.29 and 9.31), which are responsible for the PP and PE signals, respectively [1]. Equations 9.29–9.32 are very similar to their SE analogues (9.14 and 9.15). The difference arises from the presence of commutators in Eqs. 9.30 and 9.32 and the substitution $X^{\dagger} \leftrightarrow X$ where appropriate.

In the following, it is convenient to consider the PP and PE signals separately.

9.3.3 PP Signal

The transient transmittance PP signal is defined through the difference polarization $\tilde{P}_{\text{PP}}(t) = P_{\text{PP}}(t) - P_{\text{PP}}^{\text{off}}(t)$ [1]. Here $P_{\text{PP}}^{\text{off}}(t)$ is the pump-off polarization induced solely by the probe pulse, which is obtained from Eqs. 9.29 and 9.30 with $\lambda_1 = 0$. Within the RWA and the slowly varying envelope approximation, the integral (int) and dispersed (dis) transient transmittance PP signals can be evaluated via the following expressions [1]:

$$S_{\mathrm{PP}}^{\mathrm{int}}(\tau_2,\omega_2) = \mathrm{Im}\int_0^{\infty} dt\, E_2(t-\tau_2)e^{-i\omega_2 t}\tilde{P}_{\mathrm{PP}}(t) \tag{9.33}$$

$$S_{\mathrm{PP}}^{\mathrm{dis}}(\tau_2,\omega_2,\omega) = \mathrm{Im}E_2(\omega-\omega_2)\tilde{P}_{\mathrm{PP}}(\omega) \tag{9.34}$$

Here

$$\tilde{P}_{\mathrm{PP}}(\omega) = \int_{-\infty}^{\infty} dt \exp\{-i\omega t\}\tilde{P}_{\mathrm{PP}}(t) \tag{9.35}$$

$$E_2(\omega) = \int_{-\infty}^{\infty} dt \exp\{i\omega t\}E_2(t-\tau_2) \tag{9.36}$$

so that $S_{\mathrm{PP}}^{\mathrm{int}}(\tau_2,\,\omega_2) = \int_{-\infty}^{\infty} d\omega\, S_{\mathrm{PP}}^{\mathrm{dis}}(\tau_2,\omega_2,\omega)$.

Analogously to the case of SE, we introduce the quantity

$$A_{\mathrm{PP}}(t,\omega_2) = i\int_0^t dt' e^{-\omega_2(t-t')}\{C_{\mathrm{PP}}(t,t') - C_{\mathrm{PP}}^{\mathrm{off}}(t,t')\} \tag{9.37}$$

which is proportional to the difference polarization induced by the pump pulse and a (fictitious) continuous-wave (CW) probe pulse. The term $C_{\mathrm{PP}}^{\mathrm{off}}(t,t')$ is defined via Eq. 9.30 with $\lambda_1 = 0$. The imaginary part of Eq. 9.37 can be regarded as the ideal dispersed PP spectrum which would be measured with a δ-function probe pulse, or as the integral PP spectrum (9.33), which would be obtained with ideal time and frequency resolution, that is, when the probe pulse would be short on the vibrational dynamics time scale but long on the time scale of the optical dephasing.

Let us further introduce the equations of motion for the auxiliary density matrices

$$\partial_t\sigma^{\mathrm{on}}(t) = -i[H + \mathcal{H}_1(t) - \tilde{\upsilon}_2^{<}(t), \sigma^{\mathrm{on}}(t)] - D\sigma^{\mathrm{on}}(t) \tag{9.38}$$

$$\partial_t\sigma^{\mathrm{off}}(t) = -i[H - \tilde{\upsilon}_2^{<}(t), \sigma^{\mathrm{off}}(t)] - D\sigma^{\mathrm{off}}(t) \tag{9.39}$$

[$\tilde{\upsilon}_2^{<}(t)$ is given by Eq. 9.17]. As in the case of SE (Eq. 9.16), there is no h.c. in the definition of $\tilde{\upsilon}_2^{<}(t)$, but this term enters Eqs. 9.38 and 9.39 through the commutator. Solving Eqs. 9.38 and 9.39 perturbatively in λ_2, one obtains $\sigma^b(t) = \sigma_0^b(t) + \lambda_2\sigma_1^b(t) + O(\lambda_2^2)$, $b = \mathrm{on}$, off. Noting that $\sigma_0^{\mathrm{on}}(t) \equiv \rho(t)$ (Eq. 9.12) and $\sigma_0^{\mathrm{off}}(t) \equiv \rho(0)$ (Eq. 9.11), one gets [21]

$$A_{\mathrm{PP}}(t,\omega_2) = \mathrm{tr}\{\tilde{\upsilon}_2^{>}(t)[\sigma^{\mathrm{on}}(t) - \rho(t) - \sigma^{\mathrm{off}}(t)]\} + O(\lambda_2^3) \tag{9.40}$$

Having computed the ideal PP spectrum, one can immediately calculate the PP spectra for laser pulses of finite duration. The integral PP spectrum is determined by the expression

$$S_{\mathrm{PP}}^{\mathrm{int}}(\tau_2,\omega_2) \sim \mathrm{Im}\int_{-\infty}^{\infty} d\omega' \int_{-\infty}^{\infty} dt'\, \Phi_{\mathrm{PP}}^{\mathrm{int}}(\tau_2 - t', \omega_2 - \omega') A_{\mathrm{PP}}(t',\omega') \tag{9.41}$$

where the joint gate probability function for the integral pump–probe signal, $\Phi_{\mathrm{PP}}^{\mathrm{int}}(\tau,\Omega) \equiv E_2(-\tau)\Phi(\tau,\Omega;0)$, is explicitly given via Eqs. 9.22, 9.23, and 9.25. Thus $\Phi_{\mathrm{PP}}^{\mathrm{int}}(\tau,\Omega)$ coincides with its TFG SE counterpart, $\Phi_{\mathrm{SE}}(\tau,\Omega)$, at $\gamma = 0$. The formulas for the dispersed PP signal read

$$S_{\mathrm{PP}}^{\mathrm{dis}}(\tau_2,\omega_2,\omega) \sim \mathrm{Im}\int_{-\infty}^{\infty} d\omega' \int_{-\infty}^{\infty} dt' \Phi_{\mathrm{PP}}^{\mathrm{dis}}(\tau_2 - t', \omega_2 - \omega') A_{\mathrm{PP}}(t',\omega') \tag{9.42}$$

$$\Phi_{\mathrm{PP}}^{\mathrm{dis}}(\tau,\Omega) = \tilde{E}_P(\omega - \omega_2)\exp\{i(\omega-\omega_2)\tau\}\Phi(\tau,\Omega;0) \tag{9.43}$$

Here $\tilde{E}_2(\omega)$ is defined as $E_2(\omega)$ in Eq. 9.36, but with $\tau_2 = 0$. The dispersed PP signal at $\omega = \omega_2$ is seen to be very similar to the integral PP signal.

9.3.4 Two-Pulse PE Signal

Let us introduce the quantity

$$A_{\mathrm{PE}}(t,\omega_2) = i\int_0^t dt' e^{-i\omega_2(t-t')} C_{\mathrm{PE}}(t,t') \tag{9.44}$$

which determines the ideal PE signal. Analogously to Eq. 9.19, we have

$$A_{\mathrm{PE}}(t,\omega_2) = \mathrm{tr}\{\tilde{v}_2^{<}(t)[\sigma^{\mathrm{on}}(t) - \rho(t)]\} + O(\lambda_2^3) \tag{9.45}$$

where $\tilde{v}_2^{<}(t)$, $\sigma^{\mathrm{on}}(t)$, and $\rho(t)$ are given by Eqs. 9.17, 9.38, and 9.12, respectively. The homodyne-detected PE signal is defined as [1]

$$S_{\mathrm{PE}}^{\mathrm{hom}}(\tau_2) \sim \int_0^{\infty} dt |P_{\mathrm{PE}}(t)|^2 \tag{9.46}$$

It can be evaluated via Eq. 9.46, in which one has to put

$$P_{\mathrm{PE}}(t) \sim \int_{-\infty}^{\infty} d\omega'\, \Phi(\tau_2 - t, \omega_2 - \omega'; 0) A_{\mathrm{PE}}(t,\omega') \tag{9.47}$$

The joint TFG function, $\Phi(\tau,\Omega)$, is given by Eqs. 9.23 and 9.25. If the probe pulse is short on the time scale of the wavepacket dynamics but long on the time scale of optical dephasing (the so-called impulsive limit), the TFG function $\Phi(\tau, \Omega)$ may be considered as a δ-function in both the time and the frequency domain. In this case, the homodyne-detected PE signal is simply expressed through the ideal PE spectrum as $S_{\rm PE}^{\rm hom}(\tau_2) \sim |A_{\rm PE}(\tau_2, \omega_2)|^2$.

The heterodyne-detected PE signal is defined as follows [1]:

$$S_{\rm PE}^{\rm het}(\tau_{\rm LO}, \tau_2) \sim {\rm Im} \int_0^\infty dt\, E_{\rm LO}(\tau_{\rm LO} - t) e^{i(\omega_{\rm LO} t - \varphi_{\rm LO})} P_{\rm PE}(t) \tag{9.48}$$

where $E_{\rm LO}(t)$, $\varphi_{\rm LO}$, and ω_L are the dimensionless envelope, phase, and carrier frequency of the so-called local oscillator field. Therefore,

$$S_{\rm PE}^{\rm het}(\tau_{\rm LO}, \tau_2) \sim {\rm Im} \int_{-\infty}^\infty d\omega' \int_{-\infty}^\infty dt\, \Phi_{\rm PE}^{\rm het}(\tau_2 - t, \omega_2 - \omega') A_{\rm PE}(t, \omega') \tag{9.49}$$

with

$$\begin{aligned}\Phi_{\rm PE}^{\rm het}(\tau, \Omega) &= \exp\{i[(2\mathbf{k}_1 - \mathbf{k}_2)\mathbf{r} - \varphi_{\rm LO}]\} \exp\{-i(\omega_{\rm LO} - \omega_2)(\tau - \tau_2)\} \\ &\quad \times E_{\rm LO}(-\tau + \tau_2 - \tau_{\rm LO}) \Phi(\tau, \Omega; 0)\end{aligned} \tag{9.50}$$

9.3.5 Discussion

The two-pulse EOM-PMA allows us to compute two-pulse (SE, PP, PE) time- and frequency-resolved spectra for overlapping pump and probe/gate pulses of arbitrary duration. The pump pulse is allowed to be of arbitrary strength (the effects of a strong pump pulse are discussed in ref. 19), while the probe pulse is assumed to be sufficiently weak. The calculation of spectra consists of two steps. (i) One performs N_ω propagations of several (two for SE and PE, three for PP) density matrices at different emission frequencies and calculates the ideal spectrum on the grid $N_t \times N_\omega$ (N_t and N_ω being the number of grid points in the time and frequency domain, correspondingly). Ideal SE and PP spectra are well known in the literature [1, 20–24] and the two-pulse EOM-PMA allows an efficient evaluation of these quantities via Eqs. 9.16, 9.38, and 9.39. (ii) The spectrum for the actual probe/gate pulse is calculated by a twofold numerical convolution of the ideal spectrum with the appropriate joint TFG function (9.23).

Within the two-pulse EOM-PMA, the signals are calculated without resort to several commonly used simplifications like the doorway–window approximation or the neglect of dissipation effects during the pulses [1]. The two-pulse EOM-PMA leads to a $\sim N_\omega$ scaling for the computation of time- and frequency-resolved spectra, in contrast to the $\sim N_t \times N_\omega$ scaling for the a posteriori decomposition of the total polarization in the nonperturbative approach [11] or for the method developed by Tanimura and Mukamel [25, 26]. The method of Hahn and Stock [27] also exhibits a $\sim N_\omega$ scaling, but it is efficient only for temporarily well separated pump–probe pulses (i.e., within the domain of validity of the doorway–window approach) and requires an additional numerical Fourier transform of the polarization.

By an appropriate extension of the Hamiltonian (9.4), the two-pulse EQM-PMA can be applied to a general electronic N-level system beyond the weak-pump limit. The method can thus be used, for example, for the calculation of so-called control kernels (see, e.g., ref. 28) within optimal control theory.

9.4 THREE-PULSE SPECTROSCOPIES

The basic master equation, which describes the three-pulse ($N=3$) driven evolution of a material system, reads (cf. Eq. 9.10)

$$\partial_t \rho(t) = -i[H + H_3(t), \rho(t)] - D\rho(t) \tag{9.51}$$

The total three-pulse induced polarization is defined as

$$P(t) \equiv \langle V\rho(t) \rangle \tag{9.52}$$

where the angular brackets indicate the trace.

As an example, we show how to extract the three-pulse photon echo (3PPE) polarization $P_{3P}(t)$ from Eq. 9.52. The polarization in any other phase-matching direction can be found in the same manner. Specifically, we search for the contribution to the total polarization $P(t)$ which is proportional to $\exp\{\pm i\mathbf{k}_{3P}\mathbf{r}\}$, where

$$\mathbf{k}_{3P} = -\mathbf{k}_1 + \mathbf{k}_2 + \mathbf{k}_3 \tag{9.53}$$

is the 3PPE phase-matching condition, so that

$$P_{3P}(t) = P_{3P}^{(+)}(t) \exp\{i\mathbf{k}_{3P}\mathbf{r}\} + \text{c.c.} \tag{9.54}$$

9.4.1 Three-Pulse-Induced Third-Order Polarization

To obtain $P_{3P}(t)$ of Eq. 9.54, it is sufficient to evaluate the complex polarization $P_{3P}^{(+)}(t)$. For this purpose, only the terms with the phase factors $\exp\{-i\,\mathbf{k}_1\mathbf{r}\}$, $\exp\{+i\mathbf{k}_2\mathbf{r}\}$, and $\exp\{+i\mathbf{k}_3\mathbf{r}\}$ need to be retained in the interaction Hamiltonian (9.5). The master equation obtained in this manner,

$$\partial_t \rho_1(t) = -i[H - v_1^{(-)}(t) - v_2^{(+)}(t) - v_3^{(+)}(t), \rho_1(t)] - D\rho_1(t) \tag{9.55}$$

and the original master equation (9.51) yield exactly the same complex polarization $P_{3P}^{(+)}(t)$. Equation 9.55 contains, however, only half of the Liouville pathways contributing to Eq. 9.10, which facilitates the extraction of $P_{3P}^{(+)}(t)$. Let us consider $\rho_1(t)$ in Eq. 9.55 as a function of the pulse strengths [15, 29],

$$\rho_1(\lambda_1, \lambda_2, \lambda_3; t) = \sum_{i,j,k=0}^{\infty} \lambda_1^i \lambda_2^j \lambda_3^k \rho_1^{i,j,k}(t) \tag{9.56}$$

where $\rho_1(\lambda_1, \lambda_2, \lambda_3; t)$ can be regarded as the generating function for the various Liouville pathways, which allows us to compute a particular contribution to the total polarization, obeying the necessary phase-matching condition. In our case,

$$P_{3P}^{(+)}(t) = \langle V\rho_1^{111}(t)\rangle \tag{9.57}$$

As can be proven by expanding $\rho_1(\lambda_1, \lambda_2, \lambda_3; t)$ in a Taylor series,

$$\begin{aligned} \lambda_1\lambda_2\lambda_3\rho_1^{111}(t) =& \rho_1(\lambda_1,\lambda_2,\lambda_3;t)+\rho_1(\lambda_1,0,0;t)-\rho_1(\lambda_1,0,\lambda_3;t) \\ &-\rho_1(\lambda_1,\lambda_2,0;t)-\rho_1(0,\lambda_2,\lambda_3;t)-\rho_1(0,0,0;t)+\rho_1(0,0,\lambda_3;t) \\ &+\rho_1(0,\lambda_2,0;t)+O(\lambda_1^n\lambda_2^k\lambda_3^m) \quad n+k+m>3 \end{aligned} \tag{9.58}$$

Therefore, the 3PPE polarization can be evaluated as

$$P_{3P}^{(+)}(t) = \langle V[\rho_1(t)-\rho_2(t)-\rho_3(t)+\rho_4(t)-\hat{\rho}_1(t)+\hat{\rho}_2(t)+\hat{\rho}_3(t)]\rangle \tag{9.59}$$

Here, $\rho_1(t)$ obeys Eq. 9.55 and

$$\partial_t\rho_2(t) = -i[H - \upsilon_1^{(-)}(t) - \upsilon_2^{(+)}(t), \rho_2(t)] - D\rho_2(t) \tag{9.60}$$

$$\partial_t\rho_3(t) = -i[H - \upsilon_1^{(-)}(t) - \upsilon_3^{(+)}(t), \rho_3(t)] - D\rho_3(t) \tag{9.61}$$

$$\partial_t\rho_4(t) = -i[H - \upsilon_1^{(-)}(t), \rho_4(t)] - D\rho_4(t) \tag{9.62}$$

The density matrices $\hat{\rho}_i(t)$ obey the same equations as the $\rho_i(t)$, but with the $\upsilon_1^{(-)}(t)$ term omitted. In writing Eq. 9.59, we have used the fact that $\langle V\rho_1(0,0,0;t)\rangle \equiv 0$, assuming that there exists no permanent dipole moment in the electronic ground state.

In the derivation of Eq. 9.59, we did not make use of the RWA. Equation 9.59 gives therefore the general three-pulse induced polarization up to the third order. The RWA leads to considerable additional simplifications. Since

$$(X^\dagger)^2 = X^2 = 0 \tag{9.63}$$

$\langle X\rho_4(t)\rangle \equiv 0$. Furthermore, $\langle X\hat{\rho}_1(t)\rangle = \langle X(\hat{\rho}_2(t) + \hat{\rho}_3(t))\rangle$, because only the terms linear in the laser fields contribute to $\hat{\rho}_1(t)$, $\hat{\rho}_2(t)$, and $\hat{\rho}_3(t)$. The last three terms in

Eq. 9.59 cancel each other, and we arrive at the result [15, 29]

$$P_{3P}^{(+)}(t) = \langle X[\rho_1(t) - \rho_2(t) - \rho_3(t)]\rangle \tag{9.64}$$

The 3PPE polarization can thus be evaluated within the RWA by performing propagations of just three density matrices. The analysis can straightforwardly be generalized to systems with more than two electronic states.

A few comments are in order. First, the $\rho_i(t)$ and $\hat{\rho}_i(t)$ are not true density matrixes: Since the $v_a^{(\pm)}(t)$ are complex, $\rho_i(t)$ and $\hat{\rho}_i(t)$ are not Hermitian operators. Second, Eq. 9.59 is valid in the leading order of the perturbation expansion in the optical fields involved, that is, $P_{3P}(t) \sim \lambda_1\lambda_2\lambda_3 + O(\lambda_1^n\lambda_2^k\lambda_3^m)$, $n+k+m > 3$. Thus the domain of validity of Eq. 9.59 coincides with that of the third-order perturbation expansion, Eq. 9.1. Third, Eq. 9.59 accounts for all effects due to pulse overlaps automatically.

Static inhomogeneous broadening of the 3PPE transients has to be taken into account in many applications: Due to the interaction with the environment, the frequencies of the electronic transitions are not fixed but have a certain distribution. Therefore, we have to average the 3PPE signal over this distribution. As has been shown in Cheng et al. [30], this procedure can efficiently be accomplished with the Gauss–Hermite integration method. More generally, the influence of the environment results in a time-dependent modulation of the transition frequencies. Such a dynamic broadening, which also occurs in intermediate situations between inhomogeneous and homogeneous broadening, can be described, for example, by combining the stochastic Liouville equation [e.g., 5, 18] with the EOM-PMA.

Once the three-pulse induced polarization is known, any four-wave-mixing (4WM) signal can straightforwardly be calculated. For example, the homodyne- and heterdyne-detected 3PPE signals are given by Eqs. 9.46 and 9.48, respectively, in terms of the third-order polarization. The calculation of 2D 3PPE signals for a model system is described in Section 9.5.3.

9.4.2 Discussion

To calculate the time evolution of the 3PPE polarization for particular values of the pulse delay times and in a specific phase-matching direction within the EOM-PMA, we have to perform three (with the RWA) or seven (without the RWA) *independent* propagations of density matrices. All other known methods for the calculation of the third-order polarization in a specific phase-matching direction are computationally more expensive. As has been shown in the literature [31–33], the a posteriori extraction [11] of the 3PPE polarization from the total polarization within the RWA requires the solution of a 12×12 system of linear equations. This implies that one has to determine the time evolution of 12 density matrices (in fact, one has to perform three additional time propagations to remove the linear terms from the nonlinear polarization) and to solve a 12×12 system of linear equations at each time step. Without invoking the RWA, the computational cost is even higher. The use of the phase-cycling procedure requires determining the time evolution of 16 auxiliary density matrices [34, 35]. Alternatively, it is possible to perform a discrete Fourier transform of the underlying Liouville equation with respect to the phases of the pertinent pulses [3, 4]. When the Frenkel exciton formalism is used and a certain closure approximation for the corresponding nonlinear exciton equations is adopted, this procedure results in a closed system of coupled Liouville equations corresponding to different Fourier components of the total density matrix [3, 4]. Without this system-specific simplification, the computational scaling of the method coincides with those employed elsewhere [31–33]. On the other hand, the approach of Renger et al. [3] and Mukamel and Abramavicius [4] can be extended beyond the weak-pulse limit at the expense of an increase in the number of coupled master equations [36].

Nonperturbative methods have been applied to calculate two-time fifth-order nonresonant Raman response functions [37–39], three-time third-order infrared response functions [40, 41] and (with additional approximations) three-time third-order optical response functions [42] via classical nonequilibrium molecular dynamics simulations. The method of Yagasaki and Saito [40] [see their Eq. (6)] is conceptually similar to the three-pulse EOM-PMA. We propose that the EOM-PMA can also be incorporated into nonequilibrium computer simulation schemes which are based on classical trajectories. The application of this strategy for optical 4WM signals may require additional approximations, similar to those made in Ka and Geva [42]. The application of the EOM-PMA to 2D infrared spectroscopy, on the other hand, seems to be quite straightforward. In this case, one can avoid the computation of stability matrixes, which is a bottleneck in semiclassical simulations of the response functions [43]. We have to perform three (with the RWA) or seven (without the RWA) series of (short) molecular dynamics simulations (with the initial conditions sampled according to the equilibrium distribution without external fields) in order to get the 4WM signal for particular values of interpulse delays and carrier frequencies. To obtain the signal for different values of the parameters, the simulation cycle must be repeated, which can be computationally expensive. Nevertheless, this procedure can be much cheaper than the direct simulation of the nonlinear response functions and subsequent calculation of signals by multiple time integrals. If we are interested, for example, in a 4WM transient as a function of the delay T between the second and the third pulses, we can obtain the desired signal by performing $3N$ (with the RWA) or $7N$ (without the RWA) series of simulations, N being the number of discretization intervals of the T axis. In the traditional perturbative approach, the complete three-time response function $S(t_1, t_2, t_3)$ is required for the calculation of a particular 4WM transient beyond the impulsive limit. This requires $\sim N_1 \times N_2 \times N_3$ series of simulations and N subsequent evaluations of triple-time integrals.

9.5 APPLICATION OF EOM-PMA TO MODEL SYSTEMS WITH NONTRIVIAL ULTRAFAST DYNAMICS

To illustrate the application of the EOM-PMA, we consider a series of model systems with nontrivial multilevel excited-state dynamics, which is governed by electronic–vibrational intrastate interactions, electronic interstate couplings, as well as weak vibrational dissipation and optical dephasing. We calculate TFG SE and electronic 2D 3PPE spectra for these models in order to illustrate how the vibrational wavepacket dynamics and interstate electronic couplings manifest themselves in spectroscopic observables.

9.5.1 Model Hamiltonians and Relaxation Operators

The system is described by three electronic states (the ground state $|g\rangle$ and two excited states $i = |1\rangle, |2\rangle$), which are coupled to a single harmonic vibrational mode with dimensionless coordinate Q. The system Hamiltonian is given by Eqs. 9.2–9.4 with

$$h_i = \tfrac{1}{2}\Omega\{P^2 + Q^2\} - \Omega\Delta_i Q \tag{9.65}$$

where $i = g$, 1, 2, P is the momentum conjugated to the coordinate Q, and Ω is the vibrational frequency of the harmonic mode; Δ_1 and Δ_2 are the dimensionless displacements of the excited-state equilibrium geometries from the ground-state geometry ($\Delta_g = 0$).

It is assumed that the excited state $|1\rangle$ is optically bright, while the state $|2\rangle$ is optically dark. In the Condon approximation, the operator X in Eq. 9.7 is thus defined as

$$X = |g\rangle\langle 1| \tag{9.66}$$

The relaxation operator D in Eq. 9.10 and in all subsequent master equations is taken as the sum of the vibrational relaxation operator $\Re$ and the optical dephasing operator $\wp$,

$$D\rho_i(t) = \Re\rho_i(t) + \wp\rho_i(t) \tag{9.67}$$

where $\Re$ is described by multilevel Redfield theory [44] as is detailed elsewhere [45]. Briefly, vibrational relaxation is introduced via a bilinear coupling of the system oscillator mode to a harmonic bath, characterized by an ohmic spectral function $J(\omega_b) = \eta\omega_b \exp(-\omega_b/\omega_c)$ [17], where η is a dimensionless system–bath coupling parameter and ω_c is the bath cutoff frequency, which is taken as $\omega_c = \Omega$. The optical dephasing operator is defined as

$$\wp\rho(t) \equiv \xi_{eg}\mathcal{P}_g\rho(t)\mathcal{P}_e + \text{H.c.} \tag{9.68}$$

ξ_{eg} being the optical dephasing rate, $\mathcal{P}_g \equiv |g\rangle\langle g|$ being the projector to the ground electronic state, and $\mathcal{P}_e \equiv 1 - \mathcal{P}_g$.

In all our calculations, we have assumed weak vibrational dissipation ($\eta = 0.1$) and zero temperature.

9.5.2 Time- and Frequency-Resolved Spontaneous Emission

In this section, we adopt the following numerical values for the system parameters (model I): The frequency of the harmonic mode is $\Omega = 0.05$ eV, so that the vibrational period $\tau_\Omega = 2\pi/\Omega$ is $\approx$83 fs. The dimensionless displacements of the excited-state equilibrium geometries from the ground-state geometry are $\Delta_1 = -2$ and $\Delta_2 = -1$. The vertical excitation energies are chosen as $\epsilon_1 = \epsilon_2 + 3.5\,\Omega$, and the electronic interstate coupling is $U_{12} = \Omega/5 = 0.01$ eV. The diabatic potential energy surfaces ($U_{12} = 0$) are shown in Figure 9.1. The pump pulse is weak and short ($\Gamma_1 = \Omega$) and possesses an exponential envelope (Eq. 9.24). The effects of strong pumping have been studied [19].

9.5.2.1 Ideal SE Specta The ideal SE spectrum simultaneously provides perfect time and frequency resolution and contains the maximum of information on the excited-state dynamics. The spectra calculated for a short pump pulse and progressively increasing values of the optical dephasing ($\xi_{eg} = \Omega/50, \Omega/5, \Omega/2, \Omega$) are shown in Figures 9.2*a*, *b*, *c*, and *d*, respectively. The shape of the spectra obviously is very sensitive to the value of ξ_{eg}. For very small dephasing (Figure 9.2*a*), wavepacket-like vibrational motion is reflected in the spectrum only at short times $\sim\tau_\Omega/2$. Afterward, vibrational interferences occur and the spectrum splits into narrow peaks which roughly correspond to emission from vibrational eigenstates. As the vibrational relaxation increases, one can see a quenching of the emission from the higher vibrational levels, whereas the peaks corresponding to emission from the lower

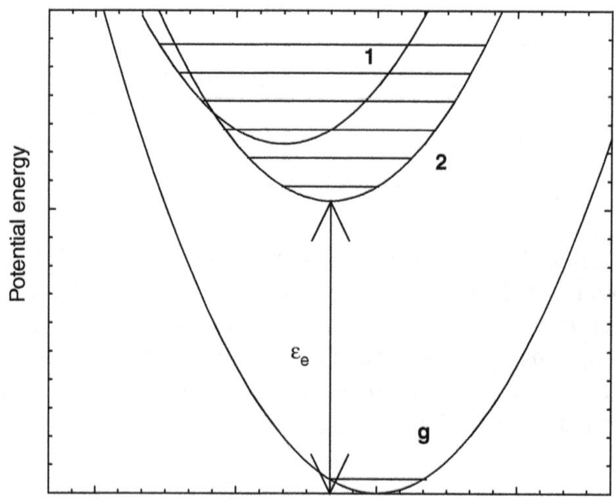

Figure 9.1 Ground-state (g) and excited-state (1, 2) diabatic potential energy surfaces of model I. Indicated are the vibrational ground level of $|g\rangle$ and the vibrational levels of the uncoupled ($U_{12} = 0$) excited states $|1\rangle$ and $|2\rangle$; ϵ_e represents the characteristic electronic excitation energy.

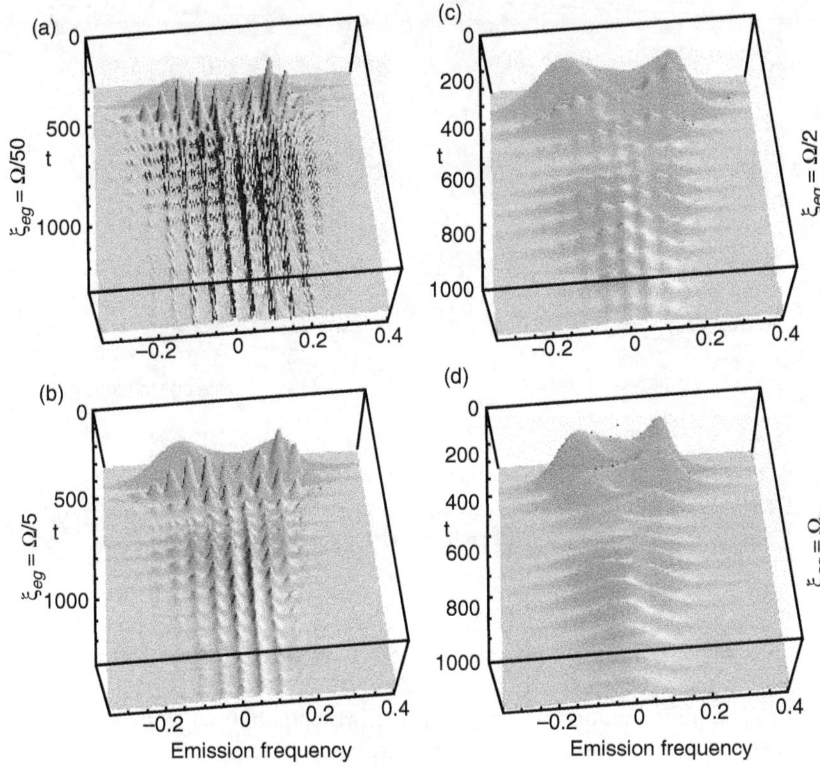

Figure 9.2 Ideal time- and frequency-resolved SE spectra $S(t, \bar{\omega}_S)$ of model I for a short pump pulse, $\tau_L = 1/\Omega$, and *(a)* $\xi_{eg} = \Omega/50$, *(b)* $\xi_{eg} = \Omega/5$, *(c)* $\xi_{eg} = \Omega/2$, and *(d)* $\xi_{eg} = \Omega$. Time and frequency units are femtoseconds and electron volts, respectively.

vibrational levels gain intensity. The pronounced interference patterns seen in Figure 9.2*a* can be interpreted as the result of interference of various coherent pathways contributing to SE at a certain frequency and at a particular time. With increasing optical dephasing, the sharp structures of the spectrum are washed out. For $\xi_{ge} = \Omega/5$ and $\xi_{ge} = \Omega/2$, one can still distinguish traces of the individual peaks (Figures 9.2*b* and *c*). If ξ_{ge} is of the order of Ω (Figure 9.2*d*), the SE spectrum essentially maps the vibrational wavepacket (see ref. 46 for further details). The recurrence of the overall intensity of emission at $t \approx 630$ fs represents a so-called electronic beating which is caused by the electronic interstate coupling U_{12}.

9.5.2.2 TFG SE Spectra The ideal time- and frequency-resolved SE spectrum would be observed with perfect time and frequency resolution. In reality, one has to sacrifice either the former or the latter when measuring two-dimensional TFG SE spectra (an increase of the temporal resolution results in a decrease of the frequency resolution and vice versa). Once the ideal time- and frequency-resolved signal is known, one can calculate the real TFG spectrum for a given gate pulse and frequency

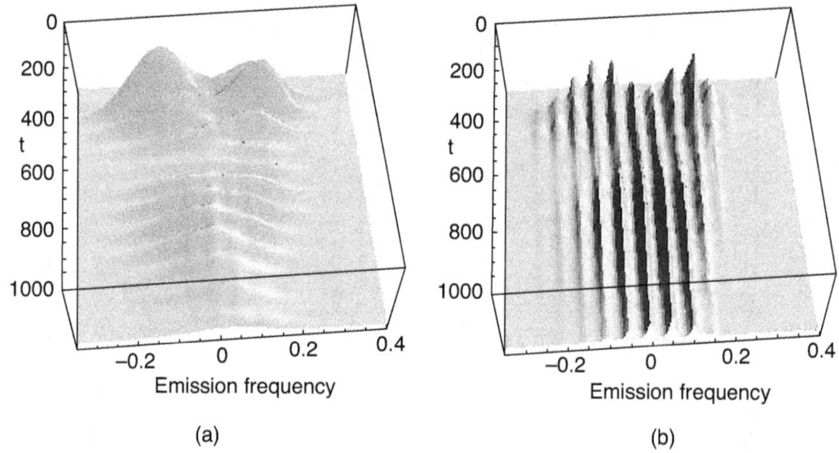

Figure 9.3 TFG SE spectra $S_{\text{TFG}}(t, \bar{\omega}_S)$ of model I calculated for a short pump pulse, $\tau_L = 1/\Omega$, weak OD, $\xi_{eg} = \Omega/50$, and good (a, $\tau_t = 1/\Omega$) as well as poor (b, $\tau_t = 5/\Omega$) time resolution. Time and frequency units are femtoseconds and electron volts, respectively.

filter by an appropriate convolution (Eq. 9.21). We assume an exponential gate pulse so that the joint TFG function is given by Eq. 9.25. There is no intrinsic physical limitation on the frequency resolution, in contrast to the time resolution, which is limited by the finite duration of the gate pulse, which is limited from below by the optical period. We thus assume the spectral resolution to be perfect in all subsequent calculations ($\gamma = 0$ in Eq. 9.25) and concentrate on the influence of the gate-pulse duration, $\tau_t \equiv 1/\Gamma$, on the TFG SE spectra.

As we have seen in Section 9.5.2.1, the ideal SE spectrum depends strongly on the rate of optical dephasing. TFG SE spectra calculated for a small dephasing ($\xi_{eg} = \Omega/50$) with good ($\tau_t = 1/\Omega$) as well as with poor ($\tau_t = 5/\Omega$) time resolution are shown in Figure 9.3. If the gate pulse is short (Figure 9.3a), the complicated interference structures of the ideal signal (Figure 9.2a) cannot be resolved and the spectrum reflects coherent wavepacket motion in the optically bright state. The vibrational motion cannot be monitored if longer gate pulses are employed ($\tau_t = 5/\Omega$, Figure 9.3b). On the other hand, the improved frequency resolution allows us to resolve the emission lines of individual vibrational levels which are clearly seen in the ideal signal in Figure 9.2a.

9.5.3 Two-Dimensional Three-Pulse Photon Echo

We adopt the following parameter values of the model Hamiltonian (model II, [47]). The frequency of the harmonic mode is $\Omega = 0.074$ eV. The dimensionless displacements of the excited-state equilibrium geometries from the ground-state geometry are $\Delta_1 = 1$ and $\Delta_2 = 3$. The vertical excitation energies are chosen as $\epsilon_2 = \epsilon_1 + 2\,\Omega$, and the electronic coupling is $U_{12} = 0.02$ eV. The corresponding diabatic and adiabatic potential energy surfaces are shown in Figure 9.4.

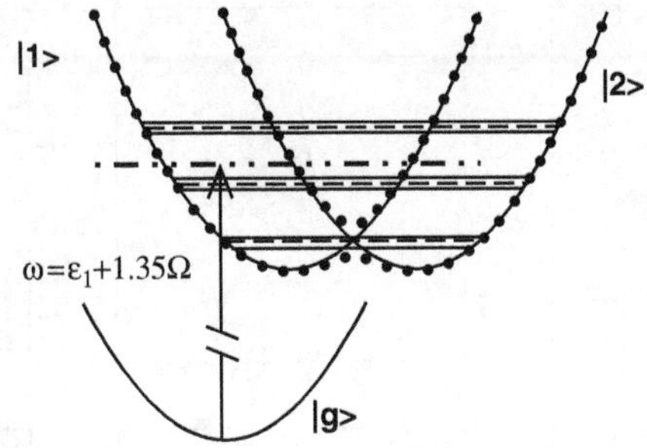

Figure 9.4 Electronic ground-state (solid line), excited-state diabatic (solid lines), and adiabatic (dots) potential energy functions of model II. Unperturbed vibrational levels (dashed lines) and vibronic energy levels (solid lines) as well as the pulse carrier frequency are indicated.

The optical dephasing rate is chosen as $\xi_{eg} = 130\,\text{fs}^{-1}$ ($\sim$0.07 Ω). All laser pulses are assumed to have Gaussian envelopes,

$$E(t) = \exp\{-(\Gamma t)^2\} \tag{9.69}$$

equal amplitudes, the same carrier frequencies ($\omega_1 = \omega_2 = \omega_3 = \omega = \epsilon_1 + 1.35\Omega$) and durations (full width at half maximum $2\sqrt{\ln 2}/\Gamma = 25$ fs).

In the limit of ideal detection, the 2D 3PPE signal is obtained by a double Fourier transformation of the nonlinear polarization $P_{3P}(t, \tau, T)$ with respect to the coherence time τ (delay between the first two pulses) and the detection time t,

$$I(\omega_\tau, \omega_t, T) \sim \int d\tau \int dt \exp(-i\omega_\tau \tau)\exp(i\omega_t t)P_{3P}(t, \tau, T) \tag{9.70}$$

where T is the population time, that is, the delay between the second and third pulses. The frequencies associated with the coherence time (ω_τ) and the detection time (ω_t) are usually referred to as absorption (or excitation) and emission (or probe) frequencies, respectively. The two dimensions in frequency space are provided by ω_τ and ω_t, and the 2D scans are recorded at a fixed population time T. The 2D signal (Eq. 9.70) is complex valued. We consider only the real part which is associated with absorption [48]. The emission and absorption frequencies (ω_t and ω_τ) are given relative to the vertical excitation energy ϵ_1. The latter is arbitrary and does not need to be explicitly defined. To account for a finite duration of the detection (LO) pulse, the signal in Eq. 9.70 is appropriately convoluted. The shape and parameters of the detection pulse are as those of the incoming pulses.

Figure 9.5 shows 2D scans calculated for various population times. The absorption frequency ω_τ reveals those transitions that are excited in the experiment. The emission frequency ω_t, on the other hand, indicates at which frequencies the emission takes place. The peaks corresponding to $\omega_\tau = \omega_t$ are located on the diagonal of each 2D scan, whereas the peaks with $\omega_\tau \neq \omega_t$ are the so-called cross peaks. Under the assumed conditions (low temperature and short pulses), the 2D spectra clearly reveal the multilevel structure of the electronic states. The most strongly populated levels are

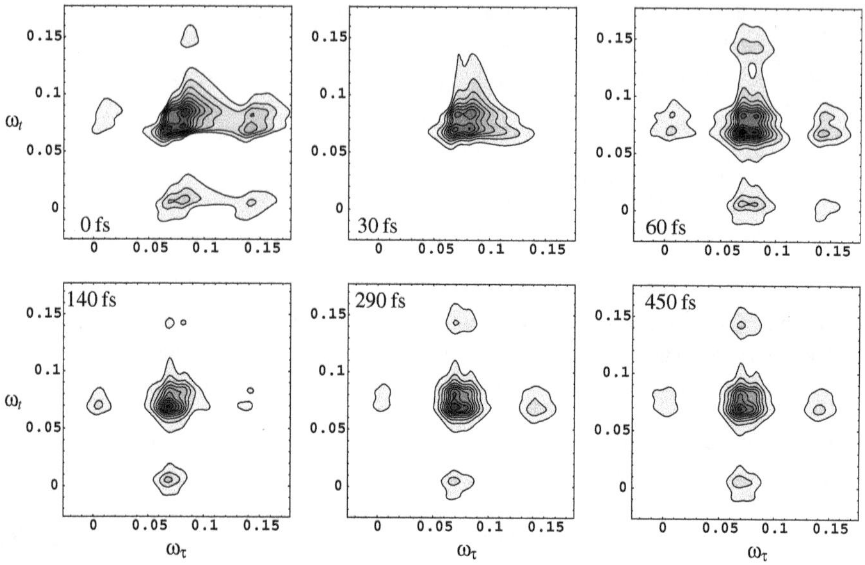

Figure 9.5 2D 3PPE scans of model II for different population times (indicated on the panels). The intensity scaling is the same in all graphs. Only the positive parts of the spectra are displayed.

those closest to the laser frequency, $\omega_\tau = 0.067$ eV and $\omega_\tau = 0.082$ eV. The transitions to the levels which are one vibrational quantum higher or lower than the most efficiently populated levels are also resolved. All the excited levels are optically coupled to the ground-state vibrational manifold. Therefore, there are many ways to satisfy $\omega_\tau \neq \omega_t$. The ground-state manifold is harmonic, and the location of the peaks in the 2D patterns reveals the system frequencies in the vibronically coupled excited state (cf. Figure 9.4). The larger peak separations are of the order of the vibrational frequency Ω. The smaller splittings arise from the electronic interstate coupling U_{12}.

A comparison of the 2D scans for various population times reveals considerable intensity modulations of the peaks. There are several mechanisms leading to these intensity modulations. One of them is population relaxation in the excited states: While absorption occurs always at the same frequencies (for given parameters), emission depends on T due to the population flow from higher to lower vibronic levels. At larger T, the 2D pattern thus loses intensity in the region of larger ω_t and gains intensity at smaller emission frequencies. Another reason for the intensity modulations with T are coherences which are created during the excitation process. If several levels are coherently excited, signal modulations with frequencies corresponding to energy differences between these levels can occur. The intensity modulations in Figure 9.5 are mostly of coherent character, since the effect of population relaxation is not yet very pronounced at the short population times considered. The two dominant time scales of oscillations in the peak intensities are determined by the characteristic system frequencies, that is, by the vibrational frequency Ω (higher frequency) and by the electronic coupling U_{12} (lower frequency). The faster time scale is revealed, for example, by comparison of the scans at T values of 0, 30, and 60 fs: The off-diagonal peaks lose their intensity at 30 fs and reappear again at 60 fs. The longer period is visible, for example, as a pronounced change in the structure of the central peak at $T = 140$ fs, compared to $T = 0$, 30, 60 fs. At early times (upper panels in Figure 9.5) one can clearly see a four-peak substructure, whereas at 140 fs essentially one peak is resolved.

A better view of the oscillations of peak intensities with T is provided by Figure 9.6, which shows the intensity modulation of cross [panel (a)] and diagonal [panel (b)]

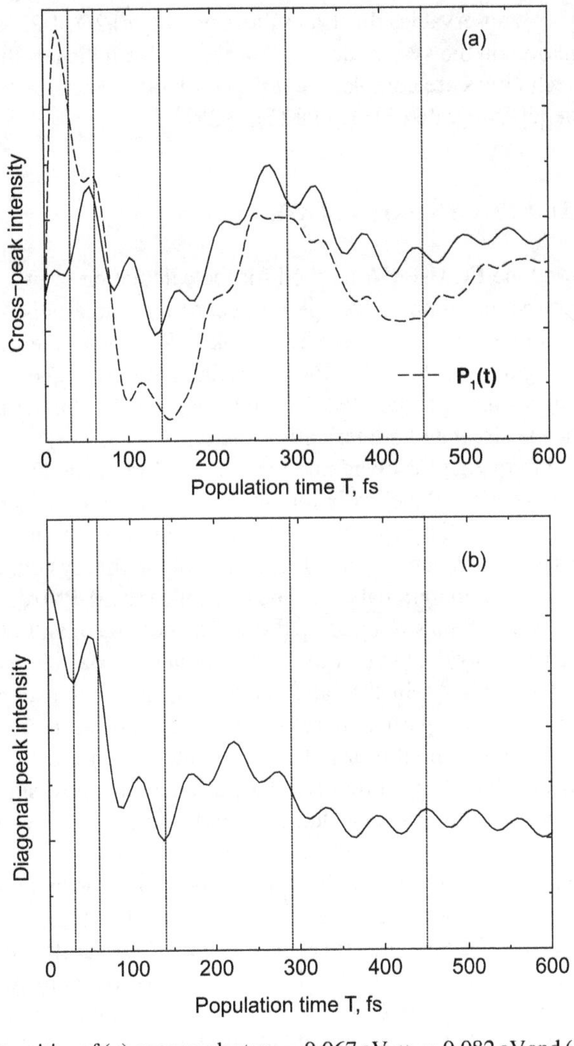

Figure 9.6 Intensities of (a) cross peak at $\omega_\tau = 0.067$ eV, $\omega_t = 0.082$ eV and (b) diagonal peak at $\omega_\tau = \omega_t = 0.082$ eV vs. population time T for model II. The dashed line in (a) represents the population dynamics $P_1(t)$ of the diabatic state $|1\rangle$. The vertical lines indicate the values of T at which the 2D spectra in Figure 9.5 are shown.

peaks. We have selected $\omega_\tau = 0.067$ eV, $\omega_t = 0.082$ eV for the cross peak and $\omega_\tau = \omega_t = 0.082$ eV for the diagonal peak. The cross peak represents the coupling between the transitions with eigenfrequencies about 0.067 and 0.082 eV, while the diagonal peak describes transitions with very similar frequencies around 0.082 eV.

Both the cross and diagonal peaks exhibit oscillations with the characteristic frequencies of the material system. In addition to these oscillations, the diagonal peak intensity (Fig. 9.6*b*) exhibits a slow decay which reflects population relaxation to lower vibronic states. Figure 9.6*a* compares the intensity evolution of the cross peak with the population dynamics $P_1(t)$ in the bright diabatic state (dashed line). The similarity of the two observables illustrates the capability of 2D 3PPE spectroscopy to provide information on the system density matrix. It also indicates that the TFG SE and 2D 3PPE techniques are complementary to each other because $P_1(t)$ essentially gives the frequency-integrated SE signal (Eq. 9.26).

9.6 CONCLUSIONS AND OUTLOOK

We have reviewed the EOM-PMA method for the calculation of two-pulse-induced (spontaneous emission, pump–probe, photon echo) and three-pulse-induced (transient grating, photon echo, coherent anti-Stokes–Raman scattering, four-wave-mixing) optical signals. In the EOM-PMA, the interactions of the system with the relevant laser pulses are incorporated into the system Hamiltonian and the driven system dynamics is simulated numerically exactly.

The time- and frequency-resolved nonlinear signals are related via a perturbation expansion (with respect to the radiation–matter interaction) to certain auxiliary density matrices which satisfy slightly modified equations of motion. The auxiliary density matrices are propagated in time together with the true system density matrix with little additional computational cost. The two-pulse EOM-PMA is not limited to weak pulses and allows for a strong pump pulse. The domain of validity of the three-pulse EOM-PMA is equivalent to that of the traditional approach based on the third-oder response functions. As in the latter approach, the nonlinear polarization is directly obtained for each particular phase-matching condition. The a posteriori decomposition of the total nonlinear polarization into the different phase-matching directions is thus avoided, which reduces the computational cost significantly. The EOM-PMA allows for arbitrary pulse durations and automatically accounts for pulse overlap effects.

The three-pulse EOM-PMA can be formulated not only in terms of density matrices and master equations but also in terms of wavefunctions and Schrödinger equations [29]. The EOM-PMA can therefore be straightforwardly incorporated into computer programs which provide the time evolution of the density matrix or the wavefunction of material systems. Besides the multilevel Redfield theory, the EOM-PMA can be combined with the Lindblad master equation [49], the surrogate Hamiltonian approach [49], the stochastic Liouville equation [18], the quantum Fokker–Planck equation [18], and the density matrix [50] or the wavefunction [14] multiconfigurational time-dependent Hartree (MCTDH) methods. When using the

RWA, just a few (two for SE and 2PPE, three for PP and 3PPE) *independent* propagations of density matrices have to be performed. The corresponding computer codes can straightforwardly be implemented on parallel computers. If necessary, averaging of the signal over a stochastic distribution of system parameters can be efficiently accomplished via a Gauss–Hermite integration.

To summarize, the EOM-PMA considerably facilitates the computation of various optical signals and 2D spectra. With slight alterations, the EOM-PMA can also be applied to compute nonlinear responses in the infrared (IR). The three-pulse EOM-PMA can be extended to calculate the N-pulse-induced nonlinear polarization [51], which opens the way for the interpretation of fifth-order spectroscopies, such as heterodyned 3D IR [52], transient 2D IR [53, 54], polarizability response spectroscopy [55], resonant-pump third-order Raman-probe spectroscopy [56], femtosecond stimulated Raman scattering [57], four–six-wave-mixing interference spectroscopy [58], or (higher than fifth order) multiple quantum coherence spectroscopy [59].

ACKNOWLEDGMENTS

This work has been supported by the Deutsche Forschungsgemeinschaft (DFG) through a research grant and through the DFG Cluster of Excellence "Munich Centre of Advanced Photonics" (www.munich-photonics.de).

REFERENCES

1. S. Mukamel, *Principles of Nonlinear Optical Spectroscopy*, Oxford University Press, New York, 1995.
2. M. F. Gelin, A. V. Pisliakov, D. Egorova, W. Domcke, *J. Chem. Phys.* **2003**, *118*, 5287.
3. T. Renger, V. May, O. Kühn, *Phys. Rep.* **2001**, *343*, 137.
4. S. Mukamel, D. Abramavicius, *Chem. Rev.* **2004**, *104*, 2073.
5. W. Zhuang, T. Hayashi, S. Mukamel, *Angew. Chem.-Int. Ed.* **2009**, *48*, 3750.
6. M. Cho, *Chem. Rev.* **2008**, *108*, 1331.
7. A. Ishizaki, Y. Tanimura, *J. Chem. Phys.* **2006**, *125*, 084501.
8. Y. Tanimura, A. Ishizaki, *Acc. Chem. Res.* **2009**, *42*, 1270.
9. Q. Shi, E. Geva, *J. Chem. Phys.* **2005**, *122*, 064506.
10. Q. Shi, E. Geva, *J. Chem. Phys.* **2008**, *129*, 124505.
11. L. Seidner, G. Stock, W. Domcke, *J. Chem. Phys.* **1995**, *103*, 3998.
12. H. Rabitz, *Theor. Chem. Acc.* **2003**, *109*, 64.
13. M. Shapiro, P. Brumer, *Phys. Rep.* **2006**, *425*, 195.
14. H. Wang, M. Thoss, *Chem. Phys.* **2008**, *347*, 139.
15. M. F. Gelin, D. Egorova, W. Domcke, *Acc. Chem. Res.* **2009**, *42*, 1290.
16. R. M. Hochstrasser, *Proc. Natl. Acad. Sci.* **2007**, *104*, 14190.
17. U. Weiss, *Quantum Dissipative System*, World Scientific, Singapore, 1999.
18. Y. Tanimura, *J. Phys. Soc. Jpn.* **2006**, *75*, 082001.

19. D. Egorova, M. F. Gelin, M. Thoss, H. Wang, W. Domcke, *J. Chem. Phys.* **2008**, *129*, 214303.
20. M. F. Gelin, D. Egorova, W. Domcke, *Chem. Phys.* **2004**, *301*, 129.
21. M. F. Gelin, D. Egorova, W. Domcke, *Chem. Phys.* **2005**, *312*, 135.
22. S. Mukamel, C. Ciordas-Ciurdariu, V. Khidekel, *Adv. Chem. Phys.* **1997**, *101*, 345.
23. S. Mukamel, *J. Chem. Phys.* **1997**, *107*, 4165.
24. F. Shuang, C. Yang, Y. J. Yan, *J. Chem. Phys.* **2001**, *114*, 3868.
25. Y. Tanimura, S. Mukamel, *J. Phys. Soc. Jpn.* **1994**, *63*, 66.
26. Y. Tanimura, Y. Maruyama, *J. Chem. Phys.* **1997**, *107*, 1779.
27. S. Hahn, G. Stock, *Chem. Phys. Lett.* **1998**, *296*, 137.
28. Y. J. Yan, R. E. Gillilan, R. M. Whitnell, K. R. Wilson, S. Mukamel, *J. Phys. Chem.* **1993**, *97*, 2320.
29. M. F. Gelin, D. Egorova, W. Domcke, *J. Chem. Phys.* **2005**, *123*, 164112.
30. Y. C. Cheng, H. Lee, G. R. Fleming, *J. Phys. Chem. A* **2007**, *111*, 9499.
31. S. Meyer, V. Engel, *Appl. Phys. B* **2000**, *71*, 293.
32. T. Kato, Y. Tanimura, *Chem. Phys. Lett.* **2001**, *341*, 329.
33. T. Mančal, A. V. Pisliakov, G. R. Fleming, *J. Chem. Phys.* **2006**, *124*, 234504.
34. D. Keusters, H.-S. Tan, W. S. Warren, *J. Phys. Chem. A* **1999**, *103*, 10369.
35. P. Tian, D. Keusters, Y. Suzaki, W. S. Warren, *Science* **2003**, *300*, 1553.
36. D. Abramavicius, Y.-Z. Ma, M. W. Graham, L. Valkunas, G. R. Fleming, *Phys. Rev. B* **2009**, *79*, 195445.
37. T. I. C. Jansen, J. G. Snijders, K. Duppen, *J. Chem. Phys.* **2001**, *114*, 10910.
38. R. DeVane, B. Space, T. I. C. Jansen, T. Keyes, *J. Chem. Phys.* **2006**, *125*, 234501.
39. S. Saito, I. Ohmine, *J. Chem. Phys.* **2003**, *119*, 9073.
40. T. Yagasaki, S. Saito, *J. Chem. Phys.* **2008**, *128*, 154521.
41. T. Hasegawa, Y. Tanimura, *J. Chem. Phys.* **2008**, *128*, 064511.
42. B. J. Ka, E. J. Geva, *J. Chem. Phys.* **2006**, *125*, 214501.
43. C. Dellago, S. Mukamel, *J. Chem. Phys.* **2003**, *119*, 9344.
44. W. T. Pollard, A. K. Felts, R. A. Friesner, *Adv. Chem. Phys.* **1996**, *93*, 77.
45. D. Egorova, M. Thoss, W. Domcke, H. Wang, *J. Chem. Phys.* **2003**, *119*, 2761.
46. D. Egorova, M. F. Gelin, W. Domcke, *J. Chem. Phys.* **2005**, *122*, 134504.
47. D. Egorova, *Chem. Phys.* **2008**, *347*, 166.
48. D. M. Jonas, *Annu. Rev. Phys. Chem.* **2003**, *54*, 425.
49. D. Gelman, G. Katz, R. Kosloff, M. A. Ratner, *J. Chem. Phys.* **2005**, *123*, 134112.
50. B. Brüggemann, P. Persson, P. H.-D. Meyer, V. May, *Chem. Phys.* **2008**, *347*, 152.
51. M. F. Gelin, D. Egorova, W. Domcke, *J. Chem. Phys.* **2009**, *131*, 194103.
52. F. Ding, M. T. Zanni, *Chem. Phys.* **2007**, *341*, 95.
53. J. Bredenbeck, J. Helbing, C. Kolano, and P. Hamm, *Chem. Phys. Chem.* **2007**, *8*, 1747.
54. H. S. Chung, Z. Ganim, K. C. Jones, A. Tokmakoff, *Proc. Natl. Acad. Sci. USA* **2007**, *104*, 14237.
55. A. M. Moran, S. Park, N. F. Scherer, *J. Chem. Phys.* **2007**, *127*, 184505.

56. D. F. Underwood, D. A. Blank, *J. Phys. Chem. A* **2005**, *109*, 3295.
57. P. Kukura, D. W. McCamant, R. A. Mathies, *Annu. Rev. Phys. Chem.* **2007**, *58*, 461.
58. Y. Zhang, U. Khadka, B. Anderson, M. Xiao, *Phys. Rev. Lett.* **2009**, *102*, 013601.
59. N. A. Mathew, L. A. Yurs, S. B. Block, A. V. Pakoulev, K. M. Kornau, *J. Phys. Chem. A* **2009**, *113*, 9261.

10

TIME-DEPENDENT APPROACHES TO CALCULATION OF STEADY-STATE VIBRONIC SPECTRA: FROM FULLY QUANTUM TO CLASSICAL APPROACHES

ALESSANDRO LAMI AND FABRIZIO SANTORO

CNR—Consiglio Nazionale delle Ricerche, Istituto di Chimica dei Composti Organo Metallici, UOS di Pisa, Area della Ricerca del CNR, Pisa, Italy

Computational Strategies for Spectroscopy: From Small Molecules to Nano Systems, First Edition. Edited by Vincenzo Barone.

In this chapter we present time-dependent (TD) eigenstate-free approaches to the computation of steady-state (continuous-wave) vibronic spectra. After introducing the theoretical framework and the TD expression of spectral lineshapes in terms of time correlation functions we review fully quantum approaches based on the direct solution of the TD Schrödinger equation with particular reference to multiconfigurational TD Hartree and its multi layer generalization, which represent, at the state-of-the-art, the most general and effective TD quantum approach to the computation of spectra also in the presence of strong nonadiabatic interactions. Special attention is given to adiabatic harmonic systems, which can also be treated by time-independent methods as described Chapter 8, showing that in these cases analytical expressions can be derived for the thermal time correlation function, making possible the study of very large systems. We briefly introduce alternative approaches based on semiclassical approximations of the time evolution propagator and discuss in which theoretical framework spectra can be obtained from pure classical dynamical simulations that span the initial-state phase space distribution. Representative examples of the different methodologies are presented and discussed.

10.1 INTRODUCTION

For many years efforts in the field of theoretical quantum chemistry have mainly focused on the task of computing eigenvalues and eigenfunctions of the molecular Hamiltonian by solving the time-independent (TI) Schrödinger equation. This occurred because the (continuous–wave) spectroscopic techniques were essentially conceived for measuring the energies as well as the intensities of each line. The realm of nonstationary states, exhibiting effects due to the movement of nuclei, was out reach since it can be accessed only by short (sub-picosecond) pulses and requires detection devices with the same time resolution. On the other hand, this attitude was also consolidated by the possibility of interpreting the behavior of a nonstationary state as due to the interference of the eigenstates in which it can be decomposed. The parallel development of pulsed lasers and ultrafast electronics has drastically changed the situation, allowing real-time investigation of photoinduced processes and providing an always increasing amount of time-resolved data [1]. The growing interest in achieving a complete analysis and understanding of these data has stimulated the development of efficient numerical methods for directly tackling the TD Schrödinger equation (TDSE), avoiding the bottleneck of the computation of eigenstates, that is, directly propagating nuclear wavepackets on single or multiple and coupled potential energy surfaces (PESs) [2]. Nowadays these methods not only are indispensable for interpreting time-resolved spectroscopic experiments but also are frequently used for computing scattering cross sections or other time-independent observables. The TD approach offers the advantage of a direct physical interpretation, since it firmly links the spectrum to the underlying dynamical process. As pointed out by Beck et al. [3], there are also important technical advantages, such as that one has to deal with square-integrable functions when dealing with continuum problems, which is not the case for the TI approach, and the wavepacket to be propagated is usually much more localized than the eigenstates.

TD methods for the computation of steady-state spectra represent probably the best and more general approach for those cases where direct calculation of eigenvalues/ eigenstates of the molecular Hamiltonian is not feasible or is computationally very demanding. These include molecular systems where some modes are strongly anharmonic or where nonlinear couplings between modes play a crucial role (the two effects usually coexist). A further important class of systems for which a TD description is best suited are those in which the electronic states involved in the electronic transition are subject to strong nonadiabatic coupling. For these cases, it is usually easier to propagate wavepackets and to extract the spectrum from time-dependent correlation function. The convenience of TD methods is even larger if the main interest is on low-resolution spectra, since they require only propagation for short time intervals. At variance, when the focus is on high-resolution spectra, TD computations become more cumbersome. Similarly, the assignment of bands in terms of quantum numbers from TD data is often more involved than it is in a TI approach, since the latter directly works on eigenstates.

As discussed in detail in Chapter 8, when nonadiabatic couplings are negligible for the electronic states involved in the optical transition and the molecular system is rigid enough to make harmonic approximation reliable, vibronic eigenstates can be computed by a simple harmonic analysis of the PESs of the initial and final electronic states, a task now feasible for both ground and excited states even in sizable systems (see Chapter 1). In these cases, the computation of the spectrum can be driven back to the evaluation of transition amplitude between known eingenstates. Nowadays, the computational challenge for these systems mainly derives from the huge number of possible transitions that must in principle be taken into account in sizable molecules (easily exceeding 10^{20} transitions in systems with several dozens of modes), thus ruling out brute-force calculations. Chapter 8 describes in detail a number of effective prescreening techniques that, selecting a priori only the relevant transitions, make TI calculations feasible in these cases [4–9]. These techniques have probably brought the capabilities of TI approaches close to the limits where prescreening strategies cannot be sufficient anymore, for the simple reason that the physically relevant transitions for the spectrum lineshape are too numerous to be taken into account. This limit can be reached, for instance, considering larger and larger molecules or focusing on high-temperature spectra, since in the latter cases also the number of initial states that must be taken into account increases steeply. In those cases TD methods are very useful, providing a complementary approach to compute fully converged low-resolution spectra, while TI methods can still be utilized for characterization and assignments of the main stick bands.

In this chapter, after briefly presenting the theoretical framework for the computation of steady-state spectra and the two alternative TI and TD expressions, we will review modern quantum and semiclassical methods, providing some examples of their application. Recent research papers, mainly focused on a proper description of the distribution of the initial molecular states, a very challenging issue when the system is in a complex environment like a solvent or a protein cavity, compute the spectrum according to more simplified classical recipes. The adopted methods are usually classified as TD methods, in the sense that they retrieve the relevant information for the initial-state distribution from molecular dynamics (MD) simulations (with classical or ab initio force fields); furthermore they are classical in the sense that quantum effects on the nuclei motion are disregarded [10, 11]. A section of this chapter is devoted to briefly rediscussing the theoretical foundations of such approaches, starting from the quantum TD description of the spectrum and analyzing the approximation necessary to achieve an expression of the spectrum utilizable in such a classical framework.

10.2 THEORETICAL FRAMEWORK

Matter–radiation interaction, at least for the cases of weak electromagnetic fields adopted in the most common spectroscopic techniques, is theoretically described in the framework of time-dependent perturbation theory. It is therefore clear that any spectroscopic signal may be written in a time-dependent formalism. Linear response perturbation theory is sufficient for one-photon transitions, and, as shown, for example, in Chapter 8, one-photon absorption (OPA) and emission (OPE) and electronic circular dichroism (ECD) can be treated in the same way by simply considering general "transition dipoles" to be specified for the given case. For the sake of simplicity, in this chapter we focus on OPA, but our results can be straightforwardly generalized also to OPE and ECD. Multiphoton spectroscopies require in principle a more complex description obtainable through higher order perturbation theory. In Chapter 8 it is described how, for some nonresonant spectroscopies, by neglecting any vibrational information on the intermediate states, it is possible to work out TI expressions where the vibrational structure only depends on the initial and final electronic states. Analogous expressions could be derived also in a TD framework at the cost of neglecting any dynamical effects on the intermediate states. These latter can be very relevant depending on, for example, the time duration of the excitation pulse [12]. A review on this subject is out of the scope of the present work. With these premises in mind in the following sections we focus on one-photon absorption.

The TD approach is the natural candidate for the interpretation of data coming from pump–probe spectroscopies [13]. If both pump and probe fields are weak, the third-order time-dependent polarization can be computed propagating wave-packets with field-free Hamiltonians [14]. A nonperturbative approach can also be followed, numerically solving the TDSE with the external fields included in the Hamiltonian [15, 16]. Chapter 9 of this book presents the equation-of-motion phase-matching approach (EOM-PMA), which can be considered a mixed perturbative–nonperturbative method.

10.2.1 Absorption Spectrum

The electronic absorption spectrum can be characterized by $W(\omega)$, the rate of energy increase of a single molecule per unit radiant energy density, as a function of the light angular frequency ω. We focus on transitions between vibronic states, referring the interested reader to Chapter 8 and references therein for a discussion of the separation between vibrational and rotational motions. As shown in Chapter 2, in the dipole approximation (when the molecule is small with respect to the light wavelength), $W(\omega)$ is given by

$$W(\omega) = 4\pi^2\omega \sum_{i,f} p_i \left| \langle i|\mu_{\xi_{if}}|f\rangle \right|^2 \cos(\theta_{if})^2 \delta(E_i + \hbar\omega - E_f) \tag{10.1}$$

where $|i\rangle$ and $|f\rangle$ are the initial and final states of the transition, E_i and E_f their energies, ξ_{if} the axis defining the direction of the transition dipole vector $\langle i|\boldsymbol{\mu}|f\rangle$, θ_{if} the angle between the electric field and ξ_{if}, and p_i the Boltzmann population of the initial state $|i\rangle$. If, as usual, the molecule is randomly oriented with respect to the electric field, the angular factor reduces to its average value $\frac{1}{3}$.

The absorption cross section (rate of photon absorption for unit radiant energy flux) is $\sigma(\omega) = W(\omega)/c$ (c being the light speed), while the molar extinction coefficient is simply $N_A\sigma(\omega)$ (N_A being Avogadro's number).

10.2.2 Time-Dependent Formulation

Let us translate Eq. 10.1 into a time-dependent language, considering the zero-temperature spectrum, just for simplicity, where only the ground vibronic state $|g\rangle$ need be considered. Consequently, in the following we will drop the indices specifying the initial state. As a first step we define the (unnormalized) doorway state $|d\rangle$ depending on the relative angle between the light electric field and the matrix element of the transition electric dipole moments:

$$|d\rangle = \sum_{f} |f\rangle\langle f|\mu_{\xi_{if}}|g\rangle \cos(\theta_f) \tag{10.2}$$

By propagating it with the unperturbed evolution operator $\exp(-i\hat{H}_0 t/\hbar)$ (where $\hat{H}_0$ is the field-free Hamiltonian), one has

$$|d(t)\rangle = \sum_f \exp\left(-\frac{iE_f t}{\hbar}\right)|f\rangle\langle f|\mu_{\xi_{if}}|g\rangle \cos(\theta_f) \tag{10.3}$$

Taking into account that for $\eta \to 0^+$ one has $\mathrm{Re}[\int_{-\infty}^{\infty} \exp(iEt/\hbar - \eta t)\, dt] = \hbar\pi\delta(E)$, Eq. 10.1 can be rewritten as

$$W(\omega) = \frac{4\pi\omega}{\hbar}\mathrm{Re}\left\{\int_{-\infty}^{\infty} \langle d(0)|d(t)\rangle \exp\left[i\left(\frac{E_g}{\hbar} + \omega\right)t\right] dt\right\} \tag{10.4}$$

Defining the dipole–dipole correlation function $C_\mu(t)$ as

$$\begin{aligned} C_\mu(t) = \langle g|\mu_H(t)\mu|g\rangle &= \exp\left(\frac{iE_g t}{\hbar}\right)\langle g|\mu \exp\left(-\frac{i\hat{H}_0 t}{\hbar}\right)\mu|g\rangle \\ &= \exp\left(\frac{iE_g t}{\hbar}\right)\langle d(0)|d(t)\rangle \end{aligned} \tag{10.5}$$

Eq. 10.4 becomes

$$W(\omega) = \frac{4\pi\omega}{\hbar}\mathrm{Re}\left[\int_{-\infty}^{\infty} C_\mu(t) \exp(i\omega)t)\, dt\right] \tag{10.6}$$

Inspection of Eqs. 10.2–10.5 clearly shows that $W(\omega)$ depends on all the $\cos^2(\theta_f)$. In the following we will refer to a freely rotating molecule (or to a nonpolarized radiation field), and after performing the average, such angular factors are placed by the unique $\frac{1}{3}$ factor:

$$W(\omega) = \frac{4\pi\omega}{3\hbar}\mathrm{Re}\left[\int_{-\infty}^{\infty} \langle d(0)|d(t)\rangle \exp[i(E_g + \hbar\omega)t]\, dt\right] \tag{10.7}$$

10.2.3 Born–Oppenheimer Separation

As shown in the previous section, the absorption spectrum can be easily evaluated as a Fourier transform of the electric dipole correlation function, whose computation is driven back to the time evolution of the doorway state (Eq. 10.2). The latter is fully determined if all the relevant eigenpairs of $\hat{H}_0$ are known, but it can be also obtained from numerical propagation through eingenstate-free approaches, which makes TD methods very appealing for spectra computations. It is worthwhile to mention here that the direct use of Eq. 10.5 is not the unique way for arriving at the absorption spectrum. The information carried out by the dipole–dipole correlation function can also be investigated by the filter diagonalization method [17], which is more targeted to obtaining precise spectra in a small energy range.

Let us now introduce the standard Born–Oppenheimer separation by which molecular states are written as a product of an electronic wavefunction depending parametrically on the nuclear coordinates $\psi_j(\mathbf{Q},\mathbf{q})$ and a purely vibrational wavefunction $\chi_{jv}(\mathbf{Q})$. This is also done in Chapter 8, but we repeat this treatment with the aim of highlighting its consequences on the system TDSE. Once more we remind the reader that, since the focus here is on condensed-phase spectroscopy, we do not consider the role of molecular rotations, whose contributions to the absorption spectra can be detected only in high-resolution experiments. It is worthwhile to note, however, that they could be included in the time-dependent approach studying the propagation of suitable rovibronic wavepackets.

With resolution of the identity written in terms of adiabatic states,

$$\sum_{j,v} |\psi_j(\mathbf{Q},\mathbf{q})\chi_{jv}(\mathbf{Q})\rangle\langle\psi_j(\mathbf{Q},\mathbf{q})\chi_{jv}(\mathbf{Q})| = 1 \tag{10.8}$$

the doorway state (Eq. 10.2), can be rewritten as

$$|d(0)\rangle = \sum_j |\psi_j(q,Q)\rangle|\phi_j(Q)\rangle \tag{10.9}$$

where

$$|\phi_j(\mathbf{Q})\rangle = \sum_{v,v'} p_{v'}|\chi_{jv}(\mathbf{Q})\rangle\langle\chi_{jv}(\mathbf{Q})|\mu_{e,jg}(\mathbf{Q})|\chi_{gv'}(\mathbf{Q})\rangle \tag{10.10}$$

$$\mu_{e,jg}(\mathbf{Q}) = \langle\psi_j(\mathbf{q},\mathbf{Q})|\mu_{\xi_j}|\psi_g(\mathbf{q},\mathbf{Q})\rangle_{\mathbf{q}} \tag{10.11}$$

and the subscript $\mathbf{q}$ indicates that an integration has been taken over the electronic coordinates only. Equation 10.9 can be rewritten in vector form as (R = row, C = column) $|d(0)\rangle = |\Psi\rangle_R|\boldsymbol{\varphi}\rangle_C$.

The TDSE for the doorway state is then ($\hbar = 1$)

$$i\frac{d}{dt}|d(t)\rangle = i|\boldsymbol{\psi}\rangle_R\frac{d}{dt}|\boldsymbol{\varphi}(t)\rangle_C = H|\boldsymbol{\psi}\rangle_R|\boldsymbol{\varphi}(t)\rangle_C \tag{10.12}$$

Multiplying by ${}_C\langle\boldsymbol{\psi}|$ gives

$$i\frac{d}{dt}|\boldsymbol{\varphi}(t)\rangle_C = H_N(\mathbf{Q})|\boldsymbol{\varphi}(t)\rangle_C \tag{10.13}$$

where

$$H_{Nij}(\mathbf{Q}) = \langle\psi_i(\mathbf{q},\mathbf{Q})|H(\mathbf{q},\mathbf{Q})|\psi_j(\mathbf{q},\mathbf{Q})\rangle_{\mathbf{q}} \tag{10.14}$$

The nuclear Hamiltonian $H_N(\mathbf{Q})$ contains diagonal and off-diagonal terms (see Chapter 8). The form of the kinetic energy operator depends on the chosen coordinates. We consider here the simplest case of normal coordinates (or obtained from them by orthogonal transformations) suitable for semirigid molecules, where the vibrational kinetic operator takes the simple diagonal form (i.e., no coupling among moments) $T_Q = -\sum_\nu (2m_\nu)^{-1}\nabla_\nu^2$, m_ν being the reduced mass associated with the normal coordinate Q_ν. We thus obtain

$$
\begin{aligned}
H_{Nij}(\mathbf{Q}) &= \sum_\nu \frac{1}{2m_\nu}\left(G_{ij}^\nu(\mathbf{Q}) + F_{ij}^\nu(\mathbf{Q})\frac{\partial}{\partial Q_\nu} + \delta_{ij}T_Q\right) \\
G_{ij}^\nu(\mathbf{Q}) &= \langle\psi_i(\mathbf{q},\mathbf{Q})\left|\frac{\partial^2}{\partial Q_\nu^2}\right|\Psi_j(\mathbf{q},\mathbf{Q})\rangle_\mathbf{q} \\
F_{ij}^\nu(\mathbf{Q}) &= \langle\psi_i(\mathbf{q},\mathbf{Q})\left|\frac{\partial}{\partial Q_\nu}\right|\psi_j(\mathbf{q},\mathbf{Q})\rangle_\mathbf{q}
\end{aligned}
\tag{10.15}
$$

The off-diagonal terms can be neglected when the adiabatic approximation holds. If this is the case, the goal is to propagate vibrational wavepackets on different (uncoupled) potential energy surfaces. The latter are described by the eigenvalues of the electronic Hamiltonian [i.e., $V_t(\mathbf{Q})$] with a small correction term $\sum_\nu (2m_\nu)^{-1} G^\nu{}_{ii}(\mathbf{Q})$, which can usually be neglected. One can then either perform separate propagations for each PES or, being only interested in the absorption spectrum for a well definite photon energy range (e.g in the $S_0 \rightarrow S_1$ range), limit himself to the propagation on a single PES (e.g. that for S_1).

The situation is different when nonadiabatic effects (conical intersections, Jahn–Teller, etc.) [18] play a role. In these cases two (or more) PESs are involved. The typical propagation problem involving two PESs can be written in matrix form (neglecting the off-diagonal G terms, which are also small):

$$
i\frac{d}{dt}\begin{pmatrix}\phi_1(\mathbf{Q},t)\\ \phi_2(\mathbf{Q},t)\end{pmatrix} = \begin{pmatrix} V_1(\mathbf{Q}) + T_Q & \sum_\nu F_{12}^\nu(\mathbf{Q})\frac{\partial}{\partial Q_\nu} \\ \sum_\nu F_{12}^\nu(\mathbf{Q})\frac{\partial}{\partial Q_\nu} & V_2(\mathbf{Q}) + T_Q \end{pmatrix}\begin{pmatrix}\phi_1(\mathbf{Q},t)\\ \phi_2(\mathbf{Q},t)\end{pmatrix} \tag{10.16}
$$

As it is well known (see, e.g., Chapter 8), the F functions diverge when two adiabatic PESs touch each other (as in conical intersections). To avoid numerical problems, it is then often convenient to look for diabatic electronic states, exhibiting a very smooth dependence on nuclear coordinates (ideally no dependence) in such a way that the off-diagonal couplings involving derivatives are very small (ideally null). They are replaced by a potential energy coupling $V_{ij}(\mathbf{Q})$. In the diabatic basis set Eq. 10.16 then becomes

$$i\frac{d}{dt}\begin{pmatrix}\phi_{1d}(\mathbf{Q},t)\\ \phi_{2d}(\mathbf{Q},t)\end{pmatrix} = \begin{pmatrix}V_{1d}(\mathbf{Q})+T_{\mathrm{Q}} & V_{12d}(\mathbf{Q})\\ V_{12d}(\mathbf{Q}) & V_{2d}(\mathbf{Q})+T_{\mathrm{Q}}\end{pmatrix}\begin{pmatrix}\phi_{1d}(\mathbf{Q},t)\\ \phi_{2d}(\mathbf{Q},t)\end{pmatrix} \quad (10.17)$$

Equation 10.17 can be easily generalized to an arbitrary number of coupled PESs.

10.3 QUANTUM METHODS OF PROPAGATION

We come to the problem of numerically propagating on a given potential energy surface an initial wavepacket, represented on a grid in the coordinate space (an analogue discussion can be done if a basis representation is adopted). The number of involved nuclear coordinates N_Q is the crucial parameter determining if a direct solution of the TDSE is affordable. In fact, using, for example, the same number n of points for each dimension of the grid, the multimode time-dependent wavefunction can be represented as a multidimensional matrix with a total number of elements of n^{N_Q} (which must be doubled if two PESs are involved). The direct method can then be applied only to small molecules (up to five or six modes). It is then necessary to switch to approximate methods to numerically solve the TDSE. Among these, a special role is played by variational methods, which share the notable property that they converge toward the exact result when the number of parameters is properly increased.

10.3.1 Variational Approaches

The variational approach to the TDSE is an old idea pursued by various authors in the past [19–24]. Here we give some general ideas focusing essentially on the multiconfiguration time-dependent Hartree (MCTDH) algorithm developed by Cederbaum and collaborators [25, 26]. An extensive description of this technique is given by Beck et al. [3], and it is implemented in a numerical code that is freely available [27]. The MCTDH revealed, in our opinion, the most powerful and flexible instrument for investigating nuclear dynamics even in the presence of strong nonadiabatic effects (i.e., in situations in which the electron and nuclear motions are strongly correlated).

Let us begin a short tour of variational methods writing the moving wavepacket as a generic function of the vectors of coordinates $\mathbf{Q}$ and of the time-dependent parameters $\boldsymbol{\alpha}: |\psi(\mathbf{Q},\boldsymbol{\alpha})\rangle$. The basic instrument is the Dirac–Frenkel TD variational principle [19, 20],

$$\langle\delta\psi(Q,\alpha)|\left(H - i\frac{\partial}{\partial t}\right)|\psi(Q,\alpha)\rangle = 0 \quad (10.18)$$

This gives

$$\sum_j \left\langle \frac{\delta\psi}{\delta\alpha_j}\right|H|\psi\rangle - i\sum_j \left|\frac{\delta\psi}{\delta\alpha_j}\right\rangle \frac{d\alpha_j}{dt} = 0 \quad (10.19)$$

The simplest approach is that of introducing a time-dependent configuration interaction (CI) representation with fixed configurations, that is, a CI in which the α's are time-dependent coefficients multiplying time-independent configurations. Each configuration is a product of f (f being the number of modes) single-mode functions (SMFs):

$$\psi = \sum_{J=1,N} A_J F_J; F_J = \prod_{v=1,f} \chi_{Jv} \tag{10.20}$$

Hence

$$\delta\psi = \sum_J \delta A_J F_J, \dot{\psi} = \sum_J \dot{A}_J F_J \tag{10.21}$$

Since each variation is independent, f separate equations are obtained, which can be recast in matrix form (orthonormal SMFs are used for any mode):

$$i\dot{\boldsymbol{\alpha}} = \mathbf{H}\boldsymbol{\alpha} \tag{10.22}$$

with

$$H_{IJ} = \langle F_I | H | F_J \rangle; \boldsymbol{\alpha}^T = (\boldsymbol{\alpha}_1 \quad \cdot \quad \boldsymbol{\alpha}_N) \tag{10.23}$$

The above linear system of differential equations can be solved by various propagation methods [2], as the time split [28, 29], Chebyshev [30], or Lanczos [31, 32] method, provided an efficient and accurate method for computing the matrix elements H_{ij} is given (see, e.g., Appendix B of ref. 3). The basic limitation of the method stays in its bad scaling properties, since the number of equations goes as n^f, where n is the dimension of the single-mode basis set (supposed here to be identical for all the modes).

In order to have an intuitive idea of how a more efficient solution can be achieved, let us consider a trivial bidimensional example in which the PES for S_1 is identical but displaced with respect to that of S_0 (notice that this model is treated in detail in Chapter 8, where it is defined as the vertical gradient, VG). We know that the solution of the problem can simply be written as a product of two time-dependent wavepackets, which are coherent states, if the two PESs are harmonic. Hence

$$\psi(Q_1, Q_2; t) = \phi_1(Q_1; t)\phi_2(Q_2; t) \tag{10.24}$$

Expanding $\phi_1(Q_1; t)$ and $\phi_2(Q_2; t)$ in the harmonic oscillator basis set yields

$$\psi(Q_1, Q_2; t) = \sum_j^{n_1} \sum_k^{n_2} c_{j,1}(t) c_{k,2}(t) \chi_{1j}(Q_1) \chi_{2k}(Q_2) \tag{10.25}$$

showing that with the traditional CI approach we can solve $(n_1 n_2)$ coupled equations [here n_1 and n_2 are the number of harmonic oscillator states needed to well represent $\phi_1(Q_1;0)$ and $\phi_2(Q_2;0)$, respectively]. In other words, an extended CI is required just because the SMFs are not allowed to have an intrinsic time dependence.

10.3.1.1 MCTDH Method We can remediate the steep increase in the number of coupled equations with the number of degrees of freedom, allowing SMFs to depend on time, that is, adopting the MCTDH method [3, 25, 26]. The variational multimode wavepacket is now written linearly combining the elements of the tensor product of the SMFs:

$$\begin{aligned} \psi(\mathbf{Q};t) &= \sum_J B_J(t) F_J(\mathbf{Q};t) \\ F_J(\mathbf{Q},t) &= \prod_v \phi_{J_v}(Q_v;t) \end{aligned} \tag{10.26}$$

Notice that, since each assigned configuration (or Hartree product) F_J corresponds a well-defined product of SMFs, the index J has been used in Eq. 10.26 to label the SMFs for the sake of simplicity. In the following, however, when we need to precisely indicate a particular SMF (appearing in several F_J's) we write it as $\phi_{k_v}^{(v)}(\mathbf{Q}_v;t)$.

The above is a powerful generalization of the time-dependent Hartree method used by Gerber and co-workers [24, 33–35] in which a single Hartree product F is used.

The variational parameters are now the coefficients B_J and the single-particle functions $\phi_{k_v}^{(v)}$. The variation of ψ due to B_J is trivially $F_J : \delta\psi/\delta B_J = F_J$. To evaluate the variation with respect to $\phi_{k_v}^{(v)}$, it is useful to introduce the single-hole function $\psi^{(k_v,v)}$, which is simply the $(f-1)$-mode wavefunction obtained from ψ, dropping out $\phi_{k_v}^{(v)}(\mathbf{Q}_v;t)$ in each F_J where it is present (which leaves a product of $f-1$ SMFs) and setting to zero all the other terms. Since, by definition,

$$\psi = \sum_{v,k_v} \psi^{(k_v,v)} \phi_{k_v}^{(v)} \tag{10.27}$$

one has

$$\begin{aligned} \delta\psi &= \sum_j \delta B_j F_j + \sum_{v,k_v} \psi^{(k_v,v)} \delta\phi_{k_v}^{(v)} \\ \psi &= \sum_j B_j F_j + \sum_{v,k_v} \psi^{(k_v,v)} \phi_{k_v}^{(v)} \end{aligned} \tag{10.28}$$

As discussed, for example, by Beck et al. [3], the solution to the variational problem is not unique due to the invariance of the expansion with respect to a linear transformation of both the coefficients B and the functions $\phi^{(k_v,v)}$. The above redundancy can be utilized to impose useful constraints. For example, we may require that the wavepackets for a given mode remain normalized and orthogonal during their evolution:

$$\left\langle \phi_{k_v}^{(v)} \middle| \phi_{m_v}^{(v)} \right\rangle = \delta_{k_v,m_v}; \left\langle \phi_{k_v}^{(v)} \middle| d\phi_{m_v}^{(v)}/dt \right\rangle = 0 \tag{10.29}$$

From Eqs. 10.18 and 10.28, one obtains two sets of equations corresponding to the variations with respect to the B_J and the $\phi_{k_\nu}^{(\nu)}$ After a few passages, using the conditions in Eq. 10.29, these become, respectively,

$$i\frac{d\mathbf{B}}{dt} = \mathbf{HB} \tag{10.30a}$$

$$\frac{id\phi_{k_\nu}^{(\nu)}}{dt} = (1 - P^{(\nu)}) \sum_{l_\nu, m_\nu} (\rho_{k_\nu l_\nu}^{(\nu)})^{-1} H_{l_\nu, m_\nu}^{(\nu)} \phi_{m_\nu}^{(\nu)} \tag{10.30b}$$

Here $P^{(\nu)}$ is the projection operator onto the space spanned by the SMFs of mode ν, $\rho^{(\nu)}$ is the single-mode density matrix,

$$\rho_{k_\nu, l_\nu}^{(\nu)} = \left\langle \psi^{(k_\nu, \nu)} \middle| \psi^{(l_\nu, \nu)} \right\rangle \tag{10.31}$$

and $H_{l_\nu, m_\nu}^{(\nu)}$ is a single-mode mean-field Hamiltonian (which is an operator in the Q_ν space),

$$H_{l_\nu, m_\nu}^{(\nu)} = \left\langle \psi^{(l_\nu, \nu)} \middle| H \middle| \psi^{(m_\nu, \nu)} \right\rangle \tag{10.32}$$

If we choose an equal number n of SMFs for each coordinate, then Eqs. 10.30a and 10.30b form a set of coupled integrodifferential equations. Those determining the coefficients B are n^f while those for the ϕ's are $n \times f$. If we take into account that each ϕ is represented on a grid of n_b points (or is written as a linear combination of n_b time-independent functions), the total number of equations is $n^f + n_b \times n \times f$. As well documented in the literature [3], the numerical effort can be strongly reduced taking into account that the time variation of the matrix H and the mean-field Hamiltonians $H_{l_\nu, m_\nu}^{(\nu)}$ in Eq. 10.32 are much slower than that of the B coefficients and the SMFs.

The MCTDH expansion usually has very good convergence properties and gives accurate results with a few SMFs.

An important generalization of MCTDH comes from the idea that it can be easily extended to many-mode ϕ's [36, 37]. For this purpose it is useful to group the real coordinates $Q_1, \ldots, Q_f$ into logical coordinates which, following the authors, will be called *particles*. For example, a four-mode problem is reduced to a two-particle problem, $q_1 = \{Q_1, Q_2\}$ and $q_2 = \{Q_3, Q_4\}$. The functions ϕ become single-particle wavepackets. Equations 10.30a and 10.30b remain unchanged, but now the density matrix Eq. 10.31 and the mean-field Hamiltonians Eq. 10.32 refer to a single particle. As before, the number of configurations can be drastically reduced allowing the SMFs to be time-dependent; here we stress that grouping coordinates into particles gives the opportunity of further reducing the number of configurations, since a part of the time-dependent correlation is already included at the one-particle level.

The MCTDH enables us to propagate wavepackets in molecules with tens of modes (some applications are reviewed in refs. [38–40]). The application to problems involving coupling to a bath of hundreds of modes can be pursued generalizing the

MCTDH to a multilayer structure [41], following the same general ideas discussed when introducing logical modes. In the next section we discuss the multilayer generalization in some detail.

10.3.1.2 Multilayer MCTDH Method Let us consider a molecule with eight modes (even tough for this system a normal MCTDH would be sufficient) and choose three time-dependent functions for each mode, indicated by $\phi^{(1,\nu)}_{k^{(1,\nu)}}$ (with $\nu = 1,\ldots,8$ and $k^{(1,\nu)}$ ranging from 1 to 3 for any ν). These functions constitute the first layer (L_1) and they are built up taking TD linear combinations of functions from a given TI basis set (e.g., harmonic oscillator functions) whose dimension n_b is identical for all the oscillators. These will be indicated by $\phi^{(0,\nu)}_{k^{(0,\nu)}}$ (with the $k^{(0,j)}$ ranging from 1 to n_b). As an example,

$$\phi^{(1,1)}_{k^{(1,1)}} = \sum_{k^{(1,2)}} A^{(1,1)}_{k^{(1,1)},k^{(0,1)}}(t)\phi^{(0,1)}_{k^{(0,1)}}(Q_1) \tag{10.33}$$

Alternatively the $\phi^{(1,\nu)}_{k^{(1,\nu)}}$ can be defined on a grid of n_b points. Now let us introduce a second layer, L_2, by considering the four logical modes $Q_{21} = \{Q_1, Q_2\}$, $Q_{22} = \{Q_3, Q_4\}$, $Q_{23} = \{Q_5, Q_6\}$, $Q_{24} = \{Q_7, Q_8\}$. Taking as an example the Q_{21} particle, we may combine linearly the nine elements of the tensor product of the L_1 functions for the modes Q_1 and Q_2 and contract them to, say, three single-particle functions,

$$\phi^{(2,12)}_{k^{(2,12)}}(Q_1, Q_2; t) = \sum_{k^{(1,1)},k^{(1,2)}} A^{(2,12)}_{k^{(2,12)},k^{(1,1)}k^{(1,2)}}(t)\phi^{(1,1)}_{k^{(1,1)}}(Q_1; t)\,\phi^{(1,2)}_{k^{(1,2)}}(Q_2; t) \tag{10.34}$$

with $k^{(2,12)} = 1, 2, 3$.

This gives a total of twelve L_2 single-particle functions. As a final step we expand the total wavefunction in the space obtained by the tensor product of the L_2 single-particle basis set:

$$\psi = \sum_{\mathbf{k}^{(2)}} B_{\mathbf{k}^{(2)}}(t)\phi^{(2,12)}_{k^{(2,12)}}(Q_1, Q_2; t)\phi^{(2,34)}_{k^{(2,34)}}(Q_3, Q_4; t)\phi^{(2,56)}_{k^{(2,56)}}(Q_5, Q_6; t)\phi^{(2,78)}_{k^{(2,78)}}(Q_7, Q_8; t) \tag{10.35}$$

with $\mathbf{k}^{(2)} = \{k^{(2,12)}, k^{(2,34)}, k^{(2,56)}, k^{(2,78)}\}$, giving eighty-one B coefficients.

Notice that the first superscript in both the coefficients A and the single-particle functions refers to the layer and the second to the particle involved. To be more explicit, instead of $Q_{11}, Q_{12}, \ldots,$ we have indicated 12, 34, 56, 78, referring to the number of actual coordinates involved. Notice also that, to avoid confusion, the subscript indexing the L_m single-particle functions also carries information on both the layer and the particle involved.

Since one has two kinds of particles, the case presented above can be defined as a two-layer MCTDH. In practice, we have to determine three sets of coefficients (B, $A^{(2)}$, $A^{(1)}$) or two sets (B, $A^{(2)}$) and the functions $\phi^{(1,\nu)}_{k^{(1,\nu)}}$ defined on a grid.

Alternatively we could decide to further increase by 1 the number of layers. As an example, we could start introducing the four L_1 logical coordinates used before and choose the same number of functions for each L_1 particle, say three. Then two L_2 particles are introduced: $Q_{21}=\{Q_{11},Q_{12}\}$, $Q_{22}=\{Q_{13},Q_{14}\}$ with four functions per particle (contracted from nine). The final step is the full CI (16 configurations) to determine the B coefficients.

In practice, in the multilayer approach we perform a series of time-dependent CIs grouping the modes in particles which contain, as the layer order increases, more and more physical coordinates. The advantage is that introducing gradually the correlation between the various modes at the layer level, one gets the flexibility necessary to finally reduce the number of configurations. As pointed out by Beck et al. [3], one must be careful and increase the number of layers only when it is strictly necessary since the job may become heavier without a real advantage.

In the following we present the working equations for multilayer MCTDH. Readers not interested in this level of knowledge of the method can skip this part and go directly to the final comments given below (Eq. 10.51).

The basic equations can be obtained as sketched previously for the standard (single-layer) MCTDH. The interested reader may refer to the original paper for a complete derivation [41] (but see also ref. 42). Here we consider the previously introduced case with eight modes and two layers in the hope that working on a specific situation may be less obscure than considering a general case, for which it is probably easier to get into trouble over the notation.

It is worthwhile to notice that at any layer and for each particle one may assume, as before, that the various single-particle functions remain orthogonal and normalized, that is,

$$\left\langle \phi_{k(m,\alpha)}^{(m,\alpha)} \middle| \phi_{l(m,\alpha)}^{(m,\alpha)} \right\rangle = \delta_{k(m,\alpha)\,l(m,\alpha)};\ \left\langle \phi_{k(m,\alpha)}^{(m,\alpha)} \middle| \phi_{l(m,\alpha)}^{(m,\alpha)}/dt \right\rangle = 0 \tag{10.36}$$

Notice that, to avoid confusion, the subscript indexing the L_m single-particle functions also contains information on both the layer and the particle.

As far as the coefficients B are concerned, the variational principle gives, as for MCTDH,

$$i\frac{d\mathbf{B}}{dt} = \mathbf{HB} \tag{10.37}$$

As above, at this stage it is useful to introduce the single-hole functions for each layer and particle, defined as what remains of the total wavefunction after annihilating all the terms that do not contain the given single-particle function and after dropping the same function from the remaining terms. Hence one can write

$$\psi = \sum_{k^{(2,12)}} \psi^{(2,12,k^{(2,12)})} \phi_{k^{(2,12)}}^{(2,12)} = \sum_{k^{(2,34)}} \psi^{(2,34,k^{(2,34)})} \phi_{k^{(2,34)}}^{(2,34)} = \ldots = \sum_{k^{(2,78)}} \psi^{(2,78,k^{(2,78)})} \phi_{k^{(2,78)}}^{(2,78)} \tag{10.38}$$

and

$$\psi = \sum_{k^{(1,1)}} \psi^{(1,1,k^{(1,1)})} \phi^{1,1}_{k^{(1,1)}} = \sum_{k^{(1,2)}} \psi^{(1,2,k^{(1,2)})} \phi^{(1,2)}_{k^{(1,2)}} = \ldots = \sum_{k^{(1,8)}} \psi^{(1,8,k^{(1,8)})} \phi^{(1,8)}_{k^{(1,8)}} \tag{10.39}$$

From Eq. 10.38, taking, for example, the variation with respect to $\psi^{(2,12,k^{(2,12)})}$, one has $\delta\psi = \phi^{(2,12)}_{k^{(2,12)}}$. The Dirac–Fenkel variational principle (Eq. 10.18), then gives

$$\left\langle \phi^{(2,12)}_{k^{(2,12)}} \middle| \hat{H} \middle| \psi \right\rangle - i \sum_{l^{(2,12)}} \left\langle \phi^{(2,12)}_{k^{(2,12)}} \middle| \frac{d\psi^{(2,12,l^{(2,12)})}}{dt} \right\rangle \phi^{(2,12)}_{l^{(2,12)}} - i \sum_{l^{(2,12)}} \left\langle \phi^{(2,12)}_{k^{(2,12)}} \middle| \psi^{(2,12,l^{(2,12)})} \right\rangle \frac{d\phi^{(2,12)}_{l^{(2,12)}}}{dt} = 0 \tag{10.40}$$

which, using Eq. 10.36, becomes

$$i \frac{d\psi^{(2,12,k^{(2,12)})}}{dt} = \left\langle \phi^{(2,12)}_{k^{(2,12)}} \middle| \hat{H} \middle| \psi \right\rangle \tag{10.41}$$

Taking instead the variation with respect to $\phi^{(2,12)}_{k^{(2,12)}}$ gives $\delta\psi = \psi^{(2,12,k^{(2,12)})}$, and then

$$\left\langle \psi^{(2,12,k^{(2,12)})} \middle| \hat{H} \middle| \psi \right\rangle - i \sum_{l^{(2,12)}} \left\langle \psi^{(2,12,k^{(2,12)})} \middle| \frac{d\psi^{(2,12,l^{(2,12)})}}{dt} \right\rangle \phi^{(2,12)}_{l^{(2,12)}} - i \sum_{l^{(2,12)}} \left\langle \psi^{(2,12,k^{(2,12)})} \middle| \psi^{(2,12,l^{(2,12)})} \right\rangle \frac{d\phi^{(2,12)}_{l^{(2,12)}}}{dt} = 0 \tag{10.42}$$

Using the result in Eq. 10.36, Eq. 10.42 becomes, after a few passages,

$$i \sum_{k^{(2,12)}} \rho^{(2,12)}_{l^{(2,12)}k^{(2,12)}} \frac{d\phi^{(2,12)}_{k^{(2)}}}{dt} = \left(1 - P^{(2,12)}\right) \sum_{m^{(2)}} H^{(2,12)}_{l^{(2,12)}m^{(2,12)}} \phi^{(2,12)}_{m^{(2,12)}} \tag{10.43}$$

where

$$H^{(2,12)}_{l^{(2,12)}m^{(2,12)}} = \left\langle \psi^{(2,12,l^{(2,12)})} \middle| H \middle| \psi^{(2,12,m^{(2,12)})} \right\rangle$$

$$\rho^{(2,12)}_{l(2,12)m(2,12)} = \left\langle \psi^{(2,12,l^{(2,12)})} \middle| \psi^{(2,12,m^{(2,12)})} \right\rangle \tag{10.44}$$

$$P^{(2,12)} = \sum_{k^{(2)}} \left| \phi^{(2,12)}_{k^{(2,12)}} \right\rangle \left\langle \phi^{(2,12)}_{k^{(2,12)}} \right|$$

In order to have only the derivatives of the single-particle functions on the left-hand side (LHS) of Eq. 10.43, it is convenient to first rewrite it in matrix form ($|d\boldsymbol{\varphi}^{(2,12)}/dt\rangle$, $|\boldsymbol{\varphi}^{(2,12)}\rangle$ being column vectors):

$$i\boldsymbol{\rho}^{(2,12)}\frac{d\boldsymbol{\varphi}^{(2,12)}}{dt} = (1-P^{(2,12)})\mathbf{H}^{(2,12)}\boldsymbol{\varphi}^{(2,12)} \tag{10.45}$$

which yields

$$\frac{id\phi^{(2,12)}_{k^{(2,12)}}}{dt} = (1-P^{(2,12)})\sum_{l^{(2,12)},m^{(2,12)}}(\rho^{(2,12)})^{-1}_{k^{(2,12)},l^{(2,12)}}H^{(2,12)}_{l^{(2,12)},m^{(2,12)}}\phi^{(2,12)}_{m^{(2,12)}} \tag{10.46}$$

Similar equations hold for the functions $\phi^{(2,34)}, \ldots$.

Starting from Eq. 10.39 and following the same procedure, the equations for the L_1 functions can be obtained. As an example, we have

$$\frac{id\phi^{(1,1)}_{k^{(1,1)}}}{dt} = (1-P^{(1,1)})\sum_{l^{(1,1)},m^{(1,1)}}(\rho^{(1,1)})^{-1}_{k^{(1,1)},l^{(1,1)}}H^{(1,1)}_{l^{(1,1)},m^{(1,1)}}\phi^{(1,1)}_{m^{(1,1)}}$$

$$P^{(1,1)} = \sum_{k^{(1,12)}}\left|\phi^{(1,1)}_{k^{(1,1)}}\right\rangle\left\langle\phi^{(1,1)}_{k^{(1,1)}}\right| \tag{10.47}$$

$$(\rho^{(1,1)})_{k^{(1,1)},l^{(1,1)}} = \left\langle\psi^{(1,1,k^{(1,1)})}\middle|\psi^{(1,1,l^{(1,1)})}\right\rangle$$

The basic ingredients of Eqs. 10.43 and 10.47 are the density matrix and the matrix elements of the mean-field Hamiltonians for a given layer and particle. In the examples above $H^{(2,12)}_{l^{(2,12)},m^{(2,12)}}$ is an L_2 operator acting on the $\{Q_1, Q_2\}$ space, while $H^{(1,1)}_{l^{(1,1)},m^{(1,1)}}$ is an L_1 operator on the Q_1 space.

Finally, from Eqs. 10.43 and 10.47 the equations for the L_2 and L_1 A coefficients can be derived. Let us start from Eq. 10.34, derive with respect to time, multiply by $\phi^{(1,1)*}_{k^{(1,1)}}(Q_1;t)\phi^{(1,2)*}_{k^{(1,2)}}(Q_2;t)$, and integrate over Q_1 and Q_2. Taking Eq. 10.36 into account yields

$$\begin{aligned} i\frac{d}{dt}A^{(2,12)}_{k^{(2,12)},k^{(1,1)}k^{(1,2)}}(t) &= \left\langle\phi^{(1,1)}_{k^{(1,1)}}(Q_1;t)\phi^{(1,2)}_{k^{(1,2)}}(Q_2;t)\right|(1-P^{(2,12)}) \\ &\quad\times\sum_{l^{(2,12)},m^{(2,12)}}(\rho^{(2,12)})^{-1}_{k^{(2,12)},l^{(2,12)}}H^{(2,12)}_{l^{(2,12)},m^{(2,12)}}\left|\phi^{(2,12)}_{m^{(2,12)}}\right\rangle \end{aligned} \tag{10.48}$$

In the same way

$$i\frac{d}{dt}A^{(1,1)}_{k^{(1,1)},k^{(0,1)}}(t) = \left\langle\phi^{(0,1)}_{k^{(0,1)}}(Q_1)\right|(1-P^{(1,1)})\sum_{l^{(1,1)},m^{(1,1)}}(\rho^{(1,1)})^{-1}_{k^{(1,1)},l^{(1,1)}}H^{(1,1)}_{l^{(1,1)},m^{(1,1)}}\left|\phi^{(1,1)}_{m^{(1,1)}}\right\rangle \tag{10.49}$$

We can now easily generalize our equations to an N_L-layer MCTDH.

For the A coefficients ($m = 1, \ldots, N_L$) one has, using a somewhat redundant notation for the sake of clearness,

$$i\frac{d}{dt}A^{(m,\alpha_m)}_{k^{(m,\alpha_m)}k^{(m-1,\alpha'_{m-1})}}(t)$$
$$= \left\langle \phi^{(m-1)}_{k^{(m-1)}} \middle| (1 - P^{(m,\alpha_m)}) \sum\nolimits_{l^{(m,\alpha_m)},p^{(m,\alpha_m)}} \left(\rho^{(m,\alpha_m)}\right)^{-1}_{k^{(m,\alpha_m)},l^{(m,\alpha_m)}} H^{(m,\alpha_m)}_{l^{(m,\alpha_m)},p^{(m,\alpha_m)}} \middle| \phi^{(m,\alpha_m)}_{p^{(m,\alpha_m)}} \right\rangle$$

$$(10.50)$$

where $k^{(m,\alpha_m)}$ is a collective vector identifying the L_{m-1} functions for all the L_{m-1} particles α' forming the L_m particle α and $\phi^{(m-1)}_{k^{(m-1)}}\prod_{\alpha}\phi^{(m-1)}_{k^{(m-1,\alpha_{m-1})}}$ (the prime indicates that the L_{m-1} particles involved are only the ones forming the L_m particle α_m). For the B coefficients,

$$i\frac{d}{dt}\mathbf{B} = \mathbf{HB} \qquad H_{JK} = \langle \phi_J | H | \phi_K \rangle \tag{10.51}$$

We conclude this section noting that the MCTDH approach is flexible enough to allow us to treat different modes with different accuracy, depending on their relevance, which is important when dealing with large systems. This can be achieved for the less relevant particles by choosing either a reduced number of functions or a simplified form of the basis functions. This is the case for the hybrid MCTDH by which the less relevant modes are treated as multidimensional Gaussian wavepackets containing variational parameter [43].

It is also interesting to mention here the possibility, discussed by Cederbaum and co-workers, of introducing a hierarchy of sequentially coupled modes in the framework of the so-called linear vibronic coupling model (described in Chapter 8) in such a way that a many-mode problem can be truncated to reproduce the required number of moments of the exact absorption spectrum [44, 45].

Other quantum mechanical approaches based on Gaussian wavepackets or coherent-state basis sets are those by Methiu and co-workers [46] and Martinazzo and co-workers [47] as well as the multiple spawning method developed by Martinez et al. [48] by which the moving wavepacket is expanded on a variable number of frozen Gaussians. Elsewhere [49] such an approach, especially conceived to be run on the fly, has been utilized for computing the ethylene spectrum by directly coupling it with electronic structure calculations.

10.3.1.3 Nonadiabatic $S_0 \rightarrow S_2/S_1$ Absorption Spectrum of Pyrazine Probably the most famous application of the MCTDH method has been in the calculation of the $S_0 \rightarrow S_2/S_1$ absorption spectrum of pyrazine in full dimensionality (24 degrees of freedom) taking into account the strong nonadiabatic interaction arising from the S_2/S_1 conical intersection. The diabatic model for this system was first developed in few dimensions [50–52] and then was generalized to include all 24 degrees of freedom and fitted to ab initio data [53]. Further refinement of the model was due to Raab

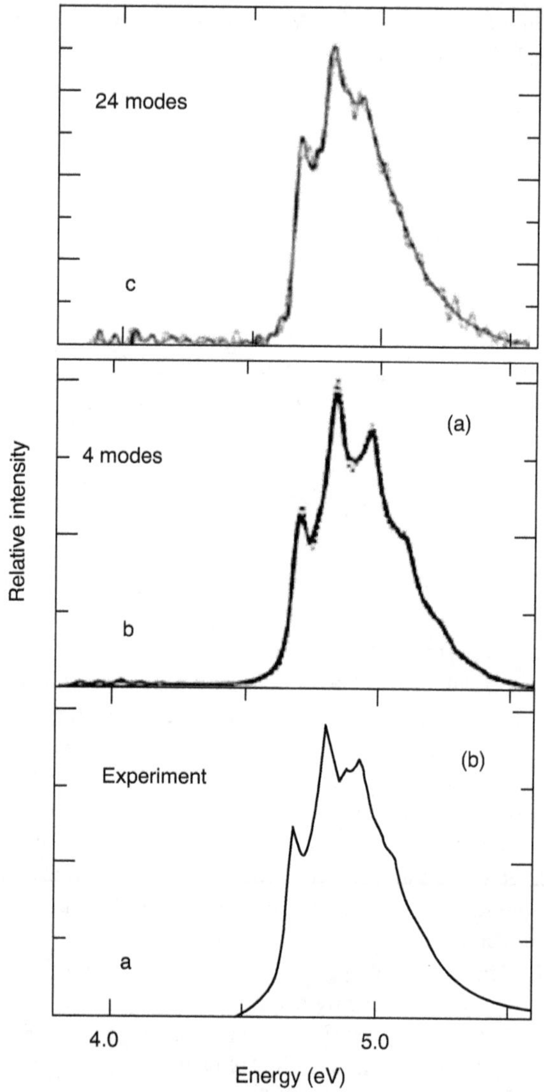

Figure 10.1 Nonadiabatic $S_0 \rightarrow S_2/S_1$ absorption spectrum of pyrazine. Computational results obtained by the MCTDH method (solid line) and by the semiclassical methods (dashed lines) for the small 4-mode model (b) and the full-dimensionality 24-mode model (c) are compared to experimental results (a). (From ref. 97, Copyright © 2004. Reproduced with permission of World Scientific Publishing Co.)

et al. [54]. The very cumbersome dynamical calculations were made affordable by adopting single-particle functions that represent more than a single physical mode. In the original paper in 1999 [54] the largest expansion included more than two million configurations. Figure 10.1 shows that the spectrum is already almost at convergence with a reduced four-dimensional (4D) model and in very good agreement with experiments. The dipole correlation function was multiplied by a damping exponential to fit the broad experimental lineshape. Interestingly, while in the reduced-dimensionality 4D model a large damping was necessary, corresponding to a lifetime of 30 fs, a 150-fs lifetime was sufficient in the full-dimensionality model, highlighting the fact that the explicit inclusion of the bath of 20 modes introduces intramolecular dephasing mechanisms that result in faster quenching of the dipole correlation function.

10.3.1.4 Application to Model Vibronic Hamiltonians: Absorption Spectrum of Adenine Stacked Dimer The improvement in terms of computational efficiency of MCTDH and multilayer MCTDH methods over traditional quantum propagation schemes has been so remarkable that the time propagation probably no longer represents the computational bottleneck of quantum dynamical studies, and the challenge is now more in the a priori computation and fit of the necessary PESs in many degrees of freedom. In this context, a feasible route, at least for some classes of systems, is the development of model Hamiltonians, with parameters fitted on accurate quantum chemical calculations, able to provide reliable PESs at moderate computational cost. Here we briefly review a recent contribution of our research group concerning the nonadiabatic absorption spectrum of oligomers of stacked DNA nucleobases [55].

DNA strongly absorbs UV solar radiation with possible mutagenic effects. Ultrafast decay channels very effectively dissipate the dangerous electronic energy into vibrational modes and finally into heat. The first step for a reliable theoretical analysis of the decay processes is clearly a reliable description of the excitation process, that is, of the absorption spectrum. Due to interaction of the electronic states of the single nucleobases, typically a dense bath of states lies in the region of the absorption band around 250 nm [56, 57]. To treat this problem, we developed a vibronic model for adenine stacked multimers and here we show its predictions concerning the absorption spectrum (we also analyzed the decay dynamics, but these results will not be reviewed here, since the focus of this Chapter is on steady-state spectra). The electronic model Hamiltonian was developed on the ground of an extended time-dependent density functional theory (TD-DFT) analysis of the excited states of the system [55]. For the case of a dimer of two stacked adenines (A) in a B-DNA orientation it is

$$\begin{aligned} H_{\mathrm{el}} &= \left(|A^{+}A^{-}\rangle\langle A^{+}A^{-}|R + \text{h.c.}\right) + \left(|A^{*}A\rangle\langle A^{+}A^{-}|t_e + \text{h.c.}\right) \\ &\quad + \left(|A^{*}A\rangle\langle A^{-}A^{+}|t_h + \text{h.c.}\right) \end{aligned} \tag{10.52}$$

where $|A^{*}A\rangle$ is a diabatic state characterized by a highest–lowest occupied molecular orbitals (HOMO $\rightarrow$ LUMO) excitation on the first adenine, $|A^{+}A^{-}\rangle$ a charge transfer (CT) state, the t_h and t_e parameters introduce the hopping between holes and electrons, and the electrostatic term R (depending on the electron–hole distance) determines the relative stability of localized exciton states and CT states.

Vibronic effects are introduced by including three normal modes for each adenine, which are the most relevant to describe the displacements among the equilibrium

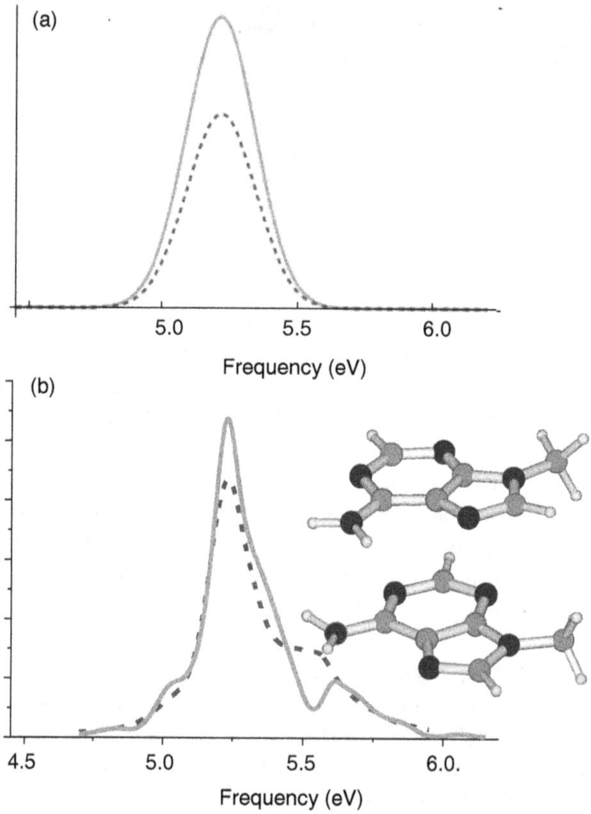

Figure 10.2 Absorption spectrum of adenine dimer (blue dashed line) and monomer (red solid line) obtained at pure electronic level (a) and at vibronic level (b) by adopting the vibronic Hamiltonian discussed in Section 10.3.1.3. It has been computed from the Fourier transform of the autocorrelation function obtained propagating a doorway state. The latter is a delocalized exciton state obtained mixing the two localized exciton states with equal weights.

geometries of the neutral ground-state (A) and the cationic (A^+), anionic (A^-), and neutral excited (A^*) states optimized at the DFT or TD-DFT level. For these degrees of freedom, harmonic approximation is invoked and frequency changes and Duschinsky mixing in the different monomer states are neglected.

The vibronic model thus includes four electronic states and six nuclear coordinates, a very challenging system not only for TI approaches but also for traditional quantum propagation schemes. Dipole time correlation was computed in the FC approximation by propagating the doorway state through the MCTDH method by a home-developed code. The computed spectrum obtained through the pure electronic Hamiltonian in Eq. 10.52 and the full vibronic Hamiltonian are reported in Figure 10.2, and they reproduce the main differences observed experimentally with respect to the monomer spectrum, namely a hypochromic effect, a slight blue shift of the maximum band and a weak wing appearing in the red part of the spectrum of the dimer. It is also worthwhile to notice that the vibrational structure observed in the monomer is partially lost in the dimer due to the coupling among nearly degenerate excitonic and charge transfer states (in the diabatic picture).

10.3.2 Analytical Expression for Thermal Time Correlation Function in Harmonic Models

When the electronic states involved in the optical transition undergo negligible nonadiabatic coupling, a remarkable simplification arises in the computation of the spectrum. In fact, the whole computation can be recast in a single-state approach, meaning that it is possible to deal with one final state at a time, even if the energy window of interest encompasses more than one electronic state. Furthermore, in semirigid molecules, when the displacements in the equilibrium positions upon electronic transition are moderate, the harmonic approximation can be invoked, at least as a reliable starting point for the description of the initial and final electronic PESs. As discussed in detail in Chapter 8, in these cases the PESs can be obtained from equilibrium geometries and harmonic analysis for both the electronic states in the so-called adiabatic Hessian (AH) approach. Final-state PESs can alternatively be constructed through a second-order Taylor expansion around the ground-state equilibrium geometry (vertical Hessian, VH). Modern electronic methods, for instance, those grounded on density functional theory and its time-dependent extension for excited state (see Chapter 1), allow us to routinely perform these calculations for sizable molecules (dozens–hundreds of normal modes), thus opening the way for the simulation of the optical spectra of systems of direct biological or technological interest. In such a single-state framework the harmonic vibronic eingenstates and energies are known analytically and the spectrum can be computed in a time-independent framework as a sum of state-to-state transitions according to Eq. 10.1. However, the number of possible vibronic transitions increases steeply with the molecular size (see Figure 8.9), ruling out brute-force calculations. Chapter 8 describes a number of effective techniques to preselect and compute only the relevant transitions, allowing, in favorable cases, for an almost blackbox calculation of fully converged spectra also for systems with hundreds of normal modes at a finite temperature [4–9]. Nonetheless, considering even larger systems or increasing the simulation temperature makes it easy to reach a physical limit where no selection scheme can be effective, due to the fact that the number of nonnegligible transitions becomes too high to be dealt with (it is not difficult to overcome the threshold of 10^{15}). In these cases, however, the interest is usually more focused on the main transitions and on the global aspect of the lineshape than on a detailed analysis of all the individual transitions, and, as discussed above, TD methods offer an effective alternative for the computation of the absorption spectrum through the Fourier transform of the dipole time correlation function. In the specific case where harmonic approximation holds for both initial and final PESs, an analytical expression can be derived for $C_{\mu}(t)$ (Eq. 10.5) also including temperature effects. While this result is rather easily obtained when Duschinsky rotation can be neglected, that is, $\mathbf{J}=\mathbf{I}$ [e.g., 58], the derivation is more involved in the more general harmonic case. Such a result was originally obtained by Mukamel [59, 60] and then derived by Tang et al. [61] and Pollak and co-workers [62, 63] at the FC level, and it has been very recently generalized to include Herzberg-Teller (HT) effects [64]. Let us consider an electronic transition from state $|\psi_i\rangle$ to state $|\psi_f\rangle$ and generalize the $C_\mu(t)$ expression to a Boltzmann ensemble of initial states at temperature T:

$$C_\mu(t,T) = Z^{-1}\,\mathrm{Tr}\left[\mu e^{-iH_0 t}\mu e^{(-\beta+it)H_0}\right] \tag{10.53}$$

where $\beta = 1/k_B T$ and k_B is the Boltzmann constant, the trace is taken over the initial vibrational states of state $|\psi_i\rangle$, and Z is their partition function. The term $C_\mu(t,T)$ is often called the thermal time correlation function.

We adopt harmonic approximation and $\mathbf{Q}_i$ are $\mathbf{Q}_f$ are the column vectors representing the sets of normal coordinates of states $|\psi_i\rangle$ and $|\psi_f\rangle$. According to Duschinsky [65], the following linear transformation holds:

$$\mathbf{Q}_f = \mathbf{J}\mathbf{Q}_i + \mathbf{K} \tag{10.54}$$

where $\mathbf{J}$ is the so-called Duschinsky matrix and $\mathbf{K}$ is the column vector of the displacements of equilibrium geometry.

As discussed in detail in Chapter 8, $\mathbf{J}$ and $\mathbf{K}$ can be obtained according to two different models, namely the VH and the AH. According to both approaches, the PES of the initial state is modeled by computing its Hessian at the equilibrium geometry. At variance, in VH, $\mathbf{J}$ and $\mathbf{K}$ are computed from the gradient and the Hessian of the final-state PES at the initial-state equilibrium geometry, while in AH it is necessary to locate the equilibrium structure of the final state and to compute directly its normal modes at that geometry.

In each of these two models one obtains, directly or indirectly, the equilibrium positions and frequencies of the final state so that the Hamitonian H_0 can be written as

$$H_0 = H_i|\psi_i\rangle\langle\psi_i| + (E_{\mathrm{ad}} + H_f)|\psi_f\rangle\langle\psi_f| \tag{10.55a}$$

$$H_i = T_N + \frac{1}{2}\mathbf{Q}_i^{\mathrm{T}}\Omega_i^2\mathbf{Q}_i = T_N + \frac{1}{2}\sum_k \omega_{ik}^2 Q_{ik}^2 \tag{10.55b}$$

$$H_f = T_N + \frac{1}{2}\mathbf{Q}_f^{\mathrm{T}}\Omega_f^2\mathbf{Q}_f = T_N + \frac{1}{2}\sum_k \omega_{fk}^2 Q_{fk}^2 \tag{10.55c}$$

where E_{ad} is the adiabatic energy difference, that is, the difference in the minima energies of the two PESs, Ω_i and Ω_f are the diagonal matrices of the vibrational frequencies of states $|\psi_i\rangle$ and $|\psi_f\rangle$, respectively, and ω_{ik} and ω_{fk} are their kth elements.

The α Cartesian component of the transition dipole moment $\mu_{if}(\alpha) = \langle\psi_i|\mu(\alpha)|\psi_f\rangle$ is in general an unknown function of the nuclear coordinates and it can be expanded in a Taylor series with respect to the normal coordinates $\mathbf{Q}_i$ around the equilibrium geometry $\mathbf{Q}_{i0}$. Retaining only the zero- and first-order terms we have

$$\mu_{if}(\mathbf{Q}_i,\alpha) = \mu_{if}^0(\alpha) + \mathbf{T}^{\mathrm{T}}(\alpha)\mathbf{Q}_i \tag{10.56}$$

where the $\mathbf{T}(\alpha)$ is the column vector of the derivatives $T_k(\alpha)$ of $\mu_{if}(\alpha)$ with respect to the normal coordinate Q_{ik} at the equilibrium position $\mathbf{Q}_{i0}$, respectively. The zero-order term gives rise to the so-called Franck–Condon (FC) contribution while the linear terms are responsible for the Herzberg–Teller (HT) effect.

For the sake of completeness, in the following we present in some detail the derivation of the analytical expression for $C_\mu(t,T)$ in the FC approximation as well as show that the two expressions reported by Tang et al. [61] and Pollak et al. [62, 63] are equivalent. (The reader not interested in the mathematical details can skip this part and go directly to the discussion below Eq. 10.75.)

In FC approximation we can simplify the expression for $C_\mu(t,T)$, obtaining

$$C_\mu(t,T) = Z^{-1}(\mu_{if}^0)^2 e^{-iE_{\mathrm{ad}}\tau}\chi(\tau_i,\tau_f) \tag{10.57a}$$

$$\chi(\tau_i,\tau_f) = Tr[e^{-iH_f\tau_f}e^{-iH_i\tau_i}] \tag{10.57b}$$

with $\tau_f = t$ and $\tau_i = -t - i\beta$. The calculation of the spectrum can therefore be driven back to the calculation of $\chi(\tau_i,\tau_f)$, and this can be done by passing to a coordinate representation where the trace is taken over the ground-state normal coordinates

$$\chi(\tau_i,\tau_f) = \int d\mathbf{Q}_i\langle\mathbf{Q}_i|e^{-iH_f\tau_f}e^{-iH_i\tau_i}|\mathbf{Q}_i\rangle \tag{10.58}$$

The coordinate representation of the evolution operator appears in Eq. 10.58 if we insert a complete set of the ground-state coordinates $I = \int d\bar{\mathbf{Q}}_i|\bar{\mathbf{Q}}_i\rangle\langle\bar{\mathbf{Q}}_i|$ and two complete sets of the excited-state coordinates $I = \int d\mathbf{Q}_f|\mathbf{Q}_f\rangle\langle\mathbf{Q}_f|$ and $I = \int d\bar{\mathbf{Q}}_f|\bar{\mathbf{Q}}_f\rangle\langle\bar{\mathbf{Q}}_f|$, where the bar is only introduced for the sake of clarity,

$$\chi(\tau_i,\tau_f) = \int d\bar{\mathbf{Q}}_f\int d\mathbf{Q}_f\int d\bar{\mathbf{Q}}_i\int d\mathbf{Q}_i\langle\mathbf{Q}_i|\mathbf{Q}_f\rangle\langle\mathbf{Q}_f|e^{-iH_f\tau_f}|\bar{\mathbf{Q}}_f\rangle\langle\bar{\mathbf{Q}}_f|\bar{\mathbf{Q}}_i\rangle\langle\bar{\mathbf{Q}}_i|e^{-iH_i\tau_i}|\mathbf{Q}_i\rangle \tag{10.59}$$

The path integral expression for the off-diagonal matrix element $\langle Q_k|e^{-iH_k\tau}|\bar{Q}_k\rangle$ is known [66]:

$$\langle Q_{fk}|e^{-iH_{fk}\tau_f}|\bar{Q}_{fk}\rangle = \sqrt{\frac{a_{fk}(\tau_f)}{2\pi i\hbar}}\exp\left\{\frac{i}{\hbar}\left[\frac{1}{2}b_{fk}(\tau)(Q_{fk}^2+\bar{Q}_{fk}^2) - a_{fk}(\tau_f)Q_{fk}\bar{Q}_{fk}\right]\right\} \tag{10.60a}$$

$$a_{fk}(\tau_f) = \frac{\omega_{fk}}{\sin(\hbar\omega_{fk}\tau_f)} \tag{10.60b}$$

$$b_{fk}(\tau_f) = \frac{\omega_{fk}}{\tan(\hbar\omega_{fk}\tau_f)} \tag{10.60c}$$

Generalizing to a molecule with N normal modes, we have, in matrix notation,

$$\langle \mathbf{Q}_f | e^{-iH_f\tau_f} | \bar{\mathbf{Q}}_f \rangle = \sqrt{\frac{\det(\mathbf{a}_f(\tau_f))}{2\pi i\hbar}} \times \exp\left\{ \frac{i}{\hbar} \left[\frac{1}{2}\mathbf{Q}_f^{\mathrm{T}}\mathbf{b}_f(\tau_f)\mathbf{Q}_f + \frac{1}{2}\bar{\mathbf{Q}}_f^{\mathrm{T}}\mathbf{b}_f(\tau_f)\bar{\mathbf{Q}}_f - \mathbf{Q}_f^{\mathrm{T}}\mathbf{a}_f(\tau_f)\mathbf{Q}_f \right] \right\} \tag{10.61}$$

where $\mathbf{a}_f(\tau_f)$ and $\mathbf{b}_f(\tau_f)$ are the diagonal matrices with elements $a_{fk}(\tau_{\mathrm{k}})$ and $b_{fk}(\tau_{\mathrm{k}})$, respectively. With these definitions (and equivalent ones for the initial-state coordinates $\mathbf{Q}_i$) we have

$$\chi(\tau_i, \tau_f) = \sqrt{\frac{\det(\mathbf{a}_f)\det(\mathbf{a}_i)}{(2\pi i\hbar)^{2N}}} \int d\mathbf{Q}_i \int d\bar{\mathbf{Q}}_i \exp\left\{ \frac{i}{\hbar} \left[\frac{1}{2}\mathbf{Q}_i^{\mathrm{T}}\mathbf{b}_i\mathbf{Q}_i + \frac{1}{2}\bar{\mathbf{Q}}_i^{\mathrm{T}}\mathbf{b}_i\bar{\mathbf{Q}}_i - \mathbf{Q}_i^{\mathrm{T}}\mathbf{a}_i\bar{\mathbf{Q}}_i - (\mathbf{K}^{\mathrm{T}} + \mathbf{Q}_i^{\mathrm{T}}\mathbf{J}^{\mathrm{T}})\mathbf{a}_f(\mathbf{J}\bar{\mathbf{Q}}_i + \mathbf{K}) \right] \right\} \exp\left\{ \frac{i}{\hbar} \left[\frac{1}{2}(\mathbf{K}^{\mathrm{T}} + \mathbf{Q}_i^{\mathrm{T}}\mathbf{J}^{\mathrm{T}})\mathbf{b}_f(\mathbf{J}\mathbf{Q}_i + \mathbf{K}) + \frac{1}{2}(\mathbf{K}^{\mathrm{T}} + \bar{\mathbf{Q}}_i^{\mathrm{T}}\mathbf{J}^{\mathrm{T}})\mathbf{b}_f(\mathbf{J}\bar{\mathbf{Q}}_i + \mathbf{K}) \right] \right\} \tag{10.62}$$

where we have integrated over the final-state coordinates $\mathbf{Q}_f$ and $\bar{\mathbf{Q}}_f$ taking into account the fact that, because of the orthonormalization condition,

$$\langle \mathbf{Q}_f | \mathbf{Q}_i \rangle = \delta(\mathbf{Q}_f - \mathbf{J}\mathbf{Q}_i - \mathbf{K}) = \prod_k \delta(Q_{fk} - \sum_j J_{kj}Q_{ij} - K_k) \tag{10.63}$$

In Eq. 10.62 we also dropped the arguments of the functions $\mathbf{a}_i(\tau_i)$, $\mathbf{b}_i(\tau_i)$, $\mathbf{a}_f(\tau_f)$, and $\mathbf{b}_f(\tau_f)$ because they are unequivocally determined by the i and f subscripts. By collecting terms with the same power dependence on $\mathbf{Q}_i$ and/or $\bar{\mathbf{Q}}_i$, we get

$$\chi(\tau_i, \tau_f) = \sqrt{\frac{\det(\mathbf{a}_f)\det(\mathbf{a}_i)}{(2\pi i\hbar)^{2N}}} \exp\left[\frac{i}{\hbar}\mathbf{K}^{\mathrm{T}}\mathbf{E}\mathbf{K} \right] \int d\mathbf{Q}_i \int d\bar{\mathbf{Q}}_i \exp\left\{ \frac{i}{\hbar} \left[\frac{1}{2}\mathbf{Q}_i^{\mathrm{T}}\mathbf{B}\mathbf{Q}_i + \frac{1}{2}\bar{\mathbf{Q}}_i^{\mathrm{T}}\mathbf{B}\bar{\mathbf{Q}}_i + \mathbf{K}^{\mathrm{T}}\mathbf{E}\mathbf{J}(\mathbf{Q}_i + \bar{\mathbf{Q}}_i) - \mathbf{Q}_i^{\mathrm{T}}\mathbf{A}\bar{\mathbf{Q}}_i \right] \right\} \tag{10.64}$$

where we have defined the matrices

$$\mathbf{G}(\tau_i) = \mathbf{b}_i(\tau_i) - \mathbf{a}_i(\tau_i) \tag{10.65a}$$

$$\mathbf{E}(\tau_f) = \mathbf{b}_f(\tau_f) - \mathbf{a}_f(\tau_f) \tag{10.65b}$$

$$\mathbf{B}(\tau_i, \tau_f) = \mathbf{b}_i(\tau_i) + \mathbf{J}^{\mathrm{T}}\mathbf{b}_f(\tau_f)\mathbf{J} \tag{10.65c}$$

$$\mathbf{A}(\tau_i, \tau_f) = \mathbf{a}_i(\tau_i) + \mathbf{J}^{\mathrm{T}}\mathbf{a}_f(\tau_f)\mathbf{J} \tag{10.65d}$$

To evaluate the integrals in Eq. 10.64, we change the variables through the orthogonal transformation $\mathbf{Z} = 2^{-1/2}(\mathbf{Q}_i + \bar{\mathbf{Q}}_i)$, $\mathbf{U} = 2^{-1/2}(\mathbf{Q}_i - \bar{\mathbf{Q}}_i)$ (notice that the Jacobian determinant is 1), thus obtaining the convenient factorization

$$\chi(\tau_i, \tau_f) = \sqrt{\frac{\det(\mathbf{a}_f)\det(\mathbf{a}_i)}{(2\pi i\hbar)^{2N}}} \exp\left[\frac{i}{\hbar}\mathbf{K}^{\mathrm{T}}\mathbf{E}\mathbf{K}\right] \int d\mathbf{Z} \exp\left\{\frac{i}{\hbar}\left[\frac{1}{2}\mathbf{Z}^{\mathrm{T}}(\mathbf{B}-\mathbf{A})\mathbf{Z} \right.\right.$$
$$\left.\left. + \sqrt{2}\mathbf{K}^{\mathrm{T}}\mathbf{E}\mathbf{J}\mathbf{Z}\right]\right\} \int d\mathbf{U} \exp\left\{\frac{i}{\hbar}\left[\frac{1}{2}\mathbf{U}^{\mathrm{T}}(\mathbf{B}+\mathbf{A})\mathbf{U}\right]\right\} \tag{10.66}$$

With the aim of highlighting the behavior of the integrand in the limits of integration and of bridging the different derivations proposed by Tang et al. and Pollak and co-workers, we make the following rearrangement:

$$\begin{aligned}\frac{i}{\hbar}\left[\mathbf{B}(\tau_i, \tau_f) + \mathbf{A}(\tau_i, \tau_f)\right] &= \frac{i}{\hbar}\left\{\mathbf{b}_i(\tau_i) + \mathbf{a}_i(\tau_i) + \mathbf{J}^{\mathrm{T}}\left[\mathbf{b}_f(\tau_f) + \mathbf{a}_f(\tau_f)\right]\mathbf{J}\right\} \\ &= -\left[\mathbf{c}_i\left(\frac{1}{2}\tau_i\right) + \mathbf{J}^{\mathrm{T}}\mathbf{c}_f\left(\frac{1}{2}\tau_f\right)\mathbf{J}\right] = -\mathbf{C}(\tau_i, \tau_f)\end{aligned} \tag{10.67a}$$

$$\begin{aligned}\frac{i}{\hbar}\left[\mathbf{B}(\tau_i, \tau_f) - \mathbf{A}(\tau_i, \tau_f)\right] &= \frac{i}{\hbar}\left\{\mathbf{b}_i(\tau_i) - \mathbf{a}_i(\tau_i) + \mathbf{J}^{\mathrm{T}}\left[\mathbf{b}_f(\tau_f) - \mathbf{a}_f(\tau_f)\right]\mathbf{J}\right\} \\ &= -\left[\mathbf{d}_i\left(\frac{1}{2}\tau_i\right) + \mathbf{J}^{\mathrm{T}}\mathbf{d}_f\left(\frac{1}{2}\tau_f\right)\mathbf{J}\right] = -\mathbf{D}(\tau_i, \tau_f)\end{aligned} \tag{10.67b}$$

$$\frac{i}{\hbar}\mathbf{E}(\tau_f) = -\,\mathbf{d}_f\left(\frac{1}{2}\tau_f\right) \tag{10.67c}$$

where for mode k (and a unspecified electronic state) we have $c_k(\tau/2) = \omega_k/\hbar \coth[i\omega_k\hbar\tau/2]$, $d_k(\tau/2) = \omega_k/\hbar \tanh[i\omega_k\hbar\tau/2]$, and we made use of the equalities

$-\coth[ix/2] = i/\tan(x) + i/\sin(x)$ and $-\tanh[ix/2] = i/\tan(x) - i/\sin(x)$. With these substitutions Eq. 10.6 becomes

$$\chi(\tau_i, \tau_f) = \sqrt{\frac{\det(\mathbf{a}_f)\det(\mathbf{a}_i)}{(2\pi i\hbar)^{2N}}} \exp\left[-\mathbf{K}^{\mathrm{T}}\mathbf{d}_f\mathbf{K}\right] \int d\mathbf{Z} \exp\left[-\frac{1}{2}\mathbf{Z}^{\mathrm{T}}\mathbf{D}\mathbf{Z} - \sqrt{2}\mathbf{K}^{\mathrm{T}}\mathbf{d}_f\mathbf{J}\mathbf{Z}\right] \times \int d\mathbf{U} \exp\left[-\frac{1}{2}\mathbf{U}^{\mathrm{T}}\mathbf{C}\mathbf{U}\right] \tag{10.68}$$

From the definitions it is easy to notice that $\mathbf{c}_f(\tau_f/2)$ and $\mathbf{d}_f(\tau_f/2)$ are diagonal matrices of pure imaginary numbers since $\tau_f = t$; on the contrary $\mathbf{c}_i(\tau_i/2)$ and $\mathbf{d}_i(\tau_i/2)$ have a real part that is always positive since $\tau_i = -t - i\beta$, and therefore the integrand always vanishes in the limit of integration.

Adopting a common trick we can put both integrands of Eq. 10.68 in diagonal form by the following change of variables:

$$\mathbf{Z}_1 = \mathbf{D}^{1/2}\mathbf{Z} + \sqrt{2}\mathbf{D}^{-1/2}\mathbf{K}^{\mathrm{T}}\mathbf{d}\mathbf{J} \tag{10.69a}$$

$$\mathbf{U}_1 = \mathbf{C}^{1/2}\mathbf{U} \tag{10.69b}$$

Noticing that

$$-\tfrac{1}{2}\mathbf{Z}^{\mathrm{T}}\mathbf{D}\mathbf{Z} - 2\mathbf{K}^{\mathrm{T}}\mathbf{d}\mathbf{J}\mathbf{Z} = -\tfrac{1}{2}\mathbf{Z}_1^{\mathrm{T}}\mathbf{Z}_1 + \mathbf{K}^{\mathrm{T}}\mathbf{d}\mathbf{J}\mathbf{D}^{-1}\mathbf{J}^{\mathrm{T}}\mathbf{d}\mathbf{K} \tag{10.70a}$$

$$-\tfrac{1}{2}\mathbf{U}^{\mathrm{T}}\mathbf{C}\mathbf{U} = -\tfrac{1}{2}\mathbf{U}_1^{\mathrm{T}}\mathbf{U}_1 \tag{10.70b}$$

The Jacobians of these new transformations are respectively det $(\mathbf{D})^{-1/2}$ and det $(\mathbf{C})^{-1/2}$. Equation 10.68 then becomes

$$\chi(\tau_i, \tau_f) = \sqrt{\frac{\det(\mathbf{a}_f)\det(\mathbf{a}_i)}{(2\pi i\hbar)^{2N}}} \exp\left[-\mathbf{K}^{\mathrm{T}}\mathbf{d}_f\mathbf{K}\right] \int d\mathbf{Z}_1 \exp\left[-\tfrac{1}{2}\mathbf{Z}_1^{\mathrm{T}}\mathbf{Z}_1\right] \int d\mathbf{U}_1 \exp\left[-\tfrac{1}{2}\mathbf{U}_1^{\mathrm{T}}\mathbf{U}_1\right] \tag{10.71}$$

We can now obtain our final expression from Eq. 10.71 by utilizing the well-known result $\int \exp(-1/2\gamma x^2) = (2\pi/\gamma)^{1/2}$:

$$\chi(\tau_i, \tau_f) = \sqrt{\frac{\det[\mathbf{a}_f(\tau_f)\det[\mathbf{a}_i(\tau_i)]}{(i\hbar)^{2N}\det[\mathbf{C}(\tau_i, \tau_f)]\det[\mathbf{D}(\tau_i, \tau_f)]}} \times \exp\left[-\mathbf{K}^{\mathrm{T}}\mathbf{d}_f(\tfrac{1}{2}\tau_f)\mathbf{K} + \mathbf{K}^{\mathrm{T}}\mathbf{d}_f(\tfrac{1}{2}\tau_f)\mathbf{J}\mathbf{D}^{-1}\mathbf{J}^{\mathrm{T}}\mathbf{d}_f(\tfrac{1}{2}\tau_f)\mathbf{K}'\right] \tag{10.72}$$

Equation 10.72 is equal to the expression given by Tang et al. [61, Eq. 12]. It can be easily proven that this expression is also equivalent to the expression given by Ianconescu and Pollak [62, Eq. 2.23]. In fact, recalling the definition of **D** and

d in Eqs. 10.67a we can write the argument of the exponential in Eq. 10.72 as

$$-\mathbf{K}^{\mathrm{T}}\mathbf{d}_f\mathbf{K}+\mathbf{K}^{\mathrm{T}}\mathbf{d}_f\mathbf{J}\mathbf{D}^{-1}\mathbf{J}^{\mathrm{T}}\mathbf{d}_f\mathbf{K}'=\frac{i}{\hbar}\left[\mathbf{K}^{\mathrm{T}}\mathbf{E}\mathbf{K}-\mathbf{K}^{\mathrm{T}}\mathbf{E}\mathbf{J}(\mathbf{B}-\mathbf{A})^{-1}\mathbf{J}^{\mathrm{T}}\mathbf{E}\mathbf{K}\right] \quad (10.73)$$

and noticing that $(\mathbf{B}-\mathbf{A})^{-1}\mathbf{J}^{\mathrm{T}}\mathbf{E}=(\mathbf{G}+\mathbf{J}^{\mathrm{T}}\mathbf{E}\mathbf{J})^{-1}\mathbf{J}^{\mathrm{T}}\mathbf{E}=\mathbf{J}^{\mathrm{T}}-(\mathbf{B}-\mathbf{A})^{-1}\mathbf{G}\mathbf{J}^{\mathrm{T}}$, we have

$$\begin{aligned}&-\mathbf{K}^{\mathrm{T}}\mathbf{d}_f(\tfrac{1}{2}\tau_f)\mathbf{K}'+2\mathbf{K}^{\mathrm{T}}\mathbf{d}_f(\tfrac{1}{2}\tau_f)\mathbf{J}\mathbf{D}^{-1}\mathbf{J}^{\mathrm{T}}\mathbf{d}_f(\tfrac{1}{2}\tau_f)\mathbf{K}\\&=\frac{i}{\hbar}\left[\mathbf{K}^{\mathrm{T}}\mathbf{E}(\tau_f)\mathbf{J}'(\mathbf{B}(\tau_i,\tau_f)-\mathbf{A}(\tau_i,\tau_f))^{-1}\mathbf{G}(\tau_i)\mathbf{J}^{\mathrm{T}}\mathbf{K}\right]\end{aligned} \quad (10.74)$$

Moreover, from Eqs. 10.68a and 10.68b and the determinant properties, the following equalities hold:

$$\det[\mathbf{C}]\det[\mathbf{D}]=\det[\mathbf{CD}]=\det\left[\left(\frac{i}{\hbar}\right)^2(\mathbf{B}-\mathbf{A})(\mathbf{B}+\mathbf{A})\right]=\left(\frac{i}{\hbar}\right)^{2N}\det[\mathbf{B}]\det[\mathbf{B}-\mathbf{A}\mathbf{B}^{-1}\mathbf{A}]$$

Therefore Eq. 10.72 can also be written as

$$\chi(\tau_i,\tau_f)=\sqrt{\frac{\det[\mathbf{a}_f]\,\det[\mathbf{a}_i]}{\det[\mathbf{B}]\,\det[\mathbf{B}-\mathbf{A}\mathbf{B}^{-1}\mathbf{A}]}}\exp\left[\frac{i}{\hbar}\left[\mathbf{K}^{\mathrm{T}}\mathbf{E}\mathbf{K}-\mathbf{K}^{\mathrm{T}}\mathbf{E}\mathbf{J}(\mathbf{B}-\mathbf{A})^{-1}\mathbf{J}^{\mathrm{T}}\mathbf{E}\mathbf{K}\right]\right] \quad (10.75)$$

which is exactly the expression obtained by Pollak and co-workers [62, 63]. From Eq. 10.75 and the definitions in Eqs. 10.57, 10.60, and 10.65, we can now write explicitly all the arguments of the thermal time correlation function $C_\mu(t, T, \Omega_i, \Omega_f, \mathbf{J}, \mathbf{K})$. It can therefore be computed analytically from the knowledge of the initial- and final-state frequencies Ω_i and Ω_f of the displacement vector **K** and the Duschinsky matrix **J**. These data can be obtained also for sizable molecules according to different electronic methods. According to Eq. 10.6, the absorption spectrum can then be calculated by Fourier transformation of $C_\mu(t, T, \Omega_i, \Omega_f, \mathbf{J}, \mathbf{K})$.

10.3.2.1 The $S_0 \rightarrow S_1$ Spectrum of* trans-*Stilbene at Room Temperature *trans*-Stilbene excited states have been the subject of very extended and detailed experimental and computational studies since it has been considered a prototypical system for the study of excited-state photoisomerization [67–71]. In the ground state the molecule belongs to C_{2h} symmetry. The relative stability of the first two excited states (B_u) has been a matter of debate for a long time [72, 73]; recently this issue has been definitively clarified by experimental results [74, 75] and accurate post-HF calculations [76]. The lowest excited-state S_1 corresponds to the stronger transition and is characterized by a HOMO → LUMO excitation. The corresponding absorption spectrum shows a remarkable sensitivity to temperature. This effect can be traced back to the role of the a_u combination of the twisting of the phenyl rings, which shows a frequency much lower in the ground state than in the excited state (experimental values are 9 and 45 cm^{-1} respectively, see Santoro et al. [7] and references therein).

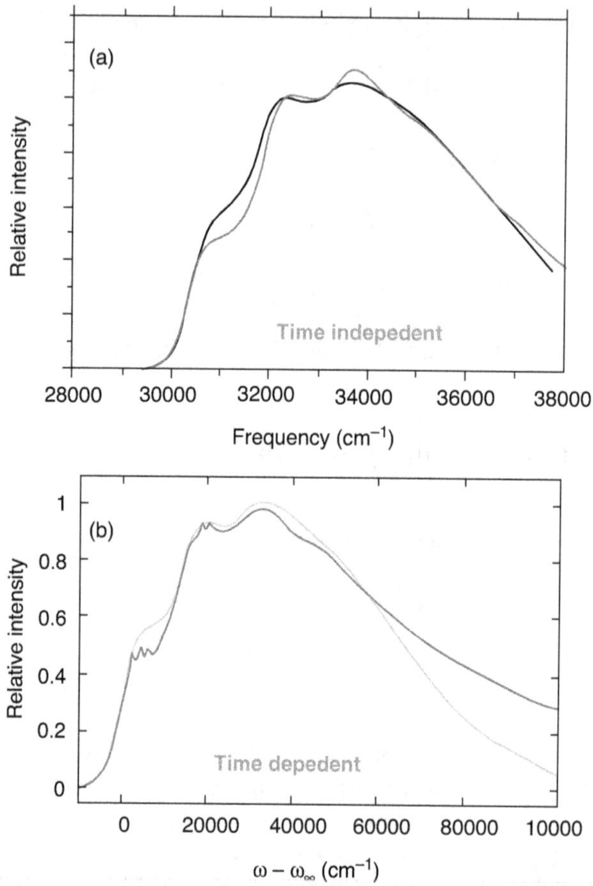

Figure 10.3 Room temperature (295 K) $S_0 \rightarrow S_1$ absorption spectrum of *trans*-stilbene computed using harmonic approximation by time-independent (upper panel [7]) and time-dependent (lower panel [63]) methods. Both computational results are compared with the experimental absorption spectrum measured at 295 K in cyclohexane by Mathies and co-workers [77]. (From J. Tatchen and E. Pollack, *J. Chem. Phys.* **2008**, *128*, 164303. Copyright © 2008. Reprinted with permission of the American Institute of Physics.)

This originates long progressions that increase with the temperature due to the population of a large number of vibrational states along this mode. Because of that, *trans*-stilbene is a typical example in which the number of relevant transitions to be taken into account becomes huge at room temperature, making the adoption of TD approaches based on the analytical computation of the thermal correlation function more suitable than TI methods based on a sum-over-state calculation. In the specific example, however, symmetry causes a block diagonalization of the Duschinsky matrix, allowing the computation of the spectrum through the convolution of spectra of independent subsystems, each of them affordable also through TI methods. Figure 10.3 reports the results of TI (upper panel) and TD (lower panel) calculations of the room temperature spectrum, performed respectively in our group [7] and by Tatchen and Pollak [63], both compared with the same experimental spectrum reported by Mathies and co-workers [77]. Both the calculations are based on harmonic analysis of the S_0 and S_1 surfaces by DFT and TD-DFT methods, respectively, performed at PBE0/6-31 + G(d,p) level (TI results) and at B3LYP/TZVP level (TD results). The two approaches deliver very similar results, both in excellent agreement with experiment. Further details can be found in the original papers where the effect of temperature was investigated also by computing and comparing with experiments, the spectrum at 77 K [7], and high-resolution dispersed fluorescence spectra [63].

10.4 MIXED QUANTUM CLASSICAL AND SEMICLASSICAL METHODS OF PROPAGATION

Despite the impressive development of efficient algorithms for the approximate solution of the TDSE, the quantum propagation of wavepackets for large systems remains a heavy task, not only for the number of coordinates involved but also because a global analytical representation of the PESs involved is needed. It is therefore desirable to develop approximate methods utilizing, at some stage, classical trajectories, that can be run on-the-fly, without requiring the knowledge of such analytical representation of the PES. This can be done according to three distinct typologies of approaches which will be briefly illustrated in this section: (a) mixed quantum classical, (b) semiclassical, and (c) classical.

10.4.1 Mixed Quantum Classical Approaches

As a promising example of the mixed quantum classical approach here we quote the possibility of coupling in a self-consistent way the MCTDH method to the classical equation of motion [78]. Typically the system modes are divided in two groups: relevant (or primary) and secondary modes. In the first group the most important modes are collected, that is, those that are expected to introduce marked quantum effects; all the other modes constitute the second group. Notice that for a molecule in the condensed phase the primary modes can include, if necessary, a few solvent modes, while some less important intramolecular modes can be considered to be secondary. The convergence may be tested varying the number of modes included in the primary group.

10.4.2 Semiclassical Approaches

The semiclassical route is an attempt to couple the simplicity and immediate interpretation of classical mechanics to the rigor of quantum mechanics. The initial step in this direction is to write down the required transition amplitude in the coordinate representation ($\hbar = 1$):

$$U_{fi}(t) = \langle \psi_f | \exp(-iHt) | \psi_i \rangle = \int d\mathbf{Q}_i \int d\mathbf{Q}_f \, \psi_f(\mathbf{Q}_f)^* \langle \mathbf{Q}_f | \exp(-iHt) | \mathbf{Q}_i \rangle \psi_i(\mathbf{Q}_i) \tag{10.76}$$

and then to use the approximate semiclassical Van Vleck–Gutzwiller propagator [79, 80]:

$$\langle Q_f|\exp(-iHt)|\mathbf{Q}_i\rangle \cong U_{\mathrm{VVG}}(\mathbf{Q}_i,\mathbf{Q}_f;t) = \sum_{\mathrm{traj}} \exp[iS_t - i\pi\upsilon(t)](2\pi i)^{-f/2}\det\left(\frac{\partial\mathbf{Q}_f}{\partial\mathbf{P}_i}\right)^{-1/2} \tag{10.77}$$

which can be derived starting from the exact Feynman path integral representation of the quantum propagator and invoking a stationary-phase approximation [81]. Notice that, for harmonic systems, the exact quantum expression of the off-diagonal element of the evolution operator in Eq. 10.76 is known, and it has been utilized in Section 10.3.2 of this book. In Eq. 10.77, $S_t = \int_0^t d\tau(\mathbf{P}\mathbf{Q} - H)$ is the classical action, and the sum is over all the trajectories starting at $t = 0$ in $\mathbf{Q}_i$ and ending in $\mathbf{Q}_f$ at time t. Using Eq. 10.77 is, however, not trivial since, for any given initial $\mathbf{Q}_i$, one must first determine for what values of the initial momentum $\mathbf{P}_i$ the trajectory $\mathbf{Q}(\mathbf{Q}_i,\mathbf{P}_i;t)$ goes to $\mathbf{Q}_f$ at a given time t. The equation $\mathbf{Q}(\mathbf{Q}_i,\mathbf{P}_i;\mathrm{t}) = \mathbf{Q}_f$ (the unknown quantity being $\mathbf{P}_i$) has in general multiple roots and the sum in Eq. 10.77 is over all such roots. The Jacobian determinant, $\det(\partial\mathbf{Q}_f/\partial\mathbf{P}_i)$ is also evaluated for the same roots, while the Maslov index $\upsilon(t)$ is the number of times it becomes zero in the time interval $(0, t)$. Due to the complication of the root-solving procedure (especially for chaotic or nearly chaotic systems), Eq. 10.77 can hardly be used for practical calculations on molecular systems. An important improvement was introduced by Miller [82, 83] with the so-called initial-value representation (IVR) by which the integral in Eq. 10.76 is performed only on the initial values Q_i, P_i (which means that the trajectory is unique). The Van Vleck–Gutziller amplitude is in such a case

$$U_{fi}^{\mathrm{VVG}}(t) = \int d\mathbf{Q}_i \int d\mathbf{P}_i \,\psi_f(\mathbf{Q}_t)^* \psi_i(\mathbf{Q}_i) \exp[iS_t(\mathbf{Q}_i,\mathbf{P}_i) - i\pi\upsilon(t)](2\pi i)^{-f/2}\det\left(\frac{\partial\mathbf{Q}_i}{\partial\mathbf{P}_i}\right)^{1/2} \tag{10.78}$$

A further numerical difficulty to be faced with for applying the method is the highly oscillatory phase space integral in Eq. 10.78. The problem can be partly circumvented by making recourse to a mixed coordinate/coherent-state representation of the propagator, as in the popular Herman–Kluk approach [84, 85]. An alternative smoothing technique was first proposed by Filinov [86, 87]. The integral in Eq. 10.78 over the initial coordinates and momenta may be computed according to (weighted) Monte Carlo techniques from the dynamics of a suitable number of classical trajectories sampling the initial conditions.

A great amount of work in the semiclassical approach to molecular dynamics can be traced back to Heller [88–91], who was also among the firsts to focus on TDSE. His basic idea, which has been used by many authors, was that of using moving Gaussian wavepackets as a basis set for the expansion of multidimensional time-dependent wavefunctions, since this establishes a natural connection with classical mechanics.

In fact, it is well known that, not only does a Gaussian wavepacket remain a Gaussian wavepacket when it moves in a quadratic potential, but also the time-dependent parameters (average position, width, phase) can be computed using classical mechanics. In the Heller approach the parameters are forced to obey classical mechanics for any potential by expanding it as a quadratic function of $\mathbf{Q}$ around the position $\mathbf{Q}_t$($\mathbf{Q}_t$ being the position along the classical trajectory at time t). It can be shown [92] that this is basically equivalent to the stationary-phase approximation invoked, as discussed before, to derive the semiclassical Van Vleck–Gutzwiller propagator, Eq. 10.77. Computational studies [43] revealed that it is sometimes better to keep fixed the width (frozen Gaussian approximation) than to allow its time variation (thawed Gaussians).

The above semiclassical methods can be used to compute absorption spectra when the propagation involves a single PES. An important advantage comes if we are able to use only local information, since this opens the route to the so-called on-the-fly methods, where the heavy electronic calculations are performed only at the points of the PES touched by the trajectories at selected time steps.

10.4.2.1 On-the-Fly Calculation of $S_0 \rightarrow S_1$ Spectrum of Formaldehyde Very recently Tatchen and Pollak computed the $S_0 \rightarrow S_1$ absorption spectrum of formaldehyde using a semiclassical propagator based on modified frozen Gaussians [93] coupled to a TD-DFT electronic structure calculation adopting the PBE functional. While the equilibrium geometry of the S_0 ground state of formaldehyde is planar, it is known both from experiments and computational studies that it is bent in the first excited state S_1 due to pyramidalization around the central C atom. The excited PES along this mode assumes a typical double-well shape that is inherently anharmonic. These cases of symmetry removal in one of the two electronic states involved in the optical transition have also been covered in Chapter 8, where it is briefly discussed how the spectra can be obtained within a vertical approach by computing the anharmonic potential energy profile along the normal modes showing an imaginary frequency at the high-symmetry saddle point structure. It is however clear that the more anharmonic is the PES of the system, the more a description grounded on a local expansion of the PES (as in harmonic approximation) is inadequate. On-the-fly semiclassical calculations avoid a priori computation of the PES and its fitting, which are necessary at the state-of-the art to perform quantum calculations (through for example MCTDH method). In this perspective, formaldehyde is a prototypical example for the calculation of the spectrum based on the semiclassical approach. Furthermore, due to its $n\pi^*$ character, the $S_0 \rightarrow S_1$ transition of formaldehyde is weak and requires a simulation at the Herzberg–Teller level, which makes this case even more interesting due to the intrinsically nonvertical quantum effects deriving by the dependence of the transition dipole on the nuclear coordinates, as, for instance, the false-origin effect. In a time-dependent perspective, the reproduction of these effects requires an explicit description of the system evolution on the excited-state PES.

Figure 10.4 reports the spectrum computed from a sample of 6000 trajectories propagated on the excited PES for 500 fs and compares it with the experiment.

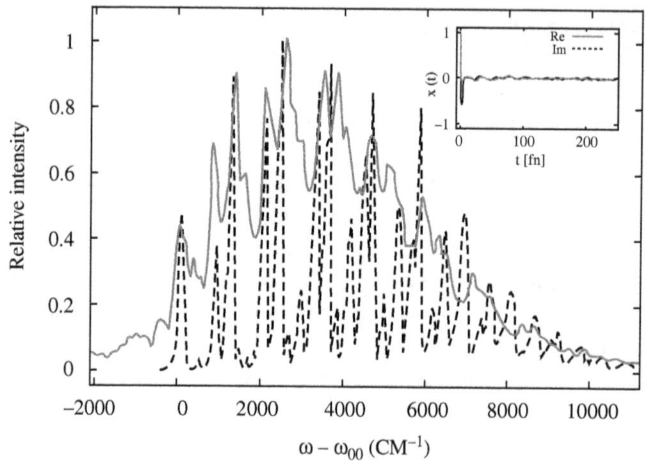

Figure 10.4 Absorption spectrum of formaldehyde computed according to a semiclassical approach by on-the-fly TD-DFT computations and its comparison with experiments. The spectrum shows remarkable Herzberg–Teller effects. (From J. Tatchen, E. Pollack, *J. Chem. Phys.* **2009** *130*, 041103. Copyright © 2009. Reprinted with permission of the American Institute of Physics.)

The results are very encouraging apart from an overestimated broadening of the spectrum, which can be attributed to the approximations introduced as well as to the short propagation time due to the computational cost of excited-state Born–Oppenheimer dynamics on TD-DFT PESs. It is easy to foresee that future developments and increase of computational power will make this approach more and more satisfactory.

10.4.2.2 Nonadiabatic $S_0 \rightarrow S_2/S_1$ Absorption Spectrum of Pyrazine Most of the semiclassical theoretical models have been developed in the limit of a dynamics taking place on a single electronic PES. In order to apply semiclassical methods to nonadiabatic dynamics, further generalization is needed and it is necessary to write down a Hamiltonian with a well-defined classical limit. In this context, the problem of the description of discrete quantum degrees of freedom (electronic states) can be tackled in two steps; the first is a mapping onto continuous variables, and the second is a semiclassical treatment of the resulting problem [94, 95]. These methods have been recently reviewed by Stock and Thoss [96].

As we discussed in Section 10.3.1.3, due to the wide range of different applied methodologies, nowadays the $S_0 \rightarrow S_2/S_1$ absorption spectrum of pyrazine is considered a benchmark system to investigate the performance of novel strategies for computing steady-state electronic spectra in the presence of remarkable nonadiabatic interactions [50–54]. Stock and Thoss [97] adopted the same model Hamiltonian developed for quantum dynamical simulations and described in Section 10.3.1.3 and performed both quasi-classical [98] and semiclassical [97] approximate calculations of the spectrum, comparing them with experiment and with the reference MCTDH results [54]. Results obtained assuming an exponential damping of the correlation function with a time constant $T = 30$ fs are reported in Figure 10.1. It is clearly seen that the semiclassical method provides results in very good agreement with the quantum predictions, thus reproducing very accurately the experimental features, both in the region of the strong absorption band (around 5 eV) and in the redwing at about 4 eV due to the absorption of the S_1 state allowed by a Herzberg–Teller vibronic borrowing mechanism. Quasi-classical results, not reported here for the sake of brevity, obtained neglecting the semiclassical phase information delivered worse results where the fine details of the spectrum were not resolved, thus demonstrating the importance of quantum interferential effects for a correct description of the spectral features. It is interesting to note that the results in Figure 10.1 were obtained without resorting to filtering techniques (like Filinov's one; see Section 10.4.2). However, due to the chaotic classical dynamics of the systems, a very large number of trajectories (10^7) were necessary to converge the spectrum calculation. In these cases semiclassical calculations become very expensive and methodological advancements are required to make such calculations feasible for more complex systems.

10.4.3 Classical Molecular Dynamics Approaches and Their Theoretical Foundation

The most economic way for investigating the dynamical behavior of large systems is to use classical MD, thus completely forgetting the quantum nature of the nuclear motion. The classical approach often gives reasonable results in computing time-dependent observables, especially at room temperature, but it is in principle completely unsatisfactory when the target is an amplitude, as for the case of the absorption spectrum, since here the phase of the wavefunction is crucial. Classical trajectories, however, are very useful in dealing with the computation of absorption spectra of large molecules in the condensed phase, especially when the interaction with the surrounding (solvent) plays an important role, since the fluctuating perturbation of the environment strongly reduces the importance of the phase information.

In this context, many authors compute electronic absorption spectra relying on a classical version of the Franck–Condon principle and use MD to span the initial-state classical phase space distribution. These methods are classified as time dependent, since they are based on the results of dynamics simulations, but here molecular dynamics (driven by parameterized or ab initio force fields) is just a technique for averaging over the *initial-state* distribution, while, according to the classical Franck–Condon principle, any dynamical effect on to the final state is neglected. The average over the initial-state distribution plays here the same role of the trace over the initial states with Boltzmann weights in the quantum description. The quantum methods discussed above are, instead, grounded on the genuine propagation of the initial wavepacket on the *final-state* PES, as required from Eq. 10.7.

In order to apply MD simulations to the calculation of absorption spectrum one needs a classical version of the absorption cross section. This has been the subject of many papers (see, e.g., ref. 99 and references therein), introducing equivalent formulations. We briefly sketch here the most direct one, focusing on a case in which the adiabatic approximation holds (i.e., only a single electronic excited state $|e\rangle$

is involved). Let us first notice that, for a thermal distribution, Eq. 10.7 can be rewritten, utilizing the density matrix, as ($\hbar = 1$)

$$W(\omega) = \frac{4\pi\omega}{3}\mathrm{Re}\left\{\mathrm{Tr}_Q\left[\int_{-\infty}^{\infty}\rho_g(\mathbf{Q},T)\langle g|\exp[i(H+\omega)t]\mu\exp[-iHt]\mu|g\rangle\,dt\right]\right\} \tag{10.79}$$

where the part of the trace involving electronic states has been explicitly performed assuming that only the electronic ground state is initially populated.

By repeatedly using the electronic resolution of the identity $|g\rangle\langle g| + |e\rangle\langle e| = 1$, Eq. 10.79 can be rewritten as

$$\begin{aligned} W(\omega) = \frac{4\pi\omega}{3}\mathrm{Re}\Bigg\{\mathrm{Tr}_Q\Bigg[\int_{-\infty}^{\infty}&\rho_g(T)\exp[i(V_g(\mathbf{Q})+T_N+\omega)t]\mu_{ge}(\mathbf{Q}) \\ &\times\exp[-i(V_e(\mathbf{Q})+T_N)t]\mu_{eg}(\mathbf{Q})\,dt\Bigg]\Bigg\} \end{aligned} \tag{10.80}$$

In order to have a classical picture we have to neglect the nuclear kinetic energy operators (fixed nuclei approximation):

$$\begin{aligned} W(\omega) &\cong \frac{4\pi\omega}{3}\mathrm{Re}\left\{\mathrm{Tr}_Q\left[\rho_g(T)|\mu_{ge}(\mathbf{Q})|^2\int_{-\infty}^{\infty}\exp[i(V_g(\mathbf{Q})+\omega-V_e(\mathbf{Q}))t]^2\,dt\right]\right\} \\ &= \frac{4\pi^2\omega}{3}\int(\rho_g(\mathbf{Q},T)|\mu_{ge}(\mathbf{Q})|^2\delta\left[V_g(\mathbf{Q})+\omega-V_e(\mathbf{Q})\right]dQ \end{aligned} \tag{10.81}$$

For a given frequency ω the integration over $\mathbf{Q}$ receives a contribution only from the value $\mathbf{Q}^*$ such that $V_g(\mathbf{Q}^*) + \omega - V_e(\mathbf{Q}^*) = 0$ (we suppose here that a single $\mathbf{Q}^*$ exists):

$$\begin{aligned} \delta\left[V_g(\mathbf{Q})+\omega-V_e(\mathbf{Q})\right] &= \delta(Q-Q^*)R(\omega) \\ R(\omega) &= \left|\left(\frac{d\left[V_g(\mathbf{Q})-V_e(\mathbf{Q})\right]}{dQ}\right)_{Q=Q^*(\omega)}\right| \end{aligned} \tag{10.82}$$

Hence one gets the final result:

$$W(\omega) \cong \frac{4\pi^2\omega}{3}R(\omega)\rho_g\left[\mathbf{Q}^*(\omega),T\right]\left|\mu_{ge}\left[\mathbf{Q}^*(\omega)\right]\right|^2 \tag{10.83}$$

The above expression, which is essentially an application of the classical Franck–Condon principle, is in many respects a severe approximation, since it determines the loss of information on the peaks corresponding to well-defined vibrational progressions. However, since it is applied when the coupling with the environment gives rise to large inhomogeneous broadening, it is in many cases an acceptable compromise between accuracy and ease of interface with MD simulations. It is instructive in this respect to notice, following Lax [100], that Eq. 10.81 can be obtained from Eq. 10.80 without invoking the drastic "frozen-nuclei approximation" $T_N = 0$ but simply by neglecting the commutators $[H_g, H_e] = [T_N, V_e(\mathbf{Q}) - V_g(\mathbf{Q})]$ and $[T_N, \mu_{ge}(\mathbf{Q})]$. By neglecting in fact the first commutator, one can write

$$\exp[iH_g t]\exp[-iH_e t] \approx \exp[i(H_g - H_e)t] = \exp[i(V_g(\mathbf{Q}) - V_e(\mathbf{Q}))t] \quad (10.84)$$

Such an alternative derivation is interesting since it allows us to connect the approximations behind the derivation of Eq. 10.83 with the dynamics on the excited state. First, it highlights that in cases where Herzberg–Teller effects are important, Eq. 10.83 is expected to work worse than in FC transitions, where $\mu_{ge}(\mathbf{Q})$ can be considered constant and hence $[T_N, \mu_{ge}(\mathbf{Q})] \approx 0$. Second, expanding the exponentials in Eq. 10.84 in powers of t, we have

$$\exp\left[iH_g t\right] = 1 + iH_g t - \tfrac{1}{2}H_g^2 t^2 + \cdots \quad (10.85a)$$

$$\exp[-iH_e t] = 1 - iH_e t - \tfrac{1}{2}H_e^2 t^2 + \cdots \quad (10.85b)$$

$$\exp\left[-i(H_g - H_e)t\right] = 1 + i(H_g - H_e)t - \tfrac{1}{2}(H_g - H_e)^2 t^2 + \cdots \quad (10.85c)$$

Then

$$\exp\left[iH_g t\right]\exp[-iH_e t] = \exp\left[i(H_g - H_e)t\right] + \tfrac{1}{2}\left[H_g, H_e\right]t^2 + o(t^3) \quad (10.86)$$

Equation 10.86 shows that the approximation leading to Eq. 10.83 becomes worse at the increase of the propagation time of the doorway state (or better of the corresponding density matrix) on the final PES. This also explains why in many cases such a classical approximation reproduces low-resolution spectra quite well, since, for this latter a short time propagation is needed (due to the fact that the dipole correlation function is assumed to decay rapidly with time), while it fails in shaping the fine vibrational details of the spectrum which arise from partial revivals of the correlation function and then need a long-time dynamics.

To apply Eq. 10.83 one needs a Monte Carlo technique to extract a weighted sample of $\mathbf{Q}^*$ values. Alternatively, invoking the ergodic principle, it is also possible to pick up a suitable number of $\mathbf{Q}^*$ frames taking snapshots of a single MD trajectory at properly chosen time intervals. The most relevant advantage of these MD-based calculations is that one does not rely on any predetermined model for the initial-state PESs while, at the quantum level, this latter is usually considered harmonic in order to have an analytical expression of the Wigner distribution. This characteristic allows us to span the true initial-state PES on the fly, which is feasible also for complex systems like molecules in the condensed phase, describing the latter at the explicit level. Chapter 11 shows a very interesting application of the sophisticated general liquid optimized boundary (GLOB) model to the simulation of the absorption spectrum of acroelin in aqueous solution [10].

As reported above, the average on the initial-state distribution can be easily taken in the harmonic approximation by proper sampling of the Wigner distribution. The latter is analytically available after a simple frequency calculation on the initial electronic state. In fact, the notation introduced in Section 10.3.2 yields

$$\rho_i(\mathbf{Q}_i, \mathbf{P}_i, T) = \frac{1}{\pi}\det[\mathbf{N}_i]\exp\left[-\mathbf{P}_i^{\mathrm{T}}\mathbf{N}_i\Omega_i^{-1}\mathbf{P}_i - \mathbf{Q}_i^{\mathrm{T}}\mathbf{N}_i\Omega_i\mathbf{Q}_i\right] \tag{10.87}$$

where $\boldsymbol{P}_i$ is the vector of the momenta associated with the normal coordinates $\mathbf{Q}_i$. and $\mathbf{N}_i$ is the diagonal matrix $\mathbf{N}_i = \hbar^{-1}\tanh[\beta\hbar\Omega_i]$.

In the following we compare the quantum and classical predictions in a simple monodimensional (1D) harmonic system with the aim of illustrating, in practice, the consequences of the approximations leading to Eq. 10.83. Figure 10.5 reports the results of this comparison in a number of representative cases. The harmonic frequency of the ground state is $\omega_g = 500\,\mathrm{cm}^{-1}$, the associated reduced mass is 6 amu, and the difference in energy of the excited V_e and ground V_g potentials is $10000\,\mathrm{cm}^{-1}$. Panel (*a*) shows the results at $T = 300$ K for a FC transition $\mu_{ge}(Q) = \mu_{ge}(Q_0)$ (Q_0 being the ground-state equilibrium position) when the oscillator in the excited state is displaced by an amount $\delta = 2$ in dimensionless coordinates and has a frequency $\omega_e = 400\,\mathrm{cm}^{-1}$. Black dotted and green solid lines report the quantum results at high and low resolution, respectively, while red dashed and blue dot-dashed lines give the classical predictions including and excluding the factor $R(\omega)$ in Eq. 10.83, since in many practical explorations in the literature this is actually omitted. With this choice of parameters, the main vibrational structure of the quantum spectrum is due to the displacement of the equilibrium position, and it can be seen that the classical approximation compares nicely with the low-resolution quantum spectrum, while of course any trace of the fine vibrational progression is lost. The inclusion of $R(\omega)$ slightly improves the quality of the classical approximation. Notice, for example, that in this case the classical vertical transition is $V_e(Q_0) - V_g(Q_0) = 12{,}000\,\mathrm{cm}^{-1}$, the exact first moment of the spectrum; that is, the average transition energy is different, being $M_1 = 11{,}226\,\mathrm{cm}^{-1}$ at $T = 300$ K ($11{,}235\,\mathrm{cm}^{-1}$ at 0 K), and the maximum of the low-resolution quantum spectrum is slightly red shifted $11{,}170\,\mathrm{cm}^{-1}$ and tends to M_1 for infinite broadening (a detailed discussion of the relationship between vertical transition energy, average energy, and absorption maximum can be found in Chapter 8). The maximum of the classical spectrum is at $11{,}330\,\mathrm{cm}^{-1}$ and $11{,}224\,\mathrm{cm}^{-1}$ neglecting or considering the factor $R(\omega)$, respectively. Panel (*b*) reports a different case where the displacement is vanishing, $\delta = 0$, and the frequency of the excited state is much lower than in the ground state, $\omega_e = 150\,\mathrm{cm}^{-1}$. While considering $\omega_e = 400\,\mathrm{cm}^{-1}$ as in panel (*a*) would still lead to good agreement between the classical spectrum and the low-resolution quantum spectrum, this case has been chosen to show a pathology in the classical approximation. In fact, it is possible to notice in panel (*b*) that while the quantum spectrum is asymmetric, with a longer wing toward the blue, the opposite occurs in the classical approximations, most of all when the factor $R(\omega)$ is included. This pathological, even if of minor relevance, behavior arises from the fact that while the difference $\omega_e - \omega_g$

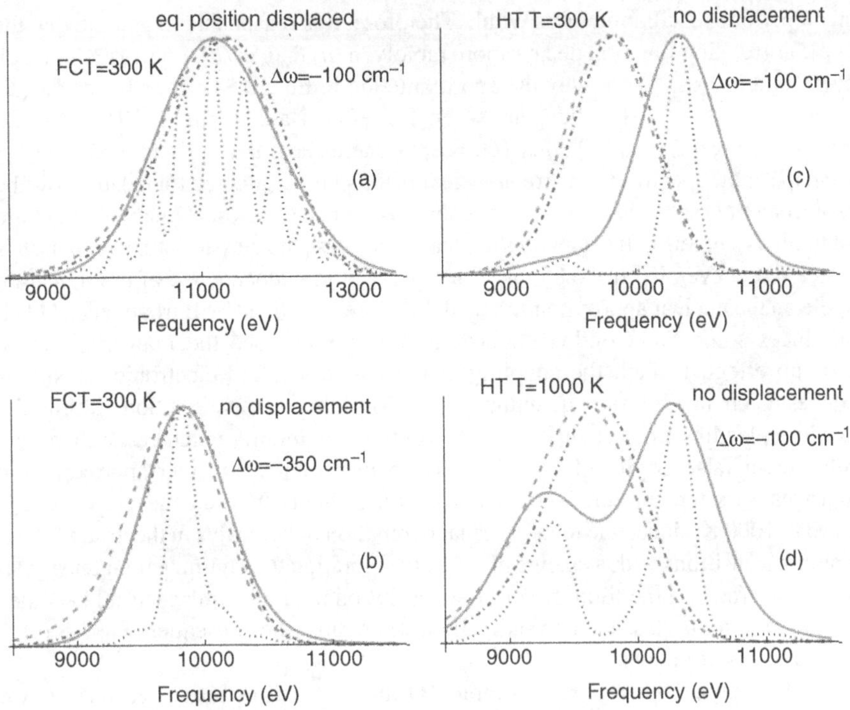

Figure 10.5 Comparison of spectrum (convoluted with Gaussian specified by standard deviation σ) of monodimensional harmonic model computed according to fully quantum methods (high-resolution spectrum, $\sigma = 100\,\text{cm}^{-1}$, black dotted line; low-resolution spectrum, $\sigma = 300\,\text{cm}^{-1}$, green solid line) and classical approximation described in Section 10.4.3. (low-resolution spectra, computed according to Eq. 10.83, red dashed line, or neglecting R function, blue dot-dashed line). (*a*) Allowed (Franck–Condon) transition at 300 K in presence of significant displacement δ of equilibrium positions and moderate change in harmonic frequency, $\Delta\omega = \omega_e - \omega_g$ (see text). (*b*) Allowed (Franck–Condon) transition at 300 K in model with no displacement of equilibrium positions and $\omega_e = 150\,\text{cm}^{-1}$. (*c*) Forbidden (Herzberg–Teller) transition at 300 K in model with $\delta = 0$ and $\omega_e = 400\,\text{cm}^{-1}$. (*d*) Forbidden (Herzberg–Teller) transition at 1000 K in model with $\delta = 0$ and $\omega_e = 400\,\text{cm}^{-1}$.

originates a vibrational progression ($n = 0, 2, 4, \ldots$) that, independently of the sign of $\omega_e - \omega_g$, elongates on the blue side (at low temperatures) of the 0–0 transition at $9650\,\text{cm}^{-1}$, in a classical approximation the vertical transition $V_e(Q_0) - V_g(Q_0) = 10{,}000\,\text{cm}^{-1}$ constitutes the maximum accessible transition energy, the energy at any other geometry being lower. Because of the definition in Eq. 10.82, inclusion of $R(\omega)$ weights more the contribution of transition energies in the red wing of the spectrum, thus worsening the agreement with quantum results. The better performance of the classical approximation in panel (*a*) with respect to panel (*b*) can be connected to the fact that, while in the former case, due to the displacement of the equilibrium position, the doorway wavepacket moves quickly away from the FC region, in panel (*b*) such a wavepacket always remains in the FC region, exhibiting only a periodic oscillation of its width. Therefore, it is reasonable to assume that the dipole correlation function decays more rapidly in (*a*) than in (*b*), thus explaining, on the ground of Eqs. 10.85, why the approximation in Eq. 10.84 works better for (*a*).

Panel (*c*) shows what happens when HT effects are dominant. The relevant parameters are μ_{ge} $(Q_0) = 0$, μ_{ge} $(Q) = \kappa Q$ (where the value of κ is irrelevant for the shape of the spectrum); the frequencies are the same as adopted for (*a*), but now the displacement is set to zero, $\delta = 0$, since, in most practical cases, HT active modes are not totally symmetric. It is clearly seen that in this case the classical approximation is poorer since, even if Eq. 10.83 takes into account the dependence of the transition dipole on the nuclear approximation (and therefore as noticed by Barbatti et al. [11] it introduces some non-Condon effects), it cannot reproduce the main nonvertical quantum effect. In fact, the coupling of the vibrational and electronic transitions induces a change of one quantum in the vibrational wavefunction during the transition, leading to a net shift of the absorption maximum (which causes, e.g., the well-known false-origin effect). The performance of the classical approximation improves as the temperature increases, as shown in panel (*d*), where the temperature is raised to 1000 K, since, due to the thermal population of hot levels in the ground state, quantum transitions with a change of -1 in the vibrational quantum gain intensity, so that, on average, in the limit of very low resolution the quantum spectrum becomes more similar to its classical approximation, where both the consequences of the $+1$ and -1 shifts are neglected.

While Figure 10.5 reports a simple 1D model for illustrative reasons, as we discussed above, the power of the classical approximation is the possibility to apply it to complex systems. Figure 10.6 reports the the absorption spectra of

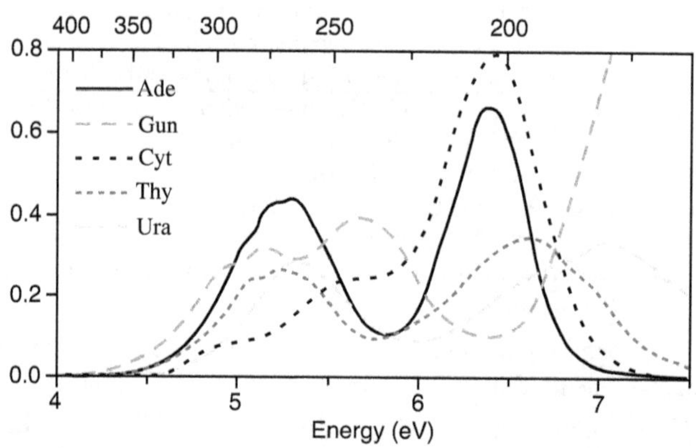

Figure 10.6 Absorption spectra of five nucleobases adenine (Ade), guanine (Gua), cytosine (Cyt), thymine (Thy), and uracil (Ura) obtained by classical sampling of ground-state distribution (in harmonic approximation), based on RI-CC2 electronic calculations. To each transition in the sampling procedure a lineshape with full width at half maximum (FWHM) = 01. eV was superimposed to remove statistical noise. (From M. Barbatti, A. J. A. Aquino, and H. Lischka, *Phys. Chem. Chem. Phys.* **2010**, *12*, 4959. Reproduced with permission of the PCCP Owner Societies.)

DNA nucleobases in an energy window encompassing several excited states computed at the classical level by Barbatti et al. [11] on the ground of resolution-of-identity approximated second-order coupled cluster (RI-CC2) electronic calculations and a sampling of the harmonic Wigner distribution in Eq. 10.87. The spectra reported have been obtained assigning to each transition, computed according to the sampling, a lineshape with full width at half-maximum of 0.1 eV to remove statistical noise. In the original paper the computed data are compared with experimental spectra showing encouraging results.

10.5 CONCLUDING REMARKS

In this chapter we reviewed different eigenstate-free time-dependent approaches to the computation of electronic spectra lineshapes. Recent methodological advancements such as the MCTDH method and its multilayer extension have boosted the potentiality of full quantum approaches and, at the state-of-the-art, should be considered the reference methods for computing spectra in strongly nonadiabatic systems. The computational bottleneck for these methods now probably lies in the difficulty to obtain a reliable description of the required multidimensional PESs in terms of analytical functions. This problem is easily solved in semirigid systems where PESs can be conveniently obtained by a Taylor expansion performed at significant nuclear configurations. For these systems and when nonadiabatic couplings are negligible, PESs can be described within the context of a harmonic approximation, and analytical expressions of the thermal time correlation function allow the fully converged computation of low-resolution spectra of large systems (hundreds of normal modes). These methods complement very nicely the time-independent ones reported in Chapter 8. With the increase of computational power, trajectory-based methods grounded in semiclassical approximations of the time evolution operator are now becoming efficient tools to obtain accurate spectra, bridging the possibility to describe quantum effects leading to fine vibrational structure with the flexibility of on-the-fly calculations that avoid the necessity to determine a priori the PESs and to constrain them to specific analytical functions, thus allowing us to take into account in principle any kind of anharmonicity. Environmental effects on the absorbing species can be explicitly described in a nonphenomenological way by exploring the free-energy hypersurface of the solute/environment system through MD simulations driven by electronic potentials computed with always increasing accuracy. In most cases, these models describe the electronic transition by applying the classical Franck–Condon principle, thus neglecting interferential effects on the dynamics of the final state that is responsible for the fine vibrational structure. It is possible that in the near future the advanced solute/environment description may be coupled with semiclassical approximations of the time correlation function providing a novel and very powerful tool for the simulation and analysis of electronic bandshapes.

ACKNOWLEDGMENTS

This work was supported by Italian MIUR (PRIN 2008) and IIT (Project Seed HELYOS). Use of the large-scale computer facilities of the CNR-VILLAGE network (http://village.pi.iccom.cnr.it) is kindly acknowledged.

REFERENCES

1. M. Kimble, W. A. W. Castleman, Jr., Eds., *Femtochemistry VII: Fundamental Ultrafast Processes in Chemistry, Physics, and Biology VII*, Elsevier Science, New York, 2006.
2. C. Leforestier, R. H. Bisseling, C. Cerjan, M. D. Feit, R. Friesner, A. Gundberg, A. Hammerich, G. Jolicard, W. Karrlein, H.-D. Meyer, N. Lipkin, O. Roncero, R. Kosloff, *J. Comp. Chem.* **1991**, *94*, 59.
3. M. H. Beck, A. Jäckle, G. A. Worth, H.-D. Meyer, *Phys. Rep.* **2000**, *324*, 1.
4. M. Dierksen, S. Grimme, *J. Chem. Phys.* **2004**, *120*, 3544; **2005**, *122*, 244101.
5. H. C. Jankowiak, J. L. Stuber, R. Berger, *J. Chem. Phys.* **2007**, *127*, 234101.
6. F. Santoro, R. Improta, A. Lami, J. Bloino, V. Barone, *J. Chem. Phys.* **2007**, *126*, 084509; **2007**, *126*, 169903.
7. F. Santoro, A. Lami, R. Improta, V. Barone, *J. Chem. Phys.* **2007**, *126*, 184102.
8. F. Santoro, R. Improta, A. Lami, J. Bloino, V. Barone, *J. Chem. Phys.* **2008**, *128*, 224311.
9. R. Improta, V. Barone, F. Santoro, *Angew. Chem. Int. Ed.* **2007**, *46*, 405.
10. G. Brancato, N. Rega, V. Barone, *J. Chem. Phys.* **2006**, *125*, 164515.
11. M. Barbatti, A. J. A. Aquino, H. Lischka, *Phys. Chem. Chem. Phys.* **2010**, *12*, 4959.
12. F. Gel'mukhanov, A. Baev, P. Macák, Y. Luo, H. Ågren, *J. Opt. Soc. Am. B* **2002**, *19*, 937.
13. S. Mukamel, *Principles of Nonlinear Optical Spectroscopy*, Oxford University Press, New York, 1995.
14. W. Domcke, G. Stock, *Adv. Chem. Phys.* **1997**, *100*, 1.
15. H. Wang, M. Thoss, *Chem. Phys.* **2008**, *347*, 139.
16. M. F. Gelin, D. Egorova, W. Domcke, *Acc. Chem. Res.* **2009**, *42*, 1290.
17. H. Beck, D. Mayer, *J. Chem. Phys.* **2001**, *114*, 2036.
18. W. Domcke, D. R. Yarkony, H. Köppel, Eds., *Conical Intersections: Electronic Strucutre, Dynamics & Spectroscopy*, World Scientific, Singapore, 2004.
19. P. A. M. Dirac, *Proc. Philos. Soc.* **1930**, *26*, 376.
20. J. Frenkel, *Wave Mechanics*, Clarendon, Oxford, 1934.
21. A. D. McLachlan, *Mol. Phys.* **1964**, *8*, 39.
22. N. Makri, W. H. Miller, *J. Chem. Phys.* **1987**, *87*, 5781.
23. A. D. Hammerich, R. Kosloff, M. A. Ratner, *Chem. Phys. Lett.* **1990**, *171*, 97.
24. P. Jungwirth, R. B. Gerber, *J. Chem. Phys.* **1995**, *102*, 6046.
25. H.-D. Meyer, U. Manthe, L. S. Cederbaum, *Chem. Phys. Lett.* **1990**, *165*, 73.
26. U. Manthe, H.-D. Meyer, L. S. Cederbaum, *J. Chem. Phys.* **1992**, *97*, 3199.
27. MCTDH code http://www.pci.uni-Heidelberg.de/tc/usr/mctdh/doc/index.html, last accessed July 11, 2011.
28. M. D. Feit, J.A. Fleck Jr., A. Steiger, *J. Comp. Phys.* **1982**, *47*, 412.

29. M. D. Feit, J. A. Fleck Jr., *J. Chem. Phys.* **1983**, *78*, 301.
30. H. Tal-Ezer, R. Kosloff, *J. Chem. Phys.* **1984**, *81*, 3967.
31. C. Lanczos, *Res. Nat. Bur. Stand.* **1950**, *45*, 225.
32. T. J. Park, J. C. Light, *J. Chem. Phys.* **1986**, *85*, 5870.
33. P. Jungwirth, R. B. Gerber, *J. Chem. Phys.* **1995**, *102*, 8855.
34. P. Jungwirth, E. Fredj, R. B. Gerber, *J. Chem. Phys.* **1996**, *104*, 9332.
35. R. B. Gerber, P. Jungwirth, E. Fredj, A. Rom, D. L. Thomson, Eds., *Modern Methods for Multidimensional Dynamics Computations in Chemistry*, World Scientific, Singapore, 1998.
36. M. Ehara, H.-D. Meyer, L.S. Cederbaum, *J. Chem. Phys.* **1996**, *105*, 8865.
37. G. A. Worth, H.-D. Meyer, L. S. Cederbaum, *J. Chem. Phys.* **1998**, *109*, 3518.
38. H.-D. Meyer, G. A. Worth, *Theor. Chem. Acc.* **2003**, *109*, 251.
39. U. Manthe, *J. Theor. Comp. Chem.* **2002**, *1*, 153.
40. F. Huarte-Larranaga, U. Manthe, *Z. Phys. Chem.* **2007**, *221*, 171.
41. H. Wang, K. Thoss, *J. Chem. Phys.* **2003**, *119*, 1289.
42. U. Manthe, *J. Chem. Phys.* **2008**, *128*, 164116.
43. I. Burghardt, K. Giri, G. A. Worth, *J. Chem. Phys* **2008**, *129*, 174104.
44. L. S. Cederbaum, E. Gindensperger, I. Burghardt, *Phys. Rev. Lett* **2005**, *94*, 113003.
45. E. Gindensperger, L. S. Cederbaum, *J. Chem. Phys.* **2007**, *127*, 124107.
46. S.-I. Sawada, R. Heather, B. Jackson, H. Methiu, *J. Chem. Phys* **1985**, *83*, 3009.
47. R. Martinazzo, M. Nest, P. Saalfrank, G. Tantardini, *J. Chem. Phys.* **2006**, *125*, 194102.
48. T. J. Martinez, M. Ben-Nun, R. D. Levine, *J. Phys. Chem.* **1996**, *100*, 7884; **1997**, *101*, 6389.
49. M. Ben-Nun, T. J. Martinez, *J. Phys. Chem. A* **1999**, *103*, 10517.
50. R. Schneider, W. Domcke, *Chem. Phys. Lett.* **1988**, *150*, 235.
51. R. Seidner, G. Stock, A. L. Sobolewski, W. Domcke, *J. Chem. Phys.* **1992**, *96*, 5298.
52. C. Woyvod, W. Domcke, A. L. Sobolewski, H.-J. Werner, *J. Chem. Phys.* **1994**, *100*, 1400.
53. G. Stock, C. Woywood, W. Domcke, T. Swinney, B. S. Hudson, *J. Chem. Phys.* **1995**, *103*, 6851.
54. A. Raab, G. A. Worth, H.-D. Meyer, L. S. Cederbaum, *J. Chem. Phys.* **1999**, *110*, 936.
55. R. Improta, F. Santoro, V. Barone, A. Lami, *J. Phys. Chem. A* **2009**, *113*, 15346.
56. F. Santoro, V. Barone, R. Improta, *Proc. Natl. Acad. Sci. USA* **2007**, *104*, 9931.
57. F. Santoro, V. Barone, R. Improta, *J. Am. Chem. Soc.* **2009**, *131*, 15232.
58. T. Petrenko, F. Neese, *J. Chem. Phys.* **2007**, *127*, 164319.
59. S. Mukamel, S. Abe, Y. J. Yan, R. Islampour, *J. Phys. Chem.* **1985**, *89*, 201.
60. Y. J. Yan, S. Mukamel, *J. Chem. Phys.* **1986**, *85*, 5908.
61. J. Tang, M. T. Lee, S. H. Lin, *J. Chem. Phys.* **2003**, *119*, 7188.
62. R. Ianconescu, E. Pollack, *J. Phys. Chem. A* **2004**, *108*, 7778.
63. J. Tatchen, E. Pollack, *J. Chem. Phys.* **2008**, *128*, 164303.
64. Q. Peng, Y. Niu, C. Deng, Z. Shuai, *Chem. Phys*, **2010**, *370*, 215.
65. F. Duschinsky, *Acta Physicochim. URSS* **1937**, *7*, 551.
66. R. P. Feynman, *Statistical Mechanics*, Benjamin, New York, 1972.

67. G. Orlandi, W. Siebrand, *Chem. Phys. Lett.* **1975**, *30*, 352.
68. J. A. Syage, P. M. Felker, A. H. Zewail, *J. Chem. Phys.* **1984**, *81*, 4685.
69. V. Molina, M. Merchán, B. O. Roos, *J. Phys. Chem. A* **1997**, *100*, 3478.
70. J. S. Baskin, L. Bañares, S. Pedersen, A. H. Zewail, *J. Phys. Chem.* **1996**, *100*, 11920.
71. D. M. Leitner, B. Levine, J. Quenneville, T. J. Martínez, P. G. Wolynes, *J. Phys. Chem. A* **2003**, *107*, 10706.
72. R. Improta, F. Santoro, C. Dietl, E. Papastathopoulos, G. Gerber, *Chem. Phys. Lett.* **2004**, *387*, 509.
73. R. Improta, F. Santoro, *J. Phys. Chem. A* **2005**, *109*, 10058.
74. G. Hohlneicher, R. Wrzal, D. Lenoir, R. Frank, *J. Phys. Chem. A* **1999**, *103*, 8969.
75. C. Dietl, E. Papastathopoulos, P. Niklaus, R. Improta, F. Santoro, G. Gerber, *Chem. Phys.* **2005**, *310*, 201.
76. C. Angeli, R. Improta, F. Santoro, *J. Chem. Phys.* **2009**, *130*, 174307/1–6.
77. A. B. Meyers, M. O. Trulson, R. A. Mathies, *J. Chem. Phys.* **1985**, *83*, 5000.
78. H. Wang, M. Thoss, W. H. Miller, *J. Chem. Phys.* **2001**, *115*, 2979.
79. J. H. Van Vleck, *Proc. Natl. Acad. Sci. USA* **1928**, *14*, 178.
80. M. C. Gutzwiller, *J. Math. Phys.* **1971**, *12*, 343.
81. L. S. Schulman, *Techniques and Applications of Path Integrals*, Wiley, New York, 1981.
82. W. H. Miller, *J. Chem. Phys.* **1970**, *53*, 3578.
83. X. Sun, W. H. Miller, *J. Chem. Phys.* **1997**, *106*, 907.
84. M. F. Herman, E. Kluk, *Chem. Phys.* **1984**, *91*, 27.
85. E. Kluk, M. F. Herman, H. L. Davis, *J. Chem. Phys.* **1986**, *84*, 326.
86. V. S. Filinov, *J. Nuc. Phys. B* **1986**, *271*, 717.
87. N. Makri, W. H. Miller, *Chem. Phys. Lett.* **1987**, *139*, 10.
88. E. J. Heller, *J. Chem. Phys.* **1981**, *75*, 2923.
89. S. Y. Lee, E. J. Heller, *J. Chem. Phys.* **1982**, *76*, 3035.
90. M. J. Davis, E. J. Heller, *J. Chem. Phys.* **1984**, *80*, 5036.
91. J. R. Reimers, K. R. Wilson, E. J. Heller, *J. Chem. Phys.* **1989**, *79*, 4769.
92. M. Baranger, M. A. M. de Aguiar, F. Keck, H. J. Korsch, B. Schellhaass, *J. Phys. A Math. Gen.* **2001**, *34*, 7227.
93. J. Tatchen, E. Pollack, *J. Chem. Phys.* **2009**, *130*, 041103.
94. X. Sun, H. W. Miller, *J. Chem. Phys.* **1997**, *106*, 6346.
95. M. Thoss, G. Stock, *Phys. Rev. A* **1999**, *59*, 64.
96. G. Stock, M. Thoss, *Adv. Chem. Phys.* **2005**, *131*, 243.
97. G. Stock, M. Thoss, in *Conical Intersections: Electronic Strucutre, Dynamics & Spectroscopy*, W. Domcke, D. R. Yarkony, H. Köppel, Eds., World Scientific, Singapore, 2004, Chapter 15.
98. G. Stock, W. H. Miller, *Chem. Phys. Lett.* **1992**, *197*, 396.
99. J. P. Bergsma, P. H. Berens, K. R. Wilson, D. R. Fredkin, E. J. Heller, *J. Phys. Chem.* **1984**, *88*, 612.
100. M. Lax, *J. Chem. Phys.* **1952**, *20*, 1752.

11

COMPUTATIONAL SPECTROSCOPY BY CLASSICAL TIME-DEPENDENT APPROACHES

GIUSEPPE BRANCATO

Italian Institute of Technology, IIT@NEST Center for Nanotechnology Innovation, Pisa, Italy

NADIA REGA

Dipartimento di Chimica "Paolo Corradini", Università di Napoli Federico II, Naples, Italy

Computational Strategies for Spectroscopy: From Small Molecules to Nano Systems, First Edition.
Edited by Vincenzo Barone.

Starting from a brief theoretical excursus, this chapter covers the modern computational repertoire for the modeling of spectroscopic measurements via classical time-dependent approaches, such as ab initio, mixed ab initio–classical, and purely classical molecular dynamics. In the first part of the chapter, we discuss important features of spectroscopic computations issuing from molecular dynamics methods, underlying both advantages and critical issues, with particular regard to the vibrational and electronic analyses. To this purpose, a sketch of the nonperiodic general liquid optimized boundary (GLOB) for molecular dynamics is provided. In the second part, key examples of applications are illustrated in some detail.

11.1 INTRODUCTION

During recent years classical molecular dynamics became an invaluable support to computational spectroscopy, ranging from magnetic and optical to X-ray diffraction/absorption techniques, for both equilibrium (steady-state) and nonequilibrium (time-resolved) experiments. In general, the different procedures by which molecular dynamics can be exploited to simulate spectra can be classified according to two main pictures. The spectrum (transition energy and cross section) can be calculated for each configurational snapshot of a molecular dynamics trajectory of the system under investigation and then averaged to account for the thermal/solvent broadening as observed in spectroscopic bands/signals. Optical absorption and emission or X-ray absorption fine-structure (XAFS) techniques are examples of spectroscopy which can be simulated by this kind of approach. A similar philosophy is adopted when the configurational sampling from molecular dynamics is exploited to estimate average spectroscopic parameters, which in turn can be used in fitting analysis of experimental spectra. Examples of this approach are the calculations of effective magnetic tensors in nuclear magnetic resonance (NMR) and electron spin resonance (ESR) or the structural parameters of XAFS and extended XAFS (EXAFS) techniques. On the other hand, the time-dependent information provided by molecular dynamics can be directly exploited to calculate spectra lineshapes, more specifically by evaluating time correlation functions of the transition moment operators (linear response theory). Examples in this case are infrared (IR) and Raman spectroscopy, and electronic spectra can be simulated as well.

The two pictures (configurational averaging and time correlation functions) share similar advantages and critical issues. A rigorous treatment of theoretical spectroscopy is unavoidably based on a quantum mechanical picture of the system and the calculation of accurate energy levels by the solution of the related rovibrational–electronic Hamiltonian $\mathbf{H}_{\text{ro-vib-elec}}$. Further, available methods mostly rely on a time-independent approach (usually, stationary points of the potential energy surfaces). Variational [1–3], self-consistent [4–8], and perturbative [9–17] methods can be applied to solve the anharmonic vibrational problem, while linear coupling [18], Duschinsky-like [19] approximations and prescreening techniques [20–25], can provide accurate and effective solutions of the vibronic problem. Semiclassical methods attempt a transfer of the spectroscopic quantum treatment into a time-dependent framework. However, these quantum mechanical methods are still computationally difficult for large systems, such as molecules in condensed phase or biomolecules. On the contrary, theoretical approaches based on molecular dynamics provide, at an acceptable cost, the structural and kinetic detail of very complex systems (macromolecules, liquids) and the mandatory statistical information for the realistic modeling of spectroscopic bands. Beside this invaluable advantage, by performing a spectroscopic analysis in terms of a molecular dynamics simulation, it is often possible to unravel indirect and subtle relationships occurring between spectroscopic observables and structural/dynamic properties.

The most critical issue when adopting molecular dynamics–based methods is the extent to which a classical sampling of the nuclear motions can be used to characterize states that are quantum mechanical in nature, regardless of whether the calculated energy potential is more or less accurately computed. In particular, when considering the classical time correlation function approach, the energy levels are determined according to a classic description. For example, it is reasonable to expect that the classically determined frequencies will underestimate anharmonic shifts, since the classical amplitudes of the motion are smaller than their quantum counterparts at the same temperature [26]. Furthermore, the classical time correlation function computed at 300 K for a stiff anharmonic mode in a single well reproduces the oscillating behavior of its quantum counterpart, but within a smaller amplitude scale [27]. Many efforts during the past decades were aimed at defining quantum corrections to the classical time correlation functions in either the time or frequency domain. Such corrections introduce frequency and/or temperature factors which take into account important symmetry properties usually absent in the classical picture. However, the corrections cannot be defined in a universal way, since they arise from comparison of several possible quantum correlation functions with the single classical counterpart. All the proposed corrections only affect the width and shape of the IR bands, while the accuracy of the calculated frequencies still relies on the ability of the classical approach to describe the vibrational modes. Comparisons of spectra computed by classical, semiclassical, and quantum approaches have been presented in a few recent papers [26–29] focusing mainly on time correlation functions [27], intensities [28], and peak positions [26–29]. While the use of quantum-corrected time correlation functions can indeed improve the accuracy of the computed intensities, the agreement between quantum and classical approaches in the frequency values is satisfactory only within the harmonic regime, that is, when molecular motions correspond to the normal modes.

It is also worth noting that eigenstate-free time-dependent methods are the main (when not the only) route to deal with systems affected by significant nonadiabatic interactions for which eigenstate calculations are unfeasible, as it is the case of conical intersections [30], or for systems propagating on highly anharmonic potential energy surfaces (PESs) [31, 32]. Such cases require that dynamical effects are properly taken into account. Furthermore, large-amplitude motions and solvent librations cannot be described by computations based on a harmonic approximation or perturbative anharmonic corrections. Then, appropriately tailored quantum mechanics/molecular mechanics (QM/MM) schemes are necessary to perform molecular dynamics (MD) simulations in order to sample the configurational space. In this respect, we have recently developed the GLOB model [33, 34], which can be successfully applied to perform QM/MM molecular dynamics simulations of complex molecular systems in solution. Then, spectroscopic observables may be computed on-the-fly or in an a posteriori step by averaging the corresponding estimators over a suitable number of snapshots. In the general case of solute–solvent systems, it is customary to carry out also simulations of the solute in the gas phase, so as to quantitatively evaluate the solvent effects. The a posteriori calculations of spectroscopic properties, compared to other on-the-fly approaches, allow us to more freely exploit different QM/MM schemes than in the MD simulations. In this way, a more accurate treatment for the more demanding molecular parameters of both first (e.g., hyperfine coupling constants) and second (e.g., electronic g-tensor shifts) order can be obtained independently of the sampling method, as far as for the latter the accuracy in reproducing reliable molecular structures and statistics is proven.

In this chapter, we focus on the most important methodological features of the vibrational and electronic treatment by a time-dependent approach. Then, we give a brief sketch of the nonperiodic GLOB model. A list of illustrative applications is discussed in Section 11.3. Regarding the IR and vibrational analysis, we choose two important benchmark systems for the polypeptides and ions in solution, namely, *N*-methyl-acetamide and Zn(II) in aqueous solution. Further, optical absorption spectra are illustrated for a solvatochromic shift prototype of the carbonyl $n \rightarrow \pi$ transition (acrolein) and for an extended system, such as liquid water. Finally, we consider the characterization of the phosphorescence emission spectroscopy involving the acetone molecule in the electronic triplet state. Concluding remarks and perspectives are sketched in Section 11.4.

11.2 SPECTROSCOPIC ANALYSIS FROM MOLECULAR DYNAMICS

Molecular dynamics allows one to found a relationship between molecular structure and properties involved in the spectroscopic event. At variance with time-independent or Monte Carlo approaches, this provides the invaluable capability to treat transient species, relaxation processes, and, more in general, time-resolved phenomena. Also, with respect to the more sophisticated quantum [35, 36] and semiclassical [37, 38] dynamics, time-dependent approaches based on classical dynamics are often the only accessible choice to bring into spectroscopic analysis the complexity of condensed-phase systems, such as solute–solvent, liquids, or macromolecules. In this context, the advent of Car–Parrinello ab initio dynamics [39, 40] paved the way for an effective and accurate simulation of realistic systems, opening a completely new scenario for spectroscopic applications [41–44]. More recently, smart ways to perform Born–Oppenheimer or single-surface dynamics have been presented, and the sampling of excited states is now also available [45, 46]. Furthermore, novel methods based on the extended-Lagrangian formalism have been presented, such as the atom centered density matrix propagation (ADMP) [47–49] method, which propagates, along with the nuclei, the one-electron density matrix described by an atomic basis set. Often, the above quantum mechanical methodologies are combined with low-level molecular mechanics calculations according to a hybrid QM/MM scheme in order to make feasible the modeling of extended molecular systems in a complex environment. In the following, we will give a general picture of the vibrational and electronic treatment from molecular dynamics. Then, a brief sketch of our GLOB QM/MM molecular dynamics model will provide the theoretical basis of the applications illustrated in the next section.

The theoretical background to both configurational averaging and time correlation approaches adopted to reproduce spectra from molecular dynamics is represented by the linear response theory as applied to the radiation absorption by a molecular system. By using the Heisenberg formalism to express the so-called golden rule of time-dependent quantum mechanical perturbation theory, the general form of the absorption lineshape $I(\omega)$ for an N-body system interacting with an electric field of frequency ω is [50]

$$I(\omega) = \frac{3}{2\pi} \int_{-\infty}^{\infty} dt\, e^{-iwt} \sum_i \rho_i \langle i|\hat{\epsilon}\mu(0)\hat{\epsilon}\mu(t)|i\rangle \tag{11.1}$$

where $\hat{\epsilon}$ is the unit vector along the monochromatic electric field, ρ_i is the probability that the system is in the initial state i when the interaction occurs, and μ is the total electric dipole moment operator:

$$\mu(t) = e^{iH_f t/\hbar} \mu e^{-iH_i t/\hbar} \tag{11.2}$$

with H_i and H_f being the Hamiltonians of the system in the initial and final states in the absence of the electric field and $\hbar$ the reduced Planck constant. Therefore, the spectroscopic band contour can be obtained by simulating the evolution over the time of the summation on i in Eq. 11.1. When the dependence on the nuclei coordinates is treated classically, a semiclassical version of the lineshape in Eq. 11.1 can be introduced, and suitable expressions for Eq. 11.2 can be obtained by an a posteriori analysis of pure classical, pure ab initio, or mixed ab initio–classical trajectories. Several approaches of this kind have been proposed to treat electronic spectra [51–53]. In the simplest treatment, expression 11.1 can be further substituted by a sampling average recurring to a spectral density of the initial electronic states, leading to the configurational averaging approach. When absorption among vibrational (nuclear) states is taken into account (IR spectroscopy), a full classical treatment of the operator integrals in Eq. 11.1 can be considered in a first approximation. In this case the sampling by molecular dynamics entirely provides the ensemble averaging to be Fourier transformed in Eq. 11.1. In the following we discuss in detail these different approaches.

11.2.1 Vibrational Analysis

The solution of the vibrational problem for polyatomic molecules is the key step in the theoretical treatments of IR, Raman, and related spectroscopic techniques [e.g., two-dimensional (2D) IR] and of vibrationally resolved techniques (absorption, emission spectra). Moreover, analysis of spectroscopic properties in terms of normal-mode contributions can be of paramount importance in several other techniques, such as XAFS, NMR, and ESR, elucidating the direct dependence of a spectroscopic parameter on the molecular structure/dynamics. The solution of the vibrational levels beyond the harmonic approximation can be obtained by variational [1–3], self-consistent [4–8], and perturbative [9–17] approaches.

An alternative route is based on time-dependent approaches, where the standard statistical mechanics formalism relies on Fourier transform of the time correlation of vibrational operators [54–57]. These approaches can provide a complete description of the experimental spectrum, that is, the characterization of the real molecular motion consisting of many degrees of freedom activated at finite temperature, often strongly coupled and anharmonic in nature. However, computation of the exact quantum dynamics evolution of the nuclei on the ab initio potential surface is as prohibitive as the quantum/stationary-state approaches. In fact, even a semiclassical description of the time evolution of quantum systems is usually computationally expensive. Therefore, time correlation methods for realistic systems are usually carried out by sampling of the nuclear motion in the classical phase space. In this context, summation over i in Eq. 11.1 is a classical ensemble average; furthermore, the field unit vector $\hat{\varepsilon}$ can be averaged over all directions of an isotropic fluid, leading to the well-known expression

$$I(\omega) = \frac{1}{2\pi} \int_{-\infty}^{\infty} dt\, e^{-iwt} \langle \mu(0)\mu(t) \rangle \tag{11.3}$$

where $I(\omega)$ can be obtained by the Fourier transform of the autocorrelation function of the dipole moment $\mu(t)$. The ensemble average $\langle \cdot \rangle$ can be calculated by molecular dynamics.

Expression 11.3 provides the whole simulated spectrum, while a detailed vibrational analysis requires the unambiguous assignment of each mode contribution. Recently, a number of methods appeared in the literature aimed at the extraction of normal-mode-like analysis from ab initio dynamics [58–63]. Some of these [58–60] refer to the quasi-harmonic model introduced by Karplus [64, 65] in the framework of classical molecular dynamics and individuate normal-mode directions as main components of the nuclear fluctuations in the NVE or NVT ensemble. The quasi-normal model relies on the equipartition of the kinetic energy among normal modes; thus problems arise when the simulation time required to obtain such a distribution is computationally too expensive, as is often the case for ab initio dynamics. Other approaches [61–63] carry out the time evolution analysis in the momenta subspace instead of the configurational space. In these approaches the basic consideration is that, at any temperature, generalized normal modes Q_i correspond to uncorrelated momenta such that [61]

$$< \dot{Q}_i(t)\dot{Q}_j(t) >= \lambda_i \delta_{ij} \tag{11.4}$$

where i and j run over the 3N generalized modes, δ is the Kronecker delta and λ_i is the average kinetics energy associated to each ith mode.

Directions of generalized modes Q_i compose the unitary transformation matrix **L**, which diagonalizes the covariance matrix **K** of the mass weighted atomic velocities $\dot{\mathbf{q}}$, with elements

$$K_{ij} = \frac{1}{2} < \dot{q}_i \dot{q}_j > \tag{11.5}$$

The frequencies associated to each mode *i* can be obtained by Fourier transform of the autocorrelation function of normal-mode velocities $\dot{\mathbf{Q}}$.

The definition in Eq. 11.4 is more general than the corresponding one provided by the quasi-normal model, because an effective quadratic shape of the potential is not assumed and equipartition of thermal energy among modes is not required.

The generalized mode approach has been adapted to ab initio molecular dynamics combined with a polarizable continuum model [66] to include the effect of a bulk solvent [63]. In this context a simple procedure for the vibrational analysis of a generic molecular property was also proposed. Moreover, comparison to Hessian-based perturbative approaches [67] to treat anharmonic frequencies validated the method with very promising results. A similar static versus *dynamic* comparison has been performed for isolated systems [68].

The method of Martinez et al. [62] introduces, beside the condition imposed by Eq. 11.4, a frequency localization procedure which involves separation of an effective atomic forces matrix and has been adopted in ab initio dynamics applications to solute–solvent systems [69, 70].

11.2.2 Electronic Analysis

Electronic spectra (UV–vis, photoelectron, X-ray etc.) can be computed from classical molecular dynamics in both equilibrium and nonequilibrium regimes. Here, the main assumptions correspond to the Born–Oppenheimer approximation and the Franck–Condon principle leading to the separation of the electronic and nuclear degrees of freedom. While nuclear motions are treated classically, the electronic transitions are computed quantum mechanically by determining the transition energies and dipole moments. While more sophisticated semiclassical and fully quantum mechanical treatments of the electronic transitions have been proposed, which are able to account for nuclear quantum effects in vibronic couplings and conical intersections, in the present discussion we consider only the case where nuclei can be safely modeled at the classical level. While this appears as a severe limitation, it is often indeed the only practical way to simulate electronic spectra of large and complex systems in the condensed phase, which represent most of the experimentally recorded spectra. It is worth noting, however, that the present approach is also based on a sound theoretical framework as described elsewhere [53].

Because we will focus especially on molecular liquids, it is of primary importance to define an accurate model for the treatment of solvent effects. In this respect, we have adopted a discrete/continuum model, which is well suited for solute–solvent systems and is nicely consistent with the time-dependent approach described in the following section. Moreover, such a model allows us to investigate the subtle solvent effects induced on a solute, which is in general a challenging task. In particular, it is often difficult to elucidate such effects occurring in polar H-bonding solvents, like water, where solute–solvent interactions are the result of a delicate balance between specific interactions, such as hydrogen bonds, and long-range effects due to solvent polarization [42]. While the electrostatic response of the solvent is generally well described by continuum models, this is not the case for solute geometry changes, which in turn may affect appreciably spectroscopic parameters.

In this context, the QM/MM scheme represents perhaps the only viable route for the accurate and extended studies of the electronic properties of large, flexible systems, provided electronic excitations are relatively well localized and the corresponding system portion can be treated at the full QM level. Recent developments make the use of time-dependent density functional theory (TD-DFT) methods and corresponding TD-DFT/MM schemes among the most reliable and effective methods in order to compute both electronic absorption and emission spectra, that is, allowing also the study of molecular excited states at equilibrium. In particular, the combination of molecular dynamics sampling and TD-DFT/MM calculations is often able to account for basically all the relevant features of the electronic spectra, meaning peak positions, band broadening, and relative intensities. Especially, the spectral broadening as due to thermal and solvent effects is usually very well captured by the present approach. At most, the effect of the transition finite lifetime can be further added by an ad hoc Gaussian or Lorentzian parameter. In the simplest case, such as the study of a single chromophore in solution, the basic required information to simulate an electronic spectrum is the transition energies and oscillator strengths, which are eventually averaged over a relatively large number (>100) of representative molecular configurations. However, if the system under consideration is a more electronically complex system, for example, there are quite a large number of chromofores and, consequently, excited states, the observed electronic spectra can be better reproduced by a sum-over-states technique, which is easily adapted to obtain, for example, the frequency-dependent dielectric polarization [71].

11.2.3 Time-Dependent Approach for Study of Complex Systems in Solution: GLOB Model

As far as large, complex, and flexible molecular systems are considered, an effective computational treatment is represented by the use of a hybrid QM/MM methodology that allows us to combine two or more computational methods for different portions of the system in such a way that only the chemical and physical interesting region is modeled at the highest level of accuracy. As an example, the well-known ONIOM [72–74] scheme allows the combination of a variety of quantum mechanical, semiempirical, and molecular mechanics methods, providing an accurate and well-defined Hamiltonian.

In this framework, a generalization of a hybrid explicit/implicit solvation model for the treatment of polarizable molecular systems at different levels of theory has been recently proposed by our group, the so-called GLOB model [33, 34]. Such a

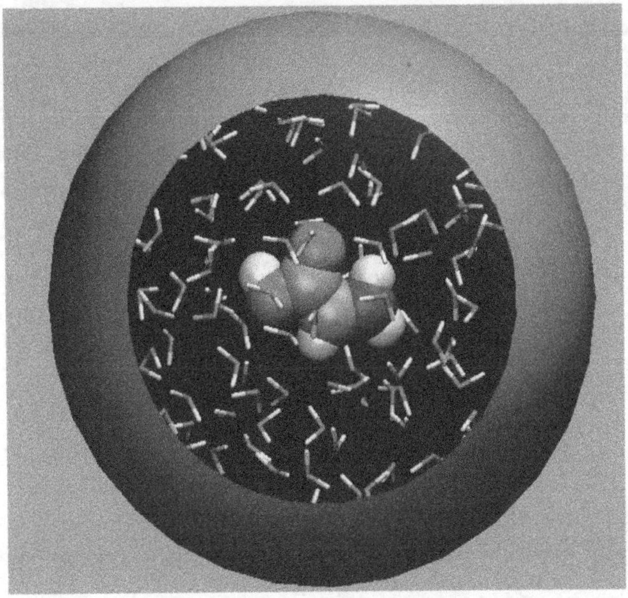

Figure 11.1 Graphical representation of solute–solvent system simulated using GLOB models. The explicit system is embedded into a spherical cavity of a dielectric continuum.

model is particularly well suited to perform ab initio or QM/MM molecular dynamics simulations of solute–solvent systems under non periodic boundary conditions and using localized basis sets within an extended-Lagrangian formalism. Thanks to an effective procedure, a complex solute with a few explicit solvation shells can be reliably modeled, ensuring solvent bulk behavior at the boundary with the continuum medium. In the following, we sketch the general features of the GLOB model, pictorially represented in Figure 11.1 (see refs. 33, 34, 75, and 76 for more details).

First, let us consider a simple subdivision of a molecular system in two portions or layers according to the ONIOM partitioning scheme, where the region of interest is treated at the QM level and the remaining system at the MM level. It is worth noting that the layers do not have to be inclusive. In this case, each energy evaluation requires three different calculations according to the expression

$$E^{\mathrm{QM/MM}} = E^{\mathrm{QM}}_{\mathrm{model}} + E^{\mathrm{MM}}_{\mathrm{real}} - E^{\mathrm{MM}}_{\mathrm{model}} \tag{11.6}$$

where the *real* system is the entire molecular system under consideration and the *model* is the core region to be modeled at the highest level of theory, which does include the influence of the remaining system treated at the MM level as a distribution of embedding charges. Such a decomposition provides a well-defined, single-valued, and differentiable potential well suited to perform QM/MM calculations and simulations. Within the framework of formally monoelectronic QM methods (e.g., Hartree–Fock or Kohn–Sham models), $E^{\mathrm{QM/MM}} = E^{\mathrm{QM/MM}}(\mathbf{P}_0, \mathbf{x})$ represents the QM/MM gas-phase energy of the whole explicit system expressed as a function of the nuclear coordinates, $\mathbf{x}$, and the unpolarized (no implicit solvent effects) one-electron density matrix, $\mathbf{P}_0$.

According to the GLOB model, the explicit system (solute plus solvent) is embedded into a suitable cavity of a dielectric continuum possibly with a regular and smooth shape, such as a sphere, an ellipsoid, or a spherocylinder. In combination with molecular dynamics techniques, such a cavity could be kept fixed, corresponding to *NVT ensemble* conditions, or allowed to change volume, according to *NpT ensemble* simulations. In this case, the solvation free energy, $\Delta A_{\mathrm{sol}}(\mathbf{x})$, of the system at a given molecular configuration can be written as the sum of the internal energy plus the so-called mean-field (or potential of mean force) contribution that accounts for the interactions with the environment (solvent) minus the gas-phase energy:

$$\Delta A_{\mathrm{sol}}(\mathbf{x}) = [E^{\mathrm{QM/MM}}(\mathbf{P},\ \mathbf{x})] - E^{\mathrm{QM/MM}}(\mathbf{P}_0,\ \mathbf{x}) \tag{11.7}$$

where $\Delta A_{\mathrm{sol}}(\mathbf{x})$ is the free energy of the system and $W(\mathbf{P}, \mathbf{x})$ is the mean field term. The density matrix, $\mathbf{P}$, is determined by a self-consistent calculation in the first two terms on the right-hand side (RHS), that is, the mean-field response is always considered at equilibrium. Such a mean-field contribution is introduced as a modification of the ONIOM [72–74] scheme for the isolated systems as described elsewhere [33, 34]. It should be pointed out that a large number of discrete/continuum models have been proposed in the literature that differ in the way W is approximated. Here, we assume that the mean-field potential is composed of conceptually simple terms, according to Ben-Naim's definition of the solvation process [77], namely a long-range electrostatic contribution, due to the linear response of the polarizable dielectric continuum, and a short-range dispersion–repulsion contribution, which accounts effectively for the interactions in proximity of the cavity boundary, that is, $W = W_{\mathrm{elec}} + W_{\mathrm{disp\text{-}rep}}$.

The W_{elec} term is modeled by means of the conductor-like version [78–80] of the polarizable continuum model (PCM) [81], which is one of the most refined boundary element methods successfully used in many applications ranging from structure and thermodynamics to spectroscopy in both isotropic and anisotropic environments [81–83]. According to the C-PCM method, the continuum medium that mimics the response of liquid bulk is completely specified by a few parameters, for example, the dielectric permittivity (ϵ_r), generally depending on the nature of the solvent and on the physical conditions, for example, density and temperature. In C-PCM, the reaction field potential, Φ_{RF}, due to the dielectric-induced polarization is exerted by a number of "apparent" surface charges (q_{asc}) centered on small tiles or *tesserae*, which are the result of a fine subdivision of the cavity boundary surface into triangular area elements (tesserae) of about equal size, and computed by a self-consistent calculation with respect to the solute electronic density [84]. The determination of q_{asc}'s requires the solution of a system of N_{tes} linear equations, with N_{tes} the number of tesserae:

$$\mathbf{D} \cdot \mathbf{q}_{\mathrm{asc}} = -\mathbf{\Phi}_I \tag{11.8}$$

where $\mathbf{q}_{\mathrm{asc}}$ is the array of the apparent surface charges, $\mathbf{\Phi}_I$ is the electrostatic potential evaluated at the center of each tessera due only to the charge distribution of the system, and $\mathbf{D}$ is a matrix that depends only on the surface topology and the dielectric constant [79, 80],

$$D_{ii} = \frac{\epsilon}{\epsilon - 1} 1.0694 \sqrt{\frac{4\pi}{a_i}} \tag{11.9}$$

$$D_{ij} = \frac{\epsilon}{\epsilon - 1} \frac{1}{|\mathbf{s_i} - \mathbf{s_j}|} \tag{11.10}$$

where $\mathbf{s_i}$ and a_i are respectively the position vector and the area of the ith tessera and ϵ is the continuum dielectric constant. Hence, for a given molecular configuration of the explicit system, **x**, the q_{asc}'s are determined from Eq. 11.8 and the electrostatic potential, $\Phi_{\text{RF}}(\mathbf{r})$, and the corresponding free energy, W_{elec}, are given by

$$W_{\text{elec}} = -\tfrac{1}{2}\Phi + \mathbf{D}^{-1}\boldsymbol{\Phi} \tag{11.11}$$

Note that if we neglect any cavity deformations, for convenience, the energy derivatives with respect to a generic coordinate assume a quite simple form with respect to the general case [79]. Moreover, in this case Eq. 11.8 can be solved by matrix inversion by computing and storing $\mathbf{D}^{-1}$ only once at the beginning of the simulation and using an inexpensive matrix–vector product in the following steps. In our implementation of the model, we use an improved GEPOL procedure [85, 86] in order to partition the cavity surface enclosing the explicit molecular system.

On the other hand, the dispersion-repulsion contribution, $W_{\text{disp-rep}}$, which accounts for the short-range solvent (explicit)–solvent (implicit) interactions, has been introduced to remove any possible source of physical anisotropy in proximity of the cavity surface, that is, deviation from bulk behavior. In the same spirit as other methodologies [87–94] developed in the framework of continuum models, we have also treated $W_{\text{disp-rep}}$ as a purely classical potential, so not perturbing the electronic density of the system. In particular, $W_{\text{disp-rep}}$ is obtained from an effective empirical procedure parametrized on structural and thermodynamic properties originally presented [75] and further developed [34] by Brancato et al. (see also refs. [76] and [33] for applications in the context of MM and QM/MM molecular dynamics simulations, respectively). The basic assumptions that have been made in the derivation of such a potential are the following: (1) $W_{\text{disp-rep}}$ can be represented by an effective potential acting on each explicit solvent molecule irrespective of the others; (2) $W_{\text{disp-rep}}$ depends only on the molecule distance and possibly orientation with respect to the cavity surface; and (3) $W_{\text{disp-rep}}$ can be expanded in a series of terms corresponding to increasing levels of approximation as $W_{\text{disp-rep}} = W^0_{\text{disp-rep}} + W^1_{\text{disp-rep}} + \cdots$. As an example, the first term, $W^0_{\text{isp-rep}}$, which depends only on the distance of the center of mass of the solvent molecule from the cavity surface, does ensure an isotropic density distribution of the liquid at the interface with the continuum, so avoiding artifacts in the simulations due the presence of a physical boundary as observed in other continuum-based methodologies [95–97]. Analogously, higher order terms are introduced, if needed, to prevent other possible physical deviations from liquid bulk behavior, as the solvent polarization effect that may appear by using discrete/continuum models. Hence, $\text{W}_{\text{dis-rep}}$ can be expressed in a simple general form as

$$W_{\text{dis-rep}} = \sum_{i}^{N} \lambda(\mathbf{r}_i) \quad (11.12)$$

where $\lambda(\mathbf{r}_i)$ is the potential acting on the ith molecule and the sum is extended over the total number of explicit solvent molecules. In practice, the dispersion–repulsion free-energy term is obtained "on the fly" from a test simulation of a neat liquid by discretization of the distance from the cavity boundary with a set of equally spaced Gaussian functions whose height is adjusted after a certain time interval on the basis of the local density [75]. It is worth noting that the so-obtained $W_{\text{disp-rep}}$ term is parametrized for a given solvent at specific physical conditions (e.g., density and temperature), but we can reasonably assume that it is constant for any solution of the same solvent irrespective of the cavity size and shape, providing that the boundary surface is smooth and the number of explicit solvent molecules is sufficiently large [e.g., 34, 76].

The QM/MM scheme briefly sketched above can be directly applied to the spectroscopic studies performed within the time-independent framework. Such an approach is suitable for large and semi rigid molecules when nonadiabatic couplings are negligible, harmonic approximation reliable, and spectroscopic properties can be evaluated considering only a small conformational region close to equilibrium.

11.3 ILLUSTRATIVE APPLICATIONS

As anticipated in the introduction, the methodological machinery presented in the above sections can be successfully applied to different computational spectroscopic studies ranging from ESR, IR/Raman, low-resolution UV–vis up to rovibronic spectra, and a large variety of physicochemical systems from small molecules in solution to macromolecules and extended systems. The following examples, which are chosen to illustrate the flexibility of the present approaches, focus on IR and vibrational analysis, optical absorption, and phosphorescence spectra.

11.3.1 IR Spectra: *trans*-*N*-Methylacetamide in Aqueous Solution

The characteristic amide modes are well-known fingerprints of polypeptides and proteins in IR and related spectroscopies. They are usually referred to as AI, AII, and AIII and roughly correspond to the CO stretching and two different combinations of NC stretching and CNH bending, respectively. The AI mode in particular is of great interest due to the peculiar resonant off-diagonal coupling which can provide plenty of information on the structures and dynamics of polypeptides (i.e., distances and orientations of the groups involved in the vibrational coupling and their time evolution).

Here we consider a well-studied prototype of the peptide moiety, namely the *N*-methylacetamide in the trans form (NMA), and illustrate the spectroscopic analysis of a NVE QM/MM simulation of the molecule in aqueous solution. More specifically, a molecule of *trans*-NMA (solute) has been solvated by 134 water molecules and simulated for about 25 ps (including 4 ps of equilibration) by the GLOB/ADMP molecular dynamics. The hybrid density functional B3LYP along with the N07D basis set and the TIP3P model have been employed for the solute and the water molecules, respectively. QM/MM parameters (van der Waals radii of NMA) were previously calibrated to reproduce structural arrangement and energy binding of NMA–water clusters optimized at the full quantum mechanical level. Further, a comparison with results obtained by more usual Hessian-based approaches is very helpful to the present discussion: therefore, we considered as reference the frequencies for the NMA minimum structure (B3LYP/N07D level) calculated by a second-order perturbative anharmonic treatment (PT2). Remarkably, the levels of theory adopted in the time-dependent and time-independent pictures are the same.

In Figures 11.2, 11.3, and 11.4 we draw the general mode directions corresponding to the AI, AII, and AIII modes as obtained on average by analysis of the GLOB/ADMP trajectory according to the time-dependent approach (general mode

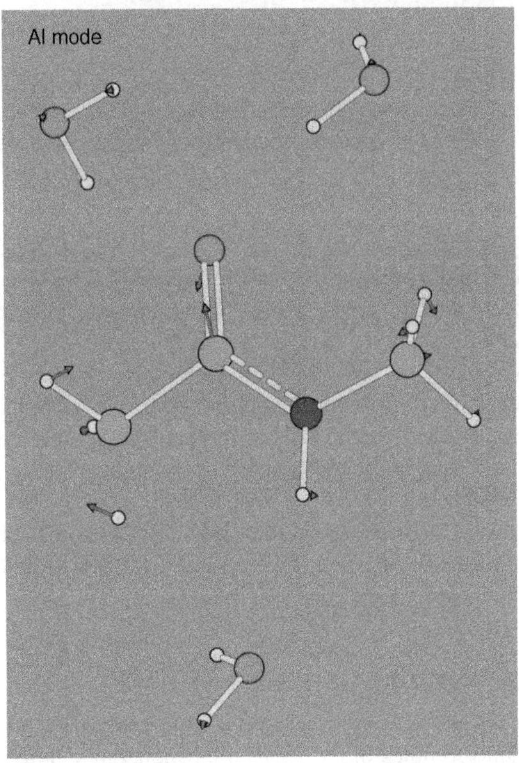

Figure 11.2 AI general mode of *trans-N*-methylacetamide in aqueous solution obtained by GLOB/ADMP dynamics.

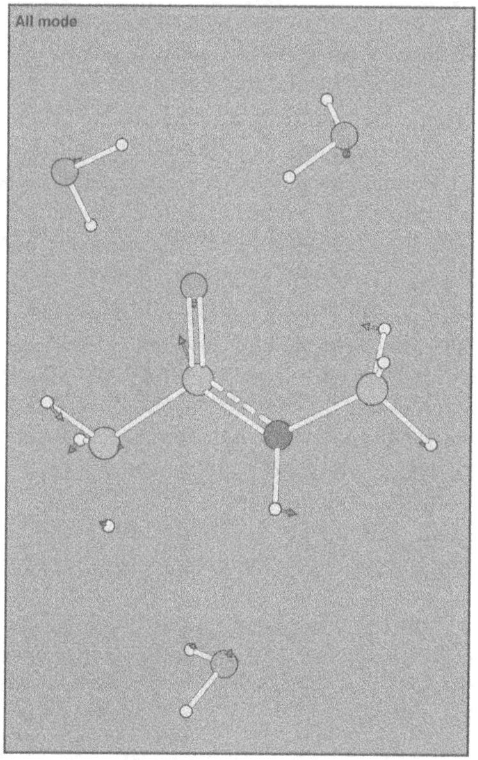

Figure 11.3 AII general mode of *trans-N*-methylacetamide in aqueous solution obtained by GLOB/ADMP dynamics.

definition) described in the previous section. In Figure 11.5 we also report the power spectra of the three mode autocorrelation functions. We note that the three modes are well separated and correspond to frequency values of 1625, 1613, and 1272 cm^{-1}, respectively. A summary of the NMA characteristic frequencies is reported in Table 11.1. Solvent shifts are evaluated with respect to anharmonic frequencies (PT2, first column) calculated for the isolated molecule in the minimum structure. The nice agreement with the experimental values points out the reliability of the proposed computational approach.

Finally, in Figure 11.6 we report the NMA IR spectrum calculated by the GLOB/ADMP trajectory as the power spectrum of the dipole–dipole autocorrelation function.

11.3.2 Vibrational Analysis: Zn(II) Hexa Aqua Ion

As a challenging example of vibrational analysis via a time-dependent approach, we choose the hexa–aquo complex of the Zn(II) ion in aqueous solution, which has been extensively studied both theoretically [99–102] and experimentally [101]. Several methodologies aimed to analyze EXAFS spectra include molecular dynamics results,

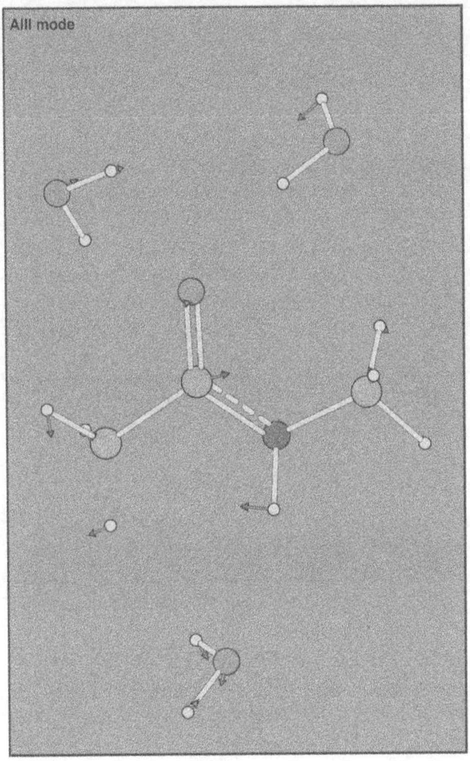

Figure 11.4 AIII general mode of *trans-N*-methylacetamide in aqueous solution obtained by GLOB/ADMP dynamics.

taking advantage of the structural and kinetic detail of the ion solvation provided by the simulation. In particular, a challenging benchmark is the connection of EXAFS parameters representing the thermal disorder and the vibrational density of states (VDOS) of ion complexes. In this context the actual benefit from theory strictly depends on the accuracy of the simulation, which must be capable of properly accounting for the solute–solvent structure on average and, at a comparable accuracy, for the molecular dynamics modulating the scattering events. This is particularly crucial for the vibrational analysis of ions in solution, where the Gaussian assumption of the atomic motion or the harmonic treatment for the normal modes must be abandoned.

The Zn(II) hexa–aquo complex is kinetically stable over the picosecond time scale, which is sufficient to obtain an exhaustive sampling of the complex motion in aqueous solution. This feature allows one to safely adopt a QM/MM description by which the zinc hexa–aquo ion is treated at the ab initio level of theory. The vibrational analysis illustrated here refers to a NVE GLOB QM/MM simulation of the zinc ion in aqueous solution performed for about 20 ps, including 4 ps of equilibration, using the ADMP methodology [47–49]. Regarding the QM potential for the zinc complex we adopted

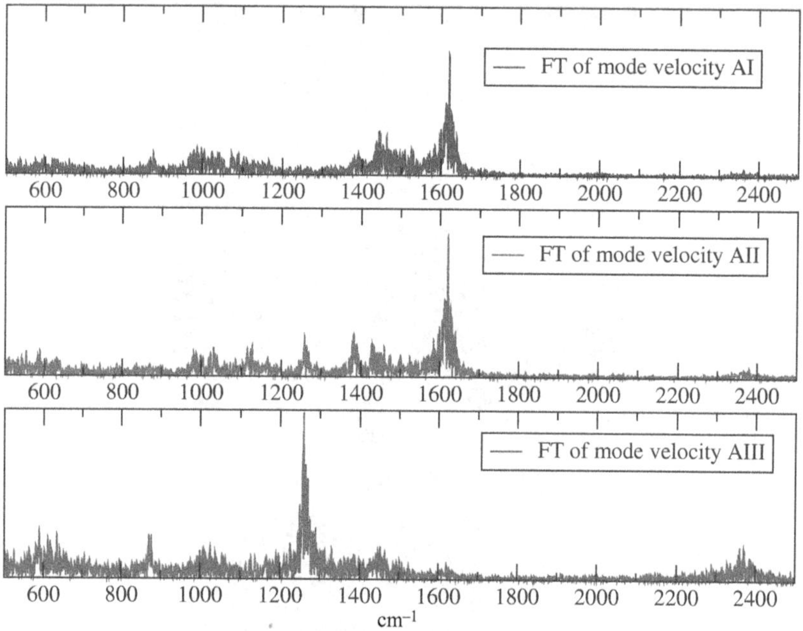

Figure 11.5 Power spectra of velocity autocorrelation functions for AI, AII, and AIII modes of *trans-N*-methylacetamide in aqueous solution obtained by GLOB/ADMP dynamics.

the hybrid density functional B3LYP and the new triple-ζ basis set (N07T), well suited for a good performance of ab initio dynamics, along with the Stuttgard/Dresden effective core potentials for the metal ion.

In Figures 11.7 and 11.8 we draw the radial distribution functions (RDFs) of zinc–oxygen and zinc–hydrogen distances obtained by analyzing the GLOB/ADMP simulation. These are characterized by a well defined first peak with variance $\sigma^{(2)}$ of 0.009 and 0.0146 Å ^{2} in the zinc–oxygen and zinc–hydrogen cases, respectively. Computed values are generally in very nice agreement with their experimental EXAFS counterparts [101], namely 0.0087 and 0.016 Å^2. The peak position R's are also simulated comparably to the experiment, the zinc–hydrogen maximum being at 2.75 Å (EXAFS value 2.73 Å), while the zinc–oxygen is located at 2.12 Å

Table 11.1 Vibrational Frequencies (cm^{-1}) of *trans-N*-Methylacetamide

	PT2 Gas Phase	GLOB/ADMP Solution	Experimental Solution[a]
AI	1723	1625 ($\Delta = -98$)	1625 ($\Delta = -98$)
AII	1533	1613 ($\Delta = +80$)	1582 ($\Delta = +82$)
AIII	1244	1272 ($\Delta = +28$)	1316 ($\Delta = +51$)

Note: Solvent shifts are given in parentheses.
[a] From ref. 98.

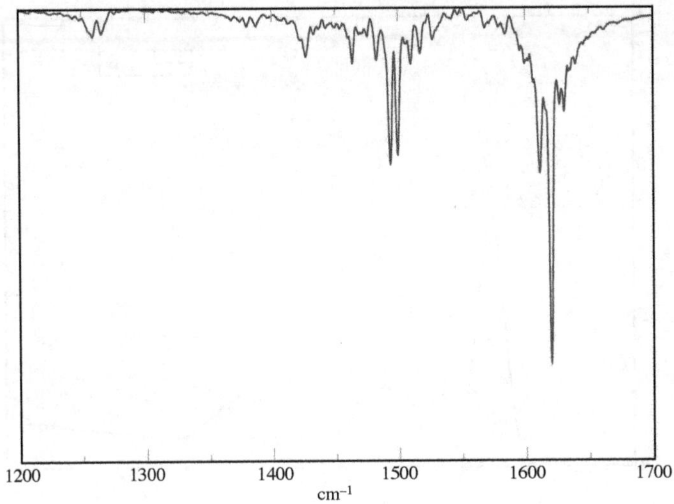

Figure 11.6 IR spectrum of *trans-N*-methylacetamide in aqueous solution obtained by GLOB/ADMP dynamics.

(EXAFS value 2.08 Å). Overall, these results suggest that the thermal motion of water molecules around the ion is well reproduced by the GLOB/ADMP simulation.

The hexa–aqua complex in aqueous solution is characterized by a tilted arrangement of the six water molecules. In fact, both the minimum structure calculated by a solvent continuum model and the structure obtained on average

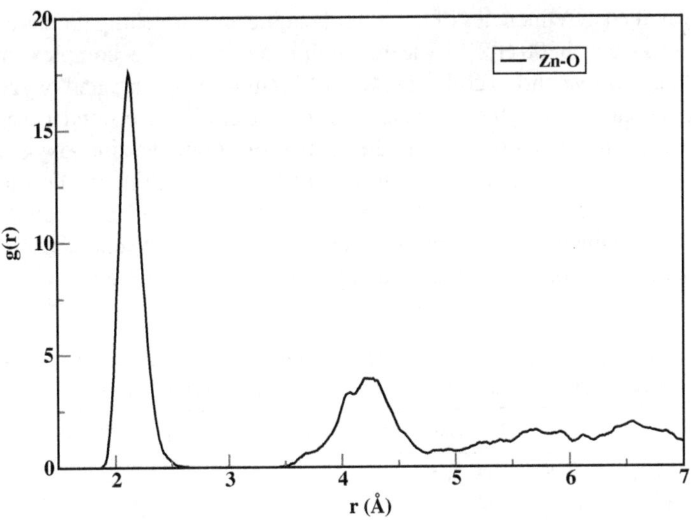

Figure 11.7 Radial distribution functions of Zn–oxygen distances obtained by ADMP/GLOB ab initio dynamics for Zn(II) in aqueous solution.

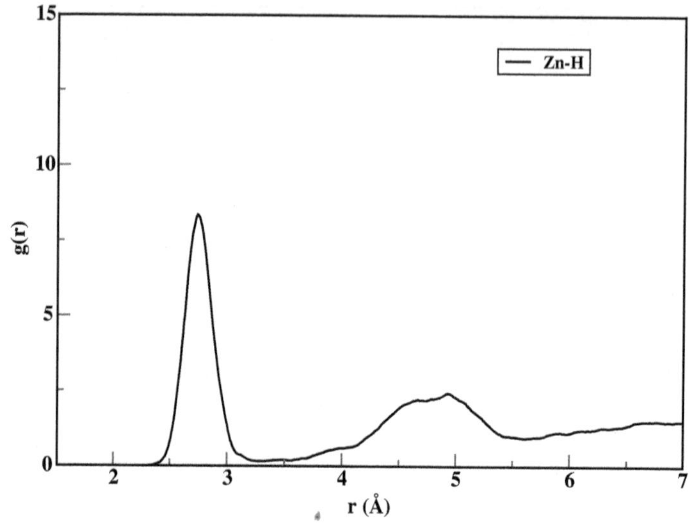

Figure 11.8 Radial distribution functions of Zn–hydrogen distances obtained by ADMP/GLOB ab initio dynamics for Zn(II) in aqueous solution.

from the GLOB/ADMP trajectory involve an angle θ between the oxygen–zinc axis and the water dipole moment of 138°. Therefore, the complex normal modes are characterized by a low degree of symmetry ($C1$ group).

In Table 11.2 we report a summary of the frequencies obtained for the hexa–aqua complex by analysis of the GLOB/ADMP trajectory along with harmonic and anharmonic (PT2) values calculated at the B3LYP/N07T/CPCM level. Experimental value of the A_g Raman band (about 390 cm^{-1}) is also reported [103]. According to the harmonic analysis, frequencies of the water bending and stretching modes are spread between 1600 and 3800 cm^{-1}. The remaining modes of the complex are below 610 cm^{-1} and correspond to collective water librations, wagging, and oxygen–metal stretching. In particular, values between 222 and 306 cm^{-1} correspond to asymmetric oxygen–metal stretching modes, with the symmetric collective zinc–oxygen stretching at 334 cm^{-1}. These modes also partially represent the tilting of the water plane with respect the oxygen–metal axis. As a consequence, collective modes below $<$610 cm^{-1} obtained by anharmonic analyses (PT2) are characterized by a strong coupling beyond the harmonic approximation. Asymmetric metal–oxygen

Table 11.2 Low Frequencies (cm^{-1}) of Hexa–Aqua Zinc Complex in Aqueous Solution Obtained by Harmonic, Anharmonic, and Molecular Dynamics Analysis

Mode	Harmonic	Anharmonic (PT2)	ADMP/GLOB	Experiment
Zn–O asymmetric stretch	222–306	/	345,370	
Zn–O symmetric stretch	334	394	400	390[2]

Note: In all cases the hexa–aqua ion is represented at the B3LYP/N07T level of theory. Experimental values are also shown for comparison.

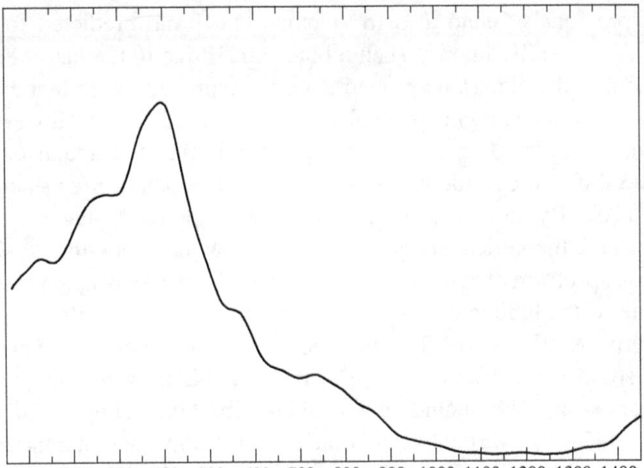

Figure 11.9 Symmetric Zn–oxygen stretching power spectra obtained by analysis of ADMP/GLOB ab initio dynamics for Zn(II) in aqueous solution.

stretching modes in particular are characterized by a high value (>300 cm^{-1}) of the extradiagonal cubic force constants involving the low frequencies of water librations and wagging. The zinc–oxygen symmetric stretching is less affected by the anharmonic coupling, and the PT2 result (394 cm^{-1}) is in excellent agreement with experiment. However, the anharmonic analysis by a perturbative approach is not the method of choice in the present case.

On the other hand, the performance of the vibrational analysis from dynamics is not affected by the harmonic nature of the modes, and asymmetric zinc–oxygen stretchings are easily assigned. In this case the five asymmetric modes are localized at about 370 and 345 cm^{-1}. All of them are mixed to the librations modes spread in the range of frequencies from 70 up to about 250 cm^{-1}. The symmetric band is peaked at about 400 cm^{-1} (Figure 11.9), in a very good agreement with the experimental value of 390 cm^{-1}.

11.3.3 Optical Absorption Spectra: Acrolein in Gas Phase and Aqueous Solution

Acrolein is a small organic molecule with some very interesting features. It can be considered as a prototype of molecular systems containing two conjugated chromophores, namely C=C and C=O, a common feature for many natural molecules. Also, the presence of two characteristic functional groups at the same time has important consequences in the chemistry and photochemistry of acrolein. For these reasons, its UV absorption spectrum has been extensively studied in different solvents [104–110] as well as in the gas phase [111–113], and a solvatochromic blue shift of the $n \rightarrow \pi^*$ transition of the C=O group has been observed in going from the gas phase to aqueous solution, as previously found also for acetone [114] and predicted for formaldehyde [115]. As generally known, such a blue shift is due to the larger extent of the dipole moment in the electronic ground-state as compared to the first excited state, which leads to a larger energy gap in polar solvents, such as water. However, a deeper investigation, as reported in the following, reveals that the actual extent of the observed blue shift is the result of different subtle and opposite effects, not only polar ones [e.g., 116]. By applying a time-dependent approach, namely the GLOB model [42], to the theoretical study of acrolein, it has been possible to simulate the UV absorption spectrum of acrolein both in the gas phase and in aqueous solution and uncover some of the hidden effects behind the observed blue shift.

To this purpose, NVT QM/MM simulations of acrolein both in the gas phase and in aqueous solution (acrolein solvated with 134 TIP3P water molecules) were performed for about 24 ps, including 4 ps of equilibration, using the GLOB/ADMP methodology [47–49]. Vertical excitation energies and oscillator strengths have been computed within the TD-DFT formalism employing the B3LYP functional and the 6–311++G(2d,2p) basis set on selected molecular configurations. Note that the consistency of such a basis set for spectroscopic calculations was validated in previous work [117].

Here, we will discuss the nature of the solvent effects on the UV $n \rightarrow \pi^*$ transition energy of acrolein in terms of the direct and indirect contributions, where the former are due to solvent polarization and H–bonding formation and the latter to solute structural changes. First, it is interesting to note that acrolein in aqueous solution represents a good example for time-dependent approaches due to the fluctuational nature of its microsolvation shell at room temperature, which is hard to represent satisfactorily by a simple cluster model. In fact, the carbonyl moiety of acrolein is surrounded, on average, by a noninteger number of water molecules, forming hydrogen bonds with the C=O group. Also, the distribution of water molecules is nonsymmetrical considering the two sides of the molecular plane divided by the carbonyl axis. Such a peculiar solvent distribution is quite different, for example, with respect to the one observed for acetone, and it can be easily revealed only by applying a finite-temperature molecular simulation technique.

The $n \rightarrow \pi^*$ vertical transition energies of acrolein issuing from the gas-phase and condensed-phase MD simulations at room temperature are reported in Table 11.3, and

Table 11.3 UV $n \rightarrow \pi^*$ Transition Energies of Acrolein in Gas Phase and Aqueous Solution Computed at TD-B3LYP/6-311++G(2d,2p) Level of Theory

	Energy	Shift
Gas phase	3.58	
Solution		
Acrolein	3.49	−0.08
Acrolein + 2 H_2O^{QM}	3.68	+0.10
Acrolein + 2 H_2O^{QM} + 132 H_2O^{MM} + PCM	3.84	+0.26

Note: Values are in eV, standard error is 0.01 eV.

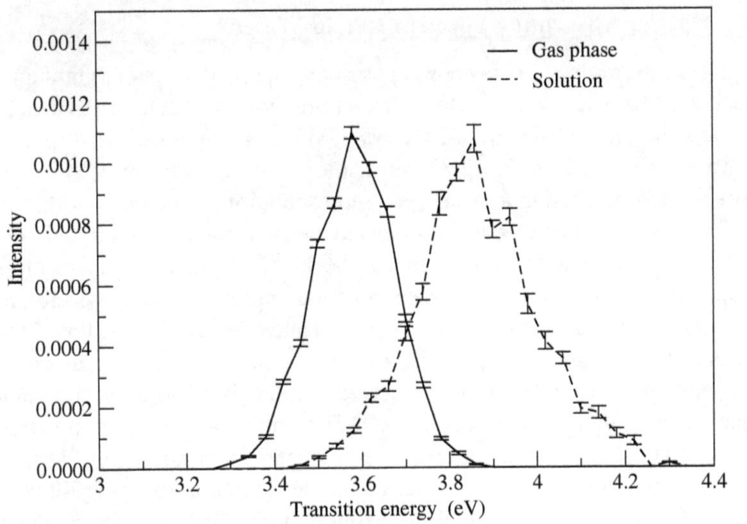

Figure 11.10 Optical absorption spectra of acrolein issuing from gas-phase and aqueous solution MD simulations computed at TD-DFT B3LYP/6-311+G(2d,2p) level of theory. (Adapted from ref. 42.)

the corresponding spectra are depicted in Figure 11.10. Overall, we have observed a blue shift of 0.26 ± 0.01 eV (last line of Table 11.3), in good agreement with experiments (0.20–0.25 eV). Moreover, from the aqueous solution simulations, we have performed different calculations by considering the acrolein molecule fully solvated, that is, acrolein and the two closest water molecules to the C=O group treated at the QM level and the remaining water molecules at the MM level (acrolein + $2H_2O^{QM}$ + $132H_2O^{MM}$ + PCM) only partially solvated, that is, including only the two closest water molecules forming hydrogen bonds (acrolein + $2H_2O^{QM}$) and without solvent molecules. In such a way, we have been able to evaluate the separate contributions to the blue shift coming solely from the solute structural changes (second line of Table 11.3) and from the carbonyl first solvation shell. Interestingly, the acrolein geometry distortion in solution leads to a nonnegligible red shift (−0.08 eV), in contrast to the overall observed transition energy shift (+0.26 eV). Hence, the direct solvent effects on the present spectroscopic property are responsible for a significant 0.34-eV blue shift. In particular, more than half of such a shift is induced by the first two water molecules surrounding the C=O group (0.18 eV). Therefore, we may conclude that the contributions from H–bonding and bulk effects are roughly the same. Moreover, from a theoretical point of view, it is worth noting that the evaluation of the solvatochromic shift does not change if we treat all the water molecules as embedded charges (acrolein + all H_2O^{MM} + PCM), including also the two hydrogen-bonded water molecules, with a blue shift of 0.25±0.01 eV. In other words, the nature of the solvent effects on the $n \rightarrow \pi^*$ vertical transition is essentially electrostatic, as also observed in the case of acetone [118].

11.3.4 Optical Absorption Spectra: Liquid Water

In the present example, we show how a physically sound discrete–continuum solvent approach can be successfully used, in combination with TD-DFT, to study the electronic structure and absorption spectra of an extended molecular system, taking liquid water as a test case. Both MD simulations, and a posteriori QM calculations have been performed using a mean–field representation of the interactions between the explicitly simulated molecular system and the environment, which is described as a structureless polarizable continuum via the PCM [118]. Contrary to the single-molecule case, optical absorption spectra of a complex extended system are more properly computed according to the sum-over-states (SOS) [71] method. Moreover, for the accurate modeling of nonlocalized electronic states, the inclusion of long-range effects in the calculation of valence excitations is of primary importance, for example by including such effects in the DFT exchange–correlation functionals.

A classical GLOB MD simulation has been carried out in the canonical ensemble at normal conditions of 241 SPC/E water molecules [119], keeping rigid the internal degrees of freedom, by embedding the explicit system into a spherical cavity of a dielectric continuum with a radius of 12.0 Å. The induced polarization effects of the continuum medium have been treated with C-PCM, plus dispersion–repulsion interactions modeled, as usual, by an effective potential. From an equilibrated 2-ns trajectory selected configurations have been extracted and QM/MM calculations have been performed by partitioning the explicit system into a small quantum core region of a few water molecules, n, and a surrounding classical region ($N - n$ molecules) modeled as embedding effective charges.

Optical transitions and spectra have been computed at the TD-DFT level using the LDA, PBE, long-range corrected LCPBE-TPSS, and hybrid M052X functionals with the N07D basis set, including also a diffuse s function on hydrogen atoms. It is worth noting that while the correlation part of the last two functionals is free of self-interaction, the long-range exchange is exact in LCPBE-TPSS, whereas both self-interaction and long-range effects are partially accounted for in M052X by inclusion of the Hartree–Fock exchange. The optical absorption spectra of liquid water have been obtained by computing a large number of excited states (up to 280) and by performing the SOS calculations of the dielectric susceptibility. The calculated spectra have been obtained from an average over 10 uncorrelated molecular configurations. A broadening due to the excited-state finite lifetime has also been considered. In Figure 11.11, we report the low-energy (6–12-eV) computed and experimental [120] optical absorption spectra of liquid water. Contrary to LDA and GGA (PBE) functionals, LCPBE-TPSS (not shown) and M052X are able to reproduce very well the positions and intensities of both characteristic peaks, with an excellent and unprecedented agreement with experiments. At higher energy, the intensity of the calculated spectra decreases with respect to the experimental counterpart due to the limited number of states included in the SOS expansion. Hence, such results do support the combination of molecular dynamics and a posteriori TD-DFT calculations for the study of the electronic properties of molecular liquids, even within the adiabatic approximation, provided that a proper DFT functional is chosen.

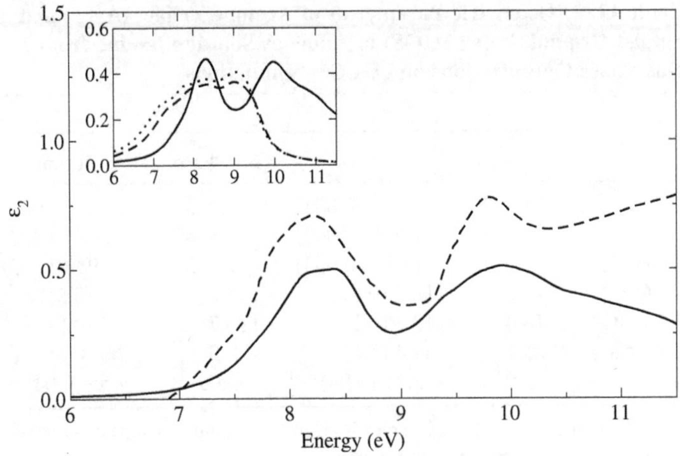

Figure 11.11 Computed (solid line) and experimental (dashed line) optical absorption spectra of liquid water. Inset: calculated optical spectra for one MD configuration. Dotted line, LDA; dashed line, PBE; solid line, M052X. A broadening term of 0.2 eV was used in the computed spectra. (Adapted from ref. 2.)

Remarkably, the present study on water has also allowed us to obtain relevant physical insights about the controversial nature of the first optical band of liquid water due to the sound treatment of the electronic density rearrangements upon excitation.

11.3.5 Optical Emission Spectra: Acetone Triplet in Aqueous Solution

As an illustrative example of the use of a time-dependent approach to the study of a transient excited-state molecule in the condensed phase, we consider the case of the $(n \rightarrow \pi^*)$ triplet state of acetone. Despite the large number of experimental [116, 122] and theoretical [118, 123–125] works on acetone, certainly one of the most studied organic compounds, its triplet state has been less investigated and mainly in the gas phase [126–128]. However, due to its relatively long lifetime in solution (about 20–50 μs), [129] triplet acetone has a quite rich and interesting chemistry and photochemistry. For example, it is commonly used as a kinetic probe for the formation of contacts with suitable quenchers [130] and it undergoes photoreduction through direct hydrogen abstraction [131, 132] from C–H or N–H bonds. Furthermore, triplet acetone is involved also in some biological processes, where it is formed by enzymatic chemiexcitation [133], and shows a different selectivity toward H abstraction reactions from amines with respect to the structurally similar excited singlet state. From an electronic viewpoint, the $^3(n \rightarrow \pi^*)$ triplet state of acetone is generated by an electron promotion from the oxygen lone pair $(\pi)$to the π^* antibonding molecular orbital of the carbonyl moiety, corresponding respectively to the highestoccupied (HOMO) and lowest unoccupied (LUMO) molecular orbitals of the ground-state acetone, and, as a consequence, the different behavior between the triplet state and the ground state can be interpreted basically in terms of this electronic structural change.

Table 11.4 Geometric Parameters of Acetone Triplet (ActT) and Singlet Ground State (ActGS) in Aqueous Solution Issuing from Gas-Phase Optimizations and GLOB Simulations

	ActT		ActGS	
	Gas Phase	Solution	Gas Phase	Solution
C=O	1.322	1.324 (0.003)	1.213	1.226 (0.003)
C–C	1.519	1.526 (0.003)	1.516	1.508 (0.003)
C–H	1.096	1.099 (0.003)	1.095	1.098 (0.003)
C–C=O	113.3	113.3 (0.5)	121.6	121.0 (0.5)
C–C–C	118.4	118.7 (0.5)	116.7	117.7 (0.5)
ϕ(COCC)	25.1	24.4 (0.6)	0.0	0.0 (0.6)
μ	1.92	2.71 (0.04)	3.09	4.55 (0.04)

Note: Bond distances are in Å, angles in degrees, and dipole moments in debye. Standard error are reported in parentheses.

First, we note that the gas-phase structural parameters, as issuing by geometry optimizations at the B3LYP/N07D level of both acetone ground and triplet states, reveal a significant distortion of the typical planar conformation (e.g., considering the heavy atoms) upon excitation (see Table 11.4), as shown by the ϕ(COCC) dihedral angle (25.1°). Moreover, the C=O bond distance is somewhat elongated [Δd(C=O) = 0.11 Å], along with a reduction of the molecular dipole moment ($\Delta\mu$). Such a result appears consistent with the acquired single bond character of the C=O group as well as a partial sp^3 hybridization of the carbon atom. Also, quantum mechanical calculations of the acetone–H_2O complex in vacuo have shown that a hydrogen-bonded water to the carbonyl group has a lower binding energy in the case of the triplet state (triplet state: -3.1 kcal mol^{-1}; ground-state: -5.7 kcal mol^{-1}). Considering the dipole moment reduction after the promotion of an electron, a solvatochromic blue shift can be predicted in going from the gas phase to the aqueous solution as a result of the enhanced stabilization energy of the acetone ground state with respect to the triplet state.

In aqueous solution, the microsolvation of the triplet acetone, as compared to the ground state, has been investigated by means of QM/MM GLOB molecular dynamics simulations [134, 135]. The average structural parameters are reported in Table 11.4. As noted above, the C_2C=O group is distorted with respect to the planar conformation, with the oxygen atom going out of the C–C–C plane [ϕ(COCC) = 24.4°], and no inversion of such a pyramidal conformation has been observed during the dynamics. Quite apparent is the change in the hydrogen-bonding pattern, which shows just one water molecule H bonded to the carbonyl in the case of the excited species, as shown by the RDF and of the acetone oxygen with respect to the water oxygen and hydrogen atoms (see Figure 11.12) and by the spatial distribution function (SDF) of the water molecules (considering here only the hydrogen atoms) around the carbonyl group, as reported in Figure 11.13. It is worth recalling that acetone in its ground state forms, on average, two hydrogen bonds in aqueous solution. Here, a similar analysis leads to an average number of hydrogen bonds with water of only 0.8 (see ref. 118 for the criteria

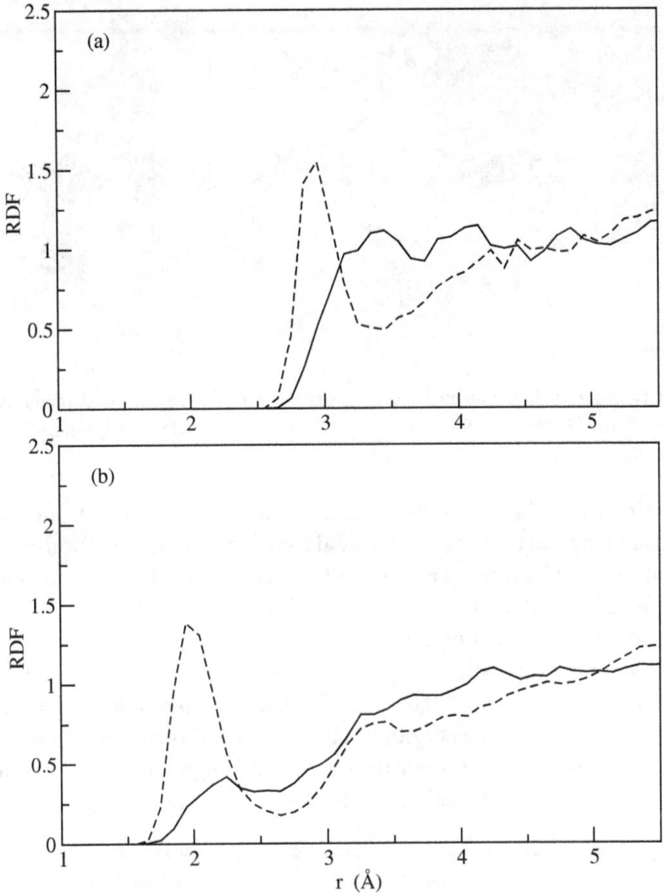

Figure 11.12 (*a*) O. . .Ow and (*b*) O. . .Hw radial distribution functions of acetone in aqueous solution as obtained from GLOB simulations. Solid line, ActT; dashed line, ActGS. (Adapted from ref. 136.)

of hydrogen bond definition used in this work). Such differences are also reflected in the photophysical behavior in solution.

By means of DFT and TD-DFT calculations, we have evaluated theoretically both the $S_0 \rightarrow S_1$ (optical absorption) and the $T_1 \rightarrow S_0$ (phosphorescence) vertical transition energies in aqueous solution of acetone. From experiments, we know that $\Delta E(S_0 \rightarrow S_1) = 4.69$ eV [137] and $\Delta E(T_1 \rightarrow S_0) = 2.72$–$2.82$ eV, [132, 138] with a Stokes shift of 1.87–1.97 eV. The same physical model used in the simulations has been retained in the QM/MM spectroscopic calculations, where the QM core region was represented by the solute and the two closest water molecules to the carbonyl group and the electrostatic response of the environment was modeled by the explicit solvent molecules treated as embedding point charges plus the dielectric continuum. Optical transition energies and spectra have been computed at DFT and TD-DFT

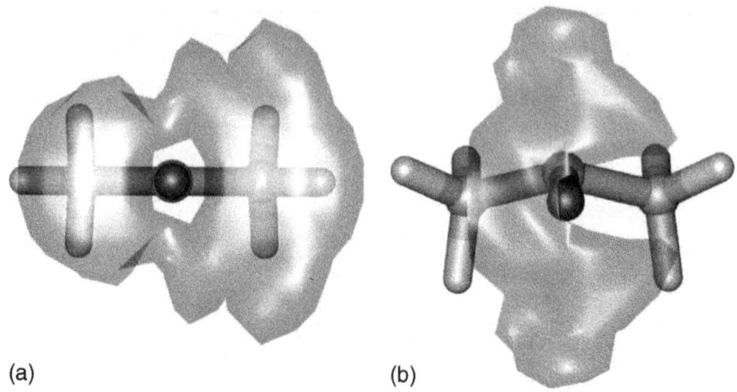

Figure 11.13 Spatial distribution functions of water molecules around (a) ActGS and (b) ActT issuing from GLOB simulations. Grey cloud, water hydrogen atoms. (Adapted from ref. 136.)

levels by extracting a large number of molecular configurations from the sampled trajectories and by performing QM calculations on both excitation energies and oscillator strengths. For consistency with the dynamics, the B3LYP method was used in the spectroscopic calculations in combination with a well-tested basis set [117], 6-311 + G(2d,2p). For the computation of the $^3(\pi^* \rightarrow n)$ emission spectrum, which requires in principle a sophisticated spin–orbit coupling calculation, we have assumed that the transition intensity is the same for each configuration. Hence, only the information on the excitation energies has been used in this case, as obtained from the electronic energy difference between the triplet and (singlet) ground states. Moreover, due to the known underestimation of the $S_0 \rightarrow T_1$ vertical transition energies of

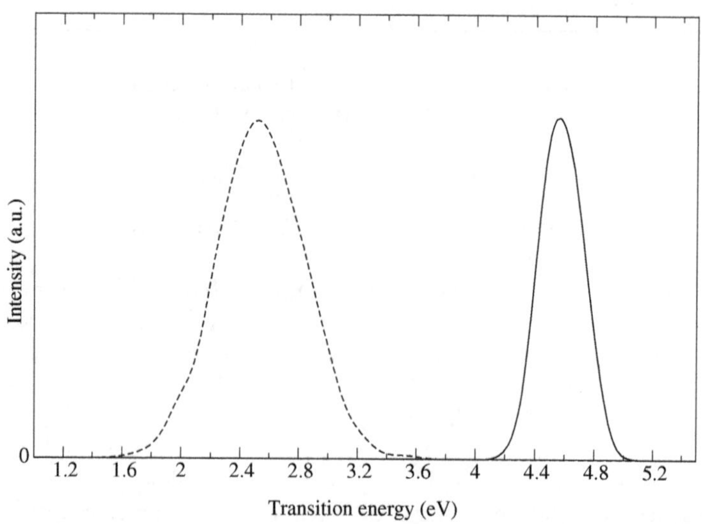

Figure 11.14 Acetone $^1(n \rightarrow \pi^*)$ absorption (solid line) and $^3(\pi^* \rightarrow n)$ emission (dashed line) spectra as computed from GLOB simulations, (Adapted from ref. 136.)

carbonyl molecules by DFT methods [e.g., 127], we have corrected the excitation energy based on a coupled-cluster calculation on a previously optimized gas-phase geometry of the triplet acetone computed at the B3LYP/N07D level [$\Delta E(S_0 \rightarrow T_1)$: 2.51 eV, CCSD(T)/aug-cc-pVTZ; 2.21 eV, B3LYP/6-311 + G(2d,2p)]. Therefore, the $\Delta E(T_1 \rightarrow S_0)$ excitation energies computed at the B3LYP level have been systematically shifted by 0.3 eV. In Figure 11.14, the computed optical absorption and emission spectra are reported, showing a peak position at about 4.57 and 2.53 eV, respectively, and consequently a Stokes shift of 2.04 eV, in good agreement with experiments.

11.4 CONCLUSIONS

Computational spectroscopy can take great advantage from well-calibrated molecular dynamics techniques, which can provide a direct and controlled interpretation of results and a deep understanding of the structure/spectroscopic relationship. Such theoretical approaches can be very helpful in several spectroscopic techniques ranging from magnetic, optical, to X-ray diffraction/absorption techniques for both equilibrium (steady-state) and nonequilibrium (time-resolved) experiments. Results obtained for challenging prototype applications encourage us to improve and further develop connections to more sophisticated full-quantum mechanical approaches.

ACKNOWLEDGMENT

The invaluable scientific and professional support of Professor Vincenzo Barone (Scuola Normale Superiore in Pisa) is gratefully acknowledged.

REFERENCES

1. S. Carter, S. J. Culik, J. M. Bowman, *J. Chem. Phys.* **1997**, *107*, 10458.
2. J. Koput, S. Carter, N. C. Handy, *J. Chem. Phys.* **2001**, *115*, 8345.
3. P. Cassam-Chenai, J. Lievin, *J. Quant. Chem.* **2003**, *93*, 245.
4. J. M. Bowman, *Acc. Chem. Res.* **1986**, *19*, 202.
5. G. M. Chaban, J. O. Jung, R. B. Gerber, *J. Chem. Phys.* **1999**, *111*, 1823.
6. N. J. Wright, R. B. Gerber, D. J. Tozer, *Chem. Phys. Lett.* **2000**, *324*, 206.
7. N. J. Wright, R. B. Gerber, *J. Chem. Phys.* **2000**, *112*, 2598.
8. S. K. Gregurick, G. M. Chaban, R. B. Gerber, *J. Phys. Chem. A* **2002**, *106*, 8696.
9. D. A. Clabo, W. D. Allen, R. B. Remington, Y. Yamaguchi, H. F. Schaeffer III, *Chem. Phys.* **1988**, *123*, 187.
10. W. Schneider, W. Thiel, *Chem. Phys. Lett.* **1989**, *157*, 367.
11. A. Willets, N. C. Handy, W. H. Green, D. Jayatilaka, *J. Phys. Chem.* **1990**, *94*, 5608.
12. S. Dressler, W. Thiel, *Chem. Phys. Lett.* **1997**, *273*, 71.
13. J. Neugebauer, B. A. Hess, *J. Chem. Phys.* **2003**, *118*, 7215.

14. O. Christiansen, *J. Chem. Phys.* **2003**, *119*, 5773.
15. T. A. Ruden, P. R. Taylor, T. Helgaker, *J. Chem. Phys.* **2003**, *119*, 1951.
16. K. Yagi, T. Tatetsugu, K. Hirao, M. S. Gordon, *J. Chem. Phys.* **2000**, *113*, 1005.
17. V. Barone, *J. Chem. Phys* **2005**, *122*, 14108.
18. P. Macak, Y. Luo, H. Ågren, *Chem. Phys. Lett.* **2000**, *330*, 447.
19. F. Duschinsky, *Acta Physicochim. URSS* **1937**, *7*, 551.
20. M. Kemper, J. Van Dijk, H. Buck, *Chem. Phys. Lett.* **1978**, *53*, 121.
21. R. Berger, C. Fischer, M. Klessinger, *J. Phys. Chem. A* **1998**, *102*, 7157.
22. M. Dierksen, S. Grimme, *J. Chem. Phys.* **2005**, *122*, 244101.
23. F. Santoro, R. Improta, A. Lami, J. Bloino, V. Barone, *J. Chem. Phys.* **2007**, *126*, 084509.
24. F. Santoro, R. Improta, A. Lami, J. Bloino, V. Barone, *J. Chem. Phys.* **2007**, *126*, 184102.
25. H.-C. Jankowiak, J. L. Stuber, R. Berger, *J. Chem. Phys.* **2007**, *127*, 234101.
26. M. Schmitz, P. Tavan, *J. Chem. Phys.* **2004**, *121*, 12247.
27. R. Ramirez, T. Lopez-Ciudad, P. Kumar, D. Marx, *J. Chem. Phys.* **2004**, *121*, 3973.
28. A. L. Kaledin, X. Huang, J. M. Bowman, *Chem. Phys. Lett.* **2004**, *384*, 80.
29. M. Schmitz, P. Tavan, *J. Chem. Phys.* **2004**, *121*, 12233.
30. H. Koeppel, W. Domcke, L. Cederbaum, *Adv. Chem. Phys.* **1984**, *57*, 59.
31. H.-D. Meyer, U. Manthe, L. S. Cederbaum, *Chem. Phys. Lett.* **1990**, *171*, 97.
32. A. Raab, G. Worth, H.-D. Meyer, *J. Chem. Phys.* **1999**, *110*, 936.
33. N. Rega, G. Brancato, V. Barone, *Chem. Phys. Lett.* **2006**, *422*, 367.
34. G. Brancato, N. Rega, V. Barone, *J. Chem. Phys.* **2008**, *128*, 144501.
35. M. H. Beck, A. Jackle, G. A. Worth, H. D. Meyer, *Phys. Rep.* **2000**, *324*, 1.
36. E. Gindensperger, L. Cederbaum, *J. Chem. Phys.* **2007**, *127*, 124107.
37. M. Thoss, G. Stock, *Adv. Chem. Phys.* **2005**, *134*, 243.
38. J. C. Tully, *J. Chem. Phys.* **1990**, *93*, 1061.
39. R. Carr, M. Parrinello, *Phys. Rev. Lett.* **1985**, *55*, 2471.
40. D. Marx, J. Hutter, Ab initio molecular dynamics: Theory and implementation, in *Modern Methods and Algorithms of Quantum Chemistry*, Vol. 1, John von Neumann Institute for Computing, Julich, 2000, p. 301.
41. M. Sulpizi, U. F. Roehrig, J. Hutter, U. Rothlisberger, *Int. J. Quant. Chem.* **2005**, *101*, 671.
42. G. Brancato, N. Rega, V. Barone, *J. Chem. Phys.* **2006**, *125*, 164515.
43. M. Cascella, I. Tavernelli, M. Cuendet, U. Rothlisberger, *J. Phys. Chem. B* **2007**, *111*, 10248.
44. G. Brancato, N. Rega, V. Barone, *J. Am. Chem. Soc.* **2007**, *129*, 15380.
45. I. Tavernelli, U. F. Roehrig, U. Rothlisberger, *Mol. Phys.* **2005**, *103*, 963.
46. E. Tapavicza, I. Tavernelli, U. Rothlisberger, *J. Phys. Chem. A* **2009**, *113*, 9595.
47. H. B. Schlegel, J. M. Millam, S. S. Iyengar, G. A. Voth, A. D. Daniels, G. E. Scuseria, M. J. Frisch, *J. Chem. Phys.* **2001**, *114*, 9758.
48. S. S. Iyengar, H. B. Schlegel, J. M. Millam, G. A. Voth, G. E. Scuseria, M. J. Frisch, *J. Chem. Phys.* **2001**, *115*, 10291.
49. H. B. Schlegel, S. S. Iyengar, X. Li, J. M. Millam, G. A. Voth, G. E. Scuseria, M. J. Frisch, *J. Chem. Phys.* **2002**, *117*, 8694.

50. D. A. McQuarrie, *Statistical Mechanics*, 2nd ed.,University Science Books, Mill Valley, CA, 2000.
51. S. A. Egorov, E. Rabani, B. J. Berne, *J. Chem. Phys* **1998**, *108*, 1407.
52. J. R. Reimers, K. R. Wilson, *J. Chem. Phys* **1983**, *79*, 4749.
53. J. P. Bergsma, P. H. Berens, K. R. Wilson, D. R. Fredkin, E. J. Heller, *J. Phys. Chem.* **1984**, *88*, 612.
54. S. S. Iyengar, *Int. J. Quant. Chem.* **2009**, *109*, 3798.
55. X. Li, D. T. Moore, S. S. Iyengar, *J. Chem. Phys.* **2008**, *128*, 184308.
56. S. S. Iyengar, X. Li, I. Sumner, *Adv. Quant. Chem.* **2008**, *55*, 333.
57. I. Sumner, S. S. Iyengar, *J. Phys. Chem. A* **2007**, *111*, 10313.
58. R. A. Wheeler, H. Dong, S. E. Boesch, *Chem. Phys. Chem.* **2003**, *4*, 382.
59. M. Schmitz, P. Tavan, *J. Chem. Phys.* **2004**, *121*, 12233.
60. M. Schmitz, P. Tavan, *J. Chem. Phys.* **2004**, *121*, 12247.
61. A. Strachan, *J. Chem. Phys.* **2004**, *120*, 1.
62. M. Martinez, M.-P. Gaigeot, D. Borgis, R. Vuilleumier, *J. Chem. Phys.* **2006**, *125*, 034501.
63. N. Rega, *Theor. Chem. Acc.* **2006**, *116*, 347.
64. M. Karplus, J. N. Kushick, *Macromolecules* **1981**, *14*, 325.
65. B. R. Brooks, D. Janezic, M. Karplus, *J. Comput. Chem.* **1995**, *16*, 1522.
66. M. Cossi, G. Scalmani, N. Rega, V. Barone, *J. Comput. Chem.* **2003**, *24*, 669.
67. V. Barone, *J. Chem. Phys.* **2005**, *122*, 014108.
68. P. Carbonniere, A. Dargelos, I. Ciofini, C. Adamo, C. Pouchan, *Phys. Chem. Chem. Phys* **2009**, *11*, 4375.
69. M.-P. Gaigeot, M. Martinez, R. Vuilleumier, *Mol. Phys.* **2007**, *105*, 2857.
70. A. Cimas, P. Maitre, G. Ohanessian, M.-P. Gaigeot, *J. Chem. Theor. Comp.* **2009**, *5*, 2388.
71. P. N. Butcher, D. Cotter, *The Elements of Nonlinear Optics*, Cambridge University Press, Cambridge, 1990.
72. S. Dapprich, I. Komaromi, K. S. Byun, K. Morokuma, M. J. Frisch, *J. Mol. Struct. Theochem.* **1999**, *462*, 1.
73. T. Vreven, K. Morokuma, O. Farkas, H. B. Schlegel, M. J. Frisch, *J. Comp. Chem.* **2003**, *24*, 760.
74. N. Rega, S. S. Iyengar, G. A. Voth, H. B. Schlegel, T. Vreven, M. J. Frisch, *J. Phys. Chem. B* **2004**, *108*, 4210.
75. G. Brancato, A. Di Nola, V. Barone, A. Amadei, *J. Chem. Phys.* **2005**, *122*, 154109.
76. G. Brancato, N. Rega, V. Barone, *J. Chem. Phys.* **2006**, *124*, 214505.
77. A. Ben-Naim, *Solvation Thermodynamics*, Plenum, New York, 1987.
78. A. Klamt, G. Schüurmann, *J. Chem. Soc. Perkin 2 Trans.* **1993**, *220*, 799.
79. V. Barone, M. Cossi, *J. Phys. Chem. A* **1998**, *102*, 1995.
80. M. Cossi, N. Rega, G. Scalmani, V. Barone, *J. Comp. Chem.* **2003**, *24*, 669.
81. J. Tomasi, M. Persico, *Chem. Rev.* **1994**, *94*, 2027.
82. J. Tomasi, B. Mennucci, R. Cammi, *Chem. Rev.* **2005**, *105*, 2999.
83. C. Benzi, M. Cossi, V. Barone, R. Tarroni, C. Zannoni, *J. Phys. Chem. B* **2005**, *109*, 2584.
84. M. Cossi, G. Scalmani, N. Rega, V. Barone, *J. Chem. Phys.* **2002**, *117*, 43.

85. J.-L. Pascual-Ahuir, E. Silla, I. Tũnon, *J. Comput. Chem.* **1994**, *15*, 1127.
86. G. Scalmani, N. Rega, M. Cossi, V. Barone, *J. Comput. Meth. Sci. Eng.* **2002**, *2*, 469.
87. F. M. Floris, J. Tomasi, *J. Comput. Chem.* **1989**, *10*, 616.
88. F. M. Floris, J. Tomasi, J. L. Pascual-Ahuir, *J. Comput. Chem.* **1991**, *12*, 784.
89. J. D. Thompson, C. J. Cramer, D. G. Truhlar, *Theor. Chem. Acc.* **2005**, *113*, 107.
90. C. P. Kelly, C. J. Cramer, D. G. Truhlar, *J. Chem. Theory Comp.* **2005**, *1*, 1133.
91. I. Tuñón, M. F. Ruiz-López, D. Rinaldi, J. Bertrán, *J. Comp. Chem.* **1996**, *17*, 148.
92. C. Curutchet, M. Orozco, F. J. Luque, *J. Comp. Chem.* **2001**, *22*, 1180.
93. C. Curutchet, A. Bidon-Chanal, I. Soteras, M. Orozco, F. J. Luque, *J. Phys. Chem. B* **2005**, *109*, 3565.
94. A. Klamt, V. Jonas, T. Burger, J. C. W. Lohrenz, *J. Phys. Chem. A* **1998**, *102*, 5074.
95. D. Beglov, B. Roux, *J. Chem. Phys.* **1994**, *100*, 9050.
96. J. W. Essex, W. L. Jorgensen, *J. Comput. Chem.* **1995**, *16*, 951.
97. G. Petraglio, M. Ceccarelli, M. Parrinello, *J. Chem. Phys.* **2005**, *123*, 044103.
98. J. Kubelka, T. A. Keiderling, *J. Phys. Chem. A* **2001**, *105*, 10922.
99. M. Q. Fatmi, T. S. Hofer, B. R. Randolf, B. M. Rode, *J. Chem. Phys.* **2005**, *123*, 054514.
100. A. M. Mohammed, H. H. Loeffler, Y. Inada, K. Tanada, S. Funahashi, *J. Mol. Liq.* **2005**, *119*, 55.
101. P. D'Angelo, V. Barone, G. Chillemi, N. Sanna, W. Meyer-Klaucke, N. V. Pavel, *J. Am. Chem. Soc.* **2002**, *124*, 1958.
102. G. Chillemi, P. D'Angelo, N. V. Pavel, N. Sanna, V. Barone, *J. Am. Chem. Soc.* **2002**, *124*, 1968.
103. H. Kanno, *J. Phys. Chem.* **1988**, *92*, 4232.
104. A. M. Buswell, E. C. Dunlop, W. H. Rodebush, J. B. Swartz, *J. Am. Chem. Soc.* **1940**, *62*, 325.
105. A. D. Walsh, *Trans. Faraday Soc.* **1945**, *41*, 498.
106. G. Mackinney, O. Temmer, *J. Am. Chem. Soc.* **1948**, *70*, 3586.
107. K. Inuzuka, *Bull. Chem. Soc. Jpn.* **1961**, *34*, 6.
108. R. S. Becker, K. Inuzuka, J. King, *J. Chem. Phys.* **1970**, *52*, 5164.
109. E. J. Bair, W. Goietz, D. A. Ramsay, *Can. J. Chem.* **1971**, *49*, 2710.
110. G. A. Osborne, D. A. Ramsay, *Can. J. Chem.* **1973**, *51*, 1170.
111. A. Luthy, *Z. Phys. Chem.* **1923**, *107*, 284.
112. F. E. Blacet, W. G. Young, J. G. Roof, *J. Am. Chem. Soc.* **1937**, *59*, 608.
113. K. Inuzuka, *Bull. Chem. Soc. Jpn.* **1960**, *33*, 678.
114. N. S. Bayliss, E. G. McRae, *J. Phys. Chem.* **1954**, *58*, 1006; A. Balasubramanian, C. N. R. Rao, *Spectrochim. Acta* **1962** *18*, 1337; N. S. Bayliss, G. Wills-Johnson, *Spectrochim. Acta A* **1968**, *24*, 551.
115. J. T. Blair, K. Krogh-Jespersen, R. M. Levy, *J. Am. Chem. Soc.* **1989**, *111*, 6948. J. Kongsted, A. Osted, T. B. Pedersen, K. V. Mikkelsen, O. Christiansen, *J. Phys. Chem. A* **2004**, *108*, 8624.
116. N. S. Bayliss, E. G. McRae, *J. Phys. Chem.* **1954**, *58*, 1002.
117. O. Crescenzi, M. Pavone, F. D. Angelis, V. Barone, *J. Phys. Chem. B.* **2005**, *109*, 445.

118. M. Pavone, G. Brancato, G. Morelli, V. Barone, *Chem. Phys. Chem.* **2006**, *7*, 148.
119. H. J. C. Berendsen, J. R. Grigera, T. P. Straatsma, *J. Phys. Chem.* **1987**, *91*, 6269.
120. G. D. Kerr, R. N. Hamm, M. W. Williams, R. D. Birkhoff, L. R. Painter, *Phys. Rev. A* **1972**, *5*, 2523.
121. G. Brancato, N. Rega, V. Barone, *Phys. Rev. Lett.* **2008**, *100*, 107401.
122. B. Tiffon, J. E. Dubois, *Org. Magn. Reson.* **1978**, *11*, 295.
123. K. Aidas, J. Kongsted, A. Osted, K. V. Mikkelsen, O. Christiansen, *J. Phys. Chem. A* **2005**, *109*, 8001.
124. U. F. Röhrig, I. Frank, J. Hutter, A. Laio, J. VandeVondele, U. Rothlisberger, *Chem. Phys. Chem.* **2003**, *4*, 1177.
125. Y. Lin, J. Gao, *J. Chem. Theory Comput.* **2007**, *3*, 1484.
126. E.W.-G. Diau, C. Kötting, A. H. Zewail, *Chem. Phys. Chem* **2001**, *2*, 273.
127. C. Angeli, S. Borini, L. Ferrighi, R. Cimiraglia, *J. Chem. Phys.* **2005**, *122*, 114304.
128. K. Aidas, K. V. Mikkelsen, B. Mennucci, J. Kongsted, *Int. J. Quant. Chem.* **2011**, *111*, 1511.
129. G. Porter, R. W. Yip, J. M. Dunston, A. J. Cessna, S. E. Sugamori, *Trans. Faraday Soc.* **1971**, *67*, 3149.
130. G. L. Indig, L. H. Catalani, T. Wilson, *J. Phys. Chem.* **1992**, *96*, 8967.
131. G. Porter, S. K. Dogra, R. O. Loutfy, S. E. Sugamori, R. W. Yip, *J. Chem. Soc. Faraday Trans.* **1973**, *69*, 1462.
132. U. Pischel, W. M. Nau, *J. Am. Chem. Soc.* **2001**, *123*, 9727.
133. W. Adma, G. Cilento, Eds., *Chemical and Biological Generation of Excited States*, Academic, New York, 1982.
134. G. Brancato, V. Barone, N. Rega, *Theor. Chem. Acc.* **2007**, *117*, 1001.
135. G. Brancato, N. Rega, V. Barone, *Chem. Phys. Lett.* **2009**, *481*, 177.
136. G. Brancato, N. Rega, V. Barone, *Chem. Phys. Lett.* **2008**, *453*, 202.
137. N. S. Bayliss, E. G. McRae, *Spectrochim. Acta A* **1968**, *24*, 551.
138. R. F. Borkman, D. R. Kearns, *J. Chem. Phys.* **1966**, *44*, 945.

12

STOCHASTIC METHODS FOR MAGNETIC RESONANCE SPECTROSCOPIES

ANTONINO POLIMENO

Dipartimento di Scienze Chimiche, Università degli Studi di Padova, Padova, Italy

VINCENZO BARONE

Scuola Normale Superiore di Pisa, Pisa, Italy

JACK H. FREED

Department of Chemistry and Chemical Biology, Cornell University, Ithaca, New York

Computational Strategies for Spectroscopy: From Small Molecules to Nano Systems, First Edition.
Edited by Vincenzo Barone.

Physicochemical properties of molecules in solution depend on the action of different motions at several time and length scales, and information on multiscale dynamics can be gained, in principle, by a variety of spectroscopic techniques. In this work we review theoretical tools for the investigation of "slow" molecular motions, such as solvent cage effects in liquids and liquid crystals, global and local dynamics in proteins, reorientation dynamics, and internal (conformational) degrees of freedom. Spectroscopic techniques which are most sensitive to such motions are electron spin resonance and nuclear magnetic resonance and they require ad hoc theoretical treatment. In particular, we discuss the definition of multidimensional stochastic models and their treatment to interpret magnetic resonance spectroscopic data of rigid and flexible molecules in isotropic media, liquid crystals, and biosystems.

12.1 INTRODUCTION

It is natural for a chemist to consider molecules as dynamical systems. Thermal effects and interactions with other molecules influence both internal and global molecular degrees of freedom. Macroscopic chemical and physical properties of molecules depend on their dynamics to varying degrees, based upon the physical observable considered. Examples of dynamical physical chemistry are numerous: Collision theory is based on the assumption that molecules move (in order to collide) to react; temperature is the macroscopic physical observable which is related to the average square velocity of particles; osmotic pressure in biological cells is kept at a fixed point value by the action of Na/K pumps, which are molecular machines that carry out their function due to internal dynamics; many enzymes can react and transform a substrate because of change of conformation that occurs in bonding, and this serves to create the right chemical environment around the substrate.

Thus interpretation of structural properties and dynamic behavior of molecules in solution is of fundamental importance to understand their stability, chemical reactivity, and catalytic action. Great interest exists in the development of new materials and the study of biological macromolecules. In general, one has to treat complex systems in which motions are present over a wide range of time scales encompassing global dynamics (microseconds), domain dynamics (nanoseconds), and localized fluctuations involving selected chemical groups (picoseconds to femtoseconds).

Given that a key role of theoretical chemistry is to interpret macroscopic observations in terms of physicochemical properties of molecules, dynamics is a fundamental ingredient as well as structure. This is especially true for models designed to interpret processes occurring in large biomolecules or complex ("soft") materials. In this work our main purpose is to review integrated theoretical/computational approaches for interpreting motions typically in the range 10^{-9}–10^{-6} s in complex molecular systems. We will refer to this range as slow molecular motions or just "slow motions". The main objective is the study of the dynamics (mobility) of complex systems, mainly of biomolecular interest, by means of the interpretation of spectroscopic data for obtaining information on their dynamics [1]. Indeed, information on dynamics can be inferred, in most cases, only indirectly from experiments. A theoretical framework is therefore required to link macroscopic observations to molecular dynamics. A sensible plan of action is then to (i) choose a reference experimental technique which is particularly sensitive to the type of motions we are interested in; (ii) set up a framework for describing the dynamics and its influence on the chosen physical observable; and (iii) select model systems which serve to build and test theoretical models.

Experimental determination of dynamical properties of molecular systems is often based on sophisticated spectroscopic techniques. Given that the properties of molecules in solution result from motions at several time and length scales, insight on multiscale dynamics can be gained, in principle, by a range of spectroscopic techniques: magnetic [nuclear magnetic resonance (NMR) and electron spin resonance (ESR)] and optical [fluorescence polarization anisotropy (FPA), dynamic light scattering (DLS), and time-resolved Stokes shift (TRSS)]. In this review we focus on slow molecular motions (e.g., dynamic solvation effects, reorientation dynamics, conformational dynamics) monitored by magnetic spectroscopies, both ESR and NMR. In the case of ESR, this means that slow-motion processes have characteristic time scales that are comparable to those of electronic spin relaxation.

This contribution reviews the basic tools which are currently employed for interpreting ESR and NMR observables in condensed phases, with an emphasis on stochastic modeling as key for the prediction of continuous-wave ESR (cw-ESR) lineshapes and NMR relaxation times of proteins. Section 12.2 is therefore devoted to the definition of reduced (effective) magnetic Hamiltonians and the stochastic (Liouville) approach to spin/molecular dynamics in order to clarify the basic stochastic approach to cw-ESR observables. Section 12.3 provides a short overview of rotational stochastic models for the evaluation of relaxation NMR data in biomolecules. Conclusions are briefly summarized in Section 12.4.

12.2 MODELING A cw-ESR EXPERIMENT

Magnetic resonance spectroscopies and theoretical chemistry have always been linked. On the one hand, the rich and detailed information hidden in ESR and NMR spectra has been a challenge for physicochemical interpretations and computational models. On the other hand, magnetic resonance spectroscopists have been looking for better tools to interpret the spectra.

12.2.1 ESR: Modeling and Observables

The intrinsic resolution of ESR spectra together with the unique role played by paramagnetic probes in providing information on their environment makes ESR one of the most powerful methods of investigation of electronic distributions in molecules and the properties of their environments. The theoretical tools needed by ESR spectroscopists come from quantum chemistry to provide the parameters of the spin Hamiltonian appropriate for room temperature (experiments usually can supply them but for frozen solutions at low temperatures) and from molecular dynamics and statistical mechanics for the spectral lineshapes.

Because of their favorable time scales, ESR experiments can be very sensitive to the details of the rotational and internal dynamics. In particular, with the advent of very high field ESR corresponding to frequencies at and above 140 GHz, the rotational dynamics of spin-labeled molecules observed by ESR is more commonly found to be in the so-called slow-motion regime than is the case at conventional ESR frequencies (e.g., 9.5 GHz) [2]. For this regime, the spectral lineshapes take on a complex form which is found to be sensitive to the microscopic details of the motional process [2]. This is to be contrasted with the fast-motion regime, where simple Lorentzian lineshapes are observed, and only estimates of molecular parameters (e.g., diffusion tensor values) are obtained independently from the microscopic details of the molecular dynamics. The interpretation of slow-motion spectra requires an analysis based upon sophisticated theory, as will be emphasized in the next section. ESR spectroscopy is applied extensively to materials science and to biochemistry. Great interest is focused particularly on the study of the dynamics of biological molecules, such as proteins and, in particular, ESR studies of proteins via site-directed spin labeling (SDSL) with stable nitroxide radicals [3–6]. The wealth of dynamic information which can be extracted from a cw-ESR or an electron–electron double-resonance (ELDOR) spectrum with nitroxide labels is at present limited experimentally by the challenge of obtaining extensive multifrequency data [6] and theoretically by the necessity of employing computationally efficient dynamic models [2, 7–9]. The review of Borbat et al. [10] provides a discussion of modern ESR techniques for studying basic molecular mechanisms in proteins and membranes by using nitroxide spin labels. These include the direct measurement of distances in biomolecules and unraveling the details of complex molecular dynamics. These studies can, for instance, provide information on phospholipid membranes [11–14] which can be described via augmented stochastic models. Since the relationship between ESR spectroscopic measurements and most molecular properties can be obtained only indirectly via modeling and numerical simulations one may utilize the spectroscopic data as the "target" of a fitting procedure of molecular, mesoscopic, and macroscopic parameters entering the model.

An intrinsic limitation of this approach is the difficulty of avoiding uncertainties due to multiple minima in the fitting procedure and the difficulty, in many cases, to reconcile best-fit parameters with more general approaches or known physical trends (e.g., temperature dependence).

A more refined methodology is based on an integrated computational approach, that is, the combination of (i) quantum mechanical (QM) calculations of structural parameters and magnetic tensors possibly including average interactions with the environment (by discrete–continuum models) and short-time dynamical effects; (ii) direct feeding of calculated molecular parameters into dynamic models based on molecular dynamics and coarse-grain dynamics; and, above all, (iii) stochastic modeling. Fine tuning of a limited set of molecular or mesoscopic parameters via limited fitting can still be employed. In particular, ESR measurements are becoming particularly amenable to an integrated approach, due to increasing experimental progress, advancement in computational methods, and refinement of available dynamical models.

Nitroxide-derived paramagnetic probes allow in principle the detection of several types of information at once: secondary-structure information, interresidual distances, if more than one spin probes is present, and large-amplitude protein motions from the overall ESR spectrum shape [15–19]. An ab initio interpretation of ESR spectroscopy needs to take into account different aspects regarding the structural, dynamical, and magnetic properties of the molecular system under investigation, and it requires, as input parameters, the known basic molecular information and solvent macroscopic parameters. The application of the stochastic Liouville equation formalism integrates the structural and dynamic ingredients to give directly the spectrum with minimal additional fitting procedures in the presence of internal dynamics, anisotropic environments, and so on [2, 20–27] Notice that alternative computational treatments of multifrequency ESR signals are nowadays emerging. In particular, standard molecular dynamics–based approaches [28] have been employed recently, and novel augmented treatments are being developed [29].

Properties of liquid crystals as order parameters, dynamics, and cage effects have been studied by several authors using ESR spectroscopy of dissolved spin probes and a stochastic Liouville equation (SLE)–based approach for interpretation. For instance, Sastry and co-workers [7, 8] studied two-dimensional Fourier transform (2D-FT) ESR of the rigid rodlike cholestane (CSL) spin label in the liquid crystal solvent butoxy benzylidene octylaniline (4O,8) and the small globular spin probe perdeuterated tempone (PDT) in the same solvent. Experimental spectra were collected over a wide range of temperatures in such a way as to include isotropic, nematic, smectic A and B, and crystal phases of 4O,8. 2D-FT-ESR was chosen because it provides greatly enhanced sensitivity to rotational dynamics over cw-ESR analysis. For both the CSL and PDT spin probes, experimental spectra were interpreted via the slowly relaxing local structure (SRLS) model [30] in which the dynamic of the system is described with two coupled relaxing processes which are interpreted as a fast global tumbling of the probe and a slow relaxation of the solvent cage collective motions. Zannoni and co-workers [31] used the ESR spin probe technique to study the changes in phase stability, orientational order, and dynamics of the nematic 5-cyanobiphenyl (5CB) doped with different cis/trans *p*-azobenzene derivatives. CSL was again adopted as the spin probe to monitor the order and the dynamics of the liquid crystal system, owing to its size, rigidity, and rodlike shape analogous to that of the 5CB [32–34]. Interpretation of the experimental spectra was carried out by simulations with the one-body model implementation by Freed [35] by assuming the probe as a rigid rotator that reorients under the action of a second-rank potential.

The theoretical approach to the interpretation of ESR spectra is based on the solution of the SLE. This is essentially a semiclassical approach based on the Liouville equation for the magnetic probability density of the molecule augmented by a stochastic operator which describes the relevant relaxation processes that occur in the system and is responsible for the broadening of the spectral lines [2]. The SLE approach can be linked profitably to density functional theory (DFT) evaluation of geometry and magnetic parameters of the radical in its environment. Dissipative parameters, such as rotational diffusion tensors, can in turn be determined at a coarse-grained level by using standard hydrodynamic arguments. The combination of the evaluation of structural properties, based on quantum mechanical advanced methods, with hydrodynamic modeling for dissipative properties and, in the case of multilabeled systems, determination of dipolar interaction based on the molecular structures beyond the point approximation are the fundamental ingredients needed by the SLE to provide a fully integrated computational approach (ICA) that gives the spectral profile. A number of parameters enter in the definition of the SLE and customarily a multicomponent fitting procedure is employed. ICA attempts to replace fitting procedures as much as possible with the ab initio evaluation of parameters in order to give them a sound physical interpretation, and fitting may be retained as a "refining" step. The calculation of ESR observables can in principle be based on the complete solution of the Schrödinger equation for the system made of paramagnetic probe + explicit solvent molecules. The system can be described by a "complete" Hamiltonian $\hat{H}(\mathbf{r}_i, \mathbf{R}_k, \mathbf{q}_\alpha)$, which can be written in the form

$$\hat{H}(\mathbf{r}_i, \mathbf{R}_k, \mathbf{q}_\alpha) = \hat{H}_{\mathrm{probe}}(\mathbf{r}_i, \mathbf{R}_k) + \hat{H}_{\mathrm{probe-solvent}}(\mathbf{r}_i, \mathbf{R}_k, \mathbf{q}_\alpha) + \hat{H}_{\mathrm{solvent}}(\mathbf{q}_\alpha) \quad (12.1)$$

where probe and solvent terms are separated. The Hamiltonian $\hat{H}(\mathbf{r}_i, \mathbf{R}_k, \mathbf{q}_\alpha)$ contains (i) electronic coordinates $\mathbf{r}_i$, of the paramagnetic probe (where index i runs over all probe electrons), (ii) nuclear coordinates $\mathbf{R}_k$ (where index k runs over all rotovibrational nuclear coordinates), and (iii) coordinates $\mathbf{q}_\alpha$, in which we include all degrees of freedom of the solvent molecules, each labeled by index α. The basic object of study, to which any spectroscopic observable can be linked, is given by the density matrix $\hat{\rho}(\mathbf{r}_i, \mathbf{R}_k, \mathbf{q}_\alpha)$, which in turn is obtained from the Liouville equation

$$\frac{\partial \hat{\rho}}{\partial t} = -\frac{i}{\hbar}\left[\hat{H}, \hat{\rho}\right] = -i\hat{\mathcal{L}}\hat{\rho} \quad (12.2)$$

Solving Eq. 12.2 in time—for instance, via an ab initio molecular dynamics scheme—allows in principle the direct evaluation of the density matrix and hence calculation of any molecular property [29]. However, significant approximations are possible which are basically rooted in time-scale separation. The nuclear coordinates $\mathbf{R} \equiv \mathbf{R}_k$ can be separated into fast-probe vibrational coordinates $\mathbf{R}_{\mathrm{fast}}$ from slow-probe coordinates, that is, rotational and intramolecular "soft" torsional degrees of freedom, $\mathbf{Q}$, relaxing at least in a picosecond time scale. Then the probe Hamiltonian is averaged on (i) femtosecond and subpicosecond dynamics pertaining to probe electronic coordinates and (ii) picosecond dynamics pertaining to probe internal vibrational degrees of freedom. The averaging over the electron coordinates is the usual implicit procedure for obtaining a spin Hamiltonian from the complete Hamiltonian of the radical. In the frame of a Born–Oppenheimer approximation, the averaging over the picosecond dynamics of nuclear coordinates allows one to introduce in the calculation of magnetic parameters the effect of vibrational motion. In this way a probe Hamiltonian is obtained characterized by magnetic tensors. By taking into account only the electron Zeeman and the hyperfine interactions, for a probe with one unpaired electron and N nuclei, we can define an averaged magnetic Hamiltonian $\hat{H}(\mathbf{Q}, \mathbf{q}_\alpha)$:

$$\hat{H}(\mathbf{Q}, \mathbf{q}_\alpha) = \frac{\beta_e}{\hbar}\mathbf{B}_0\mathbf{g}(\mathbf{Q}, \mathbf{q}_\alpha)\hat{\mathbf{S}} + \gamma_e \sum_n \hat{\mathbf{I}}_n\mathbf{A}_n(\mathbf{Q}, \mathbf{q}_\alpha)\hat{\mathbf{S}} + \hat{H}_{\text{probe-solvent}}(\mathbf{Q}, \mathbf{q}_\alpha) + \hat{H}_{\text{solvent}}(\mathbf{q}_\alpha) \tag{12.3}$$

The first term is the Zeeman interaction depending upon the $\mathbf{g}(\mathbf{Q}, \mathbf{q}_\alpha)$ tensor, external magnetic field $\mathbf{B}_0$, and electron spin operator $\hat{\mathbf{S}}$; the second term is the hyperfine interaction of the nth nucleus and the unpaired electron, defined with respect to hyperfine tensor $\mathbf{A}_n(\mathbf{Q}, \mathbf{q}_\alpha)$ and nuclear spin operator $\hat{\mathbf{I}}_n$. Additional terms are $\hat{H}_{\text{probe-solvent}}(\mathbf{Q}, \mathbf{q}_\alpha)$ to account for interactions between the probe and the medium which do not affect directly the magnetic properties (e.g., solvation energy) and $\hat{H}_{\text{solvent}}(\mathbf{q}_\alpha)$ for solvent-related terms. Here tensors $\mathbf{g}$, $\mathbf{A}_h$ are diagonal in local (molecular) frames GF, A_nF; operators $\hat{\mathbf{S}}$, $\hat{\mathbf{I}}_n$ are defined in the laboratory or inertial frame (LF). An explicit dependence is left in the magnetic tensor definition from slow-probe coordinates (e.g., geometric dependence upon rotation) and solvent coordinates. The averaged density matrix becomes $\hat{\rho}(\mathbf{Q}, \mathbf{q}_\alpha, t) = \langle\hat{\rho}(\mathbf{r}_i, \mathbf{R}_k, \mathbf{q}_\alpha)\rangle_{\mathbf{r}_i, \mathbf{R}_{\text{fast}}}$ and the corresponding Liouville equation, in the hypothesis of no residual dynamic effect of averaging with respect to subpicosecond processes, can be simply written as in Eq. 12.2 with $\hat{H}(\mathbf{Q}, \mathbf{q}_\alpha)$ instead of $\hat{H}(\mathbf{r}_i, \mathbf{R}_k, \mathbf{q}_\alpha)$. The next step, that is, the projection or "elimination" of solvent/bath coordinates to obtain an effective time evolution equation depending just on the relevant set of coordinates $\mathbf{Q}$, is not a trivial passage and in truth can be addressed only in terms of a semiphenomenological, albeit very effective, theoretical approach. In essence, one assumes that averaging the density matrix with respect to solvent variables is tantamount to (1) redefining the variables as a Markov stochastic process. A simplified modified time evolution equation for $\rho(\mathbf{Q}, t)$ is defined assuming that (2) the stochastic process is not affected by the system (absence of back-reaction) and therefore that an independent equation for the conditional probability $P(\mathbf{Q}, t)$ describing the stochastic process is given by $\partial P/\partial t = -\hat{\Gamma}P$, where $\hat{\Gamma}$ is the stochastic (Fokker–Planck or Smoluchowski) operator modeling the time evolution of the reduced density matrix on relaxing processes described by stochastic coordinates $\mathbf{Q}$, with an equilibrium solution $\hat{\Gamma}P_{\text{eq}}(\mathbf{Q}) = 0$. A time evolution equation for $\rho(\mathbf{Q}, t)$ is then defined according to the so-called stochastic Liouville equation (SLE) formalism by the direct inclusion of $\hat{\Gamma}$ in the (effective) Liouville equation [2]

$$\frac{\partial\hat{\rho}}{\partial t} = -\frac{i}{\hbar}\left[\hat{H}(\mathbf{Q}), \hat{\rho}(\mathbf{Q}, t)\right] - \hat{\Gamma}\hat{\rho}(\mathbf{Q}, t) = -i\hat{\mathcal{L}}\hat{\rho} \tag{12.4}$$

where the reduced Liouvillian is defined with respect to the effective Hamiltonian

$$\hat{H}(\mathbf{Q}) = \frac{\beta_e}{\hbar}\mathbf{B}_0\mathbf{g}(\mathbf{Q})\hat{\mathbf{S}} + \gamma_e \sum_n \hat{\mathbf{I}}_n\mathbf{A}_n(\mathbf{Q})\hat{\mathbf{S}} \tag{12.5}$$

and $\mathbf{g}(\mathbf{Q})$, $\mathrm{A}_n(\mathbf{Q})$ are now averaged tensors with respect to all solvent coordinates. The inclusion of relevant variables within a phenomenological semiclassical time evolution equation for the reduced density matrix operator of a molecular system is at the basis of the SLE, originally proposed by Kubo [36, 37] to describe the dynamics of a quantum system perturbed by a Markovian stochastic process. Formal justification of the SLE has been proposed by several authors and is reviewd by Schneider and Freed [2], and it should be clear that in the absence of a coherent theory of stochastic quantum systems, it remains a phenomenological ansatz (but see, e.g., Wassam and Freed [38, 39]). A comprehensive review of recent theoretical development of the SLE formalism is given, for instance, by Tanimura [40]. Here we point out that this is a general scheme which allows for additional considerations and further approximations. First, the average with respect to picosecond dynamic processes is carried out, in practice, together with averaging with respect to solvent coordinates to allow the QM evaluation of magnetic tensors corrected for solvent effects. Second, time separation techniques can also be applied to treat approximately relatively faster relaxing coordinates included in the relevant set $\mathbf{Q}$, such as restricted (local) torsional motions. Third, complex solvent environments (e.g., highly viscous fluids) can be described by an augmented set of stochastic coordinates, to be included in $\mathbf{Q}$, which describes slow-relaxing local solvent structures, or in other words to maintain the generalized Markovian nature of $\mathbf{Q}$.

12.2.2 Setting Up the SLE

From the spin Hamiltonian it is clear that a number of parameters are required, that is, the $\boldsymbol{g}$ tensors of the unpaired electron and the $\boldsymbol{A}$ hyperfine coupling tensors for all nuclei. All these quantities are purely quantum mechanical properties and their evaluation can be carried out via a first-principles treatment (see below). The choice of the stochastic operator, $\hat{\Gamma}$, is a basic step in the methodology. Here we comment on two canonical cases frequently occurring in standard applications: (i) rigid-body model, where the probe is seen as a rigid rotator diffusing and the stochastic variables are $\mathbf{Q} = \Omega$, the set of Euler angles which give the relative orientation of the molecule with respect to the inertial laboratory frame; (ii) "flexible"-body model, where the molecule is described as a rotator with one internal degree of freedom represented by a torsional angle, so the stochastic variables, $\mathbf{Q} = (\Omega, \theta)$, are the set of angles Ω (for the global rotation) and the torsional angle θ. In both models the stochastic variables are considered as diffusive processes and the stochastic operator has the general form

$$\hat{\Gamma} = -\hat{\nabla}_{\mathrm{Q}}^{\mathrm{tr}} \boldsymbol{D}(\mathbf{Q}) P_{\mathrm{eq}}(\mathbf{Q}) \hat{\nabla}_{\mathrm{Q}} P_{\mathrm{eq}}^{-1}(\mathbf{Q}) \tag{12.6}$$

where $\hat{\nabla}_{\mathrm{Q}}$ is the vector operator of partial derivatives over the stochastic variables, $\boldsymbol{D}(\mathbf{Q})$ is the diffusion tensor of the system (which in general may depend on the stochastic variables), and $P_{\mathrm{eq}}(\boldsymbol{Q})$ is the Boltzmann equilibrium distribution probability

$$P_{\mathrm{eq}}(\mathbf{Q}) = \frac{\exp[-V(\mathbf{Q})/kT]}{\langle \exp[-V(\mathbf{Q}) - kT] \rangle} \tag{12.7}$$

Here, $V(Q)$ is the potential acting on the stochastic coordinates and $\langle \cdot \rangle$ represents the integration over **Q**. Assumptions can be made by requiring that the potential has separated contributions, for example, an "external" term acting on the global orientation (e.g., ordering effects in liquid crystals) and an internal term acting on the torsional angle (if present) which is the torsional potential, that is,

$$V(\mathbf{Q}) = V_{\text{ext}}(\Omega) + V_{\text{int}}(\theta) + V_{\text{coupling}}(\mathbf{Q}) \approx V_{\text{ext}}(\Omega) + V_{\text{int}}(\theta) \tag{12.8}$$

Mesoscopic parameters, such as the full-diffusion tensor and potential V, are usually determined phenomenologically or from complementary approaches. For instance, dissipative properties described by the diffusion tensor can be obtained on the basis of hydrodynamic modeling (see below). The internal potential can be evaluated as a potential energy surface scan over the torsional angle θ. For small molecules this operation can be easily conducted at the DFT level, while for big molecules such as proteins, mixed quantum mechanical/molecular mechanics (QM/MM) methodologies can be employed.

12.2.3 Magnetic Tensors

The introduction of the DFT is a turning point for the calculations of the spin Hamiltonian parameters [41–44]. Before DFT, ab initio calculations of the magnetic parameters of spin Hamiltonians were either prohibitively expensive already even for medium-size radicals [45–47] or less reliable than semiempirical methods. These latter were based on the approaches introduced by McConnell [48, 49] and Stone [50] for the calculations of the hyperfine coupling and the $\boldsymbol{g}$ tensors, respectively. Based on semiempirical parameters taking into account separately the spin density on the singly occupied molecular orbital (SOMO) and that due to spin polarization [51], the method for the evaluation of hyperfine tensors has been an invaluable tool for understanding the correlation between the magnetic parameters of the spin Hamiltonian, the spin distribution, the conformation of radicals, and the molecular properties in general. However, the reliability of the method was very restricted, as being limited to predictions within groups of similar radicals for which the same set of semiempirical parameters were sound, and the parameters to be calculated were only the SOMO spin densities [51]. Within these limits the calculated hyperfine tensors were quite reliable. On the other hand, the agreement between calculated and experimental values for $\boldsymbol{g}$ tensors were in general much worse. To this end, it should be noted that the recently improved methods of calculating reliable $\boldsymbol{g}$ tensors by DFT on the one hand [52–55] and to measure them by high-frequency ESR on the other has provided a new largely unexplored source of information on the molecular properties attainable by ESR analysis.

Today, the agreement between experimental and calculated parameters of the spin Hamiltonian by DFT is outstanding [41–44, 52, 56]. Both the vibrational averaging of the parameters [57–59] and the interactions of the probe with the environment [60–65] are taken into account, thus providing a set of tailored parameters that can be used confidently for further calculations. It should be noted that this approach is a step forward with respect to the traditional starting point, that is, the use of a set of experimental hyperfine and $\boldsymbol{g}$ tensors generally obtained for a different system and extrapolated to the case of interest.

The $\boldsymbol{g}$ tensor can be dissected into three main contributions [52–56],

$$\boldsymbol{g} = g_e \mathbf{1}_3 + \Delta \boldsymbol{g}^{\mathrm{RMC}} + \Delta \boldsymbol{g}^{\mathrm{GC}} + \Delta \boldsymbol{g}^{\mathrm{OZ/SOC}} \tag{12.9}$$

where g_e is the free-electron value ($g_e = 2.002319$) and $\mathbf{1}_3$ is the 3 unit matrix; Δg^{RMC} and Δg^{GC} are first-order contributions which take into account relativistic mass (RMC) and gauge (GC) corrections, respectively. The last term, $\Delta g^{\mathrm{OZ/SOC}}$, is a second-order contribution arising from the coupling of the orbital Zeeman (OZ) and the spin–orbit coupling (SOC) operators. The SOC term is a true two-electron operator, but here it will be approximated by a one-electron operator involving adjusted effective nuclear charges [66]. This approximation has been proven to work fairly well in the case of light atoms providing results close to those obtained using more refined expressions for the SOC operator [52–54]. In our general procedure, spin-unrestricted calculations provide the zero-order Kohn–Sham (KS) orbitals and the magnetic field dependence is taken into account using the coupled-perturbed KS formalism described by Neese but including the gauge including/invariant atomic orbital (GIAO) approach [52–54]. Solution of the coupled-perturbed KS equation (CP-KS) leads to the determination of the OZ/SOC contribution.

The second term is the hyperfine interaction contribution, which in turn contains the so-called Fermi contact interaction (an isotropic term), is related to the spin density at the corresponding nucleus n by [67]

$$A_{n,0} = \frac{8\pi}{3} \frac{g_e}{g_0} g_n \beta_n \sum_{\mu,\nu} P_{\mu,\nu}^{\alpha-\beta} \langle \varphi_\mu | \delta(r_{kn}) | \varphi_\nu \rangle \tag{12.10}$$

and an anisotropic contribution which can be derived from the classical expression of interacting dipoles [68],

$$A_{n,ij} = \frac{g_e}{g_0} g_n \beta_n \sum_{\mu,\nu} P_{\mu,\nu}^{\alpha-\beta} \langle \varphi_\mu | r_{kn}^{-5} (r_{kn}^2 \delta_{ij} - 3 r_{kn,i} r_{kn,j}) | \varphi_\nu \rangle \tag{12.11}$$

The $\boldsymbol{A}$ tensor components are usually given in gauss (1 G = 0.1 mT); to convert data to megahertz one has to multiply by 2.8025.

Magnetic tensors evaluated at this level do not give sufficiently accurate estimates of experimental values, especially if one considers a molecule in a solvent with high polarity and/or a solvent that can form hydrogen bonds. Environmental effects (e.g., solvent) need to be taken into account and the most promising general approach to the problem can be based on a system–bath decomposition. Calculations can be performed on the system, including the part of the solute where the essential part of the process to be investigated is localized together with, possibly, the few solvent molecules strongly and specifically interacting with it. This part is treated at the electronic level of resolution and is immersed in a polarizable continuum, mimicking the macroscopic properties of the solvent. So, the solution process can then be dissected into the creation of a cavity in the solute process requiring an energy E_{cav}, and the successive switching on of dispersion/repulsion, with energy $E_{dis\text{-}rep}$, and electrostatic, with energy E_{el}, interactions with surrounding solvent molecules. All of these contributionsm, for both isotropic and anisotropic solutions, are included into the so-called polarizable continuum model (PCM) [69–72]. Taking into account solvent effects gives the corrections required in order to predict values of the tensors very close to the experimental ones (see Tables 2 and 7 of ref. 73).

While in some cases considering the environment is sufficient to reproduce experimental values of the **g** and hyperfine tensors, there are molecules presenting fast motions in the neighborhood of the unpaired electron. Dependence of the magnetic parameters on these small geometric variations can be very significant [57, 74–76]. These motions are usually too fast with respect to the ESR time scale window so the effective contribution is a correction that can be calculated as an average over short-time dynamics calculated at a QM level [77, 78].

12.2.4 Friction and Diffusion Tensors

We review in this section a coarse-grained (hydrodynamic-based) recipe for evaluation of friction and diffusion tensors of flexible molecular systems. Let us consider a molecule made of N_A atoms which has been partitioned into N_P fragments. The ith fragment is composed of N_i atoms and its orientation relative to the $(i+1)$ th fragment is defined by the torsional angle θ_i. We limit our discussion to noncyclic molecules, so that a generic molecular system is considered in general as a sequence of N_F fragments, and the total number of torsional angles is $N_T = N_F - 1$. Notice that $\sum_{i=1}^{N_F} N_t = N_A$.

We define a molecular frame (MF) fixed on a chosen fragment v (hereafter referred to as the *main fragment*) which is placed for convenience in the center of mass of the main fragment itself (see Figures 12.1 and 12.2). The atoms in the main fragment are characterized by translational and rotational motions, while atoms belonging to the other fragments have also additional internal motions. We define the set of generalized coordinates $\mathcal{R} = [\boldsymbol{r}, \Omega, \dot{\boldsymbol{\theta}}]$ for describing the translational and rotational coordinates of the main fragment and internal torsional motions. Associated with $\mathcal{R}$ is the set of velocities $\mathcal{V} = [\boldsymbol{v}, \boldsymbol{\omega}, \dot{\boldsymbol{\theta}}]$ (where the dot stands for time derivative) and also the total force consists of three contributions $\mathcal{F} = [\boldsymbol{f}, \boldsymbol{\tau}, \boldsymbol{\tau}_i]$ corresponding, respectively, to the translational force and the global torque and internal torque moments. Forces and velocities are related by the friction tensor ξ, which is defined as a $(6 + N_T) \times (6 + N_T)$ matrix

$$\begin{pmatrix} \boldsymbol{f} \\ \boldsymbol{\tau} \\ \boldsymbol{\tau}_i \end{pmatrix} = -\xi \begin{pmatrix} \upsilon \\ \omega \\ \dot{\theta} \end{pmatrix} \tag{12.12}$$

or simply $\mathcal{F} = -\xi\mathcal{V}$. If one considers the system without constraints (bonds), that is, the position of each atom is independent of the positions of the other atoms, the friction tensor Ξ, of the N_A independent atoms is represented as a $3N_A \times 3N_A$ matrix.

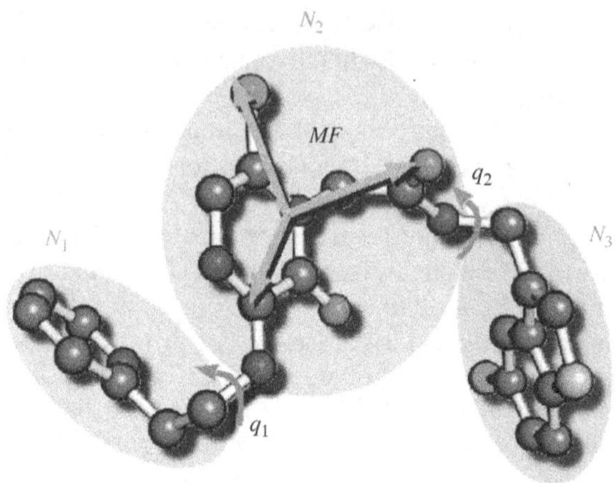

Figure 12.1 Partitioning of generic molecule into linear chain of three fragments and two torsional angles; MF is set on second fragment.

If $\boldsymbol{F}_i$ and $\boldsymbol{V}_i$ are, respectively, the translational force and velocity of the ith atom, we can write

$$\begin{pmatrix} \boldsymbol{F}_1 \\ \vdots \\ \boldsymbol{F}_{N_A} \end{pmatrix} = -\Xi \begin{pmatrix} \boldsymbol{V}_1 \\ \vdots \\ \boldsymbol{V}_{N_A} \end{pmatrix} \tag{12.13}$$

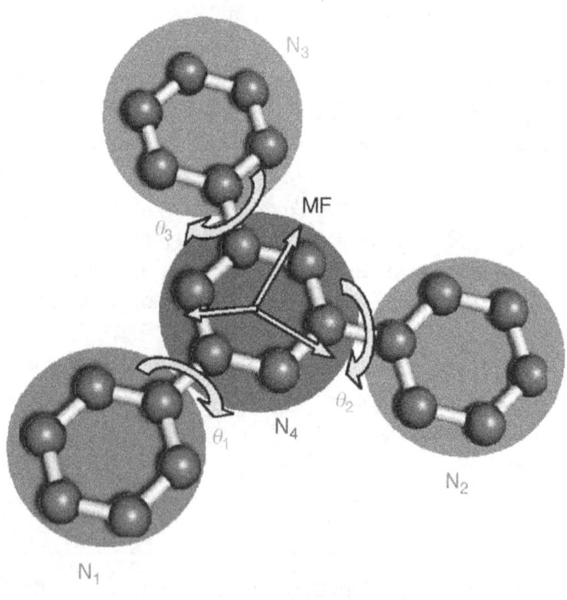

Figure 12.2 Partitioning of generic molecule in branched chain of four fragments and three torsional angles; MF is set on central fragment.

or $\boldsymbol{F} = -\Xi \boldsymbol{V}$. Following standard geometric arguments [79], one can show that $\mathcal{F} = \boldsymbol{A}\boldsymbol{F}$ and $\boldsymbol{V} = \boldsymbol{B}\mathcal{V}$, where $\boldsymbol{A}$ and $\boldsymbol{B}$ are $(6 + N_T) \times 3N_A$ and $3N_A \times (6 + N_T)$ matrices which depend on the molecular geometry; additionally, $\boldsymbol{B} = A^{\text{tr}}$. It follows that

$$\xi = \boldsymbol{B}^{\text{tr}} \Xi \boldsymbol{B} = \begin{pmatrix} \xi_{TT} & \xi_{TR} & \xi_{TI} \\ \xi_{RT} & \xi_{RR} & \xi_{RI} \\ \xi_{IT} & \xi_{IR} & \xi_{II} \end{pmatrix} \tag{12.14}$$

where the subscripts stand for $T =$ translational, $R =$ rotational, and $I =$ internal. The diffusion tensor is obtained from Einstein relation as the inverse of ξ,

$$\boldsymbol{D} = k_B T \xi^{-1} = \begin{pmatrix} \boldsymbol{D}_{TT} & \boldsymbol{D}_{TR} & \boldsymbol{D}_{TI} \\ \boldsymbol{D}_{RT} & \boldsymbol{D}_{RR} & \boldsymbol{D}_{RI} \\ \boldsymbol{D}_{IT} & \boldsymbol{D}_{IR} & \boldsymbol{D}_{II} \end{pmatrix} \tag{12.15}$$

where k_{B} is the Boltzmann constant and T the absolute temperature. The friction tensors are linked to the diffusion tensors $\boldsymbol{D}$ (constrained spheres) and $\boldsymbol{d}$ (unconstrained spheres) via the generalized Einstein relations $\boldsymbol{D} = k_B T \xi^{-1}$ and $d = k_B T\ \Xi^{-1}$. It follows that the molecular diffusion tensor for the joint translation, rotation, and internal conformational motion for the molecule, that is, D, is obtained as

$$\boldsymbol{D} = \left(\boldsymbol{B}^{\text{tr}} \boldsymbol{d}^{-1} \boldsymbol{B}\right)^{-1} \tag{12.16}$$

The main ingredients for the calculation of the diffusion tensor are the geometric matrix $\boldsymbol{B}$ and the unconstrained diffusion tensor d. Let us first consider the calculation of the geometric matrix. We define r_j^i as the vector of the ith atom in the jth fragment, u_n as the unitary vector defining the rotation θ_n, taken to be parallel to the nth torsional angle and pointing away from the main fragment, and $r_{j,k}^i$ the distance vector between the jth atom of the ith fragment and the atom at the origin of the unitary vector u_k. Atoms in the main fragment are characterized only by the translational and global rotational velocity

$$\upsilon_j^{\nu} = \upsilon + \omega \times r_j^{\nu} \tag{12.17}$$

while for the remaining fragments ($i \neq \nu$) the torsional contributions must be included,

$$\upsilon_j^i = \upsilon + \omega \times r_j^i + \sum_k \dot{\theta}_k u_k \times r_{j,k}^i \tag{12.18}$$

where the summation is taken over the angles that link the main fragment to the ith fragment. Equations 12.17 and 12.18 can be rewritten in matrix form as

$$\upsilon_{ij}^i = {}^T\boldsymbol{B}^i \upsilon + {}^R B_j^i \omega + \sum_k {}^I B_{j,k}^i \dot{\theta}_k \tag{12.19}$$

where ${}^{T}\boldsymbol{B}_{j}^{i} = 1_3$, ${}^{R}\boldsymbol{B}_{j}^{i} = \boldsymbol{r}_{j}^{i\times}$, and ${}^{I}\boldsymbol{B}_{j,k}^{i} = \boldsymbol{r}_{j,k}^{i\times}\boldsymbol{u}_k$ or 0 depending on k and i and $r_{\alpha\beta}^{\times} = r_k \xi_{\alpha\beta\gamma}$, where $\xi_{\alpha\beta\gamma}$ is the Levi-Civita tensor with α, β, $\gamma = 1,2,3$. For a linear chain of fragments, numbered sequentially from the first to the last one, the general form of the $\boldsymbol{B}$ matrix is

$$B = \begin{pmatrix} 1_3 & \boldsymbol{r}_j^{1\times} & \boldsymbol{r}_{j,1}^{1\times}\boldsymbol{u}_1 & \cdots & \boldsymbol{r}_{j,\nu-1}^{1\times}\boldsymbol{u}_{\nu-1} & 0 & 0 & \cdots & 0 \\ 1_3 & \boldsymbol{r}_j^{2\times} & 0 & \cdots & \boldsymbol{r}_{j,\nu-1}^{2\times}\boldsymbol{u}_{\nu-1} & 0 & 0 & \cdots & 0 \\ \vdots & \vdots & \vdots & \vdots & \vdots & \vdots & \vdots & \vdots & \vdots \\ 1_3 & \boldsymbol{r}_j^{\nu-1\times} & 0 & \cdots & \boldsymbol{r}_{j,\nu-1}^{\nu-1\times}\boldsymbol{u}_{\nu-1} & 0 & 0 & \cdots & 0 \\ 1_3 & \boldsymbol{r}_j^{\nu\times} & 0 & \cdots & 0 & 0 & 0 & \cdots & 0 \\ 1_3 & \boldsymbol{r}_j^{\nu+1\times} & 0 & \cdots & 0 & \boldsymbol{r}_{j,\nu}^{\nu+1\times}\boldsymbol{u}_\nu & 0 & \cdots & 0 \\ \vdots & \vdots & \vdots & \vdots & \vdots & \vdots & \vdots & \vdots & \vdots \\ 1_3 & \boldsymbol{r}_j^{N_F-1\times} & 0 & \cdots & 0 & \boldsymbol{r}_{j,\nu}^{N_F-1\times}\boldsymbol{u}_\nu & \boldsymbol{r}_{j,\nu+1}^{N_F-1\times}\boldsymbol{u}_{\nu+1} & \cdots & 0 \\ 1_3 & \boldsymbol{r}_j^{N_F\times} & 0 & \cdots & 0 & \boldsymbol{r}_{j,\nu}^{N_F\times}\boldsymbol{u}_\nu & \boldsymbol{r}_{j,\nu+1}^{N_F\times}\boldsymbol{u}_{\nu+1} & \cdots & \boldsymbol{r}_{j,N_F-1}^{N_F\times}\boldsymbol{u}_{N_F-1} \end{pmatrix} \tag{12.20}$$

The form of the geometric matrix $\boldsymbol{B}$ is dependent on the topology and also on the numbering scheme chosen for the fragments.

Evaluation of $\boldsymbol{d}$ can be carried out at the simplest possible level assuming the model of noninteracting spheres in a fluid, or one can include hydrodynamic interactions, for example, based on the Rotne–Prager (RP) approach [80, 81], which ensures a satisfactory albeit not too cumbersome treatment of sphere–sphere hydrodynamic interactions. The resulting elements of $\boldsymbol{D}$ depend upon a purely geometric tensorial component $\overline{\boldsymbol{D}}$ and the translational diffusion coefficient for an isolated sphere D_0, that is,

$$\boldsymbol{D} = D_0\overline{\boldsymbol{D}} \tag{12.21}$$

where $D_0 = k_BT/CR_e\eta\pi = k_BT/\Xi_0$: here C is a constant depending upon hydrodynamic boundary conditions, R_e is the average radius for the spheres, and η is the local viscosity. The RP unconstrained diffusion tensor is given as

$$d_{ii} = \frac{k_BT}{\Xi_0} 1_3$$

$$= \begin{cases} \dfrac{k_BT}{\Xi_0}\dfrac{3R_e}{4r_{ij}^3}\left[\left(r_{ij}^2 + \dfrac{2}{3}R_e^2\right)1_3 + \left(1 - 2\dfrac{R_e^2}{r_{ij}^2}\right) r_{ij} \otimes r_{ij}\right] & \text{if } r_{ij} > 2R_e \\ \dfrac{k_BT}{\Xi_0}\left[\left(1 - \dfrac{9}{32}\dfrac{r_{ij}}{R_e}\right)1_3 + \dfrac{3}{32}\dfrac{r_{ij} \otimes r_{ij}}{r_{ij}^2}\right] & \text{if } r_{ij} < 2R_e \end{cases} \tag{12.22}$$

where i and j are two generic atoms, $\boldsymbol{r}_{ij} = \boldsymbol{r}_i - \boldsymbol{r}_j$, and $\otimes$ indicates the dyadic product. Notice that the general methodology reported above can be applied with minor changes to other types of internal motions, such as stretching of bonds, bending of bond angles, domain and loop motions.

12.2.5 Solving the SLE

Once magnetic, structural, and dissipative parameters have been estimated, the SLE is completely defined. At this point, physical properties can be calculated, with the knowledge of $\hat{\Gamma}$ and P_{eq}, either directly from the conditional probability $P(\boldsymbol{X}, t)$ or in terms of time correlation functions, which are defined, for two correlated observables $f(\mathbf{Q}, t)$ and $g(\mathbf{Q}, t)$, as

$$G(t) = \langle f(\mathbf{Q}, t) | \exp(-\hat{\Gamma} t) | g(\mathbf{Q}, t) P_{eq}(\mathbf{Q}) \rangle \tag{12.23}$$

from which it is possible to calculate the spectral density, that is, the Fourier–Laplace transform of G(t), as

$$J(\omega) = \frac{1}{\pi} \int_0^\infty d\omega \, G(t) e^{-i\omega t} = \frac{1}{\pi} \langle f(Q, t) | \left(i\omega + \hat{\Gamma} \right)^{-1} | g(\mathbf{Q}, t) P_{eq}(\mathbf{Q}) \rangle \tag{12.24}$$

The formalism for evaluating cw-ESR spectra is now easily interpreted in terms of spectral densities. In the SLE framework, the stochastic operator $\hat{\Gamma}$ is part of the generic stochastic Liouvillian $\hat{\mathcal{L}}$ and the cw-ESR spectrum is given by

$$I(\omega - \omega_0) = \frac{1}{\pi} \mathcal{R}\mathrm{e} \left\{ \left\langle \left\langle \upsilon \middle| \left[i(\omega - \omega_0) + \hat{\mathcal{L}} \right]^{-1} \middle| \upsilon P_{eq} \right\rangle \right\rangle \right\} \tag{12.25}$$

that is, as the real part of the spectral density for the autocorrelation function for the observable, usually called the *starting vector*, corresponding to the X component of the magnetization as well as P_{eq}.

It is convenient to transform the Liouvillian with the symmetrization

$$\tilde{\mathcal{L}} = P_{eq}^{-1/2} \hat{\mathcal{L}} P_{eq}^{1/2} = i\hat{\mathcal{H}}^{\times} + P_{eq}^{-1/2} \hat{\Gamma} P_{eq}^{1/2} = i\hat{\mathcal{H}}^{\times} + \tilde{\Gamma} \tag{12.26}$$

where the density matrix is $\tilde{\rho}(\boldsymbol{Q}, t) = \rho(\boldsymbol{Q}, t) / \rho_{eq}(\boldsymbol{Q})$ and the equilibrium probability density is $\tilde{P}_{eq} = P_{eq}^{1/2}$. The spectral density becomes

$$I(\omega - \omega_0) = \frac{1}{\pi} \mathcal{R}\mathrm{e} \left\{ \left\langle \left\langle \upsilon P_{eq}^{1/2} \middle| \left[i(\omega - \omega_0) + i\hat{\mathcal{H}}^{\times} + \tilde{\Gamma} \right]^{-1} \middle| \upsilon P_{eq}^{1/2} \right\rangle \right\rangle \right\} \tag{12.27}$$

The definition of the starting vector depends on the radical studied. Consider as an example the case of a monoradical in which the unpaired electron is coupled to a nucleus of spin I: The starting vector takes the form

$$\left| \upsilon P_{eq}^{1/2} \right\rangle \rangle = (2I + 1)^{-1/2} \left| \hat{S}_X \times 1_I \times P_{eq}^{1/2} \right\rangle \rangle \tag{12.28}$$

The cw-ESR spectrum is obtained by numerically evaluating the spectral density defined in Eq. 12.27, and here we adopt the standard methodology of spanning the Liouvillian over a proper basis set defined by the direct product

$$\left|\Sigma\right\rangle\!\!\right\rangle = \left|\sigma\right\rangle\!\!\right\rangle \otimes \left|\Lambda\right\rangle \tag{12.29}$$

The basis set for the spin coordinates, $|\sigma\rangle\rangle$ is the space of spin transitions and is defined elsewhere [2, 9, 82]. For the stochastic part we make the standard choice of employing Wigner rotation matrices for the global rotation and complex exponentials for the internal torsional angle, that is, $|\Lambda\rangle = |LMK\rangle \otimes |n\rangle$ with,

$$|LMK\rangle = \sqrt{\frac{2L+1}{8\pi^2}} D^L_{Mk}(\Omega) \tag{12.30}$$

$$|n\rangle = \frac{1}{\sqrt{2\pi}} e^{-in\theta} \tag{12.31}$$

To obtain the spectral density, usually iterative algorithms such as Lanczos [83, 84] or conjugate gradients [2] are employed. In particular, we make use of the Lanczos algorithm, a recursive procedure to generate orthonormal functions which allow a tridiagonal matrix representation of the system Liouvillian. Assuming as a first function the normalized zero-average observable, $|1\rangle\rangle = |\upsilon P_{\mathrm{eq}}^{1/2}\rangle\rangle/\langle\langle\upsilon|P_{\mathrm{eq}}|\upsilon\rangle\rangle 1/2$, the following functions are obtained recursively:

$$\beta_{n+1}|n+1\rangle\rangle = (\tilde{\mathcal{L}} - \alpha_n)|n\rangle\rangle - \beta_n|n-1\rangle\rangle \tag{12.32}$$

$$\alpha_n = \langle\langle n|\tilde{\mathcal{L}}|n\rangle\rangle \tag{12.33}$$

$$\beta_n = \langle\langle n|\tilde{\mathcal{L}}|n-1\rangle\rangle \tag{12.34}$$

Coefficients α_n and β_n actually form the first and second diagonal of the tridiagonal (complex) symmetric matrix representation of the symmetrized Liouvillian, and the spectrum can be written in the form of a continued fraction [84]

$$I(\omega) = \cfrac{1}{i\omega - \alpha_1 - \cfrac{\beta_2^2}{i\omega - \alpha_2 - \cfrac{\beta_3^2}{i\omega - \alpha_3 - \cdots}}} \tag{12.35}$$

Evaluation of Eqs. 12.32–12.34 is carried on in finite arithmetic by projecting the symmetrized Liouvillian and the starting vector on the basis set 12.29, defining the matrix operator and starting vector elements

$$\mathcal{L} = \left\langle\!\!\left\langle \Sigma \right| \tilde{\mathcal{L}} \left| \Sigma' \right\rangle\!\!\right\rangle \tag{12.36}$$

$$\upsilon = \left\langle\!\!\left\langle \Sigma \right| 1 \right\rangle\!\!\right\rangle \tag{12.37}$$

so that the matrix–vector counterparts of Eqs. 12.32–12.34 become

$$\beta_{n+1}\mathbf{v}_{n+1} = (\mathcal{L} - \alpha_n)\mathbf{v}_n - \beta_n\mathbf{v}_{n-1} \tag{12.38}$$

$$\alpha_n = \mathbf{v}_n \cdot \mathcal{L} \cdot \mathbf{v}_n \tag{12.39}$$

$$\beta_n = \mathbf{v}_n \cdot \mathcal{L} \cdot \mathbf{v}_{n-1} \tag{12.40}$$

Symmetry arguments can be employed to significantly reduce the number of basis function sets required to achieve convergence, together with numerical selection of a reduced basis set of functions based on "pruning" of basis elements with negligible contributions to the spectrum [2]. New strategies for reducing matrix dimensions in densely coupled spin systems are being investigated [85].

12.2.6 Case Study: Interpretation of cw-ESR Spectra of Tempo-Palmitate in 5CB

In the following we perform a complete a priori simulation of the ESR spectra of the prototypical nitroxide probe 4-(hexadecanoyloxy)-2,2,6,6-tetramethylpiperidine-1-oxy (usually referred to as Tempo-palmitate, TP) in isotropic and nematic phases of 5CB, for which detailed cw-ESR data are available in the literature [86]. The system is described as a flexible body reorienting under the influence of an external field, which favors its orientation along the nematic director, which is assumed parallel to the external magnetic field along the Z axis of the inertial laboratory frame (LF). We shall adopt a number of simplifying hypotheses aimed at keeping the required computational effort at a reasonable level. The molecule is considered as split into two fragments, the alkyl chain and the paramagnetic probe (Tempo) (Figure 12.3). Geometry and dynamics are described by two stochastic variables: (i) the set of Euler angles (Ω) which describes the orientation between the LF and a molecule fixed frame (MF) and (ii) an internal angle (θ) which defines the relative orientation between the Tempo fragment and the alkyl chain.

Structural properties were obtained by means of quantum mechanical calculations performed to find the minimum energy geometry of the molecule, evaluate the magnetic tensors, and calculate the internal potential [44]. On the basis of a previous study [87], the alkyl chain of TP was replaced by an ethyl group. Internal torsional potentials and magnetic tensors were then evaluated by the PBE0 hybrid functional [88] and the purposely tailored N06 basis set. Solvent effects were taken into account by our anisotropic version of the polarizable continuum model [87].

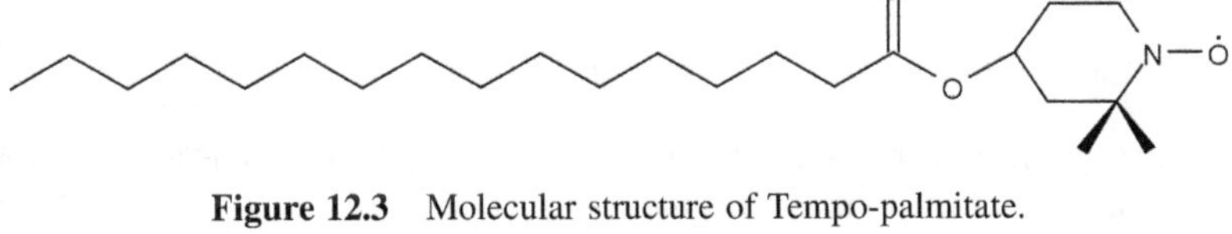

Figure 12.3 Molecular structure of Tempo-palmitate.

Of course, the diffusion tensor was evaluated for the true TP radical using the geometry optimized for the all-trans conformer.

The MF is fixed on the alkyl chain, which is considered as a rigid entity in the all-trans conformation; the MF is chosen in such a way that the rotational part of the diffusion tensor (see below) is diagonal. Magnetic tensors are diagonal in the same reference frame (μF) fixed on paramagnetic probe. The total potential energy of the system is defined according to Eq. 12.8, that is, we neglect potential coupling terms $\boldsymbol{V}_{\mathrm{coup}}(\Omega, \theta)$ between internal (θ) and external (Ω) variables (Figure 12.5). The external potential is chosen according to the simple Maier–Saupe form [89–91]

$$U_{\mathrm{ext}} = \frac{V_{\mathrm{ext}}}{k_B T} = -\epsilon \mathcal{D}^2_{0,0}(\Omega) \tag{12.41}$$

This is the simplest potential which assures the presence of an energy minimum when the alkyl chain is parallel to the nematic director.

An accurate evaluation of the internal deg potential is obtained directly by QM calculations. An energetic barrier is observed corresponding to $\theta = 180°$. In general, we may define the potential via the expansion $V_{\mathrm{int}}/k_B T = -\sum \chi_n e^{-in\theta}$, where $\chi_n = \chi^*_{-n}$ ensure that the potential is real. In practice terms up to $n = 1$ have been retained to fit the potential to the shifted cosine form

$$U_{\mathrm{int}} = \frac{V_{\mathrm{int}}}{k_B T} \approx A(1 - \cos\theta) \tag{12.42}$$

To summarize, energetics is defined by the simplified expression

$$U = U_{\mathrm{ext}} + U_{\mathrm{int}} = -\epsilon \mathcal{D}^2_{0,0}(\Omega) + A(1 - \cos\theta) \tag{12.43}$$

defined by parameters ϵ and A.

In the case under investigations, which includes nematic (anisotropic) phase environments, we shall assume the usual approximation of considering isotropic local friction, and the macroscopic local viscosity is taken equal to half of the fourth Leslie–Ericksen coefficient η_4 [92–95]. The diffusion tensor of the system is obtained, neglecting translational contributions, as a 4×4 matrix, that is,

$$\boldsymbol{D} = \begin{pmatrix} \boldsymbol{D}_{RR} & \boldsymbol{D}_{RI} \\ \boldsymbol{D}_{IR} & \boldsymbol{D}_{II} \end{pmatrix} \tag{12.44}$$

where the 3×3 $\mathbf{D}_{RR}$ block is the purely rotational contribution, the $\boldsymbol{D}_{IR} = \boldsymbol{D}^{\mathrm{tr}}_{RI}$ blocks describe the rotoconformational interaction, and $\boldsymbol{D}_{II}$ is the conformational diffusion coefficient. The general outcome of the elements of the 4×4 diffusion tensor shows, as expected, a weak dependence upon the internal angles. We express the tensor as

$$\boldsymbol{D}(T) = D(T)\overline{\boldsymbol{d}} \tag{12.45}$$

in order to separate the purely geometric tensorial component $\overline{\boldsymbol{d}}$ and the translational diffusion coefficient for an isolated sphere $D(T)$, that is, $D(T) = k_B T/CR\pi\eta(T)$: Here

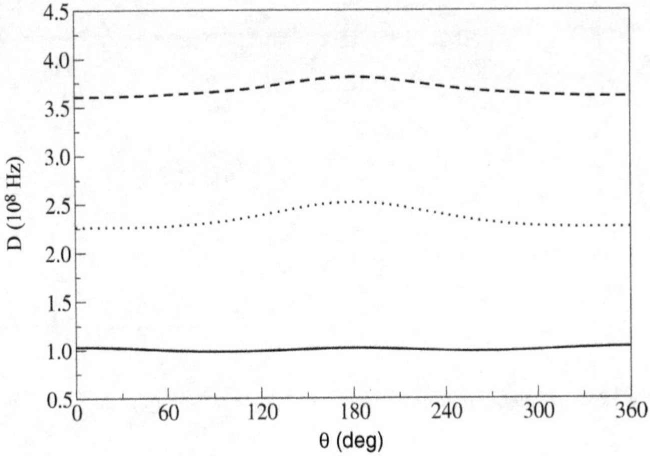

Figure 12.4 Values of $\mathrm{Tr}\,\boldsymbol{D}_{RR} \times 10^{-7}$ s (full line), $|D_{RI}| \times 10^{-7}$ s (dashed line), and $D_{II} \times 10^{-7}$ s (dotted line) for $T = 316.09$ K plotted vs. conformation angle θ.

C is a constant depending on hydrodynamic boundary conditions, R is the average radius for the spheres, and η is the local viscosity.

Selected tensor functions of the diffusion tensor, namely $Tr\{D_{RR}\}$, $|D_{RI}|$, and D_{II}, are shown for $T = 316.92$ K in Figure 12.4 as function of θ: Variation is indeed minimal; therefore we assume the diffusion tensor calculated for the minimum energy configuration ($\theta = 0$).

Next we need to define the form of the time evolution operator (Liouvillian) for the density matrix described by the SLE. The molecule being partitioned in two fragments, as described above, we have (i) two local frames respectively fixed on the palmitate chain (CF) and on the tempo probe (PB): these are chosen with their respective z axes directed along the rotating bond, for convenience; (ii) the molecular frame (MF), fixed on the palmitate chain: this is the frame which diagonalizes the

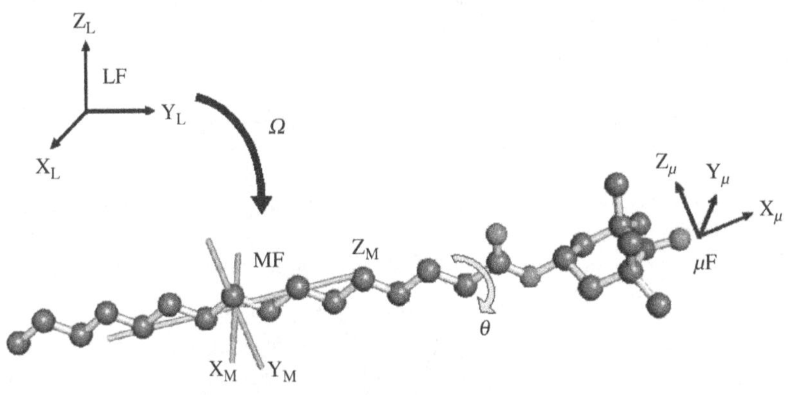

Figure 12.5 Relevant stochastic coordinates.

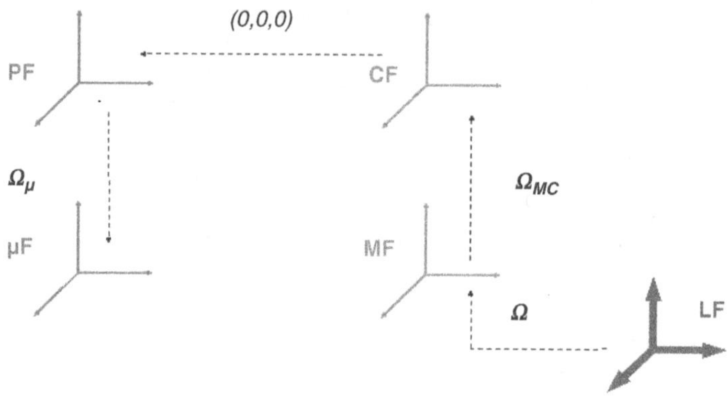

Figure 12.6 Molecular frames and Euler angle sets employed in the model.

rotational part of the diffusion tensor $\boldsymbol{D}_{RR}$; the magnetic frame, fixed on the probe (μF) where magnetic tensors are diagonal (Figure 12.6). Several sets of Euler angles are defined: Ω_{MC} is the set of Euler angles that transforms MF to CF, which has the z axis parallel to the rotating bond, Ω_μ is the set of Euler angles that transforms from PF to μF; the set $(0, 0, \theta)$ is the rotation from CF to PF; finally Ω transforms from the laboratory frame LF to MF. Following the established methodology [2, 30, 82, 84] the general form of the spin super-Hamiltonian is cast in the contracted tensorial form

$$\hat{H}^{\times} = \sum_{\mu} \omega_\mu \sum_{l=0,2} \sum_{m=-l}^{l} F_{\mu,\mathrm{LF}}^{(l,m)*} \hat{A}_{\mu,\mathrm{LF}}^{(l,m)\times} \tag{12.46}$$

where $\mu = g, A$ runs over the magnetic interactions, that is, the Zeeman interaction between the electron and the external field (g) and the hyperfine interaction between the electron and the ^{14}N nucleus (A). Parameters ω_μ, with $\mu = g, A$ are defined as $\beta_e B_0$ Tr $\boldsymbol{g}/3\hbar$ and γ_{e}Tr $\boldsymbol{A}/3$, respectively. Notice that for the generic irreducible spherical tensor F one can write

$$F_{\mu,\mathrm{LF}}^{(l,m)*} = \sum_{m',m''} D_{m,m'}^{l}(\Omega) e^{-im''\theta} G_{\mu}^{(l,m',m'')} \tag{12.47}$$

with

$$G_{\mu}^{(l,m,m')} = \mathcal{D}_{m,m'}^{l}(\Omega_{\mathrm{MC}}) \sum_{m''} D_{m',m''}^{l}(\Omega_\mu) F_{\mu,\mu\mathrm{F}}^{(l,m'')*} \tag{12.48}$$

Explicit forms for $F_{\mu,\mu\mathrm{F}}^{(l,m)*}$ and superoperators $\hat{A}_{\mu,\mathrm{LF}}^{(l,m)\times}$ are provided in the literature [82].

Finally, we define the form of the diffusion operator. In a symmetrized form (vide supra) we write

$$\tilde{\Gamma} = -P_{\mathrm{eq}}^{1/2} \begin{pmatrix} \hat{M} \\ \dfrac{\partial}{\partial\theta} \end{pmatrix}^{\mathrm{tr}} \mathbf{D} P_{eq} \begin{pmatrix} \hat{M} \\ \dfrac{\partial}{\partial\theta} \end{pmatrix} P_{\mathrm{eq}}^{-1/2} = \tilde{\Gamma}_{RR} + \tilde{\Gamma}_{II} + \tilde{\Gamma}_{RI} \tag{12.49}$$

where $\tilde{\Gamma}$ acts on $\boldsymbol{X} = (\Omega, \theta)$, the set of relevant variables; $\hat{\boldsymbol{M}}$ is the infinitesimal rotation operator. Finally, for the explicit evaluation of matrix elements, it is convenient to define

$$\begin{aligned}
\tilde{\Gamma}_{RR} &= -P_{\mathrm{eq}}^{-1/2}\hat{\boldsymbol{M}}^{\mathrm{tr}}\boldsymbol{D}_{RR}P_{\mathrm{eq}}\hat{\boldsymbol{M}}P_{\mathrm{eq}}^{-1/2} \\
\tilde{\Gamma}_{II} &= -D_{II}P_{\mathrm{eq}}^{-1/2}\frac{\partial}{\partial\theta}P_{\mathrm{eq}}\frac{\partial}{\partial\theta}P_{\mathrm{eq}}^{-1/2} \\
\tilde{\Gamma}_{RI} &= -P_{\mathrm{eq}}^{-1/2}\left(\hat{\boldsymbol{M}}^{\mathrm{tr}}\boldsymbol{D}_{RI}P_{\mathrm{eq}}\frac{\partial}{\partial\theta} + \frac{\partial}{\partial\theta}\boldsymbol{D}_{IR}P_{\mathrm{eq}}\hat{\boldsymbol{M}}\right)P_{\mathrm{eq}}^{-1/2}
\end{aligned} \tag{12.50}$$

The detailed forms of the rotational, internal, and rotational–internal operators are reported elsewhere [2, 30, 82, 84]. Although rather cumbersome, the whole algebraic derivation is straightforward. The numerical solution is based on the standard methodology described above.

Let us first report on the calculated set of parameters obtained from the QM calculations for structural and magnetic properties and the hydrodynamic modeling for diffusion properties. The principal values of the magnetic tensors minus the isotropic part are $g_{xx} = 0.00221$, $g_{yy} = 0.00020$, $g_{zz} = -0.00240$, $A_{xx} = -9.19$ G, $A_{yy} = -8.98$ G, and $A_{zz} = 18.18$ G. The orientations of the internal frames of reference are specified by angles $\Omega_{\mathrm{MC}} = (90, 35, 0)$ degrees and $\Omega_{\mu} = (0, 55, 180)$ degrees. The isotropic values of the hyperfine and gyromagnetic tensors are significantly different for different phases and are taken from experiment (see Table 12.1). A comparison with QM computed values is discussed in the next section. The computed torsional barrier of 1.8 kcal/mol^{-1} for the θ angle leads to a potential parameter $A = 453$ K/T. The diffusion tensor is expressed by Eq. 12.45 with

$$\overline{d} = \begin{pmatrix} 2.387\times10^{-3} & 0.0 & 0.0 & 1.560\times10^{-2} \\ 0.0 & 2.989\times10^{-3} & 0.0 & 1.313\times10^{-2} \\ 0.0 & 0.0 & 4.513\times10^{-2} & -3.071\times10^{-2} \\ 1.560\times10^{-2} & 1.313\times10^{-2} & -3.071\times10^{-2} & 5.884\times10^{-2} \end{pmatrix} \mathring{A}^2 \tag{12.51}$$

Table 12.1 Parameters Employed in Simulations

T/K	A_{iso}/Gauss	g_{iso}	ϵ	η/mPa s
316.09	15.5	2.00615	0.0	18.89
309.03	15.5	2.00629	0.0	23.80
308.72	15.7	2.00659	0.0	25.78
307.88	14.7	2.00679	0.9	26.80
299.02	13.5	2.00706	1.0	31.70

and

$$D(T) = D(T_0)\frac{\eta(T_0)}{\eta(T)}\frac{T}{T_0}$$

where $D(T_0)$ is the translational diffusion coefficient for a sphere of radius R at reference temperature T_0 given by

$$D(T_0) = \frac{k_B}{RC}\frac{T_0}{\pi\eta(T_0)}$$

Choosing $R = 2.0$ Å, $C = 6$, $T_0 = 316.92$ K (as reference temperature), and, $\eta(T_0) = 18.89 \times 10^{-3}$ Pa s, one gets $D(T_0) = 6.12 \times 10^9$ Hz.

We can now simulate the cw-ESR spectra of the Tempo-palmitate in 5-cyanobiphenyl in the range of temperatures from 316.92 K (isotropic phase) to 299.02 K (nematic phase). In Figure 12.7 simulated spectra are reported superimposed on experimental spectra taken from the literature [86]. The spectra at different temperatures and in different phases are reproduced with a very limited number of fitting parameters (ordering potential and isotropic parts of the magnetic tensors).

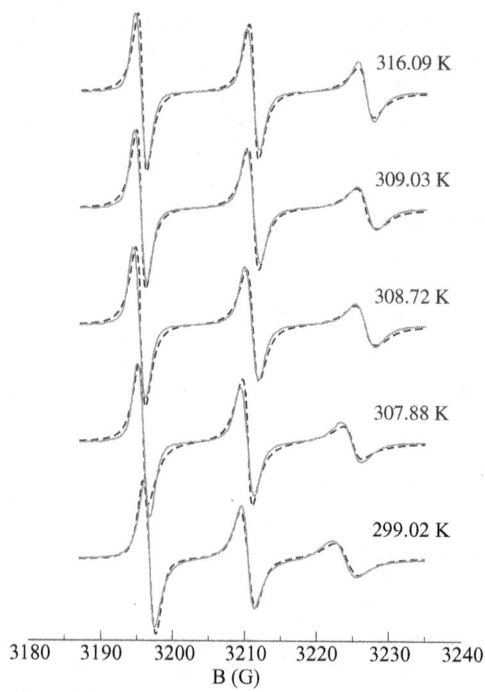

Figure 12.7 Experimental (full line) and simulated (dashed line) cw-ESR spectra of Tempo-palmitate in 5-cianobiphenyl at 316.09, 309.03 K (isotropic phase), 308.72 K (isotropic–nematic transition) and 307.88, 299.02 K (nematic phase).

12.3 INTERPRETING NMR RELAXATION DATA IN MACROMOLECULES

Spectroscopic techniques, both magnetic and optical, are widely used in structural and dynamical investigation of microscopic parameters of biomolecules [96], and, in particular, nuclear magnetic resonance (NMR) spectroscopy showed to be an important and powerful experimental technique in the interpretation of the micro-dynamics of proteins. The macroscopic physical observables are the T_1, T_2 and NOE relaxations of ^{15}N, ^{2}H, and ^{13}C nuclei, which have been found to be very sensitive to dynamics. The interpretative potential of the methodology comes from the fact that isotopic enrichment can be targeted to single residues of the protein, leading to the possibility of understanding localized dynamics (e.g., studying conformational motions specifically in the active site of the protein) and, moreover, comparison of data coming from different residues of the same protein permits us to make spatial (structural) considerations.

NMR relaxation data depend on dipolar (^{15}N and ^{13}C) and quadrupolar (^{2}H) interactions on chemical shift anisotropy and cross-correlation effects. It is well known that the NMR relaxations can be written as functions of the spectral densities of the magnetic interactions, and this is the intersecting point between macroscopic and microscopic descriptions: The spectral densities are calculated within the theoretical framework describing the dynamics of the system.

The most challenging part of the work is the introduction of the theoretical model. An early approach was proposed by Lipari and Szabo [97, 98] with their "model free" (MF) analysis. This approach is based on considering the presence of two uncoupled motions in the system: the global tumbling of the protein and the local motion of the probe. The assumption of decoupling leads to an easy formulation for the spectral density as the sum of spectral densities calculated from the two different motions. Simple mathematical expressions and fast calculations come from this approach, but also a number of limitations, leading to a restricted range of validity. The two most important limitations of their approach are: (i) MF considers isotropic global tumbling of the protein so that it works well with globular proteins but not with other molecules the geometry of which is not well approximated by a sphere (in later versions anisotropy was introduced); (ii) it fails to reproduce NMR data when the time scales of the motions are similar, that is, where the decoupling approximation cannot be assumed a priori.

An advanced approach to the modeling of two coupled dynamical processes was introduced by Polimeno and Freed [9, 30], originally in the interpretation of the electron spin resonance (ESR) of probes in ordered phases such as liquid cystals and glasses [8, 99]. The model is known as the slowly relaxing local structure (SRLS) model, which is a two-body Smoluchowski equation describig the coupled motion of two rigid rotors. This model has been applied by Meirovitch et al. [100–102] to the interpretation of NMR data. Due to the fact that coupled relaxation is taken into account rigorously and because the interaction potential can be interpreted in terms of local ordering imposed by the protein to the probe, the SRLS model has been shown to be useful, yielding good fitting to experiment even in cases that are out of the range of validity of the MF approach.

12.3.1 Two-Body Stochastic Modeling

Magnetic relaxation times T_1, T_2 and NOE of ^{15}N, ^{13}C, and ^{2}H nuclei depend on dipolar (^{15}N and ^{13}C) and quadrupolar (^{2}H) interactions, chemical shift anisotropy, and cross-correlation effects. In particular, we consider here as a spin probe the ^{15}N$-^{1}$H bond for which, following standard theory [103], it is possible to express the NMR relaxation times as functions of the spectral densities $J^{D}(\omega)$ (dipolar interaction) and $J^{C}(\omega)$ (chemical shift anisotropy):

$$\begin{aligned}
\frac{1}{T_1} &= d^2[J^D(\omega_H - \omega_N) + 3J^D(-\omega_N) + 6J^D(\omega_H - \omega_N)] + c^2 J^C(-\omega_N) \\
\frac{1}{T_2} &= d^2[4J^D(0) + J^D(\omega_H - \omega_N) + 3J^D(-\omega_N) + 3J^D(\omega_H) + 6J^D(\omega_H - \omega_N)] \\
&\quad + \frac{1}{3}c^2\left[3J^C(-\omega_N) + 4J^C(0)\right] \\
\mathrm{NOE} &= 1 + d^2\frac{\gamma_H}{\gamma_N}T_1\left[6J^D(\omega_H + \omega_N) - J^D(\omega_H - \omega_N)\right]
\end{aligned} \tag{12.52}$$

where $d = \mu_0\gamma_H\lambda_N/4\pi r_{NH}^3$, $c = \sqrt{2/15}\omega_N/\delta_{CSA}$, δ_{CSA} is the anisotropy of the chemical shift tensor, and ω_A is the Larmor frequency of nucleus A.

Spectral densities are calculated within the framework of the theoretical model for the dynamical evolution of the system. In the SRLS approach a two-body Smoluchowski equation describes the time evolution of the density probability of two relaxation processes (at different time scales) coupled by an interaction potential. In the application of this model to the description of protein dynamics, the two relaxing processes are interpreted as the slow global tumbling of the whole protein and the relatively fast local motion of the spin probe, the local motion of the ^{15}N$-^{1}$H bond in our case. Both processes are described as rigid rotators the motion of which is coupled by a potential correlating their reorientation, and it is interpreted as providing the local ordering that the molecule imposes on the probe.

Figure 12.8 gives a complete overview of the relevant coordinates and frames which are invoked in the model:

LF is the fixed inertial laboratory frame.

M_1F is the protein fixed frame where the diffusion tensor of the protein, ${}^{M_1}\boldsymbol{D}$, is diagonal.

M_2F is the protein fixed frame where the diffusion tensor of the probe, ${}^{M_2}\boldsymbol{D}$, is diagonal.

VF is the protein fixed frame having the z axis aligned with the director of the orienting potential.

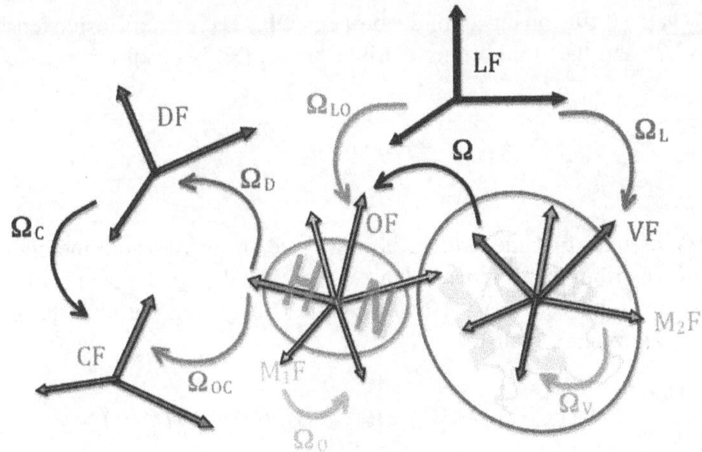

Figure 12.8 Definition of frames and Euler angles in SRLS model applied to NMR.

OF is the probe fixed frame the *z* axis of which tends to be aligned to the director of the potential.
DF is the probe fixed frame where the dipolar interaction is diagonal.
CF is the probe fixed frame where the chemical shift tensor is diagonal.

To complete the picture, we have to define the set of Euler angles that transform among the different frames:

Ω_L transform from LF to VF, while Ω_{LO} transform from LF to OF.
Ω tranform from VF to OF.
Ω_V tranform from M_1F to VF.
Ω_O transform from M_2F to OF.
Ω_C transform from OF to DF, while Ω_{OC} tranform from OF to DF.
Ω_C transform from CF to DF.

The system is fully described with two sets of stochastic Euler angles, and in particular our choice is on the set of Euler angles Ω_L, giving the orientation of the protein relative to the laboratory frame, and Ω, which represents the relative orientation of the probe and the protein. Using this set of stochastic variables, $\mathbf{Q} = (\Omega, \Omega_L)$, the diffusion operator describing the time evolution of the density probability of the system is

$$\hat{\Gamma}(\mathbf{Q}) = {}^{O}\hat{\mathbf{J}}^{\dagger}(\Omega)\,{}^{O}\mathbf{D}P_{\mathrm{eq}}(\mathbf{X})\,{}^{O}\hat{\mathbf{J}}(\Omega)P_{\mathrm{eq}}^{-1}(\mathbf{Q})$$
$$+ \left[V\hat{\mathbf{J}}(\Omega) - {}^{V}\hat{\mathbf{J}}(\Omega_L)\right]^{\dagger V}\mathbf{D}_1 P_{\mathrm{eq}}(\mathbf{Q})\left[V\hat{\mathbf{J}}(\Omega) - {}^{V}\hat{\mathbf{J}}(\Omega_L)\right]P_{\mathrm{eq}}^{-1}(\mathbf{Q}) \quad (12.53)$$

where ${}^{0}\mathbf{D}_2$ is the diffusion tensor of the probe in OF, ${}^{V}\mathbf{D}_1$ is the diffusion tensor of the protein in VF, and the equilibrium distribution P_{eq} ($\mathbf{X}$) is given by

$$P_{eq}(\mathbf{Q}) = \mathcal{N} \exp\left[-\frac{V(\Omega, \Omega_L)}{k_B T} \right] \tag{12.54}$$

We may assume that the protein is immersed in an isotropic medium, so the equilibrium distribution is independent of Ω_L and the total potential is just the interaction potential between the two processes for which we take the following expansion over Wigner matrices:

$$\begin{aligned} -\frac{V(\Omega)}{kT} &= c_0^2 \mathcal{D}_{0\,0}^2(\Omega) + c_2^2[\mathcal{D}_{0\,-2}^2(\Omega) + \mathcal{D}_{0\,2}^2(\Omega)] + c_0^4 \mathcal{D}_{0\,0}^4(\Omega) \\ &\quad + c_2^4[\mathcal{D}_{0\,-2}^4(\Omega) + \mathcal{D}_{0\,2}^4(\Omega)] + c_4^4[\mathcal{D}_{0\,-4}^4(\Omega) + \mathcal{D}_{0\,4}^4(\Omega)] \end{aligned} \tag{12.55}$$

Observables are expressed as spectral densities, that is, Fourier–Laplace transforms of correlation functions of Wigner functions of the absolute probe Euler angles, $\Omega_{LO} = \Omega + \Omega_L$

$$j_{k,k'}(\omega) = \langle \mathcal{D}_{mk}^{j}(\Omega_{LO}) P_{eq}(\Omega_{LO}) | \left(i\omega - \hat{\Gamma} \right)^{-1} | \mathcal{D}_{m'k'}^{j'}(\Omega_{LO}) P_{eq}(\Omega_{LO}) \rangle \tag{12.56}$$

Considering the symmetry of the magnetic interactions (dipolar and chemical shift anisotropy) contributing to the spin Hamiltonian of the system for the ^{15}N–^{1}H probe, only physical observables with $j = j' = 2$ and $m = m' = 0$ have to be considered.

From these spectral densities it is possible to calculate the spectral densities for every magnetic interaction, μ (dipolar, CSA), as

$$J^{\mu}(\omega) = \sum_{k,k'=-2}^{2} \left[\mathcal{D}_{k\,0}^{2\,*}(\Omega_\mu) \mathcal{D}_{k'0}^{2}(\Omega_\mu) \right] jk, k'(\omega) \tag{12.57}$$

where Ω_μ is the set of Euler of angles transforming from OF to the frame where the μth magnetic tensor is diagonal.

Calculation of spectral densities $j_{k,k'}$ (ω) is achieved by spanning the diffusive operator over a proper basis set, as usual. An obvious choice is the direct product $|\Lambda\rangle = |\lambda_1\rangle \otimes |\lambda_2\rangle = |L_1 M_1 K_1\rangle \otimes |L_2 M_2 K_2\rangle$, where

$$|L_1 M_1 K_1\rangle = \sqrt{\frac{2L_1 + 1}{8\pi^2}} \mathcal{D}_{M_1 K_1}^{L_1}(\Omega_L) \tag{12.58}$$

$$|L_2 M_2 K_2\rangle = \sqrt{\frac{2L_2 + 1}{8\pi^2}} \mathcal{D}_{M_2 K_2}^{L_2}(\Omega) \tag{12.59}$$

It is simpler to work with autocorrelation functions so instead of directly calculating spectral densities in Eq. 12.56, we define the function $2C_{k,k'} = \mathcal{D}^2_{0k} + \mathcal{D}^2_{0k'}$, and calculate the symmetrized spectral densities

$$j^S_{k,k'}(\omega) = \langle C_{k,k'}(\Omega_{LO})P_{\text{eq}}(\Omega_{LO})|\left(i\omega - \hat{\Gamma}\right)^{-1}|C_{k,k'}(\Omega_{LO})P_{\text{eq}}(\Omega_{LO})\rangle \tag{12.60}$$

and then obtain the $j_{k,k'}(\omega)$ functions as linear combinations of the symmetrized spectral densities:

$$j_{k,k'}(\omega) = \frac{1}{10}\left[2(1+\delta_{k,k'})j^S_{k,k'}(\omega) - j^S_{k,k}(\omega) - j^S_{k',k'}(\omega)\right] \tag{12.61}$$

Using the closure relation for the basis $|\Lambda\rangle$, the integral in Eq. 12.60 can now be rewritten in matrix form as

$$j^S_{k,k'} = \boldsymbol{v}^t(i\omega 1 - \Gamma)^{-1}\boldsymbol{v} \tag{12.62}$$

where

$$(\Gamma)_{i,j} = \langle \Lambda_i|\hat{\Gamma}|\Lambda_j\rangle \tag{12.63}$$

$$(\boldsymbol{v})_i = \langle \Lambda_i|C_{k,k'}(\Omega_{LO})P_{\text{eq}}(\Omega_{LO})\rangle \tag{12.64}$$

Details on the evaluation of eqs. 12.63 and 12.64 are reported elsewhere [100].

12.3.2 Case Study: AKeko Protein

A set of residues of the *Escherichia coli* adenylate kinase (AKeco) protein has been selected in order to illustrate and test the application of the methodology to real experimental data. In Figure 12.9 are highlighted the chosen residues with different colors. The color scheme is: yellow for the AMPbd domain, red for the CORE domain, blue for the LID domain, and green for the small P-loop. We followed the standard definition in dividing the protein into those domains [100]. For the experimental values see the supporting information in Shapiro et al. [100].

The diffusion tensor of the protein, in water, was evaluated with slip boundary conditions, effective radius of the spheres of 2.0 Å, and room temperature and viscosity of 0.9 cP. With this parameters we obtained ${}^1D_{XX} = 1.11 \times 10^7$ Hz, ${}^1D_{YY} = 1.20 \times 10^7$ Hz, and ${}^1D_{ZZ} = 1.65 \times 10^7$ Hz. Because of the near axiality of the tensor, in the calculations we assumed the average values ${}^1D_{XX} = {}^1D_{YY} = 1.15 \times 10^7$ Hz. We imposed an axial orienting potential coupling the two bodies.

As outlined above, the first body describes the motion of the protein, while we interpret the second body as the (collective) local motions in the neighborhood of the magnetic probe, the ^{15}N–^{1}H bond. In this picture we assume, for the second body, a diffusion tensor which is diagonal in a frame having the Z axis parallel to the

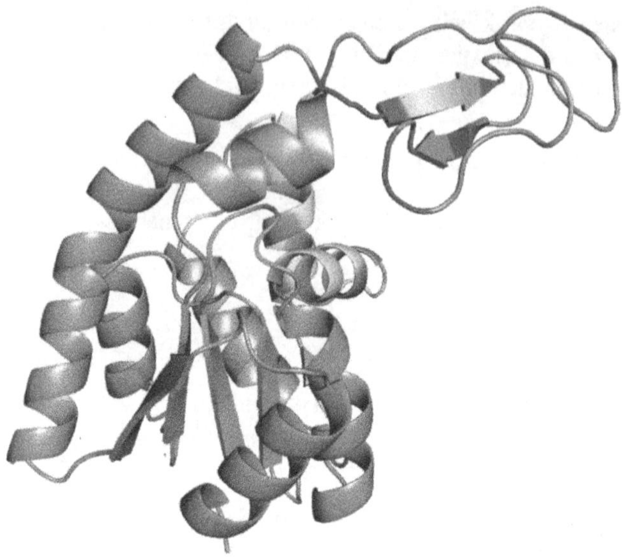

Figure 12.9 Pictorial overview of distribution of residues chosen for calculations.

^{15}N–^{1}H bond and the X axis perpendicular to the peptide bond plane. Moreover, we consider the tensor to be axially symmetric in such a frame. To interpret data, we make the simplifying assumption that the coupling potential tends to align the Z axis of the second body (i.e., of the OF frame), parallel to the direction containing the ^{15}N=^{1}H bond in the equilibrium geometry of the protein. This is reproduced by defining a frame VF having the Z axis parallel to the ^{15}N–^{1}H bond, which in general is tilted from the M_1F, where the diffusion tensor of the protein is diagonal. So, for every residue, we extracted from the geometry of the protein the set of Euler angles that transform from M_1F to VF, Ω_1. We assume that the magnetic tensors are diagonal in the same frame, that is, $\Omega_C = (0.0, 0.0, 0.0)$ degrees, and a constant tilt with respect to the OF, $\Omega_D = (0.0, 18.0, 90.0)$ degrees, following Meirovitch et al. [101].

A set of four parameters were considered adjustable and obtained via fitting: the parallel and perpendicular components of the diffusion tensor of the second body, ${}^{O}D_1$ and ${}^{O}D_{\|}$, the strength of the axial potential, c_0^2, and a parameter called rate of exchange, R_{ex}, which gives a correction due to a very slow change in configuration of the protein [101]. Table 12.2 summarizes the values obtained for the 37 residues considered.

Figures 12.10–12.12 show the experimental and theoretical values of the T_1, T_2 and NOE at 600.0 MHz. The overall agreement is good: All the relative errors between theoretical and experimental values are within 5%. Figure 12.13 plots the values of the order parameters obtained with the standard formula

$$S = \langle \mathcal{D}^2_{0\,0}(\Omega_O) P_{eq}(\Omega_O) \rangle \tag{12.65}$$

Table 12.2 Values of Model Parameters Obtained from Fitting

Domain	Residue	${}^{O}D_{\perp}$ (10^7 Hz)	${}^{O}D_{\parallel}$ (10^{10} Hz)	R_{ex} (Hz)	c_0^2	S
AMPbd	32	1.69	10.5	2.95	2.64	0.55
	33	2.04	13.0	1.51	3.65	0.68
	36	1.55	12.7	1.38	2.82	0.58
	41	2.56	5.42	0.277	4.81	0.77
	42	2.54	4.23	0.873	4.80	0.77
	46	2.09	7.02	1.16	4.32	0.74
	48	1.38	7.27	1.30	2.50	0.53
	50	2.23	6.84	1.09	4.69	0.76
	52	2.12	7.09	0.118	4.34	0.74
	53	1.93	5.34	0.882	4.06	0.72
	55	2.36	6.55	1.01	5.13	0.78
	56	2.25	6.29	0.427	4.40	0.74
	60	2.29	5.27	0.196	5.01	0.78
CORE	2	1.29	17.7	1.60	1.77	0.39
	3	1.40	35.1	1.24	2.24	0.49
	16	1.83	11.4	4.21	3.40	0.65
	77	1.32	20.9	1.75	2.06	0.45
	86	1.69	15.6	2.38	2.70	0.56
	97	1.72	19.5	0.54	4.00	0.71
	107	1.33	28.1	2.14	1.83	0.41
	117	1.42	31.5	2.36	2.37	0.51
	170	1.63	8.95	0.898	3.47	0.66
	191	2.20	5.21	0.000	4.27	0.73
	210	1.35	25.4	1.51	3.15	0.62
LID	122	1.70	25.4	6.05	4.16	0.72
	123	1.70	12.4	2.90	3.58	0.67
	126	1.84	15.4	0.000	4.28	0.73
	132	2.58	6.59	0.000	5.38	0.80
	136	2.05	6.87	1.54	5.06	0.78
	137	2.25	6.77	0.000	5.73	0.81
	145	1.64	9.07	1.42	3.53	0.67
	151	1.35	14.7	1.20	3.09	0.62
	158	2.30	3.96	1.49	4.37	0.74
	159	1.79	8.82	0.458	4.28	0.73
P loop	8	1.84	15.4	0.161	4.33	0.74
	11	1.46	13.5	2.30	2.96	0.60

Analysis of NMR relaxation data applied to the investigation of microscopic dynamics is very promising, and a wealth of experimental measures are just waiting for advanced interpretative tools. The SRLS model is a first somewhat primitive but systematic approach which attempts to combine simplified but clearly defined physical hypotheses with a reliable physical interpretation of both dynamical and structural (through the interaction potential) properties.

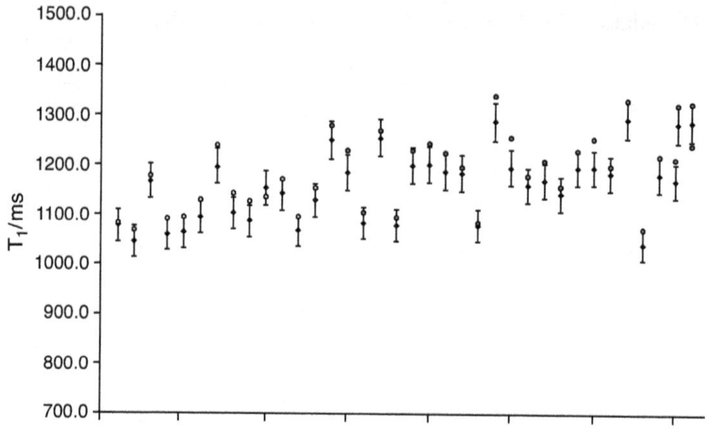

Figure 12.10 Experimental (rhombi) and theoretical (circles) T_1 values at 600.0 MHz.

12.4 CONCLUSIONS

Stochastic models are a comprehensive and mature tool for interpreting molecular relaxation phenomena observed from magnetic resonance spectroscopies. Modern implementations [104, 105] allow one to exploit the modularity of numerical algorithms to obtain highly efficient software tools which can tackle diverse molecular systems, especially in connection with QM determination of structural and dynamical properties of complex molecular systems. The future of stochastic approaches can be thought of in connection with the proper development of

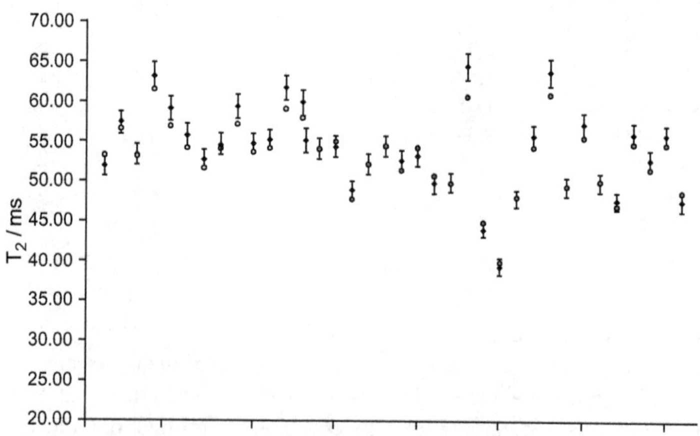

Figure 12.11 Experimental (rhombi) and theoretical (circles) T_2 values at 600.0 MHz.

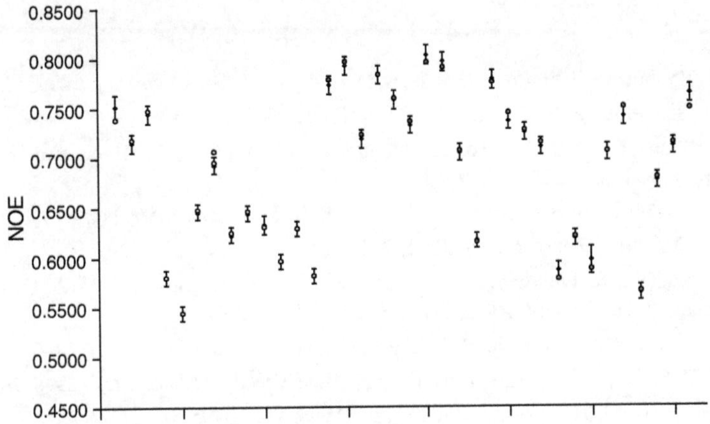

Figure 12.12 Experimental (rhombi) and theoretical (circles) NOE values at 600.0 MHz.

multiscale approaches. Indeed, in the near future one can envision integrated mesoscopic–atomistic methods which combine stochastic modeling of slow, or "soft," variables and appropriate treatment (at a molecular dynamics level) for fast, or "hard," degrees of freedom. This methodology would be ideal to treat large flexible biomolecules, allowing an economical computational treatment. Moreover, foundations of stochastic many-body approaches can be based on atomistic-derived descriptions, rendering these augmented treatments predictive in nature; data fitting could then be seen as a refining step geared toward overcoming errors in parameter evaluation implied by the approximations inherent in the various components of the protocol.

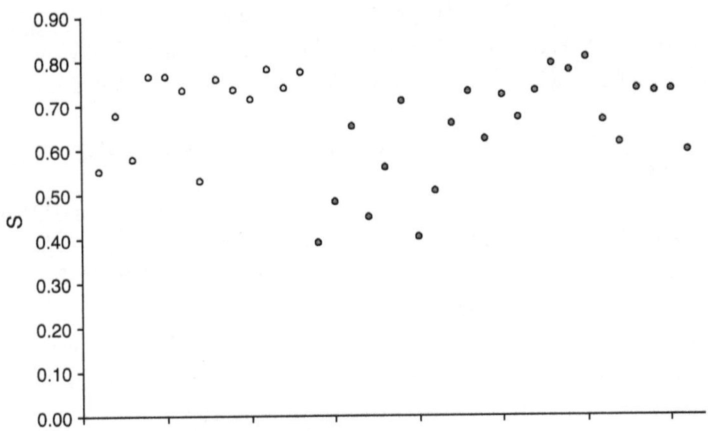

Figure 12.13 Order parameters obtained from fitting.

REFERENCES

1. Y. E. Ryabov, C. Geragthy, A. Varshney, D. Fushman, *J. Am. Chem. Soc.* **2006**, *128*, 15432.
2. D. J. Schneider, J. H. Freed, *Adv. Chem. Phys.* **1989**, *73*, 387.
3. M. Balog, T. Kálai, J. Jeko, Z. Berente, H. J. Steinhoff, M. Engelhard, K. Hideg, *Tetrahedron Lett.* **2003**, *44*, 9213.
4. W. L. Hubbell, H. S. Mchaourab, C. Altenbach, M. Lietzow, *Structure 1996 4*, **1996**.
5. H. J. Steinhoff, *Front. Biosci.* **2002**, *7*, 2002.
6. Z. Zhang, M. R. Fleissner, D. S. Tipikin, Z. Liang, J. K. Moscicki, K. A. Earle, W. L. Hubbell, J. H. Freed, *J. Phys. Chem. B* **2010**, *114*, 5503.
7. V. S. S. Sastry, A. Polimeno, R. H. Crepeau, J. H. Freed, *J. Chem. Phys.* **1996**, *107*, 5753.
8. V. S. S. Sastry, A. Polimeno, R. H. Crepeau, J. H. Freed, *J. Chem. Phys.* **1996**, *107*, 5773.
9. A. Polimeno, J. H. Freed, *J. Phys. Chem.* **1995**, *99*, 10995.
10. P. P. Borbat, A. J. Costa-Filho, K. A. Earle, J. K. Moscicki, J. H. Freed, *Science* **2001**, *291*, 266.
11. M. Ge, J. H. Freed, *Biophys. J.* **2009**, *96*, 4925.
12. Y.-W. Chiang, A. J. Costa, J. H. Freed, *J. Phys. Chem. B* **2007**, *111*, 11260.
13. M. J. Swamy, L. Ciani, M. Ge, A. K. Smith, D. Holowka, B. Baird, J. H. Freed, *Biophys. J.* **2006**, *90*, 4452.
14. S. Van Doorslaer, E. Vinck, *Phys. Chem. Chem. Phys.* **2007**, *9*, 4620.
15. O. Schiemann, T. F. Prisner, *Q. Rev. Biophys.* **2007**, *40*, 1.
16. P. P. Borbat, J. H. Freed, Measuring distances by pulsed dipolar ESR spectroscopy: spin-labeled histidine kinases, in *Two-Component Signaling Systems, Part B*, M. Simon, B.R. Crane, A.B. Crane, Eds., Methods in Enzymology 423, Ch. 3, 2007, pp. 52–116.
17. P. Borbat, H. Mchaourab, J. Freed, *J. Am. Chem. Soc.* **2002**, *124*, 5304.
18. A. J. Costa-Filho, Y. Shimoyama, J. H. Freed, *Biophys. J.* **2003**, *84*, 2619.
19. G. Jeschke, Y. Polyhach, *Phys. Chem. Chem. Phys.* **2007**, *9*, 1895.
20. F. Tombolato, A. Ferrarini, J. H. Freed, *J. Phys. Chem. B* **2006**, *110*, 26248.
21. F. Tombolato, A. Ferrarini, J. H. Freed, *J. Phys. Chem. B* **2006**, *110*, 26260.
22. A. Polimeno, M. Zerbetto, L. Franco, M. Maggini, C. Corvaja, *J. Am. Chem. Soc.* **2006**, *128*, 4734.
23. V. Barone, M. Brustolon, P. Cimino, A. Polimeno, M. Zerbetto, A. Zoleo, *J. Am. Chem. Soc.* **2006**, *128*, 15865.
24. M. Zerbetto, S. Carlotto, A. Polimeno, C. Corvaja, L. Franco, C. Toniolo, F. Formaggio, V. Barone, P. Cimino, *J. Phys. Chem. B* **2007**, *111*, 2668.
25. P.-O. Westlund, H. Wennerstrom, *Phys. Chem. Chem. Phys.* **2010**, *12*, 201.
26. S. K. Misra, *J. Mag. Res.* **2007**, *189*, 59.
27. A. Borel, R. B. Clarkson, R. L. Belford, *J. Chem. Phys.* **2007**, *126*, 054510.
28. M. Lindgren, A. Laaksonen, P.-O. Westlund, *Phys. Chem. Chem. Phys.* **2009**, *11*, 10368.
29. D. Sezer, J. H. Freed, B. Roux, *J. Am. Chem. Soc.* **2009**, *131*, 2597.
30. A. Polimeno, J. H. Freed, *Adv. Chem. Phys.* **1993**, *83*, 89.
31. A. Arcioni, C. Bacchiocchi, L. Grossi, A. Nicolini, C. Zannoni, *J. Phys. Chem. B* **2002**, *206*, 9245.

32. A. Nayeem, S. Rananavare, V. Sastry, J. Freed, *J. Chem. Phys.* **1992**, *96*, 3912.
33. J. Moscicki, Y. Shin, J. Freed, *J. Chem. Phys.* **1993**, *99*, 634.
34. A. Arcioni, C. Bacchiocchi, I. Vecchi, G. Venditti, C. Zannoni, *Chem. Phys. Lett.* **2004**, *396*, 433.
35. J. H. Freed, *Electron Spin Relaxation in Liquids*, Plenum, New York, 1972.
36. R. Kubo, *J. Math. Phys.* **1963**, *4*, 174.
37. R. Kubo, *J. Phys. Soc. Jpn. Suppl.* **1969**, *26*, 1.
38. W. Wassam, J. Freed, *J. Chem. Phys.* **1982**, *76*, 6133.
39. W. Wassam, J. Freed, *J. Chem. Phys.* **1982**, *76*, 6150.
40. Y. Tanimura, *J. Phys. Soc. Jpn.* **2006**, *75*, 082001.
41. V. Barone, *J. Chem. Phys.* **1994**, *101*, 10666.
42. V. Barone, *Theor. Chem. Acc.* **1995**, *91*, 113.
43. V. Barone, *Advances in Density Functional Theory. Part I,* Vol. 287, World Science, Singapore, 1995.
44. R. Improta, V. Barone, *Chem. Rev.* **2004**, *104*, 1231.
45. D. Feller, E. R. Davidson, *J. Chem. Phys.* **1988**, *88*, 5770.
46. S. A. Perera, L. M. Salemi, R. J. Bartlett, *J. Chem. Phys.* **1997**, *106*, 4061.
47. A. R. Al Derzi, S. Fau, R. J. Bartlett, *J. Phys. Chem. A* **2003**, *107*, 6656.
48. H. M. McConnell, *J. Chem. Phys.* **1963**, *39*, 1910.
49. H. M. McConnell, *Proc. R. A. Welch Found. Conf. Chem. Res.* **1967**, *11*, 144.
50. A. J. Stone, *Mol. Phys.* **1964**, *7*, 311.
51. C. Adamo, V. Barone, R. Subra, *Theor. Chem. Acco. Theory Comput. Model. (Theor. Chim. Acta)* **2000**, *104*, 207.
52. F. Neese, *J. Chem. Phys.* **2001**, *115*, 11080.
53. R. Ditchfield, *Mol. Phys.* **1974**, *27*, 789.
54. J. R. Cheeseman, G. W. Trucks, T. A. Keith, M. J. Frisch, *J. Chem. Phys.* **1996**, *104*, 5497.
55. O. L. Malkina, J. Vaara, B. Schimmelpfennig, M. Munzarova, V. G. Malkin, M. Kaupp, *J. Am. Chem. Soc.* **2000**, *122*, 9206.
56. I. Ciofini, C. Adamo, V. Barone, *J. Chem. Phys.* **2004**, *121*, 6710.
57. V. Barone, R. Subra, *J. Chem. Phys.* **1996**, *104*, 2630.
58. F. Jolibois, J. Cadet, A. Grand, R. Subra, N. Rega, V. Barone, *J. Am. Chem. Soc.* **1998**, *120*, 1864.
59. V. Barone, P. Carbonniere, C. Pouchan, *J. Chem. Phys.* **2005**, *122*, 224308.
60. J. A. Nillson, L. A. Eriksson, A. Laaksonen, *Mol. Phys.* **2001**, *99*, 247.
61. M. Nonella, G. Mathias, P. Tavan, *J. Phys. Chem. A* **2003**, *107*, 8638.
62. J. R. Asher, N. L. Doltsinis, M. Kaupp, *Magn. Reson. Chem.* **2005**, *43*, S237.
63. M. Pavone, C. Benzi, F. De Angelis, V. Barone, *Chem. Phys. Lett.* **2004**, *395*, 120.
64. M. Pavone, P. Cimino, F. De Angelis, V. Barone, *J. Am. Chem. Soc.* **2006**, *128*, 4338.
65. M. Pavone, A. Sillanpaa, P. Cimino, O. Crescenzi, V. Barone, *J. Phys. Chem. B* **2006**, *110*, 16189.
66. S. Koseki, M. W. Schmidt, M. S. Gordon, *J. Phys. Chem.* **1992**, *96*, 10768.
67. E. Fermi, *Zeits. Phy. A Hadrons Nuclei* **1930**, *60*, 320.

68. R. A. Frosch, H. M. Foley, *Phys. Rev.* **1952**, *88*, 1337.
69. J. Tomasi, B. Mennucci, R. Cammi, *Chem. Rev.* **2005**, *105*, 2999.
70. V. Barone, M. Cossi, J. Tomasi, *J. Chem. Phys.* **1997**, *107*, 3210.
71. C. Benzi, M. Cossi, R. Improta, V. Barone, *J. Comp. Chem.* **2005**, *26*, 1096.
72. M. Cossi, G. Scalmani, N. Rega, V. Barone, *J. Chem. Phys.* **2002**, *117*, 43.
73. V. Barone, A. Polimeno, *Phys. Chem. Chem. Phys.* **2006**, *8*, 4609.
74. V. Barone, A. Grand, C. Minichino, R. Subra, *J. Chem. Phys.* **1993**, *99*, 6745.
75. V. Barone, C. Adamo, Y. Brunel, R. Subra, *J. Chem. Phys.* **1996**, *105*, 3168.
76. V. Barone, *Chem. Phys. Lett.* **1996**, *262*, 201.
77. M. Pavone, C. Benzi, F. D. Angelis, V. Barone, *Chem. Phys. Lett.* **2004**, *395*, 120.
78. O. Crescenzi, M. Pavone, F. de Angelis, V. Barone, *J. Phys. Chem. B* **2005**, *109*, 445.
79. G. Moro, *Chem. Phys.* **1987**, *118*, 181.
80. H. Yamakawa, *J. Chem. Phys.* **1970**, *53*, 436.
81. J. Rotne, S. Prager, *J. Chem. Phys.* **1969**, *50*, 4831.
82. E. Meirovitch, D. Igner, E. Igner, G. Moro, J. H. Freed, *J. Chem. Phys.* **1982**, *77*, 3915.
83. G. Moro, J. H. Freed, *J. Chem. Phys.* **1981**, *74*, 3757.
84. G. Moro, J. H. Freed, *Large-Scale Eigenvalue Problems, Mathematical Studies Series*, Vol. 127, Elsevier, New York, 1986.
85. H. J. Hogben, P. J. Hore, I. Kuprov, *J. Chem. Phys.* **2010**, *132*, 174101.
86. M. A. Morsy, G. A. Oweimreen, J. S. Hwang, *J. Phys. Chem.* **1996**, *100*, 8331.
87. C. Benzi, M. Cossi, V. Barone, *J. Chem. Phys.* **2005**, *123*, 194909.
88. C. Adamo, V. Barone, *J. Chem. Phys.* **1999**, *110*, 6158.
89. S. Chandrasekhar, *Liquid Crystals*, 2nd ed., University Press, Cambridge, 1992.
90. P. G. De Gennes, J. Prost, *The Physics of Liquid Crystals*, 2nd ed., Oxford University Press, New York, 1993.
91. G. Vertogen, W. H. de Jeu, *Thermotropic Liquid Crystals, Fundamentals*, Springer-Verlag, Berlin, 1993.
92. F. M. Leslie, *Quart. J. Mech. Appl. Math.* **1966**, *19*, 357.
93. F. M. Leslie, *Adv. Liq. Cryst.* **1979**, *4*, 1.
94. J. L. Ericksen, *Trans. Soc. Rheol.* **1961**, *5*, 23.
95. J. L. Ericksen, *Adv. Liq. Cryst.* **1976**, *2*, 233.
96. A. G. PalmerIII, *Annu. Rev. Biophys. Biomol. Struct.* **2001**, *20*, 129.
97. G. Lipari, A. Szabo, *J. Am. Chem. Soc.* **1982**, *104*, 4546.
98. G. Lipari, A. Szabo, *J. Am. Chem. Soc.* **1982**, *104*, 4559.
99. K. A. Earle, J. K. Moscicki, A. Polimeno, J. H. Freed, *J. Chem. Phys.* **1997**, *106*, 9996.
100. Y. E. Shapiro, E. Kahana, V. Tugarinov, Z. Liang, J. H. Freed, E. Meirovitch, *Biochemistry* **2002**, *41*, 6271.
101. E. Meirovitch, A. Polimeno, J. H. Freed, *J. Phys. Chem. B* **2006**, *110*, 20615.
102. E. Meirovitch, A. Polimeno, J. H. Freed, *J. Phys. Chem. B* **2007**, *111*, 12865.
103. A. Abragam, *The Principles of Nuclear Magnetism*, Oxford University Press, London, 1961.
104. M. Zerbetto, A. Polimeno, V. Barone, *Comput. Phys. Comm.* **2009**, *180*, 2680.
105. M. Zerbetto, A. Polimeno, E. Meirovitch, *J. Phys. Chem. B* **2009**, *113*, 13613.

INDEX

Computational Strategies for Spectroscopy: From Small Molecules to Nano Systems, First Edition. Edited by Vincenzo Barone.